Rock Mechanics and Engineering

Rock Mechanics and Engineering

Volume 5: Surface and Underground Projects

Editor

Xia-Ting Feng
Institute of Rock and Soil Mechanics, Chinese Academy of Sciences,
State Key Laboratory of Geomechanics and Geotechnical Engineering,
Wuhan, China

CRC Press
Taylor & Francis Group
Boca Raton London New York

CRC Press is an imprint of the
Taylor & Francis Group, an **informa** business
A BALKEMA BOOK

Published by:
CRC Press/Balkema
Schipholweg 107C, 2316 XC Leiden, The Netherlands

First issued in paperback 2023

ISBN-13: 978-1-138-02763-3 (hbk)
ISBN-13: 978-1-03-265227-6 (pbk)
ISBN-13: 978-1-315-70811-9 (ebk)

Typeset by Integra Software Services Private Ltd

Library of Congress Cataloging-in-Publication Data

Names: Feng, Xia-Ting, editor.
Title: Rock mechanics and engineering / editor, Xia-Ting Feng, Institute of Rock and Soil Mechanics, Chinese Academy of Sciences, State Key Laboratory of Geomechanics and Geotechnical Engineering, Wuhan, China.
Description: Leiden, The Netherlands ; Boca Raton : CRC Press/Balkema, [2017]– | Includes bibliographical references and index. Contents: volume 1. Principles
Identifiers: LCCN 2016053708 (print) | LCCN 2017004736 (ebook) | ISBN 9781138027596 (hardcover : v. 1) | ISBN 9781315364261 (ebook : v. 1)
Subjects: LCSH: Rock mechanics.
Classification: LCC TA706 .R532 2017 (print) | LCC TA706 (ebook) | DDC 624.1/5132–dc23
LC record available at https://lccn.loc.gov/2016053708

Cover photo: Mining equipment
Copyright: deadmeat243
Courtesy of: www.shutterstock.com

Contents

Foreword

Although engineering activities involving rock have been underway for millennia, we can mark the beginning of the modern era from the year 1962 when the International Society for Rock Mechanics (ISRM) was formally established in Salzburg, Austria. Since that time, both rock engineering itself and the associated rock mechanics research have increased in activity by leaps and bounds, so much so that it is difficult for an engineer or researcher to be aware of all the emerging developments, especially since the information is widely spread in reports, magazines, journals, books and the internet. It is appropriate, if not essential, therefore that periodically an easily accessible structured survey should be made of the currently available knowledge. Thus, we are most grateful to Professor Xia-Ting Feng and his team, and to the Taylor & Francis Group, for preparing this extensive 2017 "Rock Mechanics and Engineering" compendium outlining the state of the art—and which is a publication fitting well within the Taylor & Francis portfolio of ground engineering related titles.

There has previously only been one similar such survey, "Comprehensive Rock Engineering", which was also published as a five-volume set but by Pergamon Press in 1993. Given the exponential increase in rock engineering related activities and research since that year, we must also congratulate Professor Feng and the publisher on the production of this current five-volume survey. Volumes 1 and 2 are concerned with principles plus laboratory and field testing, *i.e.*, understanding the subject and obtaining the key rock property information. Volume 3 covers analysis, modelling and design, *i.e.*, the procedures by which one can predict the rock behaviour in engineering practice. Then, Volume 4 describes engineering procedures and Volume 5 presents a variety of case examples, both these volumes illustrating 'how things are done'. Hence, the volumes with their constituent chapters run through essentially the complete spectrum of rock mechanics and rock engineering knowledge and associated activities.

In looking through the contents of this compendium, I am particularly pleased that Professor Feng has placed emphasis on the strength of rock, modelling rock failure, field testing and Underground Research Laboratories (URLs), numerical modelling methods—which have revolutionised the approach to rock engineering design—and the progression of excavation, support and monitoring, together with supporting case histories. These subjects, enhanced by the other contributions, are the essence of our subject of rock mechanics and rock engineering. To read through the chapters is not only to understand the subject but also to comprehend the state of current knowledge.

I have worked with Professor Feng on a variety of rock mechanics and rock engineering projects and am delighted to say that his efforts in initiating, developing and seeing through the preparation of this encyclopaedic contribution once again demonstrate his flair for providing significant assistance to the rock mechanics and engineering subject and community. Each of the authors of the contributory chapters is also thanked: they are the virtuosos who have taken time out to write up their expertise within the structured framework of the "Rock Mechanics and Engineering" volumes. There is no doubt that this compendium not only will be of great assistance to all those working in the subject area, whether in research or practice, but it also marks just how far the subject has developed in the 50+ years since 1962 and especially in the 20+ years since the last such survey.

John A. Hudson, Emeritus Professor, Imperial College London, UK
President of the International Society for Rock Mechanics (ISRM) 2007–2011

Introduction

The five-volume book "Comprehensive Rock Engineering" (Editor-in-Chief, Professor John A. Hudson) which was published in 1993 had an important influence on the development of rock mechanics and rock engineering. Indeed the significant and extensive achievements in rock mechanics and engineering during the last 20 years now justify a second compilation. Thus, we are happy to publish 'ROCK MECHANICS AND ENGINEERING', a highly prestigious, multi-volume work, with the editorial advice of Professor John A. Hudson. This new compilation offers an extremely wide-ranging and comprehensive overview of the state-of-the-art in rock mechanics and rock engineering. Intended for an audience of geological, civil, mining and structural engineers, it is composed of reviewed, dedicated contributions by key authors worldwide. The aim has been to make this a leading publication in the field, one which will deserve a place in the library of every engineer involved with rock mechanics and engineering.

We have sought the best contributions from experts in the field to make these five volumes a success, and I really appreciate their hard work and contributions to this project. Also I am extremely grateful to staff at CRC Press / Balkema, Taylor and Francis Group, in particular Mr. Alistair Bright, for his excellent work and kind help. I would like to thank Prof. John A. Hudson for his great help in initiating this publication. I would also thank Dr. Yan Guo for her tireless work on this project.

Editor
Xia-Ting Feng
President of the International Society for Rock Mechanics (ISRM) 2011–2015
July 4, 2016

Slopes

Chapter 1

Discontinuity controlled slope failure zoning for a granitoid complex: A fuzzy approach

Z. Gurocak[1], S. Alemdag[2], H.T. Bostanci[3] & C. Gokceoglu[4]
[1]*Department of Geological Engineering, Firat University, Elazig, Turkey*
[2]*Department of Geological Engineering, Gumushane University, Gumushane, Turkey*
[3]*Department of Geomatics Engineering, Gumushane University, Gumushane, Turkey*
[4]*Department of Geological Engineering, Hacettepe University, Ankara, Turkey*

Abstract: Kinematic analysis is one of the most used methods to evaluate potential failures such as planar, wedge and toppling in rock slopes. To analyze possible failure types, orientations of slope and discontinuity geometry with the friction angle of the discontinuities planes are used. In this study, kinematical analyses were evaluated to obtain discontinuity controlled potential failure types in the Gumushane Granitoid Complex which is exposed in Gumushane City and the surroundings. Unstable slope orientations for planar, wedge and toppling failure types were determined. The unstable orientations obtained from kinematic analyses were used to create Geographic Information System (GIS) based instability risk maps. The maps generated by the GIS were evaluated using the Fuzzy Interference System (FIS) method to produce instability risk maps for the study area.

1 INTRODUCTION

Stability analyses in rock slopes are carried out routinely in many engineering studies such as mining, construction and geotechnical projects. The main objectives of these stability analyses can be listed as follows:

- Determination of rock slope stability conditions,
- Researching the potential failure mechanisms,
- Determination of the factors that affect the stability of slopes,
- Carrying out optimum and safe slope designs,
- Determination and testing of different support and enhancement methods.

Various methods such as kinematic analyses, limit equilibrium method and numerical analyses are used today for evaluating rock slope stability. However, the potential failure type, terrain conditions as well as the weak and strong aspects of the preferred method should be taken into account when selecting the analysis method to be used.

In this study, kinematic analyses have been carried out to determine the possible failure types that might occur due to discontinuities in the area where the Gumushane Granitoid Complex is exposed (Figure 1) and to determine the orientations that such failures can occur. The results obtained from kinematic analyses have been evaluated

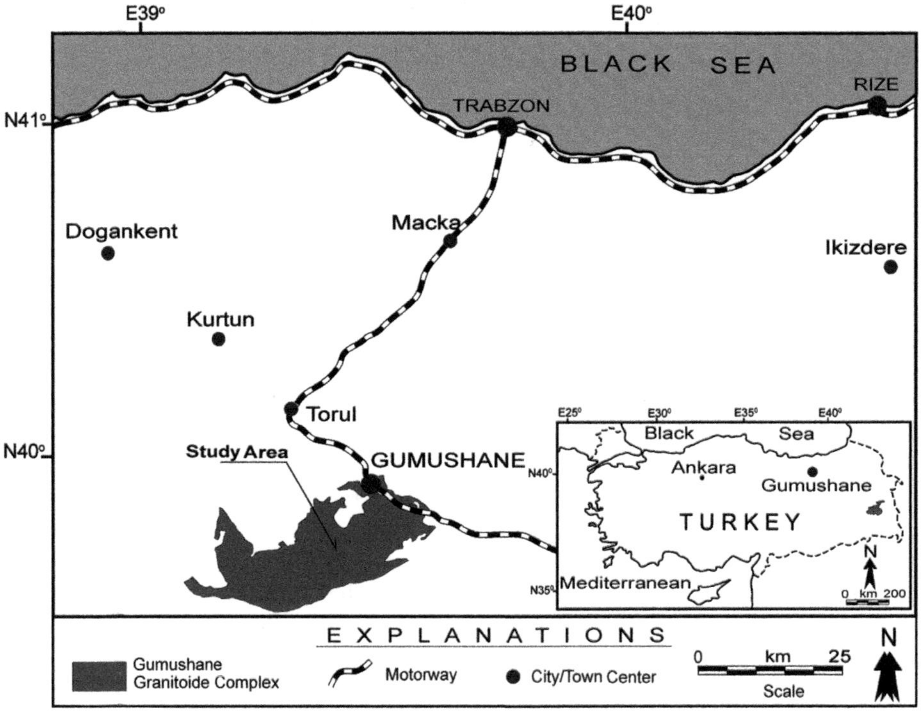

Figure 1 The location map of study area.

via Geographical Information System (GIS) and Fuzzy Inference System (FIS) methods and risk maps have been generated for the study area.

The studies have been carried out in three stages. In the first stage, the main orientations and the shear strengths of joints have been determined. To this end, orientation measurements were obtained from the joints and these joint orientations were evaluated using the Dips v5.1 (Rocscience, 2002) software after which the main orientations of joints were determined. Whereas the shear strength of the joints was determined according to the method suggested by Barton & Choubey (1977).

In the second stage, the results obtained from field and laboratory works were evaluated using Dips v5.1 (Rocscience, 2002) software and kinematic analyses were performed for planar, wedge and toppling failures to determine the orientations of unstable orientations of slope. In the final stage, the results obtained from kinematic analyses were used to generate GIS based instability risk maps and the obtained maps were evaluated according to the FIS method after which instability risk map was generated for the study area.

2 GEOLOGY OF STUDY FIELD

The Pontide Orogenic Belt of Turkey is located in the middle sections of the Alpine-Himalaya Orogenic Belt. The eastern portion of it, which is also known as the

Eastern Pontides, has been divided into Northern Zone and Southern subzones by researchers according to the lithological and structural properties of the exposed units. In the Southern Zone including the study area, metamorphic, magmatic and sedimentary rocks that have formed during time intervals varying from Late Paleozoic to Quaternary crop out in the study area (Figure 2).

The oldest rocks in the region are the Early Carboniferous Pulur and Kurtoglu Metamorphic Complexes aged Early Carboniferous (Topuz *et al.*, 2004, 2007). These metamorphic complexes include many different metamorphic rocks that vary in origin from continental rocks to oceanic rocks (Dokuz *et al.*, 2011, 2015).

Middle Carboniferous Kose-Gumushane Granitoid complexes that settle by cutting the Pulur-Kurtoglu Metamorphic Complex mostly consist of continental crust based granodiorite and granite and less of quartz diorite, quartz monzonite, spherulitic dacite

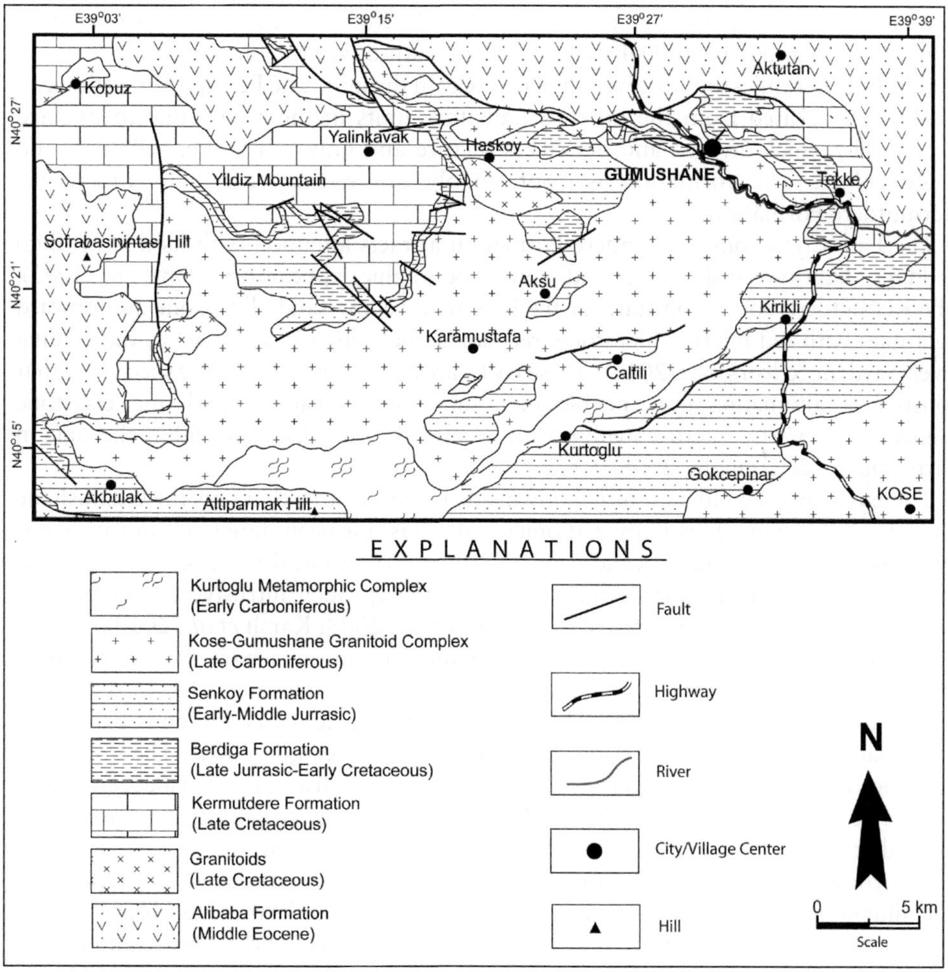

Figure 2 The geological map of study area.

and riolites (Topuz *et al.*, 2010; Dokuz, 2011). Aplitic dikes are also observed in granitic complexes. This study is focused on the Late Carboniferous Gumushane Granitoid Complex, which includes at least 4 or 5 well developed joint sets.

The Early-Middle Jurassic Senkoy Formation unconformably covers the Carboniferous basement rocks. It starts at the base with the basal conglomerate consisting of quartzite and granite pebbles and the carbonous intermediate level sandstones that are on top of them. Red limestones with macro-fossil (Ammonitico Rosso facies) lie on top of these (Kandemir & Yilmaz, 2009). Overlying is a sequence consisting of volcanic-clastic pebble, sandstone, siltstone and marl interbedded with various basic volcanic rocks (tuff, tuffite, andesite, spilitic basalt and pyroclastics) (Dokuz & Tanyolu, 2006). In places, cherty limestones and basalts are also observed within the unit (Dokuz *et al.*, 2006; Sen, 2007).

The Late Jurassic – Early Cretaceous aged Berdiga Formation that creates a key stratigraphic horizon in the Eastern Pontides conformably covers the Senkoy Formation unconformably. The unit begins at the base with medium to thick bedded and partly massive dolomitic limestones (Vörös & Kandemir, 2011). Upward it contains pebble, sandstone and siltstone and ends with dolomitic limestone (Koch *et al.*, 2008). Chert nodules are observed in places within the unit. The Berdiga Formation can easily be separated from the other sedimentary units in the region due to its hard topography, thick layered and massive structure and gray, off-white, yellowish and beige colors. The formation is conformably covered by the Late Cretaceous Kermutdere Formation.

The Late Cretaceous Kermutdere Formation attracts attention with its thin-medium bedding, light greenish color and soft topographic appearance (Saydam, 2002). The Kermutdere Formation starts at the base with yellow sandy limestone and upward continues with red micritic limestone, red sandstone, siltstone and marl as the guiding level (Saydam & Korkmaz, 2011). These rocks are covered by a turbiditic succession that consists of alternating gray pebblestone, sandstone, siltstone, marl and claystone sequences. In parts, micro conglomeratic levels are also observed within the unit in parts (Dokuz & Tanyolu, 2006).

There is also a second a granitic body inside the study area that cuts the Gumushane Granitoid Complex. This body radiometrically has not been dated yet, but its age is given as Late Cretaceous-Eocene based on the stratigraphical position of the similar intrusive bodies to the further north (Guven, 1993). Such intrusive bodies become dominant in the areas to the north (Kaygusuz *et al.*, 2008; Karsli *et al.*, 2010a).

The Middle Eocene is represented by Alibaba Formation, which starts at the base with sandstone and tuffite interbedded with nummulitic limestones. Upward the unit continues with andesite, basalt and associated pyroclastic rocks and ends with limestone, sandstone, marl tuff alternation (Arslan & Aliyazicioglu, 2001). The formation non-conformably covers the Kermutdere Formation as well as the Late Paleocene-Early Eocene adakitic stocks (Karsli *et al.*, 2010b, Dokuz *et al.*, 2013). Younger rocks are located at the northern zone of the Pontides (Aydin *et al.*, 2008).

The youngest units of the region are composed of Quaternary aged hill slope wash and alluviums. Slope wash has sprung forth from the older units that have been exposed and consists of elements with different dimensions and is exposed especially on the hills. Whereas alluvium consists of well-rounded elements at river beds with clay, sand, pebble and block dimension.

3 ORIENTATION AND SHEAR STRENGTH OF DISCONTINUITIES

Orientation of discontinuities and shear strength are two parameters that play important roles in the development of instabilities that develop due to discontinuity surfaces. These two parameters are used as input parameters in rock slope analyses such as kinematic analysis, limit equilibrium analysis. Hence, the main orientations of the discontinuities in the rock as well as shear strengths have to be determined.

3.1 Orientation

Discontinuity orientation is expressed as the position in space of a discontinuity surface and is defined by the dip and dip direction of the surface. The measurements taken directly from the rock mass using geologist compass is evaluated via the contour diagrams prepared to test this method using stereographic projection method and the main orientations of the discontinuity sets contained within the rock mass are determined. Generally, stereonets with lower hemisphere and equal-angle are used.

In this study, a total of 1797 orientation measurement taken from the joints of Gumushane Granitoide Complex were evaluated in accordance with the methods suggested by ISRM (2007) using Dips v5.1 software (Rocscience, 2002) and main orientations of the joint sets were determined (Figure 3, Table 1). As can be seen

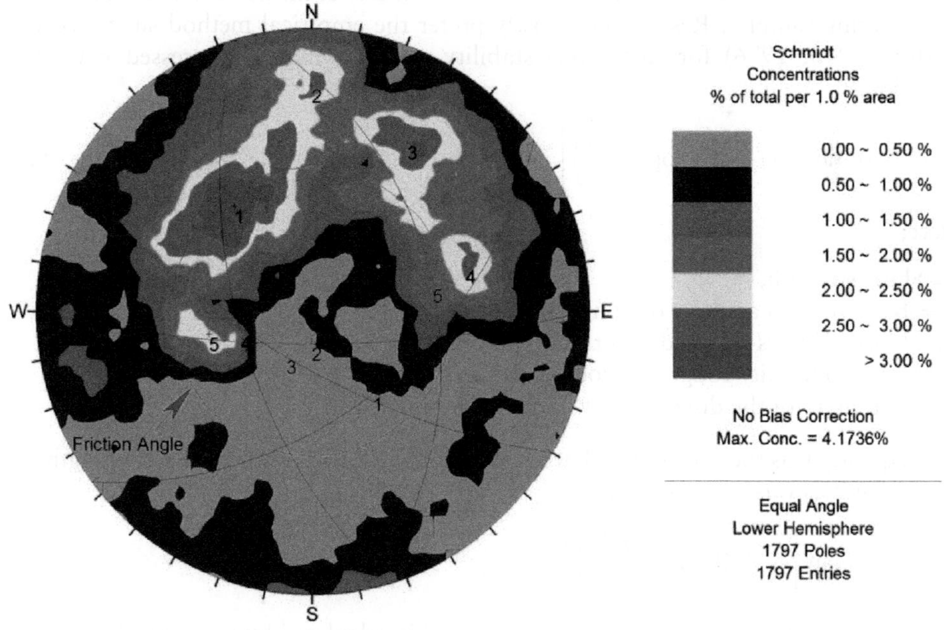

Figure 3 Stereographic projections of joint sets.

Table 1 The major orientations of joint sets.

Type of discontinuity	*Dip direction / Dip*
Joint-1	143/50
Joint-2	180/77
Joint-3	210/69
Joint-4	254/60
Joint-5	78/42

from Figure 3, Gumushane Granitoide Complex includes 5 joint sets in general and has joints with random orientation.

3.2 Shear strength

The shear strength of the discontinuity planes in rock masses as well as the stability are among the most important factors that are controlled and the attributes that control this parameter are roughness, undulation, weathering degree and the normal stress applied to the discontinuity plane. This parameter is determined in the laboratory via the deformation controlled direct shearing experiment suggested by ISRM (2007) and CANMET (1977b) or the empirical failure criterion suggested by Barton (1973, 1976) (Equation 1). However, deformation controlled direct shearing experiment is not preferred by researchers due to the crushing of the filler under high normal stresses that occur during the determination of the shear strengths of surfaces with high roughness and undulated discontinuity, the failure to provide a standard loading rate, the failure to read the shear movement based on the property of the device and the difficulty of preparing samples. Researchers mostly prefer the empirical method suggested by Barton (1973, 1976) for rock slope stability studies which is expressed with the equation:

$$\tau = \sigma_n \tan\left[\phi_b + \mathrm{JRC}\log\left(\frac{\mathrm{JCS}}{\sigma_n}\right)\right] \tag{1}$$

Here;

τ = Shear strength,
ϕ_b = Basic friction angle of discontinuity,
σ_n = Normal stress on the discontinuity plane,
JRC = Discontinuity roughness coefficient, and
JCS = Strength of the discontinuity surface.

This equality was then revised by Barton & Choubey (1977) and the shear strength of discontinuities was expressed as;

$$\tau = \sigma_n \tan\left[\phi_r + \mathrm{JRC}\log\left(\frac{\mathrm{JCS}}{\sigma_n}\right)\right] \tag{2}$$

ϕ_r in Equation 2 is the residual friction angle and is calculated by Equation 3.

Table 2 Used parameters for determining of internal friction angle of discontinuities.

Parameter	ϕ_b (°)	ϕ_r (°)	*JCS* (*MPa*)	*JRC*	*r*	R	γ (kN/m^3)	h (m)	ϕ (°)	ϕ_b (°)
Min	25.00	19.21	33.00	6.22	27	38	25.20	–	29.87	25.00
Mean	29.00	23.00	34.00	8.00	28	40	25.50	15	36.43	29.00
Max	32.00	29.04	48.00	9.15	46	54	26.00	–	45.17	32.00
Std. dev.	2.81	4.01	6.35	1.34	5	4.6	0.400	–	4.38	2.81

$$\phi_r = (\phi_b - 20) + 20(^r/_R) \tag{3}$$

Where,

ϕ_b = Basic friction angle of discontinuity,
R = Schmidt rebound hammer hardness value of the fresh surface, and
r = Schmidt rebound hammer hardness value of the weathered surface.

In this study, Equation 2 suggested by Barton & Choubey (1977) has been used to determine the shear strength of the discontinuities in Gumushane Granitoid Complex. Nine block samples collected from the Gumushane Granitoid Complex were cut in the dimensions of (6 × 6 × 1cm) in order to determine the basic friction angle (ϕ_b) which were then smoothed out and three shear experiments were carried out for each sample set under different normal strain. The basic friction angle (ϕ_b) has been determined using the Normal stress (σ) – Shear stress (τ) graphs drawn using the data of shear experiments.

Discontinuity roughness coefficient (JRC) has been determined as a result of the comparison of the roughness profiles obtained during field studies via Barton's comb with the standard roughness profiles suggested by Barton & Choubey (1977).

Whereas, the strength of the discontinuity surface was determined using the Schmidt hardness values obtained from discontinuity planes and unit weight of the rock with the chart suggested by Deere & Miller (1966).

The unit weight of the rock was determined in accordance with ISRM (2007) standards at the laboratory using shaped samples. Another parameter used in the calculation is slope height (H). Slope height is an important parameter of slope design and varies according to the engineering work that will be carried out. However, the H value in this study has been accepted as 15 m by assuming that the maximum slope height that can be applied in practice is about 15 m.

The values of parameters obtained from field and laboratory studies were used as input parameter for the RocData computer software (Rocscience, 2004) and the internal friction angle of the discontinuities was calculated on average as 36.43°. The parameters used in the calculation can be seen in Table 2.

4 KINEMATICAL ANALYSES

When the failures in the rock slopes are considered, it is observed that the failures occur mostly due to structural elements in the rock mass that are defined as discontinuities such as joint, layering, fault, shear fractures. Kinematic analysis method is widely preferred to analyze these instabilities controlled by discontinuities. This method was first defined by Hoek & Bray (1981), developed by Goodman (1989) and revised by Wyllie & Mah (2004) and it is used to evaluate stability analyses for failures such as planar, wedge and toppling by taking into consideration the discontinuity orientations, slope direction and the internal friction angles of discontinuity surfaces.

Kinematic analysis method has certain limitations as well even though it is the most preferred method and provides many advantages. The main advantages of kinematic analysis method are that the parameters used in the analysis are easy to determine, it provides an idea about the failure potential, it is related with limit equilibrium analyses and it can be used together with statistical analyses. Due to its advantages, several researchers have used the kinematic analysis method to evaluate potential rock slopes failures (Gokceoglu *et al.*, 2000; Yoon *et al.*, 2002; Kentli & Topal, 2004; Gurocak *et al.*, 2008; Yilmaz *et al.*, 2012; Tudes *et al.*, 2012; Alemdag *et al.*, 2014).

However, this method also has significant limitations in addition to its advantages. The main limitations of this method is that it is only suited for pre-design, that it is obligatory to determine critical discontinuities accurately and that the cohesion of the failure surface, external loads, pore water pressure, weight of the sliding mass and dynamic loads are not taken into account.

Planar failure

According to Hoek & Bray (1981) and Norrish & Wyllie (1996) the following geometrical conditions must be provided for planar failure on a single plane to occur (Figure 4):

- The dip direction of discontinuity (α_p) must be within ±20 degrees of the dip direction of slope face (α_f),

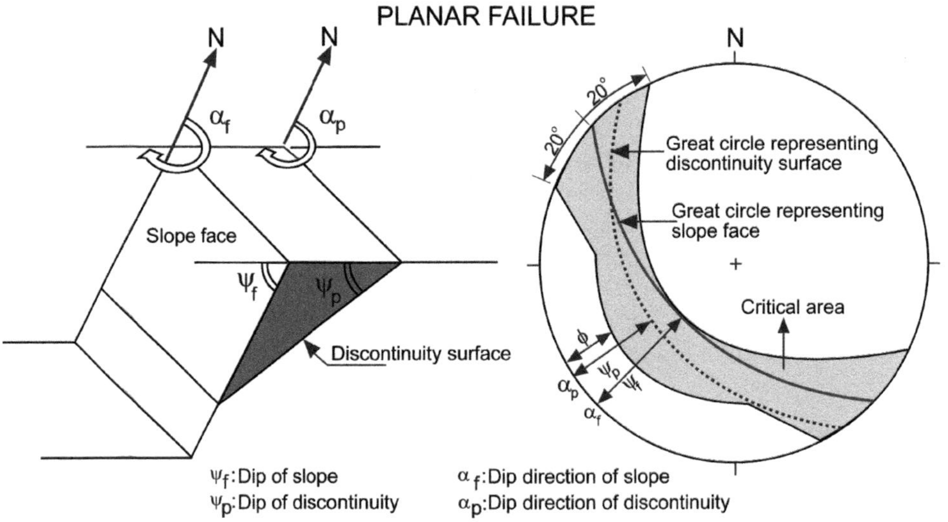

Figure 4 Kinematic and geometric conditions for planar failure (Norrish & Wyllie, 1996).

- The dip of discontinuity (Ψ_p) must be less than the dip of slope face (Ψ_f), and
- The dip of discontinuity (Ψ_p) must be greater than the friction angle of failure plane ϕ.

Also, the lateral extent of the potential failure mass must be isolated by lateral release surfaces which free a block for sliding. This is the requirement that reduces the likelihood of planar failure occurrence.

Wedge failure

Wedge failures result when a rock mass slides along two intersecting discontinuities both of which dip out of the cut slope at an oblique angle to the cut face, forming a wedge-shaped block. According to Hoek & Bray (1981) and Norrish & Wyllie (1996), three conditions are required for wedge failures to occur, as shown in Figure 5:

- The azimuth of line of intersection (α_i) must be similar to the dip direction of slope face (α_f),
- The plunge of line of intersection (Ψ_i) must be less than the dip of slope face (Ψ_f),
- The plunge of line of intersection (Ψ_i) must be greater than the friction angle of failure plane (ϕ).

On the stereographic projection, the point of intersection of the two great circles representing plane A and plane B must plot within the critical area, which is called the daylight zone, and lies on the convex side of the cut slope.

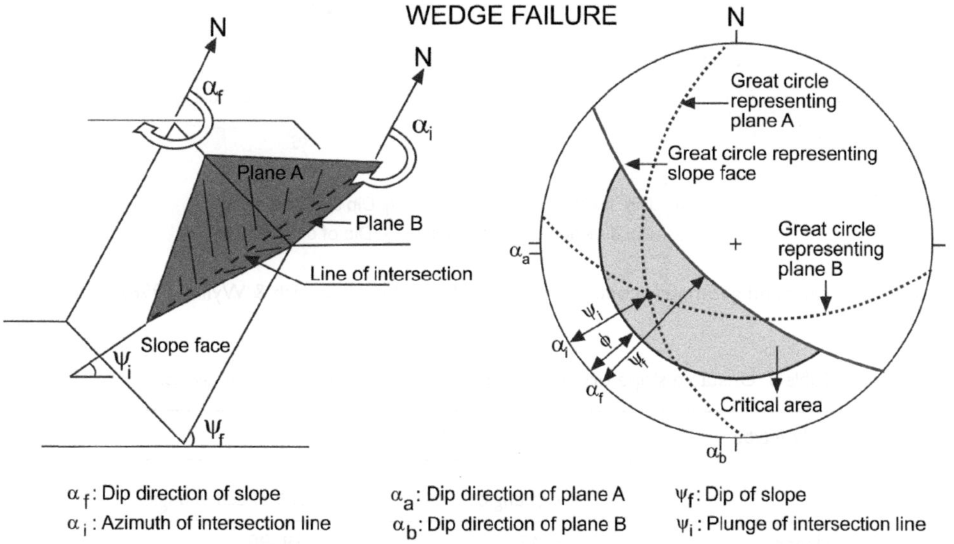

Figure 5 Kinematic and geometric conditions for wedge failure (Norrish & Wyllie, 1996).

Toppling failure

According to Goodman & Bray (1976) and Goodman (1980), the necessary conditions for toppling failure are as follows (Figure 6):

- The difference between the strikes of discontinuity and slope face must be + 30° or −30°,
- The pole of the discontinuity must plot within the critical area, and
- The following equation must be provided;

$$(90 - \Psi_p) \leq \Psi_f + \phi \tag{4}$$

In this study, kinematic analyses were carried out to determine the slope orientations for which planar, wedge and toppling type failures might occur due to the joints of Gumushane Granitoid Complex and the slope orientations for which these failures might occur have been given in Table 3.

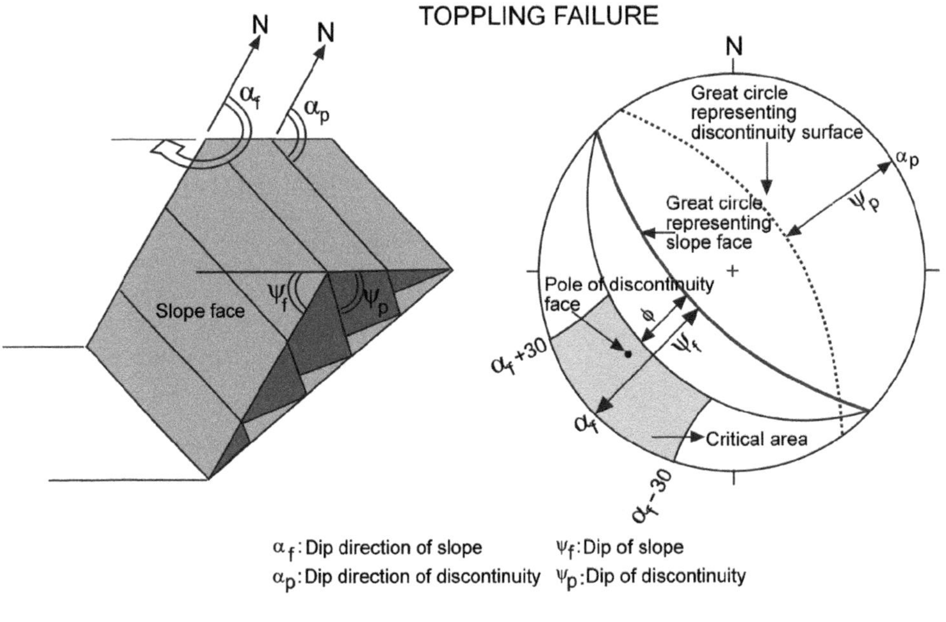

Figure 6 Kinematic and geometric conditions for toppling failure (Norrish & Wyllie, 1996).

Table 3 Unstable slope orientations obtained from kinematical analyses.

Failure type	*Slope orientation*	
	Dip angle	*Dip direction angle*
Planar	> 42	58–98 123–274
Wedge	> 39.5	11–344
Toppling	> 49	104–228

5 MAPPING BASED ON GIS FOR DISCONTINUITY CONTROLLED SLOPE FAILURE

The requirement to manage large amounts of spatial data effectively has made it obligatory to get computer support and thus Geographical Information Systems (GIS) have come to the forefront. Even though there are many different definitions according to different perspectives (Maguire, 1991) the GIS can be defined with a general approach as "the total of software and hardware elements designed for the management of a spatial database" (Lee & Zhang, 1989). The GIS was first developed during the beginning of the 1960s which started to gain popularity after the 1980s following the advancements in computer technology (Malczewski, 2004). Today, it is used as an effective tool in many fields from engineering and planning activities to tourism and marketing.

Using the extensive analysis options provided by the GIS, it is possible to make rapid and practical deductions for much wider areas with the same or similar properties based on the data obtained from certain locations. While this opportunity is generally used in geology engineering discipline to prepare landslide susceptibility and risk maps (Carrara *et al.*, 1991, 1995; Gokceoglu & Aksoy, 1996; Gokceoglu *et al.*, 2000; Irigaray *et al.*, 2003; Guzzetti *et al.*, 2004; Ayalew & Yamagishi, 2004; Yilmaz, 2007; Akgun & Bulut, 2007; Pradhan, 2010; Gao *et al.*, 2011; Yilmaz *et al.*, 2012), in this study it was used to generate rock slope instability maps.

The objective of this study was to generate instability maps for toppling, planar and wedge type failures in the area at which the Gumushane Granitoid Complex spreads by taking the limit values determined via kinematic analysis method as basis. Since the input for instability maps are slope and aspect maps, first the digital elevation model (DEM) of the study area was generated using 1/25000 scale digital topographic maps (Figure 7), after which slope (Figure 8) and aspect (Figure 9) maps were generated from this DEM.

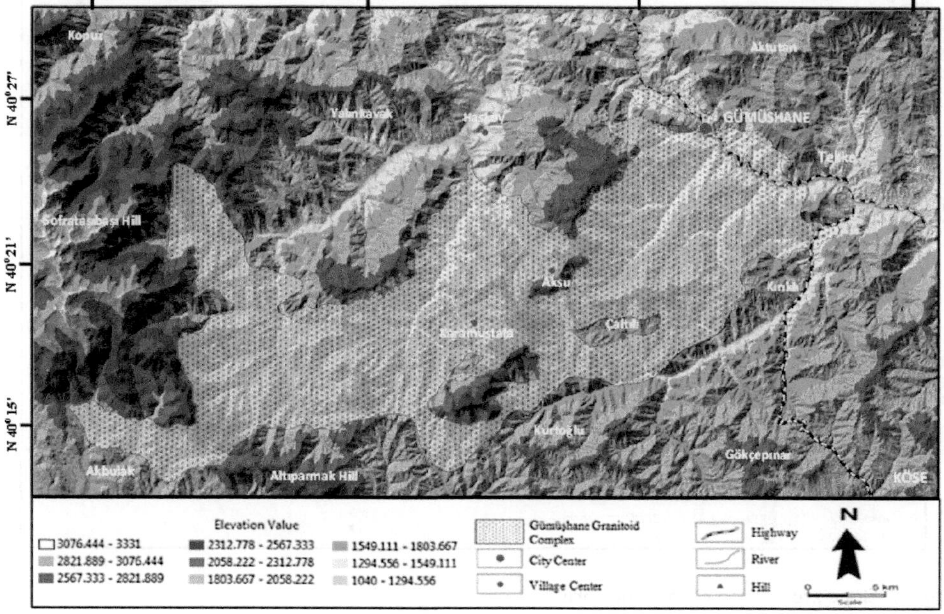

Figure 7 Digital Elevation Model (DEM) of the study area.

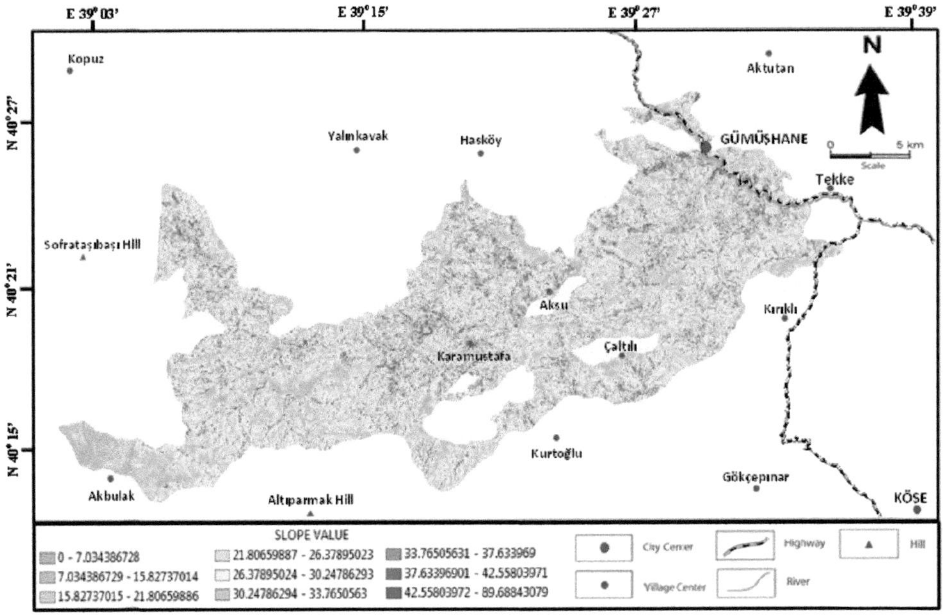

Figure 8 Slope map of the study area.

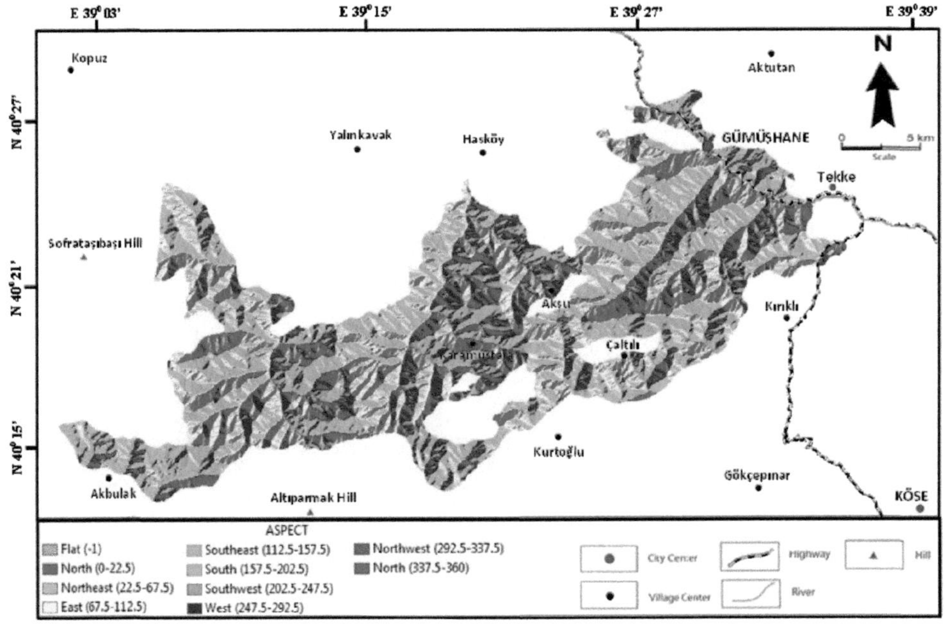

Figure 9 Slope aspect map of the study area.

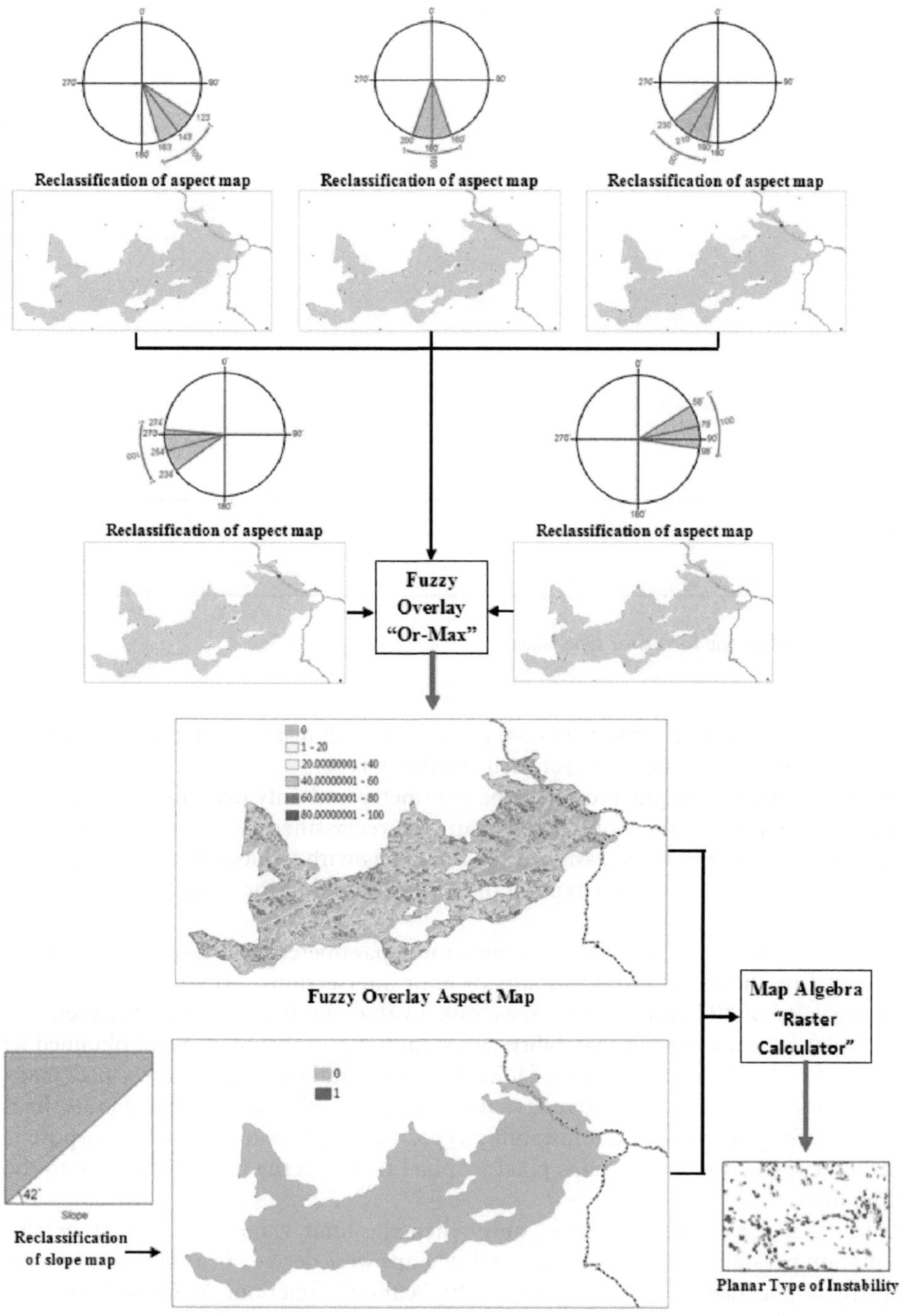

Figure 10 Production procedure of the planar-type instability map.

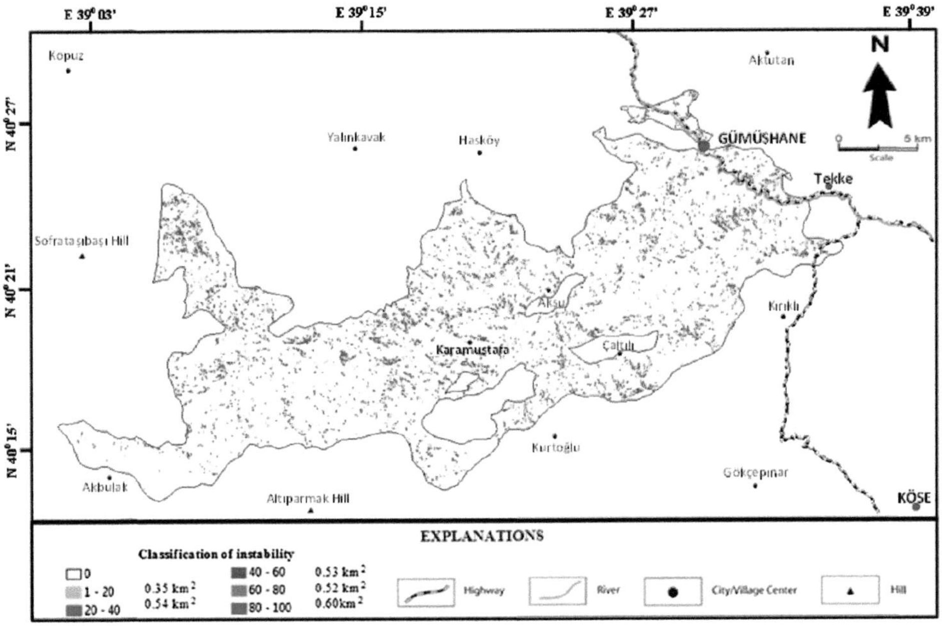

Figure 11 Instability map for planar type failure.

However, since different limit values are present for each instability type, the slope and aspect maps were reclassification for each instability type.

No weighting was applied to the slope parameter and only instability was considered. For this reason, the slope index map was reclassified for different instability types; on the map, the value assigned for the pixels with values lower than the limit value of the relevant instability type (stable) is "0" and that for other pixels (instable) is "1".

Since weighting was applied for orientation parameters, the aspect maps were reclassified on the basis of sets of joints or joint intersections so that the stable areas took the value of "0" and instable areas took a value that linearly varies between "1" and "100" for each type of instability. Since multiple aspect maps were obtained as a result of reclassifications by sets of joint for each instability type, these aspect maps were overlaid through fuzzy overlay analysis and "or" method to generate final weighted aspect maps for each instability type. The analysis used for this purpose is based on comparison of the maps pixel by pixel to assign the maximum value to the output map.

Given that the instability occurs only when the limit values for both slope and orientation parameters concur, the instability maps were derived from slope and aspect maps by application of "map algebra multiplication" function. As a result of these functions the instability maps weighted for orientation parameter that indicate instability areas according to slope and orientation parameters were obtained. The procedure followed for generation of the instability maps as well as the instability maps obtained are given in the Figures 10–15.

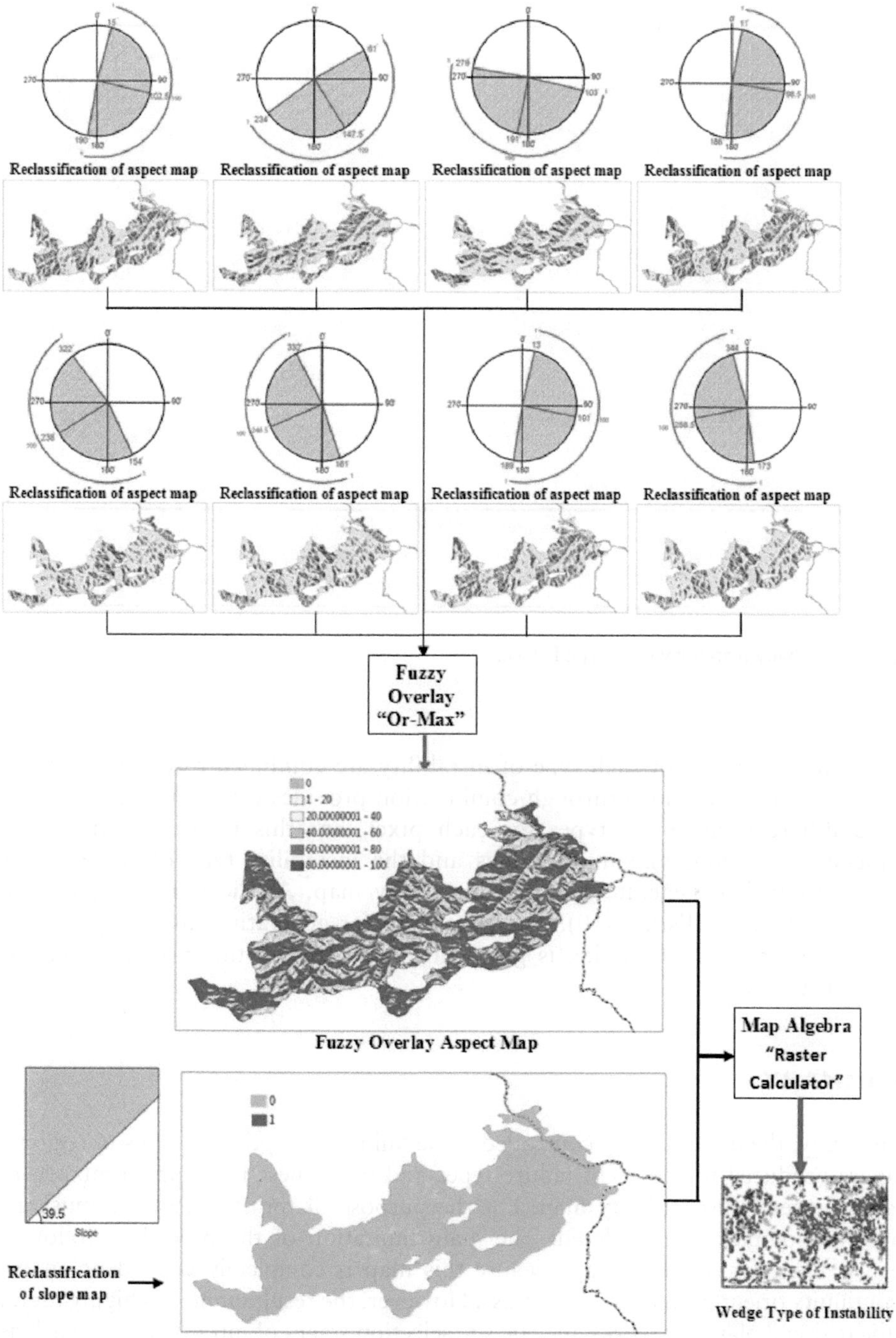

Figure 12 Production procedure of the wedge-type instability map.

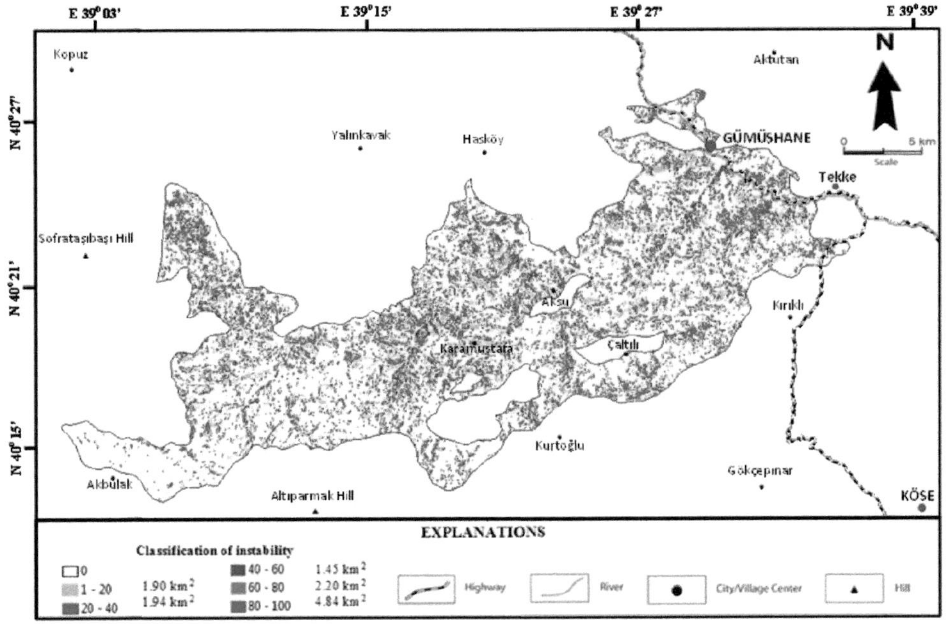

Figure 13 Instability map for wedge type failure.

The maps obtained for each type of instability are combined through combined analysis. The map obtained through combination provides different values derived from 3 different instability types for each pixel. For this reason, by means of membership function, maximum values and the instability type of these values were determined to generate the final instability map. Applications were carried out in ArcGIS 10.0 (Esri, 2010). The procedure for generating instability map of Gumushane Granitoid Complex is given in Figure 16 and the instability map is given in Figure 17.

6 SUMMARY

In this study, the discontinuity controlled slope failure susceptibility maps in regional scale are produced for different failure types and then, these maps are combined to obtain a general susceptibility map. For the purpose of the study, the Gumushane Granitoid Complex is considered. The main limitation of the procedure followed herein is valid for one lithology because this map is completely dependent on the discontinuity orientation in the rock mass. However, the resultant map is highly useful for the regional planning works and the site selection stages of infra structures such as tunnels, roads, railroads, deep slope cuts.

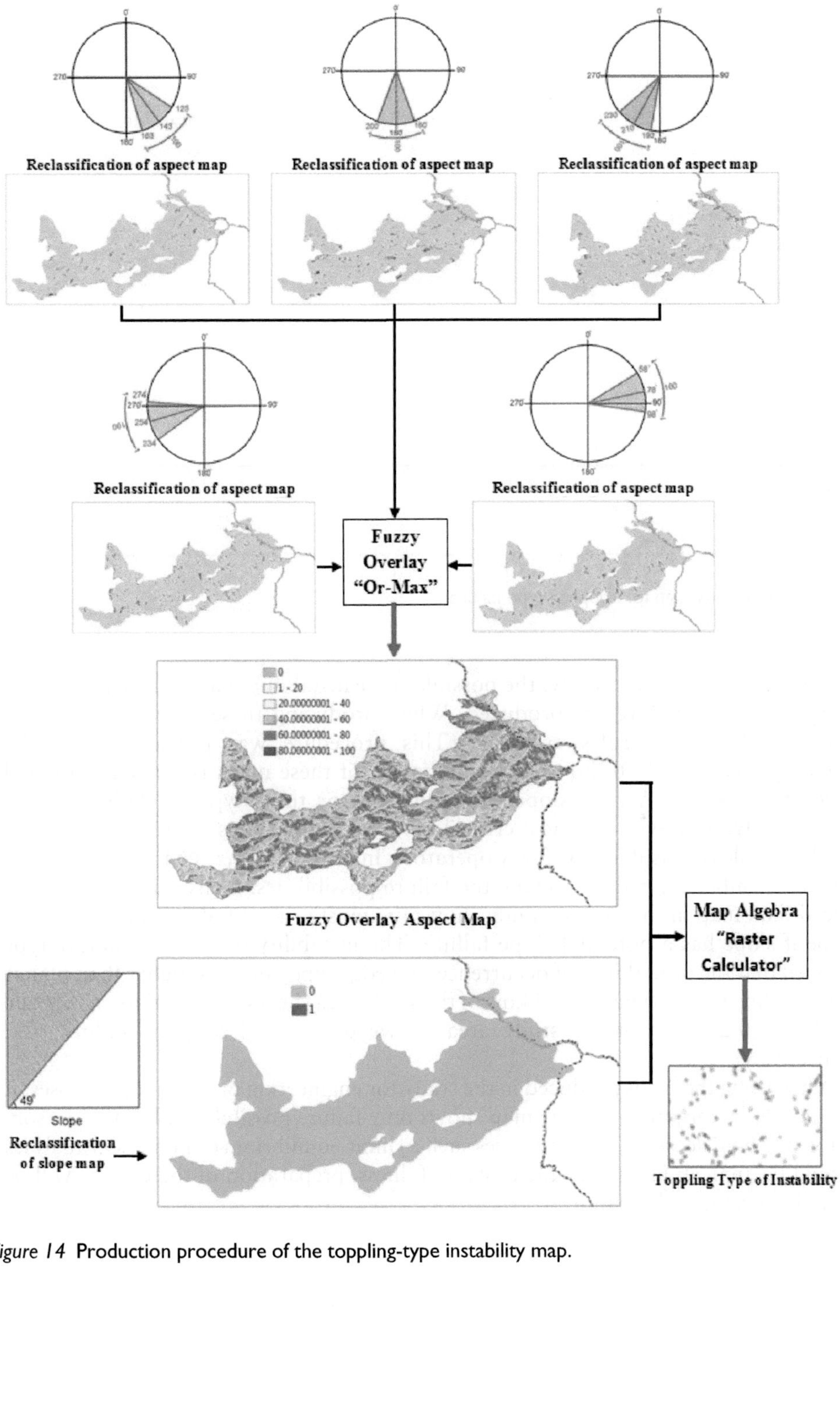

Figure 14 Production procedure of the toppling-type instability map.

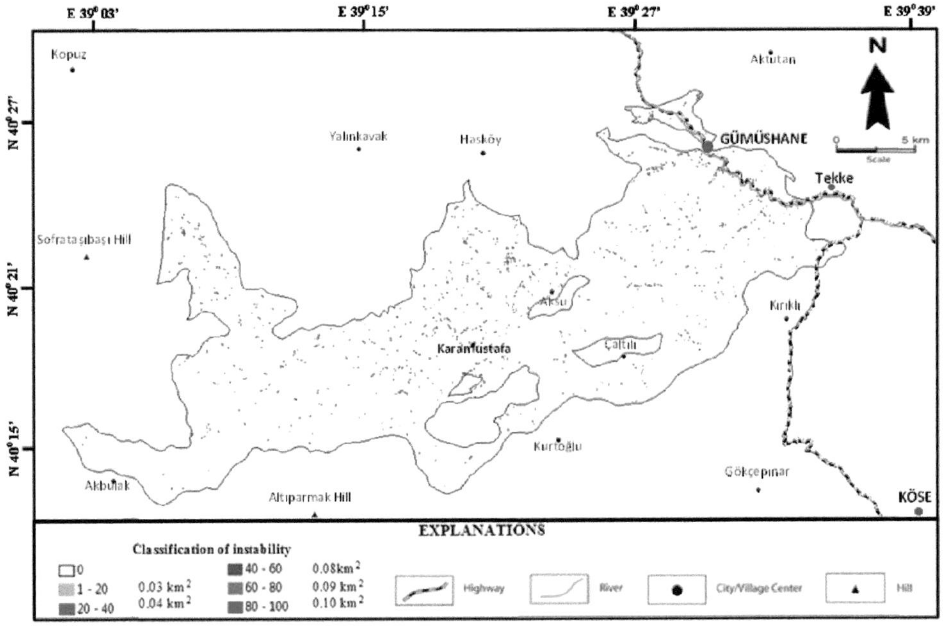

Figure 15 Instability map for toppling type failure.

In the first stage of the study, the possible slope instability maps for each discontinuity controlled failure are produced. When producing these maps, the classical kinematical analysis rules are used. This procedure was previously used by Gokceoglu *et al.* (2000), however combination of these maps is the main original part of this study. A rock slope can show two or three types of failure modes theoretically. However, the most critical mode should be considered. In this study, this basic rule is considered by fuzzy operators. In other words, each failure mode has a fuzzy membership value showing the failure possibilities. Consequently, the produced final map in the present study shows the most critical failure modes for each slope if slope has a potential slope failure. The instability map for all failure type showed that the probability of occurrence of wedge type failure is higher than planar and toppling failure types. 11.84 km^2 of the study area shows wedge failure risk while 0.42 and 0.23 km^2 of the study area shows planar and toppling failure risk, respectively.

When using the maps produced in the study for engineering or planning purposes, it should not be forgotten that the map shows only failure possibilities. For this reason, before the final decision, extra analyses such as limit-equilibrium or numerical analyses should be applied. However, due to its usefulness, preparation of these maps will be more important in the future.

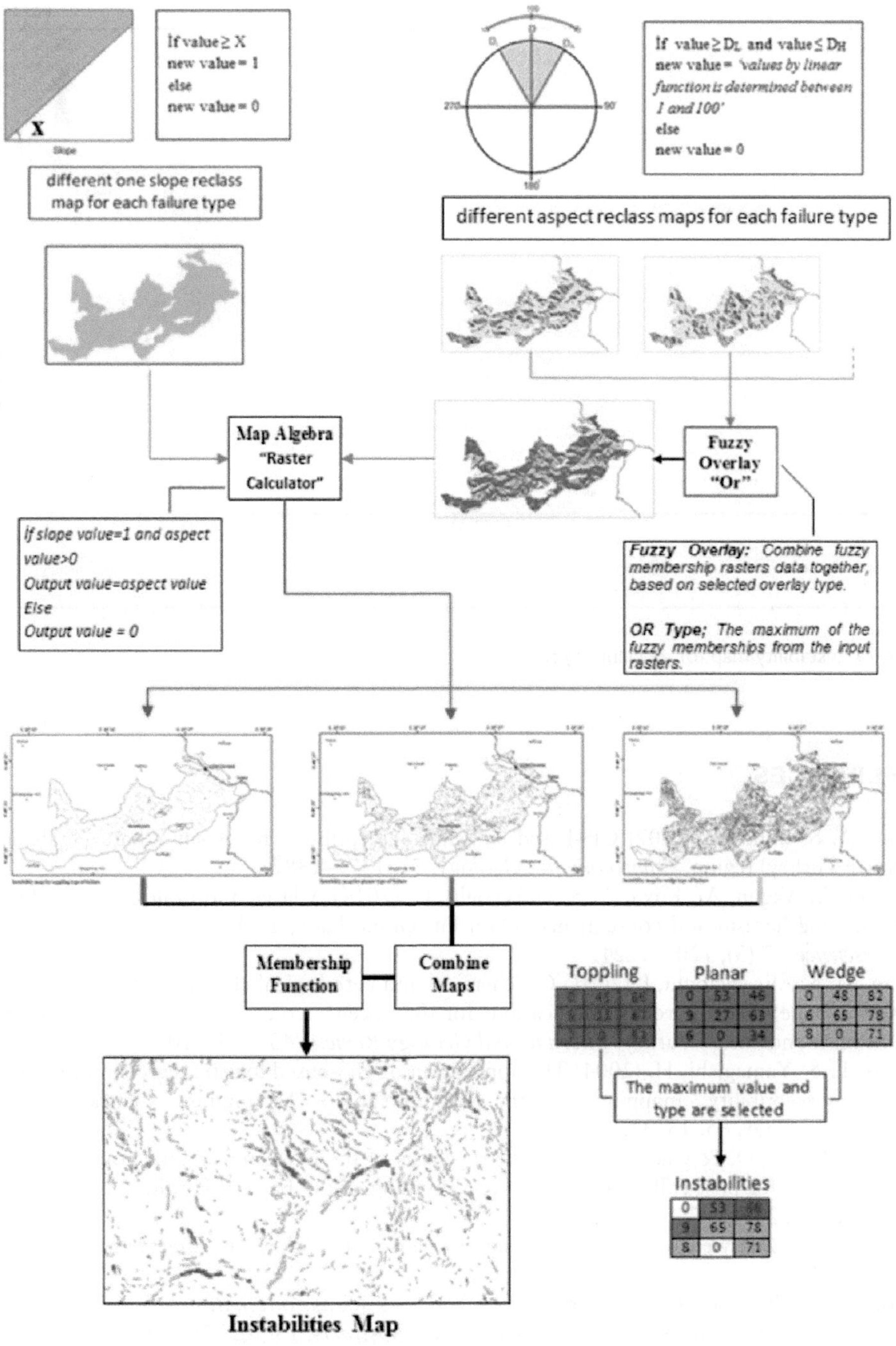

If value ≥ X
new value = 1
else
new value = 0
X
different one slope reclass map for each failure type
If value ≥ DL and value ≤ DH
new value = 'values by linear function is determined between 1 and 100'
else
new value = 0
different aspect reclass maps for each failure type
Map Algebra
"Raster Calculator"
Fuzzy Overlay
"Or"
If slope value=1 and aspect value>0
Output value=aspect value
Else
Output value = 0
Fuzzy Overlay: Combine fuzzy membership rasters data together, based on selected overlay type.
OR Type; The maximum of the fuzzy memberships from the input rasters.
Membership Function
Combine Maps
Toppling
Planar
Wedge
The maximum value and type are selected
Instabilities
Instabilities Map

Figure 16 Production procedure of the instabilities map.

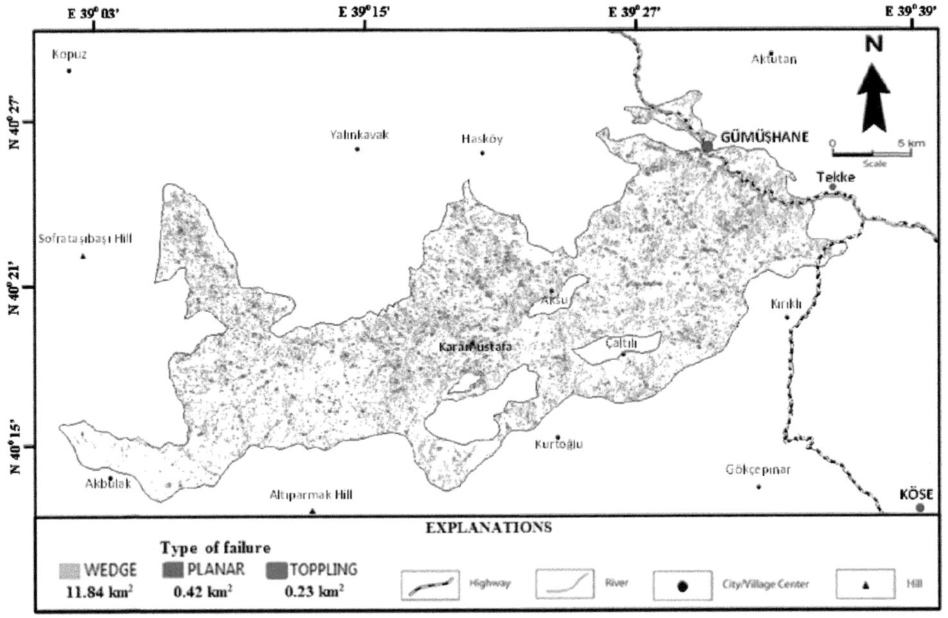

Figure 17 Instability map for all failure type.

REFERENCES

Akgun, A. & Bulut, F. (2007) GIS-based landslide susceptibility for Arsin-Yomra (Trabzon, North Turkey) region. *Environmental Geology*, 51, 1377–1387.

Alemdag, S., Akgun, A., Kaya, A. & Gokceoglu, C. (2014) A large and rapid planar failure: causes, mechanism and consequences (Mordut, Gumushane, Turkey). *Arabian Journal of Geosciences*, 7 (3), 1205–1221.

Arslan, M. & Aliyazicioglu, I. (2001) Geochemical and petrological characteristics of the Kale (Gümüşhane) volcanic rocks: Implications for the Eocene evolution of eastern Pontide arc volcanism, northeast Turkey. *International Geology Review*, 43, 595–610.

Ayalew, L. & Yamagishi, H. (2004) The application of GIS-based logistic regression for landslide susceptibility mapping in the Kakud-Yahiko Mountains, Central Japan. *Geomorphology*, 65, 15–31.

Aydin, F., Karsli, O. & Chen, B. (2008) Petrogenesis of the Neogene alkaline volcanics with implications for post-collisional lithospheric thinning of the Eastern Pontides, NE Turkey. *Lithos*, 104, 249–266.

Barton, N. (1973) Review of a new shear strength criterion for rock joints. *Engineering Geology*, 7, 287–332.

Barton, N. (1976) The shear strength of rock and rock joints. *International Journal of Rock Mechanics and Mining Sciences & Geomechanics Abstracts*, 13 (9), 255–279.

Barton, N. & Choubey, V. (1977) The shear strength of rock joints in theory and practice. *Rock Mechanics*, 1 (2), 1–54.

CANMET (1977) *Pit Slope Manual: Supplement 5–1, Plane Shear Analysis*. Canmet Report 77–16.

Carrara, A., Cardinalli, M., Detti, R., Guzetti, F., Pasqui, V. & Reichenbach, P. (1991) GIS techniques and statistical models in evaluating landslide hazard. *Earth Surface Process Landforms*, 16 (5), 427–445.

Carrara, A., Cardinali, M., Guzzetti, F. & Reichenbach, P. (1995) GIS-based techniques for mapping landslide hazard. In: Carrara, A. and Guzzetti, F. (eds.), *Geographical Information Systems in Assessing Natural Hazards*. Kluwer Publications, Dordrecht, The Netherlands, pp. 135–176.

Deere, D.U. & Miller, R.P. (1966) *Engineering Classification and Index Properties for Intact Rock*. Air Force Weapons Laboratory Technical Report, AFWL-TR, 65–116.

Dokuz, A. & Tanyolu, E. (2006) Geochemical constraints on the provenance, mineral sorting and subaerial weathering of Lower Jurassic and Upper Cretaceous clastic rocks from the Eastern Pontides, Yusufeli (Artvin), NE Turkey. *Turkish Journal of Earth Sciences*, 15, 181–209.

Dokuz, A., Tanyolu, E. & Genc, S. (2006) A mantle- and a lower crustderived bimodal suite in the Yusufeli (Artvin) area, NE Turkey: trace element and REE evidence for subduction-related rift origin of Early Jurassic Demirkent intrusive complex. *International Journal of Earth Sciences*, 95, 370–394.

Dokuz, A. (2011) A slab detachment and delamination model for the generation of Carboniferous high-potassium I-type magmatism in the Eastern Pontides, NE Turkey: The Köse composite pluton. *Gondwana Research*, 19, 926–944.

Dokuz, A., Uysal, I., Kaliwoda, M., Karslı, O., Ottley, C.J. & Kandemir, R. (2011) Early abyssal- and late SSZ-type vestiges of the Rheic oceanic mantle in the Variscan basement of the Sakarya Zone, NE Turkey: Implications for the sense of subduction and opening of the Paleotethys. *Lithos*, 127, 176–191.

Dokuz, A., Uysal, İ., Meisel, W., Turan, M., Duncan, R. & Akçay, M. (2013) Post-collisional adakitic volcanism in the eastern part of the Sakarya Zone, Turkey: evidence for slab and crustal melting. *Contributions to Mineralogy and Petrology*, 166, 1443–1468.

Dokuz, A., Uysal, I., Dilek, Y., Karsli, O., Meisel, T. & Kandemir, R. (2015) Geochemistry, Re-Os isotopes and highly siderophile element abundances in the Eastern Pontide peridotites (NE Turkey): Multiple episodes of melt extraction-depletion, melt-rock interaction and fertilization of the Rheic Ocean mantle. *Gondwana Research*, 27 (2), 612–628.

ESRI. (2010) ArcGIS 10: www.esri.com

Gao, H. X. & Yin, K. L. (2011) GIS-based spatial prediction of landslide hazard risk. *Journal of Natural Disasters*, 20 (1), 31–36.

Goodman, R. E. & Bray, J. W. (1976) Toppling of rock slopes. *Proceedings of the Specialty Conference Rock Engineering for Foundations and Slopes*. Boulder, CO: American Society of Civil Engineers, ASCE Press, Reston, VA. Vol. 2, pp. 201–234.

Goodman, R.E. (1980) *Introduction to Rock Mechanics*. Wiley and Sons, New York. https://books.google.com.tr/books?id=KKFU8WYUemkC&dq=Introduction%20to%20Rock%20MechanicsC&source=gbs_similarbooks [Accessed 07 July 2015].

Goodman, R.E. (1989) *Introduction to Rock Mechanics*. 2nd edition. Wiley and Sons, New York. https://books.google.com.tr/books?isbn=8126525665 [Accessed 07 July 2015].

Gokceoglu, C. & Aksoy, H. (1996) Landslide susceptibility mapping of the slopes in the residual soils of the Mengen region (Turkey) by deterministic stability analyses and image processing techniques. *Eng Geol*, 44, 147–161.

Gokceoglu, C., Sonmez, H. & Ercanoglu, M. (2000) Discontinuity controlled probabilistic slope failure risk maps of the Altındağ (settlement) region in Turkey. *Engineering Geology*, 55, 277–296.

Gurocak, Z., Alemdag, S. & Zaman, M. (2008) Rock Slope Stability and Excavatability Assessment of Rocks at the Kapikaya Dam Site, Eastern Turkey. *Engineering Geology*, 96 (1–2), 17–27.

Guzzetti, F., Reichenbach, P. & Ghigi, S. (2004) Rockfall hazard and risk assessment along a transportation corridor in the Nera Valley, Central Italy. *Environmental Management*, 34, 191–208.

Guven, İ.H. (1993) General Directorate of Mineral Research and Exploration (MTA) of Turkey. *Geology and Compilation of the Eastern Pontide*, 1:250000.

Hoek, E. & Bray, J. W. (1981) *Rock Slope Engineering*. Third edition. The Institution of Mining and Metallurgy, London. Available from: http://www.beknowledge.com/wpcontent/uploads/2010/10/c4ca4Rock_Slope_Engineering_Civil_and_Mining.pdf.

Irigaray, C., Fernández, T. & Chacón, J. (2003) Preliminary rock-slope-susceptibility assessment using GIS and the SMR classification. *Natural Hazards*, 30, 309–324.

ISRM (2007) *The Complete ISRM Suggested Methods for Rock Characterization, Testing and Monitoring: 1974–2006*. In: Ulusay, R. and Hudson, J.A. (eds.), Suggested methods prepared by the commission on testing methods, International Society for Rock Mechanics, compilation arranged by the ISRM Turkish National Group, Kozan Ofset, Ankara, Turkey.

Kandemir, R. & Yilmaz, C. (2009) Lithostratigraphy, facies, and deposition environment of the Lower Jurassic Ammonitico Rosso type sediments (ARTS) in the Gümüşhane area, NE Turkey: Implications for the opening of the northern branch of the Neo-Tethys ocean. *Journal of Asian Earth Sciences*, 34, 586–598.

Karsli, O., Dokuz, A., Uysal, I., Aydin, F., Kandemir, R. & Wijbrans, J. (2010a) Relative contributions of crust and mantle to generation of Campanian high-K calc-alkaline I-type granitoids in a subduction setting, with special reference to the Harsit Pluton, Eastern Turkey. *Contributions to Mineralogy and Petrology*, 160, 467–487.

Karsli, O., Dokuz, A., Uysal, I., Aydın, F., Kandemir, R. & Wijbrans, J. (2010b) Generation of the Early Cenozoic adakitic volcanism by partial melting of mafic lower crust, Eastern Turkey: implications for crustal thickening to delamination. *Lithos*, 114, 109–120.

Kaygusuz, A., Siebel, W., Sen, C. & Satir, M. (2008) Petrochemistry and petrology of I-type granitoids in an arc setting: the composite Torul pluton, Eastern Pontides, NE Turkey. *International Journal of Earth Sciences*, 97, 739–764.

Kentli, B. & Topal, T. (2004) Assesment of rock slope stability for a segment of the Ankara-Pozanti motorway, Turkey. *Engineering Geology*, 74 (1–2), 73–90.

Koch, R., Bucur, I.I., Kirmaci, M.Z., Eren, M. & Tasli, K. (2008) Upper Jurassic and Lower Cretaceous carbonate rocks of the Berdiga Limestone-Sedimentation on an onbound platform with volcanic and episodic siliciclastic influx. Biostratigraphy, facies and diagenesis (Kircaova, Kale-Gümüşhane area; NE-Turkey). *Neues Jahrbuch für Geologie und Paläontologie – Abhandlungen*, 247 (1), 23–61.

Lee, Y. C. & Zhang, G. Y. (1989) Developments of geographic information systems technology. *Journal of Surveying Engineering*, 115 (3), 304–323.

Maguire, D. J. (1991) An overview and definition of GIS. *Geographical Information Systems: Principles and Applications*, 1, 9–20.

Malczewski, J. (2004) GIS-based land-use suitability analysis: a critical overview. *Progress in Planning*, 62 (1), 3–65.

Norrish, N.L. & Wyllie, D.C. (1996). *Rock Slope Stability Analysis*. In: Turner, A.K. and Schuster, R.L. (eds.), Landslides Investigation and Mitigation. Transportation Research Board National Research Council, National Academy Press, Washington DC, Special Report 247.

Pradhan, B. (2010) Remote sensing and GIS-based landslide susceptibility analysis and cross-validation using multivariate logistic regression model on three test areas in Malaysia. *Advances in Space Research*, 45, 1244–1256.

Rocscience (2002) *Dips v5.1 Graphical and Statistical Analysis of Orientation Data*. Rocscience Inc. Toronto, Ontario, Canada. https://www.rocscience.com/rocscience/products/dips.

Rocscience (2004) *User's Guide of RocData: Strength Analysis of Rock and Soil Masses Using the Generalized Hoek–Brown, Mohr–Coulomb, Barton–Bandis and Power Curve Failure*

Criteria, Version 3.0 ed. Rocscience Inc., Toronto, Ontario, Canada. https://www.rocscience.com/rocscience/products/rocdata.

Saydam, C. (2002) *Sedimentary Petrographical and Organic Geochemical Properties of Late Cretaceous Aged Clastic Deposites in Eastern Pontides.* PhD. Thesis, Karadeniz Technical University, 320p.

Saydam Eker, C. & Korkmaz, S. (2011) Mineralogy and whole-rock geochemistry of Late Cretaceous sandstones from the eastern Pontides (NE Turkey). *Neues Jahrbuch fur Mineralogie – Abhandlungen,* 188 (3), 235–256.

Sen, C. (2007) Jurassic volcanism in the Eastern Pontides: Is it rift related or subduction related? *Turkish Journal of Earth Sciences,* 16 (4), 523–539.

Topuz, G., Altherr, R., Kalt, A., Satir, M., Werner, O. & Schwarz, W.H. (2004) Aluminous granulites from the Pulur complex, NE Turkey: A case of partial melting, efficient melt extraction and crystallization. *Lithos,* 72, 183–207.

Topuz, G., Altherr, R., Schwarz, W.H., Dokuz, A. & Meyer, H.P. (2007) Variscan amphibolitefacies rocks from the Kurtoglu metamorphic complex. Gümüşhane area, Eastern Pontides, Turkey. *International Journal of Earth Sciences,* 96, 861–873.

Topuz, G., Altherr, R., Siebel, W., Schwarz, W.H., Zack, T., Hasözbek A, Barth, M., Satır, M. & Sen, C. (2010) Carboniferous high-potassium I-type granitoid magmatism in the Eastern Pontides: The Gümüşhane pluton (NE Turkey). *Lithos,* 116, 92–110.

Tudes, S., Ceryan, S. & Bulut, F. (2012) Geoenvironmental evaluation for planning: an example from Gumushane City, close to the North Anatolia Fault Zone, NE Turkey. *Bulletin of Engineering Geology and the Environment,* 71 (4), 679–690.

Vörös, A. & Kandemir, R. (2011) A new early Jurassic brachiopod fauna from the Eastern Pontides (Turkey). *Neues Jahrbuch für Geologie und Paläontologie Abhandlungen,* 260, 343–363.

Wyllie, D.C. & Mah, C.W. (2004) *Rock Slope Engineering Civil and Mining.* Spon Press, Taylor and Francis e-library.

Yilmaz, I. (2007) GIS based susceptibility mapping of karst depression in gypsum: A case study from Sivas basin (Turkey). *Engineering Geology,* 90, 89–103.

Yilmaz, I., Marschalko, M., Yildirim, M., Dereli, E. & Bednarik, M. (2012) GIS-based kinematic slope instability and slope mass rating (SMR) maps: Application to a railway route in Sivas (Turkey). *Bulletin of Engineering Geology and the Environment,* 71, 351–357.

Yoon, W.S., Jeong, U.J. & Kim, J.H. (2002) Kinematic analysis for sliding failure of multi-faced rock slopes. *Engineering Geology,* 67 (1–2), 51–61.

Chapter 2

Risk management of rock slopes in a dense urban setting

K.K.S. Ho, D.O.K. Lo & R.W.H. Lee

Geotechnical Engineering Office, Civil Engineering and Development Department, Hong Kong SAR Government

Abstract: The acute slope safety problems in Hong Kong can be attributed to high seasonal rainfall together with intense urban development on a steep hilly terrain. The dense urban environment gives rise to a vulnerable setting in which a relatively small landslide or rockfall can result in casualties or significant socioeconomic impact. Since 1977, a comprehensive Slope Safety Management System has been developed and implemented by the Geotechnical Engineering Office to combat the slope safety problems in a holistic manner. This paper outlines the characteristics of rock slope failures in Hong Kong, the landslide risk management framework adopted and the prevailing rock slope engineering practice. The key lessons learnt from studies of notable rock slope failures are presented and the experience in managing some 9,000 man-made rock cut slopes, as well as the hazards associated with rockfalls and boulder falls from natural hillsides over the past 40 years or so, is consolidated.

1 INTRODUCTION

Hong Kong has a small land area of about 1,100 m^2 with a population of over 7 million. The terrain is hilly, with about 75% of the land steeper than 15° and over 30% steeper than 30°. The scarcity of flat land has resulted in intense urban development encroaching on the hillside. Combined with the high seasonal rainfall (hourly and annual rainfall can exceed 100 mm and 3,000 mm respectively), this has resulted in severe landslide problems leading to significant loss of life and damage to property in a dense urban setting.

Following a series of multiple-fatality incidents in the 1970s, the Hong Kong Government in 1977 set up a central policing body, the Geotechnical Control Office (GCO, renamed Geotechnical Engineering Office (GEO) in 1991), to regulate and manage slope safety. A comprehensive Slope Safety Management System was developed and implemented to combat the acute landslide problems in a holistic manner (Wong, 2009a). The system has evolved and undergone continuous improvement over the years in response to lessons learnt from landslide studies and rising public expectations. With concerted efforts to improve slope safety over the years, the landslide risk in Hong Kong has been significantly reduced.

This paper outlines the characteristics of typical rock slope failures, the landslide risk management framework adopted by the GEO and the prevailing rock slope engineering practice in Hong Kong. The key lessons learnt from notable rock slope failures and the application of novel technology in rock slope engineering practice are also presented.

2 ROCK SLOPES IN HONG KONG

2.1 Background and setting

Since the 1950s, a large number of man-made slopes were formed in the course of site formation works as part of the infrastructure and building development to cope with the rapid increase in population. Before 1977, the vast majority of the man-made slopes were formed with no or very limited geotechnical input (referred to as 'old slopes' in this paper) and they generally do not comply with the current safety standards. Some disused quarry sites had also been used for building development. However, poorly controlled drilling and blasting at these old quarries had resulted in the formation of numerous unsatisfactory rock slopes with inherent instability potential, *e.g.* undulating faces with overhangs and obvious damage to the cut faces.

Hong Kong has a high seasonal rainfall with intensities over 100 mm per hour being not uncommon. Together with the intense urban development on a hilly terrain with steep old slopes in close proximity to buildings and roads, this has resulted in acute slope safety problems. On average, some 200 to 300 landslides, including rockfalls and boulder falls, are reported to the Government every year and most of these failures occurred on man-made slopes.

At present, some 60,000 sizeable man-made slopes (about 65% being old slopes) have been registered by the GEO in the Catalog of Slopes (an inventory providing a principal source of comprehensive and up-to-date information on the sizeable man-made slopes in Hong Kong). Rock cut slopes and soil/rock cut slopes make up about 15% (*i.e.* about 9,000) of these registered man-made slopes. In the present context, these 9,000 rock and soil/rock cut slopes are collectively referred to as 'rock cut slopes' for simplicity. These are typically about 10 m high but the larger ones can reach over 80 m in height. About 5,000 of these affect facilities of moderate to high consequence-to-life (primarily buildings and roads) in the event of a slope failure, of which about 3,000 have been engineered (*i.e.* formed or upgraded to the required safety standards).

2.2 Geology

Figure 1 presents a simplified geological map showing the distribution of the principal rock types in Hong Kong. The predominant rock types in the urban areas comprise granites and volcanics. These rocks have generally been subjected to severe in situ weathering with a heterogeneous and highly variable weathering profile (which can be in excess of 100 m locally). A six-grade weathering-based system, ranging from fresh rock to residual soil, has been adopted for rock mass classification (GCO, 1988). The bottom two grades of rock (*i.e.* completely decomposed rock and residual soil) are regarded as soil based on their engineering properties. The joints in granites are generally widely spaced with typical joint spacing between 0.5 m and 2 m as compared to the generally more closely jointed volcanics with typical joint spacing between 0.2 m and 1 m. Sheeting joints often exist in near-surface granites and these usually govern the engineering behavior of the rock mass. Hencher *et al.* (2011) gave a detailed account of the typical engineering characteristics of sheeting joints encountered in Hong Kong.

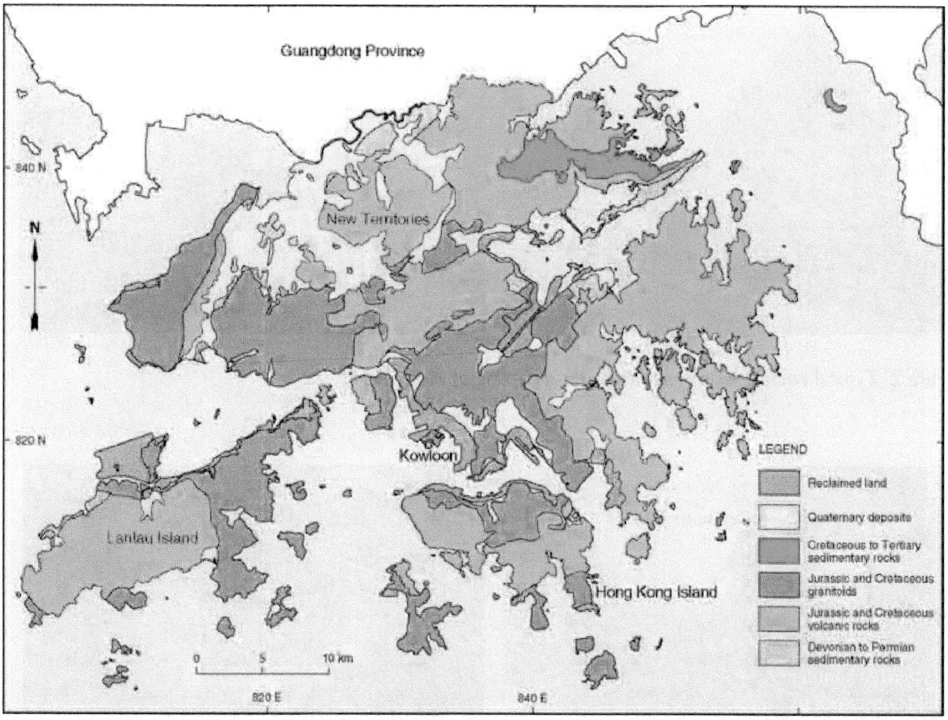

Figure 1 Simplified geological map of Hong Kong.

Natural hillsides are typically strewn with isolated boulders or boulder clusters (or fields), particularly in granitic terrain. Boulders of granitic origin are typically 5 m^3 to 25 m^3 in size, being generally larger than those of volcanic origin which are typically in the order of 0.5 m^3 to 2.5 m^3. Some of the hillsides may locally comprise natural rock cliffs.

2.3 Rock slope failures

2.3.1 Triggers and modes of typical rock slope failures

Rock slope failures in Hong Kong generally comprise rockfalls and rockslides occurring on man-made rock cuts. In addition, boulder falls, involving the detachment and downhill movement of isolated boulders from natural hillside, and failures of natural rock cliffs can occur. Typical settings of registered rock cut slopes within the densely urbanized areas in Hong Kong are shown in Figure 2. Figure 3 shows the natural rock cliffs (known as the Seymour Cliffs) and extensive boulder fields on the hillside in the Mid-levels area flanking high-rise residential buildings.

While human influence could have a significant bearing on the stability of man-made rock cut slopes, failures of natural rock cliffs are mostly part of the natural degradation or mass wasting process. The vast majority of slope failures in

Figure 2 Typical rock cut slopes in the urban setting of Hong Kong.

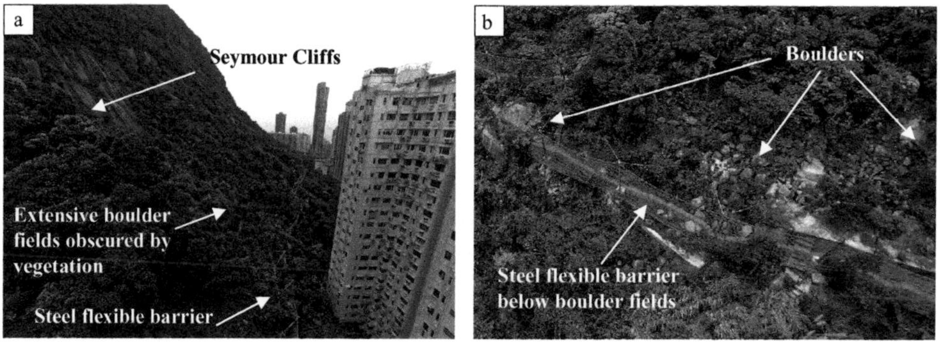

Figure 3 (a) Natural rock cliffs and boulder fields overlooking high-rise residential development at Mid-levels; (b) Close-up view of the boulder fields.

Hong Kong are rain-induced. Ingress of rainwater leading to rise of groundwater pressure and/or build-up of cleft water pressure in joints constitutes the most common trigger. Other factors such as wedging action by tree roots or unplanned vegetation growing in adversely orientated rock joints, inadequate drainage provisions leading to uncontrolled and concentrated surface water flow (*e.g.* uncontrolled discharge above slopes, overspilling of surface water at road bends due to vulnerable site setting, etc.), inadequate maintenance of slope or road drainage and localized deterioration can constitute adverse predisposing factors that would contribute to making a slope more vulnerable to failure or detachment of rock blocks or slabs. Rock slope failures and boulder falls due to blasting at a nearby site, or leakage or bursting of water-carrying services, have also been observed, although these are relatively rare.

For rock cut slopes, failure modes involving planar sliding, wedge failure, toppling and raveling (viz. detachment of loosened rock fragments or blocks locally from the slope face due to progressive deterioration and possible tree root action, which can be exacerbated by typhoons) are commonly observed. Failures are in most cases caused by the build-up of cleft water pressure in rock joints and in some cases involving tree root action for raveling.

Some boulders are perched precariously on hillsides. The manner in which individual boulders are naturally supported ranges from surface friction or soil embedment to interlocking with other boulders. Boulder falls occur often as a result of loss of support when the embedding soil is washed away during rainstorms, and sometimes due to external forces (*e.g.* overland surface water flow, landslides, etc.), or natural deterioration of the boulder itself by continued weathering.

2.3.2 *Failure statistics – insights from landslides*

The GEO has been keeping systematic records of reported landslides in Hong Kong since 1984. Apart from reported landslides, an inventory of natural terrain landslides has been compiled from the review of aerial photographs taken since 1924. The failure statistics and observations in respect of rockfalls, rockslides and boulder falls are presented below. The strategic use of the landslide records for landslide risk management is discussed in Sections 3.3.3 and 3.4.3.

2.3.2.1 Failures on man-made rock cut slopes

Since 1997, all reported landslide incidents have been examined by the GEO under the systematic landslide investigation (LI) program (see Section 3.5 for details). Between 1997 and 2014, about 360 rockfall or rockslide incidents (*i.e.* on average about 20 incidents per year) were reported to have occurred on registered rock cut slopes. Figure 4 presents the volume distribution of these 360 incidents. About 97% of these incidents are 'minor' failures with a failure volume of less than 50 m^3 and the remaining 3% involved 'major' failures with a failure volume exceeding 50 m^3. It is noteworthy that a fairly large proportion of the failures (about 65%) involved detachment of small rock blocks with a volume of less than 1 m^3.

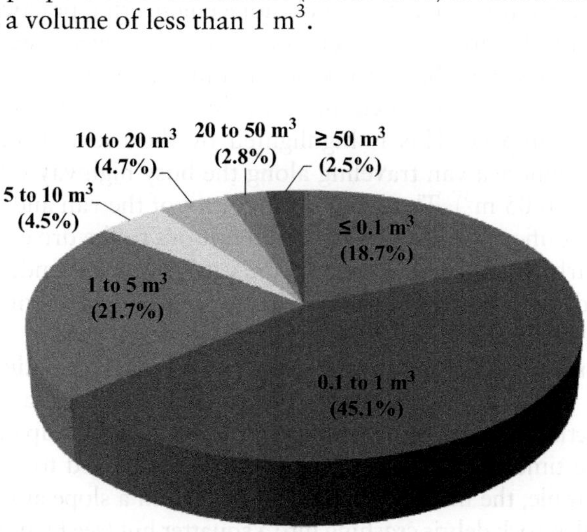

Figure 4 Volume distribution of reported rockfall and rockslide incidents on registered rock cut slopes between 1997 and 2014.

Figure 5 (a) Failure of rock cut slope resulting in serious property damage; (b) Minor rockfall resulting in significant vehicle damage and injuries.

As mentioned above, most failures are primarily triggered by rainfall. In respect of the common contributory causes of failure, diagnosis of systematic landslide investigation revealed that almost all the major failures (taken as 50 m^3 to 500 m^3) are kinematically-controlled due to the presence of adverse geological features or rock joints, which are sometimes infilled with weak weathered materials. For minor failures (< 50 m^3), root-wedging action due to unplanned vegetation (which plays a role in about 60% of the minor failures) often gives rise to a vulnerable setting. Among these 360 incidents, about 80 (*i.e.* 22%) were associated with engineered rock cut slopes. Major failures involving engineered rock cut slopes are relatively rare and there has been thus far no massive failure (≥ 500 m^3) on engineered rock cut slopes. The findings that some 22% of the reported rock cut failures involving engineered slopes (*i.e.* slopes that were accepted as conforming to the required safety standards) is pertinent which emphasizes the need to further enhance rock slope engineering practices.

Damage due to the failures of rock cut slopes can be severe (see Figure 5(a) for example) and casualties have been reported in some cases. In a vulnerable densely urbanized setting, even a minor rockfall can result in casualties and/or significant economic loss/social impact. This is highlighted by the rockfall incident at Castle Peak Road in 2001 where a van traveling along the busy highway was hit by a small piece of rock debris (0.05 m^3). The driver lost control of the van and crashed into the slope resulting in significant vehicle damage and injuries to the driver and a passenger (Figure 5(b)). In early 2015, small rockfalls (0.05 m^3) occurred under dry weather at Stanley Gap Road, which punched through the window of a tour bus and four passengers aboard suffered cuts from the broken glass fragments.

From time to time, there are reported incidents where damage and/or causalities are narrowly avoided. These 'near-miss' cases (*e.g.* landslides affecting a busy road at off-peak hours or affecting a flimsy residential structure when the occupants are not at the affected area at the time of failure; see Figure 6) could have led to more serious consequences. For example, the minor rockfalls (0.5 m^3) from a slope at Cha Kwo Ling in 2014 resulted in fallen rock debris crashing into a squatter hut (see Figure 6(b)). Although it was fortunate that the incident did not result in any casualty, the affected squatter hut had to be permanently evacuated given the potential risk of further instabilities.

Figure 6 Typical examples of 'near-miss' rockfall cases.

2.3.2.2 Failures on natural rock cliffs and boulder falls from natural hillside

Failures on natural rock cliffs are considered as natural terrain landslides. The corresponding hazards are rockfalls (detachment of isolated rock blocks) and rockslides with a fairly large volume (*e.g.* up to about several hundred cubic meters). In general, the available landslide inventory based on aerial photograph interpretation (API) precludes the differentiation of rock cliff failures from other natural terrain landslides (*e.g.* open hillside failures).

Where the setting of the natural hillside is adverse, this would be liable to lead to an increase in the landslide debris mobility (*i.e.* the runout distance of the debris). For instance, the channelized debris flow which demolished some vacant squatter structures and caused temporary road closure at Lei Pui Street in 2001 (Figure 7) originated from the failure of rock slabs above a natural rock cliff (source volume about 250 m^3) along some adversely orientated sheeting joints at the source area, with the debris discharged into the incised drainage line below (MGSL, 2004). A maximum active debris volume of about 780 m^3 (due to entrainment of loose materials along the drainage line) was mobilized along the debris trail. The debris traveled a plan distance of about 320 m with a travel angle (Wong & Ho, 1996) of about 23°.

The number of reported boulder fall or rockfall incidents from natural hillsides is not statistically high (on average about 10 incidents per year), as many probably occurred in remote areas and did not impact on developed areas. Notwithstanding this, the consequences are liable to be significant. For example, the boulder fall incidents at Holy Cross Path in 1976 and Quarry Bay in 1981 resulted in three fatalities plus six injuries and one fatality plus two injuries respectively. Boulder fall incidents resulting in temporary road closures and damage of structures or vehicles have been reported from time to time (Figure 8(a)). 'Near-miss' cases are sometimes noted, *e.g.* two fallen boulders (weighing about 70 tonnes in total) traveled a plan distance of over 70 m and came to rest on a slope berm just above the heavily-trafficked Tate's Cairn Tunnel Portal at Diamond Hill in 2012 (Leung *et al.*, 2013) (Figure 8(b)).

Figure 7 Channelized debris flow above Lei Pui Street in 2001.

Figure 8 (a) Boulder falls causing damage of vehicles; (b) Example of 'near-miss' boulder fall case.

3 LANDSLIDE RISK MANAGEMENT FRAMEWORK

3.1 General

In Hong Kong, a risk-based approach has been adopted to manage landslide risk. The GEO pioneered the development and application of quantitative risk assessment (QRA) methodology to quantify the landslide risk. Overall QRA for the whole of Hong Kong was undertaken to assess the societal risk arising from the different landslide hazards posed to the community. This provides a useful and invaluable basis for the formulation of landslide risk management strategy, particularly for policy-making and resources allocation consideration.

In this section, the key components of the Slope Safety Management System are highlighted, followed by the presentation of the risk management framework adopted to manage the landslide risk from man-made rock cut slopes and natural rock cliffs, and that from boulder fall hazards.

3.2 Slope safety management system

The landslide risk in Hong Kong is being managed in a holistic and multi-pronged manner through a Slope Safety Management System that has been developed by the GEO (Wong, 2009a). The system incorporates the application of fundamental risk management concepts at the policy administrative level. It has been subjected to continuous improvement over the years. It has evolved into a comprehensive slope safety regime to manage landslide risk, involving a suite of policy, legislative, administrative, technical, emergency preparedness, public education and community-based provisions. The key components of the system together with their functions are presented in Table 1.

Table 1 Key Components of the Slope Safety Management System in Hong Kong.

Components of the Slope Safety Management System	*Contribution by each component*		
	to reduce landslip risk		*to address public attitudes*
	hazard	*vulnerability*	
Policing			
• cataloging, safety-screening and statutory repair orders for slopes	✓		
• checking new works	✓	✓	
• slope maintenance audit	✓		
• inspecting squatter areas and recommending safety clearance		✓	
• input to land use planning	✓	✓	
Safety standards and research	✓	✓	✓
[e.g. *natural terrain hazard study and mitigation, landslide debris mobility, landslide risk assessment and management, slope greening, etc.*]			
Works projects			
• upgrading existing government man-made slopes	✓		
• safety-screening studies of private man-made slopes	✓		
• mitigating natural terrain landslide risk	✓	✓	
Regular slope maintenance			
• routine and preventive slope maintenance	✓	✓	
Education, information and emergency preparedness			
• maintenance campaign	✓		✓
• personal precautions campaign		✓	✓
• awareness program	✓	✓	✓
• information services	✓	✓	✓
• landslip warning and emergency services	✓	✓	✓

Note: Maintenance of registered government man-made slopes and natural terrain defense/stabilization measures is under the responsibility of designated government departments in accordance with the 'beneficiary maintains' principle.

The GEO is the de facto slope safety manager in collaborating input from various government departments, as well as promoting actions by the private slope owners and the general public. The Key Result Areas (KRA) of the system include:

1. improve slope safety standards, technology, and administrative and regulatory framework;
2. ensure safety standards of new slopes;
3. rectify substandard existing government slopes;
4. maintain all sizeable government man-made slopes;
5. ensure that owners take responsibility for slope safety;
6. promote public awareness, preparedness and response in respect of slope safety through public warnings and information services, public education and publicity campaigns; and
7. improve slope appearance.

As a result of the concerted slope safety efforts over the past years, the landslide risk in Hong Kong has been substantially reduced since 1977 as illustrated by the notional landslide risk trend given in Figure 9, which is corroborated by the rolling-average landslide fatality rate as well as the QRA findings (Ho *et al.*, 2015). The overall residual landslide risk has reached a reasonably low level that is commensurate with the international best practice in risk assessment (*i.e.* the 'As Low As Reasonably Practicable' (ALARP) level). To contain the risk within the ALARP level in the long run, continued efforts in landslide risk management are called for in order to prevent risk increase due to progressive slope degradation, encroachment of more urban development on steep hillside, population growth, potential impact from extreme weather events or climate change, etc.

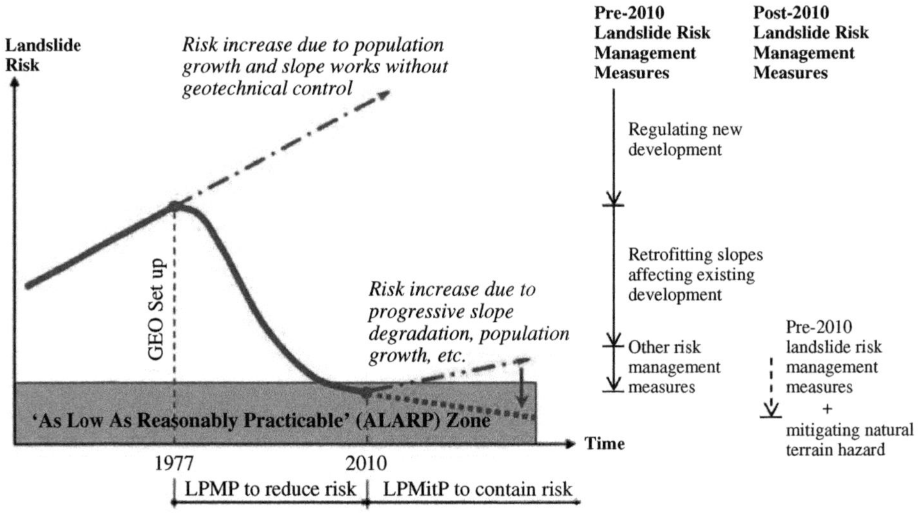

Figure 9 Notional landslide risk trend in Hong Kong.

3.3 Management of landslide hazards from man-made slopes

3.3.1 Strategy

A proactive approach has been adopted in regulating the safety of man-made slopes (including rock cut slopes). For new slopes, the Government's strategy is to contain the associated landslide risk through geotechnical control. While for substandard existing slopes, the respective landslide risk is reduced primarily by retrofitting works and slope maintenance.

3.3.2 New man-made slopes

The landslide risk from new slopes is contained by the implementation of geotechnical control on building and infrastructure developments. Under the prevailing framework, private developments are regulated under the Buildings Ordinance and public works are controlled under an administrative mandate. The GEO has been auditing the design and supervision of the private and public slope works to ensure that the new slopes formed (or old slopes upgraded) as part of the new development would meet the required slope safety standards.

3.3.3 Existing man-made slopes

The Catalog of Slopes (see Section 2.1) provides the essential information for assessment of the scale of the problem arising from existing man-made slopes, risk quantification, determination of slope ownership and planning of the retrofitting studies and works. This together with a wealth of reported landslide records (see Section 2.3.2) facilitate the formulation of the landslide risk management strategy.

The GEO has embarked on a Landslip Preventive Measures Programme (LPMP) in 1977 to proactively and systematically reduce the landslide risk from the potentially substandard man-made slopes posed to existing development in accordance with a risk-based priority ranking system. Upon the completion of the LPMP in 2010, the GEO launched a rolling Landslip Prevention and Mitigation Programme (LPMitP), to dovetail with the LPMP. The pledged annual outputs of the LPMitP are to (i) retrofit 150 government man-made slopes, (ii) conduct safety-screening studies for 100 private man-made slopes, and (iii) implement studies and necessary risk mitigation works for 30 natural hillside catchments.

Given the legacy of a large stock of potentially substandard government man-made slopes to be dealt with under the LPMP and LPMitP, a risk-based ranking system has been formulated for prioritizing the more deserving slopes for detailed studies and retrofitting. Four separate systems have been developed for the four main types of man-made slope features respectively (viz. soil cut slopes, rock cut slopes, fill slopes and retaining walls), with due consideration given to the different factors that govern the failure of each specific type of slope feature. For rock cut slopes, four different mechanisms of failures, namely planar sliding, wedge failure, toppling and raveling, are embraced by the ranking system. Each system provides a list of ranked slopes. By reference to the findings of the overall QRA (Cheng & Ko, 2010), the four lists are merged into a single list to establish their priority for follow-up action under the

LPMP and LPMitP. This risk-based approach provides a sound basis for selection of the most deserving slopes for action and hence maximizes the rate of landslide risk reduction. The overall QRA also demonstrated, by means of cost-benefit calculations, that the investment in systematic retrofitting of substandard slopes was cost effective in terms of the projected number of lives saved (Lo & Cheung, 2005).

The GEO also undertakes safety-screening studies under the LPMP (and subsequently the LPMitP) on privately owned man-made slopes. Where prima facie evidence is established indicating that a private slope is dangerous or liable to become dangerous, a statutory order would be served on the private owners requiring them to investigate and where necessary upgrade their slopes within a designated time period.

In addition to the LPMP and LPMitP, an integrated approach is adopted as part of the implementation of government projects (*e.g.* road widening projects) to upgrade existing substandard government slopes that could affect or be affected by the projects.

Regular maintenance can arrest or slow down slope deterioration and is essential to the continued stability of all man-made slopes. Assessments by Wong & Ho (1997) and Lo *et al.* (1998) revealed that although inadequate maintenance may not be a sufficient condition alone to cause slope failure, it can however greatly increase the likelihood of failure, particularly for substandard slopes. A clear demarcation of the responsibility of slope maintenance is crucial to facilitate slope owners to take responsible action. For this purpose, an information system on the maintenance responsibility of individual registered man-made slopes has been set up and made available to the public through the internet. The Government maintains, through a number of designated maintenance departments, all registered government man-made slopes. In addition to regular maintenance, the respective maintenance departments have been undertaking preventive maintenance, involving the use of prescriptive measures (refer to Section 4.3.1 for details), in reducing the landslide risk of those government man-made slopes that do not urgently require full scale upgrading under the LPMP or LPMitP. Private owners are responsible for maintaining their own slopes and upgrading those that are found to be substandard. The technical aspects of slope maintenance are described in Section 4.7.

With the concerted efforts made through the above measures, the overall landslide risk arising from old man-made slopes in Hong Kong has been reduced to less than 25% of the 1977 level by 2010 as confirmed by means of QRA (Cheng, 2014). All the high risk old man-made slopes (about 7,000), primarily those affecting buildings and busy roads, have been retrofitted by 2010. Notwithstanding this, there would still be around 33,000 old slopes, which are of a lower risk, awaiting assessment and retrofitting to current safety standards.

3.4 Management of natural terrain landslide hazards

3.4.1 Strategy

The Government's strategy in managing the natural terrain landslide hazards aims at keeping the respective risk to an ALARP level. The policy for new development or redevelopment is to contain the corresponding increase in landslide risk due to the proposed development. The policy for natural terrain affecting existing development is to deal with the risk posed to the community pursuant to the 'react-to-known-hazard' principle, in accordance with a risk-based priority ranking order. Details are given by Wong (2009b).

The above policies have been implemented for many years and so far they appear to be effective in managing the landslide risk and appropriate in discharging the due diligence of the GEO as the slope safety manager. The applications of the risk management policy to natural terrain landslides (including natural rock cliff failures) and boulder falls are elaborated below.

3.4.2 New development

Natural terrain landslide risk to new development is primarily contained through proper land use planning and development control. To ensure early geotechnical input during the land use planning stage, the GEO has been advising government town planners/land managers on land development proposals, land use zoning and special geotechnical conditions on land allocations or land disposals. Where a proposed development is located close to natural terrain, the GEO may advise, on a case-by-case basis, against development or otherwise impose the requirement on the developer (both for private and government projects) to carry out a natural terrain hazard study and the implementation of the necessary mitigation actions and/or stabilization works as part of the new development. In this respect, a set of criteria has been developed by the GEO (viz. 'In-principle Objection Criteria' and 'Alert Criteria') to facilitate the screening and the determination of the need of further studies (GEO, 2013). In applying the criteria, professional judgment is exercised, with due account taken of the nature, size and geometry of the proposed development, its proximity to the hillside, credible debris flowpath to the site, and available information on the conditions and history of the hillside. Geotechnical control is exercised by the GEO in respect of these studies and works, similar to that for man-made slopes as described above.

Under the prevailing policy, preventive action will be taken for boulders posing an *immediate and obvious danger* to the proposed development. Evaluation of boulder stability will be undertaken only when there would be a significant risk, and preventive action will be taken when considered necessary (GEO, 2004).

3.4.3 Existing development

The Government's policy in managing the natural terrain landslide risk posed to existing development is based on the 'react-to-known-hazard' principle, *i.e.* to carry out studies and mitigation actions and/or stabilization works where significant hazards become evident. A Hong Kong-wide QRA conducted by Wong *et al.* (2006) showed that the overall natural hillside landslide risk would be similar to that of man-made slopes upon the completion of the LPMP in 2010. In light of this, expanded efforts are made in the LPMitP to systematically deal with the vulnerable natural hillside catchments with known landslide hazards to existing buildings or important transport corridors (referred to as 'historical landslide catchments' (HLC)), in addition to the retrofitting of the remaining substandard man-made slopes. A risk-based ranking system was developed to prioritize the 2,700 HLC identified from interpretation of aerial photographs for action under the LPMitP (Wong *et al.*, 2006).

For boulders affecting existing development, preventive action will be taken urgently where there exists an *immediate and obvious danger*. Evaluation of boulder stability (and the implementation of preventive action where necessary) will be undertaken only when there have been persistent boulder falls or where a dangerous situation could develop (GEO, 2004).

A Hong Kong-wide QRA conducted by MGSL (2001) found that the societal risk posed by boulder falls from natural hillside is about 0.05 Potential Loss of Life (PLL) per year, which amounts to about 1% of that of natural terrain landslides. Given that the risk of boulder falls is much lower than that of natural terrain landslides, the prevailing boulder policy for existing development is considered to be appropriate (*i.e.* taking action when and only when significant hazards become known). Nevertheless, boulder hazards posed to existing development arising from the vulnerable natural hillside catchments will be addressed together with the landslide hazards (*e.g.* open hillside failure and channelized debris flow) under the LPMitP, as necessary.

3.5 Systematic landslide investigation program

The GEO has implemented a systematic LI program (Wong & Ho, 2000) since 1997. Under the program, all reported landslides are examined and cases that warrant follow-up studies are identified. Detailed studies are carried out on deserving cases to document the failures, diagnose their probable causes and failure mechanisms, and identify the key lessons learnt and necessary follow-up actions. These include forensic studies of serious landslides that may involve coroner's inquest, legal action or financial dispute. The LI program also serves as a safety net for the identification of slopes in need of early attention. Another important aspect of the LI program is to provide data for reviewing the performance of the Slope Safety Management System and identifying areas for continuous improvement. The LI program is an integral part of LPMP as well as LPMitP.

4 TECHNICAL ASPECTS OF ROCK SLOPE ENGINEERING IN HONG KONG

4.1 General

The rock slope engineering practice in the early 1980s has been summarized by Brand *et al.* (1983). The practice has been continuously enhanced to take account of the lessons learnt from studies of rock slope failures and technological development. A number of local guidelines and publications are available which cover the whole project cycle, ranging from ground investigation (GCO, 1987; GEO, 2007), design (GCO, 1984; GEO, 2000, 2007, 2009, 2014), construction control (GCO, 1984; GEO, 2000, 2006) to maintenance (GEO, 2003).

4.2 Ground investigation

A rock joint mapping is typically carried out by means of a scanline survey along the accessible slope toe/berm (and in some cases at higher levels with access provided by scaffolding or a lifting-cage) where joint sets are mapped and salient features such as infilled materials, joint opening, joint spacing, etc. and their characteristics are recorded. Rock slope surfaces may be covered either partially or entirely with shotcrete or chunam, in particular on portions that are severely weathered or with more closely-spaced joints. Slope surface stripping (*i.e.* temporary removal of a strip of the hard surface cover) is generally undertaken to expose the rock face for mapping. In addition to face mapping, acoustic borehole televiewer surveys (which essentially supplanted impression packer surveys) have also been widely used to assess the orientation of rock discontinuities. While such technique can produce a large amount of joint orientation data, not much information can be obtained on the relative persistence of joint sets which needs be duly considered in the kinematic analysis.

For the majority of routine cases, the use of generalized shear strength parameters for rock joints (that are deemed to be suitably conservative values) tends to be favored by designers instead of conducting site-specific testing. Where assessment of the shear strength of rock joints is required, the joint surface roughness and presence of infilled materials should be duly accounted for (Hencher & Richards, 1982; GEO, 1993). Weak infilled or weathered materials (*e.g.* kaolinite) in rock joints may not be recovered by conventional drilling method. Special measures, such as triple-tube core-drilling in conjunction with air-form flushing (Richards & Cowland, 1986), would enhance recovery of these materials. The fieldwork on characterization of discontinuities can also be supplemented by downhole geophysical methods, *e.g.* use of gamma density and spectral gamma ray methods to detect clay-infilled joints (Lau & Franks, 2000).

A good understanding of the site hydrogeology, including the seasonal and storm response of the groundwater regime, is also an important aspect of ground investigation. The groundwater response to rainfall is typically complex and reliable prediction can be fraught with difficulty. Even within a given discontinuity, a detailed regional study presented by Cowland & Richards (1985) observed random and discrete groundwater surges during a rainstorm where transient elevations in groundwater pressure (as high as 10 m) did not take place simultaneously over the entire surface of a large sheeting joint. This illustrates the uncertainty involved in the usual assumption of an overall rise in groundwater table generally adopted for stability analysis based on the limited knowledge on the actual groundwater regime as acquired from typical ground investigations.

4.3 Design

The design of stabilization or defensive measures for rock cut slopes, rock cliffs and boulders calls for much engineering judgment based on common sense, local experience and the application of sound engineering principles.

4.3.1 Formation and treatment of rock cut slopes

Stability assessment of a jointed rock slope is based largely on the recommendations by Hoek & Bray (1981). The factor of safety approach has generally been adopted in slope design. The stipulated safety standards are risk-based, taking due account of the consequence-to-life category of a slope (GCO, 1984). For example, the minimum required factor of safety for a rock cut slope affecting a building (*i.e.* of consequence-to-life category 1) is 1.4 for a design groundwater condition corresponding to a 1 in 10 years return period rainfall. In a dense urban setting, human factors that could aggravate slope stability should also be duly considered in the design and detailing of the corresponding stabilization measures. For instance, the routing of new water-carrying services should be kept as far away as possible from the area within which leakage could affect the slope or otherwise the effect of possible leakage may need to be accommodated in the design.

Use of prescriptive measures

Similar to standard good practice adopted in other developed countries, stabilization measures for rock cut slopes in Hong Kong typically comprise scaling/dentition and provision of rock mesh netting, rock dowels, concrete buttresses, surface channels, relief drains, etc. As a supplementary approach to the conventional means of establishing the optimal solution for the stabilization measures by design calculations and in light of the need for a more expeditious way in dealing with a large stock of substandard slopes that need to be retrofitted, the prescriptive measures approach was rationalized (Wong *et al.*, 1999). These are pre-determined, experience-based and suitably conservative modules of works that can be prescribed by qualified geotechnical professionals without the need for detailed ground investigation and design analyses. Prescriptive measures for rock cut slopes are given in GEO (2009).

The overall stability of a rock slope should be addressed based on conventional design while prescriptive rock dowels and concrete buttresses can be applied to stabilize small local unstable blocks and highly fractured zones/local overhangs respectively. Prescriptive surface protection (*e.g.* rock mesh netting) and surface/subsurface drainage measures (*e.g.* crest channels with upstand, downpipes to confine water flow on steep slopes and relief drains on hard surface cover at potential seepage sources) are strongly recommended for all rock slopes. Where appropriate, buffer zones may be prescribed to reduce the consequences of rockfalls or rockslides.

Experience has shown that minor rockfalls are exceedingly difficult to guard against in design. A pragmatic approach is to provide suitable protective and/or mitigation measures, such as rock mesh netting, rockfall fence/barrier, rock trap and buffer zone where space permits (GEO, 2014), as an integral part of the design. For high consequence roadside slopes, the provision of a rock mesh on the unprotected steep rock face could avert serious consequences and traffic disruption (see Figure 10) due to minor

Figure 10 Minor rockfalls contained by rock mesh avoiding impact on the adjoining carriageway.

failures and is strongly recommended (GEO, 2009). To further enhance the control of rockfalls, high-tensile steel wire mesh can be provided over the rock face with pretensioning to promote the confinement. In practice, the use of rock traps is relatively rare in Hong Kong mainly because of space constraints. For the design of rock traps below steep presplit rock slopes, Mak & Bloomfield (1986) presented a set of refined design charts based on the findings from a series of local boulder fall field tests.

Design of new rock cut slopes

The overall cut face profile should be determined to ensure slope stability taking due account of the geological features and the characteristics of discontinuities (including groundwater conditions) in the rock mass. In respect of slope berm provision, key factors including the rock mass characteristics, possible 'ski-jump' effect of rockfalls, berm width to contain small rockfalls, land-take, maintenance, etc. should be considered (GEO, 2000). Benefits that can be brought about by the omission of berms, such as the reduction in land take/excavation and hence the minimization of disturbance to the natural hillside (Lam *et al.*, 2002), should be properly weighed against the associated disadvantages (*e.g.* lack of safe access for future maintenance). If rock slopes are to be formed with the joints dipping into the excavation exposed, it may be safer and more economical to form these slopes on a continuous profile parallel to the joints, rather than forming steep cuts with berms which can create daylighting joints (Figure 11).

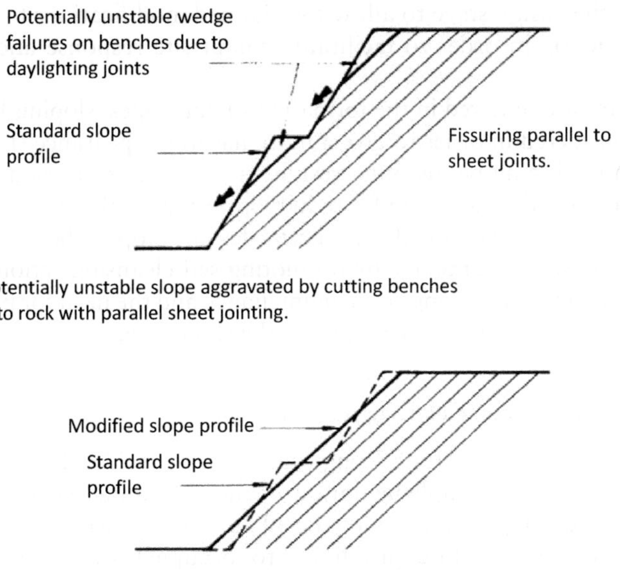

(a) Potentially unstable slope aggravated by cutting benches into rock with parallel sheet jointing.

(b) More stable slope is formed by laying back at the angle at which the sheet joint is dipping. For a reasonable range of orientations of the sheet joints, there may be little extra volume of excavation required.

Figure 11 Effect of slope profiling in adversely orientated joints (GCO, 1984).

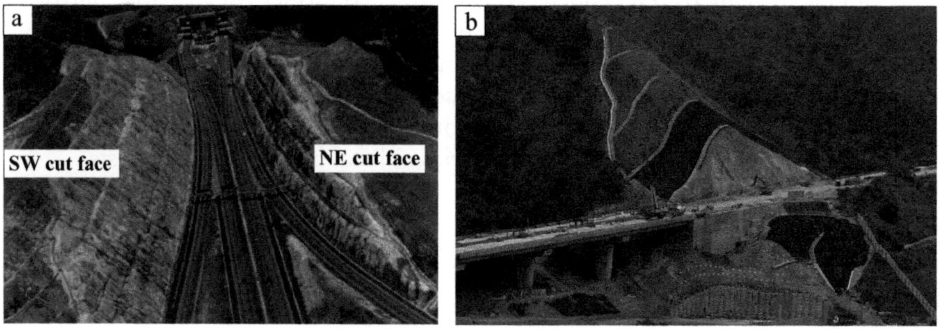

Figure 12 (a) Rock cut slopes formed on either side of the approach road to Route 3 Tai Lam Tunnel; (b) Cut slope with sloping-berms to blend in with the surrounding natural landscape.

The face mapping results obtained during construction should be critically examined for continuous updating of the design model. As an example, two rock cuts on either side of the approach road leading to the Route 3 Tai Lam Tunnel (Figure 12(a)) were originally designed to have symmetric profiles with similar slope angles. During construction, it was identified that the persistent, adversely-dipping joints on the southwest cut face had resulted in an acute planar instability problem. Consequently, the southwest slope was further cut back to form a more gentle profile which minimized daylighting of the stability-critical joints and avoided the need of extensive stabilization. The observational method (Nicholson *et al.*, 1999) should be introduced at the design stage to allow for planned modifications to be incorporated during construction in order to facilitate prompt site response and greater overall economy.

If slope berms are required in the formation of cut slopes, sloping-berms (Figure 12 (b)) can be considered in order to give a more natural appearance (Lam *et al.*, 2002). The adoption of sloping-berms often results in smoother transition between natural hillside and man-made feature and a reduction in the total volume of excavation as compared with the conventional level-berm design. Sloping-berms also provide an added benefit on surface drainage by promoting self-cleansing action of the drainage channels, hence minimizing long-term maintenance and the need for providing visually intrusive stepped channels or downpipes on the slope face.

Site constraints on upgrading existing rock cut slopes

The dense urban setting in Hong Kong often poses considerable constraints for the ground investigation and implementation of the upgrading works on existing rock slopes. Slopes requiring action under the LPMP or LPMitP are invariably steeply-inclined and located in close proximity to occupied structures or busy roads. In some cases, working space for investigation or upgrading works is practically not available, *e.g.* a slope may be obscured by existing structure at its toe (Figure 6 (b)), or a slope may abut a narrow and busy road without footpath (Figure 13). Temporary traffic arrangement can be difficult to implement and often results in severe traffic impact and major implication on the construction methods, and site

Figure 13 Typical rock cut slope abutting narrow and busy road (Ho & Wright, 2011).

logistics and site safety requirements, as well as the construction program. In addition to these constraints, problems such as slow progress due to ground conditions and public complaints arising from some inevitable social disruption or disturbance can sometimes result in delay to the works.

In practice, it is important to identify the site constraints as early as possible and take these into account throughout the project cycle, including public consultation and engagement. The upgrading works design should be tailored to suit the anticipated site setting. Factors such as limitations on access and working space, land matter (*e.g.* no works to be encroached into private lots), presence of nearby sensitive receivers, etc. should be duly considered in the design. Careful and insightful planning and consideration at the design stage can significantly enhance the efficiency of the works.

During construction, temporary road closure where required should be restricted to a bare minimum. Where working space is limited, a narrow elevated platform may be erected at the slope toe to facilitate the works. Site-specific temporary works and special construction plant (*e.g.* mobile platform and hand-held coring machine) may need to be deployed. Work sequence and logistics planning can be crucial for congested sites and 'just-in-time' delivery of materials may be warranted. Tight control on construction safety and environmental issues is also called for. Ho & Wright (2011) presented some case studies to demonstrate how some of the challenges and constraints arising from the construction of slope upgrading works along busy roads in Hong Kong can be addressed satisfactorily. Lee *et al.* (2011) highlighted some critical factors, including early liaison and close communication between the project team and various stakeholders, and the adoption of a prompt and proactive approach to resolve site problems and complaints, for the successful implementation of works on several slopes subjected to severe site constraints.

4.3.2 Boulder treatment and rock cliff stabilization measures

The investigation and treatment of boulders and rock cliffs on natural hillside is often constrained by the lack of safe access and limited space for works and stockpiling. Associated with the difficult site setting, the formation of temporary access and working platform on a hillside usually constitutes a notable portion of the overall cost of the works which should be allowed for in the design. Vegetation clearance, where required, should be minimized in order to reduce the environmental impact and avoid the aggravation of the risk of boulder falls.

For boulder fall analysis, numerical tools (Stevens, 1998; Jones *et al.*, 2000) are widely used for stochastic simulation of the probable boulder fall trajectory and runout characteristics (*e.g.* runout distance, velocity, kinetic energy and bounce height) based on a combination of factors such as the slope profile, surface material, surface roughness, coefficient of restitution, boulder shape and size, etc. Typically, a number of boulder falls are simulated to obtain a range of predicted boulder fall trajectory outcomes for probabilistic analysis of a boulder fall event. In each simulation, the values of input parameters are varied within pre-defined ranges randomly or by random sampling from pre-defined probability distributions. The reasonableness of the simulation is governed by the appropriateness of the input parameters, particularly the coefficients of restitution which are usually based on limited calibration efforts or literature on back-analysis of field trials. The runout characteristics of a boulder have to be critically assessed to evaluate the risk posed to downslope facilities.

Two approaches, namely preventive and protective methods, are typically adopted to address the risk from boulder falls (Au & Chan, 1991). The preventive method involves removal or insitu stabilization (*e.g.* surface protection to the soil embedment, buttressing, anchoring, strutting and wire lashing, see Figure 14) of the potential unstable boulders, whereas the protective method involves arresting or deflecting the

Figure 14 Typical insitu stabilization measures for potential unstable boulders (Au & Chan, 1991).

moving boulders (*e.g.* through the provision of boulder fence, rock trap, deflective barrier, etc.) before they reach the downslope facilities of concern.

Due consideration should be given to, inter alia, the size of boulders, their distribution on hillside, the facilities at risk, site constraints and cost in determining the suitable method for boulder treatment. In some cases, it can be more cost effective to adopt a combination of preventive and protective methods to mitigate the boulder fall hazard. An example is the treatment of an extensive boulder field below the Seymour Cliffs in the Mid-levels area where several thousand boulders on the hillside covering a total plan area of 61,600 m^2 were systematically inspected and assessed (Chan & Au, 1988; Smith & Yeung, 1991). The boulder treatment works comprised insitu stabilization (with typical measures as described above) of boulders greater than 3 tonnes in weight and the construction of a flexible boulder fence to catch smaller boulders in mitigating the boulder fall hazard to the downslope dense residential development. To facilitate the design of the boulder fence, field and laboratory tests were carried out as part of the study to aid assessment of the boulder velocity and boulder/fence interaction (Chan *et al.*, 1986).

While boulder barriers are normally not designed for fire resistance, the use of combustible materials should be avoided where the risk of hill fire is high. The hill fire in 1989 badly damaged a number of massive boulder barriers formed by gabions below Lion Rock Ridge. The outer geogrid sheetings of the gabions were burnt resulting in the collapse of the retained rock infill. As a protection against further hill fire, the damaged barriers were subsequently replaced using fire-resistant gabions while the undamaged ones were covered with shotcrete.

In respect of rock cliffs, the hazard from potentially unstable outcrops that are liable to suffer sizeable failures can generally be addressed through stabilization measures. Typical stabilization measures include the construction of concrete buttresses, pillars, foundations; installation of rock dowels/bolts; and cement mortar infilling of rock joints. The amount of concreting works should be minimized in view of the extreme access difficulties. An example is the use of concrete pillars instead of buttresses to support long overhanging rock sections.

For new development close to natural hillside, the relevant rockfall/boulder fall hazard may be mitigated through judicious adjustment to the development layout. Tse *et al.* (2003) described a case where a cost effective hazard mitigation strategy was adopted by locating the residential block away from the hillside and building a recreation area (*e.g.* a swimming pool) in-between to act as a buffer zone and rock trap.

4.4 Control on blasting

Blasting is the most common method for bulk excavation of rock. Uncontrolled blasting carried out in the past often resulted in highly uneven surfaces with overhangs and blast damage to the cut faces. With controlled blasting techniques involving input from qualified personnel, it is possible to create steep rock slopes and bench profiles that are intrinsically undamaged. Experience has shown that carefully controlled blasting can achieve a significant reduction in the amount of stabilization works rendering much lower overall cost of excavation and support than in the case of poorly blasted excavation (Beggs *et al.*, 1984).

Apart from the effect on the stability and integrity of the rock slopes that are formed, the disturbance caused by blasting to the nearby slopes, among other sensitive receivers, is also an important consideration given the potential safety hazard to the general public. There have been notable failures of rock slopes in the vicinity of blasting operations (see Section 5.4). The potential adverse effects of blasting on the stability of nearby rock slopes have been discussed by Nicholls *et al.* (1992). Incorporating the lessons learnt from the past blast-induced failures, enhanced methods for assessing the stability of nearby rock slopes subjected to blasting vibration and the effect of blasting gas pressure are given by Wong & Pang (1992) and Blastronics Pty Ltd. (2000) respectively.

For the sake of public safety, a blasting assessment has to be undertaken with professional input during the design stage to assess the potential adverse effects and identify the necessary precautionary or mitigation measures (GEO, 2006). The Contractor is also required to obtain a blasting permit from the Government before proceeding to use explosives for blasting works. Blasting within zones of more weathered and fractured rock deserves special attention as illustrated by a near-miss incident in 2003 where flyrocks were ejected onto a busy road up to 230 m away from the blasting site near New Clear Water Bay Road (Massey & Siu, 2003).

4.5 Landscaping and aesthetics

It has been a long-standing government policy to make slopes look as natural as possible to reduce their visual impact and improve the environment. Since the late 1990s, the Government has stepped up the promotion of slope greening to meet public expectations on safe and green slopes. A hard surface cover is used only as a last resort on slope stability grounds or as emergency repairs to landslide scars. Guidance on landscape treatment for slopes is given in GEO (2011).

The provision of rock mesh nettings on rock slopes (to protect rock face that is heavily jointed or where raveling is likely) allows the natural color and structure of the rock to be seen. Moreover, they facilitate the establishment and growth of climbers on slopes (Figure 15(a)). Colored mesh may be adopted to make it less visually intrusive. Similarly, hard surface cover and concrete buttresses, if provided, can be appropriately colored to better blend in with the surrounding environment. Soft landscaping works may also be applied on rock slopes to further enhance the greening effect. A possible approach is to provide continuous fiber reinforced soil to support the growth of vegetation on rock face or shotcreted surface (Lam *et al.*, 2002). Figure 15(b) shows an example of the satisfactory growth of vegetation on a rock slope using this technique, more than 10 years after construction.

Aesthetic design should be integrated into the geotechnical design of slope works at an early stage. For new cuttings, sloping-berm design (as discussed in Section 4.3.1) in conjunction with judicious slope face contouring can avoid monotonous planar slopes and hence achieve better visual harmony. It also gives a more natural appearance to the cut slopes. The use of toe planters and berm planters (Figure 15(c)) to support vegetation is generally warranted where feasible. They can also serve to contain minor rockfalls and aesthetically break up the 'towering effect' of a slope. Opportunities should be taken to improve the appearance of the engineering elements wherever possible. For example, the shape, form and surface finish of a buttress can be designed to reduce its apparent scale.

Figure 15 (a) Climbers growing down the wire mesh on steep rock slope; (b) Use of continuous fiber reinforced soil to support vegetation growth on rock face; (c) Berm planters on high rock slope; (d) Use of artificial rock feature to cover up concrete buttress.

Artificial rock feature can also be applied to cover up buttress in order to retain the natural appearance (*e.g.* along hiking trails, see Figure 15(d)).

4.6 Slope maintenance

The recommended standards of good practice (GEO, 2003) encompass two types of maintenance inspections to be carried out for all registered man-made slopes at regular intervals (see Table 2). Maintenance manuals are prepared for all newly formed and upgraded slopes to assist the owners or maintenance parties to appreciate the slope maintenance requirements.

During the inspection of a rock slope, the possible presence of loose blocks should be carefully examined, particularly in the vicinity of trees and shrubs where isolated rock blocks may be destabilized by root-wedging, and these should be removed or stabilized if found. Clearance of unplanned vegetation is warranted if it is deemed to be detrimental to rock slope stability, where judgment should be made with due regard to factors such as the type of vegetation, condition and orientation of rock joints, etc. Indiscriminate removal of all unplanned vegetation should be avoided. Appropriate means of effective removal of an unplanned tree, including the root

Table 2 Recommended Frequency of Maintenance Inspections for Registered Man-made Slopes.

Consequence-to-life category of man-made slope	*Recommended frequency of*	
	Routine maintenance inspection (by a responsible person such as technical officer – professional geotechnical knowledge and qualification is not necessary)	*Engineer inspection for maintenance* (by a professionally qualified geotechnical engineer)
1 (e.g. affecting buildings or busy roads)	Once every year	Once every five years
2 (e.g. affecting moderately busy road or sitting-out areas)	Once every year	Once every five years
3 (e.g. affecting country parks)	Once every two years	Once every ten years

system, should be adopted to prevent it from re-establishing. Observation should also be made in the maintenance inspections for the presence of open joints and these may need to be sealed up locally to prevent water ingress. However, care must be taken to avoid blockage of drainage path (weepholes to be provided if necessary) that may lead to the build-up of cleft water pressure in joints. Moreover, attention should be paid to the occurrence of any movement along joints that may show signs of progressive failure.

4.7 Health monitoring

Monitoring of rock slopes is generally not required but it may be considered as a risk management tool before completion of upgrading works if there exist areas with signs of distress or with known or suspected movement or complex behavior. The monitoring of rock slope or joint movement can be made by simple means such as surveying the movement of targets or surface-mounted telltale gauges on slopes and using the photogrammetry technique. Annual API surveys have been tried in some local projects for health monitoring with a view to establishing any changes in the features that could lead to the development of a potential hazardous situation, *e.g.* the monitoring of sizeable outcrops on the Seymour Cliffs at Mid-levels area between the late 1980s and the mid 1990s as discussed by Lai (1996). Experience however revealed the limited accuracy of annual API surveys in detecting any impending boulder movement or forewarning rockfall incidents. With the advancement of remote sensing technology, some novel techniques are available nowadays for more reliable movement monitoring of rock slopes/boulders as a health monitoring tool (see Section 6).

Some special measures installed on slopes, in particular those active support or drainage measures whose continual functional performance is critical to long-term slope stability, require continued monitoring throughout their service life. For example, long-term monitoring of the residual load is required for permanent prestressed ground anchors (GCO, 1989).

5 KEY LESSONS LEARNT FROM NOTABLE ROCK SLOPE FAILURES

5.1 General

In this section, some notable rock slope failures in Hong Kong are presented to illustrate the range of factors that could contribute to failures. The key lessons learnt and pertinent issues that warrant special attention are also discussed.

5.2 Major failures on engineered rock slopes

The cases presented below highlight the importance of adequate ground investigation and formulation of a representative geological model to guard against major failures on engineered rock slopes.

5.2.1 The 1993 Hiu Ming Street rockslide

In September 1993, a major rockslide involving a failure volume of about 200 m^3 occurred on a 37 m high rock cut slope above the Hiu Ming Street playground in Kwun Tong during a moderately heavy rainstorm (HCL, 2001). The debris destroyed a section of a boulder fence at the slope toe and extended beyond the buffer zone inundating part of a tennis court (Figure 16).

The cut slope was initially formed in the early 1970s. Prior to the 1993 failure, the slope had been subjected to a number of geotechnical assessments in the 1980s by different parties, which resulted in three phases of stabilization works that were

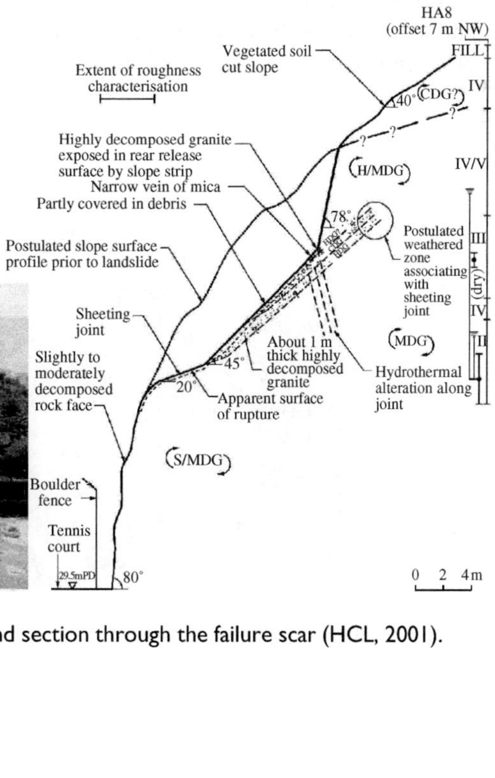

Figure 16 The 1993 Hiu Ming Street rockslide and section through the failure scar (HCL, 2001).

completed in early to mid-1980s and in 1990. The potential of minor detachments was recognized in the previous assessments and hence localized rock slope stabilization measures were carried out. In addition, a buffer zone together with a boulder fence were provided along the slope toe. However, the sizeable failure that subsequently occurred in 1993 was not envisaged by the previous assessments.

The 1993 rockslide was rain-induced and was primarily controlled by the presence of a steeply dipping and undulating basal release surface together with persistent open joints forming the side and rear release surfaces (Figure 16). The subsequent landslide investigation established that the dip angle of the sheeting joint behind the slope face was notably steeper than that exposed on the rock face. The failure was diagnosed to be caused by the build-up of transient cleft water pressure within an adversely orientated, wavy, undulating and partly open sheeting joint that daylighted across the slope. Progressive deterioration of the slope associated with the development of open, partly clay-infilled subvertical joints allowing enhanced water ingress was also a contributory factor to the failure.

This incident emphasizes the uncertainty of the characteristics of severely weathered sheeting joint which may not be readily apparent through the mapping of exposures on the rock face. It highlights the importance of proper characterization of potentially adverse and wavy sheeting joints for slope stability assessment. The need for regular maintenance to prevent slope deterioration is also reinforced.

5.2.2 The 2008 Tsing Yi Road rockslide

In June 2008, a rockslide involving a failure volume of about 250 m^3 occurred on a 30 m high rock cut slope below Tsing Yi Road (Figure 17) during a severe rainstorm (Lam & Lam, 2013). The incident resulted in temporary closure of part of an access road at the slope toe.

The slope was subjected to stability assessments in 1980 and 1999. Stabilization works comprising installation of rock bolts and shotcreting were implemented on the

Figure 17 The 2008 Tsing Yi Road rockslide.

slope. No structural support was recommended on the slope portion that failed in 2008 although a number of rock bolts were installed over the potential unstable rock blocks above it. Continuous seepage flow through the discontinuities in the rock mass over a long period of time was evident.

The 2008 rockslide was rain-induced and primarily occurred as a translational slide with a large rock mass sliding along the basal slip surface. The critical joint forming the basal slip surface was laterally-persistent and generally dipping out of the slope at about 32°. However, the joint became wavy and had a more gentle dip close to the pre-failure slope face where its critical orientation controlling the failure could not be readily appraised simply from surface mapping of the rock face exposures. The joint was mapped in the previous assessments as dipping in a direction perpendicular to the slope and hence the translational sliding mechanism in the 2008 failure was not foreseen. The failure was caused by adverse groundwater conditions, presence of adversely orientated joints and development of cleft water pressure in the joints forming the failure scarp. Other contributory factors to the failure included additional water ingress due to possible overspillage from a blocked crest channel and slope deterioration.

The incident highlights the possible waviness of some stability-critical joints which can be difficult to detect in practice by surface mapping done in some cases.

5.3 Minor failures on engineered rock slopes

The following cases illustrate the common factors contributing to minor failures on engineered rock slopes (viz. tree root-wedging action and inadequate drainage provision/slope maintenance) and highlight the importance of proper detailing of the engineering measures provided.

5.3.1 The 1998 A Kung Ngam Road rockfall

In September 1998, a small rockfall with a failure volume of about 1 m^3 occurred on a 30 m high rock cut slope above A Kung Ngam Road (FSWJV, 2003). The debris, originated from several meters above road level, blocked the footpath and partially affected a short section of the road at slope toe.

The slope was upgraded in 1985 and 1993 respectively. Past records revealed that heavy surface water flow across the slope face and seepage from rock joints in specific areas had been a long-standing issue since the late 1970s. Relatively heavy water flow over a large area of the slope in the vicinity of the failure was also noted during the landslide inspection in 1998 (see Figure 18). The water appeared to originate from near the slope crest due to subsurface flow daylighting from the slope at the soil/rock interface and seepage through rock joints.

The build-up of cleft water pressure in adversely orientated joints is considered to have contributed to the failure. In addition, poor drainage detailing and lack of maintenance were also identified at the upslope area where surface runoff was directed into the drainage system above the slope concerned. The drainage system (without any safe access for maintenance) was blocked and did not appear to have a proper engineered route to direct the flow into a designated drainage outlet. Accordingly, significant water flow overtopped the surface channel and caused the failure. The isolated rock dowels installed in the areas adjoining the failure scar was not affected in the incident.

Figure 18 The 1998 A Kung Ngam Road rockfall.

The incident highlights the importance of providing adequate drainage measures together with proper detailing.

5.3.2 The 2001 King's Road rockfall

In June 2001, a minor rockfall with a failure volume of about 0.1 m^3 occurred on a 15 m high subvertical rock cut slope above King's Road (FMSWJV, 2004). The fallen rock block, dislodged from about 10 m above road level, landed on the footpath and narrowly missed King's Road (a heavily-trafficked road). The footpath and the adjoining carriageway lane of King's Road were temporarily closed following the incident.

The detached rock block originated from a location at the margin of a large wedge (stabilized in 1984) where an unplanned small tree was growing on the slope face (Figure 19). Persistent seepage at the joints delineating the base of the wedge had promoted the growth of unplanned vegetation and the progressive tree root growth had penetrated into joint apertures, thereby wedging the joints open. Localized build-up of cleft water pressure behind the affected rock block as a result of surface infiltration into the open joints is considered to have contributed to the rockfall. The unplanned tree, which probably played a key role in the rockfall incident, was previously identified for removal during routine slope maintenance inspection. The tree was subsequently trimmed back to the main stump but it apparently regenerated rapidly, resulting in the continued penetration of trees roots into the joint apertures.

This incident highlights the potential hazard of tree root-wedging action, which could lead to the opening up of rock joints and make them more susceptible to water ingress and progressive deterioration.

Figure 19 The 2001 King's Road rockfall.

5.3.3 The 2012 Mansion Street rockfall

In July 2012, a minor rockfall with a failure volume of about 0.07 m^3 occurred on a 20 m high rock cut slope near Mansion Street. The rock debris, dislodged from about 2.5 m above ground, was deposited on an open area at the slope toe (Figure 20).

The slope was upgraded in 2002 and the irregular rock cut face was covered with a rock mesh. The source area of the 2012 incident was located close to the edge of the rock mesh. The incident was probably caused by an elevation of cleft water pressure behind the affected rock blocks. The rooting action of the unplanned vegetation in its

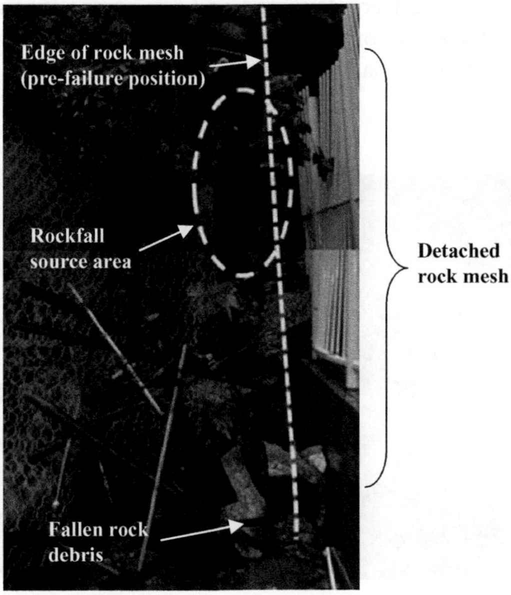

Figure 20 The 2012 Mansion Street rockfall.

vicinity could also have had a detrimental effect on the stability of these rock blocks. Post-landslide inspection revealed that the rock mesh was largely detached from the slope surface and did not closely follow the profile of the rugged rock cut surface. It was also found that the design spacing of the mesh anchorage points at the rockfall location was not met and there was no anchorage provided at the bottom part of the mesh. As a result, the mesh was unable to retain the dislodged rock blocks.

This incident illustrates the importance of paying due attention to the proper detailing of the rock mesh and the need for close site supervision and proper construction control. The inadequacy of the rock mesh anchorage was also not observed during subsequent maintenance inspections.

5.4 Blast-induced rock slope failures during construction

Blasting operations without proper control can lead to massive failures of the nearby rock slopes. Two notable cases are presented below to illustrate the effect of blasting (*e.g.* induced rock fracturing, vibration and gas pressure) on rock slope stability. The importance of an adequate blasting assessment is also highlighted.

5.4.1 The 1991 Shau Kei Wan rockslide

In February 1991, a massive rockslide occurred under dry weather during the site formation works to create a platform for a new housing estate in Shau Kei Wan (Evans & Irfan, 1991). It involved the detachment of about 2,000 m^3 of granitic rock from the crest of a 50 m high steep rock face of a disused quarry (Figure 21) shortly after the completion of a blast within some 15 m behind the slope crest. The rock debris was largely deposited within the construction site boundary. Six container offices and four vehicles at the old quarry base were damaged and one person was slightly injured.

The detached rock mass slid on a statistically minor, impersistent but adversely orientated joint which had apparently been disregarded in the previous stability

Figure 21 Blast-induced rockslides.

assessment. The amount of explosives used for the blast appeared to be excessive given the presence of a shear zone and decomposed seams in the blast area. Extensive blast damage was noted in the rock mass, with spalling and opening of joints due to shock waves and gas pressure. Fracturing of intact rock could have occurred progressively during the whole period of blasting works. The landslide investigation established that, inter alia, the sole reliance on a statistical approach to data collection and analysis is liable to miss out those infrequent but stability-critical discontinuities.

5.4.2 The 1997 Sau Mau Ping Road rockslide

In December 1997, a massive rockslide occurred under dry weather during the site formation works for a housing development above Sau Mau Ping Road (Leung *et al.*, 1999). It involved the detachment of about 1,000 m^3 of granitic rock on a 25 m high rock slope immediately after the completion of a blast about 3 m behind the slope crest. The rock debris blocked the entire four lanes of a 25 m long section of Sau Mau Ping Road and completely destroyed a section of the protective fence along the slope toe (Figure 21). While there was no casualty, the affected section of road had to be closed for a period of 17 days.

The detached rock blocks slid on sheeting joints dipping at about 25°. Extensive blast damage was noted in the rock mass, with dilation of joints and movement of blocks. Theoretical analysis carried out during the landslide investigation indicated that the blast-induced vibration alone could not have resulted in complete detachment of the rock blocks of such scale. With blasting carried out in close proximity, the slope failure could have been triggered by the near-field shock waves and gas pressure generated by the blast. The investigation also revealed the overcharging of explosives with reference to the allowable limit stipulated in the blasting permit.

5.5 Other notable rock slope failures during construction

Apart from blast-induced failures, other rock slope failures during construction owing to factors such as lack of proper understanding of the site geology, poor workmanship and inadequate site supervision have also been reported. Two case studies are presented below to demonstrate the importance of proper ground investigation and construction control/review.

5.5.1 The 1982 South Bay Close rockslides

Two massive rockslides occurred in May and August 1982, with failure volumes of about 3,800 m^3 (Figure 22(a)) and 700 m^3 respectively, on a partially completed 24 m high rock cut slope within the construction site for a private development at South Bay Close (Hencher, 1983). The rockslides were both triggered by heavy rainfall. Fortunately, the landslide debris was confined within the site boundaries and did not result in any casualty.

The key contributory factor to the landslides was the presence of a major, adversely orientated infilled discontinuity (both rockslides failed along the same discontinuity) which was not identified from the ground investigation and hence not taken into account in the design. The infill to the discontinuity was up to about 700 mm thick (Figure 22(b))

Figure 22 (a) General view of the May 1982 South Bay Close rockslide; (b) 700 mm thick infill along the surface of rupture.

consisting of moderately decomposed gravel fragments in a matrix of sand and silt with pockets of light brown and pink clay. Clover (1986) reported that such unexpected zone of weakness was recognized during excavation but the slope had failed before the completion of the design of additional stabilization measures. Moreover, the development of a high perched water table above the infilled zone (above which preferential flow through soil pipes was also noted; as indicated in Figure 22(b)) during heavy rainstorms might have played a role in the failures. It is noteworthy that the groundwater level measured using Halcrow buckets in piezometers following the first failure recorded much higher groundwater level than that assumed in the design, which was predicted based on manual dipping of the piezometers.

These incidents highlight the importance of good quality ground investigation for establishing a representative geological model and identification of the potential failure mechanism. In respect of groundwater monitoring, piezometers should be provided near the base of the zone underlain by a relatively impermeable layer (*e.g.* clay seams) and equipped with automatic recording devices or Halcrow buckets in order to capture the peak groundwater level. These should preferably be monitored for at least a complete wet season before the design is finalized. As ground variability and adverse geological features are often not fully revealed until excavation takes place, the importance of timely and regular construction review cannot be over-emphasized.

5.5.2 The 1995 Tuen Mun Highway rockfall

In August 1995, a rockfall incident occurred during the operation of scaling works on an overhanging rock outcrop at about 30 m above Tuen Mun Highway, under a road widening project (Sun & Tsui, 2005). This involved the detachment of a rock block (about 2 m by 2 m by 2.5 m) from the crest of the slope which fell onto the highway below outside the site boundary (Figure 23). A van traveling along the road hit the rock block killing the driver and injuring the only other passenger. Three carriageway lanes were

Figure 23 The 1995 Tuen Mun Highway rockfall.

temporarily closed for six days following the incident and the works under the road widening project at the section of the road of concern were suspended for about 11 months.

Prior to the incident, the top segment of the outcrop had not been successfully split in one go as intended and was left in place. Within minutes, the incident occurred suddenly with the detachment of the rock block from the outcrop. No precautionary measures, such as cabling or dowelling, were provided prior to the rock splitting operation. The incident highlights the critical importance of appropriate risk assessment prior to potentially hazardous operations and proper construction control and supervision toward public safety, in particular under a dense urban setting.

5.6 Key lessons learnt and enhancement initiatives

In addition to fundamental engineering principles, proper engineering judgment and local experience are of the essence in dealing with the highly variable tropically weathered profiles commonly encountered in Hong Kong. Face mapping could be difficult for rock slopes covered with hard surface which require surface stripping or window openings for mapping. Reference should also be made to photographs (*e.g.* from construction or maintenance records, where available) depicting the condition of the rock face before the hard surface cover was applied.

Sizeable failures on engineered rock slopes are relatively rare in Hong Kong but they have taken place given the adverse ground conditions which were not fully recognized or appreciated in the ground investigation and design. The waviness of a sheeting joint can involve local steepening of the joint dip angle behind the slope faces which can be difficult to detect in practice. The inherent variability of joint orientations should be recognized. Input from geotechnical professionals with adequate engineering geological knowledge and local experience would be critical in establishing the realistic probable failure mechanisms and identifying suitable stabilization measures.

Despite the geotechnical input, the failure rate of engineered rock slopes is not that low as compared with that for non-engineered rock slopes. Common contributory factors include the adverse effect of root action of unplanned vegetation, poor detailing of drainage provisions, and inadequate slope maintenance (*e.g.* damaged hard surface cover, blocked/broken drainage channels, etc.). Minor failures are also sometimes associated with statistically-rare but stability-critical joints. Designers should be cautious of the inherent limitations of applying stereographic projections for kinematic analysis of rock slope stability (Hencher, 1985; GEO, 2014). The contouring of rock joint data can overshadow the statistically-rare but stability-critical joint data. In light of this, uncontoured joint data falling within the potential unstable zones on the stereoplots should always be critically appraised. Over-reliance on kinematic analysis to identify unstable blocks should be avoided but rather more emphasis should be given to judicious field mapping to directly identify the problematic rock blocks.

In view that minor rockfalls can cause serious consequences in a dense urban setting, enhanced practice in relation to the design and detailing of upgrading works is called for, with due attention given to address the common factors leading to minor failures. Given the uncertainties involved and the practical difficulty in eliminating local minor failures, it is generally cost effective and more robust to adopt a pragmatic approach in minimizing the consequence of failure. The installation of rock mesh over the rock face can provide a simple and effective mean to contain minor rockfalls. Proper detailing (*e.g.* proper anchorages, extent of mesh coverage, etc.) is of paramount importance to ensure that the installed mesh would serve its intended function. It is noteworthy that in most circumstances, lack of space typically precludes the use of rock traps.

Rock slopes can be vulnerable to failure during the construction stage. Particular attention should be paid to the performance review and verification of original design against the actual site conditions, provision of appropriate temporary support and site drainage, consideration of blasting impact, implementation of appropriate precautionary and contingency measures, workmanship, site supervision, etc. Adequate provision should always be made in the construction program for detailed rock face mapping and review/amendment of the design and construction details of stabilization/mitigation measures.

Further advices and technical recommendations on enhancing the rock slope engineering practice taking due cognizance of the lessons learnt and findings from landslide studies are presented in GEO (2014) and Ho & Lau (2007).

6 USE OF NOVEL TECHNOLOGY IN ROCK SLOPE ENGINEERING IN HONG KONG

6.1 General

Although the use of novel technologies in rock slope engineering is yet to be fully integrated into the local practice, efforts have been made to explore the possible application and test the various novel tools. This section gives a brief overview of some of the trials conducted in Hong Kong on the use of novel technologies for the investigation and movement monitoring of rock slopes and boulders. The applicability and limitations of the novel techniques are also highlighted.

6.2 Novel technology for rock slope investigation and boulder detection

Discontinuity survey by means of non-contact measurement using the laser scanning technology may be resorted to under constrained site settings, *e.g.* where manual measurement is difficult or even impractical due to inaccessibility/presence of heavy overgrowth or potentially hazardous such as surveying of fresh landslide scar. Trial applications of terrestrial as well as mobile laser scanning conducted by the GEO revealed that these techniques can be used as a reliable and efficient supplement to manual measurement on joint orientations. Notwithstanding this, it should be cautioned that the methods only give spatial information on rock joints. While this may be sufficient for performing kinematic analysis, other important information critical for rock slope stability assessment, such as the degree of weathering and condition of seepage/infill materials, would need to be obtained by other means. A close-up inspection to collect such critical information may still be needed for conducting detailed stability assessment. It should also be noted that the application of laser scanning survey to extract discontinuity data is governed by the point spacing and accuracy of the laser scanning technique adopted. The surveys do not register discontinuities that are not in line-of-sight or dipping into the slopes. They also have limitations in measuring tight and narrow, low-persistent, rough and non-planar discontinuities.

Unmanned aerial vehicle (UAV) can also be applied to aid remote sensing given the advantage offered by the flexibility in taking photographs and/or measurements at favorable vantage points. This is particularly useful for sites with access problems. Apart from photogrammetric imaging, there has been on-going development in respect of equipping laser scanner on UAV and its application to rock slope investigation is worth exploring.

Site trials have been conducted by the GEO to explore the potential application of the image processing technique to aid boulder detection and mapping (Ng *et al.*, 2003). This technique could substantially reduce the amount of fieldwork required and facilitates the mapping of boulders on steep and difficult terrain. While the site trials indicated that this technique is capable of detecting boulders and mapping their respective sizes and shapes, the overall accuracy is not very high. Further study and improvement is warranted before this technique could be used for routine application.

6.3 Novel technology for movement monitoring

To facilitate monitoring of specific problematic rock slopes or active rockslides, the GEO carried out trials at several sites in 2009 to test the accuracy and resolution of ground-based Interferometric Synthetic Aperture Radar (InSAR) in capturing the movement of several moveable targets (concrete plates of about 1 m^2 in size) by remote monitoring at some 200 m to 500 m away. The results are generally encouraging although some limitations such as reduced accuracy under heavy rainfall, potential blind spots and uncertainty of the capability of penetrating through vegetation for certain wave frequencies were noted.

In respect of boulder movement monitoring, the GEO is currently exploring the use of the facet approach to detect changes in orientation of the pre-demarcated faces on a boulder by analyzing and comparing the annual airborne Light Detection and Ranging (LiDAR) data. The preliminary results of the study appear promising in detecting sizeable boulders with significant rotational movements. However, the point spacing and positional accuracy of the LiDAR data remain the major limitations which hinder the application of this method for detecting small boulders (*e.g.* of plan dimensions of less than 0.5 m × 0.5 m) with minor movements.

Airborne LiDAR technology with the multi-return capability can produce 'bare-earth' ground profiles or digital terrain models even in heavily vegetated terrain. The bare-earth model gives a better representation of the topography for boulder runout analysis and other types of natural terrain hazard studies.

7 DISCUSSION AND CONCLUSIONS

This paper has consolidated the experience from risk and asset management of a large stock of rock slopes in the dense urban setting in Hong Kong. The combination of the vulnerable site setting together with high public expectations calls for a very high standard of slope safety. A dense urban setting is not only vulnerable to minor failures but is also liable to lead to cascading failures due to interaction of the various components (*e.g.* landslides blocking a catchwater channel leading to uncontrolled overflow causing landslides or washout failures on the downhill area, overspilling of surface water from road bends, bursting of pressurized water mains, leakage from buried water-carrying services, etc.). The risk management framework and the related technical policies implemented in Hong Kong are described and their effectiveness has been reviewed. The risk management system is subject to regular reviews and continuous improvement. Following concerted slope safety efforts in the past years, the system has, by and large, worked well in containing the landslide risk posed to the community.

In practice, the basic and simple rock mechanics principles are generally found to be adequate for the routine problems posed by the typical range of rock slope hazards encountered locally. However, experience has shown that the application of theory to practice in managing landslide risk can sometimes be fraught with difficulty. This is reflected by a relatively high failure rate of engineered rock slopes, especially in respect of minor detachments, which can be problematic in the context of a dense urban setting. The application of the more sophisticated theories and methodologies would require more input parameters and the geotechnical practitioners may not necessarily have a good 'feel' for these more advanced analytical or numerical tools, which almost certainly will not be able to avert many of the failures observed in Hong Kong in hindsight.

The best practice is generally guided and shaped by feedback on actual slope performance and lessons learnt from local failures. Observations and lessons from systematic landslide studies have highlighted the key contributory factors to failure (*e.g.* root-wedging action associated with existing or unplanned vegetation, uncontrolled discharge of surface water flow leading to concentrated water ingress, progressive deterioration due to inadequate maintenance, inadequate geological model, etc.). Observations from past notable failures have revealed the many uncertainties involved in professional judgment (*i.e.* human uncertainties or human errors), including the identification of all stability-critical joints and the characterization of their pertinent engineering characteristics, by virtue of the variability of the tropically weathered rocks. The need to ensure rigor and adequate geotechnical input by suitably qualified and experienced professionals cannot be over-emphasized.

Potential time-dependent changes are also liable to be contributory factors to failure, such as progressive deterioration, changes in the site setting of the uphill area (*e.g.* leading to diversion of concentrated surface water flow), growth of unplanned vegetation, etc. This emphasizes the need for regular maintenance inspections (by technical staff or similar) supplemented by continued professional input by suitably qualified geotechnical personnel (through engineer inspections), with a view to facilitating early detection of slope distress and anticipating problems before they develop to an advanced stage.

A key element of good design as well as good detailing is to enhance the robustness of the engineering measures. For example, the use of rock mesh with adequate anchorage support on a jointed rock face has shown to be a simple and reasonably robust measure for managing the risk of minor detachments.

The physical constraints associated with the retrofitting of existing substandard rock slopes in close proximity to buildings and busy roads call for the judicious application of appropriate enabling construction technology. In addition, experience has shown that there is good potential for the use of novel digital technology to assist with the mapping and investigation of rock slopes, provided that their applicability and limitations are properly understood.

ACKNOWLEDGMENT

This paper is published with the permission of the Head of the Geotechnical Engineering Office and the Director of Civil Engineering and Development, Government of the Hong Kong Special Administrative Region.

REFERENCES

Au, S.W.C. & Chan, C.F. (1991) Boulder treatment in Hong Kong. *Selected Topics in Geotechnical Engineering – Lumb Volume*, pp. 39–71.

Beggs, C.J., Threadgold, L., Blomfield, D. & Chan, Y.C. (1984) Rockfall and its control. *Hong Kong Engineer*. Vol. 12, No. 4, pp. 41–43.

Blastronics Pty Ltd. (2000) *Methods of Assessment and Monitoring of the Effects of Gas Pressures on Stability of Rock Cuts due to Blasting in the Near-field*. GEO Report No. 100. Geotechnical Engineering Office, Hong Kong.

Brand, E.W., Hencher, S.R. & Youdan, D.G. (1983) Rock slope engineering in Hong Kong. *Proceedings of the 5th International Rock Mechanics Congress, Melbourne*. Vol. 1, pp. C17–C24.

Chan, T.F.C. & Au, S.W.C. (1988) *Mid-levels Boulder Field Preventive Works Pilot Scheme*. Design Study Report No. DSR 1/88. Geotechnical Control Office, Hong Kong.

Chan, Y.C., Chan, C.F. & Au, S.W.C. (1986) Design of a boulder fence in Hong Kong. *Proceedings of the Conference on Rock Engineering and Excavation in an Urban Environment*. Institution of Mining and Metallurgy, Hong Kong Branch, pp. 87–96.

Cheng, P.F.K. (2014) *Assessment of Landslide Risk Posed by Man-made Slopes as of 2010*. GEO Report No. 297. Geotechnical Engineering Office, Hong Kong.

Cheng, P.F.K. & Ko, F.W.Y. (2010) *An Updated Assessment of Landslide Risk Posed by Man-made Slopes and Natural Hillsides in Hong Kong*. GEO Report No. 252. Geotechnical Engineering Office, Hong Kong.

Clover, A.W. (1986) Slope stability on a site in the volcanic rocks of Hong Kong. *Proceedings of the Conference on Rock Engineering and Excavation in an Urban Environment*. Institution of Mining and Metallurgy, Hong Kong Branch, pp. 121–134.

Cowland, J.W. & Richards, L.R. (1985) Transient groundwater rises in sheeting joints in a Hong Kong granite slope. *Hong Kong Engineer*. Vol. 13, No. 2, pp. 27–32.

Evans, N.C. & Irfan, T.Y. (1991) *Landslide Studies 1991: Blast-induced Rock Slide at Shau Kei Wan*. Special Project Report No. SPR 6/91. Geotechnical Engineering Office, Hong Kong.

Fugro Maunsell Scott Wilson Joint Venture (2004) *Review of the 25 June 2001 Rockfall on Slope No. 11SE-A/C561 Above King's Road*. GEO Report No. 150. Geotechnical Engineering Office, Hong Kong.

Fugro Scott Wilson Joint Venture (2003) *Initial Study of the Landslide at Slope No. 11SE-B/C87 A Kung Ngam Road, Shau Kei Wan on 17 September 1998*. Geotechnical Engineering Office, Hong Kong.

Geotechnical Control Office (1984) *Geotechnical Manual for Slopes*. 2nd edition. Geotechnical Control Office, Hong Kong.

Geotechnical Control Office (1987) *Guide to Site Investigation (Geoguide 2)*. Geotechnical Control Office, Hong Kong.

Geotechnical Control Office (1988) *Guide to Rock and Soil Descriptions (Geoguide 3)*. Geotechnical Control Office, Hong Kong.

Geotechnical Control Office (1989) *Model Specification for Prestressed Ground Anchors (Geospec 1)*. Geotechnical Control Office, Hong Kong.

Geotechnical Engineering Office (1993) *Guide to Retaining Wall Design (Geoguide 1)*. 2nd edition. Geotechnical Engineering Office, Hong Kong.

Geotechnical Engineering Office (2000) *Highway Slope Manual*. Geotechnical Engineering Office, Hong Kong.

Geotechnical Engineering Office (2003) *Guide to Slope Maintenance (Geoguide 5)*. 3rd edition. Geotechnical Engineering Office, Hong Kong.

Geotechnical Engineering Office (2004) *Boulders*. GEO Circular No. 31. Geotechnical Engineering Office, Hong Kong.

Geotechnical Engineering Office (2006) *Geotechnical Control of Blasting*. GEO Circular No. 27. Geotechnical Engineering Office, Hong Kong.

Geotechnical Engineering Office (2007) *Engineering Geological Practice in Hong Kong*. GEO Publication No. 1/2007. Geotechnical Engineering Office, Hong Kong.

Geotechnical Engineering Office (2009) *Prescriptive Measures for Man-made Slopes and Retaining Walls*. GEO Publication No. 1/2009. Geotechnical Engineering Office, Hong Kong.

Geotechnical Engineering Office (2011) *Technical Guidelines on Landscape Treatment for Slopes*. GEO Publication No. 1/2011. Geotechnical Engineering Office, Hong Kong.

Geotechnical Engineering Office (2013) *Study and Mitigation of Natural Terrain Hazards*. GEO Circular No. 28. Geotechnical Engineering Office, Hong Kong.

Geotechnical Engineering Office (2014) *Enhancement of Rock Slope Engineering Practice Based on Findings of Landslide Studies*. GEO Technical Guidance Note No. 10. Geotechnical Engineering Office, Hong Kong.

Halcrow China Ltd. (2001) *Detailed Study of Selected Landslides on Slope No. 11NE-D/C45 Hiu Ming Street Kwun Tong*. Landslide Study Report No. LSR 7/2001. Geotechnical Engineering Office, Hong Kong.

Hencher, S.R. (1983) *Landslide Studies 1982, Case Study No. 4, South Bay Close*. Special Project Report No. SPR 5/83. Geotechnical Control Office, Hong Kong.

Hencher, S.R. (1985) Limitations of stereographic projections for rock slope stability analysis. *Hong Kong Engineer*. Vol. 13, No. 7, pp. 37–41.

Hencher, S.R., Lee, S.G., Carter, T.G. & Richards, L.R. (2011) Sheeting joints: Characterisation, shear strength and engineering. *Rock Mechanics and Rock Engineering*. Vol. 44, No. 1, pp. 1–22.

Hencher, S.R. & Richards, L.R. (1982) The basic frictional resistance of sheeting joints in Hong Kong granite. *Hong Kong Engineer*. Vol. 11, No. 2, pp. 21–25.

Ho, K.K.S., Cheung, R.W.M. & Kwan, J.S.H. (2015) Advances in Urban Landslide Risk Management. *Proceedings of the International Conference on Geotechnical Engineering*. Sri Lankan Geotechnical Society. (In print).

Ho, K.K.S. & Lau, T.M.F. (2007) Enhancement of rock slope engineering practice based on lessons learnt from systematic landslide studies. *Proceedings of the 16th Southeast Asian Geotechnical Conference*, Kuala Lumpur. Southeast Asian Geotechnical Society, pp. 811–818.

Ho, N.L. & Wright, M.J. (2011) Construction of slope upgrading works along busy roads in Hong Kong. *Proceedings of the 31st Annual Seminar*. Geotechnical Division, Hong Kong Institution of Engineers, pp. 82–89.

Hoek, E. & Bray, J. (1981) *Rock Slope Engineering*. 3rd edition. Institute of Mining and Metallurgy, London.

Jones, C.L., Higgins, J.D. & Andrew, R.D. (2000) *Colorado Rockfall Simulation Program: Version 4.0 (for Windows)*. Colorado Department of Transportation, the United States.

Lai, A.C.S. (1996) *Mid-levels Seymour Cliffs Improvement Works*. Stage 3 Study Report No. S3R 69/96. Geotechnical Engineering Office, Hong Kong.

Lam, A.Y.T., Lau, K.W.K. & Yim, K.P. (2002) Aesthetic geotechnical design for the Tsing Yi North Coastal Road project. *Proceedings of the Annual Seminar*. Geotechnical Division, Hong Kong Institution of Engineers, pp. 179–185.

Lam, C.L.H. & Lam, H.W.K. (2013) *Review of the 7 June 2008 Rock Slope Failure on Slope No. 10NE-B/C56 Below Tsing Yi Road*. GEO Report No. 288. Geotechnical Engineering Office, Hong Kong.

Lau, K.C. & Franks, C.A.M. (2000) *Phase 3 Site Characterisation Study – Stage 4: Evaluation of Downhole Geophysical Methods for Ground Investigation at Field Trial Sites*. GEO Technical Note No. TN 6/2000. Geotechnical Engineering Office, Hong Kong.

Lee, F.Y.K., Chu, B.S.W. & Wong, T.K.C. (2011) Challenges of landslip preventive measures works with complex site constraints: case study – four slope features at Coombe Road. *Proceedings of the 31st Annual Seminar*. Geotechnical Division, Hong Kong Institution of Engineers, pp. 90–98.

Leung, B.N., Leung, S.C. & Franks, C.A.M. (1999) *Report on the Rock Slope Failure at Cut Slope 11NE-D/C7 along Sau Mau Ping Road on 4 December 1997*. GEO Report No. 94. Geotechnical Engineering Office, Hong Kong.

Leung, J.C.W., Lee, R.W.H. & Ting, S.M. (2013) *Factual Report on Hong Kong Rainfall and Landslides in 2012*. Special Project Report No. SPR 3/2013. Geotechnical Engineering Office, Hong Kong.

Lo, D.O.K. & Cheung, R.W.M. (2005) *Assessment of Landslide Risk of Man-made Slopes in Hong Kong*. GEO Report No. 177. Geotechnical Engineering Office, Hong Kong.

Lo, D.O.K., Ho, K.K.S. & Wong, H.N. (1998) Effectiveness of slope maintenance in reducing the likelihood of landslide. *Proceedings of the Hong Kong Institution of Engineers Geotechnical Division 17th Annual Seminar on Slope Engineering in Hong Kong*. A.A. Balkema Publisher, Rotterdam, pp. 251–258.

Mak, N. & Bloomfield, D. (1986) Rock trap design for presplit rock slopes. *Proceedings of the Conference on Rock Engineering and Excavation in an Urban Environment*. Institution of Mining and Metallurgy, Hong Kong Branch, pp. 263–269.

Massey, J.B. & Siu, K.L. (2003) *Investigation of Flyrock Incident at New Clear Water Bay Road on 6 June 2003*. Geotechnical Engineering Office, Hong Kong.

Maunsell Geotechnical Services Ltd. (2001) *Territory Wide Quantitative Risk Assessment of Boulder Fall Hazards – Stage 2 Final Report*. Geotechnical Engineering Office, Hong Kong.

Maunsell Geotechnical Services Ltd. (2004) *Detailed Study of the 1 September 2001 Debris Flow on the Natural Hillside above Lei Pui Street*. GEO Report No. 154. Geotechnical Engineering Office, Hong Kong.

Ng, K.C., Li, X.C., Shi, W.Z., Zhu, C.Q. & Shum, W.L. (2003) Detection of boulders on natural terrain using image processing techniques. *Proceedings of the Conference on Intelligent Engineering Applications of Digital Remote Sensing Technology*. Civil, Environmental and Geotechnical Divisions, Hong Kong Institution of Engineers, pp. 55–63.

Nicholls, K.H., Cowland, J.W. & Chan, R.K.S. (1992) Effects of blasting on the stability of adjacent rock slopes. *Proceedings of the Conference 'Asia Pacific – Quarrying the Rim'*. Institute of Quarrying, Hong Kong, pp. 185–195.

Nicholson, D., Tse, C.M., Penny, C., O'Hana, S. & Dimmock, R. (1999) *The Observational Method in Ground Engineering: Principles and Applications*. CIRIA Report 185. Construction Industry Research and Information Association, London.

Richards, L.R. & Cowland, J.W. (1986) Stability evaluation of some urban rock slopes in a transient groundwater regime. *Proceedings of the Conference on Rock Engineering and Excavation in an Urban Environment*. Institution of Mining and Metallurgy, Hong Kong Branch, pp. 357–363.

Smith, M.J. & Yeung, T.H. (1991) *Boulder Fence Extension Area above Conduit Road*. Stage 3 Report No. S3R 7/91. Geotechnical Control Office, Hong Kong.

Stevens, W.D. (1998) *RocFall: A Tool for Probabilistic Analysis, Design of Remedial Measures and Prediction of Rockfalls*. Department of Civil Engineering, University of Toronto, Ontario, Canada.

Sun, H.W. & Tsui, H.M. (2005) *Review of Notable Landslide Incidents During Slope Works*. GEO Report No. 171. Geotechnical Engineering Office, Hong Kong.

Tse, S.H., Lo, D.O.K., Tsui, H.M. & Ng, S.L. (2003) Some aspects of mitigation measures against natural terrain landslide hazards in Hong Kong. *Proceedings of the Annual Seminar*. Geotechnical Division, Hong Kong Institution of Engineers, pp. 246–253.

Wong, H.N. (2009a) Holistic urban landslide risk management – challenges and practice (keynote lecture). *Proceedings of the 7th Asian Regional Conference*. International Association for Engineering Geology and the Environment, Chengdu, China, pp. 28–43.

Wong, H.N. (2009b) Rising to the challenges of natural terrain landslides. *Proceedings of the 29th Annual Seminar*. Geotechnical Division, Hong Kong Institution of Engineers, pp. 15–54.

Wong, H.N. & Ho, K.K.S. (1996) Travel distance of landslide debris. *Proceedings of the 7th International Symposium on Landslides*. A.A. Balkema Publisher, Trondheim, Norway, Vol. 1, pp. 417–422.

Wong, H.N. & Ho, K.K.S. (1997) Systematic investigation of landslides caused by a severe rainstorm in Hong Kong. *Transactions*. Hong Kong Institution of Engineers. Vol. 3, No. 3, pp. 17–27.

Wong, H.N. & Ho, K.K.S. (2000) Learning from slope failures in Hong Kong. *Proceedings of the 8th International Symposium on Landslides*. Thomas Telford, London.

Wong, H.N., Ko, F.W.Y. & Hui, T.H.H. (2006) *Assessment of Landslide Risk of Natural Hillsides in Hong Kong*. GEO Report No. 191. Geotechnical Engineering Office, Hong Kong.

Wong, H.N., Pang, L.S., Wong, A.C.W., Pun, W.K. & Yu, Y.F. (1999) *Application of Prescriptive Measures to Slopes and Retaining Walls*. 2nd edition. GEO Report No. 56. Geotechnical Engineering Office, Hong Kong.

Wong, H.N. & Pang, P.L.R. (1992) *Assessment of Stability of Slopes Subjected to Blasting Vibration*. GEO Report No. 15. Geotechnical Engineering Office, Hong Kong.

Tunnels and Caverns

Chapter 3

Tunnels in the Himalaya

R.K. Goel[1] *& Bhawani Singh*[2]
[1]*Chief Scientist & Professor (AcSIR), CSIR-CIMFR Regional Centre, Roorkee, India*
[2]*Former Professor, Department of Civil Engineering, IIT Roorkee, Roorkee, India*

"Learn from yesterday. Live for today. Hope for tomorrow.
The important thing is not to stop questioning."

– Albert Einstein

1 INTRODUCTION

Generally, a tunnel layout is first prepared and the tunneling operations are started after collecting adequate geological information of the area. Success of the tunneling operations depends upon the reliability of these geological predictions. It is easier to collect geological details when the tunnel is shallow, the terrain is flat and the rock mass is not much disturbed. In such regions the ideal approach would be first to make a quick geophysical exploration to identify such features as major faults, shear zones, sand pockets, water bodies etc. Once the presence or absence of such features is established, conventional geological exploration should be planned for detailing. Such an approach would optimize the time and cost of exploration efforts besides providing useful geological information and reducing the chances of surprises.

The need for the geological information becomes all the more valid in a tunnel where the cover is high, the terrain is inaccessible and the rock mass is highly disturbed tectonically below a thick forest cover. Under such conditions, drilling up to the tunnel grade is costly and sometimes impossible. Attempts to infer the geology up to the tunnel grade by extrapolating meager surface data and to plan the layout of a major tunneling project on the basis of such geological projections often lead to serious unforeseen problems. These problems sometimes lead to time and cost over-runs.

Major tunneling problems in India are encountered in the young Himalayan regions, particularly the lesser or lower Himalaya, where the geology is difficult, the rock masses are weak and undergoing intense tectonic activities resulting into major faults, folds and other discontinuities. Compared to this, in the peninsular (southern part) India where the rocks are strong and less disturbed, tunneling problems are rarely encountered.

A number of hydroelectric projects are located in the lower Himalaya. In addition, rail and road tunnels are also being constructed on a mass scale. These projects lie in the Himalayan states of Assam, Himachal Pradesh (H.P.), Jammu & Kashmir (J&K), Manipur, Uttarakhand, etc. in India. Detailed geological exploration work for all the projects throughout the country has been mainly undertaken by Geological Survey of India (GSI). Despite the best efforts of the geologists, inadequacies in the prediction of nature of the rock masses, at the tunnel grade were observed in most of the tunnels in the Himalaya. These inadequacies led to different tunneling problems like water-inrush, roof falls, cavity formation, face collapse, swelling, support failure, gas explosion etc. In addition, squeezing ground conditions in the weak rock masses of Himalaya have also created considerable construction problems.

Experience of TBM tunneling in the Himalaya, so far, is not encouraging. But, the success of TBM in a recently completed head race tunnel of Kishanganga hydroelectric project has certainly encouraged the morale of designers and engineers in favor of TBM. The key issues for the success of TBM in the Himalaya are highlighted.

Since the chapter is on 'Tunnels in Himalaya', a brief geology of the Himalaya is presented at first.

2 THE HIMALAYA

The Indian subcontinent is surrounded in the north by a lofty mountain chain known as the Himalaya. The Himalayan range with NW-SE general trend was formed, according to the theory of 'Continental Drift', by the collision of the Indian Plate with the Eurasian Plate. The Indian plate is known to be moving toward north at a rate of approximately 5cm per year. This collision, which began with the first contact about 40 million years ago, caused the sediments of the intervening Tethys Sea and the Indian Shield to be folded and faulted into the lofty peaks and outliers visible in the lesser Himalaya. Since the northward shift of the Indian plate is still continuing, the mountain building process is still continuing and the zone is still active seismically. On the basis of its average height from mean sea level (MSL), from south to north, the Himalayan range and the rock formations are divided as per Table 1 and shown in Figure 1. Similarly, geographically from west to east, the Himalaya is divided as given in Table 2.

Table 1 Rock formations and average height above mean sea level of Himalaya (Goel *et al.*, 1995).

Rock Formation (Broadly)	*Average Height from Mean Sea Level (MSL)*	*Popular Name*
SOUTH		
Indo-Gangetic Planes		
---------- ***Main Frontal Thrust*** ----------		
Soft, loose and easily erodible rocks, e.g. sandstone, siltstone, mudstones, clays and or claystones, conglomerates	Up to 1000m	Sub-Himalaya or Shiwaliks
---------- ***Main Boundary Fault*** ----------		
Sedimentary formations, e.g. slates, dolomites, quartzites, shales, claystones etc. Metamorphic formations, e.g. phyllites, quartzites, schists, gneisses etc.	1000m to 4000m	Lesser or Lower Himalaya
---------- ***Main Central Thrust*** ----------		
Weak sedimentary formations e.g. shales sandstones, siltstones, conglomerates and strong metamorphic formations e.g. gneisses, migmatites schists, marble etc.	> 4000m	Greater or Higher Himalaya
NORTH		

Table 2 Sub-divisions of Himalaya from west to east.

WEST
Punjab Himalaya
Kumaon Garhwal Himalaya
Nepal Himalaya
Sikkim Bhutan Himalaya
Nefa
EAST

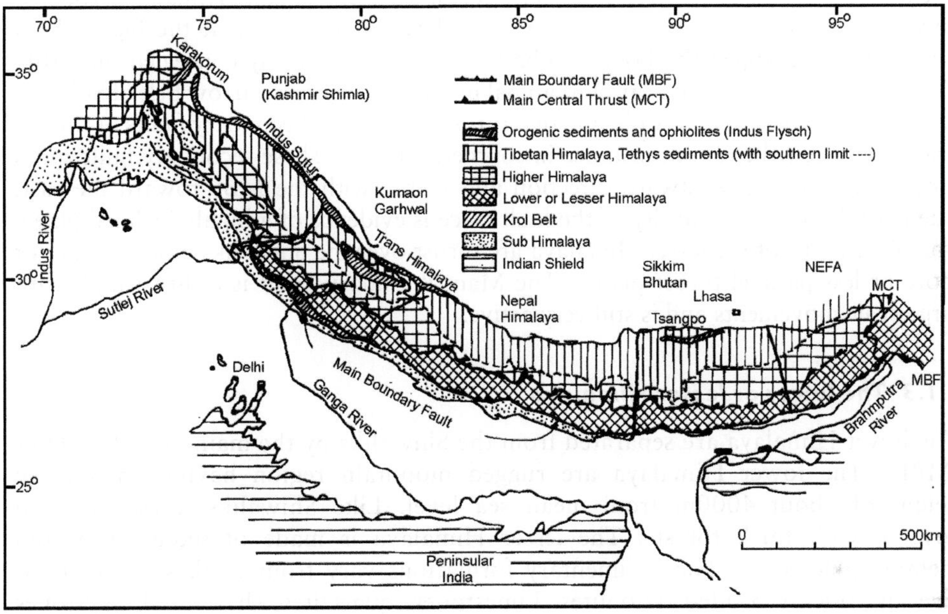

Figure 1 Longitudinal and transverse sub-divisions of the Himalaya (after Gansser, 1964).

2.1 Geology of the Himalaya

Moving from south to north, main frontal thrust (MFT) separates the Shiwaliks from the Indo-Gangetic planes (Table 1).

2.1.1 The Shiwaliks

The Shiwalik rocks constitute the southern foothills of the Himalaya. With an average height of about 1000m from mean sea level (MSL), these are generally covered with thick forests and comprise the youngest rocks in the Himalayan range. The soft, loose, and easily erodible rocks are represented by sand rocks, sandstones, siltstones, claystones, mudstones and conglomerates. Water penetrates into these rock masses along the fractures and joints and sometimes creates flowing ground conditions (*e.g.* Khara tunnel, Udhampur-Katra rail tunnel, etc.). Since the Shiwalik rocks are less resistant to weathering, the engineering behavior of these rocks is likely to vary with time.

2.1.2 The Main Boundary Fault (MBF)

Separating the Shiwalik Formations of the Sub-Himalaya from the older rocks of Lesser Himalaya lying to their north, the Main Boundary Fault is a major structural plane discernible throughout the length of the Himalaya. Hitherto regarded as a steep north dipping fault, it is more likely a thrust which flattens with depth. The MBF, originally defined as the tectonic feature separating the Shiwalik from the pre-Shiwalik Tertiaries, is exposed only in the extreme western sector in the Kumaon Himalaya, roughly between the Yamuna and Tons valleys. On East of the Yamuna, the higher Krol Thrust has overlapped the Eocene Subathu and has completely concealed the MBF. The only exception is seen near Durgapipal in the east where a narrow belt of Subathu is exposed between the Shiwalik and overthrust Krol rocks. Secondary faults or thrusts branch off the MBF, as for instance in southern Punjab. These secondary fault zones always diverge in a westward direction and merge with the MBF toward the east. The irregularity and sinuosity of the fault trace is evidence of a highly inclined plane. The older rocks of the lesser Himalaya are thrust over the Shiwaliks along a series of more or less parallel thrust planes. The Main Boundary Fault is a thrust fault with large-scale movements and is still very active.

2.1.3 The lower or lesser Himalaya

The lower Himalaya are separated from the Shiwaliks by the main boundary fault (MBF). The lower Himalaya are rugged mountain region having an average height of about 4000m from mean sea level. Like Shiwaliks, these are also covered with thick forests. The lesser Himalaya is made of sedimentary and metamorphic rocks. The sedimentary formations vary from weak slates to massive and thickly bedded dolomites. Limestones, quartzites, shales and claystones are also present. These are intensely folded and faulted. The low grade metamorphic rocks in the lesser Himalaya are phyllites, quartzites, schists and gneisses. The metamorphic formations are also folded and faulted (*e.g.* Chhibro-Khodri tunnel, Giri-Bata tunnel, Loktak tunnel, Maneri Stage I & II tunnels, Salal tunnel and Tehri tunnels, etc.).

2.1.4 The Main Central Thrust (MCT)

The main central thrust (MCT), marking the boundary between the lesser and higher Himalaya, is a zone of more or less parallel thrust planes along which the rocks of the Central Crystallines have moved southwards against, and over the younger sedimentary and metasedimentary rocks. It is a major feature.

2.1.5 The higher Himalaya

The higher Himalaya are separated from the lesser Himalaya by the Main Central Thrust (MCT). The topography is rugged and the average height above mean sea level is about 8000m. These Himalayan ranges remain covered with snow. The formations are divided into two units (a) The Central Crystallines, comprising of competent and massive high grade metamorphic rocks such as gneisses, migmatites, schists and marbles and (b) The Tibetan- Tethys Zone, composed of incompetent rocks such as shales, sandstones, siltstones and conglomerates. The rocks of higher Himalaya are also intensely folded and faulted.

The above geological description clearly indicates that the Himalayan rocks are tectonically disturbed, weak and the terrain is inaccessible. Tunnels in Himalaya have high overburden because of its great heights from MSL. Because of these features, various tunneling problems were encountered while excavating tunnels through the Himalaya.

2.1.6 Great earthquakes in the Himalaya

Many major earthquakes of differing size that have occurred during the past centuries dominate the seismicity of the Himalayan region. The major ones among them are: 1897 earthquake associated with the rupture in the south of Himalaya beneath the Shillong plateau (and the formation of the Shillong plateau, M=8.7); the 1905 Kangra earthquake (M=8.6); the 1934 Bihar–Nepal earthquake (M=8.4) and the 1950 Assam earthquake (M=8.7). In addition to these, a few more earthquakes of magnitude M > 7 have occurred during the years 1916, 1936 and 1947. During the 1991–2000 decade, three significant and damaging earthquakes with M > 6.5 have occurred in Himalaya in 1988 (M=6.6), 1991 (M=6.6), 1991 (M=6.6) and 1999 (M=6.3). An earthquake of magnitude 7.9 has stuck recently on 25th April 2015 in the north-west of Kathmandu, Nepal having widespread devastating effects.

The Himalaya is tectonically active region with number of earthquakes in the past. Weak and fragile rocks, with regional and smaller structural features have made the tunneling in the Himalaya a challenging task.

In the words of Dr. V. M. Sharma, an eminent engineer, "It is difficult to think of India, more so of Rock Mechanics in India without the great Himalaya. On the one hand, it provides an enormous source of renewable energy, and on the other, it poses some of the most difficult challenges to the Rock Engineers".

Prof. J. A. Hudson (Editor-in-Chief, Int. J. Rock Mech. Min. Science & Geomech. Abstract, 1994) once mentioned in his Editorial the importance of Rock Mechanics activities in India with these words, "To those of us who appreciate the romance and passion of Rock Mechanics, there can be no more exciting work than building structures in the Himalaya with the huge scales, the tectonic activity and the weatherability of the rocks .... Having travelled on just a few of the roads in the foothills of the Himalaya, I have experienced the romance of this work carried out at great heights, low temperatures and in adverse conditions....."

3 TUNNELING PROJECTS IN THE HIMALAYA

The Himalaya has the tunnels and other underground openings mainly for hydroelectric projects, roads and railways. Some of the important tunneling projects are listed in Table 3. The rock type, major tunneling problems and the remedial measures are highlighted against each project in Table 3.

Table 3 Major tunneling projects with problems faced and possible remedial measures.

Name of the Project, Name of Tunnel, Length, Size	*Rock Type*	*Tunneling Problems*	*Remedial Measures & Supports*
Ranganadi Hydroelectric Project, NE Himalaya: HRT – 8.5 km long, 6.8m dia.	Schist, gneiss, shiwalik sandstone besides mica chlorite/mica schist, granitic gneiss, carbonaceous shales and soft sandstone	Intake portal collapse, squeezing ground, intra-thrust zone, methane gas, chimney formation, roof falls and over breaks, crushed rock and flowing water from roof	Forepoling and drainage then tunnel driving, steel supports in squeezing grounds, shotcreting and rock bolting etc., changed tunnel alignment to cross main central thrust (MCT)
Dul Hasti Hydroelectric Project, J&K: HRT – 10.6km long, 7.5m dia. circular/ horse-shoe	Schist/gneiss on west, quartzite/phyllite on east; Kishtwar fault separating the two lithological units	Water ingress, cavity formation, TBM did not succeed in a smooth manner	Advanced probe holes and use of conventional DBM, use of 20mm wiremesh at crater location to stop the muck flow, filling of cavities with concrete, drainage
Nathpa Jhakri Hydroelectric Project, H.P.: HRT – 27.4km long, 10.15m dia, maximum tunnel depth up to 1300m.	Intrusive igneous & metamorphic rocks like gneiss/augen gneiss, quartz mica schist	No serious problem in tunneling except crossing a hot water zone of 52–53°C with large quantity of water	Aeration, ice in large quantities, ties at face; short shift operation; special precautions in concreting for lining; shotcreting and rock bolting
Uri Hydroelectric Project, J&K: HRT- 10.5 km long, 8.4m dia.	Phyllites, graphites, shales, limestones and marble; saturated condition	Highly squeezing ground	Feeler holes ahead of drilling for advance drainage; grouting and spiling in saturated horizons; steel fiber reinforced shotcrete in layers
Tehri Project, Uttarakhand: HRT (4 nos.) – 1km long each, 8.5m dia.	Thinly/thickly bedded phyllites of various grades, sheared phyllites	Minor tunneling problems generally in sheared phyllites	Steel rib supports with final concrete lining in HRTs.
Yamuna Project, Uttarakhand: Ichhari-Chhibro tunnel-6.2km long, 7.0m dia.; Chhibro-Khodri Tunnel – 5.6km long, 7.5m dia., tunnel depth > 600m	Quartzite, slate, limestone, shale, sandstone and clays; recurrence of thrust in Chhibro-Khodri tunnel	High overbreaks, water-in-rush, Squeezing conditions, high tunnel deformations, abnormal rock loads	Heading and benching and multi-drift method; shotcreting, perfo-bolting, forepoling flexible lining
Beas Sutlej Link Hydroelectric Project, H.P.: Pandoh-Baggi Tunnel- 13.12km long, 7.62m dia	Granite with schistose bands and kaolinised pockets and phyllites	Overbreak, cavity formation, flowing and squeezing ground conditions, abnormal load, twisting of steel ribs, heavy water inflow	Forepoling, distressing by drilling advance holes at heading and benching for draining rock behind the face

Table 3 (Cont.)

Name of the Project, Name of Tunnel, Length, Size	*Rock Type*	*Tunneling Problems*	*Remedial Measures & Supports*
Sundernagar-Slapper Hydroelectric project, H.P.: HRT-12.23km long, 8.5m dia.	Limestone, dolomite	Overbreak, cavity formation, flowing ground, heaving ground, heavy water inflow	Forepoling, draining water from behind the heading face, changing tunnel alignment
Giri Hydroelectric Project, H.P.: HRT-7.0km long, 3.7m dia.,	Slates with boulder beds, phyllite, shale, clay, sandstone	Overbreak, Squeezing conditions, high tunnel deformation and support pressure, twisting of steel ribs, occurrences of gases	Shotcreting with perfo-bolting, flexible lining, excavating tunnel of large diameter to allow deformation before final supporting, use of gas detectors
Maneri Stage-I Hydroelectric Project, Uttarakhand: HRT-8.56km long, 5.0m dia., circular, maximum tunnel depth 800m	Quartzite, metabasic, Chlorite schist, quartzite with minor slate; fault and recurrence of folds	Water-in-rush, cavity formation and high pressure because of squeezing condition leading to support failure	Tunneled through alternate alignment to avoid water-charged zone; formation of grouted zone around tunnel to tackle the highly jointed and crushed metabasics and quartzites in cavity area; heavy steel rib supports with steel lagging to tackle squeezing condition; secondary support of concrete lining
Maneri Stage-II Hydroelectric Project, Uttarakhand: HRT-16.0km long, 6.5m wide horse-shoe, tunnel depth > 1000m	Quartzite, gneisses, phyllites, greywackes, slates, limestone, epidiorite; Srinagar thrust and faults	The lithological contacts were sheared, squeezing, high pressure and deformation, flowing ground condition	Forepoling, grouting to tackle the crushed and weak rocks; cavity was grouted using bulkhead and inserting the bolts, steel rib supports with concrete backfill; excavation of bypass drift to release the water pressure
Loktak Project, Manipur: HRT- 6.25km long, 4.6m dia., circular, maximum tunnel depth 800m	Terrace and lake deposits, sandstone, siltstone and shale	Squeezing and swelling grounds, abnormal support pressures, methane gas explosion	Perfo-bolting, shotcreting and use of NATM; larger excavation size to allow deformation, excavation diameter in squeezing condition was 5.4m
Ranjit Sagar Hydroelectric Project, Sikkim: HRT- 3km long, 4.5m dia.	Phyllitic zone, intake portal at slope-wash /talus	Number of shear zones with flowing conditions	Cold bend rib supports, precast lagging, forepoling and backfill concrete

Table 3 (Cont.)

Name of the Project, Name of Tunnel, Length, Size	*Rock Type*	*Tunneling Problems*	*Remedial Measures & Supports*
Khara Hydroelectric Project, Uttarakhand: HRT –10.2km long, 6.0m dia.	Sandstone, clay and conglomerates	Overbreak and flowing ground conditions, saturated with water	Creation of bulkhead, grouting, excavation and heavy supports of steel ribs with final concrete lining
Parbati Stage-II Hydroelectric Project, H.P.: HRT- 31.5km long, 7.0m dia., circular, maximum tunnel depth 1300m	Granite, gneissic granite and quartzite; folded, faulted and jointed	Mild rock burst, water inundation from probe holes flooding the tunnel and TBM, work stopped from TBM side, likely to resume soon.	High capacity steel rib supports were installed in drill and blast excavated section; secondary concrete lining; rock bolts, wiremesh shotcrete and hexagonal precast concrete segments were installed in TBM excavated section
Pir Panjal Rail Tunnel, J&K: 11.2km long, 8.4m wide horse-shoe, maximum tunnel depth 1100m	Silicified limestone, andesite, basalt, quartzite, sandstone, limestone, shale and agglomerates	Squeezing in shales, high deformation, roof falls,	NATM, rock bolts, shotcrete and lattice girder supports were used; forepoles in weaker rocks; thorough monitoring to evaluate the performance of supports; secondary concrete lining
Tala Hydroelectric Project, Bhutan: HRT-22.4km long, 6.8m dia., horse-shoe, maximum tunnel depth 1100m	Gneiss, quartz mica schist, chlorite, sericite schist, quartzite; rocks are folded, faulted and highly jointed; contact of two rock types sheared	Cavity formation, flowing ground, squeezing condition, other adverse geological conditions	Face supported by bamboo bolts, shotcrete, rock bolts, self-drilling anchors, steel rib supports, forepoles; face plugged to grout the roof cavity; systematic drainage; secondary support by concrete lining
Udhampur-Katra Rail Tunnel, J&K: Tunnel T1 – 3.1km long, 6.5m wide, D-shape, maximum tunnel depth 320m	Softs sandstone, siltstone and claystone; highly jointed; claystone has swelling minerals; at places the rock mass is charged with water	Squeezing and swelling conditions; high deformation and support pressure; floor heaving; steel rib supports buckled and twisted	Double steel rib supports with invert and backfill up to inner flange; Secondary support by concrete lining
Chenani-Nashri Highway Tunnel, J&K: 9.0km long, 6.0mwide horse-shoe escape tunnel and 12m wide horse-shape main tunnel, maximum tunnel depth 1200m	Sandstone, siltstone and claystone; minor shear zones	High deformation for longer period; roof falls at places	Tunneling by NATM; longer rock bolts and additional layer of shotcrete along with lattice girders have been used as primary support with final concrete lining

Table 3 (Cont.)

Name of the Project, Name of Tunnel, Length, Size	*Rock Type*	*Tunneling Problems*	*Remedial Measures & Supports*
Rohtang Highway Tunnel, H.P.: 8.9km long, 10.0m wide horse-shoe, Maximum tunnel depth 1900m	Uniformly dipping alternate sequence of quartzites, quartzitic-schist, and quartz-biotite schist with thin bands of phyllites; Seri nala fault passes through the tunnel	Roof collapse; loose rock falls at various places; squeezing; high deformations of roof; Seri nala fault flooded tunnel with rock debris	NATM was used. Shotcrete and rock bolt supports was strengthened; longer rock bolts were used; yieldable steel rib supports are planned in poor rock conditions; DRESS technology was used to tackle the fault zone; the concrete lining will be used as final support

Notations: HRT – Head race tunnel; H.P. – Himachal Pradesh (an Indian state); J&K – Jammu and Kashmir (an Indian state); NATM – New Austrian tunneling method; DBM – Drilling and blasting method; TBM – Tunnel boring machine; DRESS – Drainage, reinforcement, excavation and systematic support

Apart from the variation in geology, the Himalayan tunnels have posed all type of challenges of tunneling, such as, face collapse, cavity formation, water-in-rush, hot water spring, gas explosion, flowing ground condition, squeezing, swelling, rock burst, etc. (Table 3). Thus, for researchers, engineers and geologists, the Himalaya provides the best field laboratory in the world where the experience and knowledge of Rock Mechanics and Tunneling Technology can be tested and established with greater confidence.

4 VARIOUS TUNNELING ISSUES AND LESSONS LEARNT

4.1 Variation in predicted and actual geology

In the Himalaya, drilling up to the tunnel grade sometimes is not possible because of high rock cover (or high tunnel depth), thick vegetation on surface, difficult and inaccessible terrain. The rocks are severely folded and faulted due to tectonic activities. Because of these reasons, the geological investigations are limited to portal areas or around the low cover zones. Hence, in number of projects, it has been observed that the geology encountered during the tunneling vary from that predicted or anticipated geology. This results in variation of excavation planning and methodology, supports type and density, etc. For example, the Chhibro-Khodri tunnel (Figs. 2a & 2b) and the Giri-Bata tunnel (Table 4).

In Chhibro-Khodri tunnel of Yamuna hydroelectric project, recurrence of Krol and Nahan thrusts have resulted in changing geology along the tunnel alignment (Figs. 2a & 2b). This resulted in the problems of water-inrush and squeezing ground conditions during tunneling through the intra-thrust zone, which delayed the project.

Table 4 shows the difference in the predicted and the actual geology along the Giri-Bata tunnel. The difference was mainly in terms of the position of faults and

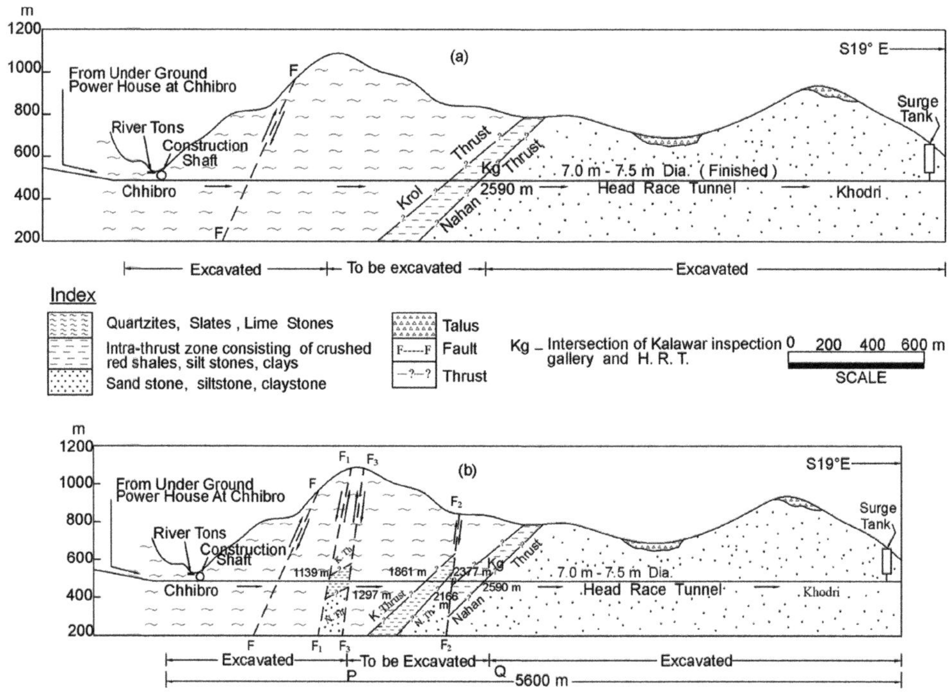

Figure 2 Geological cross-sections along Chhibro-Khodri tunnel (a) original before starting tunneling (b) actual encountered during tunneling (after Jethwa *et al.*, 1980).

Table 4 Predicted and actual geology, Giri-Bata tunnel (Dube, 1979).

Geological Features	*Predicted*	*Actual*	*Difference (m)*
1. Krol Thrust	R.D. 2780m	R. D. 3360m	580
		R. D. 2980m	200
2. Nahan Thrust	R.D. 3405m	R. D. 3520m	115
3. Sile Branch Fault	R.D. 3350m	R. D. 3196m	154
		R. D. 3236m	14
		R. D. 3266m	84
4. Marar Fault	R.D. 4959.5m	R. D. 3360m	169.5
		R. D. 2980m	89.5
5. Length of Blaini Formation	1710m	1312m	398
6. Length of Infra-Krol	1070m	1660m	590
7. Length of Dadahus	625m	384m	241
8. Length of Nahans	3710m	3593m	115

thrusts, which were struck as surprise and resulted in the delay in completion of tunnel.

Rohtang highway tunnel project in H.P. state, India is a challenging project through the higher reaches of Himalaya. The tunnel is being excavated at an altitude of more than 3000m and has the rock cover of upto 1.9km above the tunnel. While tunneling from south end, the Seri nala fault was encountered about 300m before the expected

Figure 3 Flow of clay matrix along with highly disintegrated and weathered rock mass from tunnel face at Ch. 2077m (Rao & Sharma, 2014).

location. As per the investigations, it was extrapolated to be encountered between Ch. 2.20 and 2.80km from south end. But, during the tunneling, the Seri nala fault was struck at Ch. 1.90km, about 300m before the predicted location. At Ch. 1.918km the fault line was visible on the tunnel face where left half face is weak strata charged with water and the right half of the face is undisturbed strata.

No probe hole could be drilled to ascertain the location of Seri nala fault. Generally, it is advised to have number of probe holes in different directions to know the location of such important features. In this tunnel, as the excavation from south end progressed, Seri nala fault adversely affected the tunnel excavation and created difficult conditions for tunneling as shown in Figure 3. Finally, the tunnel through the fault zone was excavated using the DRESS method, which is found to be useful to excavate tunnel through soft, weak and water charged strata (Rao & Sharma, 2014). The DRESS (**D**rainage, **R**einforcement, **E**xcavation and **S**ystematic **S**upport) method has systematic pre-drainage ahead of face, reinforcement of ground, use of forepoles to form umbrella, pre-grouting if required, excavation in small steps by mechanical means and finally the systematic supports. The DRESS method is found to be useful to excavate tunnel through such soft, weak and water charged strata (Rao & Sharma, 2014).

The variation in the predicted and actual geology sometimes makes it impossible to tunnel along the planned alignment as discussed below.

4.2 Change in tunnel alignment

In earlier hydroelectric projects through the Himalayan rocks, in absence of the geological investigations up to the tunnel grade, the straight tunnel alignment between the inlet and the outlet location were chosen. The straight alignment, quite often, had to be changed while constructing the tunnel because of difficult and adverse ground conditions. *For example*, in the head race tunnel (HRT) of Maneri stage-I project, India the tunnel had to be diverted because of the water-in-rush and chimney formation. Three alternative tunnel alignments were considered as shown in Table 5. But,

Table 5 Alternate tunnel alignments between Heena and Tiloth, Maneri stage-I project, India (Goel *et al.*, 1995b).

S.No.	*Proposed Layout*	*Total Length between Heena and Tiloth (m)*	*Tunnel Length through Water Charged Quartzites (m)*	*Increase in Tunnel Length (m)*
1	Original	5065	1200	–
2	Alternative I	5940	920	875
3	Alternative II	7170	–	2105
4	Alternative III	5535	920	470
5	New Alignment	5207	1600	142

finally alternative at S.No. 5 in Table 5, *i.e.*, 'New Alignment' was followed to complete the tunnel. This problem had led to time and cost over-runs. Thus, selection of a trouble free tunnel alignment is of great importance to complete the project within stipulated time and budget.

Almost similar problem was faced in the Chhibro-Khodri tunnel while tunneling in the intra-thrust zone (Jethwa *et al.*, 1980).

It has been experienced that a delay of about 20 per cent in time results in cost escalation by 35 to 40 per cent. Therefore, detailed geological and geo-physical investigations of the area are must and shall be carried out in the area where the geology is varying. In addition, the provision of probe holes ahead of the tunnel face shall also be mandatory in the contract.

4.3 Mixed and fragile geology

Experiences related to the Murree formation of the Himalaya is highlighted here to show the effect of mixed and fragile geology on tunneling. The Murree formation is represented by a sequence of argillaceous and arenaceous rocks that includes interbedded sandstone, siltstone, claystone/mudstone beds. These are also affected by minor shears.

The bands of sandstone, siltstone, claystone/mudstone of varying thickness have been frequently encountered during tunnel excavation. There is no fixed pattern of the bands of these rocks. Figure 4 shows an exposure of different rocks on tunnel face. In fact the bands of mixed rocks, for example, intermixed siltstone & sandstone and intermixed siltstone & claystone are also encountered frequently. The uniaxial compressive strengths of freshly obtained rock samples of sandstone, siltstone and claystone are 70–120MPa, 25–40MPa and 8–15 MPa respectively.

The claystone lies in the category of soft rocks. The claystone rock specimen, if left exposed to atmosphere, degrades and crumbles to small pieces in about a week's time. The freshly excavated claystone on tunnel face sometimes give deceptive appearance of massive rock or one or two joints, but after a day, it starts giving way. *Siltstone are also creating the problems where the joints in siltstone have erodible clay fillings, reducing the shear strength of the rock mass.*

Rock mass behavior because of tunneling in mixed rock masses is different from tunneling through only the poor rocks or through only the good rocks. In the Murree formations having mixed rocks, the sandstone layers are usually separated from each other by weaker layers of siltstone or claystone. Hence, rock-to-rock contact between

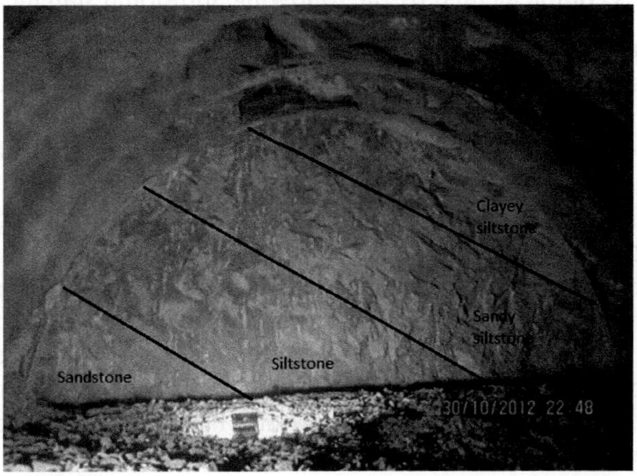

Figure 4 Photo showing exposure of different rocks on tunnel face, ch. 1546m, Chenani-Nashri main tunnel, south end.

blocks of sandstone is limited. Therefore, it is not appropriate to use the properties of the sandstone to determine the overall strength of the rock mass. On the other hand, using the 'intact' properties of the siltstone or claystone only may be conservative since the sandstone skeleton certainly contributes to the rock mass strength.

Murree formations of the Himalaya are comparable with the flysch rocks of the Alps. In order to know the uniaxial compressive strength of mixed rocks, Marinos & Hoek (2001) have proposed that a 'weighted average' of the intact strength properties of the strong and weak rock layers should be used.

Barton's rock mass quality Q (Barton *et al.*, 1974) and Bieniawski's rock mass rating RMR (Bieniawski, 1994) have wide range for different rock masses being encountered in the tunnel through layered mixed rocks. The variation in the values of Q is mainly because of the variation in RQD, J_a and SRF, whereas variation in RMR is because of variation in RQD, UCS and joint condition. In most of the cases there are three joint sets including the bedding plane plus random. In case of mixed rocks, since the rock mass behavior will vary as mentioned in above paragraphs, it is understood that the Q and RMR values shall be influenced by the per cent of different rocks. *This highlights the need of a new engineering rock mass classification for characterizing the mixed (layered) rocks.*

4.4 Squeezing ground condition

Commission on Squeezing Rocks in Tunnels of International Society for Rock Mechanics (ISRM) has published *Definitions of Squeezing* as reproduced here (Barla, 1995).

"Squeezing of rock is the time-dependent large deformation, which occurs around a tunnel and other underground openings, and is essentially associated with creep caused by (stress) exceeding shear strength (limiting shear stress). Deformation may terminate during construction or continue over a long time period".

This definition is complemented by the following additional statements:

- Squeezing can occur in both rock and soil as long as the particular combination of induced stresses and material properties pushes some zones around the tunnel beyond the limiting shear stress at which creep starts.
- The magnitude of the tunnel convergence associated with squeezing, the rate of deformation, and the extent of the yielding zone around the tunnel depend on the geological conditions, the in situ stresses relative to rock mass strength, the ground water flow & pore pressure and the rock mass properties.
- Squeezing of rock masses can occur as squeezing of intact rock, as squeezing of infilled rock discontinuities and / or along bedding and foliation surfaces, joints and faults.
- Squeezing is synonymous of over-stressing and does not comprise deformations caused by loosening as might occur at the roof or at the walls of tunnels in jointed rock masses. Rock bursting phenomena do not belong to squeezing.
- Time dependent displacements around tunnels of similar magnitudes as in squeezing ground conditions, may also occur in rocks susceptible to swelling. While swelling always implies volume increase due to penetration of the air and moisture into the rock, squeezing does not, except for rocks exhibiting a dilatant behavior. However, it is recognized that in some cases squeezing may be associated with swelling.
- Squeezing is closely related to the excavation, support techniques and sequence adopted in tunneling. If the support installation is delayed, the rock mass moves into the tunnel and a stress re-distribution takes place around it. Conversely, if the rock deformations are constrained, squeezing will lead to long-term load build-up on rock support.

Squeezing ground conditions through weak and highly jointed rock masses under high overburden pressure (in situ stress) is quite common in the fragile Himalaya. Squeezing is mainly experienced in the argillaceous rock masses, such as, phyllites, shales, clays, soft gougy material, etc. having uniaxial compressive strength < 30MPa and the overburden pressure is high (tunnel depth is more). In most of the Himalayan tunnels, the squeezing ground condition has been experienced where the tunnel floor heaving is also common (Fig. 5).

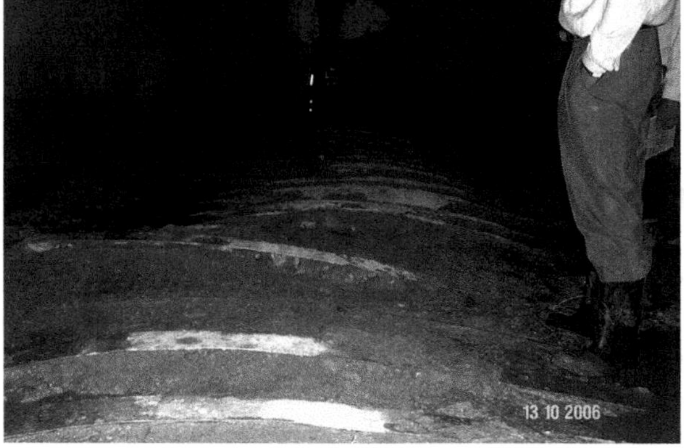

Figure 5 Floor heaving in a railway tunnel in J&K, India.

The support pressure developing far behind the tunnel face in a heavily squeezing ground depends considerably on the amount of support resistance during the yielding phase. The higher the yield of the support, the lower will be the final load. A targeted reduction in support pressure can be achieved not only by installing a support that is able to accommodate a larger deformation (which is a well-known principle), but also through selecting a support that yields at a higher pressure. Furthermore, a high yield pressure reduces the risk of a violation of the clearance profile and increases safety level against roof instabilities (loosening) during the deformation phase (Cantieni & Anagnostou, 2009).

4.4.1 Tunnel size and squeezing ground condition

In Chhibro-Khodri tunnel, in 1970s the main tunnel of 9m diameter was divided into three tunnels of 4.5m diameter each to avoid the squeezing purely on the qualitative consideration. Thus, by reducing the tunnel size, the squeezing condition was avoided. But, in Maneri Stage-II head race tunnel, in 1980s the main tunnel of 6.0m diameter experienced the squeezing ground condition. To bypass the problematic squeezing condition zone, a smaller size drift (2.5m) was excavated. But, this 2.5m wide drift had also experienced some squeezing ground condition. These two cases qualitatively showed that there is some effect of tunnel size on ground condition for tunnels and encouraged to develop approach for predicting the ground condition (see section 5.2.1).

4.4.2 Effect of tunnel depth on support pressure

According to the elasto-plastic theory, failure of the rock mass around an opening under the influence of depth pressure forms a broken zone called "coffin cover". The failure process is associated with volumetric expansion of the broken rock mass and manifests itself in the form of squeezing into the opening (Labasse, 1949; Daemen, 1975). The "characteristic line" – or the "ground reaction curve" – concept explains that the support pressure increases with depth, provided that the tunnel deformations are held constant. Further, *large tunnel deformations associated with expansion of the broken zone lead to reduced support pressures* (Fig. 7).

Higher tunnel deformations and support pressures observed in the red shales at a depth of 600 m at Chhibro-Khodri tunnel, as compared to those observed at a depth of 280m in the same tunnel, were explained by Jethwa *et al.* (1977) with the help of the elasto-plastic theory. They employed an empirical relation given by Komornik & David (1969) to estimate the swelling pressure and considered that the support pressure was the arithmetic sum of elasto-plastic (squeezing plus loosening) and swelling pressures. Later, Singh (1978) emphasized the interaction between the swelling and squeezing pressure and suggested that only the greater of the two should be considered. The average elasto-plastic pressures, estimated according to the suggestions of Singh (1978), are close to the observed values. As such, the empirical approaches, developed to estimate support pressure for tunnel support design, must be amended to include the effect of tunnel depth in order to obtain reliable results under squeezing rock conditions. The correction factor for overburden (or tunnel depth) for estimating support pressure using Q, as suggested by Singh *et al.* (1992), is accepted now. Equation 6 also shows the effect of tunnel depth (H) on support pressure.

4.4.3 Loose backfill with steel arch supports

In a deep tunnel under squeezing ground conditions, the supports are likely to attract high loads unless substantial tunnel deformations are allowed. An ideal support system for such conditions would be the one which absorbs large deformations while maintaining tunnel stability. Use of flexible supports in a slightly over-excavated tunnel provides a possible solution to such a problem. The thickness of backfill should be decided from the considerations of its compressibility and desirable tunnel deformations.

Although flexible steel arches were not used, loosely thrown tunnel muck behind steel ribs provided an element of flexibility in the Chhibro-Khodri tunnel. It was observed that the support pressure reduced to a large extent with such a loose backfill (Fig. 6a & 6b after Jethwa *et al.*, 1980).

4.5 Roof collapse and cavity formation

It has been experienced that because of frequently changing geology and presence of shear zone, support has either been inadequate or it has not been installed soon enough, which has resulted in deterioration of the rock mass quality and roof falls and cavity formation.

For example, in Maneri stage-I project head race tunnel a major cavity was formed during excavation between ch. 5038m and 5050m in highly jointed and folded quartzites. The tunnel crossed a shear zone at ch. 5050m. The crushed quartzite was also charged with water. Therefore, the crushed rock debris was continuously flowing from the roof, which formed a cavity. The total volume of the cavity was estimated as $813m^3$. The face was sealed after forepoling with rolled steel joists as shown in Figure 7. A bulkhead was constructed at the tunnel face leaving 2 to 3 pipes for grouting the muck and debris. Drainage holes were provided to release the hydrostatic pressure. A side drift was also excavated to release the water pressure. The cavity above the forepoles was then filled with concrete using the pipes inserted in the cavity for this purpose. The muck below the forepoles and behind the face was grouted using a cement water slurry to check the water flow and to consolidate the muck. The tunnel then was excavated.

In yet another incident in the same tunnel, the sheared and crushed zone between metabasics and quartzites was tackled by creating grouted plug ahead of the tunnel face all around the tunnel (Fig. 8). This was then excavated and supported leaving 5m grouted zone by following the steps shown in Figure 8.

Such collapses can be avoided by pre-grouting the rock mass ahead of the tunnel and or installing the effective supports timely close to the tunnel face. Invert supports, to complete the support ring, must be used in soft and weak rock masses, thick fault gouges and shear zones.

In the lower Himalaya, it has been observed that the contact of two rock formations invariably is sheared, which generally leads to support failure and collapse. Hence, this

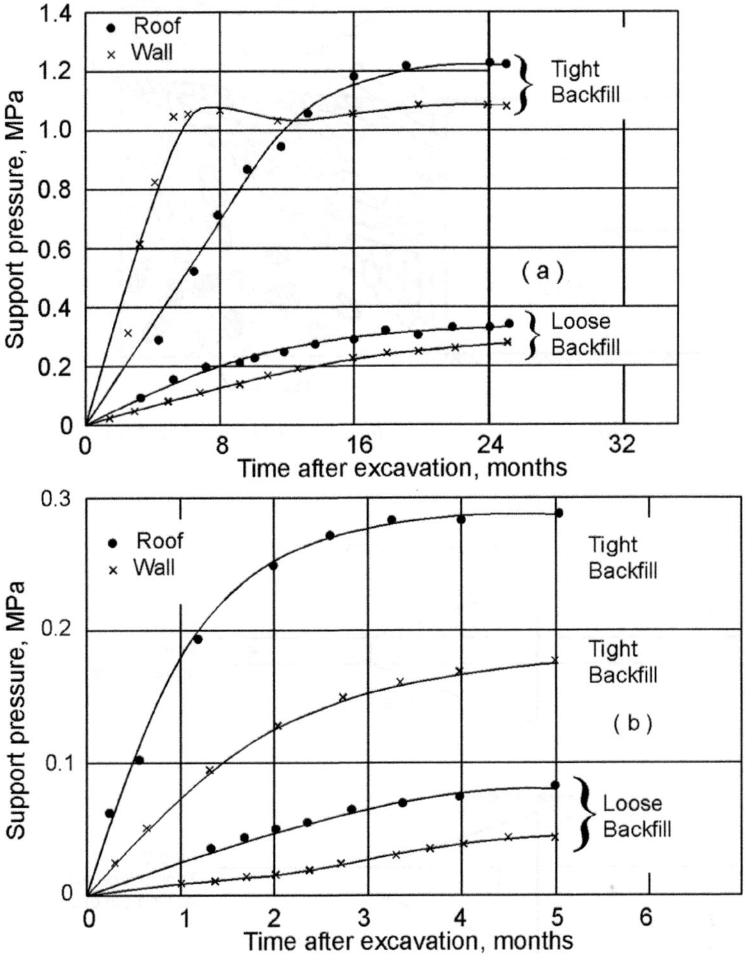

Figure 6 Effect of compressible loose backfill on support pressure in: a – black clays; b – red shales (after Jethwa *et al.*, 1980).

should be known to the geologists and site engineers so that timely preventive steps can be taken.

4.6 Shear zone treatment

There are number of small or big shear zones and faults present in the lower Himalayan rocks. These shear zones and faults are sympathetic to regional main boundary fault and main central thrust. It is generally said that in the tunnels through lower Himalayan rocks if no fault or shear is seen for 100m it means this has been missed. The contact of two rock types is also found to be generally sheared in lower Himalaya (Goel *et al.*, 1995a).

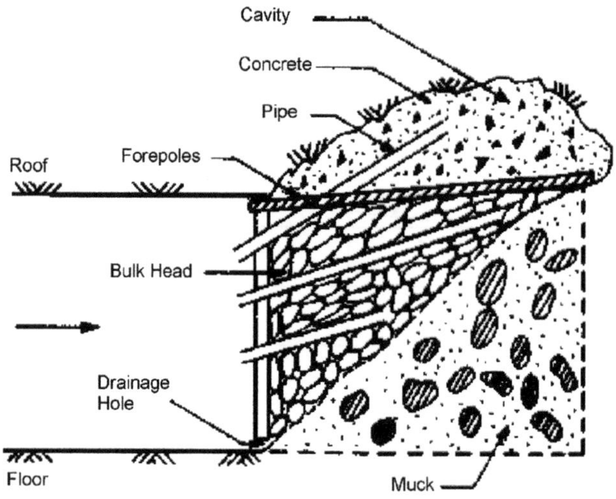

Figure 7 Method for tackling the problem of cavity formation (Goel *et al.*, 1995b).

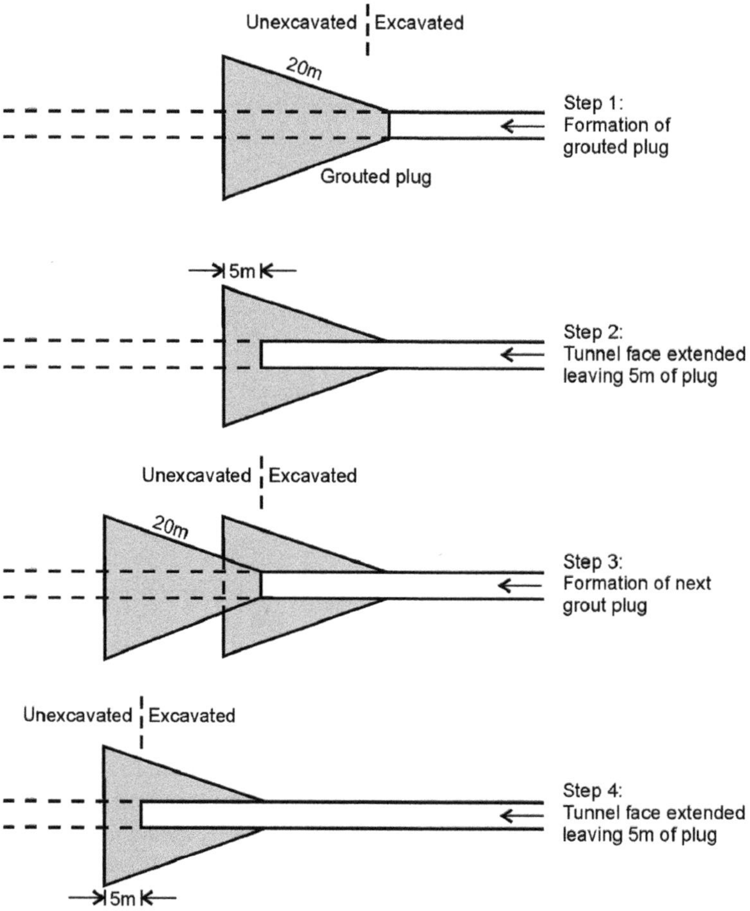

Figure 8 Steps to tackle the problem of sheared contact zone of metabasic and quartzite (Goel *et al.*, 1995b).

It is envisaged that the rock mass affected by a shear zone is much larger than the shear zone itself. Hence, the affected rock mass must be down-graded to the quality of the shear zone so that a heavier support system than a regular one can be installed. A method has been developed at NGI (Norwegian Geotechnical Institute) for assessing support requirements using the Q-system for rock masses affected by shear zones (Grimstad & Barton, 1993). This has also been used in some Himalayan tunnels in India. In this method, weak zones and the surrounding rock mass are allocated their respective Q-values from which a mean Q-value can be determined, taking into consideration the width of the weak zone/shear zone. Equation 1 may be used in calculating the weighted mean Q-value (Bhasin *et al.*, 1995).

$$\log Q_m = \frac{b.\ \log\ Q_{wz}\ +\ \log Q_{sr}}{b+1} \tag{1}$$

where,

Q_m = mean value of rock mass quality Q for deciding the support,
Q_{wz} = Q value of the weak zone/shear zone,
Q_{sr} = Q value of the surrounding rock, and
b = width of the weak zone in meters.

The strike direction (θ) and thickness of weak zone (b) in relation to the tunnel axis is important for the stability of the tunnel and therefore the following correction factors have been suggested for the value of b in the above Equation 1.

if $\theta = 90° - 45°$ to the tunnel axis then use 1b,
if $\theta = 45° - 20°$ then use 2b in place of b,
if $\theta = 10° - 20°$ then use 3b in place of b, and
if $\theta < 10°$ then use 4b in place of b.

Hence, if the surrounding rock mass near a shear zones is downgraded with the use of the above equations, a heavier support should be chosen for the whole area instead of the weak zone alone.

Figure 9 shows a typical treatment method for shear zones (Lang, 1961) in the roof of tunnel. First the shear zone is excavated with caution up to some depth. After excavation, immediately one thin layer of shotcrete with wire mesh or steel fiber reinforced shotcrete (SFRS) shall be sprayed. The weak zone is then reinforced with inclined rock bolts and finally shotcrete with wiremesh or SFRS (preferably SFRS) should be sprayed ensuring its proper thickness in weak zones. This methodology is urgently needed if NATM or NMT (Norwegian Method of Tunneling) is to be used in the tunnels of the Himalayan region, as seams/ shear zones/ faults/ thrusts/ thin intra-thrust zones are frequently found along tunnels and caverns in the Himalaya. *Stitching is perhaps the terminology that best suits this requirement.*

In case of a thick shear zone (b>>2m) with sandy gouge, umbrella grouting or rock bolting is used to enhance the strength of roof and walls in advance of tunneling. The excavation is made manually. Steel ribs are placed closely and shotcreted until the shear zone is crossed. Each (blasting) round of advance should be limited to 0.5m or

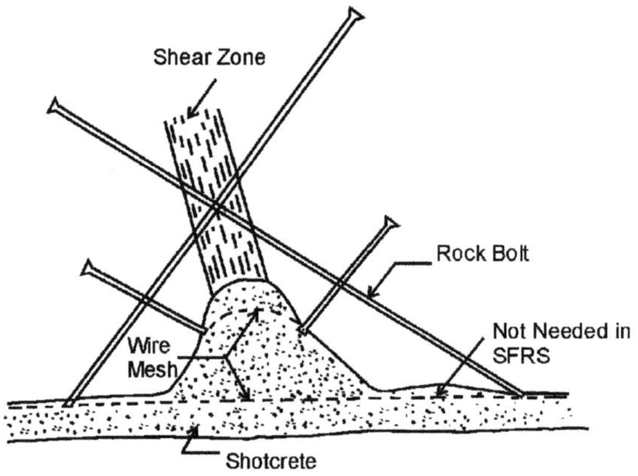

Figure 9 Shear zone treatment in an underground opening (Lang, 1961).

even smaller depending upon the stand-up time of the material and fully supported before starting another round of excavation.

4.7 Water-in-rush

In the tunnels in Himalaya, it has been experienced often that the rock mass above the shear zone is water-charged. This may be because of the presence of impermeable gouge material in the shear zone. Hence, one should be careful and be prepared to tackle the water problem in the tunnels through shear zone having impermeable gouge material.

For example, in the case of Maneri stage-I project head race tunnel in Uttarakhand state, India, because of tunneling, the trapped water in quartzites above impervious shear zone rushed in to the tunnel causing roof collapse and debris flow flooding the tunnel (Fig. 10).

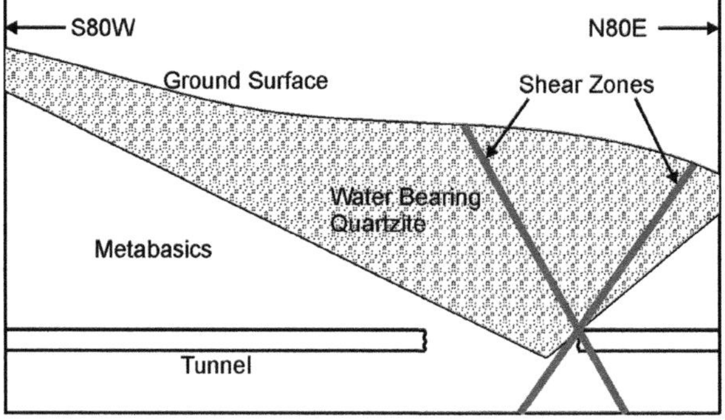

Figure 10 Schematic view of causes of water inrush in head race tunnel of Maneri stage-I project (Goel *et al.*, 1995a).

Figure 11 Water-in-rush in a rail tunnel in J&K, India.

Intersection of water-charged zones while tunneling is also a common feature. However, when the normal seepage turns into free flowing conditions, particularly with material outwash, tunneling problems attain serious dimensions. If caught unawares, these problems are capable of completely disrupting tunneling activity and influencing the time schedules involved. Such cases call for state of art tunneling techniques like ground stabilization using special grout admixtures or freezing which saves draining arrangements running into several cumec capacity, advance probing, and so on.

In a rail tunnel in J&K state, the water started flowing from the tunnel face with a discharge of more than 500 liters per minute. The tunnel was being constructed through the terrace deposits. It was thought that the discharge rate will subside with time. But, to our surprise, it remained continued with a rate of about 300 liters per minute. The tunnel was completed by pipe-roofing technique and by channelizing the water with the help of drainage holes.

It is highlighted here that once the water is stuck in the tunnel, it is difficult, costly and time consuming to tackle it and tunnel through. *Therefore, it is generally advised to take steps to probe and divert water as soon as first sign of water is seen.*

4.8 Probe holes

An important step to prevent the geological surprise is to drill the probe holes in different directions, if required, from the tunnel face to get the geological information. Having 50m deep probe holes of at least 50 to 75mm diameter ahead of the tunnel face in the regions of highly changing geology could provide valuable advance information of geology. Lesser depth of probe holes can also be drilled depending upon the requirement and availability of drilling equipment and material. The ground conditions, the support pressures, etc., can also be ascertained as per the geological details obtained from probe holes. Accordingly, the construction and supporting techniques may be modified. Such probing is now becoming popular and being used in some projects. *For example*, in Chenani-Nashri highway tunnel project, J&K probe holes were drilled to know the water condition ahead of the tunnel face from north end in the escape tunnel. The probe holes have shown that the water pressure is reducing and accordingly the tunnel activity was planned.

Probe holes sometimes can be disastrous also. For example, Parbati stage-II head race tunnel in Nov. 2005 faced a problem when approximately 12000m^3 of silt and fine sand flowing out of a probe hole buried the TBM and half of the bored tunnel with 7000 lit per min water inflow. Hence, it is essential to keep close watch on the variation of geology and water seepage with the excavation before and during drilling the probe holes.

Probing ahead of the tunnel face using geophysical means like tomographic analysis and radar is also becoming popular but is comparatively expensive and cannot be used as a regular or on a routine basis. Seismic profiling is another methodology being adopted to probe the geology ahead of tunnel face.

Probe holes planning, drilling and monitoring shall be carried out in the supervision of experienced geologists to get the desired information and results.

4.9 Tunnel Boring Machine (TBM) in the Himalaya

Use of Tunnel Boring Machine (TBM) has not been very discouraging so far in Himalayan tunneling because of varying geology and water-charged formations. To highlight, three cases of TBM are briefly presented.

The work of 6.75km long Dulhasti project head race tunnel of 8.3m excavated diameter was started with gripper type hard rock TBM. The rock mass was predominantly hard and highly abrasive quartzites. While tunneling, the TBM was inundated with a water inflow of over 1000litres/sec. This 'inrush' occurred at a minor shear zone aquifer (fractured quartzite) within impermeable interbedded phyllites and included 4,000m^3 of sand and quartzite pebbles. The inflows fell to 150 liter/sec within 100 days and five years later inflows of 100 liter/sec were still being recorded. The TBM could bore only 2.86km and finally abandoned. This experience in Himalayan geology was not encouraging. The project has subsequently been completed by conventional excavation. The project was commissioned in 2007, after a delay of about 19 years.

At the head race tunnel for Parbati Stage-II project, H.P. state of India, an incident similar to Dulhasti project tunnel occurred in May 2007 when routine probing ahead of a 6.8m diameter refurbished Jarva open TBM tunnel in sheared and faulted quartzite at 900m overburden cover punctured a water bearing horizon which resulted in inflows of water of over 120 l/sec containing about 40% sand and silt debris. The inflow was sudden and occurred at a high pressure which could not be contained. Eventually over 7500m^3 of sand and silt debris buried the TBM. The project supposed to be commissioned in 2007, was delayed for about 10 years (Sengupta *et al.*, 2008).

National Thermal Power Corporation (NTPC) is constructing the Tapovan-Vishnugad hydroelectric power project (TVHEP) with installed capacity of 520MW (4x130MW). The project has HRT of length approximately 12.1 km, of which 8.6 km has been planned to excavate using a double shield TBM by a Joint Venture (JV) of Larsen & Toubro Ltd., India, and Alpine, Austria. The remaining 3.5km of the HRT is being excavated conventionally. The tunnel passes below the steep hills of Himalaya near Joshimath, India. The tunnel depth at places is more than 1.0km. This is the first time the double shield hard rock tunnel-boring machine (ordered from Herrenknecht, Germany) has been used for a hydel power project tunnel in India. The TBM has an excavation diameter of 6.575 m for an internal finished diameter of the HRT being 5.64 m (Saxena, 2013).

Soon after TBM excavation started, it became clear that there are also groundwater-bearing, approximately NS striking, steeply inclined faults and fracture zones associated with quartz-rich lithologies such as quartzite, quartzitic gneiss and augen gneiss. These steeply inclined fracture and fault structures cut across the main foliation joint, which means that there is a high level of interconnection between the joint systems, allowing for the development of potent and high-pressure aquifers. The rock types are gneisses and quartzites (Brandl *et al.*, 2010).

During the excavation, the TBM encountered a large fault zone. A major portion of rock detached and dented the shield of the TBM and the TBM trapped. Subsequently, there was heavy ingress of ground water into the tunnel, commencing at the tail skin area of the TBM and progressing rapidly, through the ungrouted section of the annulus, some 160 m back along the tunnel. The water pressure was very high carrying the rock material and debris which resulted in more damages to TBM (Brandl *et al.*, 2010). Work remains standstill for quite some time. Subsequently, a bypass tunnel was excavated to recover the buried TBM. The TBM has been recovered, repaired and again put to use in the same tunnel.

While the excavation using TBM has been quite successful in other parts of India, *e.g.*, Delhi metro, Srisailam left bank canal tunnel and Bombay Malabar hill tunnel, the Himalaya remain a major challenge. The experience suggests that many of the problems can be avoided if sufficient advance information ahead of the face is available. Following key issues have been identified by Goel (2014) for the success of TBM in the Himalaya.

4.10 Key issues for TBM success in the Himalaya

The Himalaya pose the most challenging ground conditions for tunneling. One of the prime reasons is that they are the youngest of the mountain chains and are still tectonically active. The difficulties of tunneling at depth through high mountainous terrain pose major challenges not just for tunnel boring machines (TBM) but also for the use of drill and blast (D&B).

The big investment in a sophisticated TBM and the expectation of mostly rapid advance rates can be spoiled by the unexpected delays caused by unexpected ground. Only a few percent of the total length of a tunnel may be unexpected, yet these few percent could double the construction time in some cases.

Tunneling in adverse ground is significantly less tolerant of the limitations of the tunneling approach than in good ground. Generally, the more difficult the ground, the more flexibility is also needed. Tunneling in the Himalaya, the Andes and until recently the Alps has shied away from TBM use due to perceived inflexibility and the likelihood of the machines getting trapped by adverse ground conditions, either as a result of squeezing or spalling/bursting conditions or because of ground collapses associated with rock falls or with running or flowing ground within faults. Any of these situations can lead to problematic tunneling at best and collapses and abandonment at worst.

Following are the key issues for the success of TBM in the Himalaya.

4.10.1 Geological investigations and probe holes

The more challenging the ground, the greater the pre-planning that is required before tunneling. This challenge is not just of tackling adverse ground, stress state and/or groundwater conditions, it is also often about logistics. For deep tunnels in mountainous regions, problematic geologic zones often are at significant distance from the nearest portal and at such significant depth that surface pre-treatment is generally impractical (Carter, 2011).

Experience suggests that many of the problems associated with the TBM in the Himalaya can be avoided if sufficient geological and geohydrological information is available in advance. Faced with cost and time constraints, detailed investigations before selecting a tunnel alignment are often compromised, resulting in encountering very disturbed geological conditions. It is essential that detailed exploration work is carried out before the start of the project and exploration ahead of the face is undertaken on a continuous basis.

In particular for a TBM driven tunnels – which are not as flexible as a conventionally driven tunnels – forward probe drilling from the tunnel face is certainly not an alternative to an adequate pre-investigation. But, regular cautious probe drilling during cutter change and maintenance shifts could largely remove the unexpected; especially if performed with two slightly diverging probe holes (Barton, 2000). It must be highlighted here that while drilling the probe holes, to avoid the blow outs, the groundwater conditions shall be closely watched and an attempt shall be made to carry out geophysical investigations. It is needless to mention here that the probe holes shall be drilled under close supervision of an experienced engineering geologist.

4.10.2 Selection of TBM and add-on-features

Selection of a TBM is the key decision. Complications in the decision-making process, in general, relate to the timing when making this decision, as it needs to be made 12–18 months in advance of actually starting tunneling, so that sufficient lead time is available for building the machine. However, often detailed project site investigations are incomplete, still ongoing or sometimes not even started when this key decision is to be made. Furthermore, once the contract is awarded to the contractor, generally after a long tendering process, almost always insufficient time and/or funds have been allocated to allow the contractor any opportunity for additional customized exploration to support his own excavation technology selection procedures before initiating equipment procurement.

The choice of TBM also needs critical analysis at the planning stage. In the absence of accepted standards for the design and construction of a type of TBM and the fact that no TBM can be designed for every type of geological condition, the design and special construction characteristics of each TBM need to be carefully, project-specifically designed. The shielded TBM has a definite edge over the open TBM as it is not as sensitive to the instability phenomena of the excavation walls owing to the presence of precast concrete or steel lining inside and the protection of the shield (Saxena, 2013).

TBM can be designed with add-on-features as per the site conditions, *e.g.* probe hole drilling, forepoling, shotcrete spraying, rock bolting, pre-grouting/grouting, steel rib erection etc. As per the expected requirements, these features can be incorporated in the TBM.

In mountainous terrain, when considering a decision on whether or not to use a TBM, and which type of TBM to use for a deep tunnel, it must be appreciated that, historically, three types of ground conditions have proved to be the most problematic from the viewpoint of halting tunnel advance. In order of severity, case records suggest faults with gouge filling, heavy water and major stress, individually and/or in combination, constitute the most problematic ground conditions. The three elements which control the trouble-free tunnel excavations at significant depth are, stress state, groundwater conditions and the rock or the medium. Adverse characteristics of any of these three elements can, on its own, compromise drill and blast (D&B) or TBM tunneling, but it usually takes a combination of all three being adverse to trap a machine or halt a D&B drive to the extent that a bypass becomes necessary (Carter, 2011).

Hence, TBM shall be selected after detailed analysis of stresses, ground water and the expected rock masses and ground conditions.

4.10.3 Expert TBM crew

Even with the best possible TBM, the progress required may not be achieved. Experienced and dedicated TBM crew is very important for the success of TBM. It is the expert crew, which can take the right decision at the right time and implement it properly. Success of TBM in Kishanganga hydroelectric project in India is one such example.

Bieniawski (2007) also highlighted the influence of TBM crew and suggested an adjustment factor for the influence of TBM crew on its performance in rock mass excavatability index for TBM.

4.10.4 Timely decision and action

Extreme ground conditions present major contrasts to tunneling, so much so that they inevitably demand use of flexible rock engineering solutions for the tunnel to progress. The fact that conditions within the Himalaya can be expected to be as bad as has ever been encountered elsewhere means there has to be the ability while tunneling to allow changes in excavation procedures and in pre- and post-excavation support approaches. It has been experienced that the delays in decision have enhanced the problems. For tunnel to be completed successfully, *the rock is not going to wait.* Hence, timely decision and action is important. There may be situations where the flexibility in the designs is required. This is possible only when the engineers-in-charge are given decision and risk taking authority.

4.10.5 Risk sharing

"Engineers have to take a calculated risk, persons become wiser after an accident. If they were really wise, it was their duty to point out mistakes in the design to engineers" – Karl Terzaghi

In more difficult ground conditions, such as those encountered in the Himalaya, with minimal investigation comes more risk of the TBM getting trapped – either as a result of squeezing or spalling/bursting conditions or because of ground collapses associated with rockfalls or with running or flowing ground within faults. These cases are further complicated by heavy water inflows. To reduce these risks considerably more investment must be made in the design process in these complex mountainous regions. Significant reduction of real risk can only be achieved through more investigative effort, not through design refinement. Cost and schedule analysis of past case records suggests that for complex ground conditions, *some 5% of the engineer's estimate of capital expenditure is required to be spent on investigating ground conditions to push the process in the right direction* (Carter, 2011).

Hence, if the investigations are insufficient, whatever is the reason, various problems are bound to be encountered as a surprise (mostly not at the expected location). There should be provision of risk sharing between the client and the contractor in the contract document. Otherwise these surprises result in the time and cost over-runs and litigations.

As mentioned earlier the contract document shall have the flexibility also to accommodate the unforeseen conditions/events. The site engineers shall be given the responsibility of allowing and approving the use of newer techniques and material required for tackling the unforeseen conditions. The contractor, once allowed by the site engineer, shall get full payment after executing the job to the satisfaction of the site engineers.

The contract should include (i) clause for compensation to contractor for an unexpected geological conditions or surprises, (ii) clause on innovations by contractors and engineers on the basis of mutual agreements, (iii) clauses for first and second contingency plans for the preparedness and (iv) penalty for delays in construction. Obviously contract is not a license for injustice to any party. Injustice done should be corrected soon (Singh & Goel, 2006).

4.11 General observations

Following are the general observations from various tunneling and underground projects constructed so far.

(i) The alignments of long power tunnels have not been fixed after proper and purposeful geological exploration. The surface geological features have proved misleading. Consequently, many disastrous problems were faced in tunneling.

(ii) In squeezing grounds, the selection of size of power tunnel is important. Initially one big tunnel was excavated, as it became difficult to drive, three tunnels of smaller diameter were then driven (Mitra, 1991). This was because there were fewer supports, bridge action period was greater and heaving of the floor was limited.

(iii) The underground power house was located at one side of the river and not too high above the water level in the river. Consequently during flood, water entered into the cavern through the tail race tunnel while its excavation was going on. The rock masses in the roof and walls were submerged for a couple of weeks before water could be pumped out. Fortunately the cavern remained stable.

(iv) The seepage through dam foundation is increased after a major earthquake, as permeability of the jointed rock mass increases drastically during shearing. On the contrary permeability of micro-fissured rock is reduced due to deformation of the joints beneath the foundation due to thrust on the reservoir filling, thereby making the grout curtain redundant which led to a dam failure.

(v) The study (on the basis of 11 years of monitoring of Chhibro underground cavern of Yamuna hydro-electric project) has shown that ultimate roof support pressure for water-charged rock masses with erodible joint filling may rise up to 6 times the short-term support pressure. The damage to the support system during an earthquake of 6.3 magnitude is not appreciable except near faults with plastic gouge material (Mitra, 1991).

(vi) Very high support pressures may be generated by reduction in the modulus of deformation due to saturation of the rock mass around HRT, TRT and penstocks etc. (Verman, 1993). Mehrotra (1992) observed that the modulus of deformation is actually very low after saturation compared with that in dry conditions in the case of argillaceous rocks (claystone, siltstone, shale, phyllite etc.).

5 CONTRIBUTIONS IN THE STATE-OF-THE-ART

5.1 Squeezing ground condition

The over-stressed zone of rock mass fails where tangential stress (σ_θ) exceeds the mobilized uniaxial compressive strength (UCS) of the rock mass. The failure process will then travel gradually from the tunnel boundary to deeper regions inside the unsupported rock mass. The zone of the failed rock mass is called the 'broken zone'. This failed rock mass dilates on account of the new fractures. A support system after installation restrains the tunnel deformation and gets loaded by the support pressure.

It is evident from the ground reaction curve that the support pressure decreases rapidly with increasing tunnel deformation. *Hence, significant tunnel deformation shall be allowed to reduce the cost of support system. This is the secret of success in tunneling through squeezing ground condition.*

In case one chooses to install very stiff support system, it may be seen from Figure 12 that stiff support system will attract high support pressure as it will restrict the tunnel deformation. If a flexible support system is built after some delay, it will attract much less support pressure. This is ideal choice. However, too late and too flexible support system may attract high support pressure due to excessive loosening of rock mass in the broken zone. Yet the squeezing ground comes to equilibrium after years even in severe squeezing ground condition. Although the final deformations may be undesirable and so corrective measures are required.

The data suggests that support pressure jumps up after tunnel deformation of about 5 to 6 percent of tunnel size. Then there is sympathetic failure of entire brittle rock mass within the broken zone, rendering its residual cohesion $c_r = 0$ in the highly squeezing ground. The theoretical ground response curve is shown in Figure 12 on the basis of this hypothesis. The sympathetic failure is in fact unstable and wide spread fracture propagation in the entire failure zone, starting from the point of maximum shear strain. This brittle fracture process may be taken into account in the elasto-plastic theory by assuming $c_r = 0$ after critical tunnel deformation of 6%. *Thus, it is recommended that tunnel deformation shall not be permitted beyond 4 percent of tunnel radius to be on safe side.*

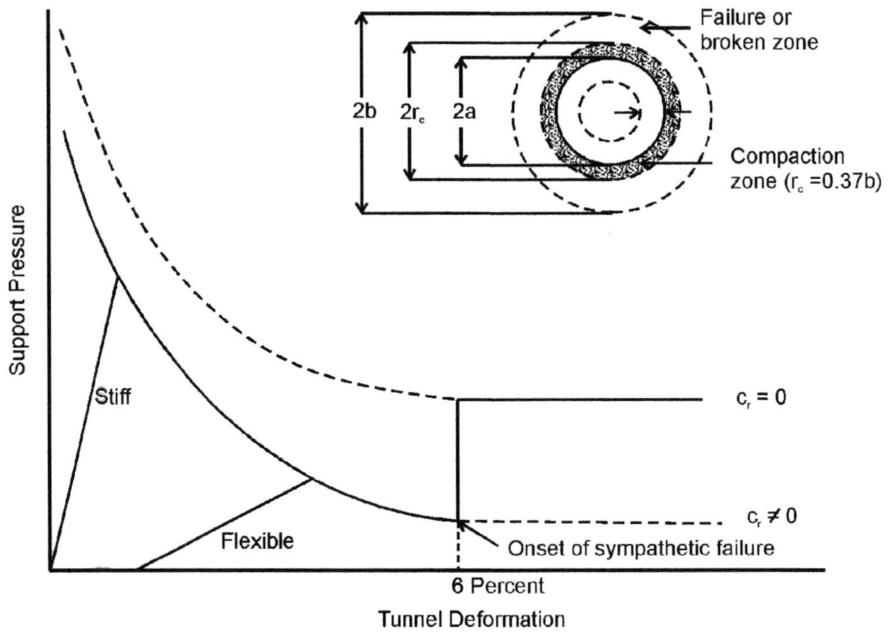

Figure 12 Effect of sympathetic failure of rock mass on theoretical ground response curve of squeezing ground condition. Support reaction curve of stiff and delayed flexible supports are superposed.

10.5.1.1 Compaction zone within broken zone

From the study of steel rib supported tunnels, Jethwa (1981) observed that the values of coefficient of volumetric expansion (K) are negative near the tunnel and increased with radius vector. Thus, he postulated the existence of compaction zone within the broken zone (Fig. 13). The radius of the compaction zone (r_c) is estimated to be approximately equal to,

$$r_c = 0.37\ b \qquad (2)$$

Thus, compaction zone will not develop where b is equal to a/0.37 or 2.7a. This is the reason why compaction zone was not observed in some of the European tunnels in the squeezing ground conditions as b was perhaps less than 2.7a.

In an ideal elasto-plastic rock mass, compaction zone should not be formed. The formation of the compaction zone may be explained as follows. A fragile rock mass around the tunnel opening fails and dilates under the influence of the induced stresses. The dilated rock mass then gets compacted due to the passive pressure exerted by the support system in order to satisfy the ultimate boundary condition that is zero rate of support deformation with time. The development of support pressure with time would reduce the deviator stresses ($\sigma_\theta - \sigma_r$) within the compaction zone which in turn will undergo creep relaxation manifested by the negative K values.

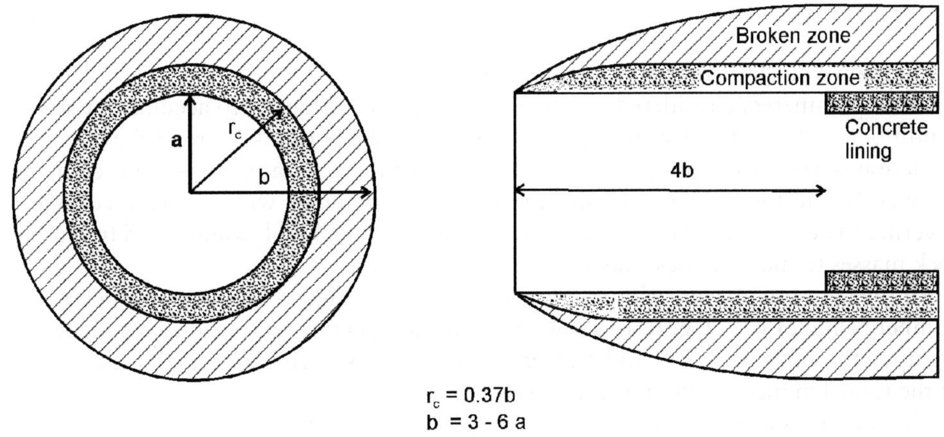

Figure 13 Compaction zone within broken zone in the squeezing ground condition (Jethwa, 1981).

Table 6 Overall coefficient of volumetric expansion of failed rock mass (K) within broken zone (Jethwa, 1981; Goel, 1994).

S.No.	*Rock Type*	*K*
1.	Phyllites	0.003
2.	Claystones / Siltstones	0.01
3.	Black clays	0.01
4.	Crushed sandstones	0.004
5.	Crushed shales	0.005
6.	Metabasics (Goel, 1994)	0.006

5.1.2 Coefficient of volumetric expansion

Jethwa (1981) has found the overall values of K from instrumented tunnels in Himalaya. These values are listed in Table 6. Actually K varied with time and radius vector. K was more in roof than in walls. So only peak overall values of K are reported and considered to have stabilized to a great extent with time (15 – 30 months). It is heartening to note that the value of K for crushed shale is the same at two different projects. It may be noted that higher degree of squeezing was associated with rock masses of higher K values.

5.2 Prediction of ground conditions

5.2.1 Empirical approach for predicting the ground conditions

To avoid the uncertainty in obtaining appropriate SRF ratings in rock mass quality Q of Barton *et al.* (1974), Goel *et al.* (1995c) have suggested rock mass number N, defined as follows, for proposing the criterion of estimating ground conditions for tunneling.

$$N = [Q]_{SRF} = 1 \tag{3}$$

Equation 3 suggests that N is Q with SRF = 1.

Other parameters considered are the tunnel depth H in meters to account for stress condition or SRF indirectly, and tunnel width B in meters to take care of the strength reduction of the rock mass with size. The values of three parameters – the rock mass number N, the tunnel depth H and the tunnel diameter or width B were collected covering a wide variety of ground conditions varying from highly jointed and fractured rock masses to massive rock masses.

All the data points were plotted on a log-log graph (Figure 14) between rock mass number N and $H.B^{0.1}$. Figure 14 shows zones of tunneling conditions/hazards depending upon the values of $HB^{0.1}$ and N. Here H is the overburden in meters, B is the width of the tunnel in meters and N is rock mass number.

The equations of various demarcating lines have been obtained as shown in Table 7. Using these equations, one can predict the ground condition and then plan the tunneling measures. These equations can also be used to plan the layout of the tunnel to avoid the squeezing ground condition.

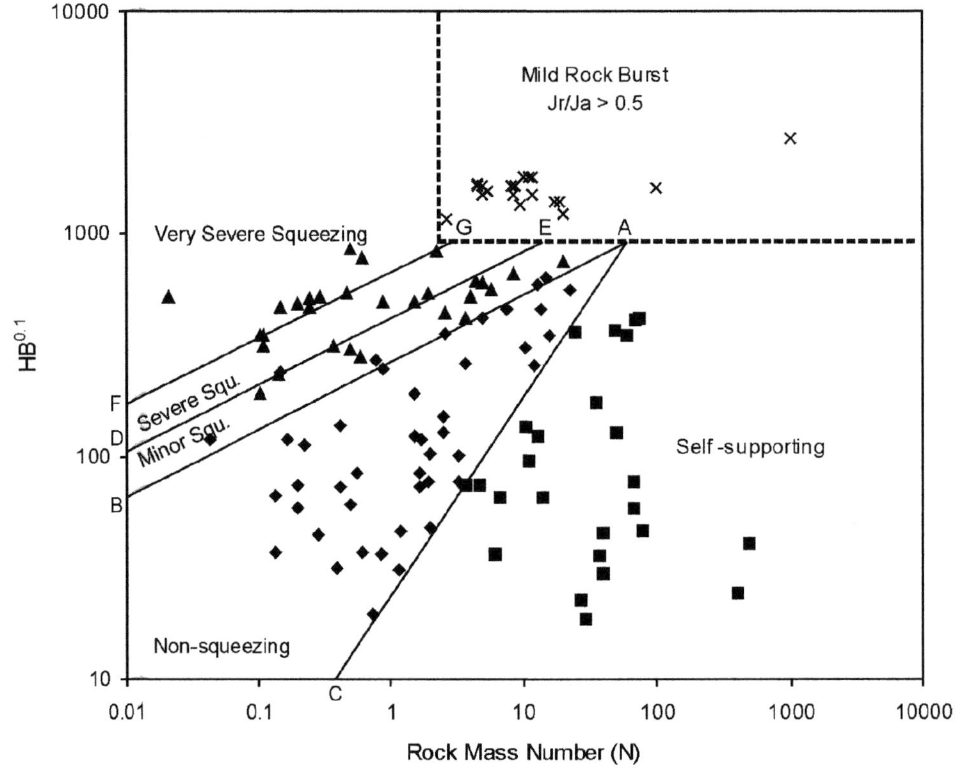

Figure 14 Plot between rock mass number N and $HB^{0.1}$ for predicting ground conditions (Singh & Goel, 2011).

Table 7 Prediction of ground condition using N (Singh & Goel, 2011).

S. No.	*Ground Conditions*	*Correlations for Predicting Ground Condition*
1.	Self-supporting	$H < 23.4\ N^{0.88}.\ B^{-0.1}$ & $1000\ B^{-0.1}$ and $B < 2\ Q^{0.4}$ m (for ESR=1) (Barton *et al.*, 1974)
2.	Non-squeezing (< 1% deformation)	$23.4\ N^{0.88}.B^{-0.1} < H < 275N^{0.33}.\ B^{-0.1}$
3.	Mild or minor squeezing (1–3% deformation)	$275\ N^{0.33}.\ B^{-0.1} < H < 450N^{0.33}.\ B^{-0.1}$ and $J_r\ /J_a < 0.5$
4.	Moderate or severe squeezing (3–5% deformation)	$450\ N^{0.33}.\ B^{-0.1} < H < 630N^{0.33}.\ B^{-0.1}$ and $J_r\ /J_a < 0.5$
5.	High or very severe squeezing (>5% deformation)	$H > 630N^{0.33}.\ B^{-0.1}$ and $J_r\ /J_a < 0.25$
6.	Mild rock burst	$H.B^{0.1} > 1000$m and $J_r/J_a > 0.5$ and $N > 2$

Notations: N = rock mass number (Barton's Q with SRF = 1); B = tunnel span or diameter in meters; H = tunnel depth in meters; J_r and J_a = Parameters of Q-system; and ESR= Barton's excavation support ratio.

Few cases of spalling and rock burst conditions have also been plotted in Figure 14. These data points are lying in the upper right hand side corner of Figure 14. The enclosure is shown by dotted line because more data points are needed before reaching to a firm conclusion. However, it may be highlighted here that spalling and rock burst have been encountered in the rock masses having $J_r/J_a > 0.5$, N>2 and $H.B^{0.1}>1000$. Hence, the rock mass having J_r and J_a ratings in category b and c in the Tables of J_r and J_a in Q-system shall not experience the rock burst condition.

Equations given in Table 7 are found to be quite useful in planning and design. In addition, the N based approach of estimating/predicting the ground condition has been found as complimentary to Q-system in respect of identifying the ground conditions and then select SRF rating to get Q-value. The, Q-value thus obtained has been used for the support design in tunnels using the Grimstad *et al.* (2003) chart.

5.2.2 SRF rating in Q-system for spalling and rock burst conditions

Kumar *et al.* (2006) have studied the spalling and rock burst condition in the Nathpa-Jhakri Project tunnels. They have experienced that the Barton's stress reduction factor (SRF) ratings in case of 'competent rock' should be different for massive rocks and for moderately jointed rocks. Accordingly they have proposed new rating of SRF for competent moderately jointed rocks as shown in Table 8. The support pressure values obtained using these new suggested SRF ratings are found to be matching with the observed values (Kumar *et al.*, 2006).

High value of J_r/J_a leads to high angle of internal friction along joints. Consequently, the intermediate principal stress (σ_2) which is the in situ stress along tunnel axis increases the rock mass strength enormously where overburden is more than 1000m. The net effect is that the rock burst condition is not serious as anticipated from Barton *et al.* (1974) and Grimstad & Barton (1993). In other words the pre-stressing effect of σ_2 on rock mass strength is similar to reduction in SRF values. Application of the polyaxial failure criterion shall be studied and Table 8 needs further research.

Table 8 Stress reduction factor (SRF) rating for jointed rock for categories L, M and N (Kumar *et al.*, 2006).

Category	*Rock Stress Problem*		q_c/σ_1	σ_θ/q_c	*SRF (Old)*	*SRF (New)*
H	Low stress, near surface open joints		>200	<0.01	2.5	2.5
J	Medium stress, favorable stress condition		200–10	0.01–0.3	0.5–2.0	1.0
K	High stress, very tight structure (usually favorable to stability, may be unfavorable to wall stability)		10–5	0.3–0.4	0.5–2.0	0.2–2.0
L	Moderately slabbing after > 1hr	Massive rock	5–3	0.5–0.65	5–9	5–50
		Moderately jointed rock			–	1.5–2.0
M	Slabbing and rock burst after a few minutes	Massive rock	3 – 2	0.65–1.0	9–15	50–200
		Moderately jointed rock			–	2.0–2.5
N	Heavy rock burst (strain burst) and immediate deformations	Massive rock	< 2	< 1	15–20	200–400
		Moderately jointed rock			–	2.5–3.0

5.3 Deformation of tunnel walls

It is common knowledge that the rock failure is associated with volumetric expansion due to creation and progressive widening of new fractures. Consequently all the points within the broken zone in a circular tunnel shift almost radially toward the opening because the expanding rock mass is kinematically free to move in radial direction only. Labasse (1949) assumed that the volume of failed rock mass increases at a constant rate (called coefficient of volumetric expansion, K).

The displacement at the boundary of broken zone (u_b) is negligible compared to that at the opening periphery (u_a). So the coefficient of volumetric expansion is defined as follows:

$$K = \frac{\pi\, b^2 - \pi\,(a-u_a)^2 - \pi\,(b^2 - a^2)}{\pi\, b^2 - \pi\, a^2} \tag{4}$$

Equation 4 can be solved as below for obtaining u_a

$$u_a = a - \sqrt{a^2 - K(b^2 - a^2)} \tag{5}$$

Using Equation 5 and the K-value from Table 6 it has become possible to estimate the tunnel wall/roof deformation.

5.4 Estimating support pressure in squeezing ground condition

First time in 1994 it has been empirically established that in the tunnels with arch roof through squeezing ground, the support pressure in steel rib supported tunnels varies with the tunnel size as per the following Equation 6 (Goel, 1994).

$$p_v(sq) = \left[\frac{f(N)}{30}\right] \cdot 10^{\left[\frac{H^{0.6}\,.\; a^{0.1}}{50.\; N^{0.33}}\right]},\quad \text{MPa} \tag{6}$$

Table 9 Correction factor f(N) for tunnel deformation in Equation 6 (Goel, 1994).

S.No.	*Degree of Squeezing*	*Normalized Tunnel Deformation %*	*f (N)*
1.	Very mild squeezing ($270\ N^{0.33}.\ B^{-0.1} < H < 360\ N^{0.33}.\ B^{-0.1}$)	1 – 2	1.5
2.	Mild squeezing ($360\ N^{0.33}.\ B^{-0.1} < H < 450\ N^{0.33}.\ B^{-0.1}$)	2 – 3	1.2
3.	Mild to moderate squeezing ($450\ N^{0.33}.\ B^{-0.1} < H < 540\ N^{0.33}.\ B^{-0.1}$)	3 – 4	1.0
4.	Moderate squeezing ($540\ N^{0.33}.\ B^{-0.1} < H < 630\ N^{0.33}.\ B^{-0.1}$)	4 – 5	0.8
5.	High squeezing ($630\ N^{0.33}.\ B^{-0.1} < H < 800\ N^{0.33}.\ B^{-0.1}$)	5 – 7	1.1
6.	Very high squeezing ($800\ N^{0.33}.\ B^{-0.1} < H$)	>7	1.7

Notations: N = rock mass number; H = tunnel depth in meters; and B = tunnel width in metres.
Note: Tunnel deformation depends significantly on the method of excavation. In highly squeezing ground condition, heading and benching method of excavation may lead to tunnel deformation > 8%.

where,

$p_v(sq)$ = short-term vertical roof support pressure in squeezing ground condition in MPa,
f(N) = correction factor for tunnel deformation obtained from Table 9, and
H & a = tunnel depth & tunnel radius in meters respectively (H < 600m).

Equation 6 has been developed using the measured value of support pressures from number of tunnels experiencing the squeezing ground conditions in the Himalaya.

Subsequently, Bhasin & Grimstad (1996) have proposed Equation 7 to estimate the support pressure in tunnels through poorer qualities of rock masses. Equation 7 also shows the effect of tunnel size on ultimate roof support pressure (p_{roof}).

$$p_{roof} = \frac{40\ D}{J_r \cdot Q^{1/3}},\ \text{kPa} \tag{7}$$

where D is the diameter or span of the tunnel in meter.

It may be highlighted here that Equation 6 has been successfully used for estimating the support pressure in some tunnels including the Udhampur-Katra rail link project tunnels in the Himalaya.

5.5 Effect of tunnel size on support pressure

Effect of tunnel size on support pressure is summarized in Table 10.

It is cautioned that the support pressure is likely to increase significantly with the tunnel size for tunnel sections excavated through the following situations:

(i) slickensided zone,
(ii) thick fault gouge,
(iii) weak clay and shales,
(iv) soft plastic clays,

Table 10 Effect of Tunnel Size on Support Pressure (Goel *et al.*, 1996).

S. No.	*Type of Rock Mass*	*Increase in Support Pressure Due to Increase in Tunnel Span or Dia. from 3m to 12m*
A. Tunnels with Arched Roof		
1.	Non-squeezing ground conditions	Up to 20 percent only
2.	Poor rock masses / squeezing ground conditions (N = 0.5 to 10)	20 – 60 percent
3.	Soft-plastic clays, running ground, flowing ground, clay-filled moist fault gouges, slickensided shear zones (N = 0.1 to 0.5)	100 percent
B. Tunnels with Flat Roof (irrespective of ground conditions)		up to 100 percent

(v) crushed brecciated and sheared rock masses,
(vi) clay filled joints, and
(vii) extremely delayed support in poor rock masses.

5.6 Principles of planning

The principles of planning tunnels and underground structures are summarized as follows:

(i) The electrical power is generated in proportion to product of the river water discharge and the hydraulic head between reservoir water level and the turbines.

(ii) So the size of machine hall depends upon the generation of electrical power and number of turbines. The machine hall should be deeper in the hill for protection from landslides. The power house may be located on a stable slope in the case of hard rocks, if it is not prone to landslides and snow avalanches. The pillar width between adjacent caverns should be at least equal to the height of any cavern or the average of widths (B) of caverns whichever is larger.

(iii) The size of the desilting chamber or settling basin (rock cover H > B) depends upon the size of particles which we need to extract from the sediment from intake. Some rivers like Ganga in Himalaya have very high percentage of feldspar and quartz particles (of hardness of 7) which damage the blades of the turbines (of weldable stainless steel) soon. In such cases desilting basins may be longer. The (pillar) spacing of desilting chambers is of critical importance. Turbines are closed when silt content is more than the design limit in the river during the heavy rains.

(iv) The rock cover should be more than three times the size of openings (power tunnels). The overburden pressure should counter-balance water head in head race tunnel and penstocks etc. However, the rock cover should be less than 350 $Q^{1/3}$ m for $J_r/J_a < 0.5$ to avoid squeezing conditions. Further the overburden should also be less than 1000m for $J_r/J_a > 0.5$ to avoid rock burst conditions.

(v) Designers have to select velocity of flow of water (say 4–6 m/sec) in head race tunnel. Higher velocity means more head loss and more abrasion of concrete lining. Less velocity, on the other hand, leads to bigger diameter of head race tunnel which increase cost and the time of construction. Thus an economical diameter of the head race tunnel is found out by trial and error approach. There is no need of hoop reinforcement in spite of high hoop tensile stresses in lining within the strong rocks as cracks (< 3mm) are self-healing. The center to center spacing of HRTs and TRTs (of width B) should be 2B in (competent) non-squeezing grounds, and preferably 6B in squeezing ground conditions for stability of the rock pillars. The width of the tunnels should preferably be limited to 6m in severe squeezing conditions, as support pressure increases with width (for $Q < 4$). Unlined HRTs and TRTs may be planned in unerodible and hard rock masses ($B < 2\ Q^{0.4}$ m and $H < 275\ N^{0.33}\ B^{-0.1}$ m, N = Q with SRF = 1, B is the tunnel size in meters), where it is more economical than the tunnels lined with PCC.

The PCC lining and the steel liner in penstocks should be designed considering actual rock sharing pressure and water hammer along with maximum head of water. The thickness of steel liner should be adequate to withstand external seepage water head when penstock is empty. The steel liner should have stiffeners to prevent its buckling when empty.

(vi) Alignment of underground structures is planned according to geology and rock mass quality. If there is one band of sound strata, we try to align the underground structures along this sound strata (with generally $Q>1$, E_d >2GPa, except in shear zones; but $H<350Q^{1/3}$ m). The penstocks may be turned along sound rock strata and need not be along the head race tunnel. The general alignment should preferably be perpendicular to shear zones/ faults/thrusts zones. The thickness of intra thrust zone should be minimized if any. The caverns should not be parallel to strike of continuous joints but be parallel to major horizontal in situ stress if possible. The displacement of the walls of machine hall should not be high at haunches (say > 6 or 7 cm) so that gantry crane can travel freely for maintenance of machines. We should also try to avoid swelling ground which may contain clay like montmorillonite etc.

(vii) In fact designers cannot plan in the beginning as a few drifts or drill holes give no idea of complex geology in tectonically active regions. We know complete geology about a project only when project is over. The important thing is to have proper documentation of each project.

(viii) A good reservoir site is located where width of the reservoir expands upstream and preferably downstream of confluence of rivers so that volume of river water storage is maximum. Steep gorges such as in Himalaya need high dams for reasonable reservoir storage (~10,000 MCM). High dam stores water in rainy season. High Tehri dam absorbed a huge flood of 3500cumecs and saved devastation of down-stream during hydrological disaster in 2013 in Uttarakhand state. Engineers are planning run-of-the-river schemes along rivers with steep bed slopes but they do not store more water for power generation but are good ecologically (in China, Himalaya). The life span and capacity of a reservoir increases with the height of a dam and so plays an important role in planning of river valley projects (Swamee, 2001).

(ix) The whole plan should be feasible geologically and economical. Several lay-outs are made and costs compared. In 21st century time of completion of project is very important as time means money (monthly profit). Cost of construction time should be added to total cost of the project for optimization of the overall cost. So engineering geology and rock mechanics studies are very important to reduce the time of construction of underground structures, especially the very long tunnels (> 5 km) under high overburden in weak rocks. In three gorges dam project (17,000 MW) in China, lot of money and construction time was saved because of properly planned rock/geotechnical investigations.

(x) Topography or slope of the hill river bed matters. The longer head race tunnel means more hydraulic head which gives more power.

(xi) Underground structures are preferred in the highly seismic area or landslide prone area or in strategic area as they are more stable, durable and invisible. Surface structures are prone to damage by landslides (such as in Himalaya and Alps).

(xii) Major tectonic features like tectonically active thrust and fault zones (*e.g.* MCT, MBT etc. in Himalaya) should be avoided, otherwise articulated or flexible concrete lining may be provided to deform without failure during slip along these active thrust zones in the entire life span of the project. It is interesting to know that the observed peak ground acceleration increases with earthquake magnitude up to 7M (on Richter's scale) and saturates at about 0.70 g up to 8 M (Krishna, 1992).

(xiii) Major shear zone should be treated by the dental concrete upto adequate depth beneath a concrete dam foundation. Abutment should be stabilized by grouted rock bolts or cable anchors if wedge is likely to slide down. In fact roller compacted concrete (RCC) dam may be more economical than conventional gravity concrete dam on weak rocks. Doubly arched dams will be economical in hard rocks and stable strong abutments. Stepped spillways are preferred now-a-days for reducing hydraulic energy of rapidly flowing water. The double rows of grout holes should be used to provide a tight grout curtain in a highly jointed rock mass below a concrete dam near heel.

(xiv) Dynamic settlement of earth or earth and rockfill dams should be less than 1m or one hundredth of dam height whichever is less. The longitudinal tensile strains (stress) should be less than permissible tensile strain (stress) in clay core. It is a tribute to the engineers that no failure of modern high earth and rockfill dams has taken place even during major earthquake (~8M) since 1975 all over the world, even when earth has passed through unprecedented peak of earthquakes. Filter upstream of clay core is also provided to drain seepage water during rapid draw down in many modern earth and rockfill dams.

(xv) The site should not be prone to formation of huge landslide dams (in unstable steep gorges subjected to large debris flow during long rains), causing the severe flash floods upon breaking. Example is temporary huge twin-reservoir created by landslide of Pute hydro-electric station, Ecuador. Nor there should be deep seated landslides near the dam site generating waves which would over-topple the dam as in Vajont dam, Italy. Further stability of villages above reservoir rim is to be looked into for safety of the people.

(xvi) Approach tunnels, hill roads and bridges should be planned and maintained well in landslide prone mountains (subjected to heavy rains or cloud bursts (> 500 mm per day) or rains for a long time) and snow avalanche prone areas for easy construction of the project. Hill roads and vibrations due to blasting in tunnels should not damage houses upon the ground, if any.

(xvii) Huge contingency funds should be made available upto 30 % of total cost for timely risk management in underground structures. Authorities should prepare risk management plan 1 and 2, in case of unforeseen geological hazards. Alternate huge diesel generators should be installed for emergency power supply in openings.

(xviii) The extensive geodynamic monitoring of displacements and seepage in to landslides along the reservoir rim and all caverns and reservoir induced seismicity etc. is undertaken along the entire hydroelectric project, and connected to the internet for global viewing, using space imageries also. The caverns are fitted with warning systems (connected to extensometers) for safety of the persons. Instrumentation is the key to success in tunneling in weak rock masses.

(xix) Portal of tunnels should be located in the stable rock slopes which should not be prone to severe landslides or severe snow avalanches.

(xx) Steel fiber reinforced shotcrete (SFRS) is found successful in tunneling in the squeezing ground conditions and weak rock masses. The invert should also be shotcreted in squeezing conditions, so that stable rings are formed.

(xxi) Invert struts should be used to enable steel ribs to withstand high wall support pressure in the squeezing ground.

(xxii) *A good bond should be ensured between shotcrete lining or concrete lining and rock masses; so that the bending stresses in the lining are drastically reduced, especially in the highly seismic areas.*

(xxiii) In highway tunnels, the thickness of shotcrete should be increased by 50mm to take care of corrosion due to toxic fumes of vehicles.

(xxiv) In the steep slopes in hard rocks, the half tunnels should be planned along hill roads.

(xxv) In the rock burst prone areas (H>1000m and J_r/J_a >0.5), full-column-resin-grouted rock bolts should be used to make the brittle rock mass as ductile mass which should be further lined by ductile SFRS to prevent rock bursts.

(xxvi) There should be international highway network in Himalaya for its rapid economic development.

6 CONCLUDING REMARKS

Tunneling in the Himalaya is a challenging task because of difficult and varying geology, the rock masses are weak and fragile undergoing intense tectonic activities resulting into major faults, folds and other discontinuities, high in situ stresses, water-charged formations, etc. Each tunnel in the Himalaya poses some problem or another. Majority of the problems are related to water-in-rush, roof falls, cavity formation, face collapse, swelling, squeezing, support failure, gas explosion etc. As such, the experiences of tunneling in the Himalaya have helped in understanding the rock behavior and also developing the state-of-the-art of tunneling.

ACKNOWLEDGMENTS

Authors are thankful to various project authorities for allowing to carry out studies presented here. They are grateful to all researchers, engineers, geologists, students and publishers whose work has been referred to in this Chapter. All enlightened engineers and geologists are requested to send their precious feedback and suggestions to the authors.

REFERENCES

Barla, G. (1995). Squeezing rocks in tunnels, *ISRM News Journal*, Vol. 2, No. 3 and 4, pp. 44–49.

Barton, N. (2000). TBM Tunnelling in Jointed and Faulted Rock, A.A. Balkema, The Netherlands, p. 173.

Barton, N., Lien, R. and Lunde, J. (1974). Engineering classification of rock masses for the design of tunnel support, Rock Mechanics, Springer-Verlag, Vienna, Vol. 6, pp. 189–236.

Bhasin, R. and Grimstad, E. (1996). The use of stress-strength relationships in the assessment of tunnel stability, Proc. Conf. on Recent Advances on Tunnelling Technology, New Delhi, pp. 183–196.

Bhasin, R., Singh, R. B., Dhawan A. K. and Sharma, V. M. (1995). Geotechnical evaluation and a review of remedial measures in limiting deformations in distressed zones in a powerhouse cavern, Conf. on Design and Construction of Underground Structures, New Delhi, India, pp. 145–152.

Bieniawski, Z. T. (1994). Rock Mechanics Design in Mining and Tunnelling, A. A. Balkema, Rotterdam, pp. 97–133.

Bieniawski, Z. T. (2007). Predicting TBM Excavatability, Tunnels and Tunnelling International, September.

Brandl, Johann, Gupta, V. K. and Millen, Bernard (2010). Tapovan-Vishnugad hydroelectric power project – experience with TBM excavation under high rock cover, *Geomechanics and Tunnelling*, Vol. 3, No. 5, pp. 501–509.

Cantieni, L. and Anagnostou, G. (2009). The interaction between yielding supports and squeezing ground, *Tunnelling and Underground Space Technology*, Elsevier, Vol. 24, pp. 309–322.

Carter, T. (2011). Successful Tunnelling in Challenging Mountainous Conditions, January [www.hydroworld.com/articles/print/volume-19/issue-3/articles/construction/successful-tunnelling-in-challenging.html], Downloaded on September 11, 2014.

Daemen, J. J. K. (1975). Tunnel support loading caused by rock failure, Ph.D. Thesis, University of Minnesota, Minneapolis, MN.

Dube, A. K. (1979). Geomechanical evaluation of tunnel stability under failing rock conditions in a Himalayan tunnel, Ph. D. Thesis, Department of Civil Engineering, IIT Roorkee, India.

Gansser, A. (1964). Geology of the Himalaya, Interscience Publishers, John Wiley and Sons, London, 289p.

Goel, R. K. (1994). Correlations for predicting support pressures and closures in tunnels, Ph.D. Thesis, VNIT, Nagpur University, India, p. 347.

Goel, R. K. (2014). Tunnel boring machine in the Himalayan tunnels, Indorock 2014, Organised by ISRMTT, November, New Delhi, pp. 54–67.

Goel, R. K., Dwivedi, R. D., Viswantahan, G. and Rathore, J. S. (2013). Monitoring of a tunnel through mixed geology in the Himalaya, World Tunnel Congress 2013, Geneva, pp. 2115–2122.

Goel, R. K., Jethwa, J. L. and Dhar, B. B. (1996). Effect of tunnel size on support pressure, Tech. Note, *International Journal of Rock Mechanics and Mining Sciences & Geomechanics Abstracts*, Vol. 33, No. 7, pp. 749–755.

Goel, R. K., Jethwa, J. L. and Paithankar, A. G. (1995a). Tunnelling in the Himalaya: Problems and solutions, *Tunnels and Tunnelling*, May, pp. 58–59.

Goel, R. K., Jethwa, J. L. and Paithankar, A. G. (1995b). Tunnelling through the young Himalaya – A case history of the Maneri-Uttarkashi power tunnel, *Engineering Geology*, Vol. 39, pp. 31–44.

Goel, R. K., Jethwa, J. L. and Paithankar, A. G. (1995c). Indian experiences with Q and RMR systems, *Tunnelling & Underground Space Technology*, Elsevier, Vol. 10, No.1, pp. 97–109.

Grimstad, E. and Barton, N. (1993). Updating of the Q-system for NMT, International Symposium on Sprayed Concrete-Modern Use of Wet Mix Sprayed Concrete for Underground Support, Fagernes, (Editors Kompen, Opsahll and Berg. Norwegian Concrete Association, Oslo).

Grimstad, E., Bhasin, R., Hagen, A. W., Kaynia, A. and Kankes, K. (2003). Q-system advance for sprayed lining, Part 1-Tunnels and Tunneling International (T&T), January 2003 and Part II – (T&T), March 2003.

Jethwa, J. L. (1981). Evaluation of Rock Pressures in Tunnels through Squeezing Ground in Lower Himalaya, Ph.D. Thesis, Department of Civil Engineering, University of Roorkee, India, p. 272.

Jethwa, J. L., Singh, B., Singh, Bhawani and Mithal, R. S. (1980). Influence of geology on tunnelling conditions and deformational behavior of supports in faulted zones – A case history of Chhibro-Khodri tunnel in India, *Engineering Geology*, Vol. 16, No. 3 and 4, pp. 291–319.

Jethwa, J. L., Singh, Bhawani, Singh, B. and Mithal, R. S. (1977). Rock pressure on tunnel lining in swelling and viscous rocks. Intern. Symp. Soil-Structure Interaction, University of Roorkee, India, January 1977, pp. 45–50, (Pub) M/s. Sarita Prakashan, Meerut, India.

Komornik, A. and David, D. (1969). Prediction of swelling pressure of clays, ASCE, *Journal of the Soil Mechanics and Foundations Division*, 95, pp. 209–225.

Krishna, Jai (1992). Seismic Zoning Maps in India, *Current Science*, India, Vol. 62, No. 122, pp. 17–23.

Kumar, N. K., Samadhiya, N. K. and Anbalagan, R. (2006). New SRF values for moderately jointed rocks, *Indian Journal of Rock Mechanics and Tunnelling Technology*, July, Vol. 12, No. 2, pp. 111–122.

Labasse, H. (1949). Les Pressions de Terrians antour des Puits, Revue Universelle des Mines, 92 e Annee, 5–9, V-5, Mars, pp. 78–88.

Lang, T. A. (1961). Theory and Practice of Rock Bolting, Transactions Society of Mining Engineers, AIME, Vol. 220, pp. 333–348.

Marinos, P. and Hoek, E. (2001). Estimating the geotechnical properties of heterogeneous rock masses such as Flysch, *Bulletin of Engineering Geology and the Environment*, 60, pp. 85–92.

Mehrotra, V. K. (1992). Estimation of engineering parameters of rock mass, Ph.D. Thesis, Civil Engineering Department, IIT Roorkee, India, p. 267.

Millen, Bernard (2014). TBM tunnelling in the Himalaya, Tapovan-Vishnugad Hydroelectric Power Plant, India, ESCI Seminar Series, March, Salzburg, Austria [www.victoria.ac.nz /sgees/about/events/esci-seminar-series-tbm-tunnelling-in-the-Himalaya,-tapovan-vishnugad -hydroelectric-power-plant,-india], Downloaded on September 11, 2014.

Mitra, S. (1991). Studies on long-term behaviour of underground power-house cavities in soft rocks, Ph.D. Thesis, Civil Engineering Department, IIT Roorkee, India, p. 194.

Naidu, B. S. K. (2001). Silt erosion problems in Hydro Power Stations and their possible solutions, 2nd Int. Conf. on Silting Problems in Hydro Power Plants, Bangkok.

Rao, R. S. and Sharma, K. K. (2014). A case study of soft soil tunnelling with DRESS methodology in Rohtang tunnel construction, Proc. National Seminar on Innovative Practices in Rock Mechanics, February, Bengaluru, India, pp. 184–190.

Saxena, P. K. (2013). Tunnelling in the Himalaya, Construction World, Downloaded on 17.9.2014, Source: www.constructionworld.in/News.aspx

Singh, Bhawani (1978). Analysis and design techniques for underground structures, Proc. Conf. Geotech. Eng., State of Art Rep. I.I.T., Delhi, 2, pp. 102–112.

Singh, Bhawani and Goel, R. K. (1999). Rock Mass Classification: A Practical Approach in Civil Engineering, Elsevier Science, UK, p. 268.

Singh, Bhawani and Goel, R. K. (2006). Tunnelling in Weak Rocks, Elsevier Ltd., UK, p. 489.

Singh, Bhawani, Jethwa, J. L., Dube, A. K. and Singh, B. (1992). Correlation between observed support pressure and rock mass quality, *Tunnelling & Underground Space Technology*, Elsevier, Vol. 7, No. 1, pp. 59–74.

Swamee, P. K. (2001). Reservoir capacity depletion on account of sedimentation, *International Journal of Sediment Research*, China, Vol. 16, No. 3, pp. 408–415.

Verman, M. (1993). Rock mass – Tunnel support interaction analysis, Ph.D. Thesis, Civil Engineering Department, IIT Roorkee, India, p. 267.

Chapter 4

Tunnels and tunneling in Turkey

N. Bilgin & C. Balci
Department of Mining Engineering, Istanbul Technical University, Istanbul, Turkey

Abstract: Recent studies revealed that tunnels were excavated under hundreds of Neolithic settlements all over Europe and the fact that so many tunnels have survived 12,000 years indicates that the original networks must have been huge from Scotland to Turkey. Some experts believe the network was a way of protecting man from predators while others believe that some of the linked tunnels were used like motorways are today, for people to travel safely regardless of wars or violence or even weather above ground (Kush, 2009). There are several underground cities from Roman Imperial Times even older in Cappadocia, Nevsehir in Turkey, the cities are linked with a network of tunnels. New details have emerged about the massive ancient underground cities discovered in Turkey's Central Anatolian province of Nevsehir. Hundreds of years ago, when the area was attacked, citizens used to flood to the underground city and stay until it was safe to re-enter the land of the living. In this chapter an attempt will be made to summarize the massive tunneling activities in Turkey from past to present focusing on the scientific contributions emerging from these activities.

1 BRIEF SUMMARY OF METRO TUNNELS

Apart from Izmir and Ankara, the biggest metro tunnel activities have been carried out in Istanbul, which is a very fast developing city with a population of almost 18 million. Tunneling activities in these cities like metro, sewerage and water tunnels are increasing tremendously day by day.

The oldest underground urban rail line in Istanbul is the Tunnel (Turkish: Tünel), which entered service on January 17, 1875. It is an underground funicular with two stations, connecting the quarters of Pera (modern day Beyoğlu) and Galata (modern day Karaköy) located at the northern shore of the Golden Horn and is about 573 meters long. It is the world's second-oldest underground urban rail line after the London Underground which was built in 1863, and the first underground urban rail line in continental Europe. The first master plan for a full metro network in Istanbul, titled Avant Projet d'un Métropolitain à Constantinople was prepared by a French engineer L. Guerby, in 1912. The plan comprised a total of 24 stations between the Topkapi and Sisli districts and included a connection through the Golden Horn.

Construction works for the first modern mass transit railway system started in 1989. The line M1 was initially called "Hafif Metro" (which literally translates as "light metro"). The first section between Taksim and 4. Levent entered service, after some delays, on September 16, 2000. This line is 8.5 km (5.3 miles) long and has 6 stations. The length of metro lines before the year 2004 was 45 km, between the years 2005–2013 it was 141 km and it is expected that the total length of metro tunnels will be 400 km between the years 2014 to 2019, while the target after 2019 is 776 km. It is expected that in the year 2017 around 20 earth pressure balance type tunnel boring machines (EPB-TBMs) will be running in Istanbul.

2 BRIEF SUMMARY OF ROAD TUNNELS

Turkey is a mountainous country necessitating continuous road tunneling activities all over the country. As seen from Table 1, the national target up to 2023 is 330,056 m of road and highway tunnels. A general outlook on the road and highway tunnels constructed and under construction is given in Table 2 which is for tunnels having a length of more than 1.5 km.

Table 1 A general outlook on the road and highway tunnels which are constructed and under construction.

Tunnel Name	*Explanation*
Mount Ovit Tunnel (under construction)	14.7 km, twin highway tunnels between Ikizdere, Rize Province and Ispir, Erzurum Province in Northeastern Turkey, the longest tunnel in Turkey.
Mount Kop Tunnel (under construction)	6,500 m, it is a road tunnel under construction located on the province border of Bayburt and Erzurum as part of the route D.915.
Sabuncubeli Tunnel (under construction)	6,480 m, is a road tunnel located on the Mount Sipylus (Turkish: Spil Dagi) in Aegean Region as part of the Manisa-Izmir highway D.565.
Eurasia Tunnel (subsea-under construction)	5.4 km, is a double-deck road tunnel under construction in Istanbul, Turkey, crossing the Bosphorus strait undersea. The project's completion is expected by October 2016. It will connect Kazlicesme on the European and Goztepe on the Asian part of Istanbul on a 14.6 km route.
Cankurtaran Tunnel (under construction)	5,228 m, is a road tunnel under construction located in Artvin Province as part of the Hopa-Borçka Highway D.010 in northeastern Turkey.
Salmankas Tunnel (under construction)	4,150 m, situated on the Mount Salmankas of Pontic Mountains is a road tunnel located on the province border of Gumushane and Bayburt connecting the provincial roads 29–26 and 69–75.
Sariyer-Cayirbasi Tunnel (2012)	4,020 m, The tunnel is part of a project of the Istanbul Metropolitan Municipality to build seven tunnels for the "City of Seven Hills", which is the nickname of Istanbul, it is a twin-tube road tunnel under the northern suburbs of Istanbul.
Ordu Nefise Akcelik Tunnel (2007)	3,820 m, originally Hapan Tunnel is a highway tunnel constructed in Ordu Province, northern Turkey, The tunnel is named in honor of the Turkish female civil engineer and earth scientist Nefise Akcelik (1955–2003)
Samanli Tunnel (under construction)	3,417 m, is a motorway tunnel under construction located at Samanli Mountains in Marmara Region as part of the Gebze-Orhangazi-Izmir Motorway O-33.
Kagithane-Piyalepasa Tunnel (2009)	3,186 m, is a twin-tube road tunnel under the inner city of Istanbul, Turkey connecting the Kagithane district and Piyalepasa Boulevard in Beyoglu district.
Mount Bolu Tunnel (2007)	3,125–3,014 m, it is a highway tunnel constructed through the Bolu Mountain in Turkey between Kaynasli, Düzce and Yumrukaya, Bolu.

Table 2 The current situation and national target for road and highway tunnels in Turkey.

Tunnels	*Numbers*		*Length*	
	>500 m	*Total*	*>500 m*	*Total*
In roads	46	160	54,300	102,390
In highways	15	25	38,380	45,350
Undergoing and project in progress	34	51	174,940	183,360
Target till 2023	95	238	287,860	330,056

Table 3 Summary of some long railway tunnels which are constructed and under construction.

Tunnel Name	*Explanation*
Ayas Tunnel (under construction)	10,064 m is a railway tunnel under construction near Ayas town of Ankara Province in Central Anatolia, Turkey. It was initially projected to shorten the railway line connecting Ankara with Istanbul.
Deliktas Tunnel (2012)	5.4 km, Kangal-Deliktas Tunnel, is a railway tunnel near Deliktas village between Ulas and Kangal districts of Sivas Province in Central Anatolia, Turkey. The construction began on November 15, 1973. Following the installation of signalization facilities, and completing the test runs with freight trains, the tunnel was put into service in late 2012.
Marmaray, subsea (2013)	1.8 km, is a partially operational rail transportation project in the Turkish city of Istanbul. It comprises an undersea rail tunnel under the Bosphorus strait, and the modernization of existing suburban railway lines along the Sea of Marmara from Halkali on the European side to Gebze on the Asian side.

3 BRIEF SUMMARY OF RAILWAY TUNNELS

Construction of the first railway line in Turkey began in 1856, by a British company that had obtained permission from the Ottoman Empire. Later, French and German companies also constructed lines – the motivation was not only economic, the region had a strategically important position as a trade route between Europe and Asia. As with other countries, rapid expansion followed; by 1922 over 8000 km of lines had been constructed in the Ottoman Empire. In 2008, Turkey had 10,991 km of railway lines. As of December 2012, total railway lines reached to 12,008 km. 888 km of this is high speed rail network. The Turkish State Railways (TCDD) owns and operates all public railways in Turkey making it the 22nd largest railway system in the world. The total length of railway tunnels is around 100km. Table 3 gives a summary of some long railway tunnels.

4 BRIEF SUMMARY OF IRRIGATION TUNNELS

DSI, the general directorate of state hydraulic works, with a legal entity and supplementary budget is the primary executive state agency for Turkish Nation overall water resources, planning, managing, execution and operation. Table 4 and 5 is the summary of major irrigation and water supply tunnels, terminated, continuing and in tendering stage.

Table 4 Summary of some major water supply and irrigation tunnels.

Tunnel Name	*Explanation*
Sanliurfa Irrigation Tunnels	26.4 km, The tunnels run in the Sanliurfa Province of Turkey. The water supply is the water reservoir of ATATURK Dam on Firat River (Euphrates). There are two parallel tunnels, the length of each being 26,400 meters. The outer diameter of each tunnel is 9.5 meters and the inner diameter is 7.62 meters. The flow rate is 328 m^3/s. With these figures, the tunnels are the largest in the world, in terms of length and flow rate.
Suruc Water Tunnel (2014)	17.185 km, The purpose of the tunnel is to provide irrigation for the Suruc Valley from ATATURK Dam. The water tunnel was commissioned by the State Hydraulic Works (DSI) on December 25, 2008. The construction of tunnels was carried out with a double shield TBM of 7.83m. The construction was finished in 1913.
Bosporus Water Tunnel (2012)	5.51 km. is an under sea waterway tunnel in Istanbul, Turkey, crossing the Bosphorus strait. It was constructed in 2012 to transfer water from the Melen Creek in Duzce Province to the European side of Istanbul. The tunnel was constructed.

Table 5 Summary of some current and future tunneling projects to be executed by DSI.

The Name of the Tunnel Project	*The name of the Tunnel*	*Excavation Diameter (m)*	*Length (m)*	*Excavation method*	*General information*
Tunnels to be opened by General Directorate of State Hydraulic Works, Diyarbakir Silvan Project I. section	Babakaya Tunnel	7.80	10,556	NATM and TBM	Continuing Irrigation tunnel
Diyarbakır Silvan Project 2. Section	İletim Tunnel	7.80	20,420	TBM	Continuing Irrigation tunnel
Kahramanmaras Kilavuzlu I.	Belpınar Tunnel	6.80	5,500	Single Shield + EPB TBM	Continuing Irrigation tunnel
Ankara domestic water transfer 2. Melen Project	Gerede Tunnel	5.3–5.2	31,592	3 TBM	Continuing
Konya Afsar (Hadimi)	Hadimi Tunnel	5.2	18,140	TBM	Waiting for TBM
Mersin Tarsus Pamukluk Project	Pamukluk Tunnel		4,809	NATM	Continuing
Kayseri Koprubasi Ekrek	–		3,075	NATM	Continuing Irrigation tunnel
Adiyaman Kocali Dam	Kocali Tunnel		5,376	NATM	To be tendered Irrigation tunnel
Adiyaman Kocali Dam	Kuyucak Tunnel		2,440	NATM	To be tendered Irrigation tunnel
Kahramanmaras Menzelet	Tunnel 2		5,450	?	To be tendered Irrigation tunnel
Malatya Yoncali Dam	Yoncali Tunnel	4.20	9,480	NATM	To be tendered Irrigation tunnel

Note: TBM; Tunnel Boring Machine, NATM; The New Austrian Tunneling method, EBP; Earth Pressure Balance

5 DIFFICULTIES IN TUNNELING, EFFECTS OF NORTH AND EAST ANATOLIAN FAULTS ON TUNNELING PERFORMANCES

Turkey is in a tectonically active region that experiences frequent destructive earthquakes. At a large scale, the tectonics of the region are controlled by the collision of the Arabian Plate and the Eurasian Plate. At a more detailed level, the tectonics become quite complicated. A large piece of continental crust almost the size of Turkey, called the Anatolian block, is being squeezed to the west. The block is bounded to the north by the North Anatolian Fault and to the south-east by the East Anatolian fault. The East Anatolian Fault (EAF) is a major strike-slip fault zone in eastern Turkey. It forms the transform type tectonic boundary between the Anatolian Plate and the northward-moving Arabian Plate. The North Anatolian Fault (NAF) is an active right-lateral strike-slip fault in northern Anatolia which runs along the transform boundary between the Eurasian Plate and the Anatolian Plate.

Six tunneling projects affected directly by the EAF and NAF faults are the Ayas, Bolu, Kargi, Dogancay, Gerede and Nurdag tunnels, which are shown in Figure 1.

5.1 Baltalimani tunnel

The first mechanized tunnel used TBM in Turkey was Baltalimani tunnel in Istanbul which was a part of the Istanbul Sewerage Project. A Robbins full face TBM was used in this tunnel. Buyukada and Trakya formations exist along the road (Figure 2). Machine utilization is found to be 28.5 % in the Buyukada Formation and less than 10% in the Trakya Formation. The Trakya Formation is closely jointed and faulted. The biggest and most important problems encountered through this formation are support

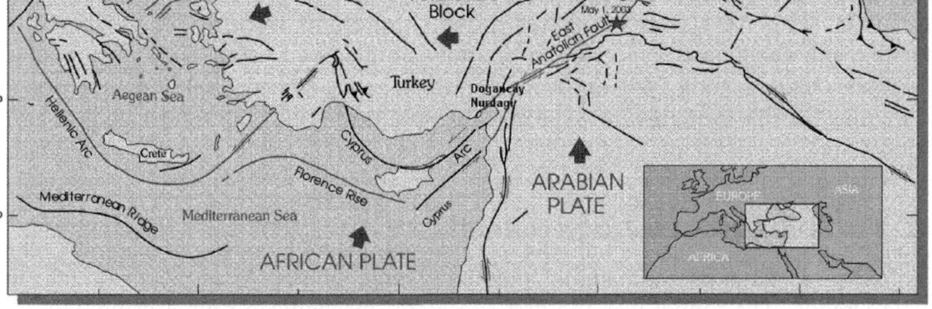

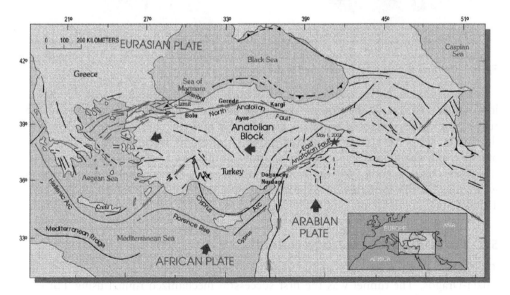

Figure 1 Tunneling operations affected by North and East Anatolian Faults (Bilgin, 2015; http://earthquake.usgs.gov/earthquakes/eqarchives/year/2003/2003_05_01_maps.php#tectonic).

Figure 2 General View of Robbins TBM used in the Baltalimani Tunnel.

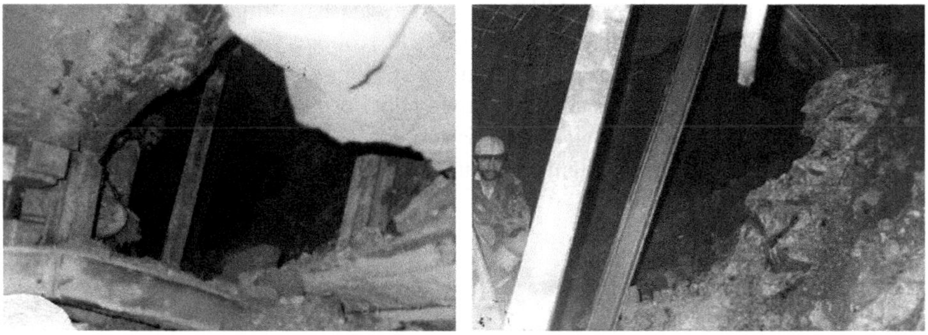

Figure 3 Collapse and support system used in the tunnel.

installation in front of the gripper pads, overbreaks at crown and sidewalls of the tunnel, sometimes collapsed especially crushed and faulted zones (Figure 3). Excessive disc wear, supporting works and bad ground conditions were the reasons for low machine utilization time. Especially, supporting of the cavern in the Trakya formation due to the collapses took 74% of the machine downtime. Machine selection and design was wrong for this formation and the machine was taken out from tunnel (Bilgin *et al.*, 1993).

5.2 Tuzla-dragos sewerage tunnel

The tunnel is situated in a highly populated area of the Asian Side of Istanbul Tuzla-Dragos. The tunnel was the first successful mechanized tunnel using TBM in Turkey. Detailed laboratory full-scale cutting tests were applied to rock samples collected from the tunnel side. The equipment used for the tests is developed under

Figure 4 General View of Robbins TBM, Model 165–162.

NATO TU-Excavation Project in Istanbul Technical University, Mining Engineering Department. Test results were used to predict the cutting performance of a TBM. Actual and predicted values are later compared using a rating system for machine utilization. Tunnels having a total length of 6,490m and final diameter of 4.5m were completed by STFA Construction Company. The tunnels between X1-X3 and K1 shafts having a length of 1670m and a depth changing between 6m and 17m were opened with Robbins TBM, Model 165–162 having a diameter of 5m. A general view of the machine is illustrated in Figure 4.

The performance of TBM and face conditions were recorded continuously during tunnel excavation. Table 6 summarizes the overall TBM performance in Tuzla Tunnel (Bilgin *et al.*, 1999). This tunnel is important in a way that it was a typical example of

Table 6 The overall TBM performance in Tuzla.

Starting date	*6th October 1997*
Finishing date	31st July 1998
Length of tunnel	1600 m
TBM diameter	5 m
Final tunnel diameter	4.5 m
Average machine utilization	10%
Machine utilization in competent rock	35%
Average net cutting rate	50 m^3/h
Average progress rate	5.1 m^3/h
Best daily advance	15.2 m/day
Average daily advance	6.2 m/day
Best weekly advance	69 m/week
Average weekly advance	33 m/week
Best monthly advance	253 m/month
Average monthly advance	135 m/month
Cutter cost	4 \$/m^3

how it was possible to predict the performance of a TBM from laboratory full scale linear rock cutting tests in Turkey.

5.3 Ayas tunnel

The construction of the railway tunnel between Beypazari and Istanbul (Arifiye-Sincan started in 1976 with a length of 10,064 km (6.253 miles). The lined inner diameter of the tunnel is 9.60 m. The tunnel's 400 m long section is constructed by cut-and-cover method while for the main part of 8,000 m the New Austrian Tunneling method was applied. 230 million US dollars were spent and still it is not completed. The main reason is highly complicated geology and difficult ground conditions. Ayas Tunnel is the longest railway tunnel in Turkey and is the sixth longest railway tunnel in the world, and constructed with no side approaches. 1,200,000 m^3 of tunneling excavation, 2,600,000 m^3 of open excavation, 450,000 tons of shotcrete, 6,700 tons of steel ribs, 6,100 tons of anchorage, 1,500 tons of wire mesh and 210,000 m^3 B225 tunneling concrete works are included within the project scope.

5.4 Bolu tunnel

The tunnel is part of the Gumusova-Gerede Highway O-4 within the Trans-European Motorway project, tunnel excavation started April 16, 1993. The total cost of the tunnel is about US $300 million. It has twin 17 m wide bores carrying three lanes of traffic in each direction. The tunnel crosses the North Anatolian Fault. The November 12, 1999 Duzce earthquake (MW = 7.2) caused substantial damage to the tunnel and viaducts, which were under construction at the time of the earthquake. The excavation of the tunnel in the Ankara-Istanbul direction was completed by the beginning of August 2005. NATM tunneling method was used in Bolu tunnel.

5.5 Kargi tunnel

The Kargi hydropower project was developed to utilize the energy potential of the Kizilirmak River between the towns of Osmancik and Boyabat. The project is a development by Kargi Kizilirmak Energy A.S, a subsidiary of Statkraft Energy A.S, which is owned 100% by the Statkraft, Europe's largest producer of renewable energy. The geology consists of Eocene aged Beynamaz volcanic mainly of agglomerate, andesite, basalt and tuff metamorphites and 2500 m of Kiraztasi-Kargi ophiolites and graphitic schist. The excavation of an 11.8 km tunnel has been recently finished. 7.8 km of the tunnel was excavated with a double shield Robbins TBM with a 9.84 m diameter, and 4 km of the tunnel from inlet part was opened with NATM. The TBM was launched in April 2012. After boring only 80m, the TBM became stuck in a section of collapsed mixed ground face of hard rock and running ground. In the first 2km of the tunnel the TBM jammed 7 times due to face collapses and bypass tunnels were required to free the TBM. After a few collapses the cutterhead drive torque capability was increased by more than 50% and the thrust capacity of TBM was also increased by hydraulic power pack cylinder jacks including overbore capabilities, shield lubrication, and belt scales. The average advance was 144.2 m/month before modifications and 407.7 m/month after modifications. The performance values of TBM and drill and blast excavations are given in Table 7 (Bilgin, 2015).

Table 7 Comparison of TBM and drill and blast operations in Kargi Tunnel.

Parameters	*TBM*	*Drill and Blast*
Boring length, km	7.8	4.0
Average advance, m/month	217.0	173.8
Best Month, m	72.3	28.5
Best Day, m	29.6	12.0
Monthly average, m	271.4	173.8
Time to mobilize	16.0	4.0

Note: TBM; Tunnel Boring Machine

5.6 Gerede tunnel

The purpose of the project is to supply drinking water to Ankara, via a tunnel with a length of 31.6 km. Tunnel excavation began in 2010 simultaneously at three points: entrance portal, shaft, and output portal, with S 690, S 691 and S 692 TBMs. TBM S 690 excavated a length of 9588 m and the first part was finished. However this is one of the most tragic TBM tunneling operations in Turkey. The first drive, downstream from Gerede to the intermediate shaft, has been already completed without any significant problems while the second drive, which started from the intermediate shaft toward Ankara, is being excavated in downstream direction by the S-691 DS TBM. TBM is currently stuck within smectite clay which has a tremendous swelling characteristic. Swelling stresses started breaking the segments and the broken segments are supported by steel arches. It is predicted that the smectite zone is behaving like a pillar zone protecting the tunnel from the high stressed pressurized water reservoir ahead of the tunnel and it is currently decided to remove the machine from the side to continue from the Ankara side to take advantage of the dip of the tunnel for water removal. The third drive, which was being excavated by the S-692 DS upstream from the Ankara side, has been continuously hindered by the complex existing geological conditions of heavily altered and weathered volcano-clastic rocks under very high water tables, which have even caused a 12 month long stoppage and required a by-pass tunnel (Ch. 27 + 582.6). Currently, the S-692 shield is trapped at chainage 24+344.86 after the tunnel suffered a collapse in July 2014. The pressure deformed the telescopic shield and about 20 meters of segmental lining, causing a huge water and material inflow which is estimated to be around 1,250 m^3 in 15 minutes (Bilgin, 2015).The tunneling activities are planned to be carried out in 2016 with another TBM which is currently being constructed.

5.7 Dogancay tunnel

The Dogancay Tunnel is located in the South East of Turkey, affected by East Anatolian Fault and it is a part of a hydroelectric Project licensed by Enerji Sa. The length of tunnel is 6,655m, the excavation started in September 2012, with excavation carried out by a double shield TBM. Carboniferous age limestone and shale, Upper Permian age limestone, Lower Permian age siltstone, claystone, sandstone and quartzite are found along the tunnel route. The overburden within 3km of the tunnel route is around 1km and only two boreholes could be done in the tunnel route. Tectonic stresses due to East

Figure 5 Traces of tectonic stresses on rock samples taken from the zones where the TBM is squeezed.

Figure 6 Traces of tectonic stresses on rock samples taken from the zones where the TBM is squeezed.

Anatolian fault squeezed the TBM several times causing considerable delays in tunnel drivage. Traces of tectonic stresses on rock samples taken from the zones where the TBM is squeezed are clearly seen in Figures 5 and 6 (Bilgin, 2015).

5.8 Nurdagi railway tunnel

The tunnel is for railway transportation and the project concerns of two tubes. Each tube has a length of 9750m. The excavation is planned to start from chainage 13+450m and to terminate in chainage 3+700m. The chainage from 13+450m to12+400 concerns Karadag Limestone of Mesozoic age, which is affected by East Anatolian Fault (EAF), fracturing the rock formation to a great extent. High water ingress is expected in

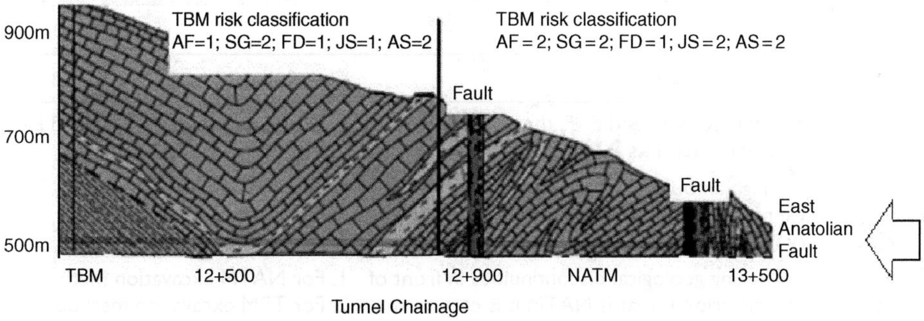

Figure 7 Geologic cross section of the tunnel between chainage 13+450m to 12+400.

this area. Karadag Limestone discharges the water at the toe of the mountain at the Nurdagi side. Several springs are available along EAF. Due to technical difficulties and time necessary to provide TBM, the first 1050m in limestone is currently being opened by NATM tunneling method. The geological cross-section of this area which is planned to be opened by drill and blast method is seen in Figure 7. The second section of the tunnel starting from chainage 12+000 is being currently excavated by a single shield hard rock TBM (Bilgin, 2015).

The geological formation from the chainage 12+400m to 4+850m concerns of middle Ordovician aged Kizlac Formation of very massive inter bedded meta-sandstone, metaquartzite and meta-mudstone having very high strength and abrasivity characteristics. This section is planned to be excavated with a single shield hard rock TBM. However the massive characteristic of the rock formation may change from chainage 4+850m to 3+700m being affected by local faults and shear zones, in this zone RQD values are very low high water ingress is also expected, this section is planned again to be excavated with NATM tunneling method. The tunnels excavated close to North and East Anatolian Faults led to the development of a risk classification method defined in Table 8. According to this table, the use of TBM in the Nurdagi tunnel between 13+500m and 12+800m was found very risky, 12+800m to12+500m was risky and favorable up to 4+850m (Bilgin, 2015). The geological formation from the chainage 12+400m to 4+850m concerns middle Ordovi.

6 THE CONTRIBUTION OF TUNNELING ACTIVITIES ON THE DEVELOPMENT OF SOME NEW CONCEPTS IN ROCK AND ROCK CUTTING MECHANICS

Drilling and blasting is the most widely used excavation method in tunneling in Turkey especially in hard rock conditions. But in recent years, the use of roadheaders and impact hammers in hard rock, especially in fractured geological formations has increased considerably in both civil and mining engineering fields. At the other hand TBMs became inevitable excavating machines for rapid mechanical excavation. However it is strongly emphasized that the prediction of the machine performance

Table 8 Risk classification system for TBM excavation in the zones close to North and East Anatolian Fault.

Factors effecting the risk of using TBMs	*Classification*
1. Distance of the tunnel to NAF and EAF, the possibility of tectonic stresses **(abbreviated as AF)**	1. Within 0.5–2 km to NAF and EAF 2. Very close to NAF and EAF
2. The possibility of high amount of water ingress into tunnel. Detailed geological reports and careful observation of drilling logs are necessary **(abbreviated as SG)**	1. Less than 100 lt/sec 2. More than 100 lt/sec
3. The possibility of seeing geological discontinuities in front of the tunnel face. The criterion is that is NATM it is easy to see and control geological discontinuities in the tunnel face **(abbreviated as FD)**	1. For NATM excavation method 2. For TBM excavation method
4. Geological discontinuities **(abbreviated as JS)**	1. Q < 1 RMR < 20 2. Q > 1.1 RMR > 21
5. The presence of anticlinal and synclinal **(abbreviated as AS)**	1. One within 1 km 2. More than one within 1 km

If the total mark obtained from the classification column from the table is between 8–10, it is very risky to use TBM; if the total mark is between 5–8, it is risky to use TBM; if the total mark is between 2–5, the risk of using TBM is medium level, if the total mark is 0–2, using TBM is not risky.

Note: TBM; Tunnel Boring Machine

plays an important role in the time scheduling and in the economy of tunneling projects and accumulated data will serve a sound basis for performance prediction models. The use of TBMs (tunnel boring machine) is found unfavorable in short tunnels and in these cases drilling and blasting method is preferred, however this method is restricted in urban areas. In favorable conditions, roadheader and impact hammer are preferred which offer many advantages over conventional methods. These include improved safety, minimal ground disturbances, elimination of blast vibration, reduced ventilation requirements and cost etc. Therefore, hydraulic impact hammers have been used widely in metro tunneling projects in Istanbul since 1990. However the uses of TBMs are also increased since 1980 in any ground conditions providing higher advance rates and more safety in many cases. Large amount of data from tunneling projects was collected continuously by researchers in the Mining Engineering Department of Istanbul Technical University (Bilgin *et al.*, 2014; Balci, 2009). This following section is a summary of models created for performance prediction of different mechanical excavators *i.e.*, impact hammers, roadheaders and TBMs.

6.1 The concept of rock mass strength index and application to impact hammers and roadheaders

The use of roadheaders and impact hammers, especially in fractured geological formations have increased considerably in several metro and utility tunnels in Istanbul and accumulated data served a sound basis for performance prediction models. Typical roadheaders used in tunneling projects are seen in Figure 8. A typical impact hammer working in a tunnel is seen in Figure 9.

Figure 8 Typical roadheaders used in Metro stations (left transverse type) and utility tunnels (right axial type).

Figure 9 Typical Impact Hammer Mounted on an excavator.

Data collected from different tunneling projects carried out lead to the development of a performance prediction model for roadheaders and impact hammers. This model is based on using the cutting power of machines used and on the concept of rock mass strength index related to rock excavatability. Equations 1 and 2 are for estimating net breaking rate (NBR) of impact hammers and Equations 2 and 3 are used for estimating net cutting rate (NCR) of roadheaders (Bilgin *et al.*, 1996, 1997, 2002, 2014).

$$\mathrm{NBR} = 4.24\mathrm{x}\mathrm{P_o}\mathrm{x}(\mathrm{RMCI})^{-0.567} \tag{1}$$

$$\mathrm{RMCI} = \mathrm{UCSx}\left[\frac{\mathrm{RQD}}{100}\right]^{2/3} \tag{2}$$

$$\mathrm{NCR} = 0.28\mathrm{xPx}(0.974)^{\mathrm{RMCI}} \tag{3}$$

Where NBR is net breaking rate of impact hammers in (m^3/h), P_O is power output of impact hammer in (HP) (it is found by multiplying impact energy by impact rate),RMCI

Table 9 A guide to estimate reduction factor K used to predict net braking rates of impact hammers related to the thickness of the dikes.

Thickness of Dikes, cm (for UCS >120 MPa)	*Reduction Factor K for Net Breaking Rate*
0–10	0.9–1.0
10–20	0.8–0.9
20–30	0.7–0.8
30–40	0.6–0.7
40–50	0.5–0.6
>50 D&B is recommended	

Note: TBM; Tunnel Boring Machine

is rock mass strength index related to rock excavatability, UCS is uniaxial compressive strength of rock in (MPa), RQD is rock quality designation in (%), NCR is net cutting rate of roadheaders in (m^3/h), and P is cutting power of roadheaders in (HP).

Equation 1 was developed for the rock formations found mainly in Istanbul where the thickness of the dikes does not exceed 10 cm. However, in some cases the thickness of the dikes varies up to several tens of meters, and it is observed that thickness of high strength non fractured dikes affects tremendously the performance of impact hammers as given in Table 9. According to this table, Equation 1 is proposed to be improved as in Equation 4 (Ocak & Bilgin, 2010).

$$NBR = Kx4.24xP_ox(RMCI)^{-0.067} \tag{4}$$

Where, K is a reduction factor when predicting net braking rates of impact hammers used in rock formations intruded by different thickness of dikes.

The other important point is that the dip and strike of the joints having an angle of 45° in favor of the cutting direction improve the net cutting rate of roadheaders tremendously about 20–25%.

Machine utilization time is as important as net breaking rate in determining the excavation efficiency and economy, since daily advance rates are directly related to these factors for a given tunnel section. The performances of mechanical excavators and drilling and blasting method are summarized in Figures 10a, 10b, 10c for a metro station tunnel in Istanbul. These figures give a unique opportunity to compare the efficiencies of roadheaders, impact hammers and drilling and blasting method for a given job. As seen from these figures machine utilization time for roadheaders is 28.2%, almost twice compared to impact hammers, and time spent for excavation in drilling and blasting which is at the same order as impact hammers with a value of 15.4 % (drilling, blast hole charging, blasting and ventilation). Mucking is another factor affecting job duration. In fractured rock formations, sometimes roadheaders excavate the rock in excess to their loading capacities with gathering arms and an extra time is necessary to load the muck. The time spent for mucking was 9.8% for roadheader, 13.3% for impact hammers and as high as 21.8% for drilling and blasting method. It is difficult to control muck size in fracture rock formations with conventional excavation methods, some big muck samples always create transportation problems and due to the disturbed zone the number of bolts

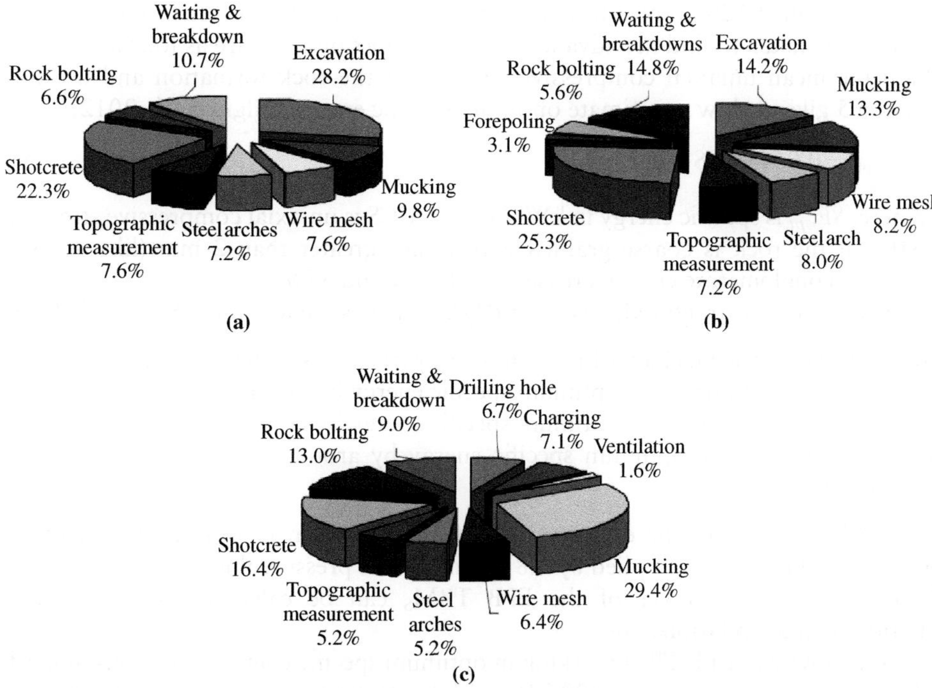

Figure 10 The general performance of a roadheader (a), impact hammers (b) drilling blasting operations (c) in a metro tunnel (Ocak & Bilgin, 2010).

used for a unit length of tunnel is always higher compared to mechanical excavation methods. Time spent for rock bolting in drilling and blasting method was found to be twice compared the other excavation methods with a value of 13%. Hydraulic impact hammers and roadheaders have gained worldwide acceptance in mining and civil engineering industries since their initial capital cost is lower. Especially hydraulic impact hammers have been widely preferred in Istanbul Metro drivages because of their initial low cost and easiness to mount in any type of available excavator. Drilling and blasting method is used in hard rock formation which roadheaders and impact hammers cannot handle.

6.2 Methodology used for predicting EPB TBM performance in hard fractured rock formations

Different TBM performance prediction methods are used in the past, like NTNU, full scale linear cutting tests (CSM) method and empirical methods (Balci, 2009; Bilgin *et al.*, 2008, 2011). However, none of them is applicable for EPB TBMs working in urban areas in challenging geology. In this work a model based on past experiences obtained in Istanbul was used. The TBM data collected from Beykoz Utility Tunnel, Cayirbasi Water Tunnel, Kartal-Kadikoy Metro Tunnel, Pendik-Kaynarca Metro Tunnel and Uluabat Power Tunnel from Bursa, and current metro Uskdar-Umraniye-Sancaktepe-

Cekmekoy projects were used to develop the model described in this section. This model was discussed in ITA2013 (Namli *et al.*, 2014). The methodology is briefly given below.

Steps to be taken to find net advance rate or production rates are as follows.

Find the mean uniaxial compressive strength of the rock formation and use the Equation 5 given below to estimate optimum specific energy (Bilgin *et al.*, 2012).

$$SE_{opt} = 0.051 \cdot UCS + 1.8 \quad (5)$$

Where, SE_{opt} is specific energy in kWh/m^3 and UCS is uniaxial compressive strength in MPa. If the rock is coarse grained with grains greater than 2 mm (like arcose, sandstone, conglomerate etc.), increase SE_{opt} by around 35%.

If the rock mass is fractured, based on *RQD* range use reduction in SE_{opt} as below:

100% ≥ RQD ≥ 70%, no change in specific energy (rock mass behaves as massive rock),
70% > RQD ≥ 50%, decrease optimum specific energy by around 15%,
50% > RQD ≥ 30%, decrease optimum specific energy by around 30%,
20% > RQD, decrease optimum specific energy by around 40% (risk of face-roof collapses).

If EPB TBM is used, the estimated specific energy from laboratory rock cutting experiments should be multiplied by 1.8 for mean face pressure fluctuating between 0. 7–1.5 Bars. Cutting power of the EPB TBM_S can be calculated the following Equations 6 and 7 relationships.

Cutting power of EPB TBM working in optimum specific energy conditions in hard and very fractured rocks up to 100 MPa of laboratory compressive strength may be calculated empirically as given below:

For TBM working in closed mode in hard fractured rock:

$$P_{cutting\text{-}EPB} = KxD \quad \text{(for EPB pressure changing between 0.7–1.5 Bar)} \quad (6)$$

K=118.8 for UCS up to 100 MPa and (for N=2–2.5 rpm, EPB pressure changing between 0.7–1.5 Bar)

K=95 for UCS up to 70 MPa and (for N=2–2.5 rpm, EPB pressure changing between 0.7–1.0 Bar)

For TBM working in open mode in hard rock

$$P_{cutting\text{-}HardRock} = 70xD \quad (7)$$

Where, $P_{cutting}$ is cutting power in kW (subscript $_{EPB}$ denotes for EPB TBMs, subscript $_{Hard\ Rock}$ denotes for hard rock TBMs) and D is cutterhead diameter in meter.

Use the following equation to estimate net production rate (Rostami *et al.*, 1994; Balci *et al.*, 2004 and Copur *et al.*, 2001).

$$NPR = k \cdot \frac{P_{cutting}}{SE_{opt}} \quad (8)$$

Where, NPR is net production rate in m^3/h, *k* is energy transfer ratio, which is around 0.8 for TBMs, $P_{cutting}$ is cutting power in kW and SE_{opt} is optimum specific energy in kWh/m^3.

Use machine utilization time as defined in Table 10 to estimate daily advance rate.

Table 10 Estimating machine utilization time.

Stoppage Type		*Stoppage Duration*
Adverse Ground	Contact zones between	Few days to 1 week
	Dikes	Few days to 1 week
	Faults	Few days to 2 weeks
	Water	few days to 1 week
TBM Breakdown	New TBM, experienced crew	2–4%
	New TBM inexperienced crew	4–6%
	Refurbishment TBM experienced crew	4–6%
	Refurbishment TBM inexperienced crew	6–8%
Cutter Replacement	Quartz content 0–20%	5%
	Quartz greater than 20%, weak and blocky ground	5–10%
	Quartz content greater than 20%, hard rock	10%
Muck Transportation	By Train: l = 0–3 km	7%
	By l greater than 3 km	10%
	By Belt Conveyor : 0–3km	5%
	By Belt Conveyor: l greater than 3 km	7%
Maintenance	Experienced contract and crew	10%
	Moderately experience contractor and crew	15%
Regripping		20–25%
TBM Mobilization at Stations		2–3 week
Other Stoppages		10–15%
MACHINE UTILIZATION		15–45%

7 CONCLUSIONS

Tunneling in Turkey has been carried out since the Stone Age. The target till 2023 for road and highway tunnels is more than 330 km and for metro tunnels in Istanbul this number rises to 800 km. Two decked Eurasia road tunnel opened under Istanbul Bosphorus with slurry TBM under hyperbaric pressure up to 11 Bars is called by many contributors the project of the century. A three deck tunnel having a diameter of 18.8 m and length of 6.5 km is already under consideration. It is expected that from 2015 to 2023 around 30 billion Euros will be spent on tunneling in Turkey. The geology of Turkey is complex leading to several problems during the construction of tunnels. The data collected by researchers leads to develop the contributions on rock and rock cutting mechanics published in different peer reviewed journals.

REFERENCES

Balci, C. (2009) Correlation of rock cutting tests with field performance of a TBM in a highly fractured rock formation: A case study in Kozyatagi-Kadikoy metro tunnel, Turkey. *Tunnelling and Underground Space Technology*, 24(4):423–435.

Balci, C. Demircin, M.A. Copur, H. & Tuncdemir, H. (2004) Estimation of optimum specific energy based on rock properties for assessment of roadheader performance. *Journal of South African Institute of Mining and Metallurgy*, 104(11) (December):633–642.

Bilgin, N. (2015) An Appraisal of TBM Performances in Turkey in Difficult Ground Conditions and Some Recommendations. *International Conference on Tunnel Boring Machines in Difficult Grounds (TBM DiGs)* Singapore, 18–20 November 2015

Bilgin, N. Balci, C. Copur, H. Tumac, D. & Avunduk, E. (2012) Rock mechanics aspects related to cutting efficiency of mechanical excavators, 25 Years of Experience in Istanbul, *EUROCK 2012*, Stockholm.

Bilgin, N. Copur, H. & Balci, C. (2014) *Mechanical excavation in Mining and Civil Industries*, CRC Press, Taylor and Francis Group, ISBN 13:978-1-4665-8474-7, p 366.

Bilgin, N. Copur, H. Balci, C. Tumac, D. Akgul, M. & Yuksel, A. (2008) The selection of a TBM using full scale laboratory tests and comparison of measured and predicted performance values in Istanbul Kozyatagi-Kadikoy Metro Tunnels. *In World Tunnel Congress*, Akra, India, pp. 1509–1517.

Bilgin, N. Copur, H. Balci, C. Tumac, D. & Avunduk, E. (2011) Experience gained in mechanized tunnelling in Istanbul and some recommendations for mining industry. In: *World Mining Congress*, Eskikaya, S.(ed), vol. 2, pp. 155–159.

Bilgin, N. Dincer, T. & Copur, H. (2002) The performance prediction of impact hammers from Schmidt hammer rebound values in Istanbul metro tunnel drivages. *Tunnelling and Underground Space Technology* 17(3):237–247.

Bilgin, N. Kuzu, C. Eskikaya, S. & Ozdemir, L. (1997) Cutting performance of jack hammers and roadheaders in Istanbul Metro Drivages. In: *Word Tunnel Congress*, Balkema, ISBN 90 5410 868 1, pp. 455–460.

Bilgin, N. Yazici, S. & Eskikaya, S. (1996) A model to predict the performance of roadheaders and impact hammers in tunnel drivages. *In Eurock 96 Symposium*, Barla, G. (ed), Torino, pp. 715–720.

Copur, H. Tuncdemir, H., Bilgin, N. & Dincer, T. (2001) Specific energy as a criterion for used of rapid excavation systems in Turkish mines. *Transactions of the Institution of Mining and Metallurgy Section A* 110:A149–157.

Kusch, H. (2009) *Tore zur Unterwelt (German) (Secrets of the Underground Door to an Ancient World)*, V.F. Samler, Graz, ISBN-10: 3853652379.

Namli, M. Cakmak, O. Pakiş, I.H. Tuysuz, L. D. Talu, D., Dumlu M., Şavk S. Bilgin, N. Copur, H. & Balci C. (2014) The performance prediction of a TBM in a complex geology in Istanbul and the comparisons with actual values, proceedings of the *World Tunnel Congress 2014 – Tunnels for a better Life*. Foz do Iguaçu, Brazil.

Namli, M. Cakmak, O. Pakis, I.H. Tuysuz, L. Talu, T. Dumlu, M. Balci, C. Copur, H. & Bilgin, N. (2013) A methodology of using past experiences in the performance prediction of a TBM in a complex geology and risk analysis. *The 39th ITA-AITES General Assembly and World Tunnel Congress*, Geneva, Switzerland.

Ocak, I. & Bilgin, N. (2010) Comparative studies on the performance of roadheaders, impact hammers and drilling and blasting method in the excavation of metro stations tunnels of Kadıköy-Kartal Metro (Istanbul). *Tunnelling and Underground Space Technology* 25 (2):181–187.

Rostami, J. Ozdemir, L. & Neil, D.M. (1994) Performance prediction: A key issue in mechanical hard rock mining. *Mining Engineering* 11:1263–1267.

US Geological Survey. (2015) USGS: Tectonic Map of Turkey. Available from: http://earthquake.usgs.gov/earthquakes/eqarchives/year/2003/2003_05_01_maps.php#tectonic/ [Accessed 20 October 2015]

Bilgin, N., Balci, C., Tuncdemir, H. Eskikaya, S. Akgul, M. Algan, M. (1999) The performance prediction of a TBM in difficult cround condition. *AFTES Conference*, p. 115–121, Paris, 25–28 October 1999.

Bilgin, N. Nasuf, E. Cigla, M. (1993) Stability problems effecting the performance of a full face tunnel boring machine in Istanbul- Baltalimani Tunnel. *Assessment and prevention of Failure Phenomena in Rock Engineeging*, Pasamehmetoglu *et al.* (eds), Balkema, Rotterdam, pp. 501–506.

Chapter 5

Tunnels in Korea

S. Jeon[1], Y.H. Suh[2], S.P. Lee[3], S.B. Lee[4] & K. Suh[5]

[1]Department Energy Systems Engineering, Seoul National University, Seoul, South Korea
[2]Infra R&D Group, Hyundai Engineering & Construction, Yongin-si, South Korea
[3]Technology Division, GS Engineering & Construction, Seoul, South Korea
[4]Research Institute of Technology, Dongbu Corporation, Seoul, South Korea
[5]Civil Engineering Research Team, Daewoo Engineering & Construction, Suwon-si, South Korea

Abstract: Over the past decades, a lot of tunnels have been constructed in Korea. In the process of fast economic growth, expressway tunnels, railway tunnels, high-speed railway tunnels, metro tunnels, and utility tunnels were in high demand. Increased amount of passenger vehicles, public transportation, and logistics will demand a greater number of tunnels in the coming decades. Korea is proud of her construction technologies, especially in tunneling and underground development. The state-of-the-art tunneling technologies acquired from the past projects have been used in many other international projects. In this Chapter, five tunneling cases are introduced that are worthy to note because of their historical and technical importance to the Korean tunneling industry.

1 INTRODUCTION

Rapid economic growth and industrialization in Korea boosted the expansion of the infrastructure. In consequence, tunneling work has enormously increased with the social demands. Lots of tunneling projects started in 1970s for the construction of the Seoul Metro system. Sophisticated site investigations, excavation, and reinforcing methods have been applied. The extension of road and railway tunnels has broadly increased in the late 1990s. Later, long and large-span tunnels were constructed in the Seoul-Busan High Speed Railway project. As the scale of tunnels increased, environmental issues were essentially considered.

Currently, Korea has tunnels of approximately 2,048 km in total length that are already completed or under construction, excluding life line tunnels such as electric power line tunnels and communication cable tunnels, as summarized in Table 1. Most tunnels were excavated in hard rock formations by the drilling and blasting method. However, mechanized tunneling methods have been widely adopted in recent decades.

The Korean Peninsula has a close affinity with the Asian continent in geology and tectonic setting, which has been regarded as stable and cratonic in nature. South Korea has mountainous areas in the eastern and central parts. More than 70% of the territory consists of mountains (having more than 5° slope angle on average). Therefore, tunnels and bridges are commonly required to construct roadways and railways with a large radius of curvature. Rock types are mostly granite and granitic gneiss, partly limestone, and complex sedimentary rocks in southeastern parts as shown in Figure 1. Granitic and gneissic rock masses are generally competent. The major horizontal stress has an orientation in the east-west direction.

Table 1 Length of tunnels in Korea.

Category	*Total length (km)*	*Remarks*
Metro tunnel	990.74	Seoul, Busan, Daegu, Incheon, Gwangju, and Daejeon Metro Systems
Roadway tunnel	754.5	National expressway and highway, municipal roadway, and provincial roadway
Railway tunnel	302.9	High-speed and conventional railway

Note: Tunnels in operation or under construction as of 2010.

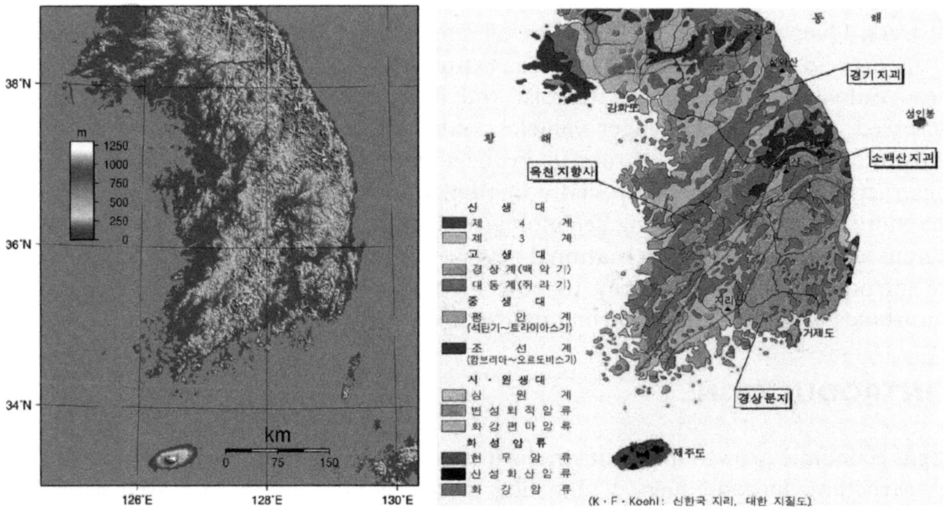

Figure 1 Topography (*left*) and geology (*right*) of Korea.

In this Chapter, five tunnels are introduced. They are worthy to note because of their technical or historical importance to the Korean tunneling industry. Among many interesting tunnels, only rock tunnels were selected. Figure 2 presents the locations of the five tunnels.

2 IN-JE TUNNEL

In-je tunnel is a 10.9 km long expressway tunnel that connects Jin-dong and Seo-rim on the route of Expressway No. 60 as shown in Figure 3. The Expressway No. 60 is newly constructed to connect the western and eastern parts of Tae-baek Mountains providing shortest distance from Seoul Metropolitan area to eastern coastal area in Korea. The tunnel excavation was completed in September, 2012 and the tunnel will be open in December, 2017 to accommodate the traffic demand for the 2018 Pyong-Chang Winter Olympic Games. When finished, the tunnel will be the longest road tunnel in Korea and the 11th longest road tunnel in the world. For reduced construction period and efficient ventilation, one inclined shaft and two vertical shafts were constructed.

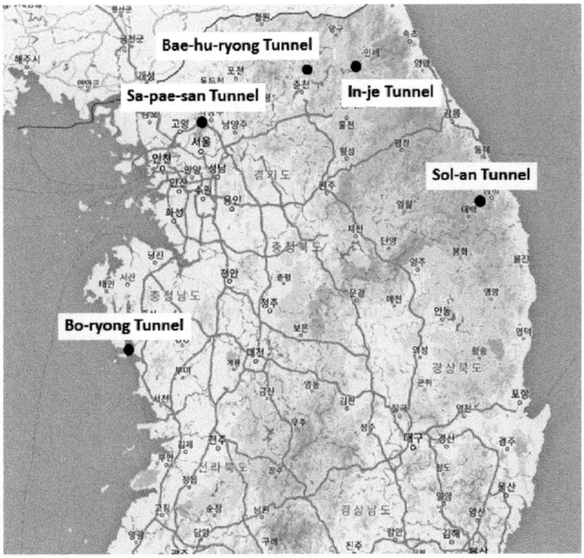

Figure 2 Locations of the five tunnels to be introduced in this Chapter.

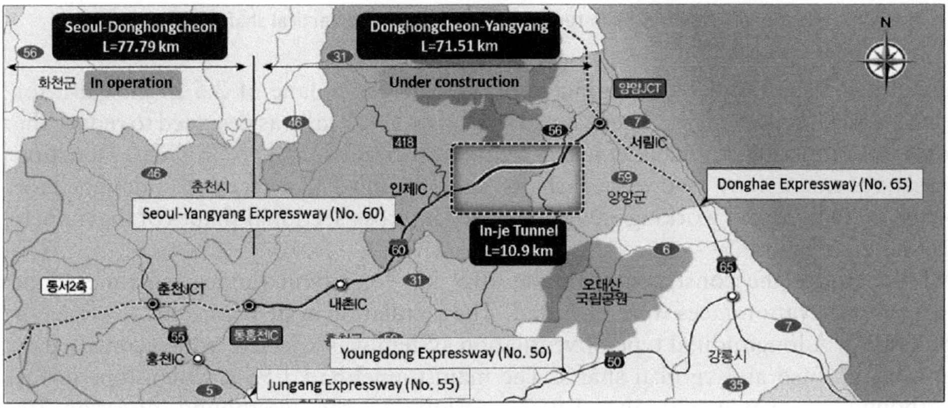

Figure 3 Route of In-je tunnel on the Expressway No. 60.

2.1 Key issues on design and construction

The geological map, satellite images, and the results of surface and borehole geological survey were used for site characterization. The geological map and the route are shown in Figure 4. The bedrock consists of Precambrian metamorphic rocks with complex geological structures and intruded Mesozoic granite. The tunnel route is mostly positioned in granitic and gneissic rock mass. The precautious measures were required to cross three fault zones.

The optimal cross-sections were selected considering the Korean design standard, workability, economy, maintenance, ventilation and disaster prevention. The main

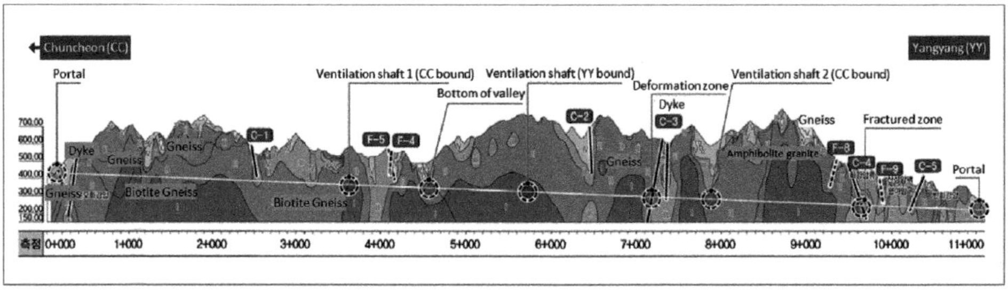

Figure 4 Geological cross-section of the route in In-je tunnel.

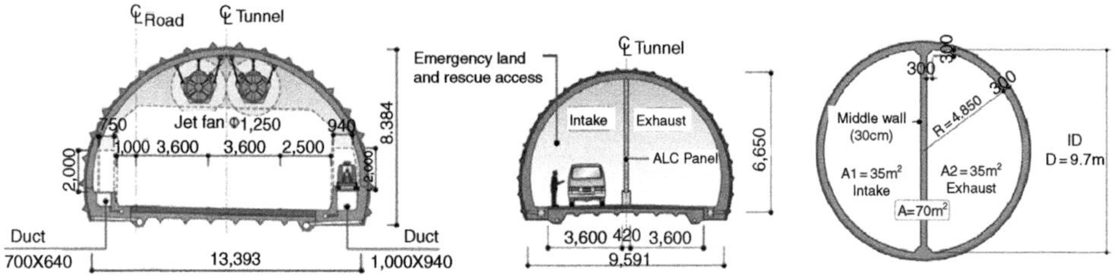

Figure 5 Cross-sectional shapes of main tunnel, inclined shaft, and vertical shaft in In-je tunnel.

tunnel has a span of 13.4 m to facilitate the emergency lane of 2.5 m in width for parking and rescue vehicles. The inclined shaft (L = 1,409 m) was designed to reduce the construction period, to make efficient ventilation circuits, and to provide evacuation passage at emergency. Two vertical shafts (L = 201 and 307 m) were also designed for better ventilation and working environments. The cross-sectional shapes of the tunnels are shown in Figure 5.

To complete the construction of the 10.9 km long twin tunnels on time, eight different excavation faces were operated. The ventilation system was the key issue of the project. A longitudinal type of ventilation system was selected, which consisted of jet fans, inclined and vertical shafts. The main tunnel had 1.95% downslope in the eastbound (to Yangyang) and 1.95% upslope in the westbound direction (to Chuncheon). Since the air pollution upslope was predicted to be 1.8 times of that downslope, the two vertical shafts were planned for the ventilation in the upslope transect and one inclined shaft in the downslope transect as shown in Figure 6.

In-je Tunnel was rated to disaster prevention Class 1. Evacuation plans and shelters were established for a rapid safe evacuation and fire suppression in particular case of fire in the large vehicles. Disaster prevention performance verification through scaled model tests and quantitative risk analysis (QRA) were carried out to establish a comprehensive disaster scenario planning for the long tunnels. As a result, 37 cross passages for passengers, 14 for passenger vehicles, and 6 for large vehicles were installed in the maximum spacing of 199 m. Also, in case of fire, inclined shafts were planned to be used as an evacuation passage and access road of the rescue vehicles.

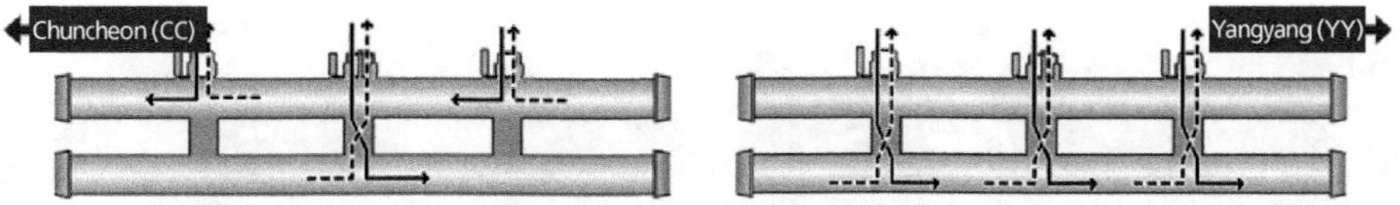

Figure 6 Ventilation system in In-je tunnel. In ordinary traffic conditions, the inclined shaft takes the ventilation in the eastbound downslope (left) and the vertical shafts take the ventilation in the westbound upslope. In congested traffic conditions, both the inclined shaft and the two vertical shafts take the ventilation.

2.2 Summary

The tunneling project is to construct the short cut expressways connecting the Seoul Metropolitan area and the east coast under the long-term plan of the national expressway network. The expected effect includes local economic growth, enhanced logistics support system, satisfying the local traffic demand, and preparation for the era of unification.

The following key design issues were considered. First, the driving safety was secured by taking a rhythmic curve route rather than a linear straight route especially considering the closed tunnel space with a great length. Second, on the consideration of the hilly topography in the project area and the grades of the tunnels, an integrated ventilation system was planned between the east and west bound tunnels. Third, the safety of the passengers were secured in the case of a fire by shortening evacuation time, placing the evacuation shelters in the middle of tunnels, and installing high capacity smoke control units. Last, comprehensive disaster prevention scenarios were established to minimize the damage and to maximize the prevention and rescue capabilities.

3 SOL-AN TUNNEL

Sol-an Tunnel is a 16.3 km long spiral railway tunnel that was completed in June, 2012 after 132 months of construction since 2001. This tunnel is the third longest railway tunnel in Korea behind the longest Yul-hyun Tunnel (in Seoul Metropolitan Express Railway) of 50.3 km in length and the second longest Keum-jeong-san Tunnel (in Seoul-Busan High Speed Railway KTX) of 20.3 km in length. Sol-an Tunnel replaced the old zig-zag type route with a switchback and sixteen tunnels between Tong-dong and Sim-po-ri where large ground deformation occurred due to thrusting. The single lane Sol-an Tunnel will be expanded to accommodate double lanes in the future.

3.1 Key issues on design and construction

Sol-an Tunnel is a spiral tunnel consisting of two inclined shafts (L_1 = 1,510 m and L_2 = 590 m) and a vertical shaft (D = 6.5 m and L = 240 m). Considering the large difference in elevation between the portals (378.7 m) and the optimal grade for locomotives, the spiral route of 1,450 m in radius of curvature was selected as shown in Figure 7.

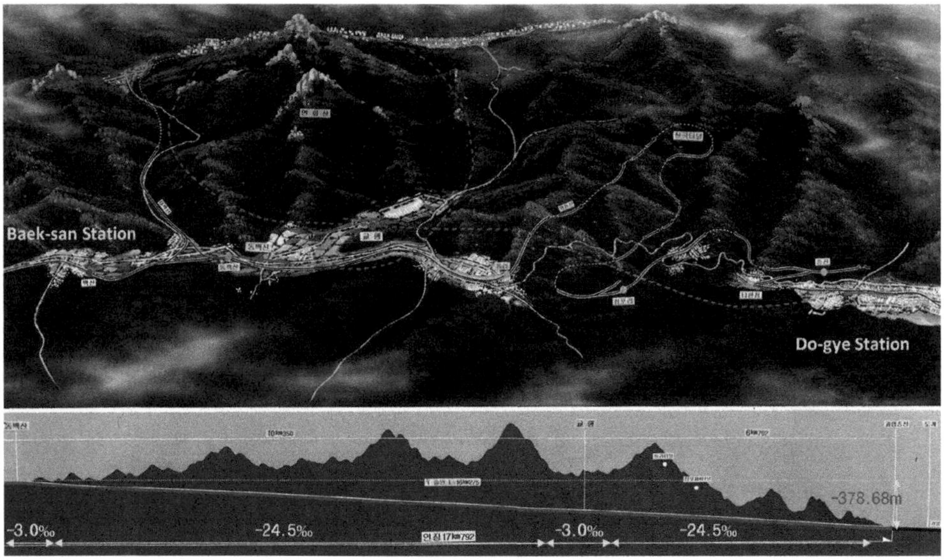

Figure 7 Planned view of Sol-an Tunnel of 16.3 km in length (top) and its cross-sectional view (bottom).

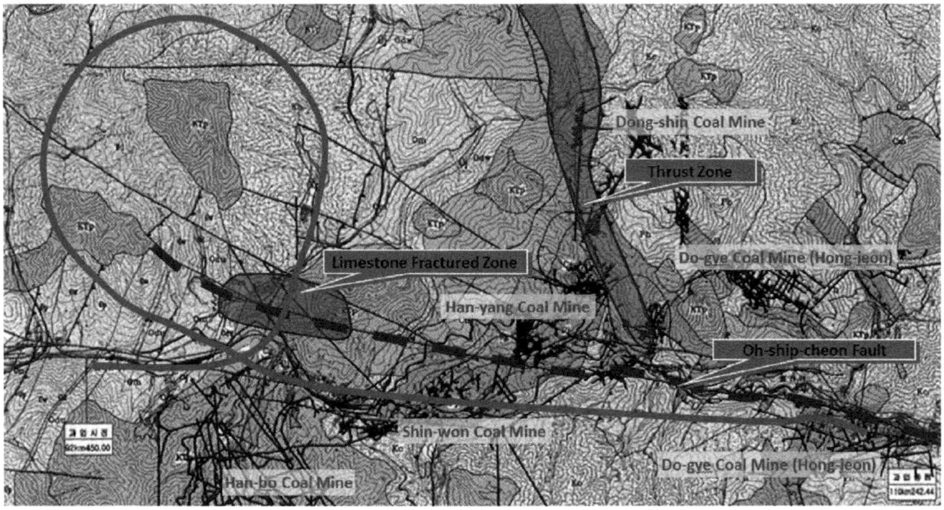

Figure 8 Geology in the Sol-an Tunnel project area (left) and the locations of abandoned mines (right).

The project area is mainly located in the sedimentary geological formation where complex interchanging layers of limestone, sandstone, shale, and coal exist. The tunnel crosses Oh-ship-cheon fault and large scale limestone fractured zones as shown in Figure 8. Lime caverns, water pockets, mined cavities at abandoned coal mines were expected to cross the tunnel as well.

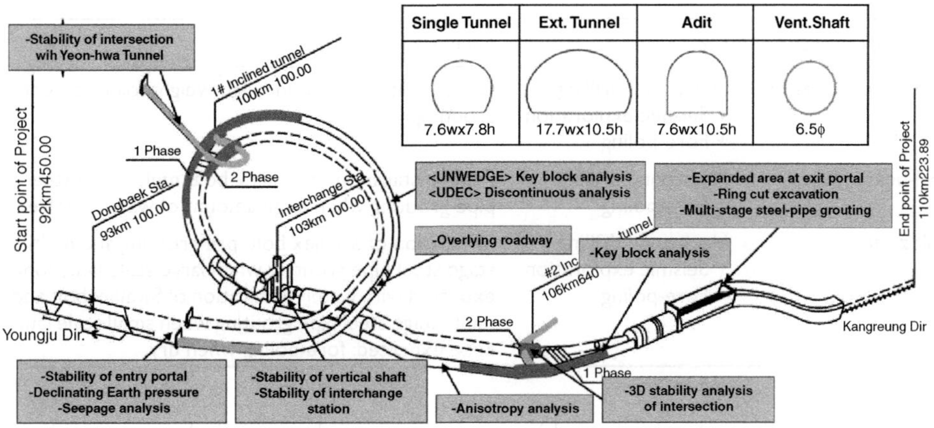

Figure 9 Spots where extensive stability analyses were made and the contents of the analyses.

The entry portal in the west was located in a limestone layer with many small scale faults, which caused unexpected water inflow. Approaching to the Inclined Shaft No. 1, mixed layers of sandstone and shale were found. When crossing the shale layer, large deformation and overbreak troubled the construction. Near the Inclined Shaft No. 2 and exit portal in the east, weak ground prevailed with shale and coal. In some areas, coal layers crossed. The mined cavities of abandoned coal mines influenced the construction as shown in Figure 8. The NATM was selected to cope with the complex geological conditions. For the vertical shaft, a raise boring machine was used followed by downward expansion drilling and blasting. Since the planned route crosses fault zones, fractures zones, intersections, enlarged sections, bifurcations, and high water pressure zones, the stability analyses were carefully made. Figure 9 presents the major issues considered and the spots where extensive analyses were made. As a disaster prevention measure, the emergency ventilation and evacuation plans were made as shown in Figure 10.

Various exploration methods and counter measures were considered in case that adverse geological conditions were met as summarized in Table 2. During the

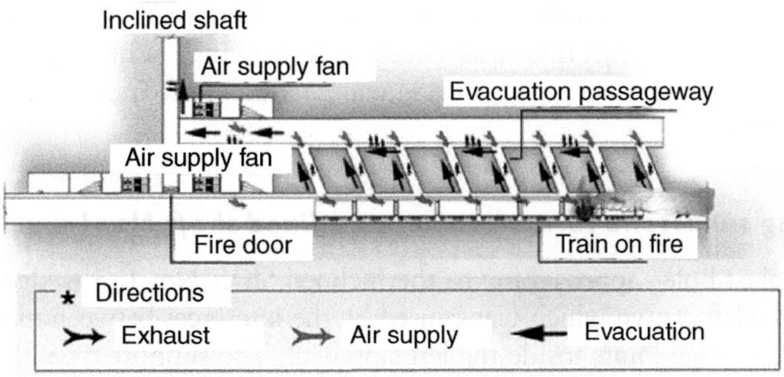

Figure 10 Emergency ventilation and evacuation plan.

Table 2 Exploration methods and counter measures for adverse ground conditions.

Geological condition	*Exploration*	*Counter measures*
Limestone formation	Horizontal drilling + Seismic exploration + Fore-poling	Cavity filling when found; lightweight foam cement grouting
Coal-bearing layer	Horizontal drilling + Fore-poling	Rapid installation of Swallex bolt; multi-stage steel pipe grouting when large amount of water involved
Fault zone	Horizontal drilling + Seismic exploration + Fore-poling	Installation of Swallex bolt, pre-grouting and multi-stage steel pipe grouting when large scale fault zone expected with water; installation of Swallex bolt and multi-stage steel pipe grouting when small scale fault zone expected; fore-poling when dry

construction, different kinds of troubles were experienced especially for vertical shaft excavation, portal opening, crossing sandstone and coal-bearing layers, and crossing water flooding sections. The trouble shooting cases are introduced in the next section.

3.1.1 Vertical shaft sinking

The typical procedure of raise boring was followed. A pilot hole was drilled downward from the surface which was enlarged to a diameter of 2.4 m by reaming with raise boring machine. The final enlargement was made by downward drill and blast.

Safety management was a key issue in this operation. Preliminary survey of neighboring structures was made. Through test blasting, the allowable amount of explosives and the optimal blast pattern was determined to minimize the blast-induce vibration, air blast, and fly rock. Because the rock waste fell on the bottom floor by gravity, the catch area was carefully restricted from access with the help of wireless communication system. The safety of the skip and the wire rope was frequently checked. The vertical shaft of 235 m in length was successfully excavated without interlocking jamming between the RBM and shaft wall in the process of reaming.

3.1.2 Entry portal opening

The entry portal was located in limestone formation where cavities were found in many spots on the crown. Lightweight foam concrete was injected into the cavities. The performance of the injection was checked by drilling. Where the performance was doubted, multi-stage steel pipe grouting was carried out to increase the stiffness of the crown and face.

3.1.3 Crossing sandstone formation near Inclined shaft No. 1

While drilling blast holes approaching to the Inclined Shaft No. 1, a wedge type caving took place on the left crown which was caused by the interfaces between major joint sets on the crown and large joints inside the left side wall. The support type of the working face was downgraded to Type 4 (1.5 m of advance) from Type 2 (2.5 m of advance) 10 m ahead of the caving spot. The caving was large in scale reaching 18 m backward

Figure 11 Coal-bearing layer at the face (left) and installation of lattice girder (right) in Sol-an Tunnel.

from the face. Indurated clay layer of 10 mm in thickness was found on the joint surfaces, which made the wedge blocks prone to move with the continuous flow of water.

To remediate the caving, heavy support was installed including rock bolts on the side wall, steel support at the spacing of 1 m, wire mesh and shotcrete. Once the roof was established, shotcrete and cement mortar was filled from the bottom to spring line. Lightweight foam concrete was injected in the rock above the spring line followed by additional rock bolt installation. The performance of the support was checked by monitoring the deformation.

3.1.4 Crossing coal-bearing layers

The planned route often crossed coal shale and coal-bearing layers as shown in Figure 11. Especially when crossing coal shale, small amount water started to inflow to the tunnel from the crown on the right. Soon small amount of coal and shale caved into the face when the labors and the machines evacuated. That was followed by caving.

After the caving, the face was closed. Lightweight foam cement was injected to fill the cavity and urethane grouting was applied to prevent water inflow. Then, a lattice girder was installed at the spacing of 0.8 m while the crown was supported with steel pipes. However, the coal slurry intruded the side wall resulting in additional cavings. In that section, urethane grouting was additionally applied to prevent water inflow and self-drilling large diameter steel pipe grouting (L = 12 m, D = 114.3 mm) was added. To protect the face, ring cut method was used, where top and bottom faces were excavated separately while leaving the core in the center of the face. Type 6 support pattern (0.8 m of advance) was selected. To obtain the full strength of shotcrete, one advance was made only per day. To prevent the face deformation, 30 mm thick shotcrete was applied to the core. The reinforcement plan of the section is shown in Figure 12.

3.1.5 Water inflow and ground subsidence

Leaving 350 m of excavation to breakthrough, a large amount of water came into the tunnel from the stream 235 m above the crown. Following the water inrush, surface subsidence was observed along the stream. Detailed site investigation was carried out to find the geological and geotechnical reasons for the inrush and subsidence.

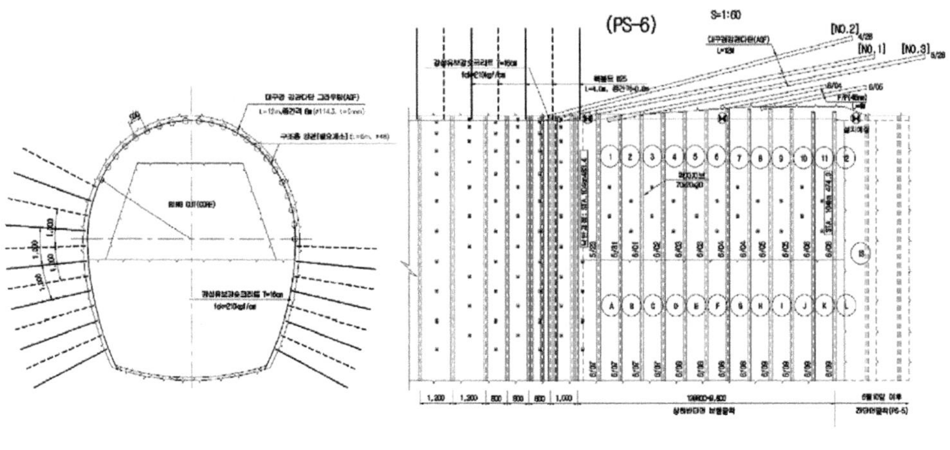

Figure 12 Reinforcement plan of the caved section in coal-bearing layer in Sol-an Tunnel.

As a result, surface grouting and JSP barrier installation were carried out to prevent water inflow and to prevent additional subsidence, respectively. In the tunnel, additional grouting was made. According to the deformation measurement, the shallow soil layer at the surface showed an elastic subsidence at the beginning but it was stabilized after grouting.

3.2 Summary

In this project, the ground conditions tuned out to be quite different from what were expected in the design stage. That caused significant increases in time and cost of the project. The 16.3 km long tunnel crossed faults, fractures zones, coal-bearing layers, weak sandstone formation, lime cavities, water pockets, and other unexpected ground conditions. The state-of-the-art techniques of forward prediction, monitoring and analysis, grouting, and reinforcement were successfully used to finish this projects in 132 months.

4 SA-PAE-SAN TUNNEL

Sa-pae-san Tunnel, an expressway tunnel, is located in the region of North Gyeong-gi Province connecting Il-san and Tae-gye-won in Seoul Beltway as shown in Figure 13. The length of twin tube tunnel is 3,993 m (Il-san direction, west bound) and 3,997 m (Tae-gye-won direction, east bound), respectively. The cross sectional area of the tunnel is 170.04 m^2 accommodating four lanes. Since the tunnel is long and very wide, the tunnel cross-section was selected based on the economic feasibility study.

In the project, the NATM was considered to be the most viable. The total excavation volume was about 1,370,000 m^3. For safe operation after completion, 5 cross passages and 12 emergency parking lots were designed. Thirty jet fans 1,530 mm in diameter were installed for ventilation and three electric smoke and dust collector

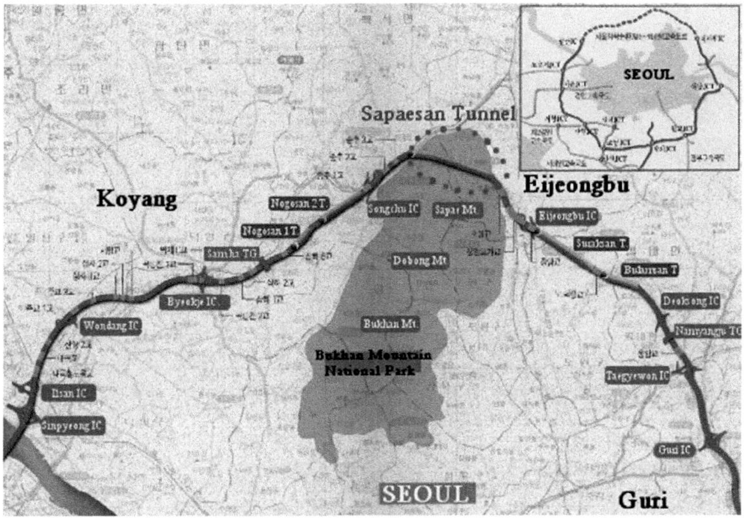

Figure 13 Route map and location of Sa-pae-san Tunnel.

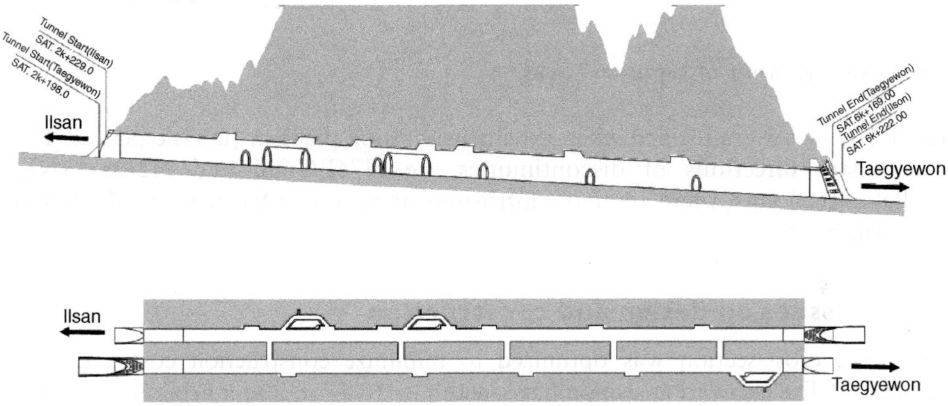

Figure 14 Cross-sectional and plan view of Sa-pae-san Tunnel.

systems were introduced considering the tunnel length. Figure 14 presents them on the cross-sectional and plan views. There were many plans considered to minimize the damage to natural environment because the Sa-pae-san tunnel passes through the Bukhan Mountain National Park. The project was planned to be completed in 54 months from January 2002 to June 2006, but it was delayed until December 2007 due to civil claims on environmental protection.

4.1 Geology

Almost the entire length of the tunnel crosses Seoul granite as presented in Figure 15, which intruded Precambrian banded biotite gneiss. Seoul granite contains many

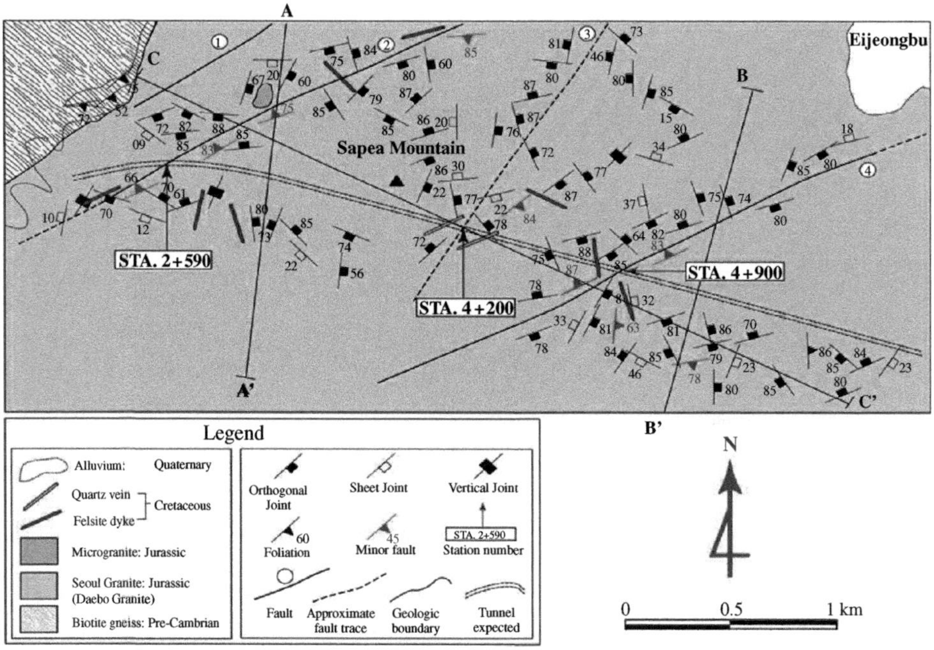

Figure 15 Geological map of Sa-pae-san Tunnel.

xenoliths which were formed when the intrusion took place in Jurassic period. The predominant directions of discontinuities are N76E/86SE and N26W/81NE. Parallel to the joints, there are many intrusions of quartz or felsite veins of 20–300 cm in width.

4.2 Key issues on design and construction

The tunnel cross-section was optimized to minimize construction cost and time (Figure 16). The excavation section has a width of 18.77 m and a height of 10.50 m (Area = 170.04 m^2).

In the design stage, the support type and excavation sequence were categorized according to Q values and four divided sections of excavation were considered in support Type II, III, IV, and V as shown in Table 3. However, it was simplified to the same sequence as in Type I during the construction stage to catch up with the construction schedule after a two-year delay due to civil claims. In addition, bulk type emulsion explosives were applied to increase blasting efficiency. The bottom section was excavated following the top section while keeping a constant distance from the top section face using long-bench cut. According to the simplified sequence, the stability must be carefully checked. Preliminary study by three-dimensional FEM analysis, TSP (Tunnel Seismic Prediction) exploration, forward prediction using computer-controlled jumbo drilling, precise face mapping, and deformation monitoring were carefully carried out.

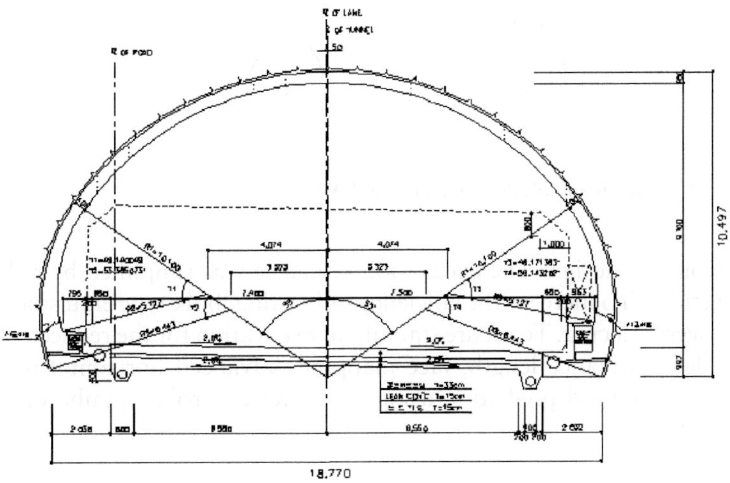

Figure 16 Standard cross-section of Sa-pae-san Tunnel.

Table 3 Support type and excavation sequence.

Type	*I*	*II*	*III*	*IV*	*V*	*VI*
Q value	> 40	10 – 40	3 – 10	1 – 3	0.2 – 1	< 0.2
Sequence	1 / 2	2 1 2 / 3	2 1 2 / 3	2 1 2 / 3	2 1 2 / 3	3 2 1 2 3 / 4
Ratio (%)	23.8	26.0	18.8	9.9	4.4	1.8

At one cross passage, eight different excavation faces were opened, *i.e.* top and bottom sections at four sides. Labor and machine transportation was made through one side of the twin tunnel while top and bottom sections were sequentially excavated in the other side. The concrete lining was installed using a movable casting wall. The effect of blast-induced vibration on the performance of the cast concrete was reviewed. The allowable level of vibration was 6.3 mm/sec and the minimum distance was 225 m. Since the distance between cross passages was 675 m, the minimum distance was well managed in the excavation sequence. With this modification, twenty one months of construction time was saved.

For safe, efficient and economic tunnel blasting, a new SAV-CUT (Stage Advance V-Cut) method was used which was developed by GS Engineering & Construction Company. Figure 17 shows the pattern of drill holes and blasting sequence of the method. Bulk type emulsion explosive were used to save time of charging.

4.3 Summary

In this large-span tunnel, the management of excavation sequence was one of the key issues. Especially to catch up with the delay schedule, several innovative modifications

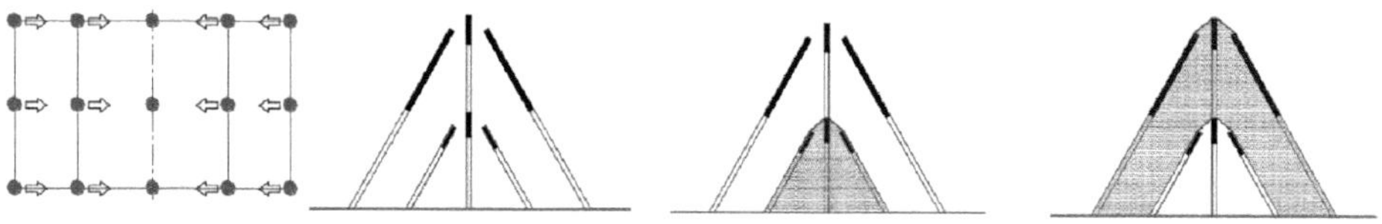

Figure 17 Drill hole pattern and blasting sequence in SAV-Cut.

were made from the design. The excavation sequence was simplified by reducing the number of sections. To increase the efficiency of blasting, bulk type emulsion explosive and SAV-cut were selected. To secure stability of excavation for the reduced number of sections and large advance length, face mapping, advanced horizontal boring, TSP exploration, and forward prediction using computer-controlled jumbo drilling were carefully carried out.

5 BO-RYEONG TUNNEL

Bo-ryeong Tunnel is the longest twin tube subsea road tunnel that links Bo-ryeong and Tae-an in the western coast of Korea as shown in Figure 18. This 6.9 km long tunnel (5.1 km subsea section) will provide a short cut of 85 km long road way. The project was started in December 2010 and will be finished by May 2018.

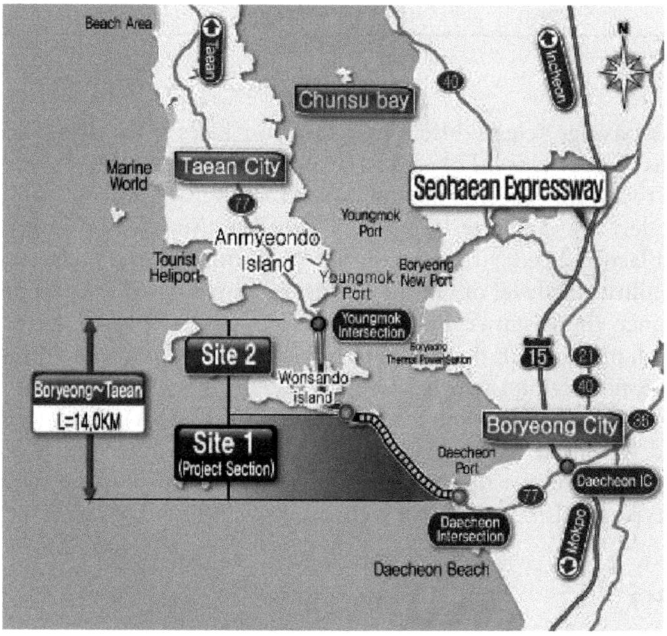

Figure 18 Geographical location of Bo-ryeong Tunnel.

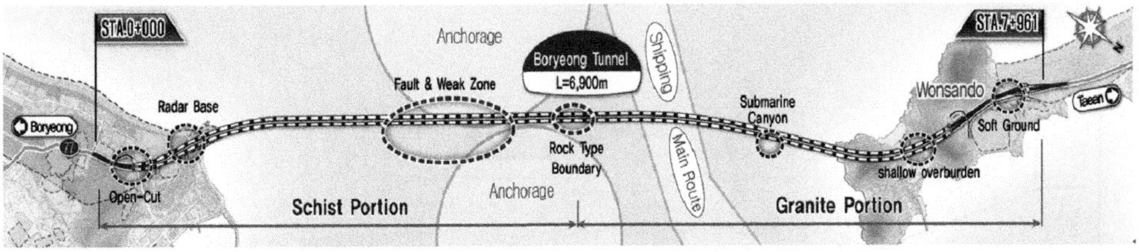

Figure 19 Major sections where special engineering considerations were made at Bo-ryeong Tunnel.

5.1 Geology

After rough geological analysis using geological mapping and satellite imaging during a pre-investigation program, an exploration campaign of multi-beam echo sound, refraction seismic profiling, vertical drillings from barges, and so on was carried out to determine the seabed topography and geological conditions. As presented in Figure 19, there are complex topography and geological characteristics in open-cut section at the entry portal (north bound), a crossing section under radar base, weak zones where faults, rupture zones are found on the seabed, interfaces between different rock types, a shallow overburden at a land section approaching to the exit portal.

The geology is mainly Precambrian metamorphic rock, penetrated by granite at the Mesozoic Jurassic and Cretaceous period. According to the results of the site investigation, metamorphic rock was made up of schist and gneiss depending on the degree of metamorphism which were penetrated by a number of Mesozoic acidic dikes.

Five lineaments, which intersect the main route, were analyzed by DEM (Digital Elevation Model) analysis on the land and the seabed, resulting in four faults and two weak zones. Among the rest, F-1 (over-thrust), F-2 (fault), F-3 (fault), L-2 (weak zone) and F-4 (fault) had a little effect on the tunnel crown. In particular, F-2 (fault) developed into the ruptured fault zone about 40 m in depth. Therefore precautions were required during the construction. At L-1 (rock type boundary), a pre-grouting method was required because of high risk of water inflow. Figure 20 shows the geological conditions at Bo-ryeong Tunnel.

5.2 Tunnel design

In the project, vertical drilling could be successfully conducted along the route of the tunnel to obtain the topography and geology of the seabed. The maximum depth of

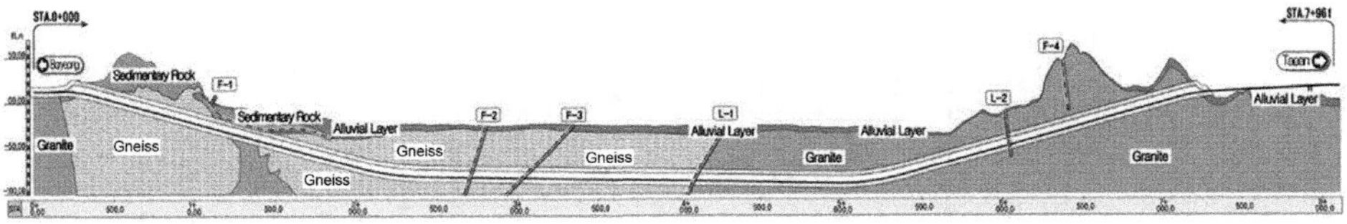

Figure 20 Geological conditions at Bo-ryeong Tunnel.

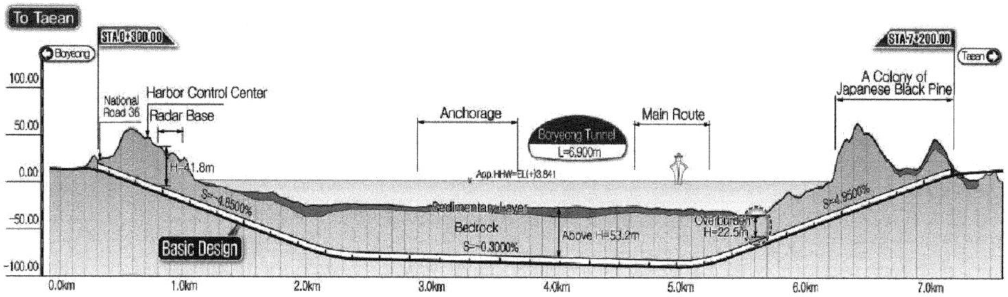

Figure 21 Longitudinal profile of Bo-ryeong Tunnel.

water and the average rock cover was about 25.0 m and 53.2 m, respectively. However, the minimum rock cover was 22.5 m where nice bedrock was covered by a thin marine sedimentary layer. The maximum gradient of the tunnel was selected to be 5% on the design speed of 70 km/h. Figure 21 presents the longitudinal profile of the tunnel.

According to the traffic demand forecast, it was reported that the Annual Average Daily Traffic (AADT) would be 15,198 vehicles requiring two lanes. Therefore, twin-bored two-lane tunnels were planned. The cross-section was designed considering 3.25 m width per lane, 1.0 m margin on both sides, installation of an inspector pathway on both sides, and jet fans of 1,030 mm in diameter as shown in Figure 22.

Water flow into a tunnel affects the stability, workability, and cost in tunnel construction and maintenance. Depending on the design rate of inflow, the capacity of drain pipes, sumps, pumps, and sewage treatment facility should be determined. Especially, at a subsea tunnel, inflow rate greatly affects the construction cost as well as maintenance cost. The two costs are inversely proportional and proportional to the designed rate of inflow, respectively. Therefore, the rate of flow should be optimized to minimize the overall cost.

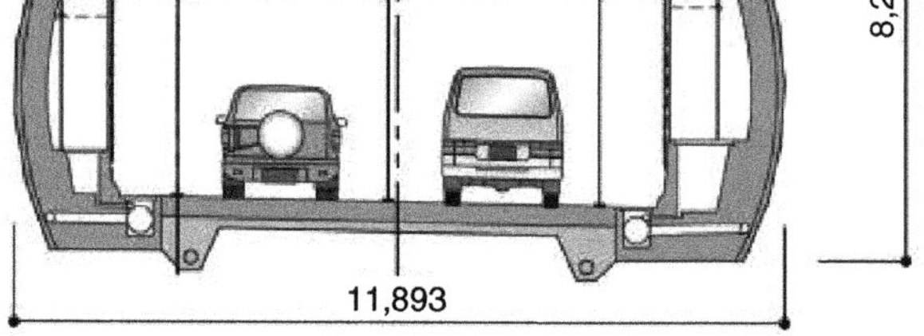

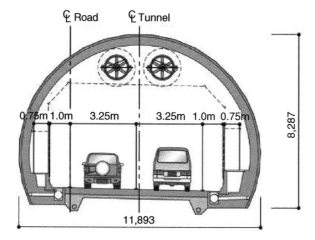

Figure 22 Cross-section of Bo-ryeong Tunnel.

In Korea, typical design inflow rates are 2–3 m^3/min/km and 0.3 m^3/min/km for Metro tunnels and expressway tunnels, respectively. But, the actual inflow rates measured at different sites are much lower than the design rates. Considering the design inflow rate at Norwegian tunnels, *i.e.* 0.3–0.5 m^3/min/km, the measured inflow rate at Seikan tunnel, *i.e.* 0.57 m^3/min/km at the beginning stage of operation, and other collected data, the design inflow rate was determined to be 0.5 m^3/min/km in this project.

5.3 Excavation and reinforcement design

Because of a limited survey of the seabed, the reliability of the geological survey is not fully secured in subsea tunnels. To overcome the uncertainty of geological conditions, a strict advanced face prediction method was planned in the project as presented in Figure 23. The face prediction methods overlapped in a certain interval to obtain required reliability.

The first step is TSP exploration. It is an advanced prediction method in which about 20 transmission boreholes are drilled on the tunnel walls and then discontinuities on the tunnel face are predicted by acquiring, processing, and analyzing the seismic reflection wave induced by explosion source. The TSP method was planned at every 200 m interval since the detective range of this method is about 200 m. In the second step, a long distance (about 60 m) horizontal drilling is planned to find unexpected geological sections and weak zones questioned by prior marine geological survey and TSP. In this step, the characteristics, locations and scales of those zones are precisely determined. In addition, by observing boreholes and collecting drilled cores, extensive prediction work can be made. In the third step, a probe drilling is executed to about 20–30 m depth from the tunnel face. By using the MWD (Measurement While Drilling) technique, detailed geological and geotechnical information ahead of a tunnel face can be obtained. Flow characteristics can be obtained as well to decide whether the pre-grouting is necessary.

Pre-grouting is done depending on the volume of inflow measured in the probe drilling (3–6 holes). If the inflow rate is greater than 4L/min/1 hole (or 8L/min/sum of 3 holes), pre-grouting is made. To decide the grouting depth and the angular range to meet the allowable inflow rate, numerical analysis was carried out. Figure 24 shows the results of the analysis. For the grouting thickness of 5–7 m all around the tunnel, the inflow rate can be reduced to the allowable value. The suggested spacing of grouting holes is 2 m for bed rock and 1.5 m for weathered rock. Ordinary Portland cement or micro-cement grouting was considered in the project to reduce water inflow.

A conservative support design was made in Bo-ryeong Tunnel. Especially to prevent the deterioration of supports by chloride attack, silica-fume was added in the shotcrete

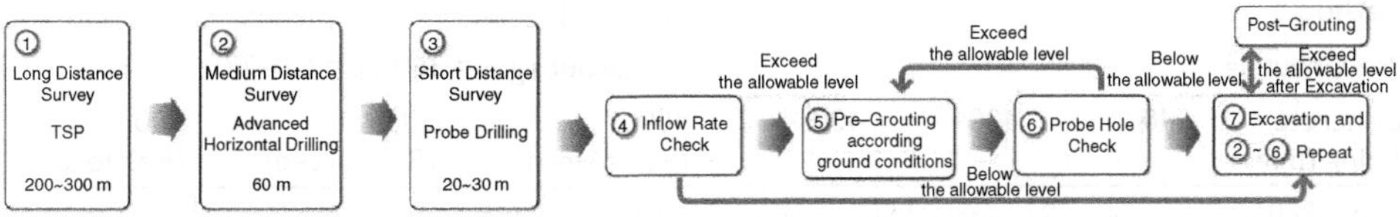

Figure 23 Flow-chart of advanced face prediction method in Bo-ryeong Tunnel.

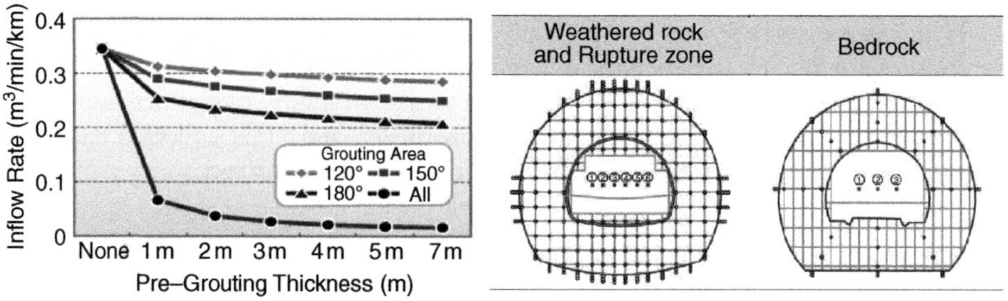

Figure 24 Pre-grouting plan in terms of grouting thickness and angle range (left) and grouting positions (right) in Bo-ryeong Tunnel.

which is known to enhance the durability of shotcrete. At the same time, a conservative minimum thickness of shotcrete, 80 mm, was selected to minimize the infiltration of water. Combi Tube type rockbolt was design to accommodate compact grout injection around the bolt. The high strength concrete lining of 40 MPa (in f_{ck}) was considered. Table 4 presents the tunnel support types and excavation sequence.

The lowest design level of the subsea tunnel is EL. −88.5m. To avoid the maximum water pressure of about 8 bars expected to act on the tunnel if undrained, a partial drainage system was selected which will reduce the water pressure and construction cost. When weak geological zones are met, pre-grouting is to be applied around the tunnel to share the water pressure with the surrounding ground. In addition, an induced drain material and a compartment waterproof system were designed to make efficient drainage and to prevent water inundation.

Table 4 Tunnel support types and excavation sequence in Bo-ryeong Tunnel.

	SP-1	*SP-2*	*SP-3*	*SP-4*	*SP-5*	*SP-6*
Outline	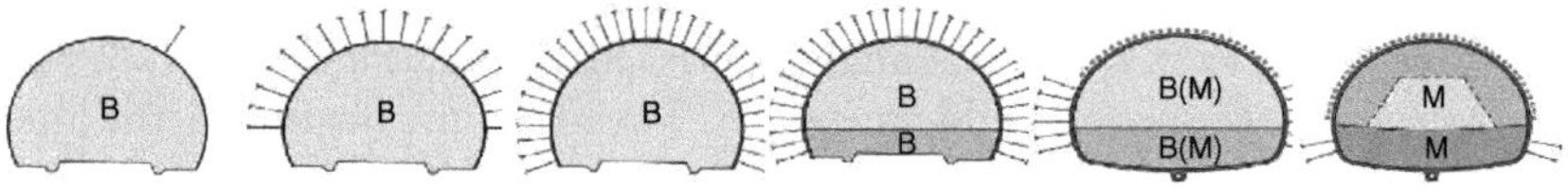					
Q	> 100	10–100	1–10	0.1–1	< 0.1	
Shotcrete (mm)	80 (SFRS)	40 (SFRS)	120 (SFRS)	160 (SFRS)	200 (SFRS)	250 (SFRS)
Rockbolt (m)	3.0	3.0	4.0	4.0	4.0	4.0
Steel Rib	–	–	–	LG 50 × 20 × 30	LG 70 × 20 × 30	H 100 × 100
Auxiliary Method	–	–	–	Fore polling if necessary	Umbrella arch reinforcement	
Pre-Grouting				Pre-Grouting θ = 360°, t = 5.0–7.0 m		
Concrete Lining (mm)	400	400	400	400 (Reinforced)	500 (Reinforced)	500 (Reinforced)

Note: SFRS = Steel Fiber Reinforced Shotcrete, LG = Lattice Girder

5.4 Summary

Considering the two major issues in subsea tunneling, limited geological and geotechnical information of seabed and the efficient measures to control potential large water flow, a careful design was made. The optimal selection of longitudinal profile, advanced face prediction methods, pre-grouting plan, tunnel support system, and waterproof design were introduced in this part. The advanced prediction of the geological condition ahead of the face is of great importance in this project. The subsequent measures to control water and reinforce the ground are equally of great importance.

6 BAE-HU-RYONG TUNNEL

Bae-hu-ryong Tunnel, which connects Chun-cheon-si and Hwa-cheon-gun in Gangwon-do, Korea, is a bi-directional two-lane single tube tunnel with a length of 5,057 m as shown in Figure 25. Its construction began in 2006 and was completed in 2012. Since the completion of construction, it has been the longest bi-directional roadway tunnel in Korea. Because the project area has a rugged topography and the tunnel is located over 400 m below the surface, the vertical shafts were costly. As a bi-directional single tube tunnel, Bae-hu-ryong Tunnel had two major disadvantages with regard to a ventilation system and evacuation plan in the case of fire. For these reasons, a service tunnel and a transverse ventilation system were selected. The service tunnel is designed with a length of 5,173 for the first time in Korea, which will be expanded for a twin tunnel system when traffic demand increases in the future.

6.1 Geology

The geology of Bae-hu-ryong Tunnel site is mainly composed of gneiss and granite as shown in Figure 26 and 27. Bae-hu-ryong fault develops parallel to the tunnel axis. Since the influenced zone of this fault is within 60 m, Bae-hu-ryong Tunnel is designed to separate from the fault with the distance of more than 100 m. Average compressive strength, total hardness, and quartz contents of rocks in the Bae-hu-ryong tunnel site are 13.5 MPa, 119 – 145, and 24 – 31%, respectively. More than 70% of rock mass in the site has the RMR value greater than 75 which is categorized into fair rock.

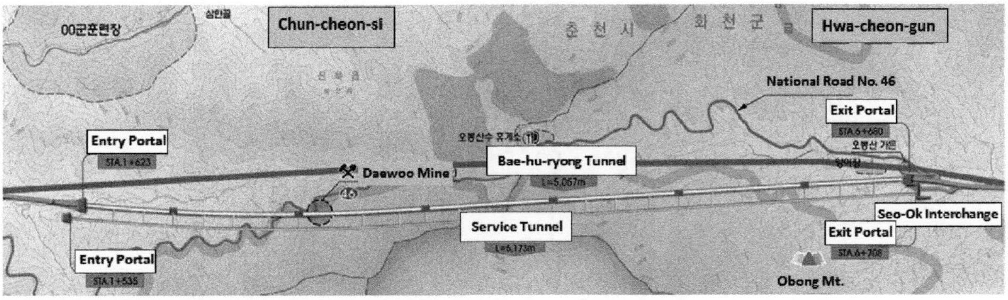

Figure 25 Plan view of Bae-hu-ryong Tunnel.

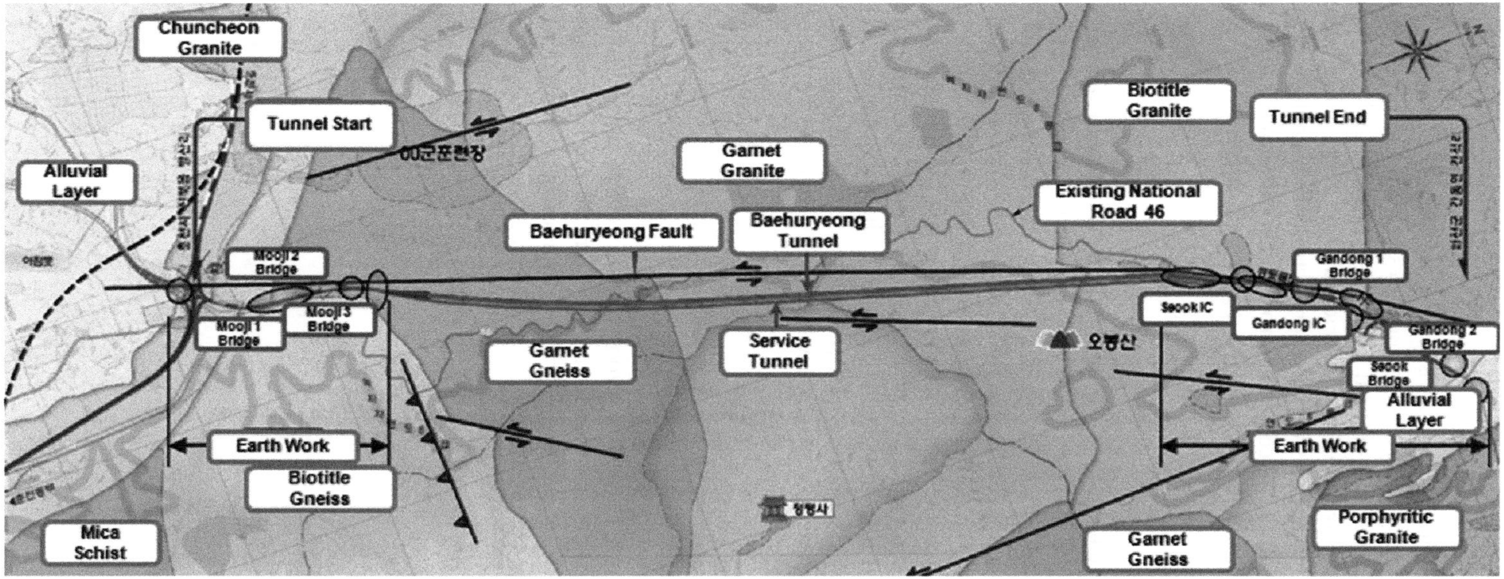

Figure 26 Geological map of Bae-hu-ryong Tunnel.

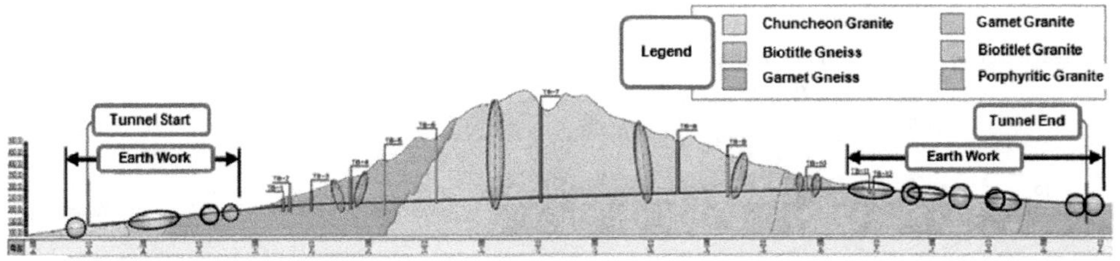

Figure 27 Longitudinal profile of Bae-hu-ryong Tunnel.

6.2 Design and construction

As a national road tunnel, the basic design specification of Bae-hu-ryong Tunnel is summarized in Table 5. The design speed limit was 80 km/hr. The traffic lane width was 3.5 m with the hard shoulder width of 1 m for east (Chun-cheon) bound and 2.5 m for west (Hwa-cheon) bound.

The main tunnel was excavated with the width of 13.7 m by the drill and blast method. The service tunnel bored ahead of the main tunnel by a TBM of 5.0 m in diameter acted as a pilot tunnel to examine the geological conditions. Figure 28 shows the typical cross section of the main tunnel and the service tunnel. In order to improve the safety of the bi-directional tunnel, a central median strip of 0.5 m in width and hard shoulders were designed for the event of emergency cases. The hard shoulder has a width of 1 m for uphill and 2.5 m for downhill. The service tunnel consists of an emergency road and a sidewalk. Various tunnel support types were incorporated according to the geological conditions. Large diameter grouted spiling method and a 50 cm thick concrete lining were applied to the portals, fault and shear zones.

Since Bae-hu-ryong tunnel is a deep bi-directional tunnel, neither a conventional longitudinal ventilation system nor a vertical ventilation system could be adopted.

Table 5 Basic design specification of Bae-hu-ryong Tunnel.

Length	*Longitudinal slope*	*Ventilation system*	*Road layout*	*Comments*
5,057 m	S = ±1.38%	Transverse	Two traffic lanes: 3.5 m × 2 Hard shoulder: 1.0 m + 2.5 m Central median strip: 0.5 m	Roadway width: 11.0 m Service tunnel: TBM ϕ5.0m

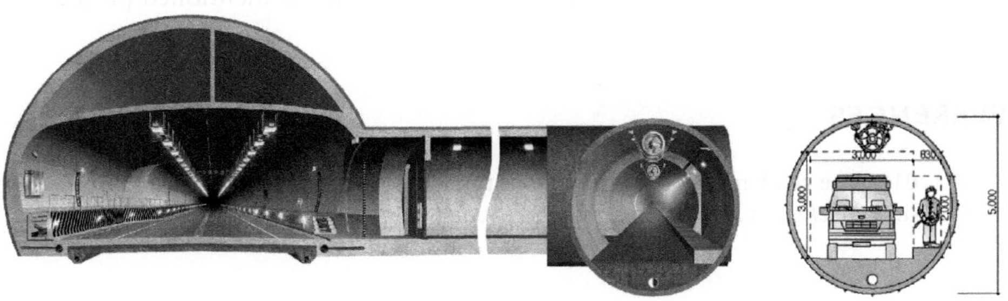

Figure 28 Cross-section of Bae-hu-ryong Tunnel (left) and service tunnel (right).

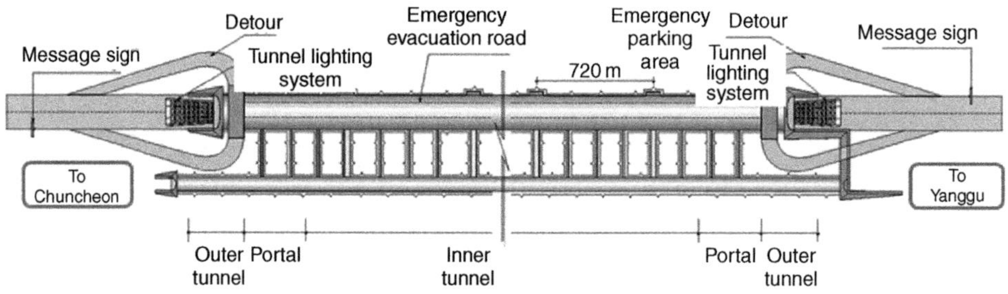

Figure 29 Emergency evacuation system in Bae-hu-ryong Tunnel.

Hence, a transverse ventilation system was selected to maintain a smoke free layer for passenger evacuation and provide excellent smoke control in case of fire. The service tunnel and the safety measure were designed to mitigate damage. The main tunnel was connected to the service tunnel via cross passages at the 180 m interval for passenger and at 720 m for vehicles, as shown in Figure 29.

The disaster prevention facilities in Bae-hu-ryong tunnel consisted of internal facilities and outside facilities. For internal facilities, an emergency exit and evacuation guidance system using radio receivers and speakers were installed. Outside the tunnel, a message sign board which shows internal traffic condition and detour information was installed. Adequate tunnel lighting system to prevent black-hole effect was installed at the entry portal. Also, a snow melting system and a frost protection system were applied.

6.3 Summary

Bae-hu-ryong tunnel is a bi-directional two-lane single tube tunnel with a length of 5,057 m. Its construction began in 2006 and was completed in 2012. After the completion of construction, Bae-hu-ryong tunnel has been the longest bi-directional roadway tunnel in Korea. For the efficient ventilation and safety measures as the bi-directional tunnel, a service tunnel and the transverse ventilation system were employed. The service tunnel bored with a TBM acted as a pilot tunnel and will be expanded to an additional two-lane tunnel in the future.

ACKNOWLEDGMENTS

The authors thank Daewoo Construction and Engineering, Hyundai Construction and Engineering, GS Construction and Engineering and Dongbu Corporation for their kind support and providing detailed information of the mentioned projects in this Chapter.

REFERENCES

Jee, W. (2012) The challenge of Boryung subsea tunnel design project, *Proceedings of the 7th Asian Rock Mechanics Symposium*, October, Seoul, Korea, 1051–1055.

Kim, H.C., Koo, I.S., Bang, S.Y., Kim, B.Y. (2010) Tunnel design history of In-je Tunnel – the longest tunnel in Korea, *Journal of Korea Geotechnical Society (Ji-ban)*, 26(9), 26–32 (in Korean)

Kim, H.C., Jang, K.Y., Kim, J.C., Lee, W.J., Kim, B.Y. (2011) History of multi-stage steel pipe grouting in shallow overburden section using automatic grout management system, *Journal of Korea Geotechnical Society (Ji-ban)*, 27(3), 29–39 (in Korean)

Korean Geotechnical Society (2013) *The Essence of Geotechnical Engineering in Korea*, Seoul, 114p.

Korea Tunnelling and Underground Space Association (2011) *Tunnels and Underground Space in Korea*, Seoul, 160p.

Lee, S.B., Je, H.C. (2005) Case study of the longest roadway tunnel in Korea – Baehuryeong Tunnel, *Tunnel and Underground Space*, 15(6), 432–440 (in Korean)

No, S.L., Noh, S.H., Lee, S.P., Kim, M.H., Seo, J.W. (2005) A case study of minimizing construction time in long and large twin tube tunnel, *Tunnel and Underground Space*, 15(3), 177–184 (in Korean)

Seo, K.W., Baek, K.H., Shin, H.S., Hong, J.S. (2009) Design and construction of Sol-an Tunnel, *Journal of Korean Association of Tunnelling and Underground Space Technology*, 11(1), 1–15 (in Korean)

Chapter 6

Siah Bisheh powerhouse cavern design modification using observational method and back analysis

M. Sharifzadeh[1], *R. Masoudi*[1] *& M. Ghorbani*[2]
[1]*Department of Mining Engineering, Curtin University, WASM Kalgoorlie, WA, Australia*
[2]*Department of Mining & Metallurgical Engineering, Amirkabir University of Technology, Tehran, Iran*

Abstract: The Siah Bisheh pumped storage powerhouse cavern with complex geometry, changeable geological formations and diverse geotechnical properties of rocks, is under construction on the Chalus River 125 km north of Tehran, Iran. The powerhouse cavern was located near the downstream (d/s) dam reservoir and its crown was more than 30 meters lower than the downstream (d/s) dam maximum lake level. After impounding of the d/s dam, the powerhouse region would be located under saturated conditions. Therefore long term stability assessment of the powerhouse cavern under saturated conditions was unavoidable. In this study displacement based direct back analysis using variable staggered grid optimization algorithm was applied and calibrated geomechanical properties of rocks, stress ratio and joints parameters were identified. The time dependent behavior of rock was tested at the laboratory and the creep test results were considered in the practical design. Numerical modeling results were in good agreement with measured displacements of extensometers which confirmed the numerical modeling accuracy and back analysis results. Then ordinary analysis of the powerhouse cavern under natural conditions using back analysis results were carried out. Results of the analysis showed that the powerhouse cavern was stable under natural conditions and existing support system had suitable efficiency and could effectively control displacements. Finally, the powerhouse cavern long term stability under saturated conditions was analyzed. Results of analysis showed that after d/s dam impounding, pore water pressure and uplift pressure in discontinuities around the powerhouse cavern would arise so the powerhouse cavern tended to have local failure around the region 2nd and 3rd instrumentation arrays in the middle of the powerhouse cavern. To obtain powerhouse long term stability, it was recommended to construct a cutoff curtain grouting around powerhouse cavern.

1 INTRODUCTION

The Siah Bisheh Pumped Storage project was located 125 km north of Tehran, Iran. The site can be reached in the vicinity of Siah Bisheh village on the main Chalus road, connecting Tehran with the Caspian Sea. Iran Water and Power Resources Development Company (IWPC) was the owner of the project. This plant was designed to produce a rated capacity of 4*260 = 1040 MW peak energy. In this project, two

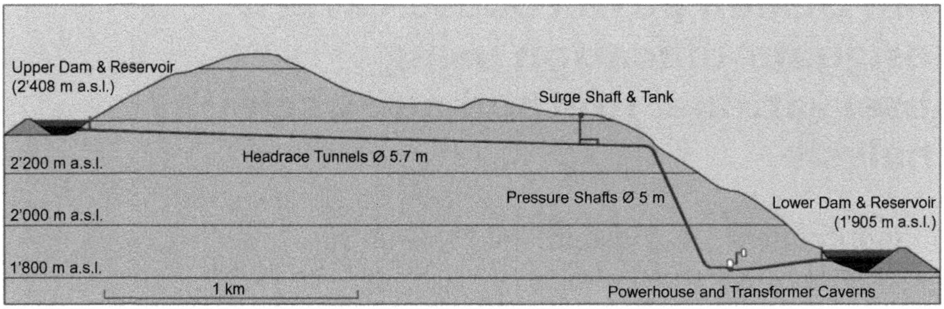

Figure 1 Schematic view of the Siah Bisheh CFRD dams and location of pumped storage powerhouse cavern (PHC) and transformer cavern (TRC).

concrete face rock fill dams (CFRD) were under construction in Chalus valley for the water storage. An underground power plant with complex geometry, changeable geological formations and diverse geotechnical properties of rocks, was under construction including powerhouse cavern, transformer cavern and guard gate cavern as well as an underground water way system in the mountain to accommodate all machinery and equipment for power generation and pumping (Figure 1).

The powerhouse cavern was located closed to the downstream (d/s) dam reservoir and its crown was more than 30 meters below the d/s dam maximum lake level. After impounding of the d/s dam, the underground powerhouse region would be located under saturated conditions. Therefore long term stability assessment of the powerhouse cavern under saturated conditions was unavoidable.

In order to do this assessment, displacement based direct back analysis using an optimization algorithm was applied and geomechanical properties of rocks, stress ratio and joints parameters were identified. Numerical modeling results were compared to actual measurements using extensometers and achieving good agreement between calculated displacements and measured displacements confirm the numerical modeling accuracy and back analysis results. Then direct analysis of the powerhouse cavern under natural conditions using back analysis results was carried out. Results of analysis showed that the powerhouse cavern was stable under natural conditions and predicted that the support system had suitable efficiency and could effectively control displacements. Finally, powerhouse cavern long term stability under saturated conditions was analyzed. Results of analysis showed that after d/s dam impounding, pore water pressure and uplift pressure in discontinuities around the powerhouse cavern would arise and had a tendency to local failure of powerhouse cavern in region 2nd and 3rd instrumentation arrays. To obtain powerhouse long term stability, it was recommended to construct a cutoff curtain (grouting) around the powerhouse cavern.

2 SIAH BISHEH POWERHOUSE CAVERN

Siah Bisheh powerhouse was constructed nearby Chalus River in the north part of Iran. The main purpose of the project was to compensate and stabilize the electricity in high and low electricity consumption period. The Powerhouse Cavern (PHC) with 131 m length, 24.5 m width and 46.5 m maximum height excavation and the Transformer Cavern (TRC) with 160.5 m length, 15.5 m width and 27 m height, were the main underground structures in this project. The other minor underground space which was constructed parallel to PHC was Guard Gate Cavern with 90.5 m length, 5.5 m width and 10.5 height. The powerhouse and transformer complex were constructed at a depth of approximately 250 m below surface. The total generating capacity of the scheme would be 1040 MW. The schematic three dimensional view of the Siah Bisheh project along with main caverns view was illustrated in Figure 2.

Siah Bisheh powerhouse cavern was located in fractured rock masses and the failure was mainly controlled by the discontinuity distribution. For cavern stability assessment, considering block size, pattern and spacing of discontinuities, three dimensional distinct element analysis was used.

3 GEOLOGY AND ENGINEERING GEOLOGY

3.1 Geology

The Siah Bisheh pumped storage project area lies in the southern part of the Paleozoic-Mesozoic Central Range of the alpine Alborz mountain chain, mainly folded and formed during the Alpine orogenic phase, with a NW-SE trend in the western parts and NE-SW in the eastern parts. Geomorphologically, Alborz is a young mountain range with deep and narrow valleys and active tectonics. The most important tectonic phenomenon of the Siah Bisheh area is the fault called the Main Thrust Fault (MTF), with a dip/dip direction of 78/028 and an almost E-W trend. The MTF has reverse mechanism. Meanwhile, the reverse fault of Chalus, which is parallel to the Chalus River in Siah Bisheh area, is another fault, which must be taken into consideration in terms of seismicity (Figure 3).

The rock sequences in the project area consist of massive limestones, detrital series (sandstones, shales) and volcanic rocks of Permian formations, Triassic dolomites and Jurassic (Lias) formations with black shales and sandstones. Several tectonic faults are crossing the project alignment. The Kandavan fault, a 15 km long and seismically active fault lies approx. 3 km south of the project area and builds the tectonic boundary between the Paleozoic-Mesozoic Central Range in the North and the Central Tertiary Zone in the South. The catchment areas of both reservoirs are of mountainous character with practically no vegetation. Based on the different strength of the geological formations, the slopes in the area of the upper dam and the headrace tunnel are generally smooth, while the lower project area lies within steep rock ridges built up by limestone and volcanic rocks.

Powerhouse and transformer caverns were constructed in the Permian Formation. Permian formations mainly consist of quartzitic sandstone, siltstone and shaly siltstone, dark and red shale and igneous rocks. Thickness of these layers varies from several centimeters to 3.5 meters (Lahmeyer & IWPC, 2005).

The attitude of the bedding planes had no considerable changes in dip and dip direction. There was uniform bedding throughout the powerhouse area with dip and dip direction of 55/195. It is noteworthy that during excavation of the powerhouse pilot gallery at chainages 40, 81 and 89 of the right wall, three shear zones, with an almost 40–50 centimeter thickness were encountered. All of these features were parallel

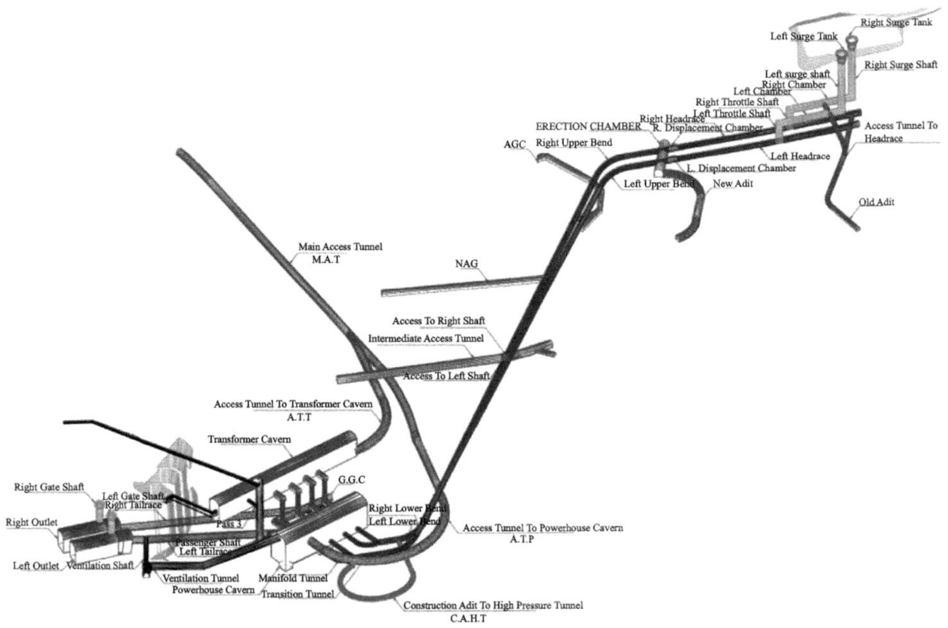

a) A 3-D model of Siah Bisheh underground excavations.

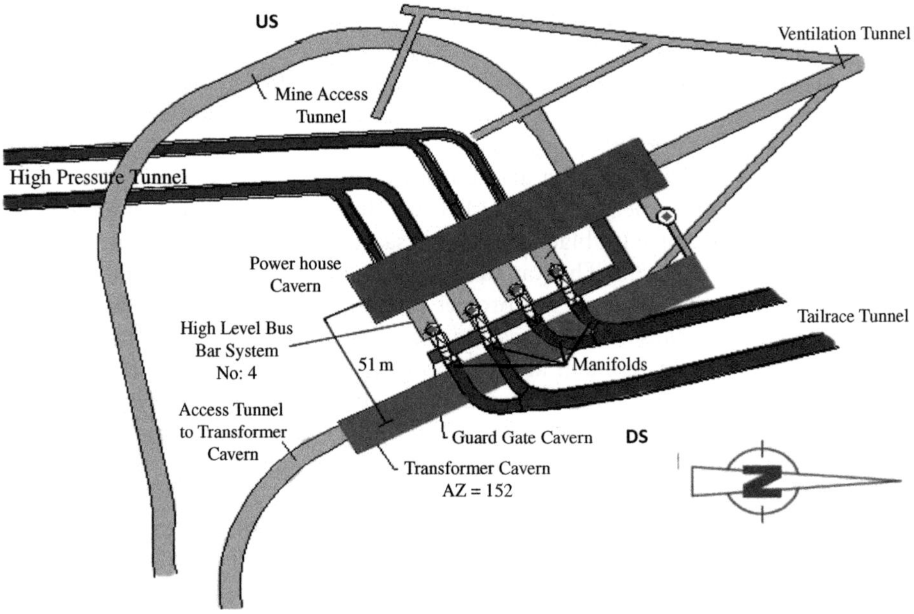

b) Plan view of the Siah Bisheh powerhouse caverns.

Figure 2 Schematic view of the Siah Bisheh pumped storage powerhouse, a) 3-D model of Siah Bisheh underground openings, b) Plan view of the Siah Bisheh powerhouse caverns.

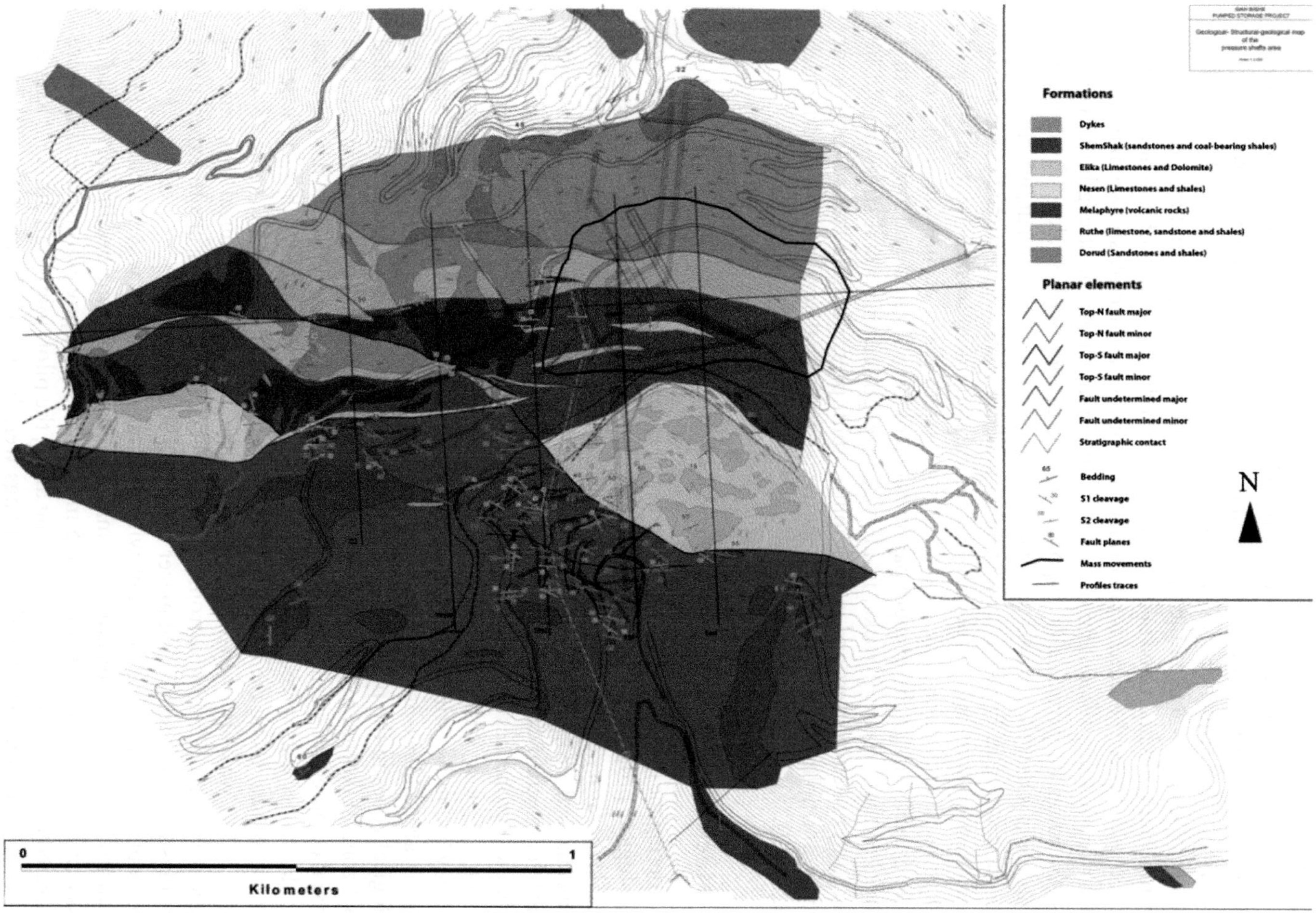

Figure 3 Geological map of Siah Bisheh Area (Lahmeyer & IWPC, 2005).

to the bedding planes. The azimuth of the powerhouse cavern was N152°E and none of the existing faults in the powerhouse area had crossed it and had an appropriate distance from it (Figure 3).

About 40 to 50 meters of the end of powerhouse cavern was completely made from igneous rock (Melaphyr) and the remaining part contained sedimentary rocks which was formed of a sequence of Quartzite Sandstone, Red Shale and Melaphyr. The influence of groundwater on the behavior of the rock mass surrounding a tunnel was very important and had to be taken into account in the estimation of potential tunneling problems. When the water is not drained it reduces the effective stresses and thus the shear strength along discontinuities and finally, in all cases, the strength of the rock mass. In addition, particularly important when dealing with shales, siltstones and similar rocks is that they are susceptible to changes in moisture content, which directly affect their strength. For long term stability analysis water effect is studied on rocks. Water effect on such rocks is mainly mechanical and pore pressure in intact rock and uplift pressure in discontinuities should be considered. Water absorption in hard rocks mainly doesn't change the strength parameters (cohesive strength and intrinsic friction angle). For these types of rocks, in all rock strength criteria, total stress should be replaced by effective stress and in rock joints, uplift pressure (u) is exerted to the joint surfaces, and uplift pressure should be subtracted from total normal stress (Sharifzadeh *et al.*, 2002).

3.2 Mechanical properties of rocks in the site

Considering the great length of the powerhouse cavern, a wide range and various types of geological properties were found as shown in the geological profile in Figure 4. Several laboratory and field tests and in situ measurements were performed to evaluate the mechanical properties of intact rock, rock joints and rock masses. The average results for mechanical and physical tests on intact rock are given in Table 1. The mechanical properties of rock joints based on test results are given in Table 2. Due to the fact that most of the geological properties could not be directly measured for this site, they had to be estimated by empirical and theoretical methods. For this purpose, the generalized Hoek-Brown failure criterion was utilized. The results showed various geological zones in the powerhouse cavern region and the area were initially divided into 2 zones. Likewise to determine the strength characteristics of the rock masses, the uniaxial compressive strength tests were carried out. Moreover the large flat jack tests and dilatometer tests were performed to determine the deformability characteristics of the rock masses. Using the field mapping the rock mass rating (RMR) value 45 at the related zones was obtained with fair rock class IV. The mechanical properties of different rock types adopted from rock mass classifications and in-situ experiments were illustrated in Table 3 (Lahmeyer & IWPC, 2005).

Discontinuity mapping program with 414 measurements was conducted in the exploratory vault adit indicating five major joint sets and one bedding plane. Rock mass consisted of bedding planes and 5 main joint sets in powerhouse area that were illustrated in Table 4. Based on surveying along the pilot, joints had different lengths of almost 3 to 10 meters and their spacings were between 200 and 600 millimeters (Lahmeyer & IWPC, 2005).

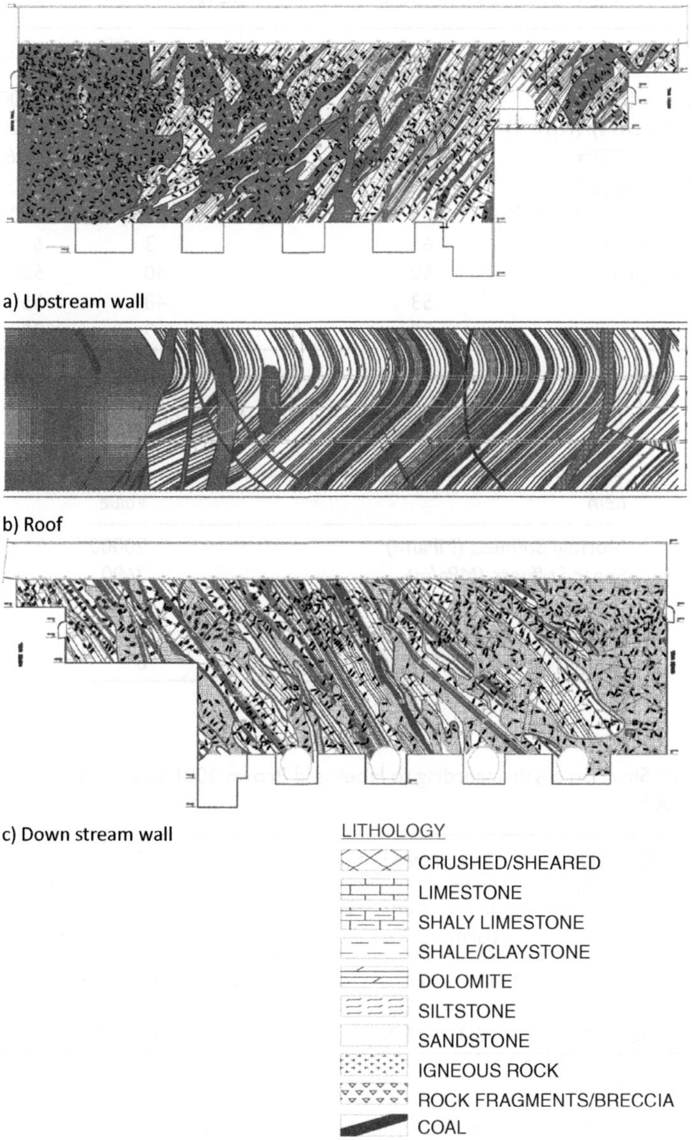

Figure 4 Geological profile of Siah Bisheh powerhouse cavern, a) Up stream wall, b) roof, and c) downstream wall (Lahmeyer & IWPC, 2005).

The shear strength parameters of $\phi = 25°$ and $c = 0$ were assumed on bedding planes. Also based on the assumption of 10 cm thick shear bands and Young's Modulus of 2000 MP, the normal and shear stiffness of rock joints were estimated to be 20,000 and 7692 MPa/m, respectively.

The value of the horizontal to vertical stress ratio (k) was estimated equal to 1.1 based on field investigation.

Table 1 Mechanical and Physical properties of intact rocks (Lahmeyer & IWPC, 2005).

Parameters	*Quartzitic Sandstone*	*Red Shale*	*Melaphyr*
Dry Density (Kg/m3)	2810	2630	2900
Saturated Density (Kg/m3)	2970	2750	2920
Bulk Modulus (GPa)	8.33	5	16.67
Shear Modulus (GPa)	6.25	3	12.5
compressive strength (MPa)	85	50	100
Tensile Strength (MPa)	6	3	6
Friction Angle (°)	50	40	50
GSI	53	48	55
Mi	20	9	25

Table 2 Mechanical properties of rock joints (Lahmeyer & IWPC, 2005).

Item	*Value*
Normal Stiffness (MPa/m)	20000
Shear Stiffness (MPa/m)	7690
Cohesion (MPa)	0.5
Friction Angle (°)	30
Tensile Strength (MPa)	0

Table 3 Rock Mass Shear Strength according to Hoek and Brown 2001 and flat jack tests (Lahmeyer & IWPC, 2005).

Rock Type	*GSI*	*UCS (MPa)*	*mi*	*Disturbance Factor = 0*				*Disturbance Factor = 0.7*				*Flat Jack Test*	
				E (GPa)	*cmσ (MPa)*	*C (MPa)*	φ *(°)*	*E (GPa)*	*cmσ (MPa)*	*C (MPa)*	φ *(°)*	*E (GPa)*	ν
Quartzitic Sandstone	53	85	20	11	22	1.6	53	7.1	14	1.1	46	15	0.2
Red Shale	48	50	9	6.3	7.9	0.98	41	4.1	4.7	0.66	32	7.5	0.25

Table 4 Discontinuity orientations in the powerhouse cavern area (Lahmeyer & IWPC, 2005).

Discontinuity	*Dip Direction [°]*	*Dip [°]*
Bedding	195	55
Joint J1	030	56
Joint J1–1	018	81
Joint J1–2	009	66
Joint J1–3	305	80
Joint J2	078	82

3.3 Time dependent behavior of rocks

The understanding of time dependent effects or creep behavior of rocks adjacent to the cavern and its influence on long-term stability is extremely important. Increasing pressure on support system due to creep behavior of rock is one of the most important issues in underground structures with weak surrounding rock mass (Barla, 2001).

The time dependent deformation of rocks has significant impact on stability of underground structures, such as nuclear waste storage facilities, tunnels and powerhouse caverns. To evaluate the stability of the underground structures and design their support systems, time dependent deformations should be highly considered (Shalabi, 2004; Tsai, 2008; Sharifzadeh *et al.*, 2013). Therefore time dependent behavior of underground structures and predicting the long-term behavior of them is assumed in special places. Predicting the time dependent behavior of underground structures is not an easy task, because it needs a reliable constitutive model which can interpret creep phenomena (Boidy & Pellet, 2000). It is also well known that rock property measurements based on laboratory tests cannot be extrapolated directly to field scale without due precaution (Boidy, Bouvard & Pellet, 2002) because the mechanical properties of jointed rock mass are strongly dependent on the properties and geometry of joints. Therefore, it is essential to use numerical analysis for simulating time dependent behavior of rock mass and compare them with measurements obtained on the monitored cavern over a long period.

Several tri-axial creep tests were performed on rock specimens of the cavern site for estimating the time dependent behavior of rock around the cavern. The Axial strain – time curves under different deviatoric stress for a typical specimen (test 1) were shown in Figure 5. The creep tests and in situ measurements were used to estimate parameters of power constitutive creep model which was able to model the primary and secondary creep regions of rock masses (Nadimi *et al.*, 2010).

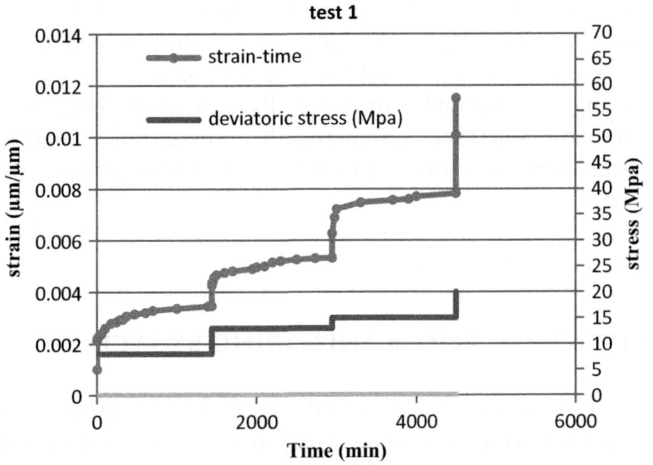

Figure 5 Axial strain – time plots of tri-axial creep test results under different deviatoric stress (Nadimi *et al.*, 2010).

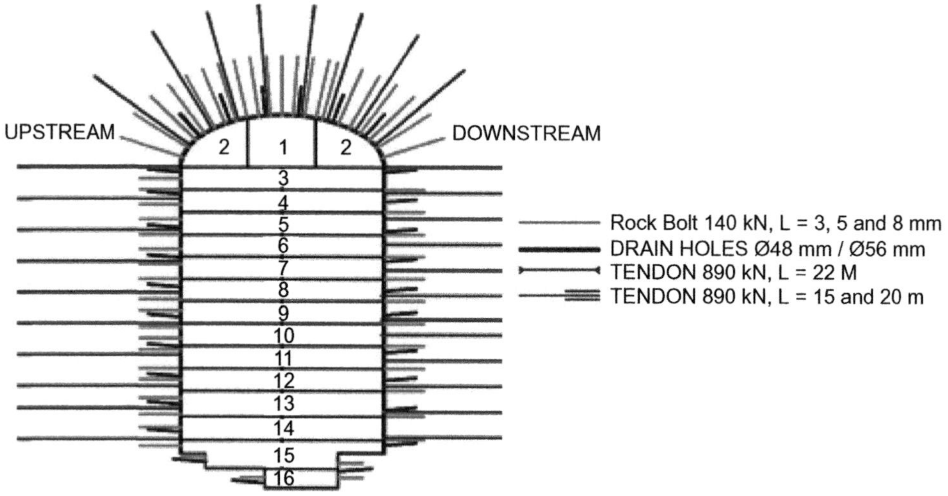

Figure 6 Excavation sequence and typical support system installed in the powerhouse cavern and excavation stages with drainage holes at roof and sidewalls (Sharifzadeh *et al.*, 2009).

4 EXCAVATION AND SUPPORT SYSTEM

All caverns were excavated using the New Austrian Tunneling Method (NATM). For excavation of the powerhouse cavern, at first a pilot was drilled at the center of the crown (sequence 1 in Figure 6) and then slashing of the crown was carried out (sequence 1 in Figure 6). After that, benching was performed with 3 meters' depth per stage which were excavated until the powerhouse floor (sequence 3 to 16 in Figure 6) (Ghorbani & Sharifzadeh, 2009).

The support system in the powerhouse cavern consists of shotcrete with wire mesh (20 cm in side walls and 25 cm in roof), fully grouted rock bolts (temporary support system) and double corrosion protected tendons (permanent support system). After each cycle of blasting, the exposed roof and walls were immediately shotcreted. Bolt installation was sometimes delayed. Systematically drainage holes 4 m in length and a 4 × 4 m spacing pattern were performed at roof and side walls of the powerhouse cavern (Figure 6) (Ghorbani & Sharifzadeh, 2009).

In Table 5 physical properties of shotcrete and interface with the rock and in Table 6 parameters of tendons were presented.

5 INSTRUMENTATION AND MONITORING SYSTEM OF CAVERN

Monitoring is the systematic collection of information as the project progresses. It is aimed at improving the safety, efficiency and design modification of a project which can be an invaluable tool to provide a useful base for parameter evaluation.

Six instrumentation arrays were set up along the axis of the powerhouse cavern at chainages of 26, 49, 67, 87,105 and 121. These arrays consist of multiple point borehole rod extensometers in the roof and sidewalls, convergence points, piezometers as

Table 5 Physical properties of the shotcrete and the interface with the rock (Ghorbani & Sharifzadeh, 2009).

Shotcrete	
Density (Kg/m3)	2400
Elastic modulus (GPa)	21
Poisson's ratio	0.2
compressive strength (MPa)	40
Tensile Strength (MPa)	20
Interface between the shotcrete and the rock	
Cohesion (MPa)	2.5
Friction Angle (°)	35
Dilation angle (°)	10
Normal Stiffness (GPa/m)	10
Shear Stiffness (GPa/m)	10

Table 6 Properties of tendons used in modeling (Ghorbani & Sharifzadeh, 2009).

Support type	*Diameter (mm)*	*Young's modulus (GPa)*	*Ultimate yield load (KN)*	*Kbond (GN/m/m)*	*Sbond (MN/m)*
Tendon	26.5	200	300	6.41	2.01
Tendon	47	200	890	6.03	3.77
Tendon	63.5	200	1540	6.79	4.59

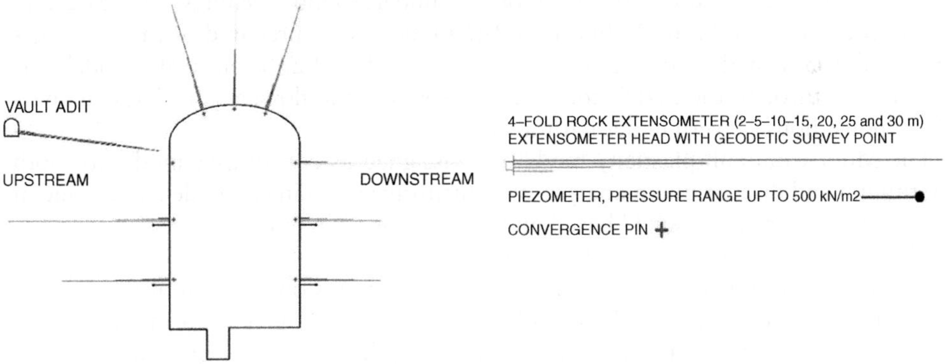

Figure 7 Typical instrumentation array installed in the powerhouse cavern (chainage 67) (Sharifzadeh *et al.*, 2009).

well as load cells on selected cable tendons. It is worth mentioning that due to delay in installation of extensometers, some part of displacement data was lost and should be considered in the calculation. A typical instrumentation section of the powerhouse cavern is illustrated in Figure 7 (Sharifzadeh *et al.*, 2009).

6 CONTINUUM-DISCONTINUUM NUMERICAL MODELING OF CAVERN

6.1 Numerical modeling of powerhouse cavern

There are two approaches available in jointed rock modeling, one is the continuum and the other is the discontinuum approach. The use of continuum modeling in tunnel engineering makes it essential to simulate the rock mass response to excavation by introducing an equivalent continuum.

The most common way to solve this problem is to scale the intact rock properties down to the rock mass properties by using empirically defined relationships such as those given by Brady and Brown (2004).

Rock joints and discontinuities in rock mass play a key role in the response of a tunnel to excavation, *i.e.* joints can create loose blocks near the tunnel profile and cause local instability; joints weaken the rock and enlarge the displacement zone caused by excavation; joints change the water flow system in the vicinity of the excavation. The use of discontinuum modeling has been gaining progressive attention in tunnel engineering mainly through the use of the UDEC and 3DEC codes, for 2D and 3D discontinuum modeling respectively (Itasca, 2007).

The Siah Bisheh powerhouse cavern was located in discontinuous media and considering low level in situ stress, the failure of rock mass was mainly controlled by the discontinuity distribution. In this study considering block size, pattern and spacing of discontinuities, three-dimensional distinct element analysis was performed.

Considering 5 joint sets, with joint spacing 12, 14 and 17cm plus bedding planes, low overburden (maximum 250 m), uniformity in monitoring data and various lithology and also bad type rock in most monitoring sections, continuum function is likely. Therefore it seemed modeling in both continuum and discontinuum was essential. In order to numerically model the Siah Bisheh underground openings, PHASE2 and 3DEC codes were utilized. At first two 2-D models were prepared in the chainages 49m and 105m of the powerhouse cavern using PHASE2. Then a 3D model was constructed through the 3DEC code. Figure 8 shows the flowchart of back analysis of the powerhouse cavern under natural conditions.

The Mohr-Coulomb plasticity model was assigned as constitutive model for both continuum and discontinuum analysis as constitutive mechanical model. The value of stress ratio (k) was determined based on field investigation to equal 1.1.

The minimization of the error function alone, does not always guarantee a correct back analysis. The qualitative trend of the displacements on the wall of the excavations should be the same in the calculation as in reality, as a confirmation of the validity of the calculation model and of the simplified assumed hypotheses. Then direct analysis of the powerhouse cavern under natural conditions (underground water table 1880 m) using these optimized parameters was implemented and stability of the powerhouse and its support system was assessed. Finally, for long term stability assessment of the powerhouse cavern under saturated conditions, the underground water table in the model was raised gradually to final elevation (1905 m). Considering instability problems especially in the area of 2nd and 3rd instrumentation array in saturated conditions, a cut-off curtain as an efficient method to guarantee long term stability was proposed.

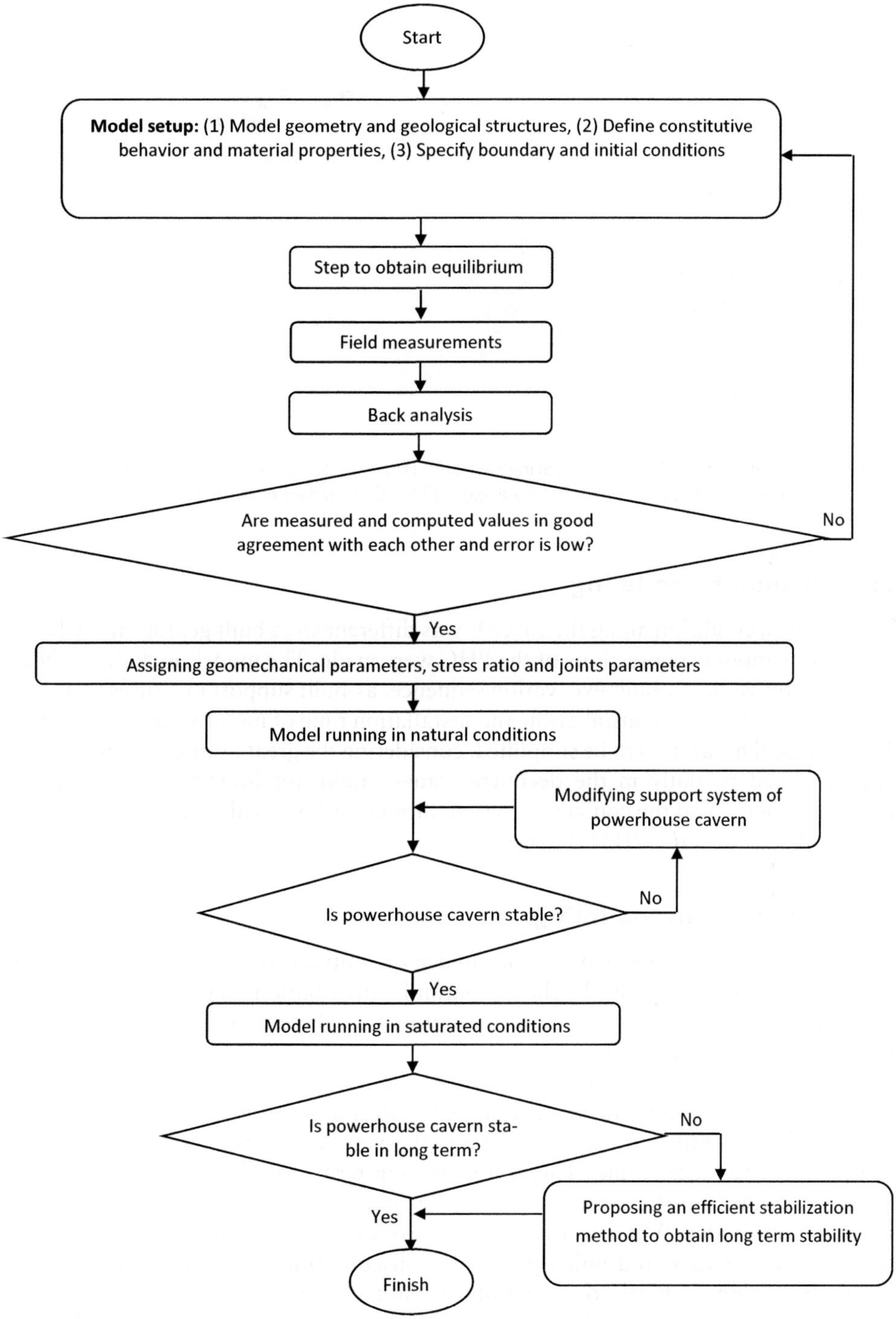

Figure 8 Flow chart of back analysis and stability analysis under natural and saturated conditions (Ghorbani & Sharifzadeh, 2009).

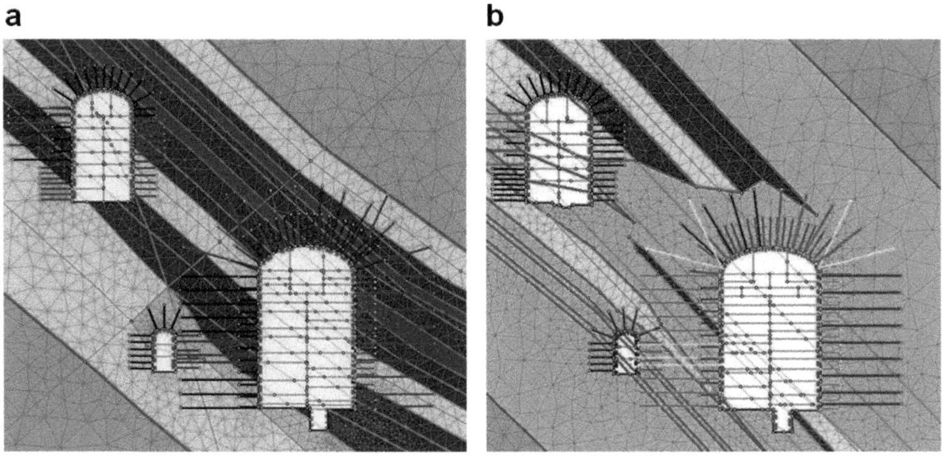

Figure 9 (a) Continuum model for monitoring section 2 (sedimentary area), and (b) Continuum model for monitoring section 5 (melaphyry area) – PHASE2 (Yazdani *et al.*, 2011).

6.2 Continuum modeling

The geological condition along the caverns was different so as built geology models for two separate monitoring sections of the PHC were made. The models include the final shape of caverns, the as-built excavation sequence, as-built support measures inclusive of their respective time of installation and installation time of monitoring instruments. Also geological model had to be simplified, considering the great number of thin layers, which changed partially in the decimeter range could not be taken over into the numerical model. Also the contacts between different lithological units were assumed, as joints (Yazdani *et al.*, 2011) (Figure 9).

6.3 Discontinuum modeling

The Siah Bisheh powerhouse cavern is located in discontinuous media and the failure of rock mass is mainly controlled by the discontinuity distribution. In this study considering block size, pattern and spacing of discontinuities, three-dimensional distinct element analysis was performed.

Siah Bisheh underground openings were under construction in rocks which are formed mainly from quartzite sandstone, red shale and igneous rocks (mainly classified as hard and competent rocks). The powerhouse cavern was constructed beneath the underground water table. Therefore for long term stability analysis the water effect was studied on these rocks and underground water table was exerted in the discontinuum model. Water's effect on such rocks is mainly mechanical and pore pressure in intact rock and uplift pressure in discontinuities should be considered. Water absorption in hard rocks mainly doesn't change the strength parameters (cohesive strength and intrinsic friction angle). For these types of rocks, in all rock strength criteria, total stress should be replaced by effective stress and in rock joints,

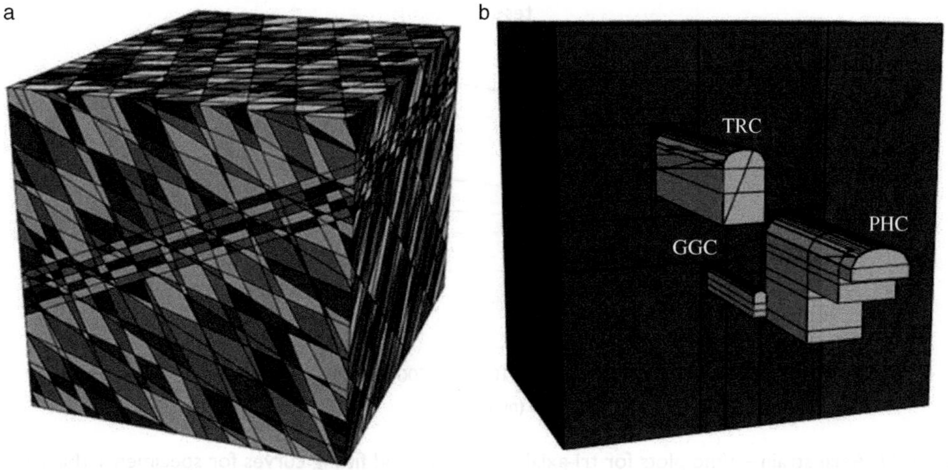

Figure 10 (a) 3D Model geometry with discontinuities, bedding planes and underground water table; and (b) location of powerhouse, transformer and guard gate caverns in discontinuum model-3DEC (Sharifzadeh *et al.*, 2007).

uplift pressure (u) was exerted on the joint surfaces, and uplift pressure was subtracted from total normal stress (Sharifzadeh, 2002) (Figure 10).

After model setup and steps to equilibrium state, direct back analysis of the powerhouse cavern using extensometer results was carried out and calibrated geomechanical properties of rocks, stress ratio and joints parameters were identified.

6.4 Time dependent numerical modeling

There are eight power models in 3DEC software for simulating time dependent behavior of structures. Based on the creep tests and in situ measurements power model was used for simulating the time dependent behavior of the cavern. The standard form of this law in 3DEC is as follow:

$$\dot{\varepsilon}_{cr} = A\overline{\sigma}^n \tag{1}$$

Where $\dot{\varepsilon}_{cr}$ is the creep rate, A and n are material properties, $\overline{\sigma} = \left(\frac{3}{2}\right)^{1/2}(\sigma_{ij}^d\,\sigma_{ij}^d)^{1/2}$ with σ_{ij}^d being the deviatoric part of σ_{ij}. The deviatoric stress increments are given by;

$$\Delta\sigma_{ij}^d = 2G(\dot{\sigma}_{ij}^d - \dot{\sigma}_{ij}^c)\Delta t \tag{2}$$

Where G is shear modulus, and $\dot{\sigma}_{ij}^d$ is the deviatoric part of the strain-rate tensor.

For time dependent analysis the system is required to be always in mechanical equilibrium, the time-dependent stress increment must not be too large compared to strain-dependent stress increment; otherwise, out of balance force will rapidly become large, and inertial effects may affect the solution. For the power law, the viscosity may be estimated as the ratio of stress magnitude $\overline{\sigma}$ to the creep rate, $\dot{\varepsilon}_{cr}$. Using Equation 1, the maximum creep timestep is;

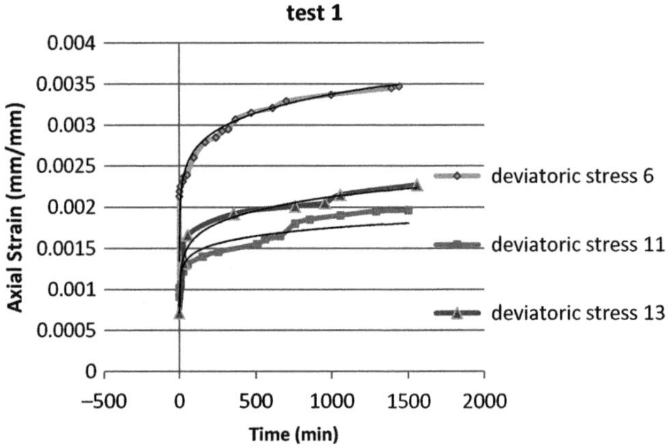

Figure 11 Axial strain – time plots for tri-axial creep tests and fitting curves for specimen 1 (Nadimi *et al.*, 2010).

$$\Delta t^{cr}_{Max} = \frac{\overline{\sigma}^{-1-n}}{AG} \tag{3}$$

Where A is power law constant; and G is elastic shear modulus (Itasca, 2007).

Triaxial creep tests were conducted on rock samples which were prepared from extensometer boreholes. The samples were red shale and quartzite sandstone and they were dry with 54 mm diameter and 110–120 mm high. The quartzite sandstone samples had high compression strength and very little creep strain; therefore triaxial creep tests of shale samples with more creep prone were used to determine power model parameters. As shown in Figure 11, the creep tests were conducted in several steps and different deviatoric stresses (Nadimi *et al.*, 2010).

7 BACK ANALYSIS OF ROCK MASS AND DISCONTINUITY PROPERTIES

Back analysis techniques as a practical engineering tool are nowadays often used in geotechnical engineering problems for determining the unknown geomechanical parameters, system geometry and boundary or initial conditions using field measurements of displacements, strains or stresses performed during excavation or construction works (Sakurai, 1993).

The direct approach employs the trial values of the unknown parameters as input data in the stress analysis algorithm, until the discrepancy between measurements and corresponding quantities obtained from a numerical analysis is minimized (Cividini *et al.*, 1981; Feng & Zhao, 2004). Trial values should be defined based on an algorithm which follows all combinations of different parameters until the optimum values of all variables are determined. This classic approach is relatively simple and suitable for parameters that are independent. While application of this method for parameters that influence or interact with one another is restricted. This method could successfully search the optimal values of parameters regardless of their initial values. Obviously it is better that variation of parameters take in a valid interval which has obtained from laboratory and field testing combined to experimental relations (Gioda & Locatelli, 1999; Oreste, 2005).

In this study displacement based direct back analysis using variable staggered grid optimization algorithm was applied. Direct formulation was very flexible and applying such a procedure for complex constitutive models was appropriate. Furthermore, development of the direct back analysis code was much less difficult than development of the code based on an inverse algorithm. The only work is appending an existing program with a module. For this purpose a Fish function was written to do the minimization of errors between measured and computed values as follows:

$$\varepsilon(p) = \sum_{i=0}^{n} \left(\frac{u_i^m(p) - u_i}{u_i} \right)^2 \tag{4}$$

Where u_i and $u_i^m(P)$, $i = 1, 2, ..., n$ were the measured and corresponding numerical results, respectively and n was the number of measured points. Obviously, $u_i^m(P)$ depends on the unknown model parameters collected in the vector P. Here we used a normalized error function to decrease the effect of measurement error.

As was mentioned before, the end part of cavern consisted of igneous rocks and for this reason to back analysis geomechanical properties of these parts two different error functions based on an equation (Swoboda *et al.*, 1999) using results of extensometers at each part were developed. The measurement results were processed before using them in back analysis. Wrong displacements due to error in installation, reading and recording of data or inaccurate performance of instruments were eliminated. Therefore after assessment of extensometer results, finally 150 displacement data among 208 displacement data were selected for back analysis. The results of back analysis for the Melaphyry section and sedimentary part are presented in Tables 7 and 8, respectively. Results showed that elastic modulus has the highest effect and Poisson's ratio, friction angle and cohesion had respectively the least effect on error function and thus on displacement values.

The relationship between the horizontal and vertical stresses in the rock mass (k) was more difficult to estimate from the preliminary investigations and it relied closely on

Table 7 Results of back analysis for Melaphyry section (Ghorbani & Sharifzadeh, 2009).

Constant Parameters	C = 2.5 MPa, ϕ = 43°, υ = 0.22, K = 1.1, β = 0°			
Young's modulus (MPa)	14	15	16	17
Error (%)	2.4410	2.1289	1.6571	1.9964
Constant Parameters	E = 16 GPa, ϕ = 43°, υ = 0.22, K = 1.1, β = 0°			
Cohesion (MPa)	2	2.5	3	3.5
Error (%)	1.7583	1.6571	1.5137	1.6852
Constant Parameters	E = 16 GPa, C = 3 MPa, υ = 0.22, K = 1.1, β = 0°			
Friction Angle (°)	40	41	42	43
Error (%)	1.3874	1.261	1.4023	1.5137

Table 8 Results of back analysis in sedimentary part (Ghorbani & Sharifzadeh, 2009).

Constant Parameters	$C = 1.5$ MPa, $\phi = 40°$, $\upsilon = 0.25$, $K = 1.1$, $\beta = 0°$				
Young's modulus (MPa)	6	7	8	9	10
Error (%)	5.5103	4.8061	4.3677	3.9664	4.4739
Constant Parameters	$E = 9$ GPa, $\phi = 40°$, $\upsilon = 0.25$, $K = 1.1$, $\beta = 0°$				
Cohesion (MPa)	1.25	1.5	1.75	2	2.25
Error (%)	4.2154	3.9664	3.6532	3.8912	4.2079
Constant Parameters	$E = 9$ GPa, $C = 1.75$ MPa, $\upsilon = 0.25$, $K = 1.1$, $\beta = 0°$				
Friction Angle (°)	37	38	39	40	–
Error (%)	3.4277	3.2238	3.4816	3.6532	–
Constant Parameters	$E = 9$ GPa, $C = 1.75$ MPa, $\phi = 38°$, $K = 1.1$, $\beta = 0°$				
Poisson's ratio	0.23	0.24	0.25	–	–
Error (%)	3.2351	3.1907	3.2238	–	–

Table 9 Results of stress ratio back analysis (Ghorbani & Sharifzadeh, 2009).

Melaphyry section parameters	$E = 16$ GPa, $\upsilon = 0.22$, $C = 3$ MPa, $\phi = 41°$, $\beta = 0°$		
Sedimentary part parameters	$E = 9$ GPa, $\upsilon = 0.24$, $C = 1.75$ MPa, $\phi = 38°$, $\beta = 0°$		
Stress ratio (k)	1.1	1.15	1.2
Total percent (%)	3.4238	3.7993	4.3486

back analysis results. For this purpose after identification of geomechanical properties for the Melaphyry section and sedimentary part, back analysis for stress ratio was carried out and results are shown in Table 9. Results also showed that stress ratio had a great effect on error function and by increasing it, values of displacements in the powerhouse wall had been increased.

Considering discontinuum modeling of powerhouse caverns and the effect of discontinuities parameters on numerical modeling results, back analysis was carried out to find strength of discontinuities and stiffness properties (Table 10). Results show that the parameters of discontinuities especially joints' normal and shear stiffness have a remarkable influence on the value of error function.

About 40 to 50 meters of the end of powerhouse cavern was igneous rock (Melaphyr) and the remainder was the sedimentary part which comprised a the sequence of Quartzite Sandstone, Red Shale, Mylonite and Melaphyr. For this reason to obtain back analysis geomechanical properties of these parts two different error functions based on formula (1) using results of extensometers installed in each part were developed in discontinuum model. But in the continuum method two different models in the chainages of 49m (sedimentary part) and 105m (melaphyry section) of the powerhouse cavern were prepared and back analysis was performed separately for these two models.

The minimization of the error function alone, does not always guarantee a correct back analysis. The qualitative trend of the displacements on the wall and vault of the

Table 10 Results of back analysis for joints parameters (Ghorbani & Sharifzadeh, 2009).

Melaphyr parameters	E = 16 GPa, υ = 0.22, C = 3 MPa, ϕ = 41°, β = 0°, K = 1.1			
Sedimentary part parameters	E = 9 GPa, υ = 0.24, C = 1.75 MPa, ϕ = 38°, β = 0°, K = 1.1			
Joint parameters	C = 0.5 MPa, ϕ = 30°			
Normal Stiffness (GPa/m)	10	20	30	40
Shear Stiffness (GPa/m)	2	7.69	10	30
Total percent (%)	5.6222	3.4238	3.0992	5.3968
Joint parameters	JKn = 30 GPa/m, JKs = 10 GPa/m, ϕ = 30°			
Cohesion (MPa)	0.4	0.5	0.6	–
Total percent (%)	2.7533	3.0992	3.6049	–
Joint parameters	JKn = 30 GPa/m, JKs = 10 GPa/m, C = 0.4 MPa			
Friction Angle (°)	25	30	35	–
Total percent (%)	2.9527	2.7533	3.1161	–

excavations should be similarly the same in the calculation as in reality, as a confirmation of the validity of the calculation model and of the simplified assumed hypotheses.

In tables 7 and 8, final results of back analysis for Melaphyry section and sedimentary part for both continuum and discontinuum models are presented. In both continuum and discontinuum models results show that elastic modulus has the highest effect and Poisson's ratio, friction angle and cohesion have respectively the least effect on error function and thus on displacement values.

Considering continuum and discontinuum modeling of powerhouse caverns and the effect of joint parameters on numerical modeling results, back analysis was carried out to find joint strength and stiffness properties (Table 8). Results in continuum models showed that friction angle had a major impact on deformations of the powerhouse cavern. Also results in the discontinuum model showed that joint parameters especially joints' normal and shear stiffness had a remarkable influence on error function values.

8 DIRECT STABILITY ANALYSIS OF POWERHOUSE CAVERN UNDER DIFFERENT CONDITIONS

8.1 Natural conditions

After finding calibrated model values for geomechanical properties of rocks, stress ratio and discontinuity parameters, direct analysis of the powerhouse cavern under natural conditions with existing underground water table (1880 m) were carried out.

In order to compare the results of analysis with measured values, deformations were utilized in several locations of the powerhouse cavern which were adjacent to extensometers of 3rd instrumentation array (Table 11). This array was very important due to presence of many shear zones in this region. Instrumentation showed large displacement and increase in the load of load cells in this array. As shown in Table 11, computed results

Table 11 Comparison between computed and measured values of displacements in 3rd instrumentation array (millimeter) (Ghorbani & Sharifzadeh, 2009).

		Measured values using Extensometers	*Computed values in natural conditions*
Upstream wall	Installed from Vault Adit (EXT.1)	18	21.2
	EL. 1858 (EXT.2)	59.06	64.2
	EL. 1847 (EXT.3)	24.5	23.17
Roof	Upstream roof (EXT.4)	11.73	16.23
	Roof center (EXT.5)	17.22	17.4
	Downstream roof (EXT.6)	4.7	14.68
Downstream wall	EL. 1858 (EXT.7)	11.3	21.18
	EL. 1858 (EXT.8)	45.6	48.1
	EL. 1858 (EXT.9)	16.98	23.26

were in good agreement with measured values. Because of delay in the installation and reading of extensometers, the first part of the deformations was lost; therefore measured data showed lower values compared to calculated results. Generally numerical modeling showed better consistency with reality. Figure 12 shows a cross section of displacement vectors in natural conditions. As seen in Figure 12, the powerhouse was stable and the existing support system had a good efficiency to control displacements. The maximum displacement of the powerhouse cavern which would occur in upstream wall equaled 6.51 cm. Transformer and guard gate caverns were both in stable condition. It is

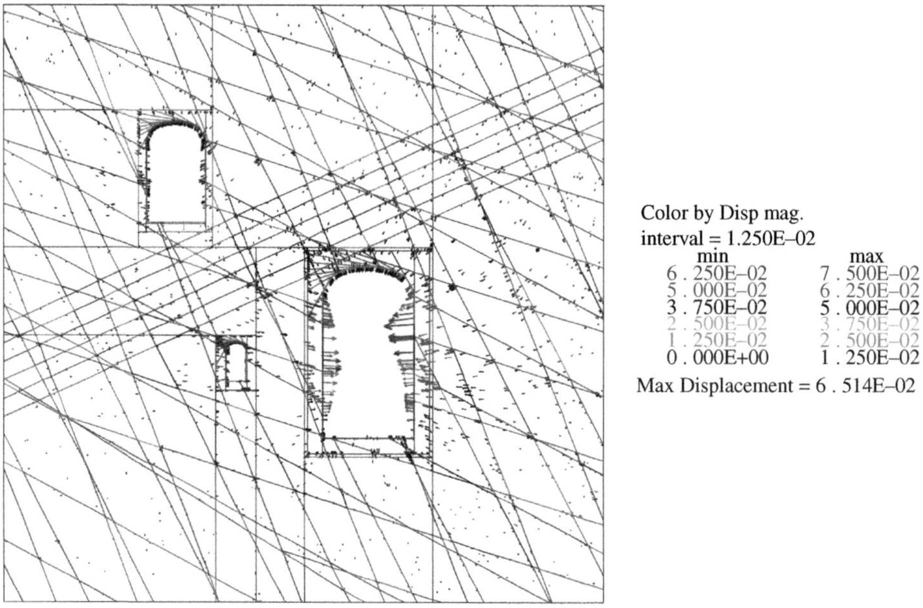

Figure 12 A cross section of displacement vectors (in m) in natural conditions (groundwater level 1880) (Ghorbani & Sharifzadeh, 2009).

noteworthy that the drainage system around the powerhouse cavern was not considered in the analysis and considering it would guarantee its stability under natural conditions.

To verify numerical simulation, displacements obtained by numerical method were compared with those obtained from direct measurements. Displacement measurements within the rock mass had been recorded in borehole extensometers installed over the periphery of the cavern. There were three types of extensometers installed in the cavern; the extensometers 30m in length, gave the displacement inside rock at 2 m, 5 m, 10 m and 30 m, the extensometers 25m in length, gave the displacement inside rock at 2 m, 5 m, 10 m and 25 m, and the extensometers 20m in length, gave the displacement inside rock at 2 m, 5 m, 10 m and 20 m from the crown of the cavern. The comparison of the measured and computed displacement-time curves showed that the power law model parameters of the data set No. 1(b) in Tab. 11 had better simulation results than another set of parameters (see Figure 11).

Figure 13 shows the failure zone around the powerhouse cavern in natural conditions. Proportionate to induced stress due to cavern excavations and rock strength, rock mass in some areas around the powerhouse cavern was in a failure condition. As seen in Figure 13, the type of failure in the powerhouse cavern was tension and the most critical situation was in the upstream wall. Depths of failure zone in upstream and downstream walls were respectively 6 m and 5 m. All design activities must be taken to prevent tension failure zone development. As shown in Figure 13, the pillars between the powerhouse and guard gate caverns were stable and their stress fields would not influence each other.

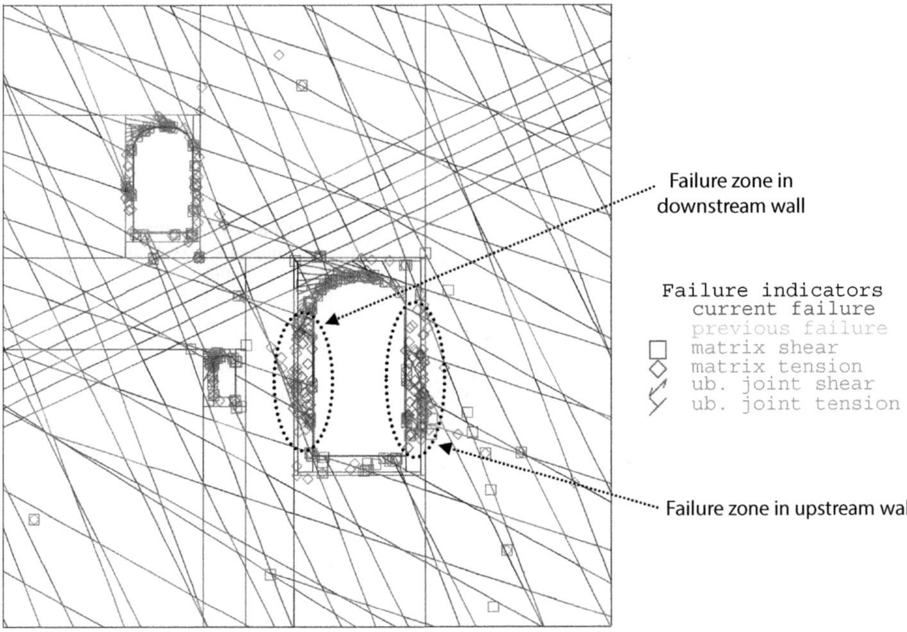

Figure 13 A cross section of failure zone around caverns in natural conditions (Ghorbani & Sharifzadeh, 2009).

8.2 Saturated conditions

After d/s dam impounding and increasing the level of the underground water table, the powerhouse cavern would be 30 m below the maximum lake level of the d/s dam. For stability analysis under such conditions the underground water table was raised gradually in five steps with 5 m intervals up to maximum level and results of analysis under fully saturated conditions were used to predict rate of displacements and efficiency of the support system. To calculate the value of uplift pressure in joints around the 3rd instrumentation array in the powerhouse cavern a Fish function was developed. This program found the nearest zone to the joint surface considering introduced points which were corresponded to extensometer installation points in the crown center, upstream and downstream walls and then draws the uplift pressure graphs based on solving time step for the model.

Figure 14 shows a cross section of displacement vectors in saturated conditions. As seen in Figure 14, powerhouse cavern walls in the chainage of the 2nd and 3rd instrumentation arrays were unstable and displacements were higher than permissible values. Powerhouse cavern roof displacements were in reasonable range and transformer and guard gate caverns were in good stable condition. The values of displacements in the downstream wall were higher than upstream wall. There were 3 reasons for this issue. First, the attitude of joints to the powerhouse cavern made some unstable blocks in the downstream wall. Second, the value of pore water pressure in the upstream wall was higher than of the downstream wall due to higher underground water table in the upstream wall. Then the value of effective stress which was the cause of displacements was higher in the downstream wall. Third, as shown in Figure 14, uplift pressure was

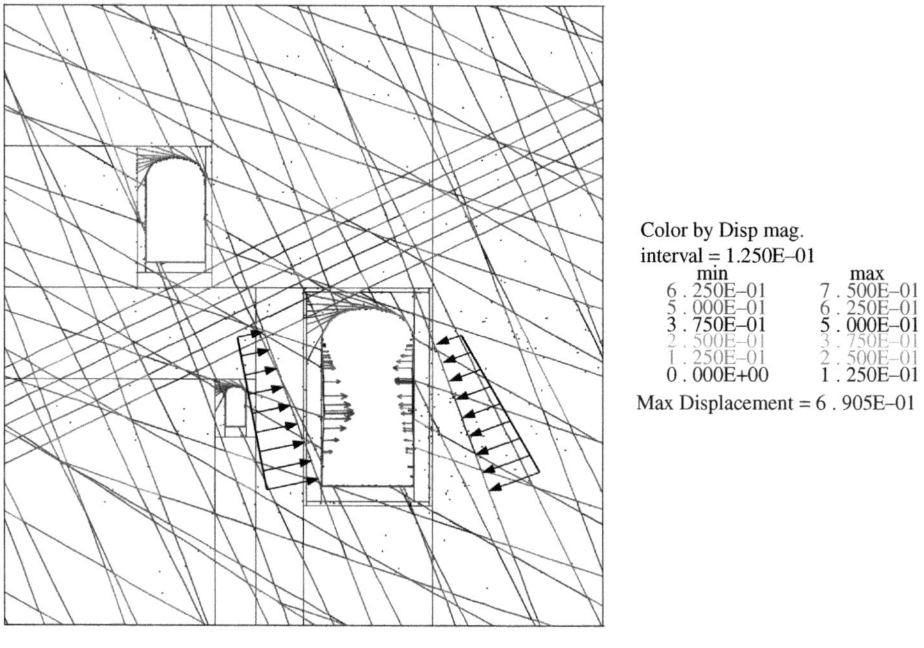

Figure 14 A cross section of displacement vectors (in m) in saturated conditions (groundwater level 1905) (Sharifzadeh *et al.*, 2008).

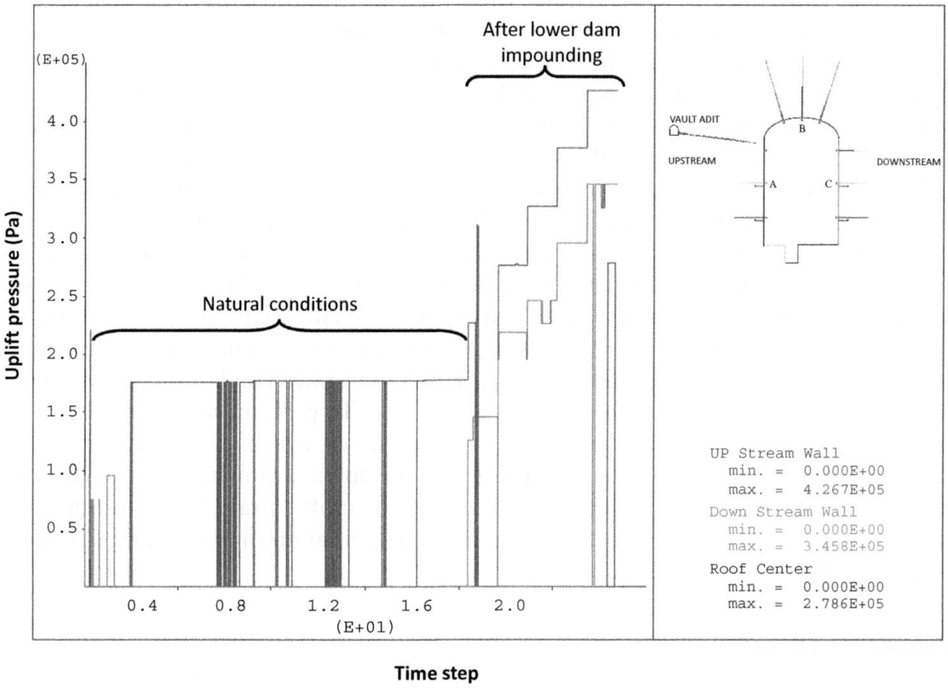

Figure 15 Histories of uplift pressure in PHC upstream wall (A), crown (B) and downstream wall (C) with increasing underground water table (Ghorbani & Sharifzadeh, 2009).

exerted on the rock mass in upstream wall joints and tended to stability but uplift pressure in downstream wall joints acted towards instability of the powerhouse cavern.

In Figure 15, the history of uplift pressures in joints surfaces of the crown center, upstream and downstream walls of the powerhouse cavern is illustrated. This figure shows increasing of uplift pressure in blocks interface correspond to 3 joints which cut the powerhouse cavern. As seen in Figure 15, with increasing of the underground water table from 1875 m (natural conditions) to 1905 m (saturated conditions) the values of uplift pressure increased in block interfaces. This is due to increasing of hydraulic pressure considering d/s dam impounding and rising underground water table.

The pressure exerted on discontinuity surfaces and called uplift pressure was computed as follows:

$$U = \gamma_w . Z \tag{5}$$

Where U is uplift pressure (Pa), γw is unit weight of water (N/m3) and z is the height of water above discontinuity surfaces (m).

With increasing uplift pressure in discontinuities, pressure on support systems would increase which tends to convergence of PHC walls and increasing the value of rock block displacements. This issue finally tends to powerhouse cavern failure in the area of the 2nd and 3rd instrumentation arrays. Therefore it was necessary to control the water pressure by an efficient stabilization method to guarantee long term stability of the powerhouse cavern.

As shown in Figure 14, the displacement plots had a similar tendency evolution; there were only some instantaneous displacements in computed plots due to shear deformation of joints or plane of layers. In addition, there were instantaneous increases of computed displacement after excavating the lower levels of the cavern. The total displacement contour after one year is shown in Figure 15. It should be implied that, by increasing the run time, the displacement in walls would be increased more than the crown.

At the time of analysis excavation of the powerhouse cavern was completed and it was impossible to modify the support system. Therefore to guarantee long term stability of the powerhouse cavern under saturated conditions, a cutoff curtain was proposed. Results of analysis under natural and saturated conditions showed that the powerhouse cavern roof was stable and there was no need to perform a cutoff curtain in the PHC roof. PHC floor concrete slab more than 5 m in height would be carried out in the future which would guarantee long term stability of this part under saturated conditions. So there was no need for a cutoff curtain for the floor of the PHC too. To perform cutoff curtain in the upstream wall of the PHC it was proposed to use vault adit which was excavated in this part of the PHC. It was recommended to perform the downstream wall's cutoff curtain from the transformer cavern (Figure 16). To perform cutoff curtain for the north wall and south wall of the PHC it was proposed to use a ventilation tunnel and transformer cavern respectively (Figure 17).

9 DISCUSSION

Back analysis is a practical engineering tool to evaluate geomechanical parameters of underground and surface structures based on field measurements of some key variables such as displacements, strains and stresses. These parameters are necessary for stability analysis and design of support system for geostructures.

Back analysis of Siah Bisheh powerhouse cavern during construction using the finite element method and distinct element method were carried out in the computer codes PHASE2 and 3DEC. Initial values of input parameters required in the both models

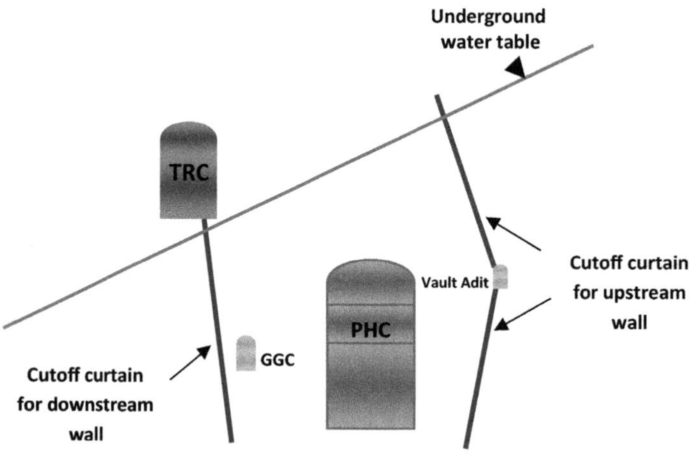

Figure 16 Proposed locations for cutoff curtain in PHC upstream and downstream walls (Ghorbani & Sharifzadeh, 2009).

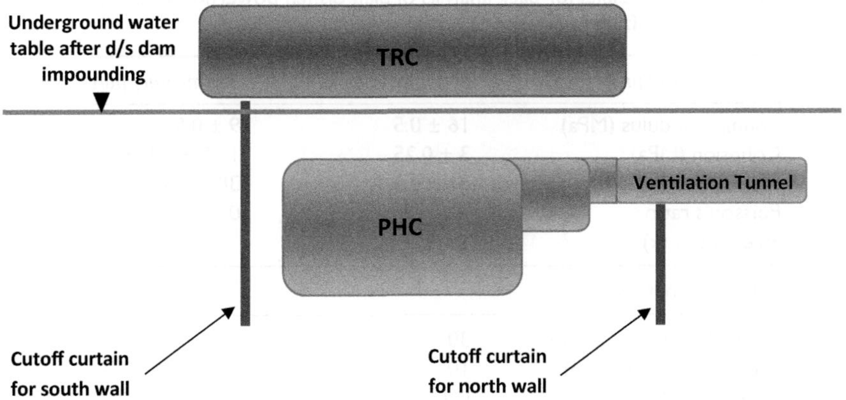

Figure 17 Proposed locations for cutoff curtain in PHC north and south walls (Sharifzadeh *et al.*, 2008).

were based on the results of geological and geotechnical investigations and estimated by empirical and theoretical methods.

The parametric studies indicated that cavern response was strongly dependent on the rock mass modulus, ratio between horizontal and vertical stresses and friction angle of joints. As could be observed from Table 7, almost all rock mass parameters resulting from back analyses in both models were in good agreement together but the elasticity module of melaphyry section and friction angle of joint parameters in both models showed discrepancy. This major difference between Young's modulus could be explained by adjacent excavation openings, shear zones and non-interference efect of rocks layers in the discontinuum model. It also seems that the difference between the values of friction angle of joint parameters was based on software performance. This study clarified that the back calculated value of Young's modulus was more representative for mechanical behavior of rock masses in a large domain. Meanwhile the results demonstrated clearly that the default assumed rock mass parameters for design powerhouse cavern were high. Eventually with reference to modeling in this practice, it seems the interest has been placed on the adoption of discontinuum models which give a more realistic and representative picture of rock mass behavior than equivalent continuum models.

It is normally considered that the creep of rock masses in situ is governed primarily by the behavior of discontinuities, *i.e.* the bedding planes, faults and joints.

Numerical simulation of time dependent behavior of the Siah Bisheh powerhouse cavern showed that the power creep model was relevant on an enlarged scale. The parameters of this model were determined on the basis of triaxial creep tests and monitoring data. Crown inward displacement increased as the time increased with decreasing rate. Although there was a scale effect on the power model parameters, the creep behavior of the small rock samples had the same character as the rock mass around the cavern. It was considered that the creep of in situ rock mass was governed by the behavior of discontinuities.

In addition, by increasing the span or scale of the cavern, the rate of the displacement increased in the first days. Also, some instantaneous displacement occurred by drilling the second and third excavation sequences but the excavation of the 4th stage had

Table 12 Final results for back analysis of Siah Bisheh powerhouse cavern (Sharifzadeh, 2008).

Geomechanical properties	*Melaphyry section*	*Sedimentary part*
Young's modulus (MPa)	16 ± 0.5	9 ± 0.5
Cohesion (MPa)	3 ± 0.25	1.75 ± 0.125
Friction Angle (°)	41 ± 0.5	38 ± 0.5
Poisson's ratio	–	0.24
Stress ratio (k)	1.1	
Joints parameters		
Normal Stiffness (GPa/m)	30	
Shear Stiffness (GPa/m)	10	
Cohesion (MPa)	0.4 ± .05	
Friction Angle (°)	30 ± 2.5	

vanishingly small effect on the strain-time curve of the crown. However, the results of power model were fairly consistent.

10 CONCLUSION

In Table 12, results of back analysis for geomechanical properties for melaphyry section and sedimentary part, stress ratio and discontinuities parameters are presented. The best way to present the final results of the back analysis is to introduce them as a mean value and its amplitude.

Results of analysis showed that powerhouse, transformer and guard gate caverns were stable under natural conditions and the existing support system had suitable efficiency and could effectively control displacements. Powerhouse cavern long term stability under saturated conditions was analyzed. Results of analysis showed that after d/s dam impounding, considering the vicinity of powerhouse cavern to d/s dam reservoir, pore water pressure and uplift pressure in discontinuities around the powerhouse cavern would arise and tend to local failure of the powerhouse cavern. The values of displacements in downstream wall under saturated conditions were higher than upstream wall values. This was due to high effective stress in this region and forming some unstable blocks considering attitude of discontinuities to powerhouse cavern. To prevent powerhouse failure and assure its long term stability, a cutoff curtain corresponding to the introduced layout was proposed.

REFERENCES

Barla, G. 2001. Tunnelling under squeezing conditions. In *Tunnelling Mechanics, Euro summer school, Innsbruck*, pp. 169–268.

Boidy, E., Pellet, F., 2000. Identification of mechanical parameters for modeling time-dependent behaviour of shales, ANDRA Workshop, Behaviour *of deep argillaceous rocks: Theory and experiment, Proc. Int. Workshop on Geomech.*, Paris, October 11–12, 2000, p. 13.

Boidy, E., Bouvard, A., Pellet, F., 2002. Back analysis of time-dependent behaviour of a test gallery in claystone. *Tunneling and Underground Space Technology*, Vol. 17, pp. 415–425.

Brady, B.H.G., Brown, E.T., 2004. *Rock Mechanics for Underground Mining* (3rd Ed.), Springer, Netherlands. ISBN 10-1-4020-2064-3 (PB).

Cividini, A., Jurina, L., Gioda, G., 1981. Some aspects of characterization problems in geomechanics. *International Journal of Rock Mechanics and Mining Sciences and Geomechanics Abstracts*, Vol. 18, pp. 487–503.

Feng, X.T., Zhao, H., Li, S., 2004. A new displacement back analysis to identify mechanical geomaterial parameters based on hybrid intelligent methodology. *International Journal for Numerical and Analytical Methods in Geomechanics*, Vol. 28, pp. 1141–1165.

Gioda, G., Locatelli, L., 1999. Back analysis of the measurements performed during the excavation of a shallow tunnel in sand. *Numerical and Analytical Methods in Geomechanics*, Vol. 23, pp. 1407–1425.

Ghorbani, M., Sharifzadeh, M., 2009. Long term stability assessment of Siah Bisheh powerhouse cavern based on displacement back analysis method. *Tunnelling and Underground Space Technology*, Vol. 24, Issue 5, September, pp. 574–583.

Hoek, E., 2001. Rock mass properties for underground mines. *Underground Mining Methods: Engineering Fundamentals and International Case Studies*. Hustrulid, W.A., Bullock, R.L. (eds), Litleton, Colorado: Society for Mining, Metallurgy, and Exploration (SME).

Itasca Consulting Group, Inc. 2007. 3DEC (3 Dimensional Distinct Element Code Manual), Version 4.1. Minneapolis: ICG.

Lahmeyer, and Iran Water & Power Resources Development Co (IWPC). 2005. Report on geology and engineering geology of powerhouse cavern.

Nadimi, S., Shahriar, K., Sharifzadeh, M., Moarefvand, P., 2010. Triaxial creep tests and back analysis of time-dependent behavior of Siah Bisheh Cavern by 3-Dimensional distinct element method. *Tunnelling and Underground Space Technology*, Vol. 24, Issue 5, September 2009, pp. 574–583.

Oreste, P., 2005. Back-analysis techniques for the improvement of the understanding of rock in underground constructions. *Tunnelling and Underground Space Technology*, Vol. 20, Issue 1, pp. 7–21.

Sakurai, S., 1993. Back analysis in rock engineering. *Comprehensive Rock Engineering*, Vol.4, pp. 543–568.

Sharifzadeh, M., Fahimifar, A., Shahkarami, A. A., Esaki, T., 2002. Classification system for evaluation of water effect on mechanical behavior of intact and jointed rocks with case study, pp. 194–204, *Proceedings of the 3rd Iranian International Conference on Geotechnical Engineering and Soil Mechanics*, December 9–11, Tehran, Iran.

Sharifzadeh, M., Ghorbani, M. Nateghi, R. Masoudi, R., 2007. Long term stability assessment of a large underground opening under saturated condition, pp. 947–950, *11th ISRM Congress*, July 9–13, 2007, Sousa, L.R., Olalla, C., Grossmann, N.F. (eds), Lisbon, Portugal.

Sharifzadeh, M., Ghorbani, M., Masoudi, R., Eslami M., 2008. Performance prediction of support system of Siah Bisheh Pumped storage Cavern under saturated condition using numerical Analysis, p. 94, *2nd National Conference of Dam and Hydro-power plants*, May 2008, Tehran, Iran.

Sharifzadeh, M., Ghorbani, M., Masoudi, R., 2009. Displacement based back analysis of Siah Bisheh pumped storage powerhouse Cavern by means of distinct element method. Sharif Journal of Science and Technology; Transaction on: *Material Science and Engineering, University Journal*, April-May 2009, Issue 47, pp. 49–57.

Sharifzadeh, M., Tarifard, A., Moridi, M. A., 2013. Time-dependent behavior of tunnel lining in weak rock mass based on displacement back analysis method. *Tunnelling and Underground Space Technology*, Vol. 38, pp. 348–356. http://dx.doi.org/10.1016/j.tust.

Shalabi, F.I., 2004. FE analysis of time-dependent behavior of tunneling in squeezing ground using two different creep models. *Tunneling and Underground Space Technology*, Vol. 20, pp. 271–279.

Swoboda, G, Ichikawa, Y, Dong, Q.X., Zaki, M., 1999. Back analysis of large geotechnical models. *International Journal for Numerical and Analytical Methods in Geomechanics*, Vol. 23, 1455–1472.

Tsai, L.S., Hsieh, Y.M., Weng, M.C., Huang, T.H., Jeng, F.S., 2008. Time-dependent deformation behaviors of weak sandstones. *International Journal of Rock Mechanics and Mining Sciences*, Vol. 45, pp. 144–154.

Yazdani, M., Sharifzadeh, M., Kamrani, K., Ghorbani, M., 2011. Displacement-based numerical back analysis for estimation of rock mass parameters in Siah Bisheh powerhouse cavern using continuum and discontinuum approach. *Tunnelling and Underground Space Technology*, Vol. 28, pp. 41–48. doi:10.1016/j.tust.2011.09.002.

Chapter 7

Construction of large underground structures in China

Xia-Ting Feng[1], *Quan Jiang*[1] & *Yong-Jie Zhang*[1,2]
[1]*State Key Laboratory of Geomechanics and Geotechnical Engineering, Institute of Rock and Soil Mechanics, Chinese Academy of Sciences, Wuhan, Hubei, PR China*
[2]*School of Civil Engineering and Architecture, Changsha University of Science & Technology, Changsha, Hunan, PR China*

Abstract: In the last twenty years, lots of large underground structures have been built in China for hydropower stations, pumped storage power stations, oil storage projects and other projects. Firstly in this chapter, large underground cavern groups constructed to house hydropower generators are mainly discussed from five aspects: the geological environmental conditions; the structural system; the construction organization and factors that influence it; the timeliness of the excavation and unloading of the surrounding rock; and the investment and stability controlling standard of large underground cavern groups. Secondly, the engineering problems of large underground cavern groups that have been built in the southwest of China are analyzed from the point of view of their layout; excavation; failures caused by rock mass structures or stress or both; deformation stability of rock anchor beam; overload or failure of anchor cables; unloading and timeliness deformation of sidewall. Regarding the layout of underground cavern groups, the longitudinal axis orientation, space between caverns, rise span ratio and layout form of different huge projects with high *in situ* stress were collected and are compared. Thirdly, the excavation and supporting design methods of large underground cavern groups, which have been used recently in China, are introduced. Then the typical research progress of large underground cavern groups – such as the intelligent analysis and optimal design method for large underground cavern groups, deformation management classification, and deformation predicting methods for the surrounding rock during excavation – are recommended. Finally, several research prospects for large underground cavern groups are proposed.

1 INTRODUCTION

There are differences between the large underground structures discussed in this chapter and highway tunnels, railway tunnels, and urban subway tunnels. The length of the type of large underground structure under consideration is smaller compared with other tunnels, but its width and height are bigger. The large underground structures include hydropower underground cavern groups, energy reserve underground cavern groups, military installation underground cavern groups, large urban underground engineering works, and so on. The first hydropower underground cavern groups have been in existence for more than sixty years. Lots of information about the hydropower underground cavern groups built in China can be found in published documents. By contrast, the energy reserve underground cavern groups have been built since the end of the twentieth century. Only a few engineering works have been built and there is little information about them. Large urban underground engineering works are usually built in the soil, which is different from the rock engineering. Therefore, the hydropower underground cavern groups and their engineering problems will be mainly discussed, with particular reference to the powerhouse project in the mountain and ravine region of southwest China.

Before 1949, only a few small, diversion type power stations had been constructed in southwest China. The first underground powerhouse was in the Tianmen River hydropower station in Guizhou Province. Its design head was 31 m, flow rate was 2.4 m^3/s, and total installed capacity was 5.76 MW. Its engineering construction was in 1939, and it was built in 1945.

The hydropower works built in the period 1949–1960 are mostly medium and small. The installed capacities of the underground powerhouses are mostly under 100 MW, and their tunnel diameters are less than 5–6 m. Their construction method was blasting excavation by hand air drill and hand haulage. The passive support was used to make the surrounding rock stable, and the construction speed was slow, working efficiency was low and security problems were serious. The first underground powerhouse built in this period is in the Gutianxi I hydropower station in Fujian Province. Its design size (width × height × length) were 12.5 m × 24 m × 85.8 m, and its total installed capacity was 62 MW. It was started in 1951 and the first unit of electricity was generated in 1956.

During 1960–1980, several large underground powerhouses were built one after another. The most representative projects are as follows.

(1) The Liujiaxia hydropower station underground powerhouse, located in Yongjing County of Gansu Province in the upper Yellow River, which was built in 1968. It has an installed capacity of 2 × 250 MW, its depth is about 55 m, its powerhouse size (width × height × length) is 24.5 m × 58.5 m × 86.1 m, and the largest excavation span is 31 m.
(2) The Baishan hydropower station underground powerhouse, located in Huadian County of Jilin Province, which is at a depth is 60–120 m. Its underground powerhouse size (width × height × length) is 25 m × 54.25 m × 121.5 m, and its installed capacity is 3 × 300 MW.
(3) The Yanshuigou hydropower station underground powerhouse, located in Yili River of Yunnan Province. Its installed capacity is 4 × 36 MW; its underground powerhouse size (width × height × length) is 12.8 m × 21.8 m × 80.3 m; its length of pressure headrace tunnel is 2,740 m, with a diameter of 3 m; and its maximum depth is 80 m. It was built in 1957, and the first unit was generated in 1966.
(4) The Yuzixi I hydropower station underground powerhouse, located in Wenchuan County of Sichuan Province. Its installed capacity is 4 × 40 MW, its length of pressure headrace tunnel is 8,429 m, and its maximum depth is 650 m. It was built in 1966 and finished in 1972.

Smooth blasting technology was first studied in 1963 to construct the spillway tunnel of Luhun reservoir. It was supported with shotcrete and bolt. After that, this technology had been used in Yuzixi I hydropower station, Baishan hydropower station, Xierhe hydropower station and so on, which promoted the progress of underground engineering excavation and its supporting technology in China.

From 1980 to 2000, the large underground engineering works of Lubuge, Ertan, Dachaoshan, and Xiaolangdi; Guangzhou pumped storage; Tianhuangping pumped storage; the Ming tombs pumped storage; and other hydropower stations in China were built, one after another. Among them, the Lubuge hydropower station was the first project constructed with a modern management concept and modern design and construction technology. The length of its pressure headrace tunnel is 9,382 m, with a diameter of 8 m. Its maximum head is 372 m, total installed capacity is 600 MW, and its underground powerhouse size (width × height × length) is 18 m × 38.4 m × 125 m. It was built with a drilling and blasting method, and rockbolting beam technology introduced from Norway. Subsequently these technologies have been used in Dongfeng, Taipingyi, Daguangba, Ertan, Dchaoshan, Xiaolangdi and other hydropower station projects. By the year 2000, more than one hundred underground powerhouses of hydropower station had been built.

After the year 2000, with the development of western hydropower construction, the amount and scale of hydropower underground cavern groups become bigger and bigger, and their geological conditions and excavating and technical support become more and more complex and difficult. For the hydropower underground cavern groups built in these years, such as the Three Gorges hydropower station, with an underground powerhouse of (width × height × length) 33.0 m × 87.0 m × 310.0 m, their unit capacities, cavern spans and construction scale are all large, and the security requirement is high.

In addition, for the surrounding rock of most underground cavern groups in the western region of China, their *in situ* stress is high, and their deformation and failure mechanism is complex. Geological and engineering disasters – such as rock burst, large deformation, collapse, water burst, surface subsidence – take place frequently. These disasters are difficult to predict, which seriously affects project construction, organization and progress. It can even cause great economic losses, sometimes bringing about adverse impacts on the environment, and other issues, such as occurred at Jinping I, Jinping II, Guandi, Goupitan, Baihetan, Houziyan and other hydropower stations. The existing technical standard and knowledge drawn form experience have been unable to satisfy the requirements of large underground engineering construction projects with complex conditions. The stability and security issues of the hydropower underground cavern groups have become more important and prominent.

Due to the high *in situ* stress, complex geological conditions and criss-cross in the space of the underground cavern groups, the caverns' stability is affected by the state of other caverns during excavation. The degree of excavation unloading of their surrounding rock is large. The deformation and failure form is complex. There are lots of loose rock blocks. The surrounding rocks are even all unstable for the not timely support, which need to stop the excavation and re-check its stability and finish the secondary system support. The deformation control of surrounding rock is difficult. Therefore, what follows first is a statistical analysis of a typical large underground engineering operation in China (Section 2). Secondly, the main engineering problems of the hydropower large underground cavern groups that have been, or are being, built in southwest China, are analyzed in detail (Section 3), and their excavation and supporting methods are also discussed (Section 4). Thirdly, the typical workings of a large underground cavern group's stability evaluation in recent years are introduced

(Section 5). Finally, the problem of large underground cavern groups in the future are put forward combined with engineering practice and the research status of such projects (Section 6).

2 AN ANALYSIS OF CHINA'S TYPICAL LARGE UNDERGROUND ENGINEERING WORKS

Since the 1990s, and especially after the year 2000, large underground cavern groups for the hydropower projects in southwest China, the pumped storage power stations in east central China, and oil storage projects become more and more important for the national economic development. These cavern groups are analyzed in this section.

2.1 An analysis of large underground engineering works used for energy reserves

The construction of underground oil storage projects began in the 1970s in China. Two water-sealed underground oil tanks, in Huangdao Shandong Province and Xiangshan Zhejiang Province, were the first to be built in a granite formation. Their storage capacities are 15×10^4 m^3 and 4×10^4 m^3, respectively, and their scale is small. The size of the Xiangshan underground tanks is 16 m × 20 m × 75 m. After 2000, attention returned to the water-sealed underground oil tanks. Four projects have been built in Shantou, Ningbo, Zhuhai and Huangdao; their capacities are 20×10^4 m^3, 50×10^4 m^3, 40×10^4 m^3 and 25×10^4 m^3, respectively. There are large oil store bases at Dalian, Huangdao, Lanshan and Aoshan. The storage capacities of the first two bases are 3×10^6 m^3, and the others are 5×10^6 m^3. The Huangdao water-sealed underground oil tank is divided into three groups, each group contains three tunnels, with widths of 20 m and heights of 30 m, as shown in Figure 1. Due to the limitation of

Figure 1 Layout of the Huangdao water-sealed underground oil tank (Web: www.wuhanins.com).

geological conditions, each tunnel is a different length. The tunnel interval of the same group is 30 m, and the tunnel interval of a different group is 58 m.

2.2 An analysis of large underground engineering pumped storage power stations

Pumped storage power station construction in China started relatively late, beginning in the 1960s. The Gangnan pumped storage power station was finished first, in 1968, then the Niyun pumped storage power station was completed in 1973. Their installed capacities are 11 MW and 22 MW, respectively. The first large pumped storage power station is the first phase of the Guangzhou pumped storage power station (1,200 MW), which was completed in 1994. The second phase (1,200 MW) began operation in 2000. This project is the biggest pumped storage power station in the world at present.

Table 1 Large pumped storage power stations built, or being built, in China.

No.	*Name*	*Main powerhouse size (length × width × height) (m)*	*Transformer chamber size (length × width × height) (m)*	*Separation distance (m)*	*Year Built*	*Lithology*	*Installed capacity (MW)*
1	Guangzhou I	146.5×21×44.5	138.1×17.2×27.4	35	1994	Granite	4×300
2	Baoquan	143×24×47.3	–	–	1994	Granite	4×300
3	Mingtan	158.4×22.4×46.5	172×13×20	–	1995	Sandstone	6×267
4	Ming tombs	145×23×46.6	136×16.6×25.5	33.6	1997	Conglomerate	4×200
5	Guangzhou II	146.5×21×47.6	–	35	2000	Granite	4×300
6	Tianghuangping I	200×21×47.7	180.9×18×27.73	33.5	2000	Tuff	6×300
7	Baishan	121.5×25×54.3	88.6×15×15	16.5	2005	Granite	2×150
8	Tongbo	182.7×25.9×60.25	162.3×18×19.5	37.3	2005	Granite	4×300
9	Pushihe	173.3×25.7×54.6	–	–	2005	Granite	4×300
10	Huhehaote	168×26.5×53.33	–	–	2006		4×300
11	Taian	180×25.9×53.7	164×17.5×18.175	35	2006	Granite	4×250
12	Langyashan	156.7×21.5×46.2			2006	Limestone	4×150
13	Yixing	155.3×23.4×52.4	134.65×17.5×20.7	40	2008	Sandstone	4×250
14	Xilongchi	149.3×22.5×49	125×16×25	45	2008	Limestone	4×300
15	Shenzhen	167×24×52.5	–	–	2008	Granite	4×300
16	Zhanghewan	145.1×23.7×52.1	–	–	2008	Andesite	4×250
17	Huizhou	152×21.5×49.4	–	–	2009	Granite	4×300
18	Bailianhe	146.4×21.9×50.88	–	–	2009	Granite	4×300
19	Heimifeng	136×27×52.7	131×20×19.5	35	2009		4×300
20	Xiangshuijian	175×26.4×55.7	167×18×20.8	35	2012		4×250
23	Xianyou	162×24×65.5	135×19.5×22	–	2013		4×300
24	Qingyuan	168.5×25.5×57.9	155×19.65×19.05	42	2015		4×320
25	Liyang	219.9×26×55.3	193.16×19.7×22	–	2016		6×250
26	Hongping	157×22×48.6	145×18×24.8	38	2016		4×300
29	Huanggou	150.5×24×47.4	114.6×20×22.7	–	2019		4×300
30	Wendeng	217.5×24.9×53	199.4×19.9×20	45	2019		6×300
31	Yangjiang	216×26×59.8	–	–	2020	Granite	4×600
33	Xianju	176×25×55	–	–	2020		4×375

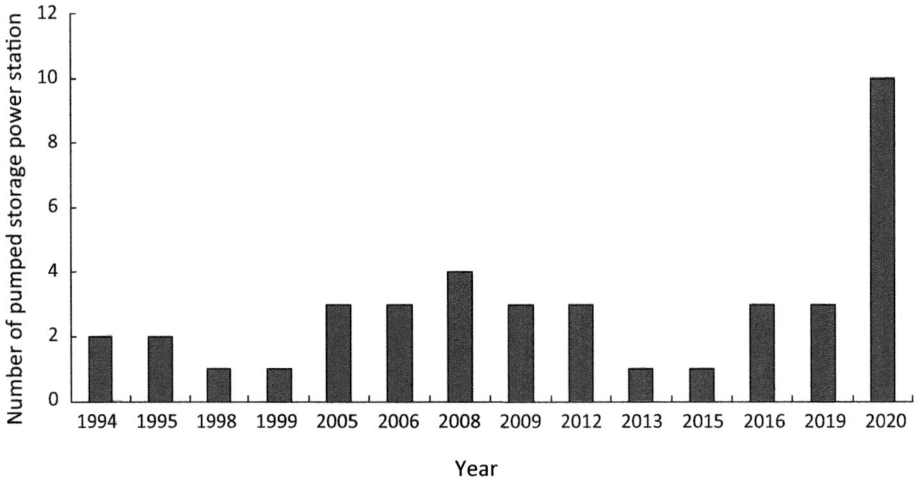

Figure 2 The current and predicted numbers of large pumped storage power stations in China.

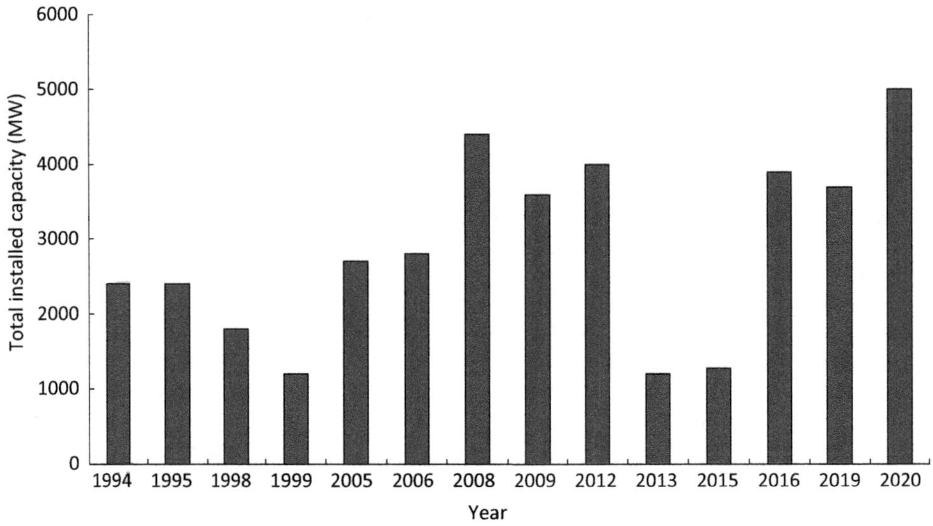

Figure 3 Large pumped storage power station installed and projected capacity in China.

The Ming tombs pumped storage power station (800 MW), built in 1997, is the first station with a high water head and capacity, which was studied earliest and with the most discussion and the longest demonstration. At present, 24 large pumped storage power stations (*i.e.* those with an underground main powerhouse span greater than 20 m) have been built in China. Nearly 20 large pumped storage power stations need to be built in the next ten years. The statistical results of large pumped storage power stations in different periods are shown in Table 1, Figure 2 and Figure 3.

2.3 An analysis of the engineering of large underground hydropower stations

An underground powerhouse as one part of hydropower engineering is an inevitable choice, because the southwest of China is an alpine and canyon region. The spans of hydropower station underground cavern groups built, or being built, in China – such as Ertan, Xiaowan, Pubugou, Guandi, Xiluodu, Jinping I, Jinping II, Houziyan, Lianghekou, Baihetan, Wudongde – are all bigger than 25 m. The largest span is nearly 34 m, and the maximum excavation height is nearly 90 m. The number of underground cavern groups with spans of more than 20 m in China is more than sixty. Among them, the number of projects with excavation spans of greater than 25 m is more than forty. More than ten projects will be finished before 2020. After that, the large hydropower underground cavern groups in China will mainly be distributed in the basins of the Jinsha River, Lancang River, the upper Nujiang River, and Yarlung Zangbo River. The statistical results of large hydropower underground cavern groups built in different years are shown in Table 2, Figure 4 and Figure 5. It is known that the number, and installed capacity, of large hydropower underground cavern groups built, or under construction, is much higher than those built in the twentieth century. In particular, the installed capacity in 2006 added nearly three times that of the first five years of the twenty-first century because many projects began being constructed before. Since then, the installed capacity has increased year by year.

Compared with the large hydropower underground cavern groups in the southwestern mountainous areas of China, the size and depth of underground oil storage cavern groups are smaller, their *in situ* stress is lower, and their geological conditions are good. The stability of surrounding rock is relatively good during excavation. The key problem is the design and construction of the underground water curtain system, so the underground cavern groups of underground oil tanks will not be discussed in this chapter. For the large underground cavern groups of pumped storage power station, although their size is large, but they are all mainly in central and eastern China. Their geological conditions are relatively good, the unloading relaxation effect is not obvious for the low *in situ* stress. Therefore, the excavation, support and stability evaluation of the large hydropower underground cavern groups in southwestern China will be mainly discussed in this paper.

The large hydropower underground cavern groups in southwestern China are seriously affected by high *in situ* stress and tectonic dynamic conditions. Due to functional requirements and limitations of terrain conditions, geological structure, hydrological conditions and so on, the large underground cavern groups with different cavern section sizes, section lengths, and complex spatial distributions must be excavated for the large hydropower projects. Their cavern scale is so huge, and bigger than existing projects. There are several types of lithology around the caverns and their distribution is complex. Meanwhile, the caverns are affected by tectonic structures, soft stratums, developed joint fissures, high *in situ* stress, strong excavation unloading action, underground water and so on. During their construction, the construction organisation requirements of underground cavern groups are high, the adoption of different construction methods and construction sequences to ensure the stability of underground cavern group surrounding rock is paramount. In conclusion, the features of large hydropower underground cavern groups in southwestern China can be shown as follows.

Table 2 Large hydropower stations built, or being built, in China (main powerhouse span bigger than 20m).

No.	*Project name*	*Main powerhouse size (length × width × height) (m)*	*Transformer chamber size (length × width × height) (m)*	*Draft tube gate chamber or pressure adjustment shaft size (length × width × height) (m)*	*Separation distance between powerhouse and transformer chamber (m)*	*Depth (m)*	*Build time*	*Lithology*	*Installed capacity (MW)*
1	Liujiaxia	86.1×24.5×59.0				63/55	1969	Quartz schist	5×225
2	Gongzui	106.0×24.5×55.0				50	1972	Granite	3×100
3	Lubuge	125.0×18.0×39.4	82.5×12.5×26.9		39.0		1988	Dolomite	4×150
4	Taipingyi	156.0×21.0×46.0			23.6	240	1994	Granite	4×65
5	Dongfeng	105.5×21.7×48.0	66.0×19.5×*		31.0	120	1995	Limestone	3×170
6	Manwan	74.0×26.6×67.8	33.5×24.0×28.0	32.0×84.0		120	1995	Rhyolite	5×250
7	Ertan	280.3×30.7×65.4	214.9×18.3×25.0	203.0×19.8×69.8	40.0	350	1998	Basalt	6×550
8	Jiangya	107.0×19.0×46.5	90.5×18.0×20.1	55.4×12.7×44.0	30.0	150	2000	Limestone	3×100
9	Dachaoshan	234×26.4×67.3	137.35×16.2×30.75	217.4×22.4×73.8	64.0	60–220	2001	Basalt	6×225
10	Xiaolangdi	251.5×26.2×57.94	174.7×14.4×17.85	175.8×10.6/6.0×20.7	32.8	110	2001	Sandstone	6×300
11	Mianhuatan	129.5×21.9×52.1	137.0×16.0×20.1	87.3×12.0×50.0	22.5	105	2002	Granite	4×150
12	Jiangkou	103.3×19.2×51.2	80.6×12.0×13.0		28.0		2003	Limestone	3×100
13	Wujiangdu	84.2×23.9×55.3			25.0	220	2003	Limestone	2×500
14	Baise	147.0×20.7×49.0	124.7×19.2×24.5		20.5	110	2005	Diabase	4×135
15	Yele	72.1×24.4×39.5	ground				2005	Diorite	2×120
16	Suofengying	135.5×24×58.4			43.0	100	2005	Limestone	3×200
17	Sanbanxi	147.2×22.7×60.1	111.0×23.0×31.8			245	2006	Sandstone	4×250
18	Xiaotiandu	77.4×20.4×40.3					2006	Granite	3×80
19	Longtan	388.5×30.7×75.4	405.5×19.8×34.2	(67+74+95)×21.6×89.7	43.0		2007	Sandstone	9×700
20	Manwan II	105.5×26.6×69.4	32.5×18.0×24.0			120	2007	Rhyolite	1×300
22	Shuibuya	168.5×23.0×68.3	ground				2007	Limestone	4×460
23	Dafa	76.5×25.4×49.5			23.6	130	2007	Limestone	2×120
24	Nuozhadu	418.0×31.0×81.7	348.0×19.0×38.6	3×38.0×94.0	45.8	184–220	2007	Granite	9×650
25	Pengshui	252.0×30.0×76.5	ground			165	2008	Shale Limestone	5×350
26	Goupitan	230.5×27.0×73.4	207.1×15.8×21.34	158.0×19.3×22.8	30.0	296	2009	Limestone	5×600
27	Laxiwa	311.8×30.0×74.8	233.6×29.0×51.5	2×32×69.3	51.0		2009	Granite	6×700

28	Pubugou	294.1×30.7×70.2	248.9×18.3×25.6	178.9×17.4×54.2	42.0		2009	Granite	6×600
29	Silin	177.8×28.4×73.5	120.0×16.3×25.4		35.0	113	2009	Limestone	4×250
30	Renzonghai	62.3×21.0×46.0					2009		2×120
31	Xiaowan	298.1×30.6×79.4	230.6×19.0×24.0	2×38.0×91.0	50.0		2010	Granite	6×700
32	Sanxia	311.3×32.6×87.3	ground			80	2011	Granite	6×700
33	Guandi	243.4×31.1×78.0	197.3×18.8×28.6	223.0×21.5×76.0	49.2	290	2012	Basalt	4×600
34	Xiangjiaba	255.0×33.0×85.5	192.0×26.0×24.4		39.0	105–225	2012	Sandstone	4×800
35	Gongguoqiao	175.0×27.4×74.5	134.8×16.5×39.0	130.0×25.0×69.7		80–275	2012	Sandstone	4×225
36	Jinping II	352.4×28.3×72.2	374.6×19.8×34.1	222.6×12.0×67.5	45.0		2012	Marble	8×600
37	Jinping I	277.0×28.9×68.8	201.6×19.3×32.7	41×80.5 / 37.0×79.5	44.9		2013	Marble	6×600
38	Ludila	267.0×29.8×77.2	203.4×19.8×29.5	184.0×24.0×75.0	45.0	140–356	2013	Sandstone	6×360
39	Yantan II	127.2×30.8×76.7	53.8×16.8×32.0		32.0		2013	Diabase	2×300
40	Xiluodu	443.3×31.9×75.6	349.3×19.8×33.2	317.0×25.0×95.0	49.5	340–480	2014	Basalt	2×9×770
41	Dagangshan	226.6×30.8×74.6	144.0×19.3×25.8	132.0×24.3×77.1	47.0	390–520	2015	Granite	4×650
42	Huangjinping	206.3×28.8×67.3	150.8×17.8×32.4		45.0	220–350	2015	Granite	4×200
43	Yingbaoliang	251.4×25.9×62.9	224.1×17.8×26.6	155.7×15×56.75			2016		6×200
44	Houziyan	219.5×29.2×68.7	139.0×18.8×25.2	140.5×23.5×75	46.7	403–655	2017	Limestone	4×425
45	Changheba	228.8×30.8×73.4	150.0×19.3×25.8	162×22.5×79.5	45.0	250–480	2017	Granite	4×650
47	Baihetan	438.0×34.0×88.7	368.0×21.0×39.5	3×50.0×100.0	60.7	260–330	2020	Basalt	2×8×1000
48	Wudongde	333.0×32.5×89.9	255.6×18.5×34.2	3×25.0×113.5		250–550		MarbleLimestone	2×6×850
49	Linxihe	93.1×20.7×47.1	31.1×10.8×23.7		28.7		2020	Limestone	2×170
51	Anning	175.8×24.9×61.8				144–281	2020	Granite	4×95
52	Lianghekou	273.0×28.7×66.7	222.0×18.8×25.6	183.0×19.5×78.4	45.0	420	2020	Sandstone	6×500
53	Shuangjiangkou	214.7×28.3×67.3	142.2×20.0×25.2	143.6×22.2×76.0	45.0	455	2020	Granite	4×500
54	Yangfanggou	228.5×27.0×75.6	210.0×18.0×22.3		45.0		2020	Diorite	4×375
55	Yagen II	220.8×30.6×67.3	188.8×18.3×25.6	194.0×24.0×76.0			2020		4×250
56	Lenggu	278.0×29.6×65.76	246.0×19.8×25.6	145.0×17.0×80.0					6×420
57	Mengdigou	227.0×30.1×70.53	182.0×19.3×25.8	150.0×20.0×80.0					4×540
58	Gangtuo	181.5×25.6×64.0	130.0×18.0×31.5						4×335
59	Yebatan	253.4×26.8×66.3	161.8×19.2×31.5	175.0×25.0×85.1					4×525

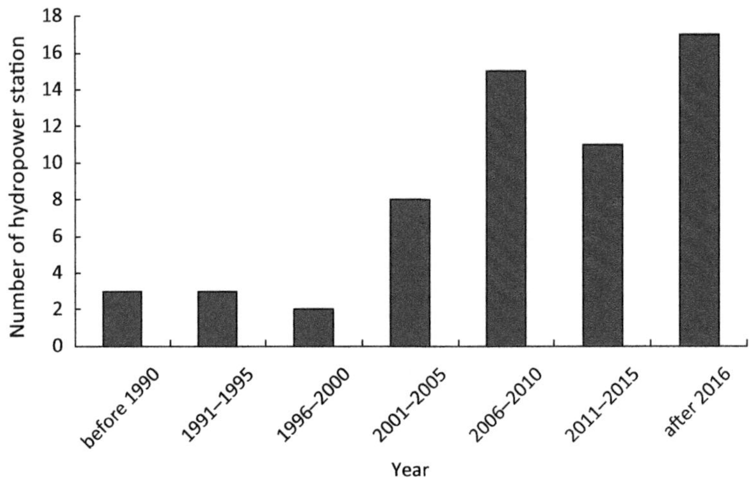

Figure 4 The current and predicted numbers of large hydropower stations in China.

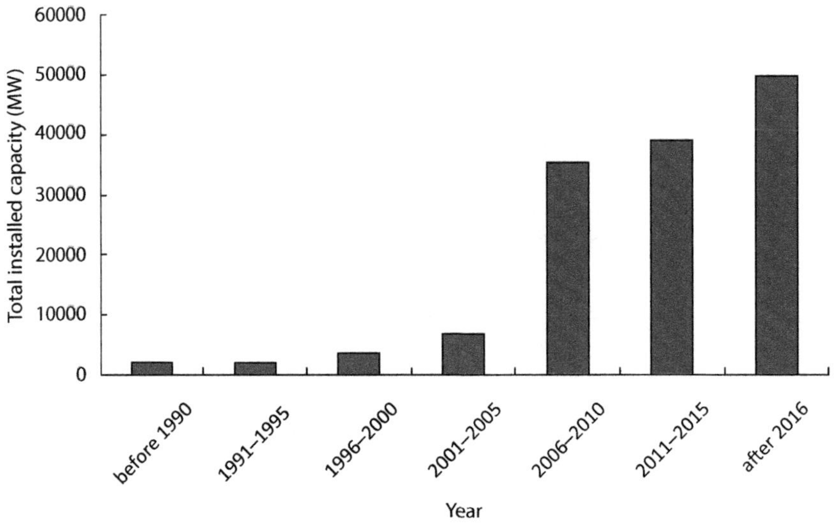

Figure 5 Large hydropower station installed and projected capacity in China.

2.3.1 Geological environmental conditions

(1) *Geological conditions are complex.* The mountain and ravine region of southwest China is affected by inner and outer geological dynamic actions, which also intensely affect the evolutionary process of a river valley. The geological actions make the faultage; bedding fault zones; and joints and fissures development much more uncertain, which make the construction risk very high. For example, the

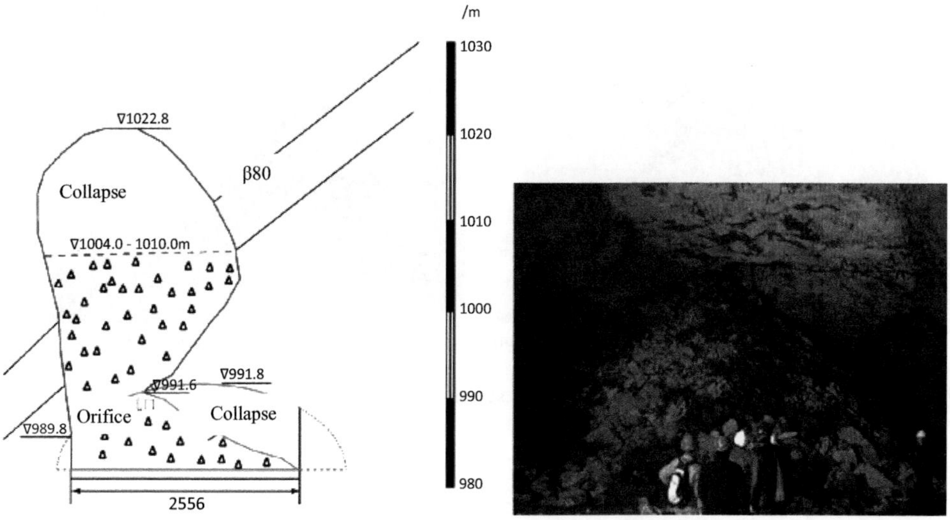

a) Transverse section b) Scene photograph

Figure 6 Geological map of the Collapse of Dagangshan main powerhouse (Zhang, 2010; Zhu *et al.*, 2013).

underground powerhouse stability of the Dagangshan hydropower station is affected by the diabase dikes $\beta 6$, $\beta 80$, $\beta 81$ and faults. Because their occurrence can be known exactly before excavation, some rock blocks near dike $\beta 80$ collapsed when the rock mass at an upstream area was blasted in December 2008, as shown in Figure 6. The volume was about 3,000 m^3, and it cost a year and a half year to dispose of the collapse.

(2) In situ *stress is high*. The *in situ* stress near the underground cavern groups in southwest China is high due to the great depth and strong geological formation movement. The rock strength-stress ratio is small and the rock mass excavation unloading degree is serious. For example, rockburst, spalling, large relaxation depth and other surrounding rock failure types happened during the construction of large underground cavern groups, such as Jinping I, Jinping II, Laxiwa, Guandi, Houziyan and other hydropower stations. These failure phenomena exceed the traditional understanding and engineering experience of hard brittle rock mass deformation and failure.

(3) *Hydrogeological conditions are complex*. There are lots of fracture aisles for groundwater seepage to occur around the underground cavern groups in southwest China, where the groundwater is rich. The groundwater seepage can affect the stability of surrounding rock. When there is karst water or karst caverns, the effect on the construction and stability of cavern groups may become much more serious. The geological conditions for engineering works within karst are difficult to explore exactly before excavation takes place, and there are some uncertainties. So the project may need increased budgets and extended construction time. For example, according to the exploration information from the Goupitan hydropower station, underground water would gush out from the #8 karst system and W24 karst system

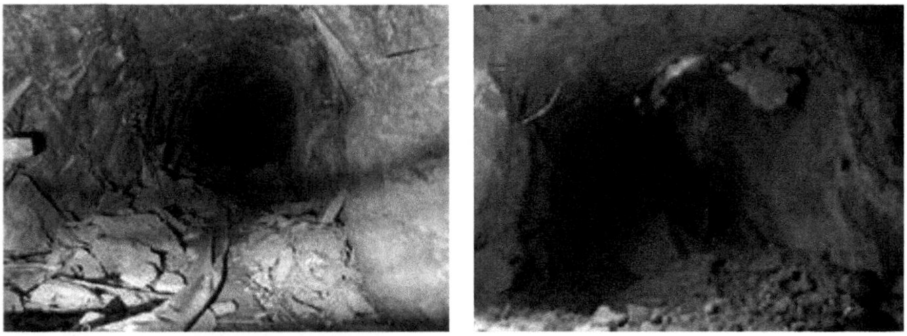

a) Spalling at drainage gallery of Houziyan (Dong *et al.*, 2014). b) Strong rockburst at one bus tunnel.

Figure 7 Rock mass failure due to the high *in situ* stress.

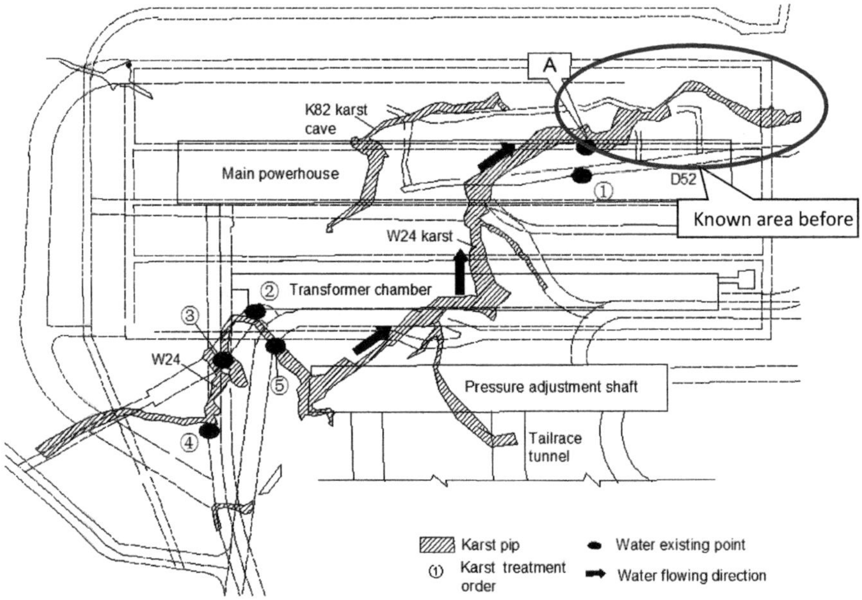

Figure 8 Water gushing out points and underground water flow direction of Pigoutan (Zi *et al.*, 2009).

during the underground powerhouse excavation. The possible water quality flowing out may be 432–7,776 m^3/d. For the transformer chamber, only a small amount of water seepage may occur in the local area. While the actual maximum water quantity was 7,000 m^3/d, and the maximum gushing water quantity was about 6,000 m^3/h, continuing for 70 minutes. The maximum gushing mud quantity was 1,000 m^3 one time, and its maximum length was up to 40 m. As the W24 karst system is connected to the surface, the gushing of water or mud occurred several dozen times. The points where water gushed out, and the underground water flow direction can be seen in Figure 8.

2.3.2 Underground cavern group structural systems

(1) *The scale of an underground cavern group is large.* In order to make good use of the abundant hydropower resources in southwest China, the unit capacity is large. So the underground cavern group scale is huge, and the problem of surrounding rock stability is very serious because of its size effect. When the width and height of the main powerhouse is big, the deformation and excavation unloading depth of the sidewalls may be greater than that of a small one. For a large excavation face, the number of unstable blocks may increase under the same geological conditions. For example, the unit capacity of Baihetan hydropower station is 1,000 MW, so its main powerhouse size (length × width × height) is 438 m × 34 m × 88.7 m, the transformer chamber size (length × width × height) is 368 m × 21 m × 39.5 m, and the size of the three pressure adjustment shafts (diameter × height) is 50 m × 100 m each. The separation distance between main powerhouse and the transformer chamber is 60.6 m. The total investment is one hundred billion yuan. The excavation scale of this underground cavern group is the biggest in China at present, and its surrounding rock stability caused by the scale of projects is the key problem, especially the effect of the bedding fault zones and columnar basalt. The underground cavern groups and main bedding fault zones are shown in Figure 9.

(2) *A lack of engineering practical experience of similar engineering projects.* The scale of large underground cavern groups are larger than existing built projects, and it is difficult to design them using the present standard, especially where the *in situ* stress is high. The excavation and supporting design methods proposed by the standards are mainly for caverns with a span smaller than 25 m. There are a few similar projects with high *in situ* stress (more than 30 MPa), and large spans (more than 25 m or 30 m) and heights (more than 70 m), which can be referenced for the stability evaluation of large underground cavern groups, such as Jinping I, Jinping II, Guandi and Houziyan hydropower stations. There is no mature engineering experience and well-defined standards for such projects.

For example, the measured maximum *in situ* stress around the underground powerhouse area of Houziyan hydropower station is 36.43 MPa, and the second principal stress is 29.82 MPa. The rock saturated uniaxial compressive strength is

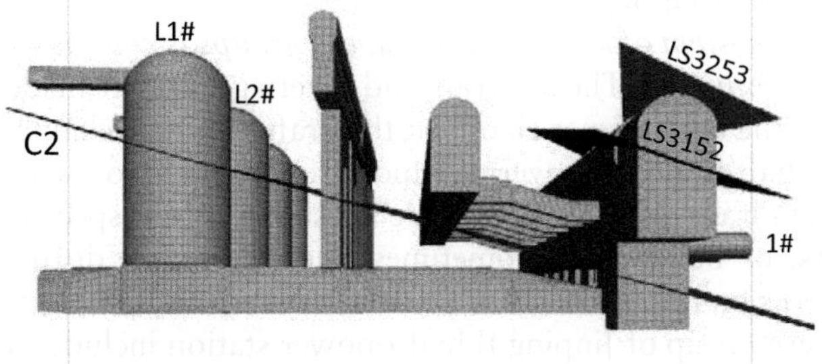

Figure 9 Schematic of the underground cavern groups and the main bedding fault zones of Baihetan hydropower station (Xu *et al.*, 2012).

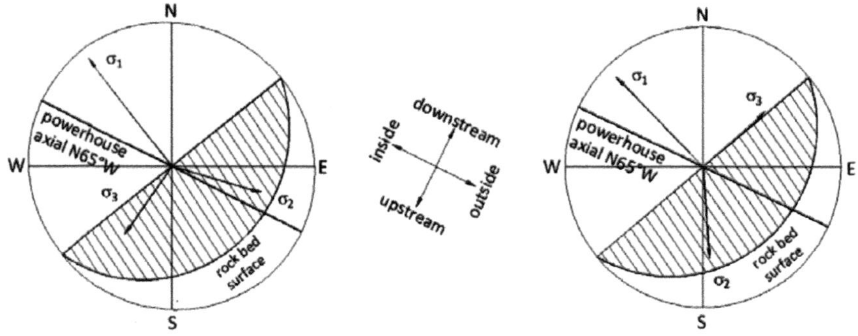

a) First group *in situ* stress. b) Second group *in situ* stress.

Figure 10 Distribution of *in situ* stress direction of Jinping I hydropower station (Huang, 2011).

between 60 MPa and 100 MPa, which belongs to medium-hard brittle rock. The rock strength stress ratio is low, about 1.65–2.74, which is relatively rare among the large underground hydropower stations built, or under construction, in China. The measured maximum and second principal stresses of Jinping I hydropower station underground powerhouse are 35.7 MPa and 20 MPa, respectively, and its rock strength stress ratio is 1.5–3.0, which is approximately equal to that of Houziyan hydropower station. But the second principal *in situ* stress of Jinping I hydropower station is lower than that of Houziyan hydropower station, and this difference may cause more problems. When designing the underground cavern group layout of Jinping I hydropower station, the adverse effects of high *in situ* stress to a high sidewall can be reduced by adjusting the intersection angle between the powerhouse axis direction and the first principal stress direction, shown in Figure 10. While for the Houziyan hydropower station, the second principal stress is also big, and its effects to a high sidewall cannot be reduced by adjusting the powerhouse axis direction, shown in Figure 11. So the stability of this cavern group is outstanding under the combined function of construction disturbance, high *in situ* stress (especially the high second principal stress), faults, compresso-crushed zone and other complex geological structures. The relaxation depth, deformation and the stress of supporting structures are all bigger than in other projects. These features bring a great challenge to the excavation and support design of the Houziyan underground cavern group.

(3) *The structural system of underground cavern groups is complex and the interaction effect is prominent.* The underground cavern groups are made up of the main powerhouse, the transformer chamber, the draft tube gate chamber, the pressure adjustment shaft and other caverns, which are connected to each other by the bus tunnels and tailrace tunnels. Their scale is usually large, especially those built in recent years, or being built. Sometimes, the excavation disturbance effect of nearby caverns on the surrounding rock is very serious. For example, the underground cavern group of Jinping II hydropower station includes more than forty caverns, shown in Figure 12. Meanwhile, there are about twenty faults among the underground cavern group of Jinping II hydropower station, shown in Figure 13.

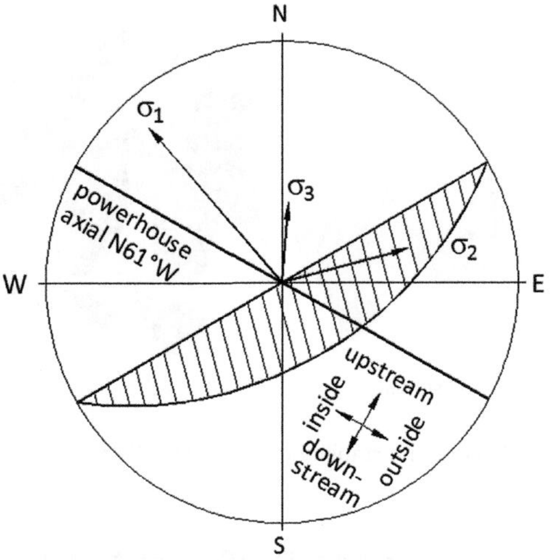

Figure 11 Relationship between the principal stress orientation, the cavern axis and the rock trend of Houziyan (Li *et al.*, 2014).

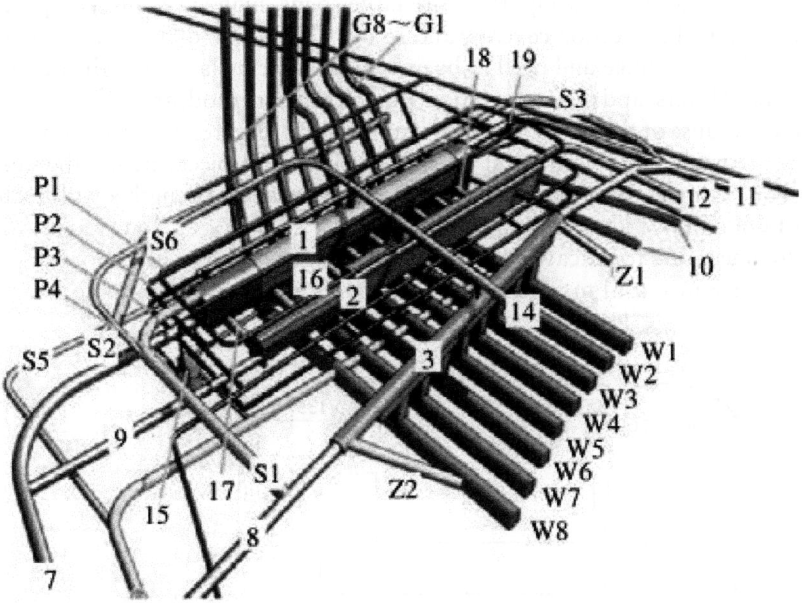

Figure 12 Schematic of the distribution of the Jinping II main caverns (Chen, *et al.*, 2008).

The complex space structure of caverns and faults bring prominent stability problems to the intersection parts between the main powerhouse and bus tunnels or transformer chamber and bus tunnels or main powerhouse and headrace tunnels, the rock middle separated wall between main powerhouse and transformer chamber or transformer chamber and draft tube gate chamber.

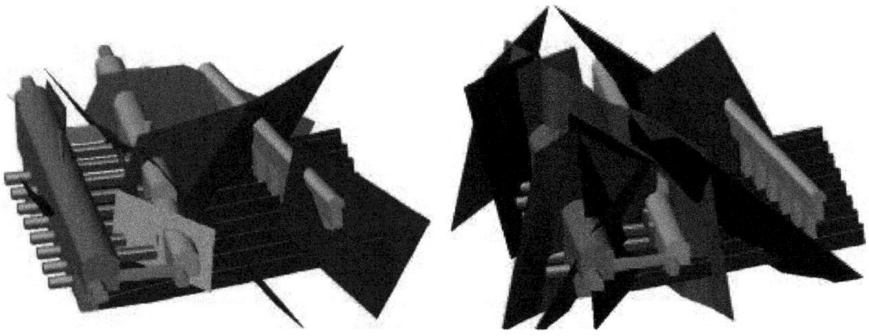

Figure 13 Schematic of faults affecting the stability of the Jinping II cavern group.

2.3.3 Construction organisation and its influence factors

(1) *The excavation construction of underground cavern groups is different from excavating tunnels.* Their excavation and arrangement are very difficult due to the large volume to be excavated and so many caverns. For a large hydropower underground cavern group, there are large spans, high sidewalls, complex structures, many intersection caverns, and so on. The section of an underground main powerhouse is huge and need to be excavated layer by layer. But the number and height of layers, and their excavation sequence and method, are different from each other because of the section size, cavern structural features, in situ conditions, rock mass structure and mechanical features of different projects are not the same.

For example, the underground main powerhouse size (length × width × height) of Jinping II hydropower station is 352.4 m × 28.3 m × 72.2 m, and it is divided into nine layers for excavation, as shown in Figure 14. The height of each layer is between 6 m and 11 m. The first to sixth layers are excavated from top to bottom,

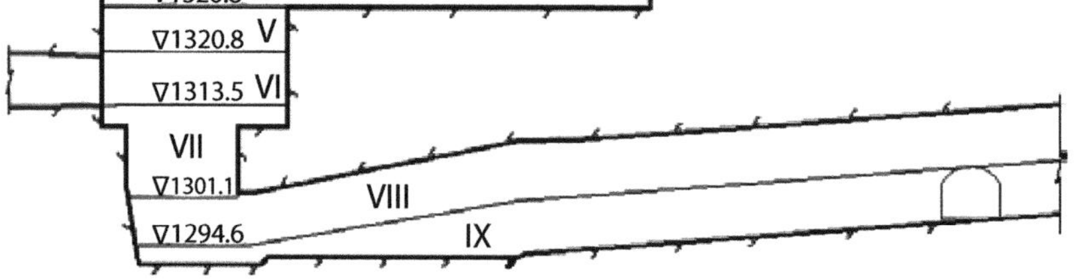

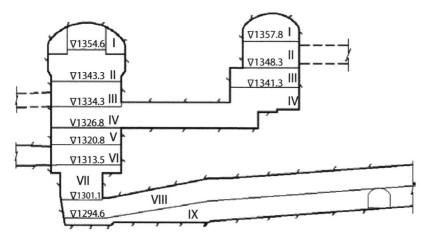

Figure 14 The excavation scheme of Jinping II underground cavern group.

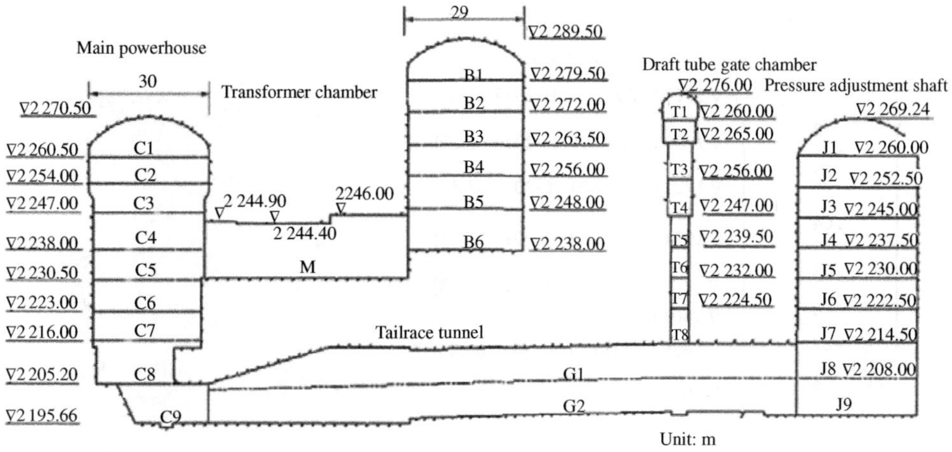

Figure 15 The excavation layers of the Laxiwa underground cavern group (Yao *et al.*, 2011).

while the seventh and eighth layers are excavated by the pilot shaft method through the tailrace tunnels excavated beforehand. For the first layer, the middle drift is first excavated, and then the two sides are excavated. For the second to sixth layer, the middle part is first excavated, then the two sides without protective layers are excavated, then the protective layers are finally excavated. The underground main powerhouse size (length × width × height) of Laxiwa hydropower station is 311.75 m × 30.00 m × 73.84 m, and it is divided into nine layers for the excavation, as shown in Figure 15. The height of each layer is between 6 m and 9 m. The underground main powerhouse size (length × width × height) of Houziyan hydropower station is 219.5 m × 29.2 m × 68.7 m, and it is also divided into nine layers for the excavation, as shown in Figure 16. The height of each layer is between 5 m and 13 m.

For the underground powerhouse of Sanbanxi, Shuibuya, Suofengying, Longtan, Xiaowan and Pubugou underground powerhouses, the average section size (length × width × height) of the latter three projects – 327.0 m × 30.7 m × 76.5 m – is bigger than that of the first three projects, which is 144.2 m × 23.2 m × 62.1 m. The mucking transport of the first three projects mainly depends on the permanent access tunnels during excavation by layer. Only one construction adit is designed in the middle part of Shuibuya underground powerhouse, while for the latter three projects, several construction adits are designed at different elevations in order to allow excavation at both ends and to accelerate the construction speed. This construction method was subsequently widely used in other large hydropower underground powerhouses. The key problem of construction organisation design is how to arrange several excavation faces, such as both ends of powerhouse; up and down layers; intersection caverns and other caverns; several construction passageways; lots of mechanics and operators; excavating; blasting; and other work.

(2) *The excavation and supporting work amount is huge and the construction cycle is long*. Subject to the constructional difficulties and space limitation, a long time

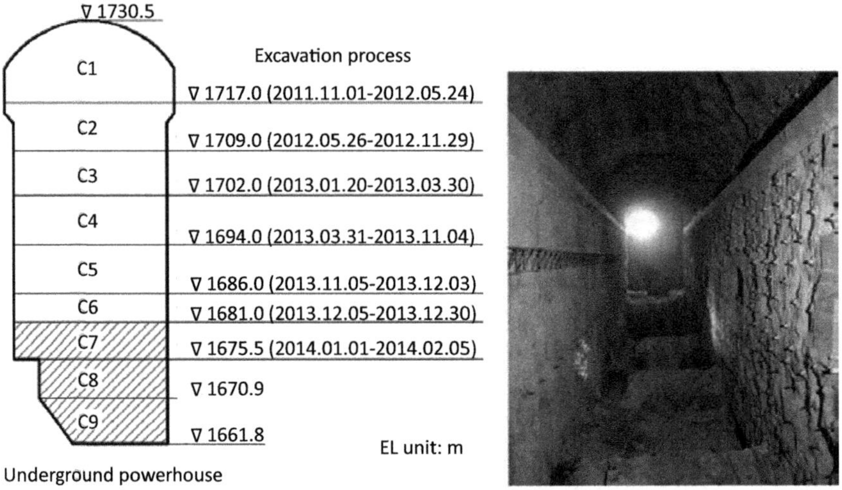

Figure 16 The excavation progress of the Houziyan underground powerhouse (Dai *et al.*, 2015 and www.zhongtianheng.com.cn).

is needed to complete the excavation and supporting work. Sometimes they can be affected by each other. For example, the underground main powerhouse size (length × width × height) of Jinping I hydropower station is 276.99 m × 28.9 m × 68.8 m. The surrounding rock is supported by steel fiber-reinforced shotcrete (Grade C30, thickness 150–250 mm), systematic bolts (diameter 32 mm; length 6 m, 9 m and 12 m; space 600–750 mm) and prestressed anchor cables (anchoring force 1,000–2,500 kN; length 20m, 25 m, 30 m and 45 m; space 3–4 mm). The geological logging, section testing and concrete spraying should be finished within four days after excavation. The short bolts, long bolts and anchor cables should be separately completed within seven days, 15 days and 30 days, respectively. The first layer was excavated from May 2006 until December 2009, when the underground powerhouse excavation was finished. The total time was nearly 44 months. For Longtan hydropower station, the underground cavern group, including 119 caverns and 30 km long, needed to be excavated within an area of 0.5 km^2. Its excavated volume for the main powerhouse is 6.6×10^5 m^3, the sprayed concrete volume is 1.0×10^4 m^3 and the number of bolts is 3.06×10^4. It was excavated from January 2001 and finished in February 2005.

(3) *The relaxation degree of the surrounding rock is affected by the excavation sequence and method.* The effects of the cavern excavation sequence and method on the surrounding rock are very large, especially for the connected caverns. The rocks surrounding the main powerhouse, the transformer chamber and bus tunnels, or between the main powerhouse and tailrace tunnels, or between the transformer chamber and the draft tube gate chamber, are unloaded in two or three directions, and their effective load-bearing areas become small. Meanwhile, the effects of small caverns on the sidewalls of large caverns are great. These actions make the damaged depth of surrounding rock bigger. In order to reduce this effect, small caverns excavated and supported first in the construction

sequence, and then the sidewall is excavated, which is useful in controlling the stability of surrounding rock and making the sidewall stable. For the nearby caverns, this method is also effective. For example, the transformer chamber and pressure adjustment shaft of the Jinping I hydropower station were excavated at the same time, which could reduce the effects of the pressure adjustment shaft excavation to another. Furthermore, the layer height and layer number are two important factors for maintaining the stability of the surrounding rock during excavation layer by layer. The excavation layer height of Jinping I underground powerhouse, with high *in situ* stress, is small and the excavation area is divided into many zones so that the degree of stress release can be reduced and controlled.

2.3.4 Excavation unloading and timeliness of surrounding rock

(1) *The stress evolutionary process of surrounding rock is complex with high* in situ *stress and low rock strength stress ratio*. The excavation unloading of surrounding rock will profoundly disturb the rock stress field. If the surrounding rock cannot bear the action of springback stress or redistribution stress, the plasticity deformation or damage – such as fractures opening and relaxed; the acoustic wave velocity of the rock mass decreasing; its permeability coefficient increasing; and so on – will happen among the surrounding rock near the free face of excavation. These phenomena usually happen from the surrounding caverns, especially from locations with focused tensile stress and compressive stress, then extend to the inside of the surrounding rock. The result is usually to form the rock loosen zone within a certain depth range of the surrounding caverns, and the stress state of surrounding rock will be adjusted due to the stress relaxation. So its development process is complex for situations of high *in situ* stress and for different rock mass structures.

For example, the measured maximum *in situ* stress of the Jinping I hydropower underground powerhouse is 35.7 MPa, and its rock strength stress ratio is 1.5–3.0. Its failure mechanics model is shown in Figure 17, which can be used to

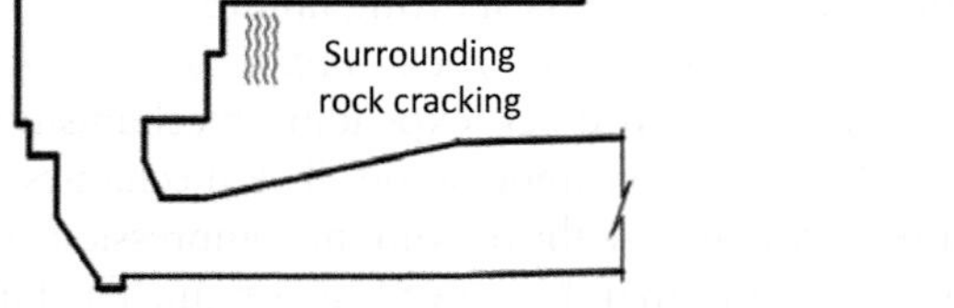

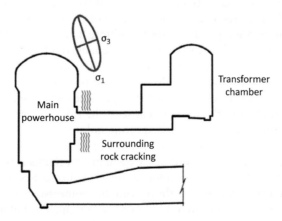

Figure 17 Mechanical model of deformation and cracking mechanism of rock surrounding the Jinping I underground powerhouse.

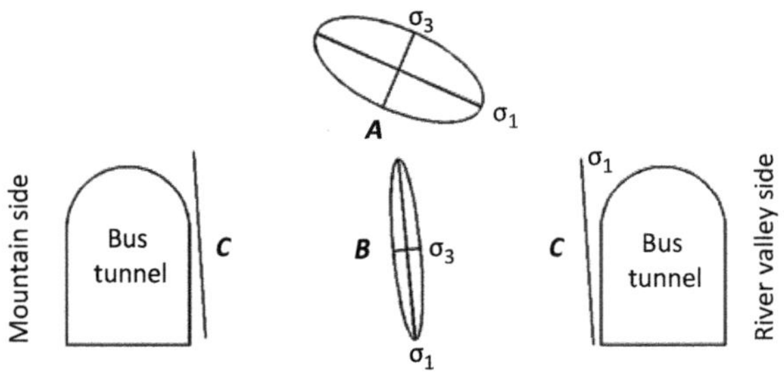

Figure 18 Mechanical analytical model of the deformation mechanism of the rock surrounding the Jinping I bus tunnels (Lu *et al.*, 2010).

reasonably explain the deformation and cracking of surrounding rock and shotcrete near the powerhouse downstream haunch. The damage to surrounding rock near the downstream roof is compressing cracking caused by the large, focused tangential stress. Meanwhile, the adverse bedding fissures open, with their deformation caused by the excavation unloading. For the downstream sidewall the stress orientation changed, affected by the excavation. Under the low rock strength stress ratio, the rock pillars between powerhouse and transformer chamber unload in three directions and the deformation and cracking occurred near the free face of the downstream sidewall. Then the features of the surrounding rock mechanics further degraded and continued adding to the surface rock mass deformation of the sidewall. This is one important reason for the large deformation of the main powerhouse downstream sidewall.

After excavation, the rock pillar between the bus tunnels was free in three directions, and it deformed into the two side bus tunnels. The two sided excavation unloading changed the undisturbed stress state of rock pillar, as shown in Figure 18. The direction of stress ellipse A changed into stress ellipses B and C. The difference between them is that the minimum principal stress of stress ellipse C, near the excavation free face of the bus tunnel, changed to zero and the minimum principal stress of stress ellipse B, near the middle of rock pillar, was small. Their two-directional stress state changed into one direction, or close to one direction. This situation happened at the downstream sidewall of the powerhouse and the upstream sidewall of the transformer chamber. The high axial stress, low uniaxial compression strength, and excavation blasting made the rock mass degrade and the primary cause of the sidewall surrounding rock and shotcrete cracking was a single axis splitting failure.

(2) *The deformation mechanism of rock mass is not known completely.* The deformation, fracture and crack expending mechanism of rock mass with high *in situ* stress has not been understood. The mechanics models, which can reflect the fracture feature before the maximum compression strength of rock, and the non-linear features after that, have not been set up. The failure phenomena of a rock mass with high *in situ* stress include spalling, rockburst, and cracking, and its mechanical actions are complex. For example, the measured maximum *in situ*

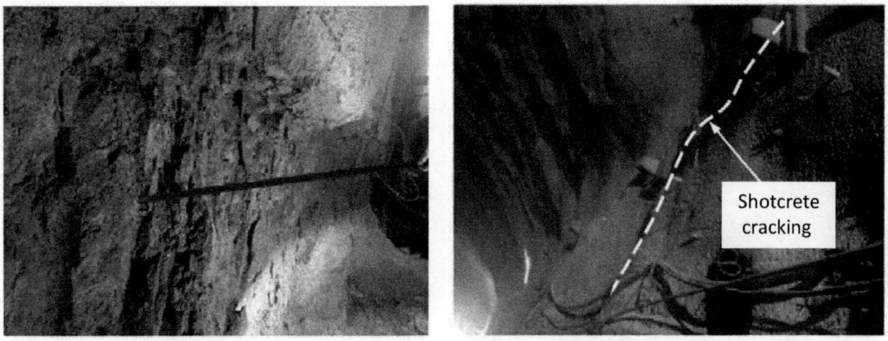

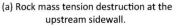

(a) Rock mass tension destruction at the upstream sidewall.

(b) Shotcrete cracking at the upstream sidewall.

Figure 19 Failure forms at the Houziyan underground powerhouse (Dai *et al.*, 2015).

stress of the Houziyan hydropower underground powerhouse is 36.43 MPa, and the rock strength stress radio is 1.65–2.74, which is similar to the Jinping I hydropower underground powerhouse. It was excavated from October 2012, and rockbursts and large deformation happened several times. The deformation of the upstream sidewall reached 10 cm by July 2013 and stopped construction for nearly two months. Although a similar situation happened in Jinping I, it could not be avoided because the complex deformation mechanism of the rock mass was not known and the exact analysis method was not set up.

(3) *The timeliness of the surrounding rock after excavation is obvious.* The deformation of large underground powerhouse with high *in situ* stress and complex geological condition is usually great, and its timeliness is obvious. Meanwhile, the anchor cables usually exceed their limited bearing capacity. So the deformation and overloaded mechanism of the surrounding rock of a high sidewall is complex, and its stability is difficult to control. There is a lack of similar

(a) Steel strand catapulting outwards

(b) Steel strand shrinking into the anchor hole

Figure 20 Failure forms of Jinping II prestressed anchor cables (Jiang *et al.*, 2013).

(a) Splitting and bending damage of rock mass.

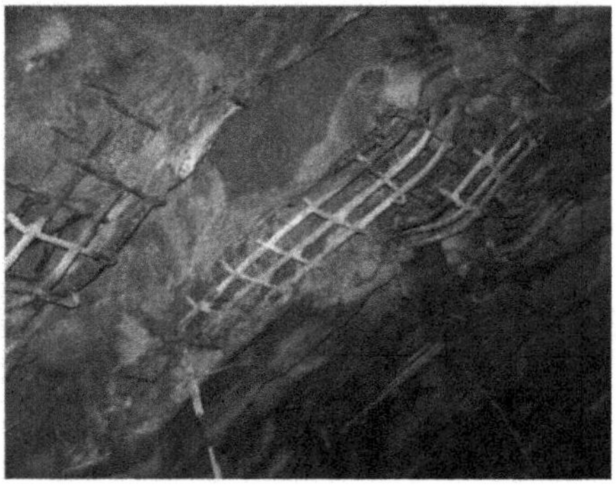

(b) Reinforced concrete arch rib bending.

Figure 21 Failure forms of Jinping I underground powerhouse downstream sidewall (Lu *et al.*, 2010).

engineering experience to draw on. Evaluation of the long-term stability of an underground powerhouse is very difficult. For example, during the construction of the Ertan, Jinping I, Jinping II and Houziyan hydropower stations, some bolts and anchor cables of large underground powerhouses overloaded, and their surrounding rock deformations were large and became more obvious over time: they were 1,24.17 mm at the downstream sidewall by December 1996, 92.9 mm at the downstream sidewall by 31 December 2009, 95.03 mm at the upstream sidewall by 30 August 2010 and 156.60 mm at the downstream sidewall by 5 March 2011. The features still cannot be explained completely by traditional engineering experience and the deformation failure rules of brittle rock. For the Jinping I underground

Figure 22 Jinping I underground powerhouse after reinforced support (Web: blog.sina.com.cn).

powerhouse, the anchor cable forces of 21.8% of the tested cables on the upstream sidewall were larger than their design values by 30 August 2011, where the number of tested cables was 55, while the anchor cable forces of 22.5% of the tested cables on the downstream sidewall were overrunning. The maximum overrunning range was 40%. For the Jinping II underground powerhouse, 38 anchor cables in 5 rows between the powerhouse and the transformer chamber failed: 16 in the first row, 18 in the second row, and 4 in the third row. For Jinping I and Houziyan in particular, the large deformation and serious cracking of the support system and surrounding rock after the support was in place caused the construction to stop and the need to reinforce the support by setting up scaffolding again, which affected the construction schedule and project investment.

2.3.5 Investment and stability controlling standard

(1) *The installed capacity, underground cavern group scale and engineering investment are all great.* The installed capacity, underground cavern group scale and engineering investment of large hydropower stations built, or being built, in southwest China are usually large. The conditions of four projects are shown in Table 3. For example, the installed capacity of Jinping I hydropower station is 6 × 600 MW, its main powerhouse size (length × width × height) is 276.99 m × 28.9 m × 68.8 m and its transformer chamber size (length × width × height) is 201.6 m × 19.3 m × 32.7 m. The space between the two caverns is 45 m. Its investment is 2.323 billion yuan.

(2) *The stability of a large underground cavern group is hard to control.* There are uncertainties about engineering geological conditions, rock unloading mechanical behavior, factors affected by construction, supporting mechanism and so on, so it is difficult to control the stability of a large underground cavern group. Before construction, it is impossible to decide the optimal excavation and supporting schemes for the underground cavern group based on the limited survey information, which is obtained through several exploration tunnels and drilling holes. The *in situ* stress and the attitude of faults and joints obtained before construction cannot be known exactly, nor do they reflect the whole condition. Then the design of the excavation and supporting schemes, based on this information, may be unsuitable for some regions of the cavern group. Taking the underground cavern group of Guandi hydropower station as an example, the

Table 3 Project conditions of large hydropower stations in southwest China.

Project name	*Main powerhouse size (length × width × height) (m)*	*Transformer chamber size (length × width × height) (m)*	*Space between the two caverns (m)*	*Installed capacity (MW)*	*Investment (billion RMB)*
Jinping I	276.99×28.9×68.8	201.6×19.3×32.7	45	6 × 600	2.323
Jinping II	352.4×28.3×72.2	374.6×19.8×34.1	45	8×600	2.498
Xiluodu	443.34×31.9×75.6	349.29×19.8×33.2	49.5	2×6×770	5.03
Baihetan	438×34×88.7	368×21×39.5	60.65	2×8×1000	10
Wudongde	333×32.5×89.9	255.6×18.5×34.2		2×6×850	5.25

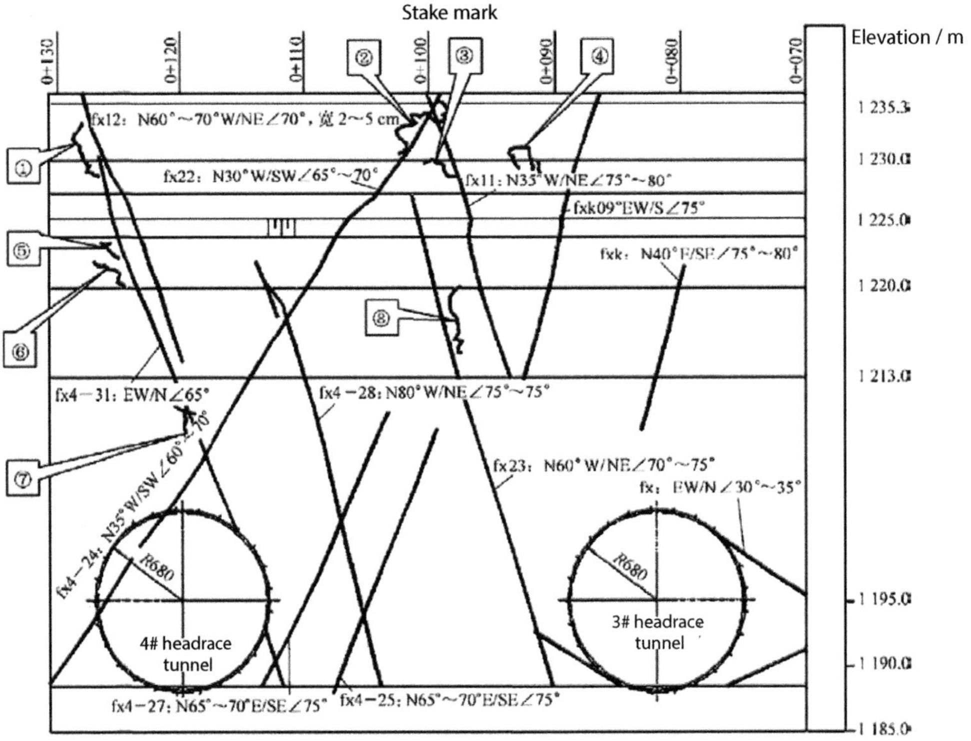

(fx4 standing for the trace of bedding fault zone 4)

Figure 23 Cracks in the shotcrete of the main powerhouse upstream sidewall between 3# and 4# turbine of the Guandi hydropower station (Lu *et al.*, 2012).

deformation of the surrounding rock was large due to the excavation blasting and rock mass structure. The surrounding rock, shotcrete and rock bolt beams cracked because of the non-uniform deformation. By October 2009, several cracks on the shotcrete of the main powerhouse's upstream sidewall between the 3# and 4# turbines were found, shown in Figure 23 and expressed with ①–⑧. There were four separate cracks on the top or bottom of the rock anchor beam, and the maximum length was 10.21 m. Most cracks distributed along the traces of bedding fault zones and were affected by them. By March 20103, 14 horizontal cracks and 16 vertical cracks were found at the upstream rock anchor beam between 0+076–088 m, their average lengths were 1.43 m and 2.07 m, respectively, and their average widths were 0.61 mm and 0.66 m, respectively. These cracks were tension fissures caused by the deformation of surrounding rock.

3 AN ANALYSIS OF THE ENGINEERING PROBLEMS ASSOCIATED WITH LARGE UNDERGROUND CAVERN GROUPS

During the construction of large underground cavern groups with high *in situ* stress in southwest China, rockburst; spalling and splitting; unloading fracturing; bucking;

structural opening; and other damage phenomena usually happen, which make their stability and safety very contentious. The deformation feature and its effect on controlling the stability of large underground cavern groups with high *in situ* stress have become the research focus of the rock mechanics and engineering fields. They are related to geological conditions and engineering problems. The effect of geological conditions has been discussed in the book *Rock Engineering Risk* by John A. Hudson and Xia-Ting Feng (2015). In this section the engineering problems, such as underground cavern group layout, excavation schemes, the stability and treatment measures of umbrella arches, rock anchor beams, high sidewalls, and so on, will be mainly discussed.

3.1 Layout of large underground cavern group

3.1.1 Choosing a suitable longitudinal axis orientation for the primary cavern

The location and longitudinal axis orientation of large underground cavern groups should be chosen according to their layout conditions, hydraulic conditions, geological structure, direction of *in situ* stress and other factors. The statistical results of different projects are shown in Table 4, and the general principles in choosing their longitudinal axes are as follows.

(1) The intersection angle between the longitudinal axis orientation of the underground powerhouse and the strike direction of the primary structural surfaces – such as rock strata, faults, joints and fractures – should be as big as possible. This angle should usually be bigger than 30° for massive rock, and 45° for layered rock. Meanwhile, the effect of minor structural surfaces on cavern group stability should also be considered.

(2) The intersection angle between the longitudinal axis orientation of the underground powerhouse and the orientation of the maximum principal *in situ* stress should be as small as possible when the *in situ* stress is high; high being usually bigger than 20 MPa. This angle should be less than 15°–30°, depending on other conditions. Its purpose is to reduce the unloading action to surrounding rock

(a) Horizontal cracks.

(b) Vertical cracks.

Figure 24 Cracks on the upstream rock anchor beam of the Guandi hydropower station (Lu *et al.*, 2012).

Table 4 The engineering and geological conditions of large hydropower stations in China.

Project name	*Axis orientation (A)*	*Strike direction of σ_I (B)*	*σ_I (MPa)*	*Angle between (A) and (B) (°)*	*Rock stratum attitude (C)*	*Angle between (A) and (C) (°)*	*σ_c (MPa)*	*Strength stress ratio*	*Rock mass structure features*
Lubuge	N61°W	N54°W	13.5–18	7	N20°W/ NE∠10°–20°	41	>80	4.4–5.9	Fault attitude: N50°E; N18°–72°W/ NW∠52°–76°.
Ertan	N6°W	N10°–30°E	17.2–38.4	16–36		34–56	100–180	2.6–5.8	Block structure, location with fractured rock mass, structure surfaces are mainly joints, their attitude: N30°–50°E/NW∠50°–70°, N40°–60° W/NE∠60°–70°.
Dachao-shan	N75°W		13–16.9		N65°–85°E/ NW∠19°–30°	20–40	62–102		Fault attitude: N72°–75°W/NE ∠70°–80°; joint attitude: N5°E–N15°W/SE, SW∠70°–90°; N60°–75°E/SE70°–85°; N75°–90°E/SE ∠70°–85°.
Xiaolangdi	N10°E	N40°E	5	30	N8°E/SE ∠9°–10°	2			F1: NW275, dip angle 85°, width 10–15cm, filled with fractured blocks and flours; F2: SW260, dip angle 80°, width 20–40 cm; joints: N20°E/NW∠84°; N60°E/ NW∠78°; N290°W/ NE∠75°; NW340/∠80°.
Longtan	N50°W	N20°–80°W	12–13	30	N5°–15°W/ NE∠57°–60°	35–45	130		Fault attitude: N30°–60°E/NW ∠60°–85°; N60°–90°W/NE∠70°–85°; N65°–85°E/ NW∠75°–85°.
Shuibuya	N64°W	N76°–80°E	5.62	36–40	N30°W	34			Fault attitude: NNE, NNW and NE three groups, high dip angles; joint attitude: NE, NW, SNT and NEE four groups, high dip angles.
Nuozhadu	N76°E	N50°–56°E	7.55–11.27	20–26					Three grade III faults, fault average distance 23.5 m, two group rigid and closed joints. F22: N0°–22°W/SW∠42°–58°, width 0.5–1.2 m, length more than 660 m, cross the whole cavern group; F20: N20°–30°E, NW∠60°–64°; F23: N20°–32°E, NW∠75°–85°.

Pengshui	N24°E	N20°E	10	4	N22°–25°E/ ∠60°–70°		60–80	Rock stratum dip direction to the sidewall at upstream and to the cavern inside at downstream. Fault attitude: NW and NNE two groups. Joints with high dip angles, length 0.5–5 m.
Goupitan	N75°E	NE43°	11–14	32	N30°–35°E/ NW∠45°–48°	40–45		Twelve faults around the main powerhouse; joint strike direction mostly NW with high dip angle; the largest and most complex karst system W24, horizontal length 420 m, pipeline length 1,000 m, elevation difference 100 m.
Laxiwa	N25°E	NS	14.6–29.7	25				Steep-inclined fault attitude: N55°W∠74°, N69°E∠68°, S65°W∠76°, S58°E∠78°; fault fracture zone width most less than 10 cm, a few 10–40 cm; several 50–70 cm.
Pubugou	N42°E	N54°–84°E	21.1–27.3	12–42		>50	>100	Fault attitude: N60°–80°E/NW∠60°–80°, N70°–75°W/SW∠45°–70°, fracture zone width less than 20 cm, joint attitude: N40°–60°W/SW∠20°–40°.
Xiaowan	N40°W	N50°–64°W	16.4–26.7	10–24	N70°–90°W/ NE∠65°–90°	30°–50°		Three faults with grade III cross the powerhouse, their attitude similar to rock stratum attitude, 1–10 cm wide compressive plane developed along with the fault.
Guandi	N5°E	N28.7°–53.0°W	25–35.2	33.7–58		>40		Fault attitude: N5°–20°W/SW∠70°–85°, N60°–80°E/SE∠70°–90°, N10°–30°W/ NE∠10°–25°, N20°–60°W/SW∠15°–40°, most width less than 10 cm, most length less than 40 m, maximum length 120 m.
Sanxia	N20.5E	N58°W	11.2–12.25	78.5		50		Fault strike direction: NNW, NE–NEE, NWW and NNE four groups, high dip angle, length 100–300 m, most width 0.3–0.5 m; joint attitude: most NNW, some NNE and NE–NEE, few NWW –EW, high dip angle.

Table 4 (Cont.)

Project name	*Axis orientation (A)*	*Strike direction of σ_1 (B)*	*σ_1 (MPa)*	*Angle between (A) and (B) (°)*	*Rock stratum attitude (C)*	*Angle between (A) and (C) (°)*	*σ_c (MPa)*	*Strength stress ratio*	*Rock mass structure features*
Xiangjiaba	N30°E	N25°–35°E	8.9–12.2	0–5	N60°–80°E/ SE∠15°–20°				Four weak intercalated rock layer cross the powerhouse, one near the crown vault with 6–42 m distance, others at the sidewall, joint attitude: N75°E/NW∠56°; N74°W/NE∠60°; N18°W/NE∠71°.
Jinping II	N35°E	N43.1°–47.4°E	10.1–22.9	8.1–12.4	N10° W–N30°E	5–40	65–80	1.7–3.5	Fault f65 strike direction along with the powerhouse axis direction, dip angle 40°–50°, its three subfaults cross the powerhouse, f16 with dip angle cross the powerhouse and transformer chamber.
Jinping I	N65°W	N28.5° W–N71°W	20.0–35.7	6–36.5	N20°W	45	60–75	1.5–3.0	Three large fault and rock dike strike direction NE, small fault strike direction NEE–EW and NW–NWW, joint attitude: N40°–60°E/NW∠25°–35°, N50°–70°E/ SE∠50°–80°, N50°–70°W/NE∠80°–90°, N25°–40°W/NE∠80°–90°.
Ludila	N50°E		10–15						Fault attitude: N40°–50°E/SW∠52°–55°, N50°–73°W/SW∠49°–52°, joint attitude: N10°–38°E/SE∠35°–40°, N30°–40°E/ SE∠60°–80°, N60°–90°E/SW∠60°–83°, fracture zone width 0.05–0.1 m.
Dagang-shan	N55°E	N44°–61°E	11.4–22.2	0–11		40–70			Rock dike β80: N25°W/SW∠65°, width 3–4 m; β81: N20°W/SW∠70°, width 1.5–2.0 m; joint attitude: N32°E/SE∠12°, N87°E/SE∠80°, N14°E/NW∠67°, N65°E/NW∠79°.
Xiluodu	Left: N24° W Right: N70°W	N60°–70°W	14.8–18.4 16–21	36–46 0–10	N20°–30°E/ SW∠5°–20°			6–10	One developed fault zone, length 20–50 m, most width 3–5 cm, some place 5–10 cm.

Huangjinping	N70°W	N80°–90°W	9.2–24.6	10–20				Steep-inclined structure surface N50°–85°W/SW∠65°–85° paralleling with the powerhouse axis, flat structure surface N20°–25°E/NW∠27° perpendicular to the powerhouse axis; joint space 0.2–0.4 m, length 3–10 m.
Houziyan	N61°W	N41°–75°W	21.53–36.43	14–20	N50°–70°E /NW∠25°–50°	60–100	1.5–3.0	Fault attitude: N60°–80°W, large fault width 1.0–1.5m, small fault with high dip angle width 1–5 cm; joint attitude: N35°–60°E/NW∠20°–55°, EW/N∠55°–80°, N30°–80°W/NE∠30°–70°, N35°–60°E/SE∠35°–45°, N20°–60°W/SW∠20°–60°; fracture zone most width 1–10 cm, maximum width 0.5–0.6 m.
Baihetan	N20°E N10°W	N30°–50°W	R: 19–23L: 22–26	10°				
Lianghe-kou	N3°E	N20.3°–57.7°E	18.09–30.44	17.3–54.7		>60		Fault attitude: NW65°–70°/SW∠65°–70°, NW70°/SW∠60°; joint attitude: NW60°–70°/SW∠60°–70°, NW0°–30°/SE∠10°–30°, NE0°–30°/SE∠40°–60°, NE0°–30°/SE∠70°–90°, space 0.2–0.6 m, length 2–3 m.
Yangfang-gou	N5°E	N61°–79°W	12.6–13	65–84				Joint strike direction: N70°–80°W, N70°–90°E, N10°–30°E, high dip angle; small scale fault and joint.
Shuang-jiangkou	N18°W	N3°–50°W	16–38	15–32				Joints undeveloped, without large fault.
Gongguo-qiao	N50°E	N5°–15°W	Self-weight stress	55–65	N5°–15°W/SW∠60°–80°	20(27)–125		Fault f2 cross through the powerhouse, N0°–20°W/SW∠60°–78°, width 0.8–1.0 m, effecting width at two sides 1–3 m; fault f15 appearing at the powerhouse vault, N57°E/NW∠75°–80°, width 15–20 cm, effecting width at two sides 1–4 m.

during excavation. For example, the intersection angles of Ertan, Pubugou, Jinping I, Jinping II, Dagangshan and Guandi hydropower stations are 16°–36°, 12°–26°, 6°–36.5°, 7.6°–11.9°, 6°–37° and 33.7°–58°, respectively, and their average angle is 15°–30°.

(3) If the orientations of the *in situ* stress and structural surfaces cannot satisfy the requirements noted above in the low *in situ* stress area, where the rock strength stress ratio is bigger than six or seven, the effect of structure surface occurrence on the surrounding rock should be considered mainly. While in the high *in situ* stress area, where the rock strength stress ratio is smaller than four or five, the effect of *in situ* stress should be considered mainly, meanwhile the structure surface occurrence is considered.

(4) The layout of the underground powerhouse's longitudinal axis should allow for a rational layout of the cavern group and a smooth flow of water. The small intersection angle between the longitudinal axis orientation of the underground powerhouse and the orientation of the maximum principal *in situ* stress can reduce the asymmetrical loading and stress concentration factor after excavation, then prevent the rockburst happening, or reduce its grade and decrease the lateral deformation of the high sidewall. These considerations can make high sidewall stable. But if the intersection angle is too small, it is not favorable to the stability of the end walls of main caverns and the rock pillars between headrace tunnels, tailrace tunnels and bus tunnels. If the maximum and second principal stresses are all high, their effects on the sidewall cannot be avoided. The higher the second principal stresses are, the more serious the rock mass fracturing is. For example, the intersection angle of Houziyan hydropower station is small – about 10° – but the second principal stresses direction is nearly perpendicular to the sidewall, finally the deformation and damage of the sidewall are very serious.

3.1.2 Choosing a suitable space between caverns

The space between the powerhouse and transformer chamber is very important to cavern group design. If the space is large, the length of bus tunnels and tailrace tunnels should all be added, and then the investment will be increased. If it is small, then the cost of engineering construction will be low and the waterpower condition will be good. When it is too small, it is also not unfavorable. The rock pillar between two caverns may easily turn plastic after excavation, and failure will happen due to rock loosening. Deformation will add the need for more supports. One such example is the Jinping I underground cavern group, the space between caverns is 45 m, the measured maximum *in situ* stress is 35.7 MPa and the rock strength stress ratio is 1.5–3.0. After excavation, the maximum deformation was 92.9 mm at the downstream sidewall near the skewback by 31 December 2009, and the two side plastic zones of rock pillar were connected. Some prestressed anchor cables were broken or overloaded and its support needed to be reinforced. So the space between caverns is usually determined by a combination of experience, numerical analysis and engineering analogy methods, according to the cavern scale, geological condition and *in situ* stress conditions.

According to the statistical results, the space (L) between nearby caverns of a large underground powerhouse is 0.45–1.1 times the height (H) of the large cavern. Among them, a value of L/H of between 0.55–0.75 occupies 77.3%, as shown in Figure 25,

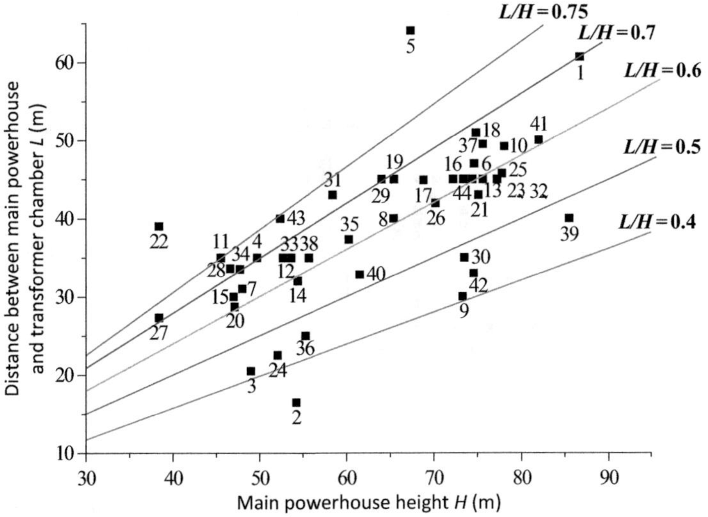

Figure 25 Distribution and boundary values of *L* and *H*.

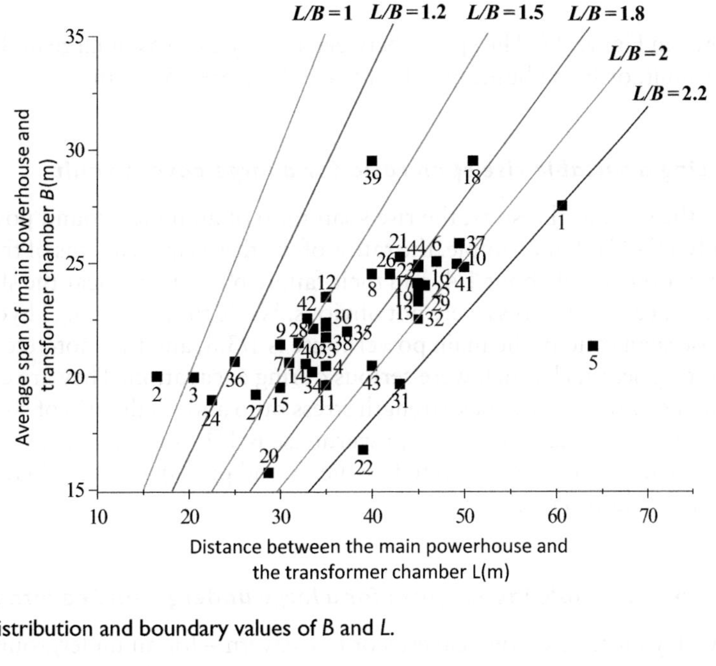

Figure 26 Distribution and boundary values of *B* and *L*.

while in high *in situ* stress areas, L/H should be more than 0.70, according to engineering experience. The space (L) is 0.80–2.0 times the average span (B) of the powerhouse and transformer chamber, as shown in Figure 26. And the space (L) is 1.29–2.57 times the span (B_1) of the main powerhouse. Among them, L/B_1 between 1.4–1.8 occupies

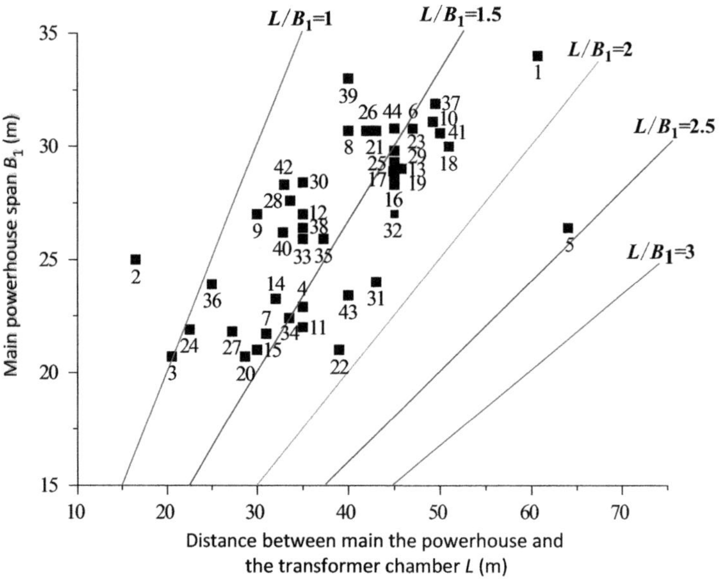

Figure 27 Distribution and boundary values of B_1 and L.

82%, as shown in Figure 27. The spaces between nearby caverns of large underground powerhouses built, or being built, in a decade in China are 35–50 m.

3.1.3 Choosing a suitable rise span ratio for a large cavern vault

According to the statistical results, the rise span ratio of an underground powerhouse vault is usually 1/3–1/5. For example, the ratios of the three large caverns of the Guandi hydropower station are all about 1/4, and their failure phenomena near the skewbacks caused by the concentrated stress were not obvious. While for the Jinping I hydropower station, the rise span ratio of the main powerhouse is 1/3.8, and the shotcrete stripping and rock splitting near its haunch were serious during excavation. They are concerned with the high *in situ* stress, low rock strength stress ratio and small ratio of rise to span. Although the ratio of Houziyan's main powerhouse is 1/4, some cracks and shotcrete stripping took place, which were affected by the second principal stress. So the ratio of rise to span of should be large.

3.1.4 Choosing a suitable layout form for a large underground cavern group

There are two layout forms – two caverns or one cavern – for an underground cavern group. The two-cavern pattern includes an independent powerhouse and transformer chamber, which is usually used for the large hydropower stations with many units. The one-cavern pattern is to arrange the main transformer in the powerhouse, which is the pattern usually used for a station with few units. The layout form of two caverns can be divided into four types, as follows.

(1) Type one: The transformer chamber is arranged at the downstream side of powerhouse and parallel to it. Their vault elevations are usually the same, or similar. The type is widely used in China, such as for the Ertan, Xiaowan, Xiluodu, Jinping I, Jinping II, Guandi, Dagangshan, Xiangjiaba, Baihetan, and Wudongde hydropower stations, as well as for the Ming Tombs, Guangzhou, Tianhuangping, Taian, Yixing, Baoquan and other pumped storage power stations.

(2) Type two: The transformer chamber is arranged at the downstream side of powerhouse and parallel to it. But the vault elevation of the transformer chamber is higher than that of the powerhouse in order to make the surrounding rock stable, such as at the Lubuge hydropower station and the Baishan pumped storage power station.

(3) Type three: The transformer chamber is arranged at the downstream side of powerhouse and perpendicular to, or intersecting with, it. This type is suitable for the station with a few units, such as the Xiaojiang, Huilongshan, and Jingbohu hydropower stations. Their lengths of powerhouse and transformer chamber are short and the spans of transformer chambers are small.

(4) Type four: The transformer chamber is arranged at the upstream side of powerhouse and parallel to it, such as at the Churchill Falls hydropower station in Canada. This type is rarely used in China.

The advantage and disadvantage of these four types can be seen in Table 5.

Table 5 Arrangement and characteristics of different types of layout for two-cavern cavern groups.

Type	Location of transformer chamber	Diagram	Advantages	Disadvantages
1	Downstream of powerhouse and with equal vault elevation	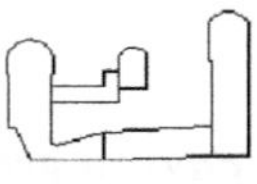	Layout compact, easy maintenance; separated arrangement can reduce degree of accident failure.	The distance between the main powerhouse and draft tube gate chamber cannot be too small, or it is detrimental to cavern stability.
2	Downstream of powerhouse and higher vault elevation	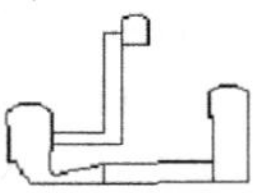	The distance between main powerhouse and draft tube gate chamber cannot be shortened; accident failure degree can be reduced.	Inconvenient operation and maintenance; longer bus tunnel; increased investment, large power loss, more equipment for ventilation and transportation.
3	Perpendicular to, or intersecting, with powerhouse	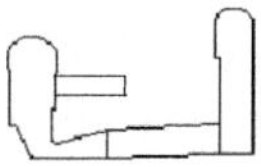	Bus tunnel short, small power loss; convenient management; tailrace tunnel can be shortened.	Only used for small hydropower stations with a few units.
4	Upstream of powerhouse	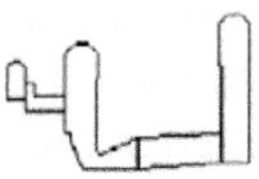	Layout compact, easy maintenance; accident failure degree can be reduced; tailrace tunnel can be shortened.	It is difficult to arrange the transformer chamber and design the drainage system. Surrounding rock stability is relatively poor.

3.2 Excavation of a large underground cavern group

3.2.1 Excavation method for a large underground cavern group

The large underground cavern groups are all excavated by blasting. The smooth excavation face can be obtained by presplitting blasting or smooth blasting. The presplitting blasting method can absorb shock and prevent cracks from being produced in the main blasting area and extending to the surrounding rock. Some rock mass stress can be also released by presplitting blasting. During the excavation of Ertan and Pubugou underground powerhouses, with high *in situ* stress, the field testing results showed that the presplitting crack quality was not good, despite the fact that the dynamite density was high without cutting blasting the middle section in advance. When the cutting blasting of the middle section was performed first, followed by expanding excavation and smooth blasting, then the excavation effect was improved. For the Jinping I and Jinping II hydropower station powerhouses, the presplitting crack quality was not perfect when presplitting blasting was performed outside the protective layer before cutting blasting middle section. Therefore, the cutting blasting of the middle section performed first is very important, regardless of whether smooth blasting or presplitting blasting is used. During the excavation of the Guandi hydropower station, the second to fourth layers of the main powerhouse were firstly excavated by cutting blasting of the middle section, followed by expanding excavation on two sides with protective layers, then presplitting blasting along the contour of the protective layers, and finally smooth blasting. Parts of the main powerhouse excavated after the fourth layer, the transformer chamber and pressure adjustment shaft, were first excavated by cutting blasting of the middle section, followed by expanding excavation on two sides with protective layers, and finally smooth blasting. The blasting excavation height was less than 6 m, and the actual blasting excavation effect was good. The blasting vibration monitoring result showed that the vibration velocity was less than the standard value.

3.2.2 Excavation sequence of large underground cavern group

The effect of the excavation sequence for large underground cavern groups on the surrounding rock stability is very large. The plastic zone, stress field and displacement field of cavern groups' surrounding rock are different with different excavation sequences, and the support parameters are also different. There are three types of excavation sequence. They are as follows.

(1) Sequence one: The transformer chamber is excavated behind the second layer excavation of the main powerhouse, and the pressure adjustment shaft is excavated behind the second layer excavation of the transformer chamber.
(2) Sequence two: The main powerhouse and pressure adjustment shaft are excavated at the same time, and the transformer chamber is excavated behind their first layer excavation, such as at the Guandi and Jinping II hydropower stations.
(3) Sequence three: The three main caverns are excavated at the same time, such as at the Jinping I and Houziyan hydropower stations.

Table 6 Large underground powerhouse excavation methods and sequences (Lu *et al.*, 2011).

Project name	*Powerhouse size (length × width × height) (m)*	*Lithology*	σ_I *(MPa)*	*Excavation sequence and contour blasting method*		
				Vault	*Sidewall*	*Rock anchor beam*
Ertan	280.3×25.5×63.9	Syenite, gabbro and basalt	15.01	AEP-II	WEP-II	BEP-II
Longtan	388.5×30.7×77.3	Sandstone and mudstone	5.25	AEP-I	WEP-III	BEP-I
Shuibuya	150.0×23.0×68.3	Limestone	2.55	AEP-II	WEP-V	BEP-I
Xiaowan	298.1×30.6×82.0	Granitic gneiss	14.92	AEP-II+ WEP-VI	WEP-III	BEP-I
Pubugou	294.1×30.7×70.2	Granite	13.50	AEP-II	WEP-II+ WEP-VI	BEP-II
Xiluodu	443.3×31.9×75.6	Basalt	10.00	AEP-II	WEP-III	BEP-I
Laxiwa	311.7×30.0×75.0	Granite	9.68	AEP-II	WEP-III	BEP-I
Sanxia	329.5×32.6×86.2	Granite and diorite	5.05	AEP-II	WEP-II+ WEP-V	BEP-I
Xiangjiaba	255.4×33.4×88.2	Sandstone	6.73	AEP-II	WEP-I	BEP-I
Jinping I	276.9×25.9×68.8	Marble	20. 0–35.7	AEP-II	WEP-V	BEP-I
Jinping II	352.4×28.3×72.2	Marble	12. 0–24.0	AEP-II	WEP-V	BEP-I
Nuozhadu	418.0×31.0×81.7	Granite	3.51	AEP-II	WEP-II	BEP-I

The analysis and testing results show that the damage depths to surrounding rock with different excavation sequences are about the same, but their damage volumes are different. Scheme two is the best because the damage volume and support work are all minimal, so it should be used preferentially. In addition, in order to reduce the unloading relaxation at the intersections of caverns, the small tunnels intersecting with the sidewall should be excavated first.

The excavation method and sequence of the large underground powerhouses of different hydropower stations in China can be seen in Table 6.

3.3 Failures caused by rock mass structures and countermeasures

The failures of large underground cavern groups caused by rock mass structure usually happen near the free face, where the structure surfaces developed. The failures may be rock block falling or sliding; collapse at a fault; a fractured zone; several free faces or the crown arch. They are not affected by the *in situ* situation and happen after excavation. When the failures happen at the cavern vault, they are usually controlled by the gently inclined structure surfaces. The failure type may be rock block falling or collapse along the gently inclined structure surface, such as the unstable rock blocks at the main powerhouse vault of the Longtan and Three Gorges hydropower stations, and the large volume collapse at Dagangshan hydropower station. When the failures happen at a cavern's upstream and downstream sidewalls, they are usually controlled by the steeply inclined structure surfaces. The failure type may be rock block falling, collapse and sliding, or a tension fracture and toppling along the steeply inclined structure surface, such as the surrounding mass blocks at the downstream sidewall of the main powerhouse of Pengshui hydropower station and the upstream sidewall of the draft tube gate chamber of the Jinping II hydropower station.

These failures usually can be easy controlled with shotcrete and bolts, and their lengths and diameters can be designed with the key block theory and discontinuous deformation analysis (DDA) method. Some controlling countermeasures are listed below.

(1) The potentially unstable rock blocks should be searched, and their stability should be analyzed in a timely manner after excavation. The stability of blocks during subsequent excavation should be judged and supported with corresponding measures.
(2) The rock block stability problem should be first considered when sudden deformations appear, which are detected by multiple position extensometers. The supports should be put in place immediately once there are analysis results.
(3) Prestressed bolts, mortar bolts with steel plates, and small tonnage anchor cables should be used to support the gently inclined structure surfaces at the cavern vault, and the steep-inclined structure surfaces at the sidewall. If the rock mass is broken, shotcrete with steel fabric can be used to prevent the rock block falling.
(4) If the rock mass is broken, or with developed structure surfaces near the rock anchor beam, preliminary grouting and advance anchor bolts can be applied before blasting to control the excavation quantity.

3.4 Failures caused by stress and countermeasures

The stress field of surrounding rock will change greatly after excavation in a high *in situ* stress area. The surrounding rock will crack under the action of secondary stress. New cracks will generate, extend and connect then damage the surrounding rock. Under this action, the damage depth to the surrounding rock will develop from outside to inside with the adjustment of stress. According to the rock strength stress ratio and structure feature, rockburst; splitting and spalling; unloading; and fracturing of the surrounding rock may happen. The failure types may be fault rockburst, strain rockburst, V-shape failure, tensile crack, splitting crack in high sidewall, and circumferential cracks in tunnels intersecting high cavern sidewalls. According to the mechanical mechanism, these failure types can be divided into three models, tension fracture (T), tension shear fracture (TS) and shear fracture (S), as shown in Figure 28.

As an example, the failures caused by stress at the Houziyan underground powerhouse usually began two hours after excavation and lasted during the whole construction. The degree and form of damage were connected to cavern location, supporting time and supporting pattern. The maximum degree of damage was at the intersection area of caverns, which was connected to the unloading in two directions. The later the supports were put in place, the more serious the damage was. The effect of damage on

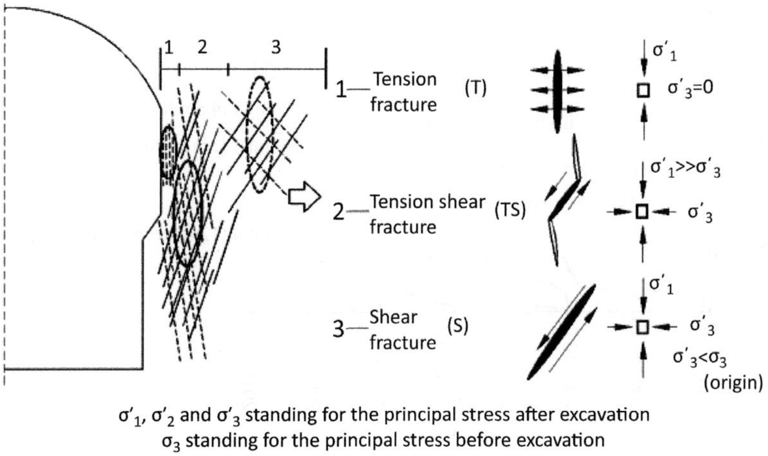

Figure 28 Rock failure modes driven by stress, and their mechanism (Wei *et al.*, 2010).

construction and supporting structure security is very large. The continued fracturing of surrounding rock is bad for the long-term security of the Houziyan powerhouse. According to the statistical results, there were more than a hundred cracks in the shotcrete on the main powerhouse sidewall. Their widths were more than 2 cm – the maximum being 11 cm – and the longest one was 13 m.

The countermeasures to control damage caused by stress are as follows.

(1) The systematic application of prestressed bolts with steel plates, shotcrete with steel fabric and shotcrete with steel fiber (5–6 cm thickness and content bigger than 50 kg/m^3) can be used to support the surrounding rock to avoid a rockburst happening. Meanwhile the pilot heading can be excavated to release the *in situ* stress.

(2) The grade of shotcrete should be changed to C30–C35; the shotcrete with steel fiber should be sprayed immediately after excavation, and then the bolts are inserted and shotcrete with steel fabric is applied.

(3) The anchor cables between the main powerhouse and the transformer chamber should be constructed before the support for the bus tunnels, and the anchor cables between bus tunnels should be put in position as soon as possible.

3.5 Failures caused by stress and rock mass structures and their countermeasures

The failures caused by stress and rock mass structures usually last for a long time after excavation. Failure phenomena include joint or bedding plane opening or splitting: joint or bedding slipping: fault sliding; bucking deformation by weak stratum squeezing or stratum bending and so on. The existing disadvantageous structure surface and geology defects are the hidden dangers to the surrounding rock stability, while high *in situ* stress activates and enlarges the controlling effect of structure surface, then the timeliness deformations are produced for the strong unloading of the geology defects with high *in situ* stress. This makes the stress field of surrounding rock adjust for a long

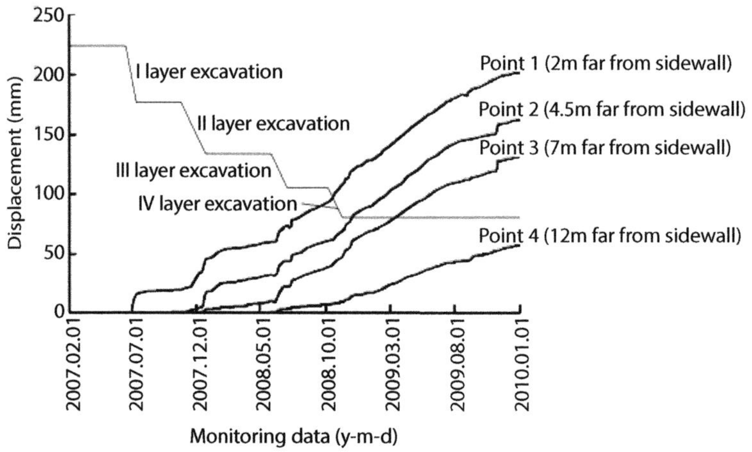

Figure 29 Deformation curves of monitoring point M^4_{PS2-8} at the downstream sidewall of the Jinping I transformer chamber (Wei et *al.*, 2010).

time and its deformation continue, such as the downstream sidewall deformation of the transformer chamber of the Jinping I hydropower station, shown in Figure 29.

The countermeasures to control the damage caused by stress are as follows.

(1) Shotcrete with steel fabric or steel fiber, systematic prestressed bolts with steel plates, and prestressed anchor cables can be used to support the surrounding rock.
(2) A suitable excavation method and sequence should be chosen.
(3) The monitoring of surrounding rock deformation, bolt stress, anchor cable tension, loosing depth, rock mass acoustic wave velocity and so on should be done and used to evaluate the stability of surrounding rock.

3.6 Deformation of rock anchor beams, and countermeasures

The deformation of the rock surrounding an underground cavern group with high *in situ* stress is large, especially at the area with developed structure surfaces. If the large deformation area is near the rock anchor beam, there may be cracks on, or dislocation of, it that will affect bridge crane working. For example, there were no cracks on the rock anchor beam of the Guandi main powerhouse by May 2009, when the templates were torn down. Then several cracks were found at the upstream rock anchor beam in March 2010, which were caused by the heterogeneous deformation of surrounding rock. The cracking area is in keeping with the large deformation area of surrounding rock and steel rail. The maximum deformation at the upstream rock anchor beam was 34 mm, and the relative deformation of the upstream and downstream rock anchor beams was 58 mm, bigger than the reserved adjustment range for deformation. Finally, the steel rail could not be adjusted and the bridge crane could not work normally. The expansion joints of the rock anchor beam of the Houziyan underground cavern group also moved; there were 18 and 12 cracks, respectively, in the upstream and downstream rock anchor beams.

One countermeasure to control the deformation of a rock anchor beam is to concrete it later, and then the steel rail of the bridge crane is installed. For example, the rock anchor beam was concreted after the excavation of bus tunnels was finished. Although the deformation of the surrounding rock was large, there were no cracks on the rock anchor beam. Another method is to reserve enough adjustment range for the steel rail of the bridge crane, according to calculations and engineering experience. In addition, the deformation of surrounding rock should also be controlled.

3.7 Overloaded or failed of anchor cables, and countermeasures

If the deformation of surrounding rock between a powerhouse and a transformer chamber is large and lasts for a long time after being supported, the prestressed anchor cables constructed after excavation may overload or fracture. These phenomena have happen in many underground powerhouses of hydropower stations, such as at Jinping I, Jinping II, Laxiwa, Houziyan. For the Guandi hydropower station, the locked loads of the prestressed anchor cables were 80% of their designed loads, and there were few overloading phenomena during construction. That is to say that the locked loads are suitable. For the Jinping I hydropower station, the locked loads of the prestressed anchor cables were 70–85% of their designed loads. However, the overloading phenomena were serious because the deformation of the surrounding rock was large, which was a consequence of the high *in situ* stress, three faults and developed joints. Their locked load should maybe have been lower. For the Jinping II hydropower station, 52 roots anchor cables at the downstream sidewall had overloaded or fractured within one year of the excavation of the underground cavern group. Of these, 23 roots were at the upside of the rock anchor beam, between units 1# and 3#, and 29 roots were at its downside between units 4# and 7#. Deformation and cracking of surrounding rock near some of the anchor heads were also found in the Houziyan underground cavern group.

According to engineering practice, the locked loads of prestressed anchor cables at the upstream and downstream sidewalls should be determined according *in situ* stress, the rock strength stress ratio, predicted deformation, structure surface development and the designed load. For common situations, their locked loads are usually 70–90% of the designed loads. The larger the deformation, the lower the locked load. If the *in situ* stress is high, the predicted deformation of surrounding rock is large and the rock strength stress ratio is low, so the locked loads of prestressed anchor cables can be 60–70% of the designed load. Where this is not the case, large locked loads should be used. But if the prestressed anchor cables are used to make the rock block stable, the locked loads should be 100% of the designed loads.

3.8 Unloading and timeliness deformation of sidewalls, and countermeasures

The deformation and failure of the rock surrounding excavations are mainly caused by rebounding stress and redistribution stress of the excavation unloading. The stability of underground caverns' surrounding rock is a contradictory problem between rock mass stress and rock strength. The effect of different factors, such as the *in situ* stress, lithology, rock mass structure, and underground to the stability of the surrounding

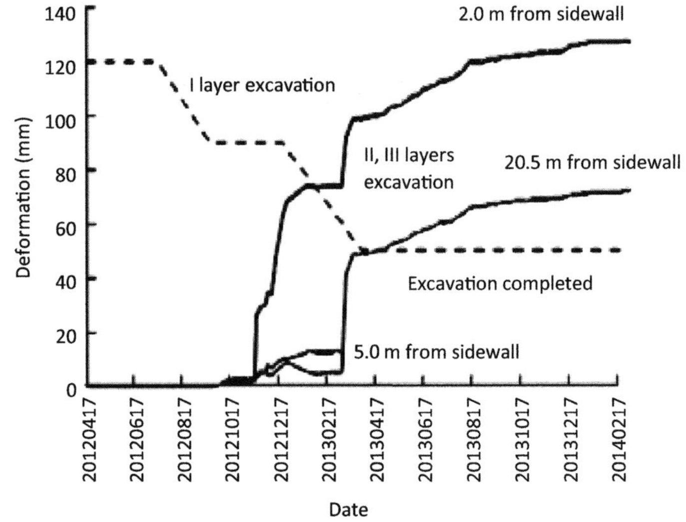

Figure 30 Deformation–time curves at upstream sidewall of the transformer chamber at the Houziyan underground cavern group: 0+83.8 m section (Li *et al.*, 2014).

rock is also reflected by these two factors. Another disadvantageous effect of the developed structure surface is to cause the unloading timeliness deformation of hard rock with high *in situ* stress. If the *in situ* stress is high, the unloading degree and the area affected after excavation are all large. And the unloading rebounding deformation is changed into the deformation of inside structure surface under the combined action of disadvantageous structure surfaces. The inside of the rock mass structure will also adjust, which will cause stress because of further adjustment in the surrounding rock. These interactions make the adjustment time of induced stress long, and the deformation large and long lasting.

For example, according to the displacement monitoring data of the Houziyan underground cavern group, the surrounding rock deformation of the high sidewall and the affected depth were all large. Some damage to the anchorage piers inset into the shotcrete and the timeliness of the deformation of the rock surrounding the high sidewall were obvious, as shown in Figure 30. By August 2014, about 17% of all the monitoring points in the surrounding rock surface were recording deformation larger than 50 mm. The maximum deformation was located at the downstream sidewall; this was 158 mm, which had been 120 mm after the fourth layer was excavated. Its deformation is bigger than that of other projects on a similar scale and with similar engineering geological conditions in China, such as the Jinping I underground cavern group, where about 11% of all testing points were recording deformations larger than 50 mm. The surrounding rock relaxation depth of the Houziyan underground cavern group was generally 6.0–12.2 m and the maximum depth was 15.0 m. The typical acoustic wave testing curves of produced by the main Houziyan powerhouse are shown in Figure 31, which show that the areas with low velocity develop inside gradually.

In general, there are three features about the distribution area of large deformation: (1) most areas are located at the upside of cavern sidewalls; (2) the deformations at

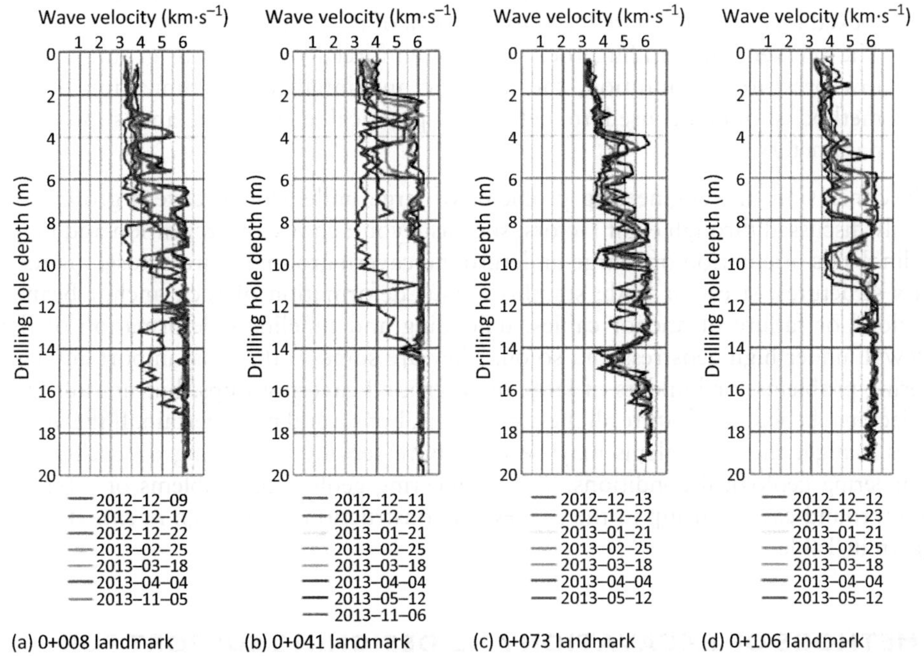

Figure 31 Acoustic wave curves from different holes at an elevation of 1,711 m of the main Houziyan powerhouse downstream sidewall (Zhang *et al.*, 2014).

downstream sidewalls are larger; (3) the deformations of rock pillars to two sides free faces are larger. The depth range affected by the deformation at an upstream sidewall of a powerhouse is 0–5.0 m, while the depths at its downstream sidewall are mostly more than 8.0 m. For the transformer chamber, the depths at its upstream sidewall are mostly greater than 5.0 m, some as much as 20.0 m, and the depths at its downstream sidewall are mostly lower than 5.0 m, some as much as 10.0 m. The depths the effect to pressure adjustment shafts are less than 8.0 m, the maximum depth being 15.0 m.

The countermeasures to control the unloading and timeliness deformation of a sidewall with high *in situ* stress are shown as follows:

(1) Each layer should be excavated with the method of deep presplitting, shallow excavation and a controlled explosive dosage, which can reduce the effect of blasting vibration on surrounding rock. Meanwhile the shallow supports, such as the shotcrete, bolts (6 m, 8 m or 9 m long) and bundles of bars (12 m long), should be done in a timely fashion, usually within five to seven days after excavation, and not more than two weeks after for bundles of bars.

(2) The placing of deep supports, such as prestressed anchor cables (15 m, 20 m, 25 m or 30 m long) and two-ended anchor cables, should be completed before the excavation of the next layer, to prevent deformation continually developing inside. Their placement should usually be finished within two weeks of completing the shallow supporting, and within not more than one month.

(3) The designed loads of the prestressed anchor cables should be increased and the locked loads should be reduced to 60–70% of the designed loads.
(4) The relaxed surrounding rock can be reinforced by concrete grouting with low pressure. In addition, the dynamic and intelligent optimization design method should be used to determine practical excavation and supporting parameters.

In conclusion, the engineering problems of large underground cavern groups are mainly discussed through eight factors, such as layout; choice of excavation and longitudinal axis orientation of a large underground cavern group; failures caused by rock mass structures, stress or both; stability versus deformation of rock anchor beams; overload or failure of anchor cables; unloading and timeliness deformation of the sidewall under high geostress and so on. They are some of the most important engineering problems for builders of large underground cavern groups, but not all. One project maybe will not feature all these problems, but several must pertain. If the *in situ* stress is high, these problems may become more serious, depending on each set of engineering geological conditions. The engineering geological problems of a typical underground cavern group in the process of construction under high *in situ* stress is shown in Table 7.

4 METHODS OF EXCAVATION AND DESIGNING SUPPORT FOR LARGE UNDERGROUND CAVERN GROUPS

4.1 The excavation method for large underground cavern groups

Underground cavern groups are excavated by the drilling and blasting method. The blasting method includes the smooth blasting method and the presplitting blasting method. During construction, the caverns are divided into several layers to be excavated. The excavating layers and their heights for existing large underground caverns are shown in Table 8. The construction sequences of different projects are not same. If the *in situ* stress is low, the order in which bus tunnels are excavated has little effect on surrounding rock stability. If the *in situ* stress is middling or high, the small caverns that intersect with the sidewalls of large caverns should be excavated and supported first, before the large caverns are excavated. The effect of the construction sequence of bus tunnels on deformation is mainly reflected in the rock anchor beams and the higher areas, especially the downstream rock anchor beam. Briefly, there are four main problems with the construction of a large underground cavern group: the excavation and support of the cavern crown; the construction of the rock anchor beam; the excavation of bus tunnels which intersect with a high sidewall; and the stability of the intersection area between tailrace tunnels and a high sidewall. In order to make the construction safe, the large underground cavern groups should be excavated as follows.

(1) The main powerhouse should be excavated first; the transformer chamber and draft tube gate chamber or pressure adjustment shaft are excavated later. They are mainly excavated from upside to downside, one layer is excavated and then supported before next layer is excavated.

Table 7 Engineering geological problems of typical underground caverns in the process of construction under high in situ stress.

Project name	*In situ stress (MPa)*	*Rock mass grade*	*Maximum displacement (mm)*	*Relaxation depth (m)*	*Engineering geological problems during construction*
Ertan	17.2–38.4	Most grade II, few grade III	185.2 at upstream sidewall of transformer chamber	Generally less than 7, location 10 for sidewalls, generally less than 2.5, location 2.5–5 at crowns	Rock cracking, rock block falling, shotcrete cracking and some anchor cables failing happened at the stress-concentrated areas, such as the upstream skewback, intersections of caverns, rock pillars and other complex locations. Large-scale deformation happened twice, caused by the excavation of 2# pressure adjustment shaft and 2# hydrogenerator pit.
Jinping II	10.1–22.9	Most grade III, few grade IV and V	95.03 at upstream sidewall of main powerhouse	1.0–3.6	Large deformation occurred at the upstream sidewall of the main powerhouse and the anchor cables overloaded at the downstream sidewall of the main powerhouse; rock blocks collapsed at the upstream sidewall of the draft tube gate chamber; spalling and cracking happened around the sidewalls and arch springs, large deformation occurred near the areas that had developed faults .
Jinping I	20.0–35.7	Most grade III, few grade IV and V	236.72 at downstream sidewall of transformer chamber	6.5–8.0 for sidewalls of main powerhouse, <2.0 for crown	Shotcrete and surrounding rock cracked at the downstream sidewall of the main powerhouse and the transformer chamber. Bending failure happened around the outside of the bus tunnels and headrace tunnels, large deformation occurred within the rock wall between the main powerhouse and the transformer chamber.
Houziyan	21.5–36.4	Most grade III, few grade IV	156.60 at downstream sidewall of main powerhouse	1.8–10 for upstream sidewall of main powerhouse, 1.0–11.8 for its downstream sidewall	Rockburst happened during the excavation of the cavern groups, the deformation of the main powerhouse downstream sidewall and transformer chamber upstream sidewall were large; cracks appeared around the rock anchor beams and the upstream sidewall of the main powerhouse and the spandrel of the pressure adjustment shaft.

Table 7 (Cont.)

Project name	*In situ stress (MPa)*	*Rock mass grade*	*Maximum displacement (mm)*	*Relaxation depth (m)*	*Engineering geological problems during construction*
Dagangshan	11.4–19.3	Most grade II and III, few grade IV and V	81.60 at sidewall of main powerhouse	<6.0	The influence of diabase dikes to the stability of the underground powerhouse and the influence of high dip angle structural surfaces to the stability of sidewalls were all considerable, the local collapse of unstable rock blocks affected the shape of the rock anchor beam.
Changheba	16.0–32.0	Most grade II and III, few grade IV	164.42 at sidewall of pressure adjustment shaft	1.2–3.6 for upstream sidewall of powerhouse, 4.2–7.6 for its downstream sidewall, 1.4–4.4 for transformer chamber, 1.0–3.0 for pressure adjustment shaft	The large deformation and failure of surrounding rock caused by high *in situ* stress were serious, as was the deformation near the intersection of the pressure adjustment shaft and 2# tailrace tunnel, and the collapse volume was nearly 200 m^3.
Huangjinping	20.1–23.7	Most grade III	87.00 at sidewall of transformer chamber	1.6–6.0 for main powerhouse, 1.8–4.6 for transformer chamber, 2.0–3.0 for pressure adjustment shaft	Some wide annular cracks appeared around the bus tunnels, and some unstable rock blocks were formed by the structural surfaces and free faces.
Guandi	25.0–38.4	Most grade II and III	66.66 at sidewall of pressure adjustment shaft	Generally 0.6–3.0, location 7.5–12.5	The controlling effects of rock mass structures were obvious, the stability of the three large cavern crowns was controlled by the gently dipping structural surfaces during excavation, and the deformation near the fracture zones was large.
Xiluodu	14.8–21.1	Most grade II and III	Generally less than 10, location 10–40	Generally 1.0–3.0, location 3.0–5.0	Several unstable rock blocks, large relaxation deformation at the cavern intersections caused by stress concentration.
Laxiwa	22–29	Most grade II, few grade III	40.2 at downstream sidewall of main powerhouse	Generally 2.0–3.0, location 4.0–7.0	Rockbursts happened at the upstream sidewall and crown of the powerhouse, the bus tunnels and the tailrace tunnels; rock blocks collapsed at the crown and downstream sidewall, shotcrete and the surrounding rock spalled and cracked at the middle-upper part of the powerhouse sidewalls and the intersection of the powerhouse and bus tunnels.

Table 8 Large underground powerhouse excavation parameters.

No.	*Project name*	*Main powerhouse size (length × width × height) (m)*	*Number of excavating layers*	*I layer height (m)*	*Other layers' heights (m)*	*Excavation time (months)*
1	Lubuge	125.0×18.0×39.4	5	7	6.5–9	21.5
2	Ertan	280.3×30.7×65.4	10	9	5.6–6.5	27
3	Dachaoshan	234.0×26.4×67.3	7	11.5	5.5–10.5	29
4	Xiaolangdi	251.5×26.2×57.9	10	9	1.3–7.7	38
5	Suofengying	135.5×24.0×58.4	7	8.3	7.4–9.2	16
6	Sanbanxi	147.2×22.7×60.1	7	8.5	4.1–9.4	14
7	Longtan	388.5×30.7×75.4	9	11.3	4.9–10.5	32
8	Manwan II	105.5×26.6×69.4	8	9.5	3–10.9	14
9	Shuibuya	168.5×23.0×68.3	8	8.5	5.5–10	26
10	Nuozhadu	418.0×31.0×81.7	10	13	4.0–9.3	30
11	Pengshui	252.0×30.0×76.5	9	9.5	6.5–10.4	33
12	Goupitan	230.5×27.0×73.4	9	9.5	6.3–11.5	24
13	Laxiwa	311.8×30.0×74.8	9	10	5.5–10.8	35.5
14	Pubugou	294.1×30.7×70.2	9	9.4	3.3–10.3	26.5
15	Xiaowan	298.1×30.6×79.4	10	12.3	2.5(4.5)–10.5	32
16	Sanxia	311.3×32.6×87.3	9	11.7	7.5–12	37
17	Guandi	243.4×31.1×78.0	11	8	4.5–8	32
18	Xiangjiaba	255.0×33.0×85.5	9	11	7–13.38	32
19	Gongguoqiao	175.0×27.4×74.5	9	10	5.3–10.9	20
20	Jinping II	352.4×28.3×72.2	9	11.2	2(4.5)–12.2	30
21	Jinping I	277.0×28.9×68.8	11	10.3	3–10	48
22	Yantan II	127.2×30.8×76.7	9	10.33	7.7–12.1	21
23	Xiluodu	443.3×31.9×75.6	10	12.1	6–9	30
24	Dagangshan	226.6×30.8×74.6	10	11.8	4–10	
25	Huangjinping	206.3×28.8×67.3	9	7.2	5–10	23
26	Houziyan	219.5×29.2×68.7	9	13.5	5–9	30
27	Changheba	228.8×30.8×73.4	10	11.4	3–9	31

(2) The large main caverns should be divided into several layers and several areas, then they can be excavated from different directions and locations at different elevations. Many construction processes can be done at one plane and at different elevations, and this speeds up construction.

(3) The middle part of the main powerhouse crown should be excavated and supported first, then the two sides are done. If there is a soft rock mass, it should be excavated and supported first, then the hard rock mass is excavated.

(4) The high caverns should be excavated layer by layer. Each layer should be presplit one at a time, then excavated and supported; this should be done through a sequence of several thin layers.

(5) For the intersection caverns, small caverns should be excavated before large caverns, and the cavern should be excavated before the sidewall. After the excavation of each layer or part, the surrounding rock should be supported immediately afterwards.

(6) The shallow supports should be constructed first, and then the deep supports are done. During construction, the monitoring of surrounding rock deformation,

bolt stress, anchor cable load and so on should be done in time, and the dynamic feedback analysis also should be done on the monitoring results in order to guide later construction.

4.2 The method for designing support for large underground cavern groups

For the underground structures, there are several support design methods, such as the experience or semi-experience method, Q-system method, and the rock mass rating (RMR) method. Among them, the convergence–constraint method is widely used, which is mainly through the deformation of surrounding rock to design the supporting parameters.

For hydropower underground cavern groups, compressive arch theory was used to design the supporting parameters in the 1950s and 1960s. The rock mass is regarded as one kind of load, and the load is determined by the loose material assumption. This method is only suitable for the broken surrounding rock. Its results are inconsistent with the practical projects in most conditions and much safer. After the 1970s, the New Austrian Tunnelling Method began to be applied gradually. Its purpose is to make good use of the surrounding rock's bearing capacity. Its main design measures are rock mass classification; optimal design; support with shotcretes and bolts; and field monitoring and feedback analysis, which should be done during the whole process of surveying, designing, constructing and operating. Base on this method, a flexible system support is adopted for the underground caverns, which includes shotcrete, system bolts and anchor cables. The whole supporting strength is generally 0.1–0.3 MPa, which is lower than the initial *in situ* stress and cannot limit the unloading deformation of surrounding rock. This has been verified by the numerical analysis and engineering practice of many projects in recent years. Good use should be made of the self-bearing capacity of the surrounding rock and it should be made basically stable before being supported.

For the large underground cavern groups in existence, the engineering exploration was usually divided into several stages. The engineering geological information obtained at different stages was not the same, so their design features are correspondingly different. Before the cavern excavation, the rock mass structure, *in situ* stress features, underground water and other factors could not be known completely and the rock mass mechanics parameters were difficult to determine exactly, so the engineering problems could not be grasped completely. The analysis results with theory method based on these parameters are mainly used for reference. These belong to the predesign works of underground structures before construction. The geology analysis method and engineering analogy method are mainly used. The rock mass grade should firstly be classified by analyzing the engineering geological conditions, *in situ* stress, and engineering factors. Secondly, the surrounding rock stability can be evaluated approximately using the direct analogy method with similar projects, or the indirect analogy method, according to the rock mass classification results. Thirdly, the supporting type and parameters can be determined. Finally, the designed support scheme should be checked by the numerical calculation methods. If necessary, the geological mechanical model experiment can be done to verify its suitability for the complex large underground cavern group.

At present, three-dimensional non-linear numerical methods – such as the finite element method, the discrete element method, and the manifold method – are widely used to analyze the stress, deformation, plastic zones and fractured zones of large underground powerhouses. But there are big differences between the calculation results and monitoring results. For the underground powerhouses with intact and hard rock mass and low *in situ* stress, the plastic zones calculated are larger than the monitored zones during the predesign stage. While for the underground powerhouses with low rock strength stress ratio, such as Jinping I and Houziyan, the monitored plastic zones, fractured zones and deformation are greater than the calculated values. One reason is that it is difficult to reflect complex geological structure in a numerical model; another is that the rock mass parameters cannot be known exactly before excavation and without feedback analysis.

In order to make the support scheme more applicable and to adjust it along with the construction, the dynamic feedback analysis method is suggested, which makes the support design method more suitable and effective. If the first layer of a main powerhouse is excavated, the deformations at different monitoring points can be used to calculate the rock mass mechanics parameters with displacement anti-analytic method. Then these parameters can be used to predict the deformation, plastic zones and fractured zones of surrounding rock after the excavation of the second layer, with the modified geological conditions by the revealed this layer. During the construction of the second layer, the rock mass mechanics parameters, support design parameters and geological conditions should be checked and adjusted according to the displacement anti-analytic method, field monitoring and geological outline. Then the support design parameters of the third layer can be presented, and the supporting effect of the second layer can be evaluated or reinforced. Finally, the caverns can be constructed with this method, which is widely used in the building of many hydropower stations at present.

The section shapes of large underground cavern groups should also be selected appropriately. These include the mushroom shape, the city-gate section channel and the ellipse. The crown of the mushroom shape cavern is supported by a rigid concrete lining. Its stress condition is the worst and it produces maximum deformation and the greatest amount of support work needed according to the study of the Mingtan pumped storage power station in Taiwan by E. Hoek. Use of this shape ceased from the 1950s to the 1980s in Japan, and before the 1970s in Italy. After that time, the New Austrian Tunnelling Method came into use gradually throughout the world. As underground powerhouses were constructed in China mainly during the 1980s, this shape was rarely used. The crowns of other shapes are usually supported by shotcrete and bolts or anchor cables. The stress and deformation conditions of caverns with an ellipse section are better than those for caverns with city-gate section channels, but there are no differences to the deformation and stress at the cavern crowns of either type. The large underground powerhouses with ellipse sections were constructed before 1999 in Japan, thereafter the city-gate section channels were widely used. In China, city-gate section channels are also chosen preferentially. For the pressure adjustment shafts, the cylinder shape and long corridor shape are the two main layout modes. The cylinder shape is usually used for single hydraulic units, such as at Taipingyi. For the large or huge hydropower stations, with several hydraulic units, either of these two modes can be used. The cylinder shape is used at Xiaowan, Jinping I, Jinping II,

Laxiwa, Nuozhadu, Baihetan, Taian and other stations, and the long corridor shape is used at Ertan, Longtan, Pubugou, Goupitan, Xiluodu and other stations.

In summary, the large underground cavern groups should be designed with the principles of engineering analogy first and numerical analysis second; flexible support first and rigid support second; and system support first and local support second. Meanwhile the dynamic feedback analysis and support design should be done according to the geological conditions revealed, and by monitoring data captured from the surrounding rock.

5 TYPICAL RESEARCH PROGRESSES RECENT YEARS

5.1 Intelligent analysis and an optimal design method for large underground cavern groups

According to the engineering features of large underground cavern groups, Xia-Ting Feng and John A. Hudson (2015) present the intelligent stability analysis, and the dynamic optimization design methods, which have been successfully used in Shuibuya, Laxiwa, Jingping II, Baihetan, Wudongde and others hydropower stations. This design method includes two stages: primary design before construction and dynamic feedback analysis during construction. The main work of primary design is as follows.

(1) The engineering conditions around the large underground cavern groups, such as the distribution of faults and joints and lithology; underground water; *in situ* stress; and rock mass mechanics parameters should be known first by engineering geological investigation, observation and analysis of the field exploration tunnels and experiment tunnels. Then the engineering rock mass classification should be done using the RMR method, the Q-system method, the HC method and BQ method in China, the GSI method and so on.

(2) The three-dimensional *in situ* stress field should be back-analyzed by the non-linear back analysis method, according to the field *in situ* testing results, meanwhile regional structural characteristics and geographic and geomorphic conditions should be considered.

(3) The deformation failure mechanism and characteristics should be revealed firstly through observations and different laboratory tests (such as uniaxial compression tests, triaxial compression tests, loading and unloading triaxial tests, and acoustic emission tests) which will indicate deformations, stresses, microseisms, elastic waves and other characteristics of the field exploration tunnels and experimental tunnels. Then the constitutive models, strength criteria and parameters of the surrounding rock can be identified by the intelligent methods, such as the hard rock mass deterioration model (RDM) (Jiang & Feng, 2008), generalized polyaxial strain energy (GPSE) strength criterion (Huang & Feng, 2008), and the parameter identification with particle swarm optimization (PSO) algorithm (Su & Feng, 2005).

(4) The number of excavation layers, the excavation sequence, layer heights, the fine excavation sequence of important areas and suitable supporting patterns and parameters should be optimized with the whole intelligent optimization method for underground cavern group excavation and their support.

(5) The potential failure modes of large underground cavern groups are identified by the failure mode classification method in Xiang & Feng (2011).
(6) The whole stability and safety factors of large underground cavern groups can be evaluated by rock mass classification, strength reduction, failure approach index (FAI), and other methods. The whole and local stability of surrounding rock can be evaluated by the numerical analysis method with the local energy release rate (LERR) and FAI, and its risk assessment can be done with the reliability analysis method.
(7) The deformation management classification of surrounding rock for the excavation of each layer of a cavern group should be defined by combining the results of engineering analogy and numerical analysis.
(8) Finally, the design scheme can be presented based on the analysis results above, and the design requirements, which will include the direction of the powerhouse axis; the excavation layer height; the excavation scheme; the support system parameters; the fine excavation sequence and support parameters for important areas; the deformation management classification of surrounding rock for excavation of each layer; and the anti-seepage design.

During construction, the dynamic feedback analysis and design optimization of a large underground cavern group should be done layer by layer until the excavation is completed. The work is mainly as follows.

(1) The geological conditions and failure modes considered during the design period should be checked according to the conditions revealed during excavation. The geological model can be adjusted dynamically according to these requirements.
(2) The distribution of a three-dimensional *in situ* stress field should be checked according to the local failure forms and the location and deformation features revealed.
(3) The failure models should be checked and identified according to the geological conditions.
(4) The mechanics parameters of surrounding rock should be back-analyzed and checked by the intelligent back analysis method based on the multi-information monitored, which can be used to evaluate the stability of caverns under their current state. Then the support parameters for caverns can be optimized and the deformation and fracture features can be predicted after the excavation of the next layer.
(5) The deformation management classification of rock surrounding the next layer of excavation can be suggested by combining the engineering analogy, numerical analysis and the experience from working the several layers excavated before.
(6) The stability of caverns during the subsequent excavation process can be estimated comprehensively based on the deformation management classification of the surrounding rock, cracked surrounding rock, overloaded anchor cables, FAI, LERR and so on.
(7) The dynamic adjustment measures needed can be identified according to practical geological conditions, and deformation and failure characteristics detected with the global optimized intelligent method. These include the optimized

excavation and support schemes for the subsequent construction, such as excavation method and time, form of support, parameters, and time.

(8) The long-term stability of large underground cavern groups should be predicted and evaluated based on the time-dependent analysis models.

5.2 Deformation management classification of surrounding rock during excavation

Concerning the construction characteristics of a large underground cavern group, both the deformation management classification and surrounding rock based on the feedback optimization analysis and its execution method were proposed according to the deformation process and failure characteristic of surrounding rock of similar projects and the numerical analysis results of this project. The deformation and its rate are regarded as the safety controlling indexes of the surrounding rock. Then safety controlling standards with three grades, *i.e.* safety grade, prediction grade and danger grade, were suggested by the dynamic feedback and adjustment based on field practice. Their corresponding engineering management methods are as follows.

(1) When the monitored data are bigger than the values for safety grade, more attention should be paid to the monitoring work and stability of surrounding rock.

(2) When the monitored data are bigger than the values for prediction grade, more attention should be paid and some controlling measures should be implemented to reinforce the surrounding rock.

(3) When the monitored data are bigger than the values for danger grade, the emergency measures should be implemented immediately to reinforce the surrounding rock.

This method has been used successfully in the underground cavern group of the Jinping II hydropower station. At present, it is being used in the underground cavern groups of Baihetan and Wudongde hydropower stations. The established deformation management classification of surrounding rock for the excavation of each layer of Jinping II's main powerhouse, and its practice, are shown in Table 9.

5.3 Methods for predicting the deformation of surrounding rock during excavation

The stability of high sidewalls is very important to the large underground main powerhouse. Their deformation, whether caused by the excavation unloading or unstable rock blocks or other factors, must be controlled with bolts or anchor cables or both. In order to design their lengths, the deformations of high sidewalls and the depths this affects should be calculated or predicted before the excavation. So Zhou & Li *et al.* calculated about one hundred examples with numerical analysis methods based on the underground cavern group model of Longtan hydropower station. The displacement changing rules of its sidewalls with four rock types, four types of buried depth, and different *in situ* stress were studied. Then the displacement predicting formula for the surrounding rock under different conditions was worked out, and the specific values of relative displacements to evaluate the stability of surrounding rock were suggested,

Table 9 The established deformation management classification and its practice at the excavation of Jinping II's main powerhouse.

Stage	Layer	Location	*Deformation management standards*						*Displacement meter with large deformation*	*Field verification*
			Safety grade		*Prediction grade*		*Danger grade*			
			Deformation increment (mm)	*Rate* $(mm{\cdot}d^{-1})$	*Deformation increment (mm)*	*Rate* $(mm{\cdot}d^{-1})$	*Deformation increment (mm)*	*Rate* $(mm{\cdot}d^{-1})$		
1	I	Crown	10	0.10	15	0.20	20	0.3		
		Upstream	8	0.15	12	0.25	16	0.4		
		Downstream	5	0.15	8	0.25	13	0.4		
2	II	Crown	4	0.10	6	0.20	10	0.3		
		Upstream	15	0.15	20	0.25	30	0.4	Mcf0+192-1 adding 20 mm	Rock blocks collapse for joints.
		Downstream	8	0.10	12	0.25	20	0.4	Mcf0+000-4 adding 15 mm	Shotcrete crack 10–12 m long.
3	III	Upstream	18	0.20	25	0.30	35	0.6	Mcf0+263.6-1 adding 29.2 mm, rate 1.1–3.4 mm/d	Soft fracture zone revealed.
		Downstream	12	0.20	18	0.30	25	0.5		
4	IV	Upstream	15	0.30	25	0.50	35	0.8	Mcf0+000-1 rate 0.4 mm/d after blasting	Shotcrete cracking, anchor cables overloading.
		Downstream	12	0.20	18	0.40	25	0.7		
	V_{up}	Upstream	15	0.30	25	0.50	35	0.8	Monitored values at EL 1,334.5 m over predicting or dangerous grade	Several cracks appearing at the third layer draining tunnel.
		Downstream	12	0.20	18	0.40	25	0.7		
5	V_{down}	Upstream	15	0.30	20	0.50	30	0.7	Average rate of section S_4 reaching 0.3 mm/d	Cracks appearing at bus tunnels, anchor cables overloading under rock anchor beam.
		Downstream	15	0.30	20	0.50	30	0.7		
	VI	Upstream	15	0.30	20	0.50	30	0.7	Monitored values at EL 1,322 m over predicting or dangerous grade	Several anchor cables overloading.
		Downstream	15	0.30	20	0.50	30	0.7		
6	VII	Upstream	12	0.30	18	0.50	24	0.7	The adding deformations of sections S_2 and S_3 at five elevations larger than 20 mm and 35 mm, respectively	Few anchor cables overloading.
		Downstream	15	0.30	20	0.50	25	0.7		
	VIII	Upstream	10	0.30	15	0.50	20	0.7		
		Downstream	12	0.30	18	0.50	24	0.7		

which can be used to judge whether anchor cables needed to be used or not. The displacement predicting formula is shown as follows.

$$u = h[a(1000\lambda\gamma H/E^2) + b(1000\lambda\gamma H/E) + c] \times 10^{-3} \quad (1)$$

where

u is the key point of the sidewall displacement (cm)
h is the height of the main powerhouse (m)
λ is the coefficient of horizontal *in situ* stress
γ is the self-weight of the rock mass (kN/m^3)
$H=H_0+h$ is the buried depth to the bottom of the main powerhouse (m)
E is the deformation modulus of the rock mass (MPa)
a, b and c are the fitting constants concerning with the project layout and characteristics.

This formula was also used to analyze the high sidewall deformation of underground cavern groups of Ertan, Sanxia, Xiluodu, Dagangshan, Shuibuya and four other hydropower projects. The predicted displacements were compared with the numerical results calculated by others, which showed that this method was suitable. The method using the relative ratios of elastic–plastic displacements was used to evaluate the degree of deterioration in surrounding rock, which can be regarded as warning values or criteria by which to judge whether long anchor cables to the high sidewall should be used or not.

Meanwhile, the predicting formulas of the maximum and average displacements of main powerhouse rock anchor beams were presented by Feng & Zhang (2015), based on the monitoring data of eight large underground cavern groups, where the cavern height to its span, and the rock strength stress ratio were mainly considered to reflect the deformation stability of the high sidewall. The statistical results of different projects can be seen in Table 10.

Table 10 Rock anchor beam displacement and strength–stress ratio.

Project name	*Powerhouse upstream sidewall*		*Powerhouse downstream sidewall*		*Average of maximum displacement (mm)*	*Average of average displacement (mm)*	*Average rock strength stress ratio*
	Rock anchor beam maximum displacement (mm)	*Rock anchor beam average displacement (mm)*	*Rock anchor beam maximum displacement (mm)*	*Rock anchor beam average displacement (mm)*			
Ludila	34.49		65.67		50.08		6.97
Jinping II	90.09	65.50	43.63	11.05	66.86	38.28	2.84
Jinping I	31.81	23.20	75.87	56.40	53.84	39.80	1.89
Laxiwa	35.60	18.60	28.40	25.70	32.00	22.15	3.70
Pubugou	65.12	39.76	77.39	35.16	71.26	37.46	3.30
Dagangshan	34.41	19.69	21.92	12.89	28.17	16.29	3.68
Guandi	36.05	18.72	23.57	12.33	29.81	15.53	4.14
Xiluoduzuoan	15.20	8.90	41.10	20.20	28.15	14.55	6.00
Xiluoduyouan	33.40	13.90	22.80	11.30	28.10	12.60	5.48

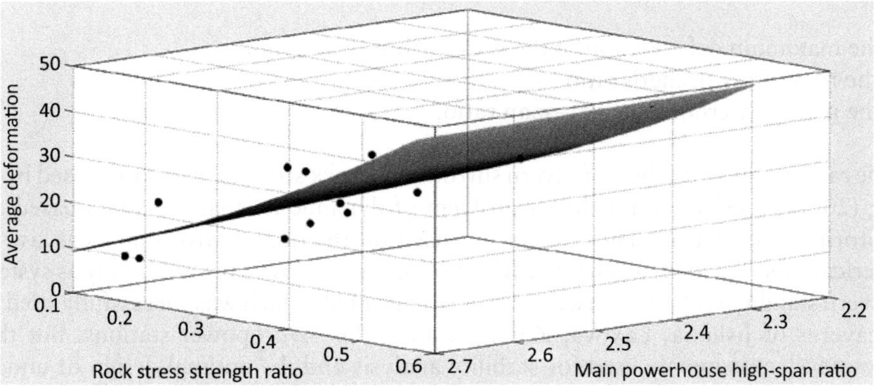

Figure 32 Statistical results of rock anchor beam average deformation.

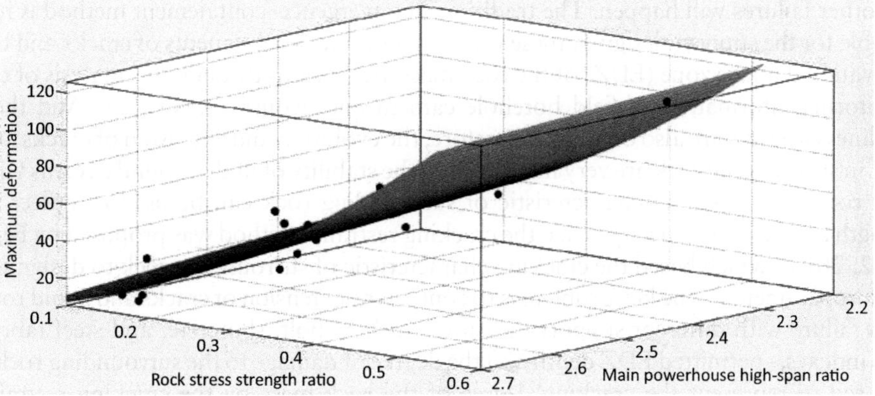

Figure 33 Statistical results of rock anchor beam maximum deformation.

The statistical relationship between monitored displacements of rock anchor beams and cavern heights, cavern spans, and rock strength stress ratios can be seen in Figures 32 and 33.

$$z = 49.42x^2 + 16.43xy - 2.817y^2 + x + y + 22.06R^2 = 0.7477 \tag{2}$$

where

z is the average deformation of the rock anchor beam (mm)
x is the rock stress strength ratio
y is the main powerhouse height–span ratio.

$$z = 278.9x^2 + 19.4xy - 6.583y^2 + x + y + 46.86R^2 = 0.8959 \tag{3}$$

where

z is the maximum deformation of the rock anchor beam (mm)
x is the rock stress strength ratio
y is the main powerhouse height–span ratio.

The rapid analysis method of cavern surrounding rock stress field was proposed by Li Ning (2015) to solve the timeliness problem of dynamic feedback analysis based on monitoring information. This method is based on the quantitative statistical law of numerical experiment results with different effecting factors. This rapid analysis system of cavern stability with one or two faults was presented, which was successfully used to the caverns of Jishixia, Laxiwa, Zipingpu and other hydropower stations. But this analysis system is mainly used for stability analysis and the optimal design of underground caverns with low *in situ* stress. For the large underground cavern groups with high *in situ* stress, its applicability still needs to be studied, and other suitable method should be proposed.

For the underground caverns with hard rock and medium or high *in situ* stress, the displacements after excavation are very small, less than 3 mm, while rockburst, buckling and other failures will happen. The traditional convergence–confinement method is not suitable for the support design of these rock masses. The developments of cracks and the excavation damage zone (EDZ) of the rock mass are obvious through the analysis of the monitoring information of field borehole camera and acoustic wave tests. And their timeliness features are also obvious. Therefore, the evolution and extension of cracks and EDZ in surrounding rock are very important to the stability of underground caverns with hard rock. The cracking characteristic of surrounding rock can be used to reflect its strength and bearing capacity. Then the cracking restraint method was proposed by Feng (2012, 2015), which based the cracking characteristic of surrounding rock to design the support parameters. The key problem is to control the extension of cracks and avoid rock mass failure with different support measures, such as bolt, shotcrete, and steel fabric. Two indexes – permitted EDZ depth and the degree of damage to the surrounding rock – are used to represent the cracking degree of the rock mass by the cracking restraint method. The purpose of the supporting design with this method is to determine suitable supporting measures, an acceptable distance from the tunnel face, and opportunity after excavation to carry out the support of the surrounding rock, which can contain the EDZ depth and degree of damage to the surrounding rock within an acceptable range.

The supporting design idea of the cracking restraint method is shown in Figure 34. If the supporting time is too late and the strength of the support system is not strong – curve a in Figure 34 – the EDZ depth and degree of damage to surrounding rock cannot be controlled effectively, as shown by curve A, meanwhile the bolt axial force is less than its design strength. If the support time is too early, shown as the curve c, the bolt axial force will exceed its ultimate strength and the EDZ depth and degree of damage to the surrounding rock will increase obviously and cannot be controlled effectively, shown as the curve C. Only when the supporting strength and supporting time are all suitable, shown as the curve b, can the EDZ depth and degree of damage to surrounding rock be controlled effectively and the bolt axial force can play an efficient role, shown as the curve B. If the EDZ depth and degree of damage to surrounding rock are too big, prestressed bolts can be used for support, shown as the curve d. This method

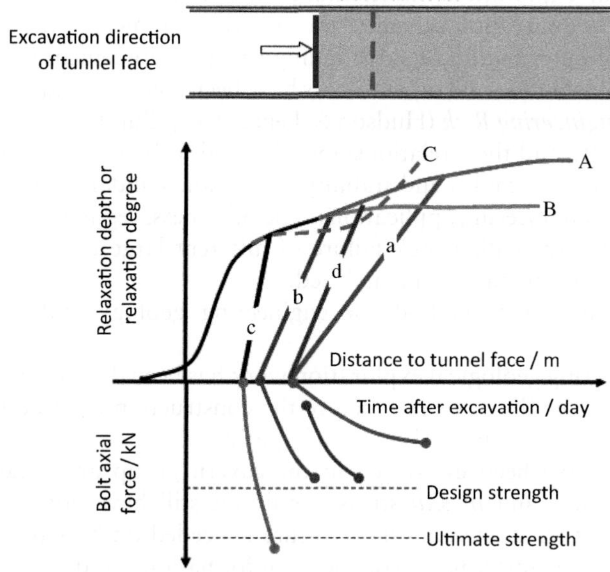

Figure 34 **Determining method of bolt supporting time based on cracking restraint method.**

has been used on the headrace tunnels of the Jinping II hydropower project and the diversion tunnels of the Baihetan hydropower project.

6 RESEARCH PROSPECTS OF LARGE UNDERGROUND CAVERN GROUPS

The engineering geological conditions, rock mass structural features, *in situ* stress, excavation unloading, and disturbance are all the key problems in the construction of large underground cavern groups with large spans and high sidewalls. Although several large underground cavern groups have been built, the suitable stability evaluation standards and measures to control the surrounding rock with high *in situ* stress and complex geological conditions. So the rapid evaluation, dynamic feedback and optimized design system for surrounding rock stability and effective support which reflect the mechanical response, deformation and fracture features of surrounding rock should be presented for the construction of cavern groups. Meanwhile, as the scale of hydropower underground cavern groups becomes larger and larger, the geological conditions become more complex. There are lots of epistemic uncertainty and aleatory uncertainty during the exploration, design and construction process of such projects. As to how to reduce the effect of different kinds of uncertainty on project construction, some work still needs to be done, some areas where research could be valuable are as follows.

(1) Statistical analysis of large underground cavern groups
Sixty large hydropower underground cavern groups with spans bigger than 20 m in China have been preliminarily analyzed by Feng & Zhang (2015).

The geological exploration information; information revealed by geological excavation; the excavation schemes; the support parameters; the failure types and degrees of deformation of surrounding rock; reinforcement measures; field monitoring information and other factors have been collected and analyzed in the book *Rock Engineering Risk* (Hudson & Feng, 2015). But these information sets are not complete and their relationships still need to be further analyzed, from which experiment formulas and stability evaluation standards can be proposed for subsequent engineering applications. The otherness, regularity and evolution mechanism of surrounding rock failure of different large underground cavern groups should be compared and analyzed.

(2) Reasonable expression method for engineering geology and hydrogeology conditions
Lots of engineering geological exploration work has been done over a long time – in some cases over thirty years – before the construction of large hydropower underground cavern groups. Although a large number of exploration tunnels and drilling holes have been excavated in the cavern group area, the rock mass structural features and *in situ* stress conditions still have not been mastered completely. Further discussion in the future is needed on how to get values to represent the engineering geological information and hydrogeological information, and then to use them to analyze the engineering stability and parameters for support design; how to reflect the effect of uncertainties on the construction of large underground cavern groups; and how to use the risk management theory during their design and construction periods.

(3) Reasonable layout of large underground cavern groups with complex conditions
Determining the axial direction of main powerhouses and the distances between caverns are usually based on the rock mass structural characteristics when the *in situ* stress is low. If the *in situ* stress is high and the second principal stress is less than the first principal stress, the axial direction is determined mainly according to the occurrence of the first principal stress and rock mass structural characteristics, at hydropower stations such as Jinping I, Jinping II, Guandi, and Laxiwa. When the second principal stress is also high, its influence should also be considered, such as at the underground cavern group of the Houziyan hydropower station. The influence of these factors on cavern group stability and their layout principle should still be researched, so as to derive rules for the future. In addition, the determining principles of distances between caverns with high *in situ* stress still needs to be discussed. Other questions that need answering are why the deformation of the rock wall between main powerhouse and transformer chamber caused by excavation unloading is so large, such as at the Jinping I and Houziyan hydropower projects which have similar first principal stresses, how to avoid the failure of prestressed anchor cables and how to optimize excavation schemes and support parameters.

(4) Rapid intelligent feedback analysis and dynamic optimization design methods
For large underground cavern groups with complex conditions, the three-dimensional numerical analysis method based on intelligent optimization dynamic feedback analysis and field monitoring information has been presented by Feng *et al.* (2013) Meanwhile, the deformation management classification to evaluate the surrounding rock stability has been established

for the underground cavern group of the Jinping II hydropower station. In addition, the cracking restrain method has been proposed for the design the support for the headrace tunnels of Jinping II hydropower project, and the diversion tunnels of the Baihetan hydropower project. But the timeliness of these methods needs to be improved and rapid intelligent feedback analysis and dynamic optimization design methods need to be established to satisfy the requirements of construction.

REFERENCES

Ali, A. (2014) Optimum depth of grout curtain around pumped storage power cavern based on geological conditions. *Bulletin of Engineering Geology and the Environment*, 73: 775–780.

Brown, E.T. (2012) Risk assessment and management in underground rock engineering-an overview. *Journal of Rock Mechanics and Geotechnical Engineering*, 4(3): 193–204.

Cai, B., Deng, Z.W. & Wu, G.Z. (2012) Stability evaluation on rockslide zone of Dagangshan underground powerhouse. *Yangtze River*, 43(22): 33–35.

Cai, D.W., Wang, D.K. & Chen, X.P. (2014) Analysis and evaluation system of rapid monitoring of large-scale underground cavern group during construction period in southwest china. *Chinese Journal of Rock Mechanics and Engineering*, 33(11): 2341–2350.

Chen, B.R., Li, Q.P. & Feng, X.T., etc. (2014) Microseismic monitoring of columnar jointed basalt fracture activity: a trial at the Baihetan hydropower Station, China[J]. *Journal of Seismology*, 18: 773–793.

Chen, B.R., Feng, X.T. & Li, Q.P., etc. (2015) Rock burst intensity classification based on the radiated energy with damage intensity at Jinping II hydropower station China. *Rock Mechanics and Rock Engineering*, 48: 289–303.

Chen, F., He, C. & Deng J.H. (2015) Concept of high geostress and its qualitative and quantitative Definitions. *Rock and Soil Mechanics*, 36(4): 971–980.

Chen, J.L., Jiang, Q. & Zhou, H. (2008) Scheme design of underground powerhouse junction for Jinping II hydropower station. *Rock and Soil Mechanics*, 29(Supp.1): 31–36.

Chen, Z.A., Sun, Z.L. & Peng, S.B, etc. (2000) *Hydropower Engineering in China (Engineering Geology Serier)*. Beijing, China Electric Power Press.

Company Standard of HydroChina Corporation (2012) Q/HYDROCHINA 009–2012. *Underground Powerhouse Design Guideline of Hydropower Station*. Beijing, Hydrochina Corporation.

Dai F., Li, B. & Xu, N.W., etc. (2015) Excavation damaged zones characteristics analysis in deep-buried underground powerhouse of Houziyan hydropower station. *Chinese Journal of Rock Mechanics and Engineering*, 34(4): 735–746.

Ding, X.L., Dong, Z.H., & Lu, B., etc. (2008) Deformation characteristics and feedback analysis of surrounding rock of large underground powerhouses excavated in steeply dipped sedimentary rock strata. *Chinese Journal of Rock Mechanics and Engineering*, 27 (10): 2019–2026.

Dong, J.X., Xu G.L. & Li Z.P., etc. (2014) Classification of failure modes and controlling measures for surrounding rock of large-scale caverns with high geostress. *Chinese Journal of Rock Mechanics and Engineering*, 33(11): 2161–2170.

Dong, Z.H., Ding X.L. & Chen S.H., etc. (2015) Monitoring and deformation mechanism analysis on rock-bolted crane girders in large underground powerhouse in high geo-stress area. *Journal of Hhydroelectric Engineering*, 34(2): 156–163.

Fan, Q.X. & Wang, Y.F. (2010) Stability analysis of layered surrounding rock mass of large underground powerhouse of Xiangjiaba hydropower Station. *Chinese Journal of Rock Mechanics and Engineering*, 29(7): 1307–1313.

Fan, Q.X. & Wang, Y.F. (2011) A Case Study of Rock Mass Engineering of Underground Powerhouse at Xiluodu hydropower Station. *Chinese Journal of Rock Mechanics and Engineering*, 30(S1): 2986–2993.

Fang, D., Chen, J.L. & Zhang, S. (2013) Stability analysis of surrounding rocks of underground powerhouse in Yangfanggou hydropower station. *Chinese Journal of Rock Mechanics and Engineering*, 32(10): 2094–2099.

Feng, X.T. (2000) *Introduction of Intelligent Rock Mechanics*. Beijing, Science Press.

Feng, X.T. & Hudson, J.A. (2011) *Rock Engineering Design*. London, CRC Press, Taylor & Francis.

Feng, X.T. & Hudson, J.A. (2015) *Rock Engineering Risk*. London, CRC Press, Taylor & Francis.

Feng, X.T., Cheng, B.R. & Zhang, C.Q., etc. (2013) *Mechanism, warning and dynamic control of rockburst development processes*. Beijing, Science Press.

Fu, B.J., Liu, Y.X. & Ma, Y.X. (1992) The development of rock caverns in hydraulic engineering of China and the case history of Lubuge project. *Chinese Journal of Rock Mechanics and Engineering*, 11(1): 79–87.

Hibino, S. (2001) Rock mass behavior of large-scale cavern during excavation and trend of underground space use. *Journal of the Mining and Materials Processing Institute of Japan*, 117(3): 167–175.

Hou, D.Q., Feng, M. & Liao, C.G., etc. (2012) Engineering practice of anchor cables supporting to large-scale underground caverns. *Chinese Journal of Rock Mechanics and Engineering*, 31 (5): 963–972.

Huadong Engineering Corporation Limited of Hydrochina Corporation. (2005) *Feasibility Study Report of Jinping II Hydropower Station along Yalong River-Engineering Geology (second volume)*. Hangzhou, China.

Huang, D. (2007) *Study on Unloading Deformation Mechanism and Stability of Excavating surrounding Rock Mass of Large Underground Caverns*. Chengdu, PhD thesis of Chengdu University of Technology.

Huang, D., Huang, R.Q. & Zhang, Y.X. (2009) Analysis on influence of fault location and strength on deformation and stress distribution of surrounding rocks of large underground openings. *Journal of Civil, Architectural & Environmental Engineering* 31(2): 68–73.

Huang, Q.X., Deng, J.H. & Su, P.Y., etc. (2011) Displacement characteristics analysis of surrounding rock in underground powerhouse chambers at Pubugou hydropower station during construction. *Chinese Journal of Rock Mechanics and Engineering*, 30(S1): 3032–3042.

Huang, Q.X., Wang, J.L. & Deng, J.H. (2013) Deformation mechanism analysis of surrounding rock at arch crown in underground powerhouse. *Chinese Journal of Rock Mechanics and Engineering*, 32(S2): 3520–3526.

Huang, Q.X., Deng, J.H. & Su, P.Y. (2013) Effects of excavation subsequence of busbars tunnel on displacement of surrounding rock mass. *Chinese Journal of Rock Mechanics and Engineering*, 32(S2): 3658–3665.

Huang, R.Q., Huang, D. & Duan, S.H., etc. (2011) Geomechanics mechanism and characteristics of surrounding rock mass deformation failure in construction phase for underground powerhouse of Jinping I hydropower station. *Chinese Journal of Rock Mechanics and Engineering*, 30(1): 23–35.

Huang, S.L., Ding, X.L. & Liao, C.G., etc. (2014) Initial 3d geostress field recognition of high geostress field at deep valley region and considerations on underground powerhouse layout. *Chinese Journal of Rock Mechanics and Engineering*, 33(11): 2210–2224.

Huang, S.L., Wang, J.M. & Ding, X.L., etc. (2011) Stability and control for underground caverns of Jinping I hydropower station based on unloading evolution of layered rock mass. *Chinese Journal of Rock Mechanics and Engineering*, 30(11): 2203–2216.

Hydro China Corporation. (2008) Q/CHECC 003–2008 *Design specification for rock bolt crane girder in underground powerhouse*. Hydro China Corporation Enterprise Standard.

Jiang, Q. (2010) *Study on model and stability of surrounding rock of large underground caverns under high geo-stress condition*. Wuhan, PhD thesis of Institute of Rock & Soil Mechanics, Chinese Academy of Sciences.

Jiang, Quan, Xiating Feng. (2011) Intelligent stability design of large underground Hydraulic caverns: Chinese method and practice. *Energies*, 4(10): 1542–1562.

Jiang, Q, Xiating Feng, & Xiang, T.B. *et al.* (2010) Rockburst characteristics and numerical simulation based on a new energy index: A case study of a tunnel at 2,500 m depth. *Bulletin of Engineering Geology and the Environment*, 69: 381–388.

Jiang, Quan, Jie Cui & Jing Chen. (2012) Time-dependent damage investigation of rock mass in an in situ experimental tunnel. *Materials*, 5, 1389–1403.

Jiang, Quan, Xia-ting Feng & Jing Chen, *et al.* (2013) Estimating in-situ rock stress from spalling veins: A case study. *Engineering Geology*, 152(1): 38–47.

Jiang, Quan, Xia-Ting Feng & Jie Cui, *et al.* (2015) Failure mechanism of unbonded prestressed thru-anchor cables: In situ investigation in large underground caverns. *Rock Mechanics and Rock Engineering*, 48: 873–878

Jiang, Quan, Jie Cui & Xiating Feng, *et al.* (2014) Application of computerized tomographic scanning to the study of water-induced weakening of mudstone. *Bulletin of Engineering Geology and the Environment*, 73(4): 1293–1301.

Jiang, Quan, Feng, X.T. & Hatzor, H.Y. *et al.* (2014) Mechanical anisotropy of columnar jointed basalts: An example from the Baihetan hydropower station, China. *Engineering Geology* 175: 35–45.

Jiang Quan, Guo-shao Su, & Xia-ting Feng, *et al.* (2015) Observation of rock fragment ejection in post-failure response. *International Journal of Rock Mechanics & Mining Sciences*, 74: 30–37.

Li, G.L. & Wu, S.H. (2011) Analysis on phased safety monitoring of underground power house in construction period. *Yangtze River*, 42(14): 59–63.

Li, J.H., Wu, W.F. & Li, J.C. (2013) Control and monitoring of rock mass engineering of super large underground powerhouse cavern groups at Xiluodu hydropower station. *Chinese Journal of Rock Mechanics and Engineering*, 32(1): 8–14.

Li, N., Sun, H.C. & Yao, X.C., etc. (2008) Cause analysis of circumferential splits in surrounding rock of busbar tunnels in underground powerhouses and reinforced measures. *Chinese Journal of Rock Mechanics and Engineering*, 27(3): 439–446.

Li, Y. (2009) *Study on the stability induced by stepped excavations and rheological effect of underground cavern groups in high in-situ stress areas*. Jinan, PhD thesis of Shandong University.

Li, Y.Z., Wang, S.C. & Sun, L.P., etc. (2014) Engineering geological research on large span underground cavern in horizontal thin layered rock mass. *Rock and Soil Mechanics*, 35(8): 2361–2365.

Li, Z.G. (1999) Deformation and supporting of surrounding rock in caverns of underground powerhouse complex at Ertan hydropower Station. *Sichuan Water Power*, 18(2): 24–26.

Li, Z.K., Zhou, Z. & Tang, X.F., etc. (2009) Stability analysis and considerations of underground powerhouse caverns group of Jinping I hydropower station. *Chinese Journal of Rock Mechanics and Engineering*, 28(11): 2167–2175.

Li, Z.P., Xu, G.L. & Dong, J.X., etc. (2014) Deformation and fracture of surrounding rock mass of underground caverns at houziyan hydropower station. *Chinese Journal of Rock Mechanics and Engineering*. 33(11): 2291–2300.

Lu, B., Ding, X.L. & Wu, A.Q., etc. (2012) Study of influence of rock structure on surrounding rock mass stability of underground caverns in hard rock region with high geostress. *Chinese Journal of Rock Mechanics and Engineering*, 31(S1): 3831–3846.

Lu, B., Wang, J.M. & Ding, X.L., etc. (2010) Study of deformation and cracking mechanism of surrounding rock of Jinping I underground powerhouse. *Chinese Journal of Rock Mechanics and Engineering*, 29(12): 2429–2441.

Lu, W.B., Chen, M. & Geng, X., etc. (2012) A study of excavation sequence and contour blasting method for underground powerhouses of hydropower stations. *Tunnelling and Underground Space Technology*, 29: 31–39.

Lu, X., Li, M.H. & Chen, Y.H., etc. (2015) Blockiness level of rock mass around underground powerhouse of three gorges project. *Tunnelling and Underground Space Technology*, 48: 67–76.

Ma, H.Q. (2006) Review and prospect of underground engineering and construction technologies in China. *Water Power*, 32(2): 52–55.

Ma, H.Q. (2011) A study on the status quo, development trend and the innovation frontline of hydraulic and hydroelectric underground projects. *Engineering Science*, 13(12): 15–19.

Pan, J.Z. & He, J. (2000) *Large Dams in China A Fifty-Year Review*. Beijing, China Water & Power Press.

Qin, J.B. & Feng, M.Q. (2007) Unloading and relaxation characteristics of Pengshui hydropower station main powerhouse downstream sidewall. *Yangtze River*, 38(9): 2254–2266.

Sa, W.Q., Zhang, S.R. & Zhang, L.M., etc. (2014) Real-time safety evaluation and early warning of large-scale underground cavern group during construction period based on internet of things. *Chinese Journal of Rock Mechanics and Engineering*, 33(11): 2301–2313.

Tang, J.F., Xu, G.Y. & Tang, X.M. etc. (2009) Cause analysis of longitudinal cracks of rock anchored beam and its development trend in underground powerhous. *Chinese Journal of Rock Mechanics and Engineering*, 28(5): 1000–1009.

Tang, J.F., Xu, G.Y. & Tang, X.M. etc. (2009) Cause analysis of longitudinal cracks of rock anchored beam and its development trend in underground powerhouse. *Chinese Journal of Rock Mechanics and Engineering*, 28(5): 1000–1009.

Wang, D.K., Peng, Q. & Tang, R. (2007) Causes analysis of cracking of rock-bolted crane girder in an underground powerhouse. *Chinese Journal of Rock Mechanics and Engineering*, 26 (10): 2125–2129.

Wang, D.K., Cai, D.W. & Dong, Y.F., etc. (2014) Analysis of deformation characteristics of surrounding rock of underground powerhouse of Houziyan Hydropower Station and its control. *Yangtze River*, 45(8): 66–69.

Wei, J.B. & Deng, J.H. (2010) Variation of excavation damaged zone and back analysis of large scale underground powerhouse with high geostress. *Rock Soil and Mechanics*, 31(S1): 330–336.

Wei, J.B., Deng, J.H. & Wang, D.K., etc. (2010) Characterization of deformation and fracture for rock mass in underground powerhouse of Jinping I hydropower station. *Chinese Journal of Rock Mechanics and Engineering*, 29(6): 1198–1205.

Xiao, H.Y. & Lin, J.C. (2014) Excavation technology of side wall of Houziyan hydropower station powerhouse in high in situ stress area. *Yangtze River*, 45(8): 74–77.

Xu, N.Q., Wang, J.S. & Li, Q.H., etc. (2014) Study of engineering geological problems in excavation of underground powerhouse cavern group. *Yangtze River*, 45: 85–88.

Xu, N.W., Li, T.B. & Dai, F., etc. (2015) Microseismic monitoring and stability evaluation for the large scale underground caverns at the Houziyan hydropower station in Southwest China. *Engineering Geology*, 188: 48–67.
Zhang, F., Zhou, S.D. & Guo, W. (2006) Layout and supporting design of Penshui hydropower station underground powerhouse. *Yangtze River*, 37(1): 35–36, 44.
Zhang, L.B., Cheng, L.J. & Hou, P., etc. (2014) Damage feature and support measures of surrounding rock of large-span cavern under high intermediate principal stress. *Journal of Yangtze River Scientific Research Institute*, 31(11): 108–113,119.
Zhang, Y., Xiao, P.X. & Ding, X.L., etc. (2012) Study of deformation and failure characteristics for surrounding rocks of underground powerhouse caverns under high geostress condition and countermeasures. *Chinese Journal of Rock Mechanics and Engineering*, 31(2): 228–244.
Zhu, W.S., Yang, W.M. & Xiang, L., etc. (2011) Laboratory and field study of splitting failure on side wall of large-scale cavern and feedback analysis. *Chinese Journal of Rock Mechanics and Engineering*, 30(7): 1310–1317.
Zhu, Z.Q., Sheng, Q. & Zhang, Y.H, etc. (2013) Research on excavation damage zone of underground powerhouse of Dagangshan hydropower station. *Chinese Journal of Rock Mechanics and Engineering*, 32(4): 734–739.
Zhu, W.S., Zhao, C.L. & Zhou, H., etc. (2015) Discussion on several key issues in current rock mechanics. *Chinese Journal of Rock Mechanics and Engineering*, 34(4): 649–658.
Zi, J.Q. (2006) *Study on the Construction Characteristic of Underground Carvens Group for Goupitan Hydropower Station*. Tianjin, Master thesis of Tianjin University.

Chapter 8

Heat transfer with ice-water phase change in porous media: Theoretical model, numerical simulations, and application in cold-region tunnels

Weizhong Chen[1,2] & *Xianjun Tan*[2]

[1]*Research Center of Geotechnical and Structural Engineering, Shandong University, Jinan, Shandong, China*

[2]*State Key Laboratory of Geomechanics and Geotechnical Engineering, Institute of Rock and Soil Mechanics, Chinese Academy of Sciences, Wuhan, Hubei, China*

1 INTRODUCTION

There are many cold region tunnels in China, Canada, Norway, Japan and Russia. In the northwest and northeast of China, because of the damage caused by frost and thaw action, some of the tunnels cannot be used for up to 8–9 months a year and transportation is negatively affected (Lai *et al.*, 1999). So the research work on reasonable cold-proof measures is of great importance to prevent their linings from being destroyed by the frost heave or thaw settlement for the cold region tunnels (Huang *et al.*, 1986; Johansen *et al.*, 1988; Qiao *et al.*, 2003). In Russia, the pipe-electric heaters heat the tunnels, and the new type and effective insulations preserve the heat of the cold region tunnels. In Norway, some of the drains in the tunnels are heated with the heated cables, some are fitted with double thermal insulation doors; and others have insulation installed. In Japan, the insulation materials are put on the lining surfaces of Shang Yu Huang tunnel at Hokkaido to prevent the local geothermal from giving out and to keep the temperatures on the lining surface from falling below the frozen temperature (Johansen *et al.*, 1986; Okada *et al.*, 1985; Sandegren, 1995; Zhang *et al.*, 2002; Guan, 2003; Lai *et al.*, 2005). In China, to keep the initial thermal situations of the surrounding rock from being broken and to prevent their linings from being destroyed, many useful methods, such as slurry injected, insulation material (such as polyphenyl plate, XPS plate), thermal insulating door, snowshed and drainage hole and so on, are proposed in cold region tunnels. Mie (1983) put forward a method for protecting tunnel from frost damage with the thermal insulation layer. Tang & Wang (2007) analyzed the effect of temperature control on a tunnel in permafrost, and the stability conditions related to the thawing depth and suitable for use during construction were established. According to the basal theories of heat transfer and seepage, considering the coupled effect of seepage field and temperature field, Zhang *et al.* (2007) studied the function of the insulation layer for treating water leakage in permafrost tunnels, three-dimensional nonlinear analysis was made either with or without the insulation layer, and the result showed that it is necessary for treating water leakage to fit the insulation layer between the two linings in Kunlun mountain tunnel. In order to research the actions of the thermal insulation material in the process of the re-frozen of permafrost Tunnel, the re-frozen analysis of Feng Huoshan tunnel with the thermal insulation material was made by Zhang *et al.* (2004). According to the observation in-situ temperature of Daban mountain tunnel in Qinghai province either with or without the insulation layer, Zhang (2000) suggested that the thermal insulating door and the snowshed were the useful method to prevent the tunnel from frost. Based on the field test of chosen material's performance of anti-freezing thermal protective layer in Daban mountain tunnel and field contrast experiments, optimum combination and construction of anti-freezing thermal protective layer materials are determined by Chen & Zan (2003). After these field experiments, a design method for anti-freezing thermal protective layers in cold area tunnels was given by Chen (2004). Lai *et al.* (2003) studied the methods for controlling frost heave in cold region tunnels. Zhou (2003) studied the constructing technology of insulation layer, water-proof and drainage water in Kunlun mountain tunnel on Qinghai–Tibet railway.

Heat transfer in freezing porous media is a complex process because of its multi-phase nature. Three important effects associated with the freezing process are: (1) effect of the latent heat during the phase change of water, (2) the nonlinear thermal properties of soil, and (3) the existence of unfrozen water in frozen porous media. With the development of economy in China and abroad, transportation infrastructure is in urgent need to extend to the poor natural conditions, remote high-altitude areas and inaccessible gorges (Jin, 2010; Cheng *et al.*, 2009). After years of engineering practice and scientific research, people gradually realized that temperature field is the foundation of frozen ground engineering, frozen soil environment, permafrost survey and forecast, and so on (Jin *et al.*, 2010; Li *et al.*, 2010; Qin *et al.*, 2009). Since the 1970s, along with the development of calculation technology, the numerical calculation method was widely used in temperature field research, which considered the influence of phase change, latent heat and so on (*e.g.*, Bonacina *et al.*, 1973; Comini *et al.*, 1974; Ershov *et al.*, 1979; Lai *et al.*, 1998, 2005; Tang *et al.*, 2007). He *et al.* (1996) analyzed and preliminarily predicted the freezing–thawing situation in the rock surrounding DabanShan tunnel. After that, He *et al.* (1999) studied the freezing–thawing situation of a rock surrounding tunnel in cold regions by a combined convection–conduction model of turbulent airflow in the tunnels and temperature field. Lai *et al.* (2002) proposed approximate analytical solution for temperature field in circular tunnels in cold regions using dimensionless and perturbative method. Zhang *et al.* (2002, 2004) gave a nonlinear analysis for the three-dimensional temperature fields in cold region tunnels based on the governing differential equations of the problem on temperature field with phase change.

As noted above, most of the studies for heat transfer mechanisms with ice-water phase change occurring in porous media assumed that the phase front is a mathematical surface separating the whole region into frozen and unfrozen zones: a two-zone model, however, it was known from experimental research (Konrad & Morgenstern, 1980; Danielian *et al.*, 1983) and theoretical analysis (Nakano, 1990; Furukawa & Shimada, 1993; Mottagy & Rath, 2006) that "two-zone" model (frozen zone and unfrozen zone) could not accurately reflect the freezing-thawing process, they thought that there is a "freezing zone" or "frozen fringe" between the frozen zone and unfrozen zone, the size of this area mainly depends on nature characteristic and temperature condition, that is so-called "three zone" theory. Bronfenbrener & Korin (1997) indicated that it is necessary to use a three-zone model in the process of freezing in fine-grained loamy soils. Although the analytical solutions (Lunardini *et al.*, 1991; Bronfenbrener, 2009) were given for the propagation of subfreezing temperatures in a porous semi-infinite, initially unfrozen media with time, t, analytic method often only suitable for simple conditions. Due to the complexity of geotechnical engineering, numerical solution is a good choice. McKenzie *et al.* (2006) successfully simulated a fully saturated, coupled porewater-energy transport process with freezing and melting porewater, using the U.S. Geological Survey's SUTRA computer code, but there were so many parameters to establish that it must contain a few assumptions before simulation.

Rock deterioration under freeze-thaw cycles is a concern in many projects such as road, railroad, pipeline, and building constructions in cold region (Grossi, 2004; Zhang, 2004). Some engineering properties of rock is severely affected by freeze-thaw cycles such as strength and compressibility (Proskin, 2010; Beier, 2009), permeability (Saad, 2010; Takarli, 2008; Bellanger, 1993), and thermodynamics (Park, 2004, Yamabe, 2001). But, the influence of freeze-thaw cycles on the mechanical properties was most remarkable in geotechnical engineering (Hale and Shakoor, 2003; Hall, 1999; Ruiz, 1999; Prick, 1995; Altindag, 2004; Seto, 2010; Karaca, 2010).

The mechanism for rock freeze-thaw damage is that: Water in micropores expands about 9% of the original volume, when rock is frozen at low temperature; this expansion induces tensile stress concentration and damages the micropores; when rock is thawed, water flows through the fractured micropores which increases the damage (Hori, 1998, Chen, 2004). Rock damage in cold regions is resulted from the number of freeze-thaw cycle, temperature, rock type, applied stress and moisture content (Takarli, 2008; Chen, 2004). A detailed graphical record about the deterioration mode for ten types of sedimentary rocks due to freeze–thaw cycles is provided by Nicholson (2000). It showed that the presence or absence of rock flaws alone does not control the deterioration mode, but rather that it is the relationship among these flaws, rock strength and textural properties which exerts the greatest influence. The effect of freeze-thaw cycles on deterioration degree was proved to be connected with the moisture content (Chen, 2004; Matsuoka, 1990).

Galongla tunnel is an important part of Zha-mo highway in Tibet which is the only connection to Mo-tuo County in China. In Galongla tunnel, the freeze period is about 8 months; the average freeze depth is about 5 to 6 m and the maximum freeze depth is about 15m; the temperature in winter is low to–30°C; the altitude is about 4300m and the oxygen content in air is only 60% of that inland. There are very few highway tunnels built in the region at so low temperature and high altitude in China and Galongla tunnel is one of them.

Based on the local geological environments and developed model, some insulation measures have been taken at entrance and exit of tunnel to prevent the surrounding rock and concrete lining from frost damage. To validate the reliability of the proposed model, here the simulated results about temperature field of surrounding rock by finite element method using the "three-zone" model are compared with the field temperature data. Then, the effect of insulation material is studied.

Figure 1 The entrance of the tunnel in winter.

2 LABORATORY INVESTIGATIONS ON THE MECHANICAL PROPERTIES DEGRADATION OF GRANITE UNDER FREEZE-THAW CYCLES

The main purpose of tests is to study the damage characteristics after the granite undergoes different number of freeze-thaw cycles, including failure pattern, strength properties, deformation characteristics and freeze-thaw degradation features.

2.1 Rock samples and properties

The rock samples were obtained from Galongla mountain in Tibet of China, where a highway tunnel will across the mountain and it's very cold in winter. The situation is shown in Fig. 1, and the samples were prepared as cylinders with a diameter of 50mm and a height of 100mm.

2.2 Test process design

The uniaxial test results are not enough to reflect the actual condition and the tri-axial tests should be adopted. Because the freeze-thaw damage of tunnel surrounding rock happens around the entrance or exit where the confining pressure is not high, two confining pressures of 5MPa and 10MPa are selected.

In order to study the freeze-thaw degradation characteristics, all the samples were provided as saturated samples and were divided into three groups. Each group had 12 samples. The first group was for uniaxial compression test, the second group was for triaxial compression test with the confining pressure of 5MPa, and the third group was for triaxial compression test with the confining pressure of 10MPa. All samples in each group were divided into four sub-groups and there were three samples in each sub-group. The 1st sub-group in each group was for the respective tests without freeze-thaw. The 2nd, 3rd and 4th sub-groups were for the respective tests with a number of freeze-thaw cycles of 50, 100,150 respectively. The uniaxial and triaxial compression tests were done on the RMT multi-function test machine.

Figure 2 The damage phenomenon in the process of freeze-thaw test.

In the freeze-thaw cycles, the conditions of THC container are set as follows: One cycle including 4 h of freezing and 4 h of thawing; the temperature varied from +40 to -40°C; the humidity was kept 100% continuously; and the number of freeze-thaw cycles was from 0 to 150.

2.3 The damage phenomenon in the process of freeze-thaw cycles test

In the whole process of freeze-thaw cycles test, 3 samples were destroyed (shown Fig. 2). It can be seen that the failure forms are different from each other. The reason for this is that the samples were taken from Qinghai-Tibetan plateau and chronically underwent alpine glacier erosion, earthquake, freezing-thaw and a series of geological tectonic movement which made them more complex and different. These three samples were broken in 150 freeze-thaw cycles. The probable reason for it may be that they had been damaged before undergoing freeze-thaw cycles.

2.4 Results and analysis for uniaxial compression test

The stress-strain relationship curves of granite after undergoing different numbers of freeze-thaw cycles are shown in Fig. 3. The name of each sample was set as following rules (take "0-50-1" for example): "0" means that the confining pressure is 0MPa, "50" means that the sample underwent 50 freeze-thaw cycles, and "1" means that this is the No. 1 sample in this group. From the stress-strain relationship curves in Fig. 3, the uniaxial compressive strength and elastic modulus for each sample can be obtained, and the results are listed in Table1.

(1) Failure form

The typical failure forms for uniaxial compressive test after different freeze-thaw cycles are shown in Fig. 4. It can be seen that failure forms are different. They can be classified into two categories:

The first category (shown in Fig. 4(a)): there are several fractures along the axial direction of sample, and one fracture is throughout the sample. The reason

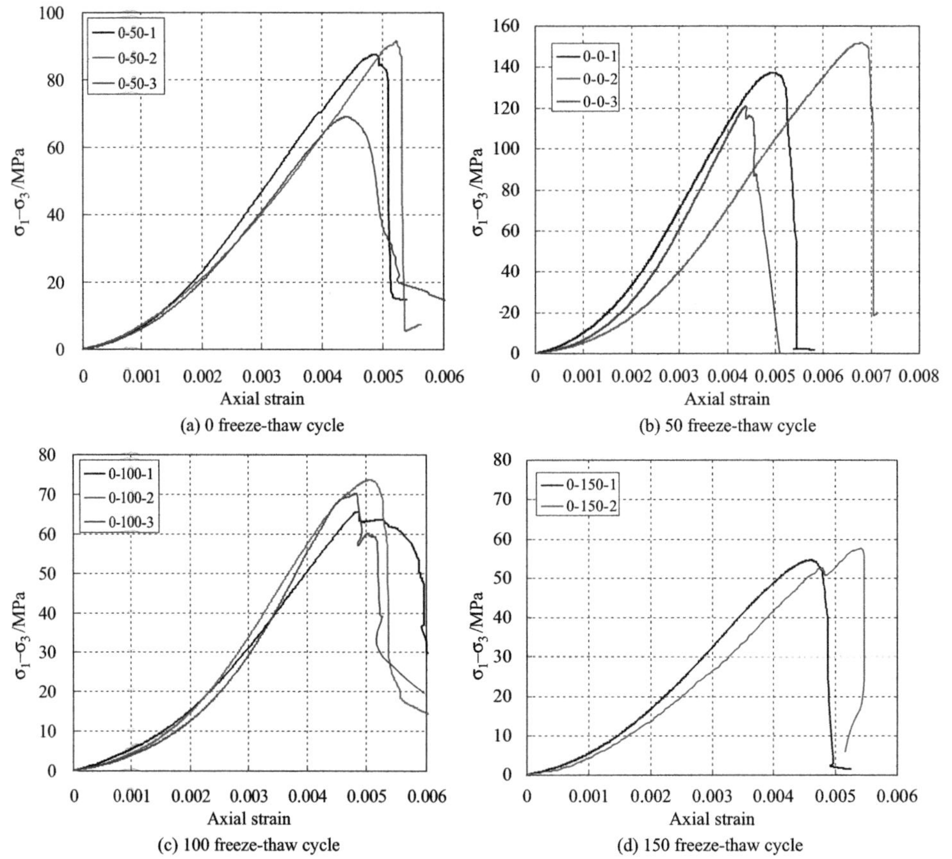

Figure 3 Stress-strain relationship for uniaxial compression tests after different number of freeze-thaw cycles.

Table 1 The uniaxial compressive test results after different number of freeze-thaw cycles.

serial numbers	*compressive strength (MPa)*				*elastic modulus (GPa)*			
	0	*50*	*100*	*150*	*0*	*50*	*100*	*150*
1	136.17	87.56	65.40	54.53	31.85	23.95	17.54	13.25
2	150.77	91.40	73.56	57.30	43.38	22.13	19.7	15.92
3	120.24	69.00	70.24	–	37.68	21.68	19.47	–
mean	135.73	82.65	69.73	55.92	37.64	22.53	18.90	14.59

for this phenomenon is that rock tensile strength is lower than the compressive strength. So, the failure surface is caused by tension.

The second category (shown in Fig. 4(b)): there are two crossing shear fractures which go through the whole sample together. In addition, there are some small

(a) Failure by tension (b) failure by shear

Figure 4 The typical failure forms for uniaxial compressive test after different freeze-thaw cycles.

shear fractures. It is concluded that some samples presented shear failure in the condition of uniaxial compression.

(2) Strength characteristics
It is shown that the uniaxial compressive strength decreases with the increasing of number of freeze-thaw cycles. The relationship between uniaxial compressive strength and the number of freeze-thaw cycles is shown in Fig. 5. An exponential function is adopted to fit the test results:

$$\sigma_b(N) = 55.28 + 80.15e^{-0.02N} \tag{1}$$

where, N is the number of freeze-thaw cycles; $\sigma_b(N)$ is the uniaxial compressive strength.

(3) Deformation characteristics
From the stress-strain curves (shown in Fig. 3), it is easy to found that the sample deformation has experienced four periods from loading to failure: compaction period, elastic growth period, plastic yield period and failure period. This reflects the commonness of rock failure process. To a certain extent, elastic modulus can reflect the deformation characteristics of rock. Therefore, the relationship between elastic modulus and the number of freeze-thaw cycles is shown in Fig. 6. An exponential function is adopted to fit the test results:

$$E(N) = 14.46 + 23.06e^{-0.02N} \tag{2}$$

where, $E(N)$ is the elastic modulus in the condition of uniaxial compression.

(4) Freeze-thaw degradation characteristics
In order to reflect the influence of freeze-thaw cycles on uniaxial compressive strength more intuitively, the stress-strain relationship for different freeze-thaw cycles are shown in one figure. Because the test results in each group have little

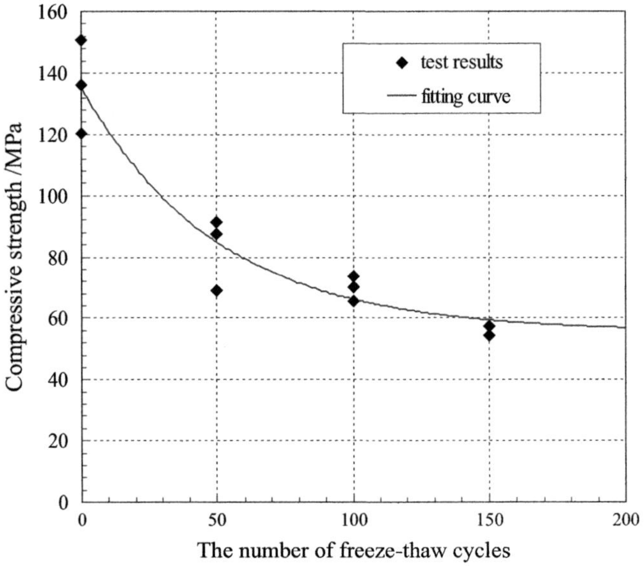

Figure 5 Relationship curve between uniaxial compressive strength and the number of freeze-thaw cycles.

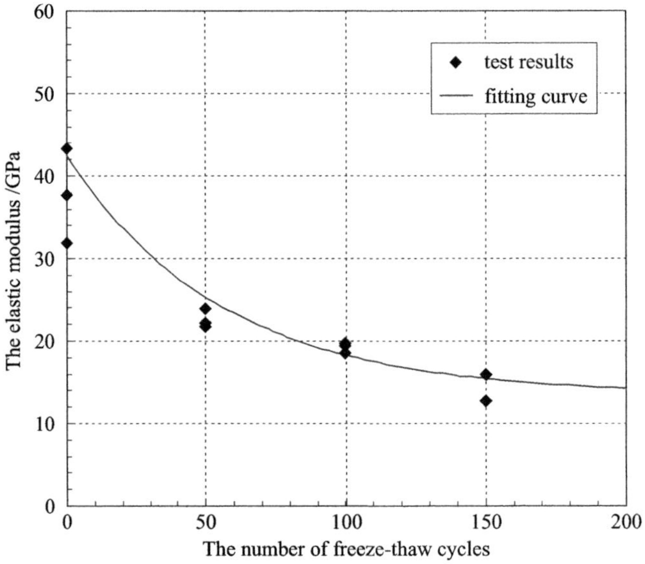

Figure 6 Relationship curve between elastic modulus and the number of freeze-thaw cycles.

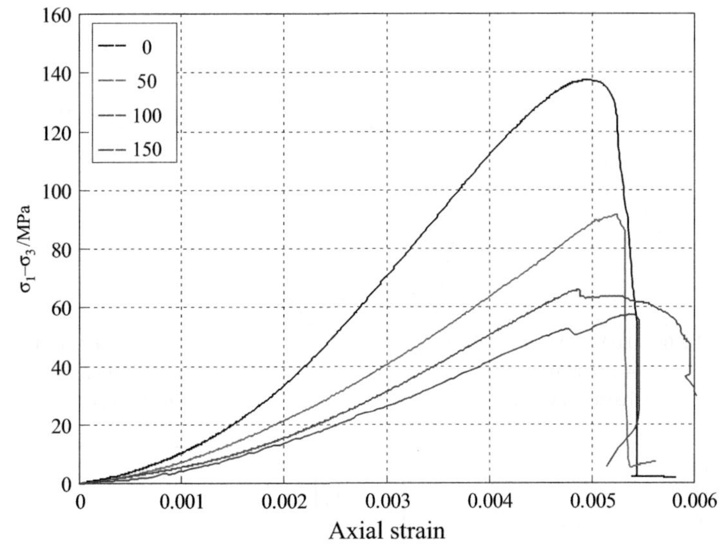

Figure 7 Stress-strain relationship for uniaxial compression tests after different number of freeze-thaw cycles.

Table 2 Triaxial compressive test results after different number of freeze-thaw cycles (confining pressure 5MPa).

serial numbers	*compressive strength (MPa)*				*elastic modulus (GPa)*			
	0	*50*	*100*	*150*	*0*	*50*	*100*	*150*
1	159.05	125.55	91.22	84.92	38.19	28.02	20.94	18.27
2	152.28	–	92.79	81.72	37.42	–	22.41	15.17
3	166.82	115.26	98.71	78.83	34.33	24.61	21.72	15.05
mean	155.67	120.41	92.01	78.83	36.65	26.32	20.07	16.16

difference, only one test in each group is shown in Fig. 7 to make graphics expression more clearly.

From Fig. 7, it is shown that the origin uniaxial compressive strength was 136.17MPa and it reduced to 57.30MPa after the sample underwent 150 freeze-thaw cycles with a loss of 58%.

2.5 Results and analysis for triaxial compression test

The stress-strain curves of granite after undergoing different number of freeze-thaw cycles are shown in Figs. 8 9 and 10. The uniaxial compressive strength and elastic modulus for each sample can be obtained, and the results are shown in Table 2 and Table 3.

(1) Failure form

The failure forms are simple in the condition of triaxial compression; all the samples are shear failure (shown in Fig. 11).

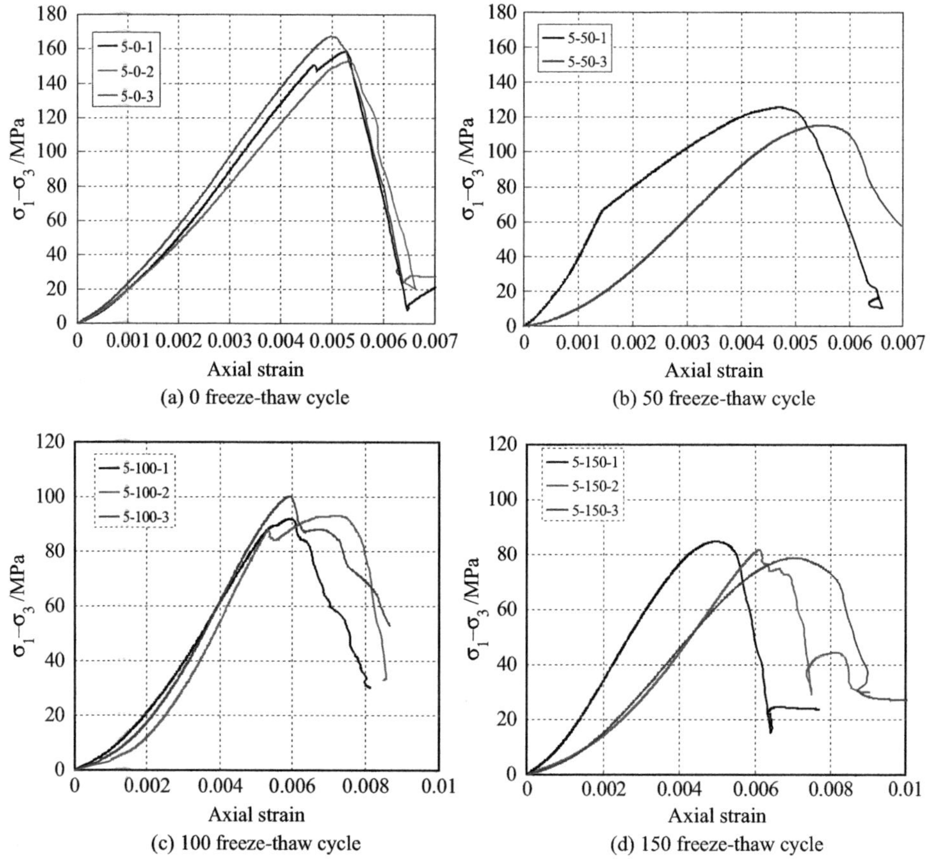

Figure 8 Stress-strain relationship for triaxial compression tests after different number of freeze-thaw cycles (confining pressure 5MPa).

Table 3 Triaxial compressive test results after different number of freeze-thaw cycles (confining pressure 10MPa).

serial numbers	*compressive strength (MPa)*				*elastic modulus (GPa)*			
	0	*50*	*100*	*150*	*0*	*50*	*100*	*150*
1	185.14	160.02	123.29	106.24	36.71	26.27	22.99	21.12
2	207.11	139.18	121.90	73.41	40.28	31.42	23.62	24.03
3	–	–	132.12	107.04	–	–	24.94	22.82
mean	196.13	149.90	125.77	95.56	38.50	28.85	23.85	22.66

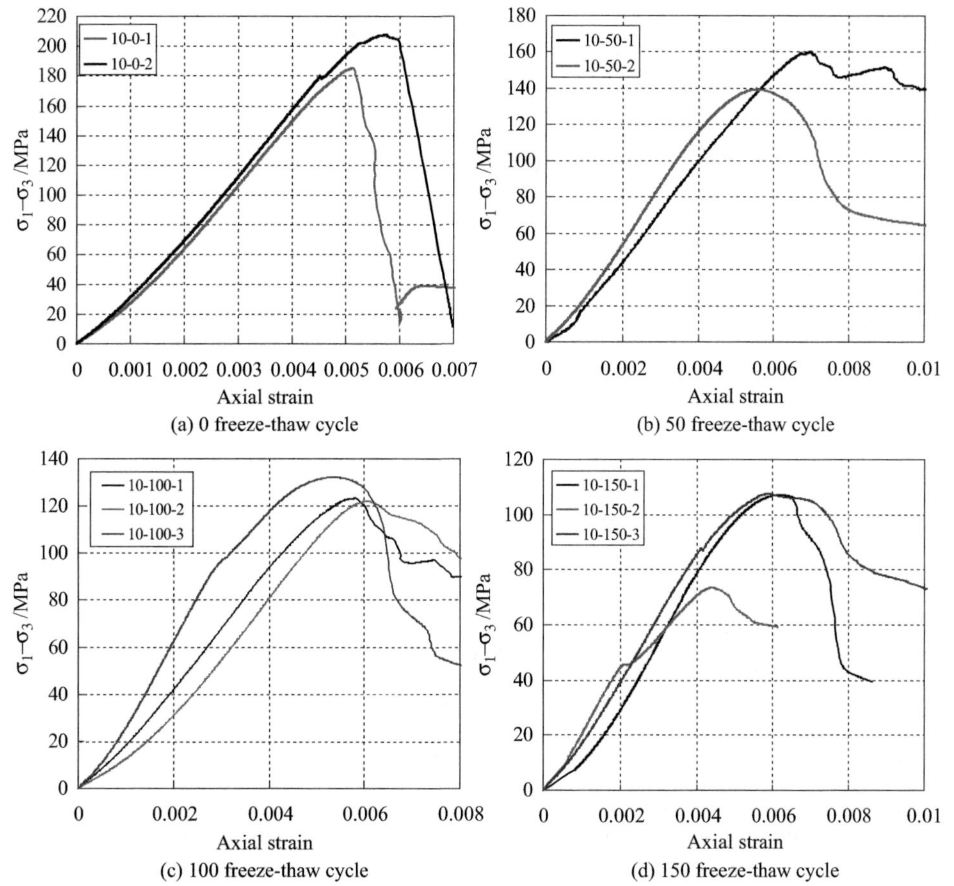

Figure 9 Stress-strain relationship for triaxial compression tests after different number of freeze-thaw cycles (confining pressure 10MPa).

(2) Strength characteristics

According to the test results, the failure forms of triaxial test are shear failure. Therefore, it is suitable to use Mohr – Coulomb strength criterion to describe the strength characteristics:

$$|\tau| = c + \sigma \tan\varphi \tag{3}$$

Where, τ and σ are the shear stress and normal stress at shear plane, respectively; c and φ are cohesive strength and internal friction angle, respectively.

The relationship between maximum principal stress and minimum principal stress can be obtained with maximum axial pressure as a vertical coordinate and confining pressure as a horizontal coordinate. The linear fit about the relationship is preceded. Combining with the following formulas, the cohesive strength c and internal friction angle φ can be calculated.

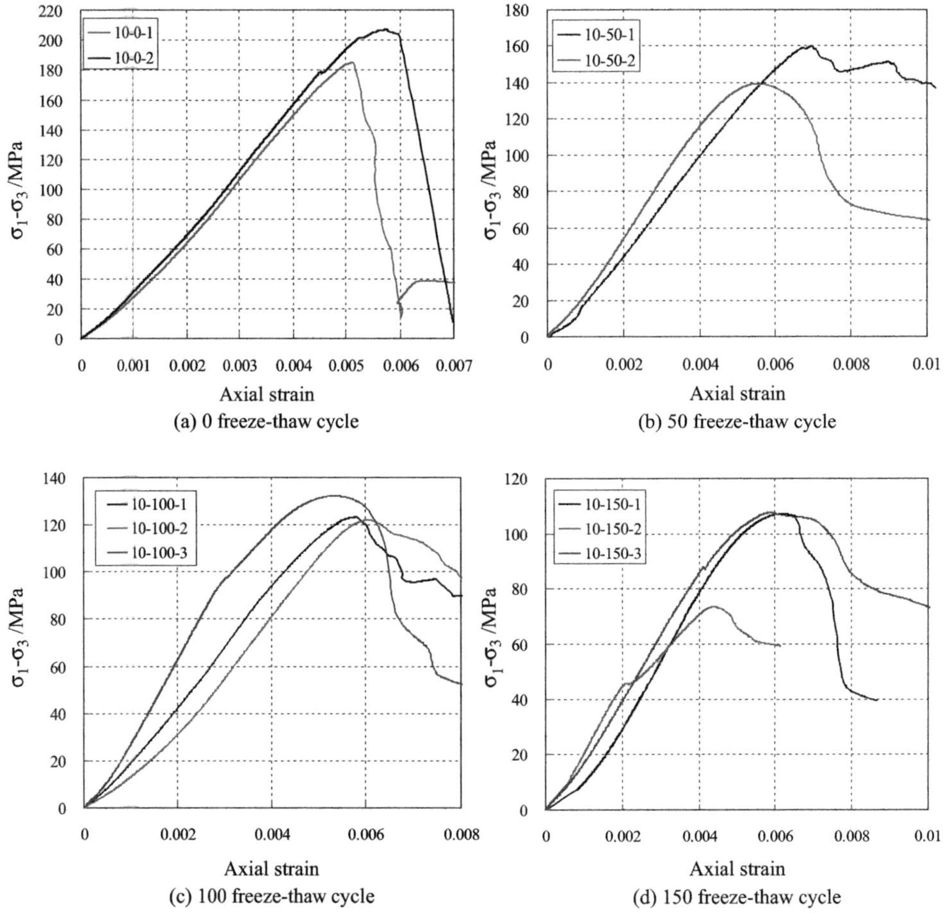

Figure 10 **Stress-strain relationship for triaxial compression tests after different number of freeze-thaw cycles (confining pressure 10MPa).**

$$c = b\frac{1 - \sin\varphi}{2\cos\varphi} \tag{4}$$

$$\varphi = \sin^{-1}\frac{m - 1}{m + 1} \tag{5}$$

Where, m is the slope of the fitted straight line and b is its intercept at vertical coordinate.

The data of compressive strength in Tables 2, 3 and 4 are extracted for drawing the scatter diagrams to describe the relationships between compressive strength and confining pressure and the experimental results are fitted into linear function (shown in Fig. 12).

According to Equation 4 and Equation 5, the cohesive strength c and internal friction angle φ with different number of freeze-thaw cycles are list in Table 4. It is shown that the cohesive strength decreases continuously with the increasing of the

Figure 11 The typical failure forms for triaxial compressive test after different freeze-thaw cycles.

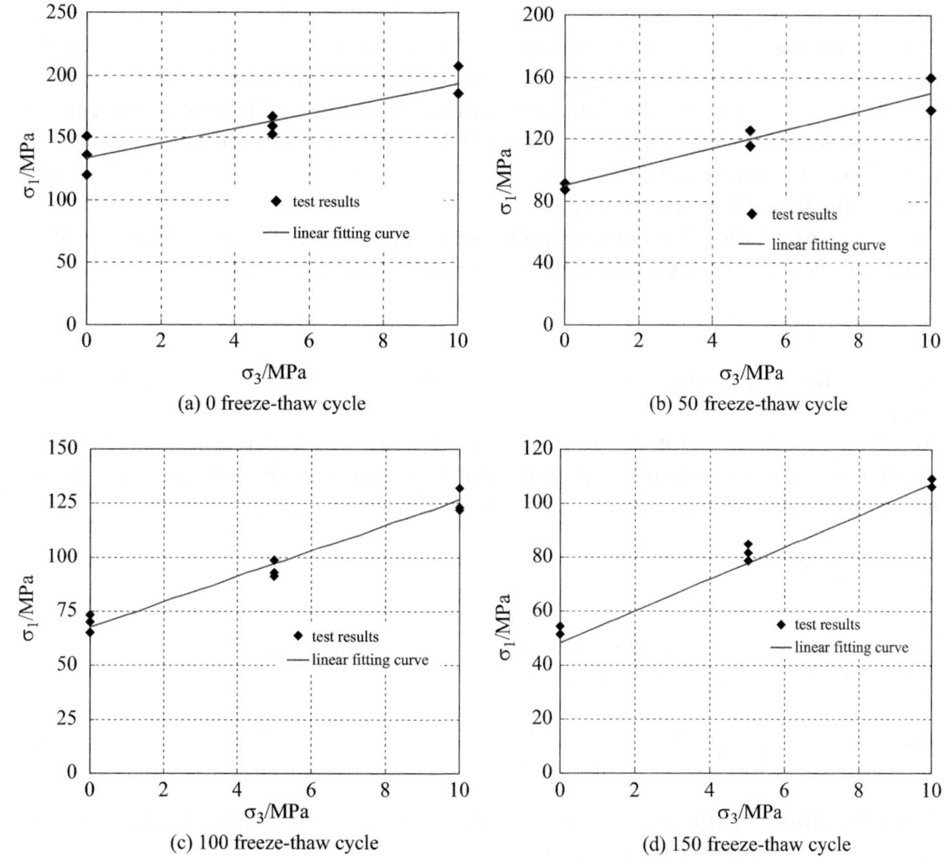

Figure 12 Relationship between maximum principal stress and minimum principal stress.

Table 4 Cohesive strength and internal frictional angle obtained by the Coulomb criterion.

	The number of freeze-thaw cycles			
	0	*50*	*100*	*150*
c(MPa)	27.43	18.31	13.90	9.91
φ(°)	45.38	45.62	45.26	45.24

Table 5 Parameters used to fit the test result by Fukuda.

n	χ^*	ρ_w(kg/m^3)	ρ_s(kg/m^3)	a	T_c(°C)
0.42	0.2	1000	2600	0.16	0

number of freeze-thaw cycles. The cohesive strength decreased by 64 % after the sample underwent 150 freeze-thaw cycles. The main reason for it is that: water in rock pore freezes and the volume expands when the temperature is below 0°C; the local tensile stress is induced among the mineral grains of rock; and these will cause partial damage to the rock at last. With the increase of the number of freeze-thaw cycles, the rock pore becomes bigger and bigger, and the cohesive reaction among mineral grains gradually declines. On the other hand, the difference of internal frictional angle in the whole process of freeze-thaw cycles is only 0.38 ° between maximum and minimum values. Thus, the internal friction angle is regarded as a constant of 45.375 ° (the average value in Table 5) in this paper.

In Fig. 13, the relationship between cohesive strength and the number of freeze-thaw cycles is shown. An exponential function is adopted to fit the test results:

$$c(N) = c_0 e^{-0.072N} \tag{6}$$

Where, c_0 is the origin cohesive strength (before freeze-thaw cycle); $c(N)$ is the cohesive strength.

According to the Coulomb Mohr – strength criterion, the relationship between maximum principal stress and minimum principal stress at the time of rock failure can be expressed with cohesive strength $c(N)$ and internal friction angle φ:

$$\sigma_1 = 2c(N)\frac{\cos\varphi}{1-\sin\varphi} + \frac{1+\sin\varphi}{1-\sin\varphi}\sigma_3 \tag{7}$$

Where, σ_1 is maximum principal stress, σ_3 is minimum principal stress.

Substituting Equation 6 into Equation 7,

$$\sigma_1 = 2c_0\frac{\cos\varphi}{1-\sin\varphi}e^{-0.072N} + \frac{1+\sin\varphi}{1-\sin\varphi}\sigma_3 \tag{8}$$

From Equation 8, the compressive strength for different confining pressure and different number of freeze-thaw cycles can be obtained.

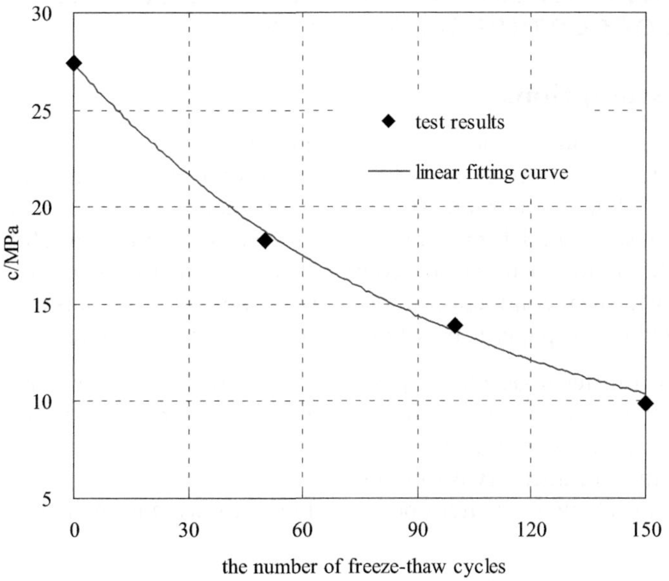

Figure 13 **Relationship curve between cohesive strength and the number of freeze-thaw cycles.**

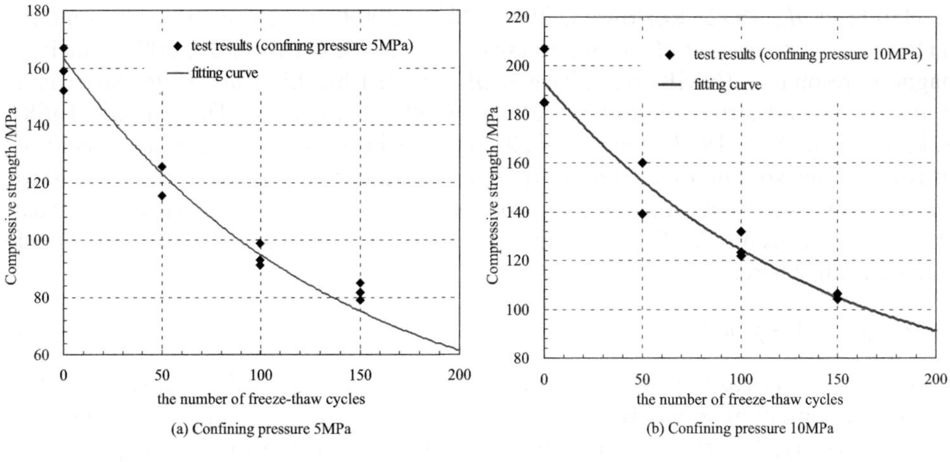

Figure 14 **Relationship between triaxial compressive strength and the number of freeze-thaw cycles.**

Assuming $c_0 = 27.43\text{MPa}$ and $\varphi = 45.375°$, the relationships between triaxial compressive strength and the number of freeze-thaw cycles calculated by Equation 8 are shown in Fig. 14, as well as the test results. From Fig. 14, it is shown that there is a good agreement between the calculation results and test results. The influence of the number of freeze-thaw cycles on the compressive strength of granite is correctly presented by the proposed formulation.

3 TRANSIENT HEAT TRANSFER MODEL WITH ICE-WATER PHASE CHANGE OCCURRING IN POROUS MEDIA

3.1 Basic assumptions

Heat is transferred from a region of higher temperature to a region of lower temperature, there are three basic modes: conduction, convection and radiation. Heat transfer by convection of liquid water and air is usually negligible in porous media (Farouki, 1981), although some models take into account (*e.g.*, Bronfenbrener, 2009; McKenzie *et al.*, 2007). Radiation heat transfer contributes less than 1% (Farouki, 1981), so in this paper, only the heat conduction is considered. Upon this, the transient heat transfer model presented in this paper is based on the following assumptions:

(1) The heat transfer takes place by conduction only, neither skeleton deformation nor water migration is associated with the freezing or thawing process,
(2) The porous media are fully saturated,
(3) The thermal conductivity is isotropic,
(4) The volume change of water upon freezing or thawing is negligible.

3.2 Unfrozen water volumetric content in porous media

The unfrozen water volumetric content plays an important role in heat transfer. It determines not only how much latent heat is released, but also the thermal properties of the porous media. Technologies such as differential scanning calorimetry (DSC) (*e.g.*, Handa *et al.*, 1992; Kozlowski, 2003, 2007, 2009), time domain reflectometry (TDR) (*e.g.*, Patterson *et al.*, 1980; Spaans *et al.*, 1995; Christ *et al.*, 2009), nuclear magnetic resonance (NMR) (*e.g.*, Tice *et al.*, 1978; Ishizaki *et al.*, 1996; Sparrman *et al.*, 2004), and ultrasonic technique (*e.g.*, Wyllie *et al.*, 1956; Timur *et al.*, 1968; Nakano *et al.*, 1972, 1973; Wang *et al.*, 2006) have been used for the measurement of unfrozen water volumetric content in frozen soils, and it was described by Blanchard *et al.*, (1985) as a power function, with parameters dependent on the specific surface area of the porous media. The following function is chosen to describe the presence of unfrozen water here:

$$\chi = \chi^* + (1 - \chi^*)e^{a(T-T_c)} \tag{9}$$

where, χ is the unfrozen water volumetric content, χ^* is the residual unfrozen water volumetric content at some lower reference temperature, a is a parameter to describe the rate of decay, T_c is the freezing point of water. A similar function, but expressed in terms of the gravimetric content rather than the volumetric content, was considered earlier by Anderson & Tice (1971, 1973). These two definitions can be transferred easily by the following expression:

$$w = \frac{n\rho_w}{(1-n)\rho_s}\chi \tag{10}$$

where, w is the unfrozen water content described by the gravimetric content.

Fig.15 shows the comparison of the unfrozen water content curve from Equations 9 and 10 with a real test result (Fukuda, 1997). The parameters are listed in Table 5.

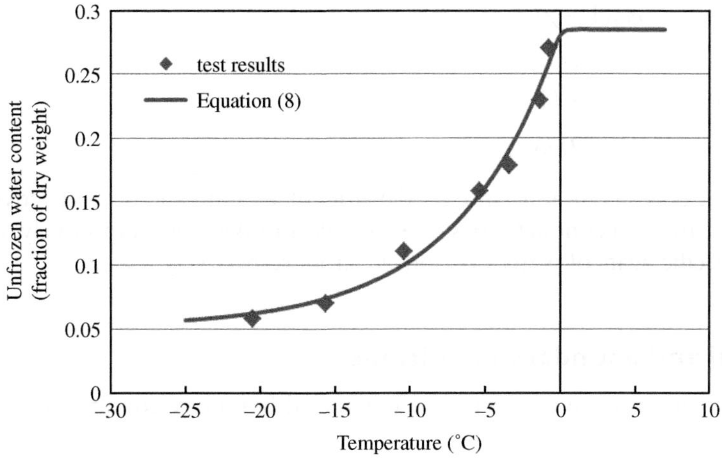

Figure 15 Unfrozen water characteristic curve.

From Fig. 15, it can be seen that the calibrated curve matches the experimental data well, which revealed that this function is flexible enough to describe the unfrozen water content and remind us that it suitable for reflecting the interval of freezing zone conveniently, in the next part of this paper, we will use different parameter a to adjust the interval of freezing zone.

3.3 Governing differential equations

According to the heat exchange theory for porous media and "three zones" theory, the governing differential equations for heat transfer can be written in the following form:

(1) Within the frozen zone:

$$C_1\frac{\partial T}{\partial t}+\nabla[-k_1\nabla T]=Q_{T1} \tag{11}$$

(2) Within the freezing zone:

$$C_2\frac{\partial T}{\partial t}+\rho_{\rm w}L_{\rm f}\frac{\partial n\chi}{\partial t}+\nabla[-k_2\nabla T]=Q_{T2} \tag{12}$$

(3) Within the unfrozen zone:

$$C_3\frac{\partial T}{\partial t}+\nabla[-k_3\nabla T]=Q_{T3} \tag{13}$$

where, $\rho_{\rm w}$ is the density of water; Q_{Ti}, k_i and C_i are the heat source, the thermal conductivity and the volumetric specific heat in the i-th zone, $L_{\rm f}$ is the latent heat of water, in which:

$$k_{1,2} = n\lambda_w\chi + n\lambda_i(1-\chi) + (1-n)\lambda_s \tag{14}$$

$$k_3 = n\lambda_w + (1-n)\lambda_s \tag{15}$$

$$C_{1,2} = n\rho_w c_w\chi + n\rho_i c_i(1-\chi) + (1-n)\rho_s c_s \tag{16}$$

$$C_3 = n\rho_w c_w + (1-n)\rho_s c_s \tag{17}$$

where, n is the porosity, ρ_i and ρ_s are the densities of ice and skeleton, respectively; λ_w, λ_i and λ_s are the thermal conductivities of water, ice and skeleton, respectively, c_w, c_i and c_s are the specific heats of water, ice and skeleton, respectively,

3.4 Initial and boundary conditions

To solve the temperature equation, the initial and boundary conditions are listed as follows:

(1) Initial condition

$$T|_{t=0} = T_0 \tag{18}$$

(2) Dirichlet boundary condition

$$T = T_0 \tag{19}$$

(3) Neumann boundary condition

$$-\mathrm{n}\cdot(-k_1\nabla T) = q_0 + \alpha(T - T_0) \tag{20}$$

3.5 Solving for the governing differential equations

Equations 9~21 are the governing equations, and initial and boundary conditions of the problem of the temperature field. Because the specific heat and thermal conductivity change with temperature, so this kind of problem is a heavily nonlinear one, hence, we obtained its solution by using a numerical analytical method. Using Galerkin's method, the following finite element formulae can be obtained:

$$[N]\frac{\mathrm{d}\{T\}}{\mathrm{d}t} + [K]\{T\} = \{F\} \tag{21}$$

$$[N] = \int_A C\cdot N_i N_j \mathrm{d}A \tag{22}$$

$$[K] = \int_A \nabla N_i \cdot k N_j \mathrm{d}A \tag{23}$$

$$[F] = \int_A N_i \cdot Q_T \cdot \mathrm{d}A \tag{24}$$

where, N_i, N_j are the shape functions.

3.6 Comparison with exact analytical solution

The proposed "three-zone" model was tested by comparison with the three-zone analytic solution presented by Lunardini *et al.* (1985), which is an exact analytical solution for the propagation of subfreezing temperature in a porous semi-infinite, and it has been used to test the validation of SUTRA computer code by McKenzie *et al.* (2007). The calculation model, initial and boundary conditions were showed in Fig. 16, where zone 1 is frozen, zone 2 is freezing and zone 3 is unfrozen. The analytic solution is given below:

$$T_1 = (T_m - T_s)\frac{erf(x/2\sqrt{\alpha_1 t})}{erf(\phi)} + T_s \tag{25}$$

$$T_2 = (T_\mathrm{m} - T_\mathrm{f})\frac{\mathrm{erf}(x/2\sqrt{\alpha_2 t}) - \mathrm{erf}(\gamma)}{\mathrm{erf}(\gamma) - \mathrm{erf}\left(\phi\sqrt{\alpha_1/\alpha_2}\right)} + T_\mathrm{f} \tag{26}$$

$$T_3 = (T_\mathrm{i} - T_\mathrm{f})\frac{-\mathrm{erfc}(x/2\sqrt{\alpha_3 t})}{\mathrm{erfc}\left(\gamma\sqrt{\alpha_2/\alpha_3}\right)} + T_\mathrm{i} \tag{27}$$

where ϕ and γ are the two solution parameters; T_1, T_2 and T_3 are the temperatures for zones 1, 2 and 3, respectively; erf and erfc are the error function and the complimentary error function, respectively; T_i is the initial temperature; T_m is the boundary temperature between frozen zone and freezing zone; T_f is the boundary temperature between freezing zone and unfrozen zone; T_s is the boundary temperature; α_1, α_2 and α_3 are the thermal diffusivity for zones 1, 2 and 3, respectively, it's the ratio of thermal conductivity and volumetric specific heat. The expression of α_2 is slightly different from α_1 and α_3, since α_2 considers the influence of latent heat in the process of phase change (McKenzie *et al.*, 2007):

$$\alpha_1 = \frac{k_1}{C_1}; \quad \alpha_2 = \frac{k_2}{C_2 + \dfrac{\gamma_\mathrm{d} L_\mathrm{f} \Delta\xi}{T_\mathrm{f} - T_\mathrm{m}}}; \quad \alpha_3 = \frac{k_3}{C_3} \tag{28}$$

Where, γ_d is the dry unit density, $\gamma_\mathrm{d} = (1-n)\rho_\mathrm{s}$, $\Delta\xi = \xi_0 - \xi_\mathrm{f}$, ξ_0 and ξ_f are the ratio of unfrozen water to soil mass for the fully thawed and frozen conditions, respectively, defined as:

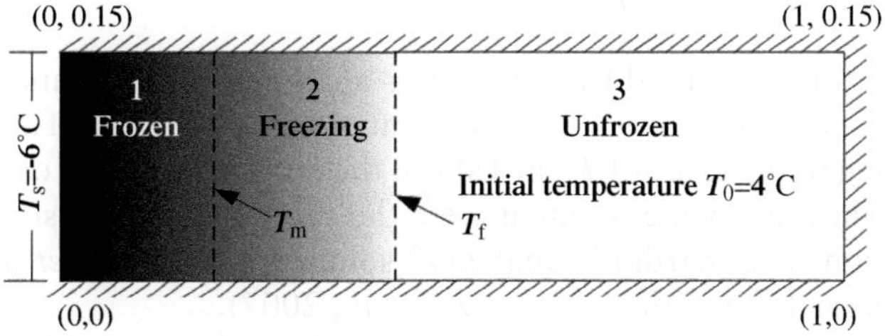

Figure 16 Calculate model and boundary condition.

Table 6 Parameters used in analytical solution by Lunardini.

Parameters	*value*
T_0(°C)	4
T_s(°C)	−6
T_f(°C)	0
T_m(°C)	−4
C_1(J/kg·K)	4187
C_2(J/kg·K)	2108
C_3(J/kg·K)	840
k_1(W/m·K)	0.58
k_2(W/m·K)	2.14
k_3(W/m·K)	2.9
γ_d(kg/m^3)	1680
L_f(J/kg)	334000
ξ_0	0.2
ξ_f	0.078
ϕ	0.0617
γ	1.395

Table 7 Parameters used in numerical simulation.

parameters	*value*
n	0.34
χ^*	0.2
a	1.5
T_c(°C)	0
c_w(J/kg·K)	4187
c_i(J/kg·K)	2108
c_s(J/kg·K)	840
λ_w(W/m·K)	0.58
λ_i(W/m·K)	2.14
λ_s(W/m·K)	2.9
ρ_w(kg/m^3)	1000
ρ_i(kg/m^3)$_s$	917
ρ_s(kg/m^3)	2600

$$\xi_0 = \frac{n\rho_w}{(1-n)\rho_s}; \quad \xi_f = \frac{n\chi\rho_w}{(1-n)\rho_s} \tag{29}$$

The calculation parameters used by Lunardini's analytical solution and finite element method are shown in Table 6 and 7. It must point out that the value of a in Table 7 was obtained according to the value of T_m in Table 6 the other parameters of to describe the characteristic of unfrozen water content were the same as Table 8, since there is not further information in Lunardini's analytical solution (Lunardini *et al.*, 1985) and McKenzie's numerical simulation (McKenzie *et al.*, 2007).

The calculation results by analytical solution and finite element method are shown in Fig. 17. It can be seen that the finite element method results successfully match the

Table 8 Parameters used to fit the test result by Fukuda

n	χ^*	ρ_w (kg/m^3)	ρ_s (kg/m^3)	a	T_c(°C)

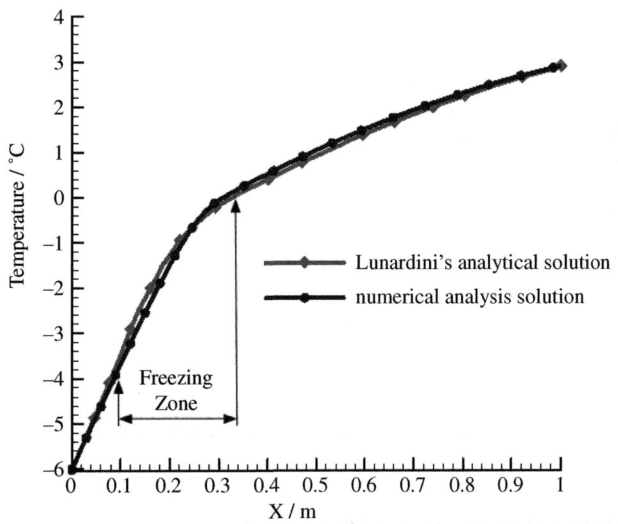

Figure 17 Comparison between numerical simulation results and classified analytical solution.

analytical solution, the maximum error between each other is 0.2°C in freezing zone, this is because Lunardini's analytical solution assumed that the thermal diffusivity of zone 2 is constant to solve problem efficiently (McKenzie *et al.*, 2007), but this value changes with temperature in the numerical analysis to simulate the real situation.

3.7 Sensitivity analyses

3.7.1 The influence of parameter a on temperature

Form the above analysis, it can be seen that the parameter a can affect the freezing zone interval, in this section we will analyse how it affects the temperature. Fig. 18 is the temperature curve in the case of different a (The other parameters are the same as Table 7), it shows that the distribution of temperature is varied with the value of a, and the maximum absolute difference was found in the freezing zone, this is because the latent heat of water is released over a range of temperatures rather than at one particular temperature, so the boundary temperature between the frozen zone and freezing zone a is important for the distribution of temperature.

3.7.2 The influence of latent heat

The results with and without latent heat are shown in Fig. 19. It can be seen that the release of the latent heat during the phase change of water has a strong influence on the

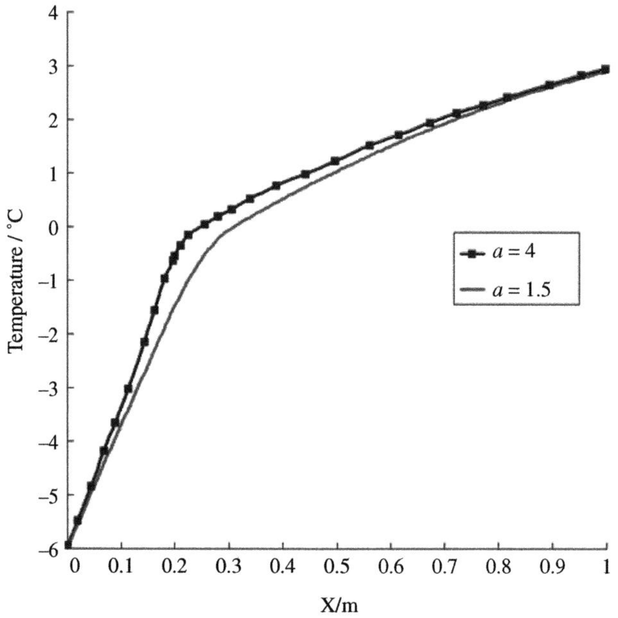

Figure 18 The distribution of temperature at different a value.

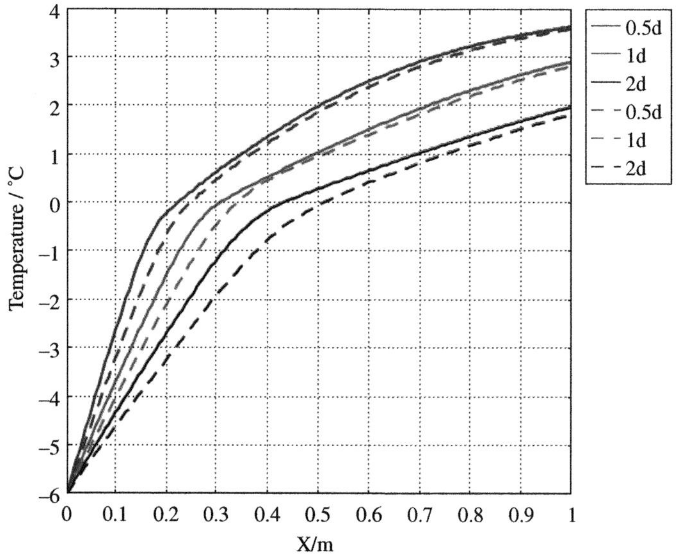

Figure 19 The distribution of temperature at different time in case of considering latent heat (line) and not considering latent heat (dotted line).

temperature profile and retards the frost penetration. As the freezing front penetrates into soil, thermal properties of the frozen soil also change. The thermal conductivity and the heat capacity of soil vary with the volumetric fractions of phases, which are dependent on the temperature.

3.8 Field temperature measurements

3.8.1 Introduction of tunnel engineering

Galongla tunnel is an important part of Zha-mo highway in Tibet which is the only connection to Mo-tuo County in China. In Galongla tunnel, the freeze period is about 8 months; the average freeze depth is about 5 to 6 m and the maximum freeze depth is about 15m; the temperature in winter is low to –30℃; the altitude is about 4300m and the oxygen content in air is only 60% of that inland. There are very few highway tunnels built in the region at so low temperature and high altitude in China and Galongla tunnel is one of them.

Based on the local temperature environment, some insulation measures must be taken at entrance and exit of tunnel to prevent the surrounding rock and concrete lining from frost damage. To valid the reliability of the proposed model, here the simulated results about temperature field of surrounding rock by finite element method using the "three-zone" model are compared with the field temperature data. Then, the effect of insulation material is studied.

3.8.2 Field work

The field temperature data used to test the proposed "three zone" model was obtained from an intensively instrumented bog in the surrounding rock of Galongla tunnel in Tibet, there are ten sections along the tunnel axis, and at each section the temperature sensors were installed at 15, 35, 50, 75, 120, 220 and 370 cm depths (in Fig. 20 and 21). The data are stored with a solar powered data logger connected to a mobile phone for data access. The data logger for temperature collection was set as follows: time 0:00; frequency: one time per day. From August 12, 2009, to April 10, 2010, this collection system worked 8 months.

3.8.3 Calculation parameters and boundary conditions

The calculation model is chosen at the entrance of Galongla tunnel (the Bo-mi side) which is about 10m far from the entrance (shown in Fig. 22).

The convective heat transfer coefficient between the air and the surface of the rock surrounding the tunnel is $\alpha = 15.0\ \mathrm{W/(m^2 \cdot K)}$ (Zhang, *et al.*, 2002), $\chi^*=0.2$, $T_c = 0$ and $a = 4$. The initial temperature is assumed to use a local geothermal gradient 1.5% under the permafrost base. The geothermal heat flux 0.06 W/m^2 flows in the computational domain from the boundary *EF*. According to the symmetry, boundaries *AB* and *DE* are diathermic boundaries. Boundary *AG* and *BCD* are in contacts with air. The boundary conditions for cave (*BCD*) and roof (*AG*) in the construction process can be expressed as follows:

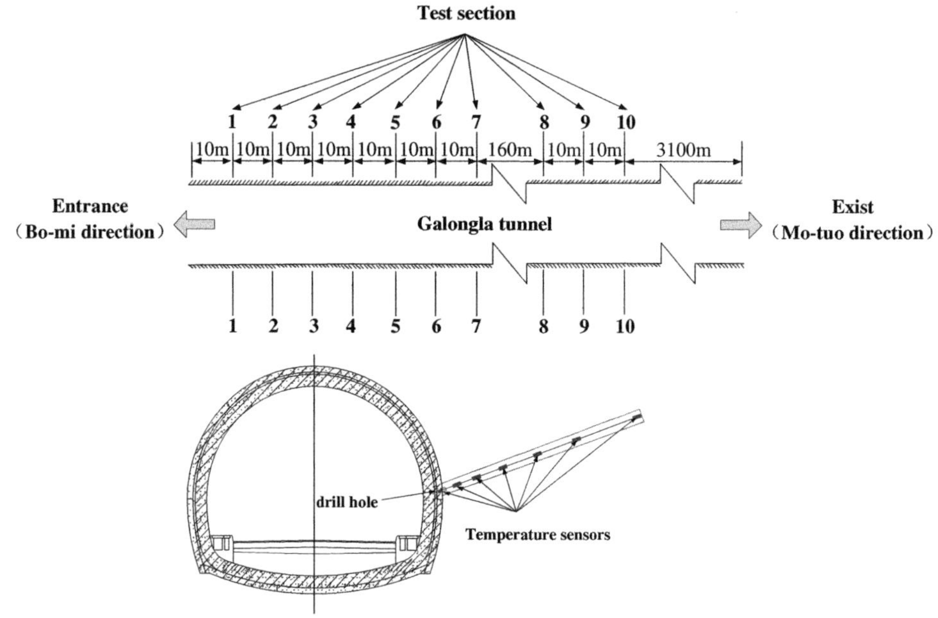

Figure 20 The arrangement of temperature sensors for the field temperature data.

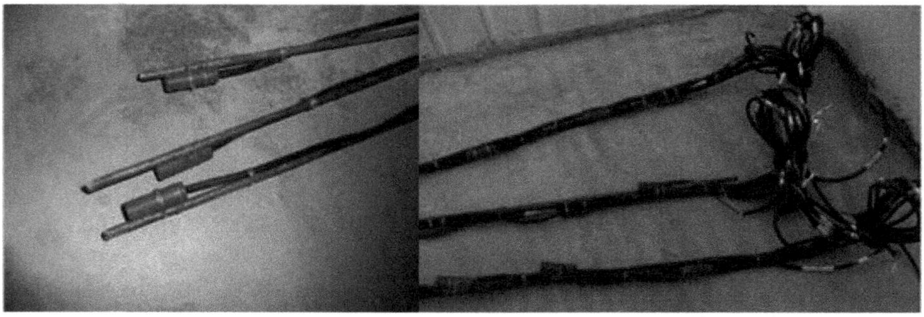

Figure 21 The field installation of temperature sensors assemblies.

$$k\frac{\partial T}{\partial y}\bigg|_{AG} = \alpha(T|_{AG} - T_{env}) \tag{30}$$

$$k\frac{\partial T}{\partial n}\bigg|_{BCD} = \alpha(T|_{BCD} - T_{env}) \tag{31}$$

where $T_{env}(t)$ is air (environment) temperature, and according to the temperature monitor results, the environment temperature can be expressed as follows:

$$T_{env}(t) = 5.85 + 6.85\sin(2\pi t/360/86400 + 7\pi/12) \tag{32}$$

where, t is time (units: s); T_{env} is environment temperature (units: ℃).

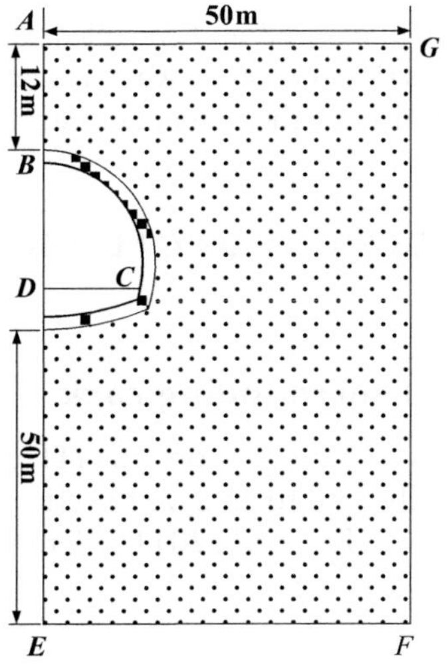

Figure 22 Calculation model for numerical simulation.

3.8.4 Comparison between FEM results and in-situ temperature measurements

Fig. 23(a)~(b) shows the results comparison between measured and modeled temperatures graphically for different depth of surrounding rock at section 1. The field measurement lasted 229 days, and the numerical analysis lasted 360 days. Comparing Fig. 23 (a) with (b), it can be seen that the temperature of the surrounding rock changed with sine function, and the changing trend was identical. With the increasing of depth, the scope of temperatures became smaller and smaller.

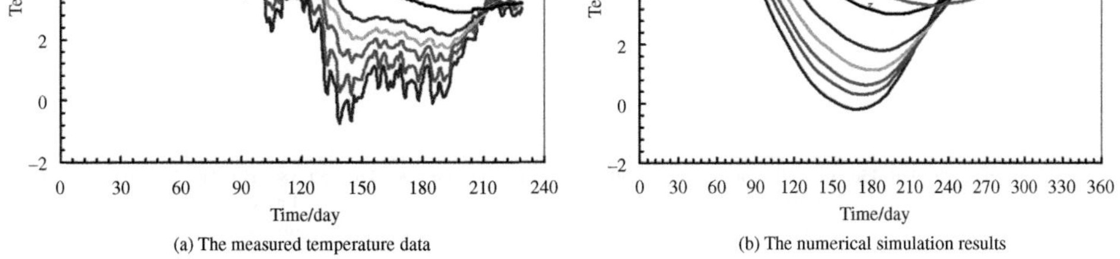

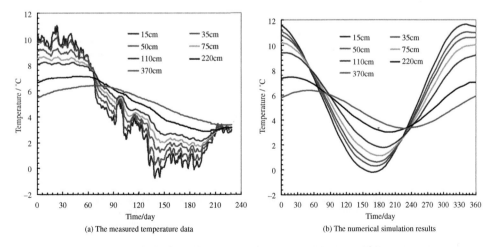

Figure 23 comparison of measured results with that of numerical simulation results at different depths.

A further comparison for individual depths was shown in Fig. 24. It can be seen that the greatest deviation from the measured and modeled temperatures occurred at the initial stage; the reason for this is that the tunnel was excavated just for a short time, and the temperature changed vigorously by the human, construction equipment and so on.

4. INSULATION LAYER DESIGN FOR GALONGLA TUNNEL IN TIBET

4.1 Temperature field of surrounding rock

In our previous work, new governing equations for HT coupling model were established, and they were successfully used to analyze the effect of insulation materials for Galongla tunnel in Tibet of China (Tan *et al.*, 2011). Here the heat transfer equation of surrounding rock is referenced directly:

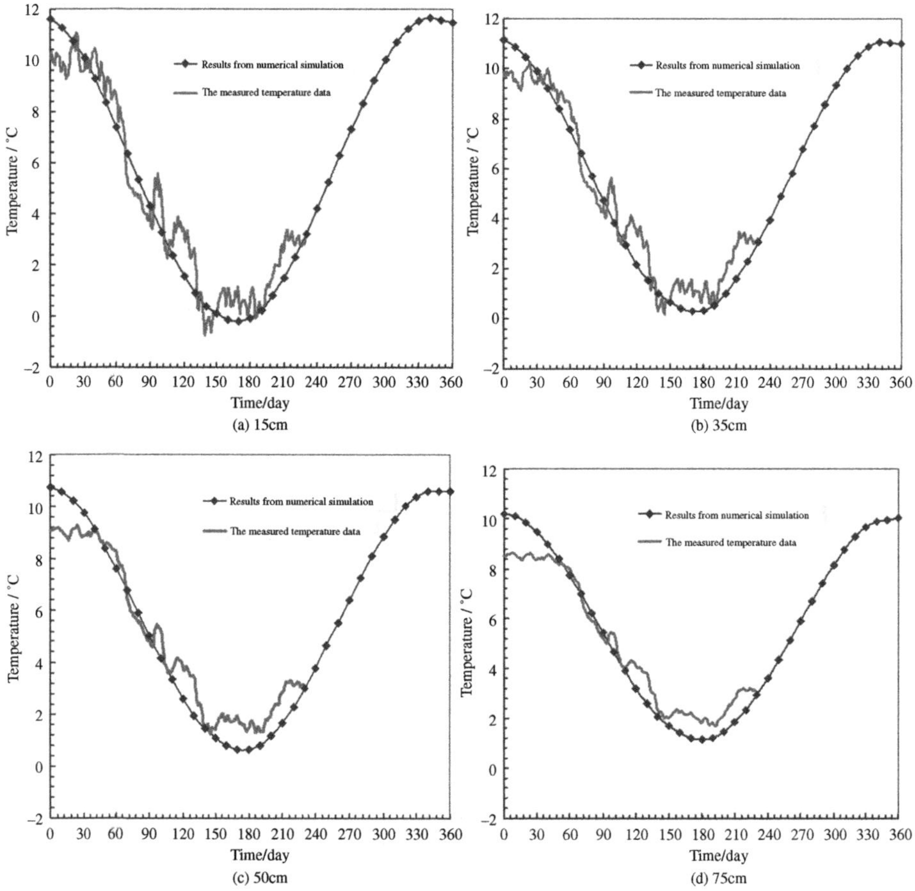

Figure 24 Comparison of measured results with numerical simulation results.

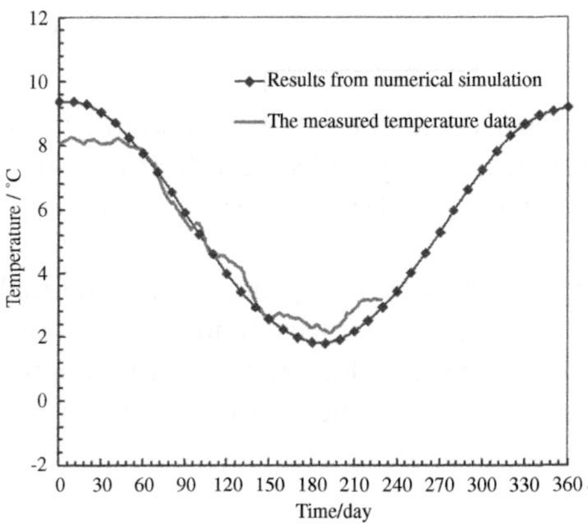

(e) 110cm

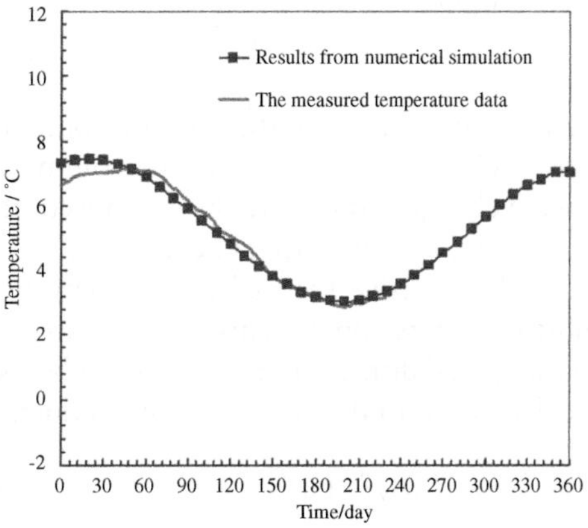

(f) 220cm

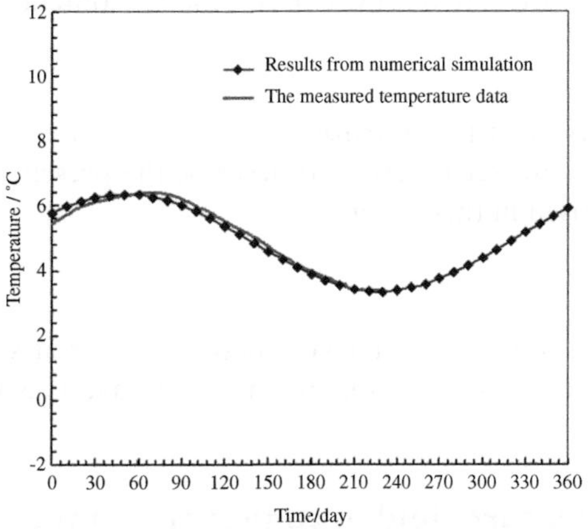

(g) 370cm

Figure 24 (cont.)

$$C_{eq}\frac{\partial T_s}{\partial t} + \nabla[-\lambda_e \nabla T_s] = Q_e \tag{33}$$

where, T_s is the temperature of surrounding rock; Q_e is the generated (consumed) heat by heating (heat release) in unit body Ω. C_{eq} is equivalent volumetric thermal capacity, it can be expressed as follows:

$$C_{eq} = \begin{cases} C_a = C + \rho_w L_f \dfrac{\partial \theta_w}{\partial T_s} & T_s \leq T_0 \\ C & T_s > T_0 \end{cases} \tag{34}$$

where, L_f is the heat latent; C is the volumetric thermal capacity, which are defined as the sum of the volumetric heat capacities of the solid, liquid and ice phases multiplied by their respective volumetric fracture, it can be expressed as follows:

$$C = \frac{\theta_s \rho_s c_s + \theta_w \rho_w c_w + \theta_i \rho_i c_i}{\theta_s + \theta_w + \theta_i} \tag{35}$$

where, θ_s, ρ_s and c_s are the volume content, density and special heat of solid grains, respectively, θ_w, ρ_w and c_w are the volume content, density and special heat of water, respectively; θ_i, ρ_i and c_i are the volume content, density and special heat of solid grains, respectively, c_i is the special heat of ice.

λ_e is the effective thermal conductivity of porous media, it depends on the thermal conductive of its components (solid, unfrozen water and ice), their volumetric fractions and the spatial distribution of the three phases. In this paper, a logarithmic law is adopted for calculations of the effective heat conductivity according to the logarithmic law

$$\lambda_e = (\lambda_s)^{\theta_s} (\lambda_w)^{\theta_w} (\lambda_i)^{\theta_i} \tag{36}$$

where, λ_s, λ_w, λ_i is the thermal conductivity of rock matrix, fluid and ice.

θ_w can be described by the unfrozen water volumetric content

$$\theta_w = n\chi \tag{37}$$

where, n is the porosity, χ is the unfrozen water volumetric content. Michalowski (1993) presented a 3-parameter function to describe the presence of unfrozen water, a similar function is chosen in this paper:

$$\chi = \chi^* + (1 - \chi^*) e^{a(T_s - T_c)} \tag{38}$$

where, χ^* is the residual unfrozen water volumetric content at some lower reference temperature, a is a parameter to describe the rate of decay, T_c is the freezing point of water.

4.2 Airflow temperature field in tunnel clearance area

The governing equations to be considered are the time-averaged continuity, momentum, and energy equations. Hence, the governing equations for the airflow region can be written as (Tan *et al.*, 2013; Nield, 1993; Kong *et al.*, 2002, 1999; Zhang *et al.*, 2012).

4.2.1 Time-averaged continuity equation

$$\nabla \cdot (\rho \mathbf{u}) = 0 \tag{39}$$

where, $\mathbf{u}$ is the velocity of airflow in tunnel; ρ is the density of airflow, which can be established by the ideal gas state equation.

4.2.2 Time-averaged momentum equation (N–S equation)

$$\nabla \cdot (\rho \mathbf{u})\mathbf{u} = \nabla \cdot \left[-p\mathbf{I} + (\mu + \mu_t)\left(\nabla \mathbf{u} + (\nabla \mathbf{u})^T - \frac{2}{3}(\nabla \cdot \mathbf{u})\mathbf{I} \right) - \frac{2\rho k}{3}\mathbf{I} \right] + S_M \tag{40}$$

where, p is the pressure; μ is the viscosity coefficient; $\mathbf{I}$ is the unit matrix; μ_t is the eddy viscosity coefficient; $S_M = \rho \mathbf{X} - \frac{\partial}{\partial x}(\mu' \nabla \cdot \mathbf{u})$, $\mathbf{X}$ is the body force along with coordinate axis orientation; μ' is the second viscosity coefficient, which is usually assumed to be equal to $\mu' = \frac{2}{3}\mu(\nabla \cdot \mathbf{u})$ for airflow.

4.2.3 Time-averaged heat transfer equation

$$\nabla \cdot (\rho c_{\mathrm{p}} \mathbf{u} T_{\mathrm{a}}) = \nabla \cdot \left[\left(\lambda_{\mathrm{g}} + c_{\mathrm{p}} \frac{\mu_{\mathrm{t}}}{\mathrm{Pr_T}} \right) \nabla T_{\mathrm{a}} \right] + Q_{\mathrm{T}} \tag{41}$$

where, T_{a} is the temperature of airflow; c_{p} is the specific heat; λ_{g} is the conductivity; $\mathrm{Pr_T}$ is the turbulence Prandtl number; and Q_{T} is the heat source.

4.2.4 $k-\varepsilon$ two equations turbulence model

The turbulence kinetic energy k and its rate of dissipation ε are governed with the following transport equations:

$$\begin{cases} \nabla \cdot (\rho k)\mathbf{u} = \nabla \cdot \left[\left(\mu + \dfrac{\mu_{\mathrm{t}}}{\sigma_{\mathrm{k}}} \right) \nabla k \right] + \mu_{\mathrm{t}} P(\mathbf{u}) - \dfrac{2\rho k}{3} \nabla \cdot \mathbf{u} - \rho \varepsilon \\ \nabla \cdot (\rho \varepsilon)\mathbf{u} = \nabla \cdot \left[\left(\mu + \dfrac{\mu_{\mathrm{t}}}{\sigma_{\varepsilon}} \right) \nabla \varepsilon \right] + \dfrac{C_{\varepsilon 1} \varepsilon}{k} \left[\mu_{\mathrm{t}} P(\mathbf{u}) - \dfrac{2\rho k}{3} \nabla \cdot \mathbf{u} \right] - \dfrac{C_{\varepsilon 2} \rho \varepsilon^2}{k} \end{cases} \tag{42}$$

where, $P(\mathbf{u}) = \nabla \mathbf{u}:[\nabla \mathbf{u} + (\nabla \mathbf{u})^{\mathrm{T}}] - \frac{2}{3}(\nabla \cdot \mathbf{u})^2$; the eddy viscosity coefficient is expressed as follows:

$$\mu_{\mathrm{t}} = \frac{\rho C_{\mu} k^2}{\varepsilon} \tag{43}$$

where, σ_{k}, σ_{ε}, $C_{\varepsilon 1}$, $C_{\varepsilon 2}$ and C_{μ} are empirical constants, the values are set as: $\sigma_{\mathrm{k}} = 1.00$, $\sigma_{\varepsilon} = 1.30$, $C_{\varepsilon 1} = 1.44$, $C_{\varepsilon 2} = 1.92$ and $C_{\mu} = 0.07$, which were recommended by Launder (1991).

4.2.5 Numerical methods

The study used commercial finite element software, Comsol Mutiphysics, to implement all the models and for all the numerical simulations. Logarithmic wall functions were used to account for solid walls subjected to turbulent flow, which assume that the computational domain begins a distance from the real wall (Ilina *et al.*, 1997; Kalilzin *et al.*, 2005), and the thermal wall function boundary condition for turbulent flow is used to model the temperature change law of solid-air interfaces Lacasse *et al.* (2004). The RANS models used the second order upwind scheme for all the variables except pressure. The discretization of pressure is based on a staggered scheme. The SIMPLE algorithm is adopted to couple the pressure and momentum equations. If the sum of absolute normalized residuals for all element become less than 10^{-6} for energy and 10^{-3} for other variables, the solution is considered converged. Adaptive finite element method and grid dependence of each case are checked using two to four different grids to ensure that grid resolution would not have a notable impact on the results (Tan *et al.*, 2013).

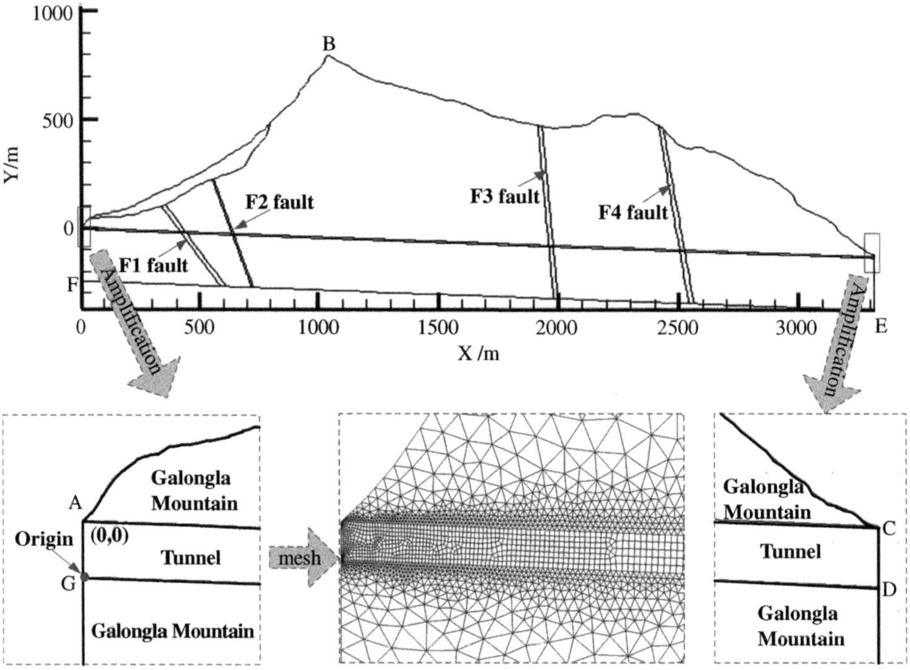

Figure 25 Calculation model.

4.2.6 Numerical analysis for Galongla road tunnel

Based on the former researches, the climate temperature in one year can be considered to change with time as a sine function (or cosine function). The extreme temperature of climate around Galongla tunnel differs with the altitude and location. According to the geological investigation, the temperature of the tunnel's entrance at point G and the tunnel's exit at point D (seen in Fig. 25) can be expressed as:

$$T_G = -4.15 + 23.85\sin(2\pi t/360/86400 + \pi/2) \tag{44}$$

$$T_d = 7.50 + 26.50\sin(2\pi t/360/86400 + \pi/2) \tag{45}$$

where, t is time(s), T_G and T_d are the climate temperature at point G and D, respectively (°C).

According to the geological survey, temperature changes very obviously in the vertical direction. Temperature decreases 0.74°C, when the location rises 100 m. So the temperature conditions of the boundary GA, AB, BC and CD can be expressed as follows (Zhang *et al.*, 2007):

$$\left.\begin{aligned} T_{GA} &= T_G - 0.74 * (H_A - H_G)/100 \\ T_{AB} &= T_F - 0.74 * (H_B - H_A)/100 \\ T_{BC} &= T_D - 0.74 * (H_B - H_C)/100 \\ T_{CD} &= T_D - 0.74 * (H_C - H_D)/100 \end{aligned}\right\} \tag{46}$$

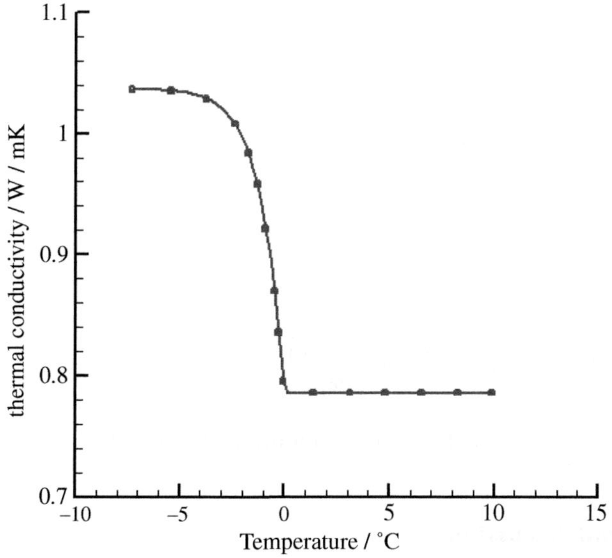

Figure 26 Inversion results for thermal conductivity.

where, T_{GA}, T_{AB}, T_{BC} and T_{CD} are the temperature of the boundary GA, AB, BC and CD, respectively (°C); H_A, H_B, H_C, H_D and H_G are the altitude of point A, B, C, D and G.

Because the calculation field is enough wide, the boundaries DE and FG are set as thermal insulation boundaries. The heat flux of the boundary EF is equal to 0.06 W/m (Zhang *et al.*, 1989). According to the in-situ surveyed data, the initial temperatures of surrounding rock are obtained by the geothermic gradient is 1.5%. The in-situ measured results for thermal conductivity is shown in Fig. 26.

Airflow boundary conditions in tunnel are specified as follows (Tan *et al.*, 2013; Lacasse *et al.*, 2004):

1) Inlet boundary conditions: $\mathbf{u} = \mathbf{u}_{in}$, its value is different with the case study; $k = (3\mathrm{I}_\mathrm{T}^2/2)(\mathbf{u}_{in} \cdot \mathbf{u}_{in})$, $\varepsilon = C_\mu^{0.75} k^{3/2}/\mathrm{L}_\mathrm{T}$, $T = T_\mathrm{in}$, its value is different with the case study.
2) Exit boundary conditions: $p = p_\mathrm{atm}$, $-\mathrm{n} \cdot \left[\left(\lambda_\mathrm{g} + c_\mathrm{p} \frac{\mu_\mathrm{t}}{\mathrm{Pr}_\mathrm{T}}\right) \nabla T_\mathrm{a}\right] = 0$.
3) The heat convective interfacial boundary between the surrounding rocks and the air conditions:

$$-\mathrm{n} \cdot \left[\left(\lambda_\mathrm{g} + c_\mathrm{p} \frac{\mu_\mathrm{t}}{\mathrm{Pr}_\mathrm{T}}\right) \nabla T\right] = h(T_\mathrm{s} - T); \quad h = \frac{\rho c_\mathrm{p} C_\mu^{1/4} k^{1/2}}{T^+}$$

where, $\mathbf{u}_{in}$ is the inlet velocity; I_T is the turbulence intensity; L_T is the turbulent length scale; p_atm is the atmosphere pressure; h is the Convective heat transfer coefficient, The dimensionless quantity $T^+ = T^+(y^+)$ is the dimensionless temperature and depends on the dimensionless wall offset, y^+, the relations are given in COMSOL as follows.

$$T^+ = \begin{cases} Pry^+ & y^+ < y_1^+ \\ 15Pr^{2/3} - 500/(y^+)^2 & y_1^+ \le y^+ < y_2^+ \\ \ln(y^+)\mathrm{Pr}/\kappa + \beta & y^+ \ge y_2^+ \end{cases} \tag{47}$$

With the following definitions:

$$\begin{cases} y_1^+ = \dfrac{10}{\mathrm{Pr}^{1/3}} \\ y_2^+ = 10\sqrt{10\dfrac{\kappa}{\mathrm{Pr_T}}} \\ \beta = 15\mathrm{Pr}^{2/3} - \dfrac{\mathrm{Pr_T}}{2\kappa}\left[1 + \ln\left(1000\dfrac{\kappa}{\mathrm{Pr_T}}\right)\right] \end{cases} \tag{48}$$

Moreover, the other parameters can be obtained in Table 9.

4.2.7 Results and discussion

4.2.7.1 Initial temperature field of Galongla tunnel

In Fig. 27(a)-(d), the initial temperature distributions of calculation area in January, April, July and October are shown. From these figures, it is concluded that the temperature in most zones of mountain is almost stable in one year, it fluctuates only in a small area of mountain surface. The maximum temperature along the tunnel axis is only about 12°C. The reason for this is that the content of rain and snow is very rich in the tunnel area and surrounding rock is fractured which decreases the feature of heat accumulation of mountain to a great extent. In addition, surface temperature on two sides of the mountain obviously differ. The main reason for this is that the temperature of the tunnel area is influenced by warm and wet flow from the Indian Ocean and southwest monsoon, the climate in Bo-Mi is different to the climate in Mo-Tuo.

After tunnel runs through the mountain, the most affected zones are entrance and exit for mountain temperature. Thus, temperature distribution at entrance and exit before excavation should be studied to compare with the temperature distribution after excavation.

Initial temperatures in the area about 120 m away from entrance (Bo-Mi) and exit (Mo-Tuo) in the tunnel axial direction in different seasons are shown in Figs. 28 and 29. It is shown that temperature in the shallow area of the mountain changed with time obviously. This phenomenon appears at a depth of 18m, especially. With a buried

Table 9 Main physical parameters for calculation.

	Water content w/%	*Thermal conductivity λ W/(m·K)*	*Density /kg·m^3*	*special heat c /J·kg^{-1}·K^{-1}*
Surrounding rock	30		2650	900
Water	–	Seen in Fig. 3	1000	4180
Ice	–		917	2090
flolic	–	0.025	300	230

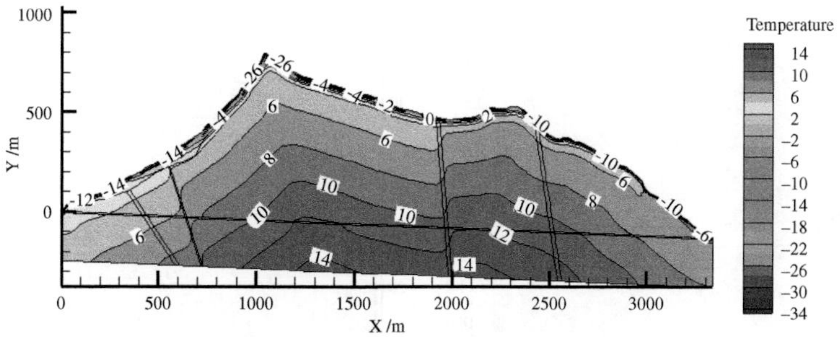

(a) January

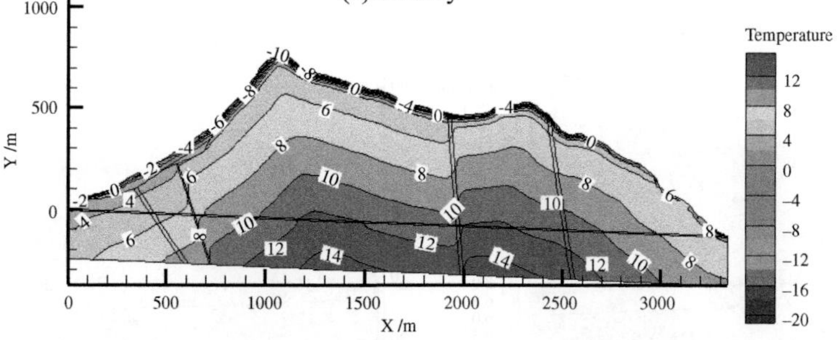

(b) April

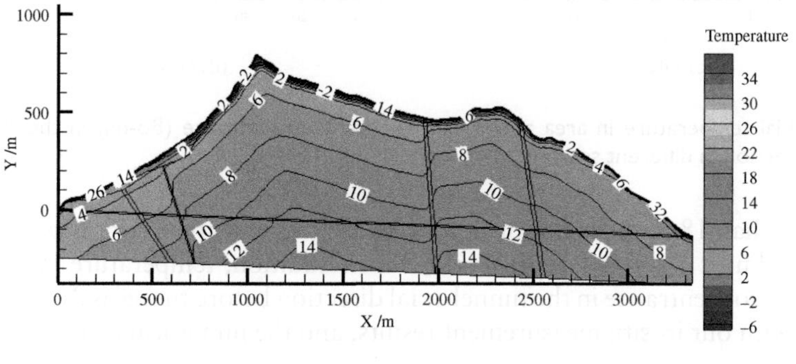

(c) July

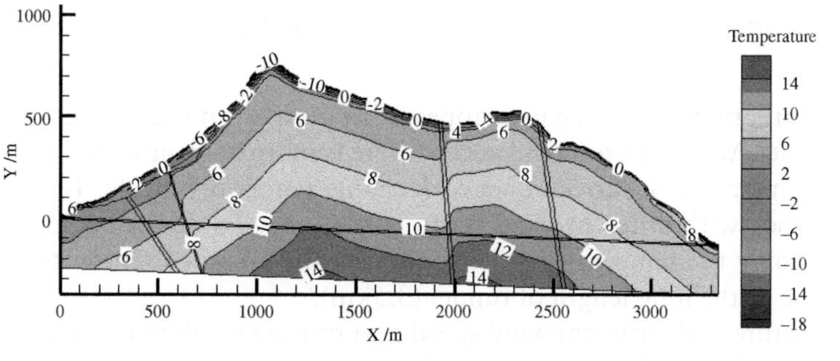

(d) October

Figure 27 The initial temperature distributions of calculation area in different seasons.

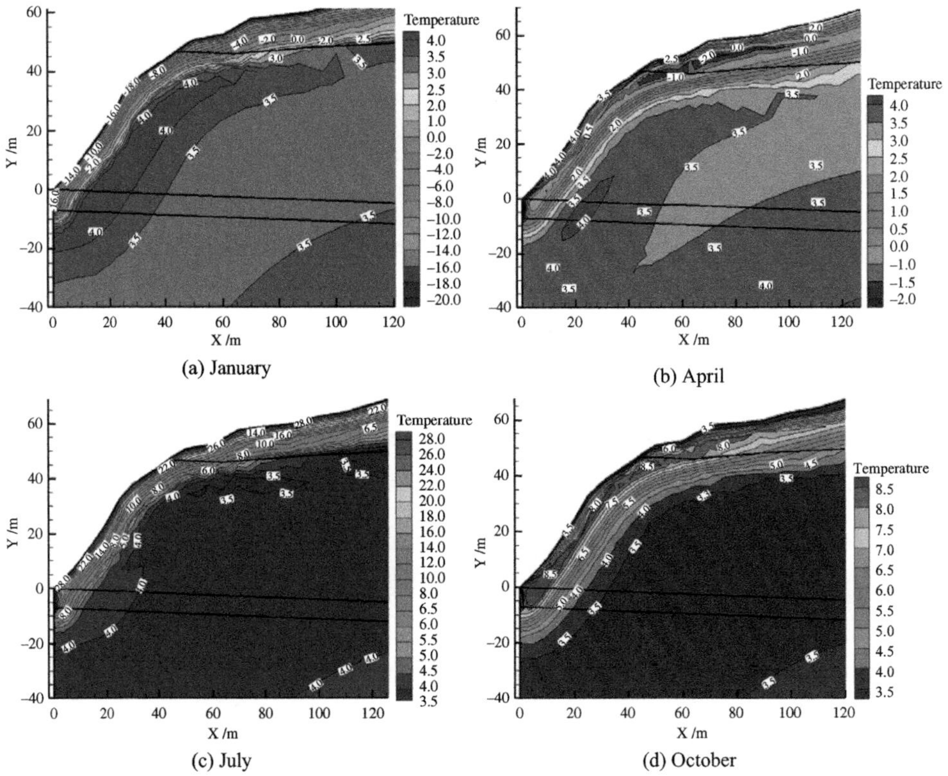

Figure 28 Initial temperature in area about 120 m away from entrance (Bo-mi) in the tunnel axial direction in different seasons.

depth larger than 18m, the variation amplitude of temperature in rock mass is less than 0.5°C. It is almost a stable temperature field. In addition, temperature in area about 120 m away from entrance in the tunnel axial direction before tunnel is 3.5~4°C. This is consistent with our in-situ measurement results, and the numerical results are verified.

4.2.7.2 Temperature distribution of surrounding rock under the condition of ventilation

Because of ventilation, heat exchange appears between air in tunnel and surrounding rock after tunnel. With a faster wind speed, there is more cold air outside entering the tunnel and temperature of surrounding rock changes more obviously. Temperature of surrounding rock with different wind speed is analyzed. To do the comparison more conveniently, only the temperature distribution in the area near entrance and exit is shown because of the long length of tunnel (3.5km).

The temperature with different wind speed at entrance of Galongla tunnel in January is shown in Fig. 30(a)-(d). The isothermal of 4°C is used for example to reveal the influence of tunnel ventilation on temperature distribution of surrounding rock. With a wind speed of 6m/s, area with a temperature of 4°C in surrounding rock is still large.

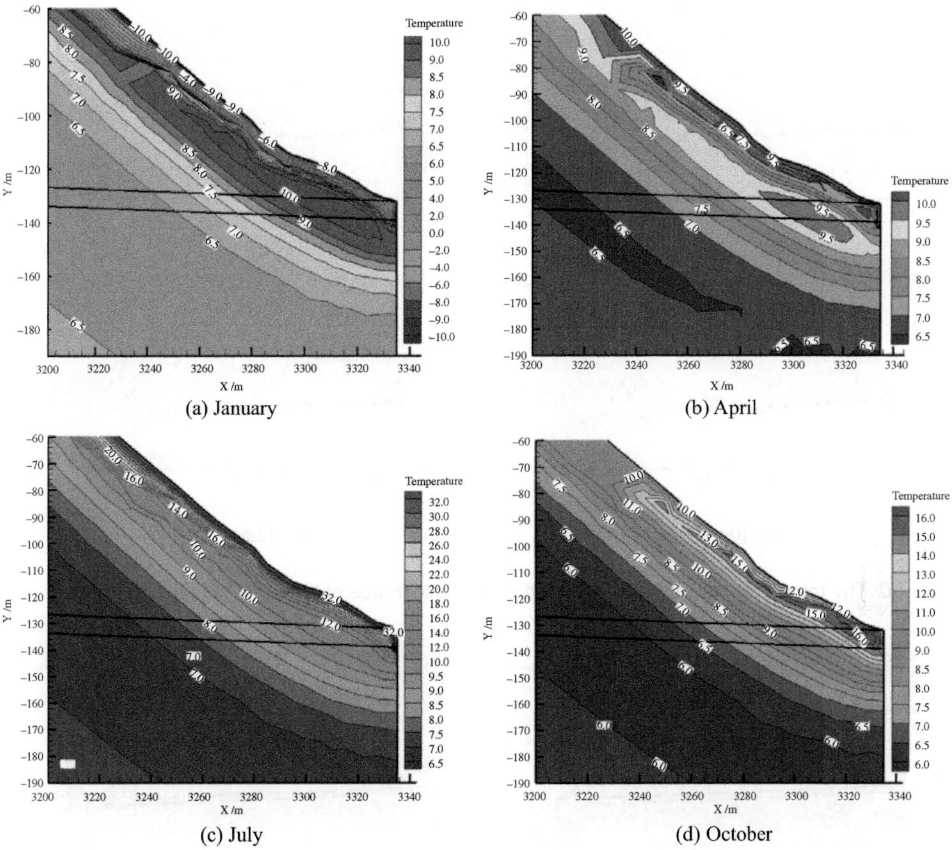

Figure 29 Initial temperature in area about 120 m away from exit (Mo-tuo) in the tunnel axial direction in different seasons.

With a wind speed of 8m/s, area with a temperature of 4°C in surrounding rock becomes smaller. When the wind speed is 10m/s, area with a temperature of 4°C in surrounding rock shrinks much more. When the wind speed is 12m/s, area with a temperature of 4°C in surrounding rock is very small. Temperature near tunnel surface becomes negative and temperature of surrounding rock with a large depth is about 2°C. Temperature distribution of air and surrounding rock at the exit of the Galongla tunnel in January with different wind speed is shown in Fig. 31(a)-(d). Just like before, the isothermal of 10°C is used for example to reveal the influence of tunnel ventilation on temperature distribution of surrounding rock. It can be seen that the area of the isothermal of 10°C is reduced with the increasing of wind speed, and this phenomenon is similar to the entrance, and temperature of surrounding rock with a large depth is about 8°C.

From Figs. 30 and 31, it is concluded that temperature of surrounding rock near tunnel entrance and exit becomes negative in January when the wind speed is 6–12m/s, It means that thermal insulation measurement should be adopted at tunnel entrance and exit.

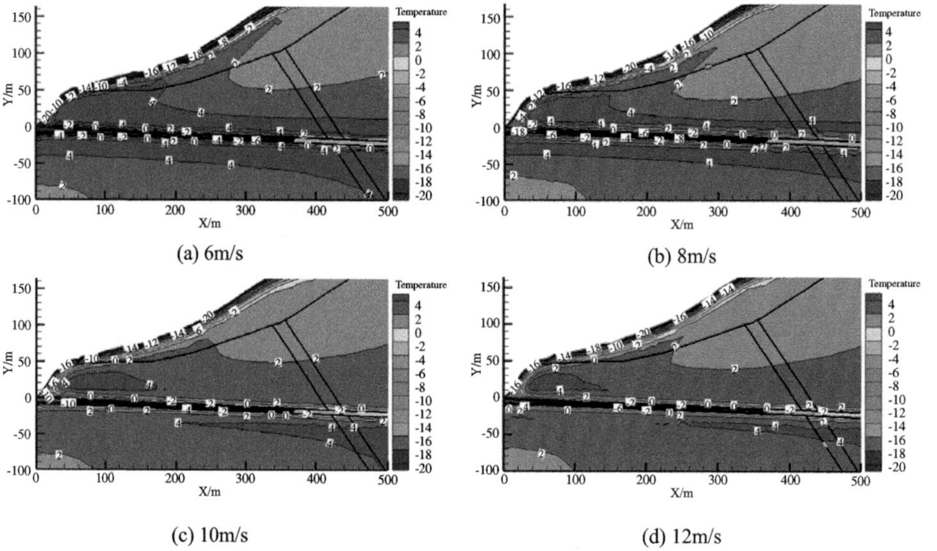

Figure 30 The temperature with different wind speed at entrance of Galongla tunnel in January.

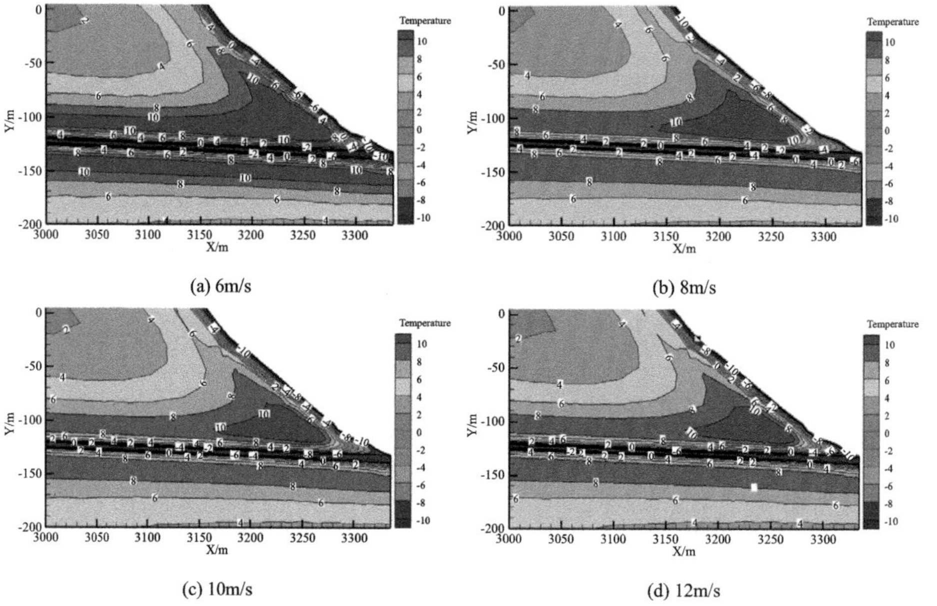

Figure 31 The temperature with different wind speed at exit of Galongla tunnel in January.

To reveal the influence of tunnel ventilation on temperature distribution of surrounding rock in different seasons, temperature distribution of air and surrounding rock at the entrance and exit of Galongla tunnel in July with different wind speeds are shown in Figs. 32 and 33.

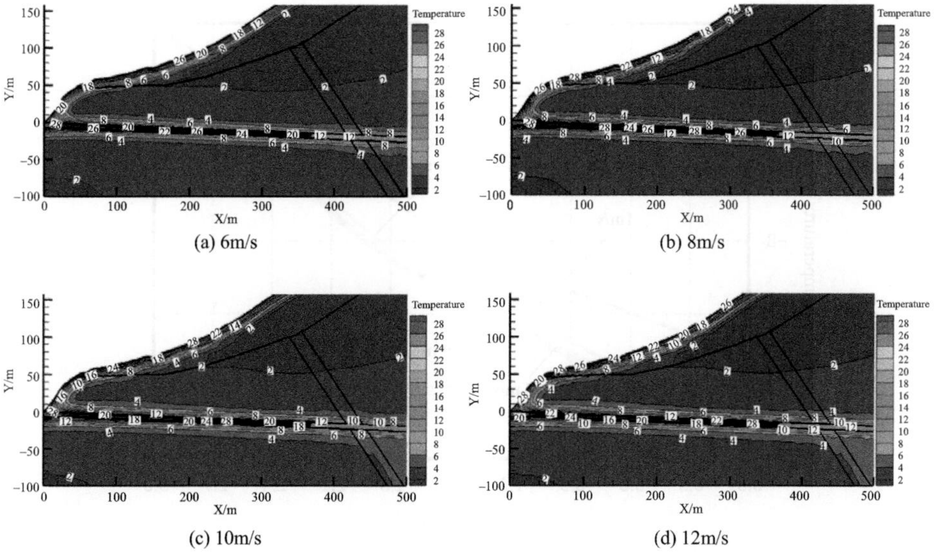

Figure 32 The temperature with different wind speeds at entrance of Galongla tunnel in July.

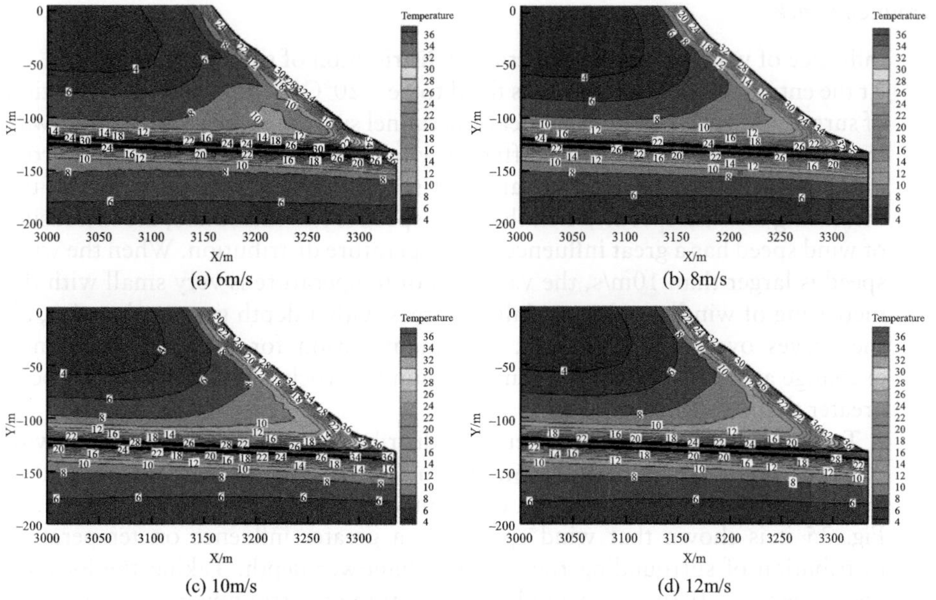

Figure 33 The temperature with different wind speeds at exit of Galongla tunnel in July.

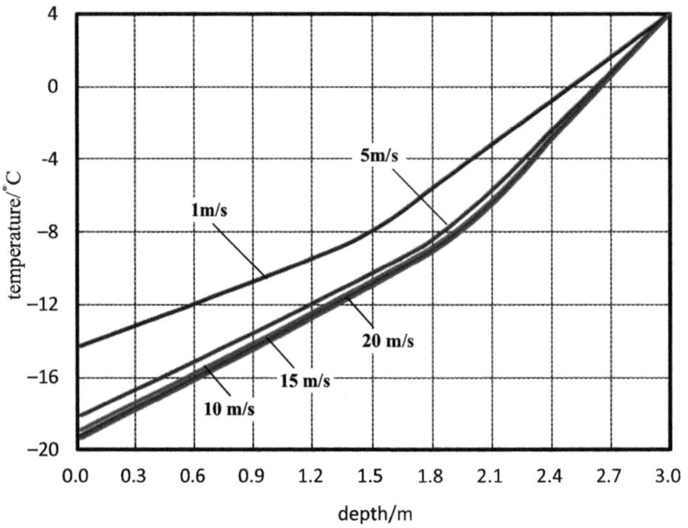

Figure 34 Temperature distribution of surrounding rock at the section 10m away from tunnel entrance with different wind speeds.

4.2.7.3 Influence of wind temperature and wind speed on temperature distribution of surrounding rock

(1) **Influence of wind speed on temperature distribution of surrounding rock**

At the entrance, air temperature is fixed to be −20°C. Temperature distribution of surrounding rock in the area near the tunnel surface at the section 10m away from the tunnel entrance with different wind speeds is shown in Fig. 34. From Fig. 34, it is shown that temperature of surrounding rock decreases with the increasing of wind speed. When the wind speed is less than 5m/s, the increasing of wind speed has a great influence on temperature distribution. When the wind speed is larger than 10m/s, the variation of temperature is very small with the increasing of wind speed, especially in areas with a depth greater than 1m, all the curves overlapped together. The main reason for this is that thermal exchange at the interface of air and surrounding rock is stable at a wind speed greater than 10m/s.

To give a further study, temperature distribution at different locations with different depths (0.2m, 0.4m, 0.6m, 0.8m and 1.0m) at the same section (10m away from tunnel entrance) with different wind speeds is shown in Fig. 35. In Fig. 35 it is shown that wind speed has a greater influence on temperature distribution of surrounding rock with a shallower depth. Taking the location with a depth of 0.2m for example, temperature of surrounding rock decreased from −13.6°C to −18.3°C, when the wind speed increased from 1m/s to 20 m/s, it decreased about 5°C. But at the location with a depth of 1.0m, temperature of surrounding rock only decreased less than 3°C.

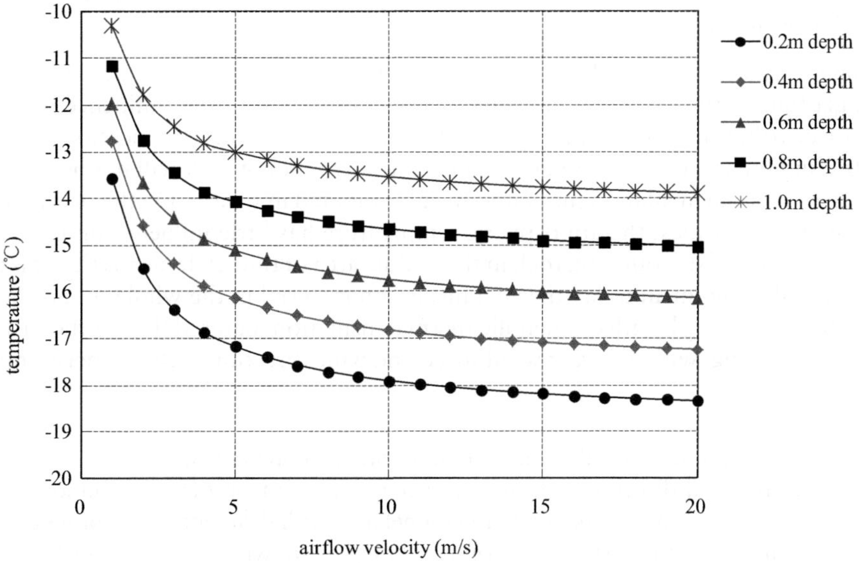

Figure 35 Temperature distribution at different locations at the section 10m away from tunnel entrance with different wind speeds.

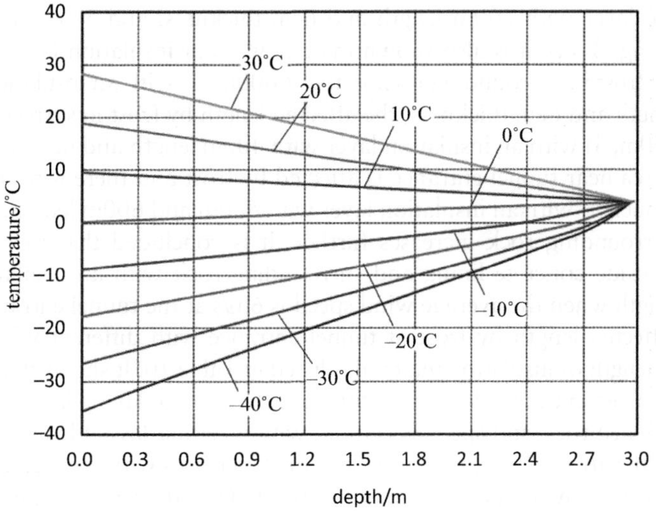

Figure 36 Temperature distribution of surrounding rock at the section 10m away from the tunnel entrance with different wind temperatures.

(2) **Influence of wind temperature on temperature distribution of surrounding rock**
At the tunnel entrance, the wind speed is fixed to be 5m/s. Temperature distribution of surrounding rock with different wind temperature in the area near tunnel surface at the section 10 m away from the tunnel entrance is shown in Fig. 36. In Fig. 36, it is shown that wind temperature has an obvious influence on temperature distribution of surrounding rock.

4.2.7.4 Determination of laying length of thermal insulation material at entrance and exit of Galongla tunnel

Through the analyses of the above section, it can be seen that at the entrance and exit of Galongla tunnel, the outside atmospheric temperature has a great influence on temperature distribution of surrounding rock. If temperature at a certain location is larger than 0°C in January, then the concrete lining and surrounding rock will not be affected by frost, so thermal insulation measures must be adopted in winter. On the other hand, because of geothermics, the atmospheric temperature has little influence on temperature distribution of surrounding rock in the middle area of tunnel. It means that it is not scientific and economical to lay thermal insulation material in the whole tunnel. So in this section the laying length of Phenolic thermal insulation material (laid at the surface of secondary lining) with different wind speed and wind direction at the tunnel entrance and exit is studied.

(1) **Laying length of thermal insulation material at tunnel entrance**
When the wind direction is from entrance to exit, temperature distribution at the surface of secondary lining along the tunnel axis with different length of insulation material (laid at the surface of secondary lining) is shown in Fig. 37 (a)~(d). Fig.37 (a) is explained in detail for an example. With an average wind speed of 6m/s at tunnel entrance, five cases are studied: 1) with an insulation layer with 300m length and 6cm thickness, 2) with an insulation layer with 400m length and 6cm thickness, 3) with an insulation layer with 500m length and 6cm thickness, 4) with an insulation layer with 600m length and 6cm thickness, and 5) without insulation layer. In Fig. 37(a), it is also shown that: 1) without insulation layer, the affected length by frost in surrounding rock is about 600m; 2) with an insulation layer with 300m length and 6cm thickness, the affected length by frost in surrounding rock is about 450m; 3) with an insulation layer with 400m length and 6cm thickness, only a small area near tunnel entrance is affected by frost and there is no frost damage inside tunnel; 4) with an insulation layer with 500m and 600m length, temperature inside surrounding rock increases further. It is concluded that there is no frost damage inside concrete lining and surrounding rock with an insulation layer of 400m length when the average wind speed is 6m/s at the tunnel entrance.
The affected length by frost at tunnel entrance with different wind speed and different length of insulation material is listed in Table 10. It shows that the affected length by frost in tunnel decreases with the increasing of laying length of insulation material when the wind speed is certain, with a certain laying length of insulation material, the affected length by frost in tunnel increases with the increasing of wind speed. When the wind speed is 12m/s, the affected length by frost is the longest (up to 850m) without insulation layer. After laying the insulation material with 600 length at tunnel entrance, only a small area near tunnel entrance is affected by frost and there is almost no frost damage inside tunnel. It is concluded that there is no frost damage inside concrete lining and surrounding rock with an insulation layer of 600m length when the average wind speed is 12m/s at the tunnel entrance.

(2) **Laying length of thermal insulation material at tunnel exit**
Similarly, when the wind direction is from exit to entrance, temperature distribution at the surface of secondary lining along the tunnel axis with different

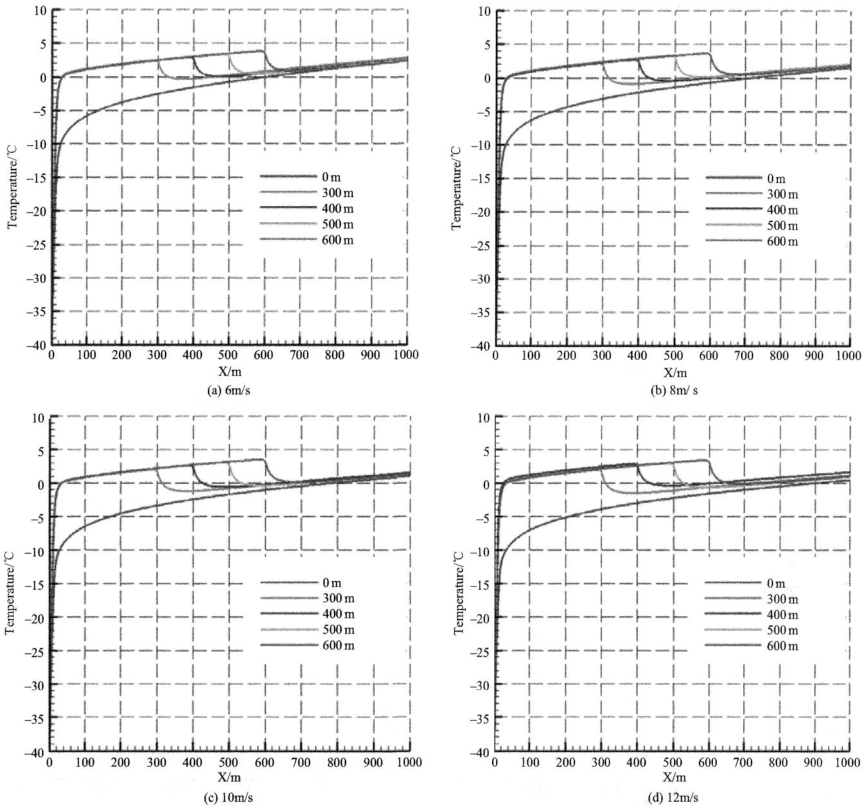

Figure 37 Temperature distribution from tunnel entrance at the surface of secondary lining with different length of insulation material.

Table 10 The affected length by frost at tunnel entrance with different wind speed and different length of insulation material.

Length of insulation material	*Wind speed*			
	6m/s	*8m/s*	*10m/s*	*12m/s*
0m	600	700	780	850
300m	450	550	630	700
400m	0	530	580	650
500m	0	0	0	600
600m	0	0	0	0

length of insulation material (laid at the surface of secondary lining) is shown in Fig. 38 (a)~(d). Fig. 38(a) is also explained in detail for an example. With an average wind speed of 6m/s at the tunnel exit, three cases are studied: 1) with an insulation layer with 300m length and 6cm thickness, 2) with an insulation layer

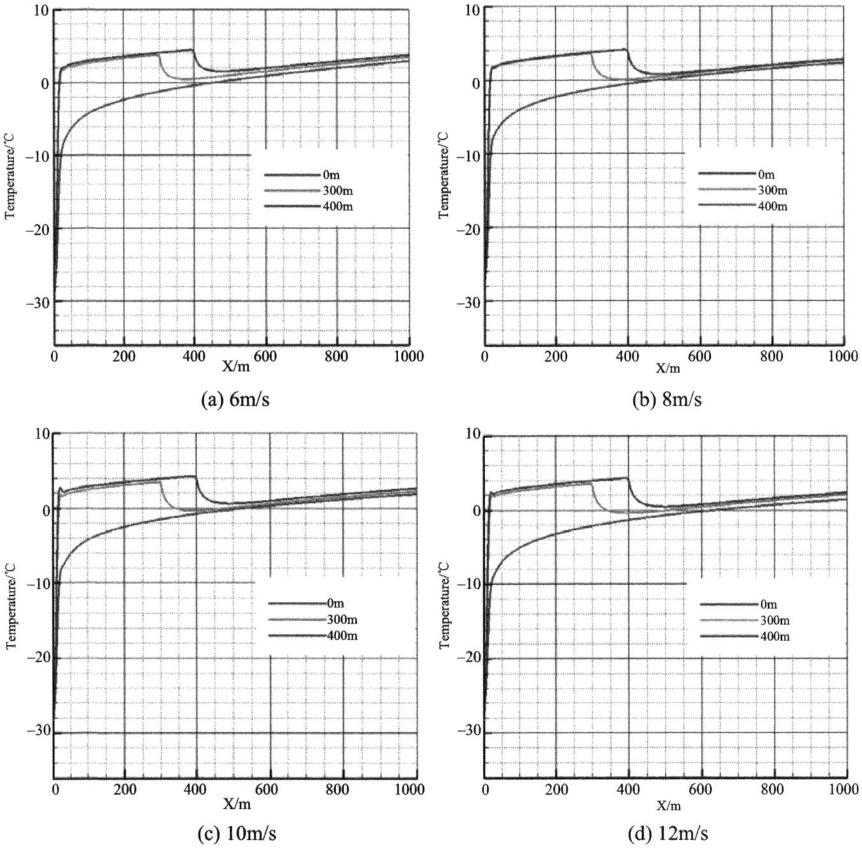

Figure 38 Temperature distribution from tunnel exit at the surface of secondary lining with different length of insulation material.

with 400m length and 6cm thickness, 3) without insulation layer. In Fig. 38(a), temperature distribution at the surface of secondary lining along the tunnel axis in three cases in January is shown. If temperature at a certain location is higher than 0°C in January, then the concrete lining and surrounding rock will not be affected by frost. Without insulation layer, the affected length by frost in surrounding rock is about 430m; with an insulation layer with 300m length and 6cm thickness, only a little area near the tunnel exit is affected by frost and there is no frost damage inside the tunnel; and with an insulation layer with 400m length, the temperature inside surrounding rock increases further. It is concluded that there is no frost damage inside concrete lining and surrounding rock with an insulation layer of 300m length when the average wind speed is 6m/s at the tunnel exit.

The affected length by frost at the tunnel exit with different wind speeds and different length of insulation material is listed in Table 11. It can be seen that when the wind speed is 12m/s, the affected length by frost is the longest (up to 600m) without insulation layer. After laying the insulation material with 400 length at the tunnel exit, only a small area

Table 11 The affected length by frost at the tunnel exit with different wind speed and different length of insulation material.

Length of insulation material	*Wind speed*			
	6m/s	*8m/s*	*10m/s*	*12m/s*
0m	430	450	500	600
300m	0	0	450	500
400m	0	0	0	0

near tunnel exit is affected by frost and there is almost no frost damage inside the tunnel. It is concluded that there is no frost damage inside the concrete lining and surrounding rock with an insulation layer of 400m length when the average wind speed is 12m/s at the tunnel entrance.

5 CONCLUSION

Based on the present study, the following conclusions are obtained:

(1) A model considering the process of freezing-thawing is presented to calculate the temperature field in porous media in terms of the "three-zone" (frozen zone, freezing zone and unfrozen zone) theory, which can simulate the formation and melting of ice. The model was successfully verified by comparing its results to an analytical solution and by matching field data.

(2) The parameter *a* was found to be very important to the distribution of temperature in the porous media. As ice forms over a range of temperatures, latent heat is released in proportion to the freezing zone. If more ice is formed over a shorter temperature interval, then more latent heat will be released over a smaller temperature range.

(3) The released latent heat has a higher energy content than the heat capacities of ice and water, and during freezing, it significantly slows further growth of ice, and this makes the distribution of temperature in porous media change greatly.

(4) Paving insulation layer on the surface of second lining is a useful method to protect tunnels against freezing-thawing damage. There has been no negative temperature on the surface of Galongla tunnel in five years with folic polyphenolic foam of 6cm thick.

(5) Air temperature and air speed are two important factors which significantly affect the temperature distribution of surrounding rock in tunnels. The simulation results indicated that 600m from the entrance of tunnel and 400m from the exit of tunnel, the application of insulation material with a depth of 6cm at the surface of the secondary lining can protect the lining and surrounding rock at Galongla tunnel from freezing-thawing damage effectively.

REFERENCES

Altindag, R., Alyildiz, I.S., Onargan, T., 2004. Mechanical property degradation of ignimbrite subjected to recurrent freeze-thaw cycles. International Journal of Rock Mechanics & Mining Sciences, 41:1023–1028.

Anderson, D.M., 1971. Tice A R. Low temperature phases of interfacial water in clay–water systems. Soil Science Society of America Proceedings, 35(1):47–54.
Anderson, D.M., Tice, A.R., 1973. The unfrozen interfacial phase in frozen soil water systems. Ecological Studies, 4:107–125.
Beier, N.A., Sego, D.C., 2009. Cyclic freeze-thaw to enhance the stability of coal tailings. Cold Regions Science and Technology, 55:278–285.
Bellanger, M., Homand, F., Remy, J.M., 1993. Water behaviour in limestones as a function of pores structure: Application to frost resistance of some Lorraine limestones. Engineering Geology, 36(1): 99–108.
Blanchard, D., Fremond, M., 1985. Soil frost heaving and thaw settlement. 4th international Symposium of ground freezing, Sapporo, 209–216.
Bonacina, C., Comini, G., Fasano, A., 1973. Numerical solution of phase-change problems. International Journal of Heat and Mass Transfer, 16(6):1852–1832.
Bronfenbrener, L., 2009. The modelling of the freezing process in fine-grained porous media: Application to the frost heave estimation. Cold Regions Science and Technology, 56 (2):130–134.
Bronfenbrener, L., Korin, E., 1997. Kinetic model for crystallization in porous media. International Journal of Heat and Mass Transfer, 40: 1053–1059.
Chen, J.X. Design method and application for anti-freezing thermal protective layers in cold area tunnels, 2004, China Civil Engineering Journal, 37 (11):85–88.
Chen, T.C., Yeung, M.R., Mori, N., 2004. Effect of water saturation on deterioration of welded tuff due to freeze-thaw action. Cold Regions Science and Technology, 38:127–136.
Cheng, G.D., Wu, Q.B., Ma, W., 2009. Innovative designs of permafrost roadbed for the Qinghai-Tibet Railway. Science in China Series E-Technological Sciences, 52(2):530–538.
Christ, M., Kim, Y.C., 2009. Experimental study on the physical–mechanical properties of frozen silt. KSCE Journal of Civil Engineering, 13(5):317–324.
Comini, G,Guidice, S.D., Iewis, R.W., 1974. Finite element solution of nonlinear heat conduction problems with special reference to phase change. International Journal for Numerical Methods in Engineering, 8(6): 613–624.
Danielian, Yu. S., Yanitcky, P.A., Cheverev, V.G., Lebedenko, Yu. P., 1983. Experimental and theoretical heat and mass transfer research in frozen soils. Engineering Geology, 3:77–83.
Ershov, E.D., 1979. Phase composition in the frozen rocks. Nauka, Moscow.
Farouki, O.T., 1981. The thermal properties of soils in cold regions. Cold Regions Science and Technology, 5(1): 67–75.
Fukuda, M., Harimaya, T., Harada, K., 1996. The study on effect of rock weathering by freezing and thawing to rock mass collapse. Gekkan Chikyu, Japan, 18(9):574–578 (in Japanese)
Furukawa, Y., Shimada, W., 1993. 3-dimensional pattern-formation during growth of ice dendrites, its relation to universal law of dendritic growth. Journal of Crystal Growth, 128:234–249.
Grossi, C.M., Brimblecombe, P., Harris, I., 2007. Predicting long term freeze-thaw risks on Europe built heritage and archaeological sites in a changing climate. Science of the Total Environment, 377(2):273–281.
Guan, B.S., 2003. Focus in Design of Tunnel. China Communications Press, Beijing, 16–30.
Hale, P.A., Shakoor, A., 2003. A laboratory investigation of the effects of cyclic heating and cooling, wetting and drying, and freezing and thawing on the compressive strength of selected sandstones. Environmental and Engineering Geoscience, 9:117–130.
Hall, K., 1999. The role of thermal stress fatigue in the breakdown of rock in cold regions. Geomorphology, 31:47–63.
Handa, Y.P., Zakrzewski, M., Fairbridge, C., 1992. Effect if restricted geometries on the structure and thermodynamic properties of ice. Journal of Physical Chemistry, 96: 8594–8599.

He, C.X., Wu, Z.W., Zhu, L.N., 1996. Preliminary prediction for the freezing –thawing situation in rock surrounding DabanShan tunnel. The Dissertations of the Fifth National Conference on Glaciology and Geocryology. Culture Press of GanSu, LanZhou, 419–425. (In Chinese).

He, C.X., Wu, Z.W., Zhu, L.N., 1999. A convection–conduction model for analysis of the freeze – Thaw conditions in the surrounding rock wall of a tunnel in permafrost regions. Science in China. Series D, 29(Supp.1):1–7. (In Chinese).

Hori, M., Morihiro, H., 1998. Micromechanical analysis on deterioration due to freezing and thawing in porous brittle materials. International Journal of Engineering Science, 36(4): 511–522.

Huang, S.L., Aughenbaugh, N.B., Wu, M.C, 1986. Stability study of CRREL permafrost tunnel. Journal of Geotechnical Engineering, 112(8):777–790.

Ilinca, F., Pelletier, D., Garon, A, 1997. An adpative finite element method for a two-equation turbulence model in wall-bounded flows, International Journal for Numerical Methods in Fluids, 24:101–120.

Ishizaki, T., Maruyama, M., Furukawa, Y., 1996. Premelting of ice in porous silica glass. Journal of Crystal Growth, 163(4):455–460.

Jin, H.J., 2010. Design and construction of a large-diameter crude oil pipeline in Northeastern China: A special issue on permafrost pipeline. Cold Regions Science and Technology, 64 (3):209–212.

Jin, H.J., Hao, J.Q., Chang, X.L., Zhang, J.M., Yu, Q.H., Qi, J.L., Lü, L.Z., Wang, S. L., 2010. Zonation and assessment of frozen-ground conditions for engineering geology along the China–Russia crude oil pipeline route from Mo'he to Daqing, Northeastern China. Cold Regions Science and Technology, 64(3):213–225.

Johansen, N.I., Huang, S.L., Aughenbaugh, N.B., 1988. Alaska's CRREL permafrost tunnel. Tunnelling and Underground Space Technology, 3(1):19–24.

Kalitzin, G., Medic, G., Iaccarino, G., Durbin, P., 2005. Near-wall behavior of RANS turbulence models and implications for wall functions. Journal of Computational Physics, 204:265–291.

Karaca, Z., Deliormanli, A.H., Elci, H., Pamukcu, C., 2010. Effect of freeze–thaw process on the abrasion loss value of stones. International Journal of Rock Mechanics & Mining Sciences, 47:1207–1211.

Konrad, J.M., Morgenstern, N.R., 1980. Effects of applied pressure on freezing soils. Can. Geotech. J., 17:473–486.

Kozlowski, T., 2003. A comprehensive method of determining the soil unfrozen water curves 1: Application of the term of convolution. Cold Regions Science and Technology, 36:71–79.

Kozlowski, T., 2007. A semi–empirical model for phase composition of water in clay–water systems. Cold Regions Science and Technology, 47:226–236.

Kozlowski, T., 2009. Some factors affecting supercooling and the equilibrium freezing point in soil–water systems. Cold Regions Science and Technology, 59: 25–33.

Lacasse, D., Turgeon, É., Pelletier, D., 2004. On the judicious use of the k-ε model, wall functions and adaptivity. International Journal of Thermal Sciences, 43:925–938.

Lai, Y.M., Liu, S.Y., Wu, Z.W., Yu, W.B., 2002. Approximate analytical solution for temperature fields in cold regions circular tunnels. Cold Regions Science and Technology, 34(1): 43–49

Lai, Y.M., Wu, Z.W., Zhang, S.J., u, W.Y., Deng, Y.S, 2003. Study of methods to control frost action in cold regions tunnels, Journal of Cold Regions Engineering, 17(4):144–152.

Lai, Y.M., Wu, Z.W., Zhu, Y.L., Zhu, L.N., 1998. Nonlinear analysis for the coupled problem of temperature, seepage and stress fields in cold-region tunnels. Tunnelling and Underground Space Technology, 13(4): 435–440.

Lai, Y.M., Wu, Z.W., Zhu, Y.L., Zhu, L.N., 1999. Nonlinear analysis for the coupled problem of temperature and seepage fields in cold-region tunnels. Cold Regions Science and Technology, 29:89–96.

Lai, Y., Zhang, X., Yu, W.B., *et al.*, 2005. Three-dimensional nonlinear analysis for the coupled problem of the heat transfer of the surrounding rock and the heat convection between the air and the surrounding rock in cold-region tunnel. Tunnelling and Underground Space Technology, 20(4): 323–332.

Lai, Y.M., Zhang, X.F., Yu, W.B., Zhang, S.J., Liu, Z.Q., Xiao, J.Z., 2005. Three-dimensional nonlinear analysis for the coupled problem of the heat transfer of the surrounding rock and the heat convection between the air and the surrounding rock in cold-region tunnel. Tunnelling and Underground Space Technology, 20:323–332.

Launder, B.E, 1991. Current capabilities for modeling turbulence in industrial flows. Applied Scientific Research, 48:247–269.

Li, G.Y., Sheng, Y., Jin, H.J., Ma, W., Qi, J.L., Wen, Z., Zhang, B., Mu, Y.H., Bi, G.Q., 2010. Development of freezing–thawing processes of foundation soils surrounding the China–Russia Crude Oil Pipeline in the permafrost areas under a warming climate. Cold Regions Science and Technology, 64(3):226–234

Lobacz, E.F., Quinn, W.F., 1966. Thermal regime beneath building constructed on permafrost. In Permafrost: Proceedings of 1st international conference, National Academy of Sciences, Lafayette, Washington, DC, 247–252.

Lunardini, V.J., 1985. Freezing of soil with phase change occurring over a finite temperature difference. Proceedings of the 4th international Offshore Mechanics and Arctic Engineering Symposium. ASM.

Lunardini, V.J., 1991. Heat transfer with freezing and thawing. Elsevier Science Publishers B.V., New-York.

Matsuoka, N., 1990. Mechanisms of rock breakdown by frost action: an experimental approach. Cold Regions Science and Technology, 17:253–270.

McKenzie, J.M., Voss, C.I., Siegel D.I., 2007. Groundwater flow with energy transport and water-ice phase change: Numerical simulations, benchmarks, and application to freezing in peat bogs. Advances in Water Resources, 30(4):966–983.

Michalowski, R. L., 1993. A constitutive model of saturated soils for frost heave simulations. Journal of Cold Regions Science and Technology, 22(1):47–63.

Mie, F.M., 1983. The frost proof problem on the drainage ditch in the cold region tunnel. Proceedings of Second National Conference on Permafrost Lanzhou Gansu People's Publish House, 405–411.

Mottagy, D., Rath, V., 2006. Latent heat effects in subsurface heat transport modeling and their impact on palaeotemperature reconstruction. Geophysical Journal International, 164:236–245.

Mutlutük, M., Altindag, R., TűrK, G., 2004. A decay function model for the integrity loss of rock when subjected to recurrent cycles of freezing-thawing and heating-cooling. International Journal of Rock Mechanics and Mining Science, 41(2):237–244.

Nakano, Y., 1990. Quasi-steady problems in freezing soils, I. Analysis on the steady growth of an ice layer. Journal of Cold Regions Science and Technology, 17:207–226.

Nakano, Y., Arnold, R., 1973. Acoustic properties of frozen Ottawa sand. Water Resources Research, 9(1):178–184.

Nakano, Y., Martin, R.J., 1972. Smith M. Ultrasonic velocities of the dilatational and shear waves in frozen soils. Water Resources Research, 8(4):1024–1030.

Nield, D.A., Bejan, A., 1999. Convection in Porous Media. Springer-Verlag, New York.

Nicholson, D.T., Nicholson, F.H., 2000. Physical deterioration of sedimentary rocks subjected to experimental freeze-thaw weathering. Earth Surface Processes and Landforms, 25 (12):1295–1307.

Okada, K., 1985. Lcile prevention by adiatic treatment of tunnel lining. Japanese Railway Engineering, 26 (3):75–80.

Park, C., Synn, J.H., Shin, D.S., 2004. Experimental study on the thermal characteristics of rock at low temperatures. International Journal of Rock Mechanics & Mining Science, 41 (Supp.1):81–86.

Patterson, D.E., Smith, M.W., 1980. The use of time domain reflectometry for measurement of unfrozen water content in frozen soils. Cold Regions Science and Technology, 3(3):205–210.

Prick, A., 1995. Dilatometrical behaviour of porous calcareous rock samples subjected to freeze-thaw cycles. Catena, 25:7–20.

Proskin, S., Sego, D., Alostaz, M., 2010. Freeze-thaw and consolidation tests on Suncor mature fine tailings (MFT). Cold Regions Science and Technology, 63:110–120.

Qiao, W.G., Li, D.Y., Wu, X.Z., 2003. Survey analysis of freezing method applied to connected aisle in metro tunnel. Rock and Soil Mechanics, 24(4):2666–2669.

Qin, Y., Zhang, J., Zheng, B., Ma, X., 2009. Experimental study for the compressible behavior of warm and ice-rich frozen soil under the embankment of Qinghai-Tibet Railroad. Cold Regions Science and Technology, 57(2–3):148–153.

Rohsenow, W.M., Hartnett, J.P., 1973. Handbook of Heat Transfer. McGraw-Hill Book Company, New York.

Ruiz, V.G., Rey, R.A., Celorio, C., *et al.*, 1999. Characterization by computed X-ray tomography of the evolution of the pore structure of a dolomite rock during freeze-thaw cyclic tests. Physics and Chemistry of the Earth, Part A: Solid Earth and Geodesy, 7(24):633–637.

Saad, A., Gue´don, S., Martineau, F., 2010. Microstructural weathering of sedimentary rocks by freeze-thaw cycles: Experimental study of state and transfer parameters. Comptes Rendus Geoscience, 342:197–203.

Sandegren, E, 1995. Insulation against ice railroad tunnels, Transportation Research Record, 1150:126–132.

Seto, M., 2010. Freeze-thaw cycles on rock surfaces below the timberline in a montane zone: Field measurements in Kobugahara, Northern Ashio Mountains, Central Japan. Catena, 82 (3):218–226.

Spaans, E.J.A., Baker, J.M., 1995. Examining the use of time domain reflectometry for measuring liquid water content in frozen Soil. Water Resources Research, 31(12):2917–2925.

Sparrman, T., Öquist, M., Klemedtsson, L., 2004. Quantifying unfrozen water in frozen soil by high–field 2H NMR. Environmental Science & Technology, 38(20):5420–5425.

Takarli, M., Prince, W., Siddique, R., 2008. Damage in granite under heating/cooling cycles and water freeze-thaw condition. International Journal of Rock Mechanics & Mining Sciences, 45:1164–1175.

Tan Xianjun, Chen Weizhong, Wu Guojun, *et al.*, 2013. Study of airflow in a cold-region tunnel using a standard $k - \varepsilon$ turbulence model and air-rock heat transfer characteristics: Validation of the CFD results. Heat and Mass Transfer, 49:327–336.

Tan, X.J., Chen, W.Z., Tian, H.M., Cao, J.J., 2011. Water flow and heat transport including ice/water phase change in porous media: Numerical simulation and application. Cold Regions Science and Technology, 68(1):74–84.

Tang, G.Z., Wang, X.H., 2007. Effect of temperature control on a tunnel in permafrost. Tunnelling and Underground Space Technology, 22:483–488.

Tice, A.R.,Burrous, C.M., Anderson, D.M., 1978. Phase composition measurements on soils at very high water contents by the pulsed Nuclear Magnetic Resonance technique. Transportation Research Record, 675:11–14.

Timur, A., 1968. Velocity of compressional waves in porous media at permafrost temperatures. Geophysics, 33(4):584–595.

Wang, D.Y., Zhu, Y.L., Ma, W., 2006. Application of ultrasonic technology for physical–mechanical properties of frozen soils. Cold Regions Science and Technology, 44 (1):12–19.

Wyllie, M.R., Gregory, A.E., Gardner, L.W., 1956. Elastic wave velocities in heterogeneous and porous media. Geophysics, 21(1):41–70.

Zhang, X.F., Xiao, J.Z., Zhang, Y.N., Xiao, S.X., 2007. Study of the function of the insulation layer for treating water leakage in permafrost tunnels. Applied Thermal Engineering, 27:637–645.

Kong, X.Y., 1999. Advanced Mechanics of Fluids in Porous Media. University of Science and Technology of China Press, Hefei, China.

Kong, X.Y., Wu, J.B., 2002. A bifurcation study of non-Darcy free convection in porous media. Acta Mechanica Sinica, 34(2):177–185.

Yamabe, T., Neaupane, K.M., 2001. Determination of some thermo-mechanical properties of Sirahama sandstone under subzero temperature conditions. International Journal of Rock Mechanics & Mining Science, 38(7):1029–1034.

Yavuz, H., Altindag, R., Sarac, S. Ugur, I., Sengun, N., 2006. Estimating the index properties of deteriorated carbonate rocks due to freeze-thaw and thermal shock weathering. International Journal of Rock Mechanics & Mining Sciences, 43:767–775.

Zhang, M.Y., Lai, Y.M., Li, D.Q., Tong, G.Q., Li, J.B., 2012. Numerical analysis for thermal characteristics of cinderblock interlayer embankments in permafrost regions, Applied Thermal Engineering, 36:252–259.

Zhang, S.J., Lai, Y.M., Zhang, X.F., Pu, Y.B., Yu, W.B., 2004. Study on the damage propagation of surrounding rock from a cold-region tunnel under freeze-thaw cycle condition. Tunnelling and Underground Space Technology, 19:295–302.

Zhang, X.F., Lai, Y.M., Yu, W.B., Wu, Y., 2004. Forecast analysis for the re-frozen of Feng Huoshan permafrost tunnel on Qing-Zang railway. Tunnelling and Underground Space Technology, 19(1):45–56.

Zhang, X.F., Lai, Y.M., Yu, W.B., Zhang, S.J., 2002. Nonlinear analysis for the three-dimensional temperature fields in cold region tunnels. Cold Regions Science and Technology, 35(3):207–219.

Zhang, X.F., Lai, Y.M., Yu, W.B., Zhang, S.J., Xiao, J.Z, 2004. Forecast analysis of the refreezing of Kunlun mountain permafrost tunnel on Qing–Tibet railway in China. Cold Regions Science and Technology, 39:19–31.

Zhang, X.F., Xiao, J.Z., Zhang, Y.N., Xiao, S.X, 2007. Study of the function of the insulation layer for treating water leakage in permafrost tunnels. Applied Thermal Engineering, 27:637–645.

Zhang, Y, 2000. The synthesis technology of treating the water leakage of the tunnel in the cold and high altitude regions, The West China Exploration Engineering, 67(6):95–97.

Zhang, Z.R., 1989. Heat Transfer. High Educational Press, Beijing.

Zhou, J.Z, 2003. The constructing technology of the drainage system and the thermal insulation material in Kunlun mountain. Journal of Glaciology and Geocryology, 6:45–50.

Chapter 9

Key problems in the design of diversion tunnel in Jinping II hydropower station

C.S. Zhang, N. Liu & W.J. Chu
HydroChina Huadong Engineering Corporation, Hangzhou, Zhejiang, China

Abstract: The construction of 4 diversion tunnels in Jinping II Hydropower Station is currently China's deepest underground engineering project. The general depth is 1500–2000 m, and the maximum depth reaches 2525 m. The highest external water pressure is 10.22 MPa. Previously unresolved issues, such as the stability of the surrounding rock of the deep buried tunnels, support design, and soft rock large deformation, are directly related to the success of the project construction. During construction, a large amount of innovative scientific and technological research was carried out to develop key design technologies in deep buried tunnel engineering. Jinping II Hydropower Station was put into operation in 2014 and has since been running in a safe and stable condition. The research achievements obtained in the construction of the project promote advances in the technology of hydraulic tunnel construction and risk control, and are of milestone significance to underground constructions in China and worldwide.

1 INTRODUCTION

Jinping II Hydropower Station is located at Jinping Bend on the Yalong River, which is at the junction of 3 counties: Muli, Yanyuan, and Mianning, in the Liangshan Yi Autonomous Prefecture of Sichuan Province, China. The power station uses the natural fall in the 150-km long river bend downstream of the Yalong River to generate power through diversion tunnels about 16.67-km long. It has a total installed capacity of 4800 MW, and is the hydropower station with the highest water head, largest power generation, and highest profit among the 21 cascade hydropower stations on the Yalong River. The power station consists of 3 main parts: the sluice in the head, the diversion system, and the underground powerhouse at the tail (Figure 1), and is a low-gate, long-tunnel, and large-capacity diversion-type power station.

The total excavation volume of underground projects involved in Jinping II Hydropower Station is more than 13.2 million m^3. Diversion tunnels, powerhouse chambers, transportation tunnels, adits, drainage tunnels, and various other tunnels cross with each other to form an enormously large-scale tunnel group, which is the only one of its kind in the history of hydropower development. The size of the deep buried tunnels and the difficulties of the accompanying technical problems are widely known.

The 4 diversion tunnels, 2 auxiliary tunnels and 1 construction drainage tunnel in Jinping II Hydropower Station are 7 parallel deep buried tunnels that form a large-scale tunnel group. The total length of the deep buried tunnel group is 118 km. For the excavation of 1# and 3# diversion tunnels, and the construction drainage tunnel, a

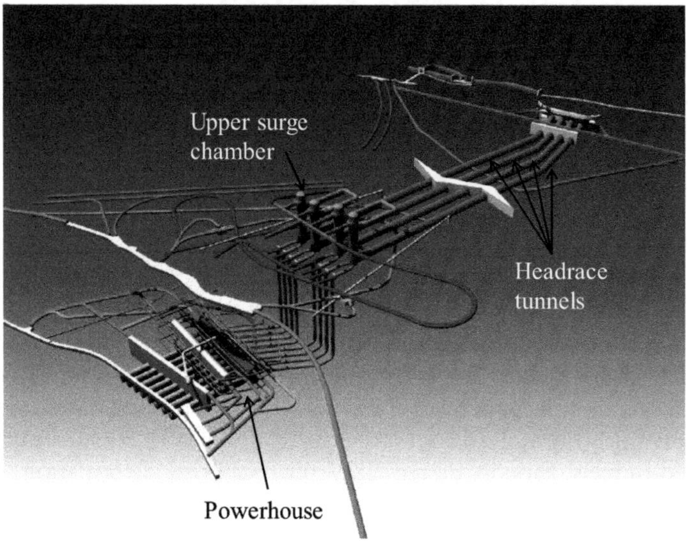

Figure 1 Three-dimensional schematic diagram of Jinping II project.

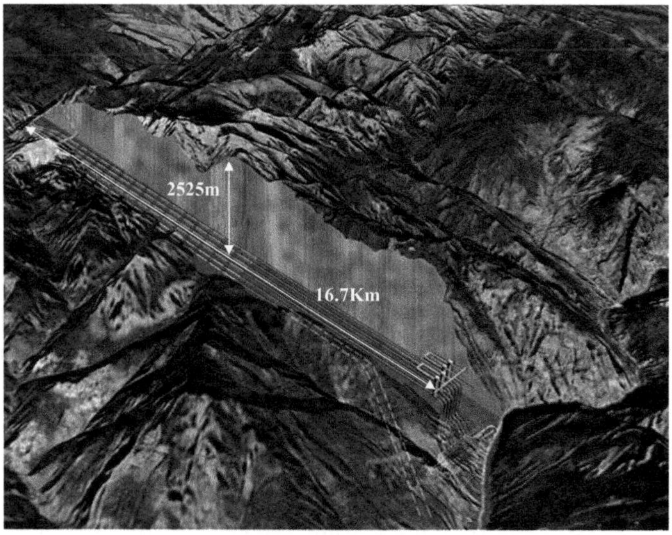

Figure 2 Diagram of the section of Jinping II deep tunnel.

tunnel boring machine (TBM) was used; for the excavation of 2# and 4# diversion tunnels, and the 2 auxiliary tunnels, the drill-and-blast method was applied. The axes of the 4 diversion tunnels are nearly orthogonal to the ridge axis of Jinping Mountain. Along this axis the mountain is steep and thick: 76.7% of the depths are greater than 1500 m, with a maximum depth of 2525 m (Figure 2). This is greater than the maximum depth of the world-famous Simplon Tunnel (2135 m), and close to that of

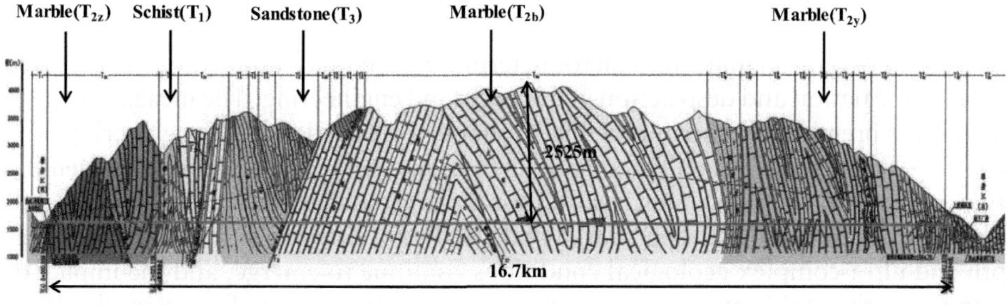

Figure 3 Simplified geological profile along the tunnel.

the world's deepest tunnel, the diversion tunnel for the Sarrans Hydroelectric Power Station in France (2619 m). The excavation diameter of the tunnel group in Jinping II Hydropower Station is 12.4–14.6 m, far greater than that of the hydraulic tunnels in the Sarrans power station at 5.8 m (Zhang, 2007: 41–44).

Along the axes of Jinping II deep buried tunnels, the hydrogeological conditions are complex, and the rock are predominantly marbles. The main stratum along the tunnel is Triassic marble, followed by sand slate, and hundreds of meters green schist (Figure 3). According to the auxiliary tunnel, there is a difference between marble of different age, and the Baishan group has better completeness and brittle characteristics. The measured maximum geostress is 113.87 MPa, and the minimum strength-stress ratio of the surrounding rock is 0.8. The ultrahigh pressure and large volume karst groundwater in the alpine gorge region underwent centralized development, and consequently the maximum external water pressure in this region is 10.22 MPa, with a maximum single-point gushing water inflow of 7.3m^3/s.

The technical difficulties in rock mass mechanical characteristics, supporting design, soft rock large deformation presented challenges far beyond the scope of existing technology both in China and in the world at that time, and the problem of construction safety was very prominent. The possibility of successful completion was therefore highly questioned by many at the beginning of the project.

To solve these problems, starting from 1991, HydroChina Huadong Engineering Corporation began the excavation of a 5 km long test tunnel. By 1995, the final length of this test tunnel was 4168 m. During the excavation a large number of targeted tests were carried out, and a large amount of first-hand information was obtained. Meanwhile, at the elevation of the diversion tunnel line, 2 parallel Jinping auxiliary tunnels were excavated. These were used for transportation, prospecting, conducting scientific experiments, and also as auxiliary construction tunnels.

Through early prospecting and experiments conducted in the 5 km long pilot tunnel and Jinping auxiliary tunnels constructed in advance, the understanding of deep buried tunnel engineering was deepened, and the involved safety risks were gradually revealed. During the construction of the diversion tunnels, a combination of research and engineering practice made a series of innovative developments, which provided a strong technical support for the safe construction and stable operation of the power station, and also a reference for future deep buried tunnel engineering projects.

2 MEASUREMENT OF SUPER-HIGH GEOSTRESS

Geostress is not only an important factor determining regional stability, but also a force causing deformation and destruction in underground engineering. The measurement of geostress is a premise for determining the mechanical properties of the engineering rock mass, performing stability analysis on the surrounding rock, and realizing the scientific design and decision-making in excavation for underground construction. As the depth of the constructions goes deeper, large-scale underground constructions with ultra-depth and ultra-complex geological conditions continue to emerge, and the impact of geostress on the stability of the construction keeps increasing. For example, large deformation of the surrounding rock, spalling and other geological disasters are all related to geostress. In addition, the inversion regression analysis of the geostress field also depends on accurate geostress measurement results. Therefore, geostress measurements have attracted much attention in engineering.

To solve the problem of testing the ultra-high geostress in Jinping, we spent three years and succeeded in developing an ultra-high pressure geostress measurement system. The course of the development included the following: (1) studies on a variety of testing methods and techniques; (2) exclusion of various geostress testing methods and cause analysis; (3) the final selection of the ultra-high pressure hydraulic fracturing measurement system, corresponding research and development, testing, feedback, and successful application.

2.1 Suitability analysis of conventional geostress testing methods

Before 2007, all geostress measurements in Jinping II Hydropower Station were completed using the conventional hydraulic fracturing method. After 2008, as the excavation of auxiliary tunnels in Jinping progressed deeper, the vertical depths gradually increased and exceeded 1500–1800 m. The geostress of the rock masses continued to increase, and the geostress test system that used the conventional hydraulic fracturing method was unable to meet the testing requirements. Subsequently, a variety of methods for geostress measurements were tested, including the aperture deformation method, the hole-bottom strain method, the surface strain method, the acoustic emission method and the local borehole wall stress relief method. The majority of the methods tested primarily involve indirect measurements instead of directly assessing the stress values. These methods are restricted by a rather large number of subjective and objective factors, which makes it difficult to obtain true on-site geostress measurement values, especially in ultra-high geostress regions such as Jinping II.

2.2 The ultra-high pressure hydraulic fracturing measurement system

The goal of the above tentative studies on the various methods for geostress measurements in Jinping II was to determine the characteristics and the order of magnitude of in-situ geostress in the rock masses. Our results show that none of the aforementioned alternative methods can be used to objectively determine the geostress characteristics of rock masses in Jinping II under the condition of ultra-high stress. The data obtained using the hydraulic fracturing method can reflect the order of magnitude and direction of the stress in the most intuitive and reliable manner. Therefore, we decided to use the hydraulic fracturing method, and made innovative improvements to solve the main problems of conventional hydraulic fracturing test systems (deficiency of the pressure supply system, insufficient pressure endurance of the piping systems, insufficient pressure endurance of the packer system and the impression system).

A preliminary estimate based on the weight of rock at a certain depth revealed that in the auxiliary tunnels with a depth exceeding two kilometers in Jinping II, the geostress will be higher than 70 MPa. The key to performing hydraulic fracturing geostress tests in rock masses under such ultra-high pressure stress is the use of suitable equipment. Currently, there is no available equipment made in China that meets such requirements. To solve this problem, HydroChina Huadong Engineering Corporation independently developed an ultra-high pressure hydraulic fracturing geostress measurement system. The development of the pressure supply system was conducted in two stages: the ultra-high pressure oil pump pressure supply system and the ultra-high pressure water pump pressure supply system.

To meet the requirements of ultra-high pressure measurements in Jinping II, the ultra-high pressure oil pump pressure supply system was improved. First, the maximum pressure supply capacity of the ultra-high pressure oil pump exceeds 120 MPa (Figure 4). Second, the pressure supply system is stable, facilitating manual control of the pressure. Lastly, under the premise of ensuring sufficient pressure supply capacity, the equipment is as lightweight as possible, making it easy to move the equipment manually. The development system includes the ultra-high pressure oil pump, the fuel saving device, the dispensing valve and the check valve. All parts can consistently endure high pressure, ensuring the safety of the test.

The ultra-high pressure oil pump pressure supply system has the following drawbacks: the pressure is insufficient; a relatively long time is needed for oil scavenging; and the oil consumption is large (when the packer ruptures during the test). To solve these problems, the pressure supply system was redesigned, and an ultra-high water pump pressure

Figure 4 The ultra-high pressure oil pump.

Figure 5 A photograph of the ultra-high pressure water pump.

supply system was developed. The maximum working pressure of this system is 150 MPa, and the flow is 10 L/min. An ultra-high-pressure check valve and a pressure relief valve were also added to the water pump; see Figure 5 for details. The development of the new ultra-high pressure water pump pressure supply system made a breakthrough in 2010, and excellent test data were recorded (Figure 6): the rock fracture pressure was 92.47 MPa; the maximum horizontal principal stress was 113.87 MPa; the minimum horizontal principal stress was 62.87 MPa, and the gravity stress at this point was 54.88 MPa. Because the stress at this test point is the highest in all Jinping deep buried tunnels, the measurement success at this point indicates that the problem of generating an ultra-high pressure supply system for the entire Jinping Mountain has been completely overcome.

3 THE MECHANICAL PROPERTIES OF DEEP MARBLES

In deep buried tunnel projects, the rock masses are often in a high-stress environment. Under such high stress, the rock masses will exhibit complex mechanical characteristics and responses.

3.1 The brittle-ductile-plastic transition characteristics of deep marbles

Figure 7 shows the triaxial stress–axial strain curves of marbles in Jinping Baishan Group which is buried in 2000m depth, and the confining stress covers 2–50MPa (Lau & Chandler, 2004: 1427–1445; Zhang *et al.*, 2010: 1999–2009).

In Figure 7 we can see that as the confining pressure increased, the difference between the peak intensity and residual strength decreased. When the confining pressure exceeded 10 MPa, after the stress–strain curve reached the yield strength it did not fall rapidly, but instead showed notable features of ductility. When the confining pressure was relatively

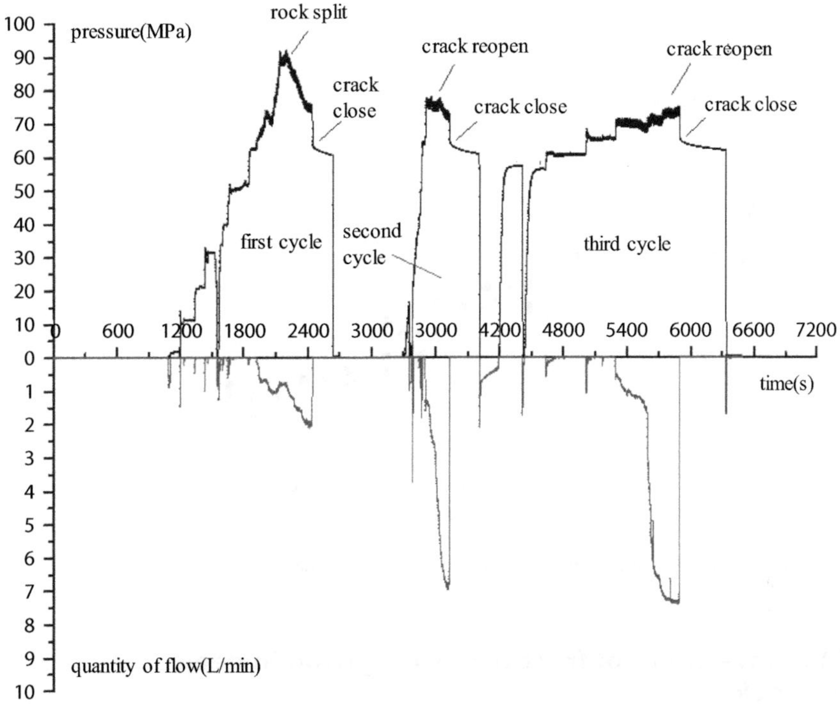

Figure 6 The pressure and flow vs. time relationship in a geostress test.

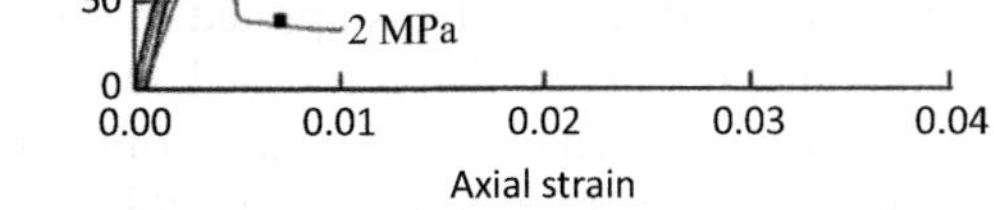

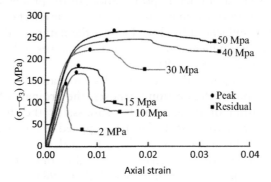

Figure 7 Triaxial test results on Jinping T_{2b} marble.

high, the slope of the curve falling from the peak intensity to the residual strength slowed down substantially. The brittle–ductile transition of the marble was highly sensitive to the level of the confining pressure. Generally, when the confining pressure exceeded 6MPa, features of ductility could be observed. At a relatively high level of confining pressure, beyond the yield strength of marble, the stress–strain curve was close to the mechanical properties of ideal plastic materials, and there was no notable stage of residual strength.

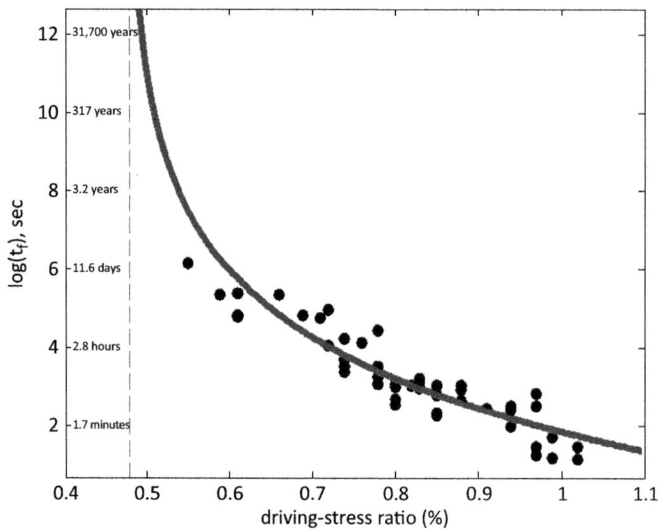

Figure 8 Static-fatigue test data for marble group with exponential fitting.

3.2 The time effect of fracture propagation in deep marble rock

In hard rock tunnels under high geostress, the scope of surrounding rock fracture and relaxation will continue to expand over a considerable period of time. This phenomenon is called the time effect of fracture propagation, or the problem of the long-term strength of hard rock. A previous study has shown that time effect of surrounding rock fracture propagation is influenced by three factors: the nature of the rock, the stress level, and changes in environmental factors, such as changes in humidity and water environment (Potyondy, 2007: 677–691).

Figure 8 shows the fatigue failure test results on marble rock collected from Jinping Yantang Group. For illustration purposes, the ordinate of the time-failure curve uses the logarithm of time, and the abscissa is the driving-stress ratio $\sigma/\sigma_c = (\sigma_1 - P_c)/(\sigma_f - P_c)$, where σ_1 is the stress applied in the axial direction, P_c is the level of the applied confining pressure, and σ_f is the peak intensity. From the figure it can be seen that as the stress condition changes, the failure of the rock samples exhibit time effect of fracture propagation (Schmidtke & Lajtai, 1985: 461–465).

4 DAMAGE EVOLUTION CHARACTERISTICS IN DEEP-BURIED TUNNEL

The in-situ rock mass properties and behaviors around the underground opening would change due to stress redistribution, blasting, moistness and temperature variation, etc. Knowledge of the degree and extent of the EDZ is important for the design and construction of Jinping deep buried tunnels. EDZ could be mechanically unstable, which in turn requires a rock support system. EDZ could also form a permeable pathway for groundwater flow, which would threaten the safety of Jinping deep buried tunnels (Cai & Kaiser, 2005: 301–310).

Both fast Lagrangian analysis of continua and particle flow code (PFC) method were used for the calibration of the ground EDZ profile measurement, followed by prediction of the maximal EDZ depth in different tunnel sections. The prediction results are helpful for support design.

4.1 Description of brittle-ductile-plastic transition behavior of Jinping marble

The Hoek-Brown constitutive model was used to describe the brittle-ductile-plastic transition behavior of Jinping marble. The EDZ depth of the headrace tunnels was predicted after model calibration. The technical details of the Hoek-Brown model properties in describing the brittle-ductile-plasticity were put forward by Zhang *et al.* (2010: 1999–2009). The model response of brittle-ductile-plastic transition is shown in Figure 9, and the Hoek-Brown strength envelop curves of peak and residual strengths are also shown in Figure 9. It can be seen that the numerical results duplicate the brittle-ductile-plastic transition with the increasing confining pressure obtained from laboratory test. In Figure 9, as the confining pressure increases to 30 MPa, the ductile response will turn into a perfect-plastic response, while under low confining pressures, for example, $\sigma_3 = 0$ or 5 MPa, the stress-strain curve drops down immediately before it reaches the peak strength. The difference between the peak and the residual strengths decreases with the increasing confining pressure. This character can also be reproduced in numerical simulations.

Figure 10 presents the numerical results of stress path near the headrace tunnel opening. The rock mass quality in simulations is of typical class II (GSI =70) according to the geological strength index (GSI). The peak strength of Hoek-Brown model is obtained from Hoek-Brown experimental strength criterion (UCS = 140 MPa, GSI = 70, m_i = 9, m_i is the Hoek-Brown constant for marble), the residual strength of rock

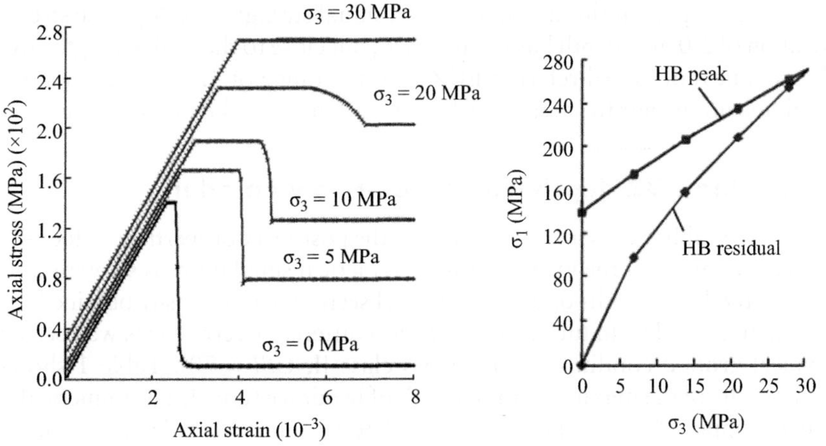

Figure 9 The description of marble brittle-ductile-plastic transition with different confining pressure.

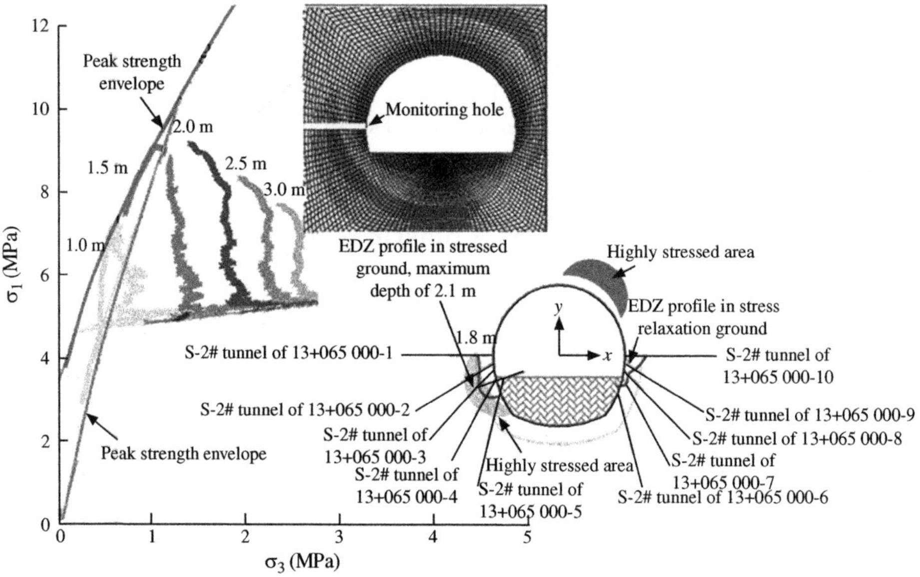

Figure 10 Measured EDZ profile on the headrace tunnel section and numerical calibration (the overburden depth is about 1733 m).

mass in the tunnel section is back calculated based on the measured ground EDZ profile.

Figure 9 also shows the calibration of post-peak behavior with Hoek-Brown constitutive model. The right-hand side in Fig.9 presents the measured EDZ profile at the two toe corners of tunnel. The measured depth of EDZ at the left corner ranges from 1.8 to 2.1 m.

The curves in the left side of Figure 9 represent simulated stress paths at the locations of monitoring hole with different depths from the opening. The stress path at the location of 1 m deep from the opening shows a typical brittle response and severe damage, and stress path at this depth reaches residual strength envelop. The stress path at the location of 2.0 m at depth has not touched, but close to the peak strength envelop. Thus the thickness of model-referred EDZ is in the range of 1.5 and 2.0 m along the horizontal hole in the left toe area, which is consistent with the measurement.

4.2 Predicting EDZ depth with Hoek-Brown model

According to the calibrations mentioned above, the post-peak properties of Hoek-Brown model can be obtained. If the in-situ stress of numerical simulations is set with different overburdens, the EDZ depths of different tunnel sections can be easily obtained.

The prediction of EDZ in tunnel sections with different overburdens was performed under typical ground conditions (rock mass class II, GSI = 70). Table 1 shows the prediction results of the maximum EDZ depth of headrace tunnel. The numerical result is helpful for support design, especially for the determination of the length of rock bolts.

In the depths of 1 800 to 2 200 m from ground surface, the maximum EDZ prediction of rock class II by Hoek-Brown model is 3.4 m. According to the principle that bolts should

Table 1 The EDZ profile prediction of rock class II under different depths by $FLAC^{3D}$.

Depth (m)	*EDZ depth (m)*
1 500	2.2
1 800	2.6
2 000	3.0
2 200	3.3

cover the damaged zone depth of surrounding rock, the length of system bolts is 4.5 m at least. Accounting for time-dependent character of EDZ in deep buried tunnels, the rock bolt of 6 m in length was selected in the tunnel section with overburden over 1 800 m.

4.3 Predicting EDZ profile using PFC

The EDZ is associated with crack initiation and propagation. Some numerical researches focus on simulation of crack initiation and propagation directly. Since the research by Griffith, many subsequent researchers have been using shearing or sliding crack to model the initiation of brittle failure (Horii & Nemat-Nasser, 1986: 337–374; Lajtai, 1990: 59–74). When using the Hoek-Brown model to calibrate parameters for relatively poor ground rock mass of class III with *GSI* = 55, it has been found that the model always intends to overestimate the depth of EDZ. This finding leads to the application of PFC to prediction of EDZ. The bonded particle model based PFC provides a numerical method that can reproduce qualitatively almost every mechanical mechanism and phenomenon that occurs in rocks, although adjustments and modifications are necessary to obtain quantitative matches in model property (Potyondy & Cundall, 2004: 1329–1364). This model also has been used for calibration and prediction of EDZ under poor ground conditions of Jinping II project.

The numerical calibration was performed by comparison of simulated EDZ and field measurement (Malmgren *et al.*, 2007: 1–15; Read, 2004: 1251–1275). Figure 11 shows a simulated EDZ profile of TBM excavated headrace tunnel section, which is well consistent with the field measurement. Table 2 shows the prediction of the maximum EDZ depth in tunnel sections with different overburdens.

Apparently, both the Hoek-Brown model based on PFC and the bonded particle model based on PFC have the same capability to deal with the EDZ prediction.

5 SUPPORT DESIGN AND OPTIMIZATION OF DEEP BURIED TUNNELS

During the construction of the long deep buried tunnels, we followed the principle that the stability of the surrounding rock should primarily rely on the self-supporting capacity of the surrounding rock itself. The surrounding rock was the main load-bearing structure, and measures including anchor support and secondary high pressure consolidation grouting were applied to reinforce the surrounding rock, so that the surrounding rock and the anchor support become a unified complex, and served as a combination load bearing structure (Figure 12). This ensured the stability of the

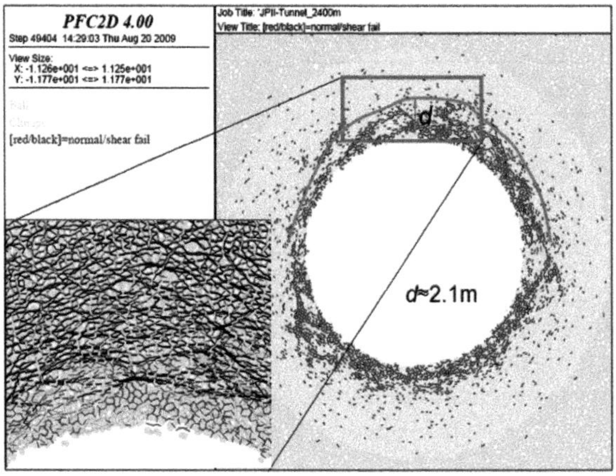

Figure 11 EDZ profile simulated by PFC (overburden 1500 m).

Table 2 Prediction of the EDZ under different ground conditions (GSI = 55).

Depth (m)	*EDZ depth (m)*
1 350	1.9
1 500	2.1
1 800	2.4
2 000	2.6
2 200	2.8

surrounding rock in the loose circle on the internal surface of tunnel, and provided triaxial confining pressure on the surrounding rock. In this way, the surrounding rock under the triaxial confining pressure itself was used to bear the geostress caused by tunnel excavation, as well as the high groundwater pressure, thereby ensuring the safety and stability of the deep buried tunnels under high external water pressure.

In practice, when designing the support, the method of dynamic design was followed. According to the geological conditions and construction conditions revealed in the actual tunnel excavation, the pre-designed rock classification system and support parameters were used as a reference only. Combined with on-site prototype monitoring and geophysical exploration data, field test results on new materials and new technologies, and analysis of surrounding rock fracture and damage in deep depth, the support structure was actively optimized.

5.1 Selecting the support unit

5.1.1 Selecting the bolt

Selecting the appropriate type of support unit required full understanding of its mechanical properties, as well as the advantages and disadvantages of the support

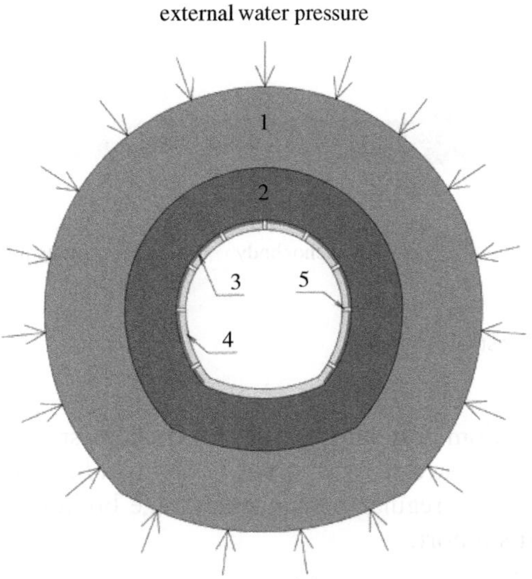

1: Impervious grouting circle; 2: High pressure consolidation grouting circle; 3: shotcrete layer; 4: reinforced concrete lining; 5: pressure-reducing hole

Figure 12 **Combination load bearing structure.**

unit itself. For specific modes of rock instability, the choice of support should be made to fully utilize its advantages and avoid its disadvantages, in order to avoid conditions unfavorable to the stability of surrounding rock. For example, the water swelling bolt belongs to friction-type anchorage bodies. Namely, it relies on the friction between the tunnel surrounding and the anchorage body to provide support. Its greatest advantage is that it can be quickly installed, which suits the requirement of rapid support in the excavation of Jinping II tunnel segments with high rock burst risk and a large cross section. However, these tunnel segments also had the prominent problem of surrounding rock fracture. Under this condition, if a water swelling bolt is used, the surface support and the anchoring force at the deep water-swelling segment need to work together to maintain the stability of fractured rock in the shallow layer. This poses challenging demands on the strength of the surface support, the bearing plate, and the anchoring force of the water-swelling bolt. These 3 aspects are the key steps to stability control of surrounding rock in deep, long, large-diameter tunnels.

For the diversion tunnel project of Jinping II Hydropower Station, after comparing the form and properties of different anchorage bodies, taking factors such as the speed, difficulty, and efficiency of the on-site construction into consideration, and through continuous testing, the water swelling bolt, the regular mortar bolt and the expanding-shell prestressed bolt were selected as the main anchorage bodies to be used on the construction sites (Figure 13). To choose among these 3 types of anchorage bodies in principle depends on the mechanical properties of marbles under high stress, and the requirement for stability control of the on-site surrounding rock, and is also affected by actual application conditions of the sites. Through field tests and assessment on the overall effects, it was ultimately determined that the water swelling anchorage bodies

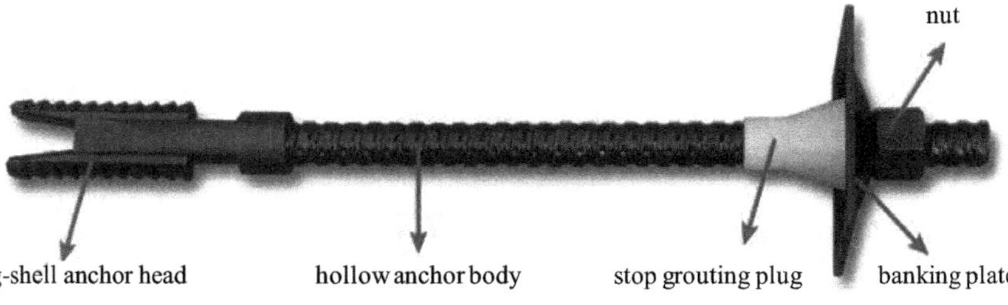

Figure 13 Expanding-shell prestressed bolt.

would be used as the random bolt, and the bolt for rock burst prevention and control; the local expanding-shell prestressed anchorage bodies would be used as the immediate permanent support; and the regular mortar anchorage bodies would be used as the lagged local permanent support.

5.1.2 Selecting the shotcrete

Due to the prominent surrounding rock stability problem, the high risk of rock burst under high geostress, and other issues in deep, long, large-diameter hydraulic tunnels, we carried out on-site tests on various new types of shotcrete. From the results of these tests, we determined to use silica fume and steel fiber double reinforced concrete, nano steel fiber concrete, nano organic imitation steel fiber reinforced concrete, and plain hanging shotcrete as the main types of supporting shotcrete (Figure 14, 15). For the general high-geostress tunnel segments, by adding nanomaterials to the shotcrete, the thickness of the primary projection was improved, and the shotcrete hardened quicker. In this way, the deformation, relaxation, and failure of the surrounding rock could be controlled. In tunnel segments with rock burst, powerful supporting measures needed to be taken to control rock failure. Under the premise of using the anchor support system, the shotcrete with the best performance, namely the silica fume and steel fiber

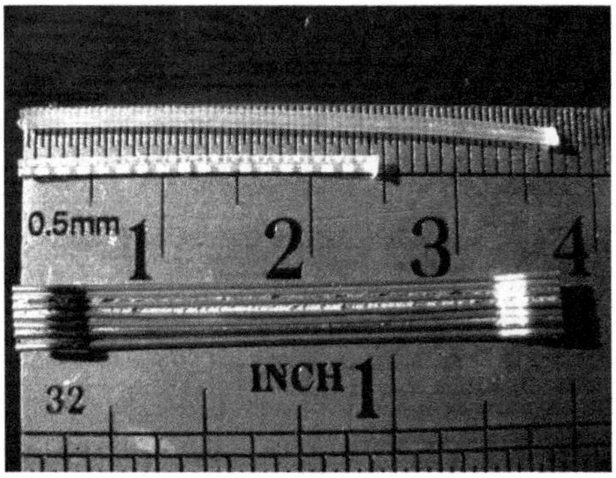

Figure 14 Nano steel fiber.

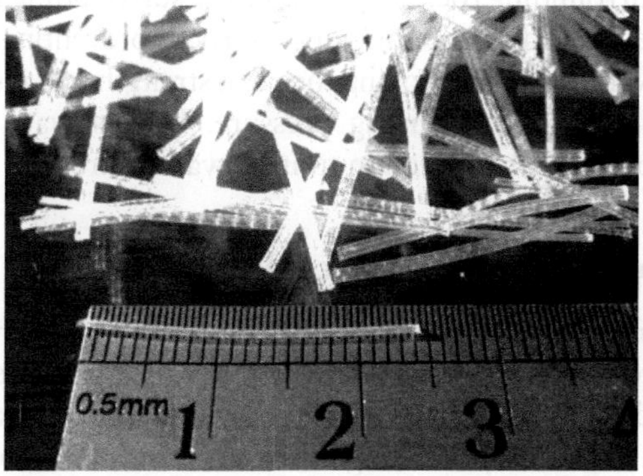

Figure 15 Nano organic imitation steel fiber.

double reinforced concrete, was used as the main supporting material, in order to ensure surrounding rock stability and construction safety. In addition, plain hanging shotcrete was subsequently applied to strengthen the support.

5.2 High-pressure seepage-proof consolidation grouting

In the process of Jinping II diversion tunnel construction, based on the actual geological conditions and the engineering requirements, we carefully examined the consolidation grouts for different surrounding rock and classified them into 4 types. The first type, consolidation grouting for fractured surrounding rock, belongs to regular grouting. Its purpose is mainly to improve the mechanical properties of the rock mass, increase its load bearing capacity, and improve the impermeability of the surrounding rock. The second type is consolidation grouting for karst tunnel segments, used for preventing internal water from leaking out after completing the backfilling of concrete or mortar in the cavern. The third type is high-pressure seepage-proof consolidation grouting. It is the primary means of counteracting the high external water pressure around the tunnel, controlling the stability of the seepage and reducing the amount of seepage. The fourth type is shallow consolidation grouting, which is designed specifically to target the widespread occurrence of relaxation, fracture, swelling, and failure of the shallow layer of brittle marble tunnels under high stress. Shallow consolidation grouting combined with secondary reinforced concrete lining provides a triaxial pressure condition to the deep surrounding rock. This increases the load bearing capacity of the surrounding rock, and meanwhile improves their seepage-proof performance.

5.3 The lining structure

Given the fact that under high geostress, fracture induced damages to surrounding rock are widespread, and have a time effect, the long-term stability and safety of anchor-plate retention cannot be guaranteed. Thus, in deep buried tunnel segments with high geostress, reinforced concrete lining was combined with shallow consolidation grouting to reinforce the surrounding rock. This ensured that the supporting structure and the surrounding rock bore forces synergistically and effectively transferred the load, and internal and external water pressure on the surrounding rock. Meanwhile the surrounding rock was closed, and high confining pressure applied to the surrounding rock, preventing the development of internal rock mass relaxation, and improving the peak strength and residual strength of the surrounding rock under triaxial confining pressure. In this way, the stability of the surrounding rock surrounding the diversion tunnels was ensured.

5.4 External water pressure reduction technology

The ability of deep, large-diameter hydraulic tunnels to bear external water pressure is rather weak. In order to ensure the long-term safe operation of these tunnels, a drainage tunnel was constructed between the diversion tunnel and auxiliary tunnel for long-term discharge of groundwater, thereby reducing the overall external water pressure along the axis of the diversion tunnel. In extreme conditions such as rainstorms, external water pressure rises sharply over a short amount of time. To handle this problem, consolidation grouting was applied to the entire length and the entire cross section of the diversion tunnel to form an impermeable ring. In addition, system decompression holes were arranged in the lining structure in order to rapidly balance and discharge external water, so that the external water pressure on the outer margin of the lining was maintained within the allowable range of its design. This ensured that the lining structure would not lose stability and collapse due to excessive external water pressure (Figure 16).

5.5 Safety analysis of combination load bearing system during its operation

As the construction approached its end, the declined water table would be gradually restored to its original level as the final closing was carried out, and the tunnels put into

Figure 16 Diversion tunnel face before operation.

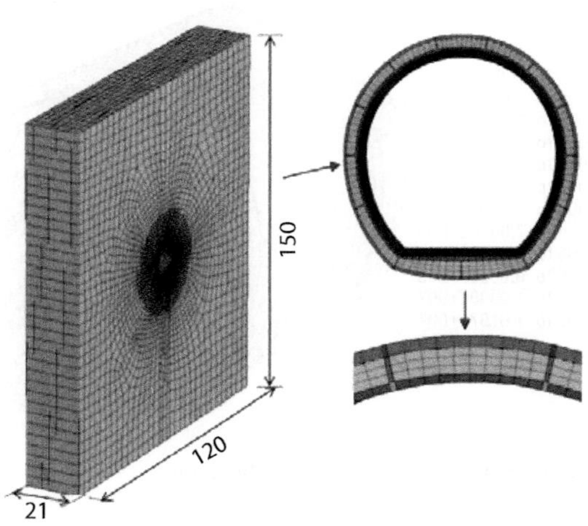

Figure 17 Calculation model.

operation. This meant the distribution of external water pressure along the tunnel would also change, posing a challenge to the long-term stability of the surrounding rock. Especially for the diversion tunnels in Jinping II Hydropower Station, the combined effects of high geostress, high external and internal water pressures, and long-term mechanical properties of the surrounding rock were completely different from that during construction. This inevitably led to complexity in the assessment of long-term safety.

To solve this problem, numerical analysis was performed. The computational model is shown in Figure 17. Within a range of 270° of the arch crown and side, decompression holes were arranged, with a row distance of 3 m. The layout of the decompression holes is illustrated in Figure 18.

The calculation results in the condition of 2500 m in depth and 10MPa of external water pressure are shown in Figures 19–21. Under high external water

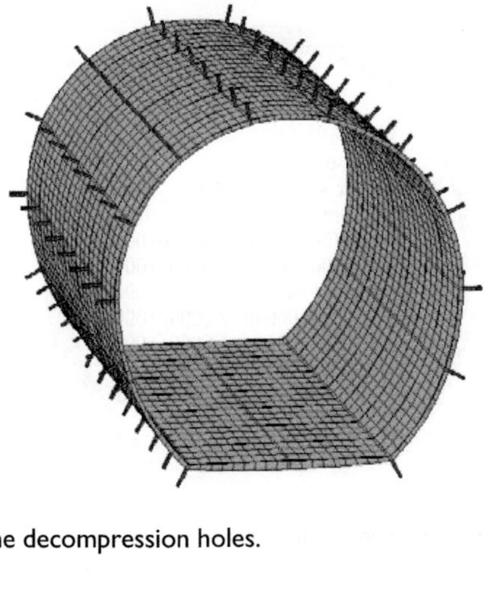

Figure 18 Figure layout of the decompression holes.

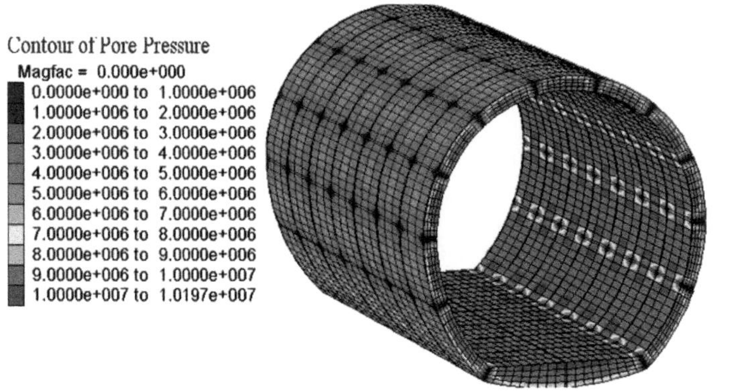

Figure 19 Distribution of pore water pressure around the lining.

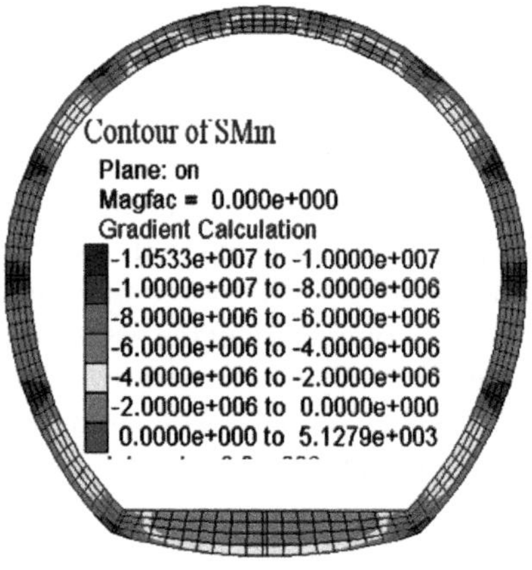

Figure 20 Distribution of compression stress in the lining.

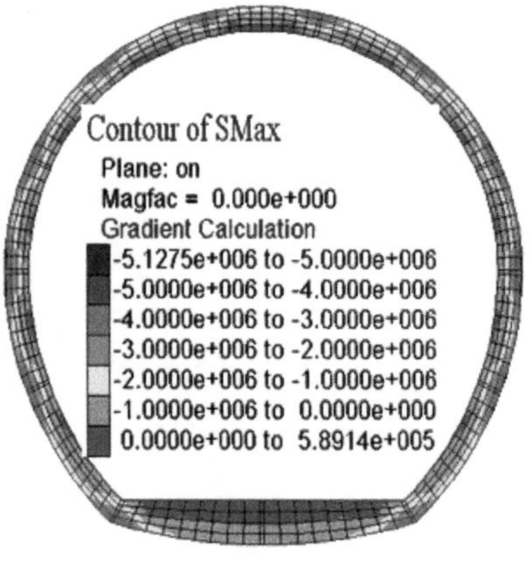

Figure 21 Distribution of tension stress in the lining.

pressure, the decompression holes laid out on the lining show substantial decompression effect, and greatly optimize the force distribution, maintaining the values of all indicators within the designed range. Thus, the combination load bearing system composed of the concrete lining structure, the surrounding rock, and the high-pressure consolidation grouting ring is safe and reliable. This simulation result is consistent with the on-site monitoring result and the actual running status of the load bearing system.

6 DEFORMATION AND STABILITY OF LARGE-DIAMETER SOFT ROCK DEEP BURIED TUNNEL

During the excavation of diversion tunnels for Jinping II hydropower station, a chlorite schist formation was encountered. The total length of this formation in the tunnel is 400 m, and the depth is 1550 m to 1850 m. Based on the gravity stress, the maximum geostress at this site should be 42–50 MPa, and hence it is a high geostress region. During the excavation of the tunnel segment with the chlorite schist formation, due to the impact of high geostress and the properties of chlorite schist, massive landslides, continuous deformation of the surrounding rock after the initial surrounding rock support, and damages to the support structure occurred, bringing a rather large negative impact throughout the entire process of diversion tunnel construction. The stability of the surrounding rock in the chlorite schist segment mainly depends on whether the lining structure can withstand the extrusion effect caused by two factors, namely the rheological effect and the softening of the surrounding rock in the presence of water (Figure 22). Thus, the stability of the surrounding rock is mainly evaluated from the perspective of the direction of the force on the lining structure.

Figure 22 Large Squeeze deformation of tunnel arch feet.

Table 3 Softening deformation parameters of rock mass.

Index	*Softening coefficient /%*	
	65	*80*
E(GPa)	1.44	2.36
v	0.28	0.27

6.1 The effect of surrounding rock softening in the presence of water

According to monitoring data, the depth at which the chlorite schist may become a clearly permeable channel is about 2 m in the vicinity of the left spandrel, and shallower in other parts. The direct result of the surrounding rock softening in the presence of water is degradation of the mechanical properties of the surrounding rock, such as reduction in deformation, and strength parameters. Indoor experimental data have shown that in the saturated condition the uniaxial compressive strength of rock is reduced by 50% compared to that in the dry condition. Taken all the above information together, assuming that at the depth 2 m the surrounding rocks are filled with water, then the rock strength and the deformation modulus at depth 1m will be decreased to 65% of the initial value, and at 2 m to 80%. Table 3 summarizes the ratios of the mechanical parameters of the surrounding rock after softening in the presence of water to their original values.

Figure 23 illustrates the distribution of the maximum tensile stresses in the lining. As can be seen, when the surrounding rock softens in the presence of water, except the arch feet on both sides, all other parts are in compression. The maximum tensile stress at the arch feet of the two sides is 1.41 MPa, which is less than the standard tensile strength of concrete, 2.01MPa, suggesting that the structure is in a safe condition.

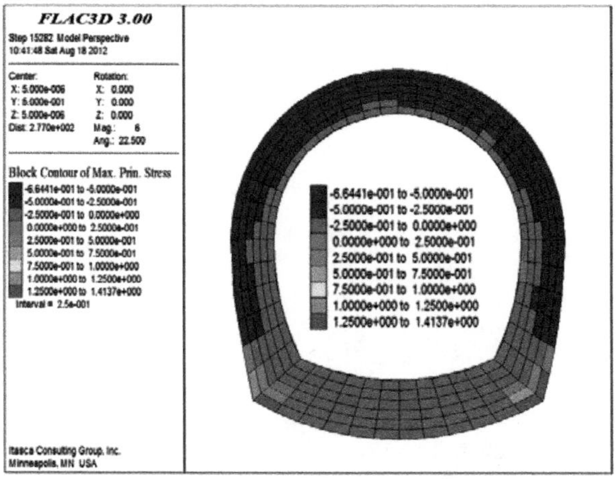

Figure 23 Maximum tensile stress in lining structure under the condition of softening.

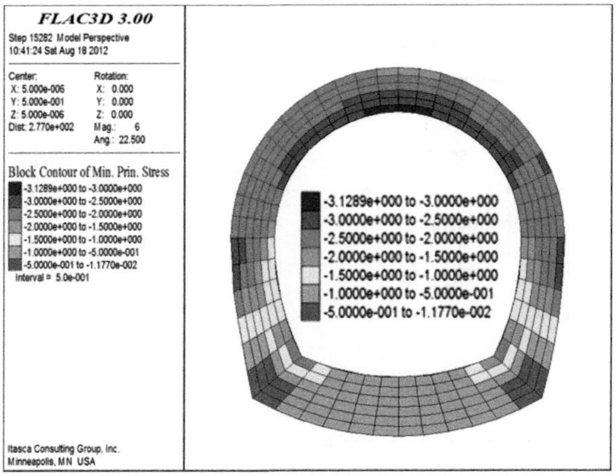

Figure 24 Maximum compression stress in lining structure under the condition of softening.

The distribution of maximum compressive stresses in the lining structure is shown in Figure 24. The maximum compressive stress is the highest at the inside of the arch top, which is 3.12 MPa, far less than the standard compressive strength of concrete C30, 20.1 MPa.

6.2 The rheological effect

The forces on the lining during operation need to be examined and the timing of the construction of the lining need to be taken into consideration. In the present calculation, the timing for constructing the lining of the diversion tunnels in Jinping II was used as a reference, and the lag between the lining construction and the excavation was set to 1.5 years. An elastic constitutive model was used for the calculation. Thus, whether the lining structure is in a safe state after it is constructed can be evaluated by the increase in the value of its compressive stress. In this calculation, a degenerated Burgers model was used to describe the attenuated rheological behavior of the chlorite schist (Figure 25). The model contains a total of seven parameters, four of which describe the elastic behaviors, *i.e.*, the spring in the Kelvin body and the spring in the Maxwell body. These four parameters are G_K, K_K, G_M, K_M, namely the shear modulus of the spring in the Kelvin body, the bulk modulus of the spring in the Kelvin body, the shear modulus of the spring in the Maxwell body, and the bulk modulus of the spring in the Maxwell body, respectively. Two parameters describe the plastic behavior of the structure: the cohesion c and friction angle φ. The latter parameter describes the mechanical behavior of the dashpot in the Kelvin body, namely the viscosity coefficient of the Kelvin dashpot η_k.

Specifically, the steps in the calculation are as follows.

- Using the elastic–plastic constitutive model, the stress state as well the deformation characteristics after excavation of the diversion tunnel are calculated. In this step, the rheological behavior is not considered.

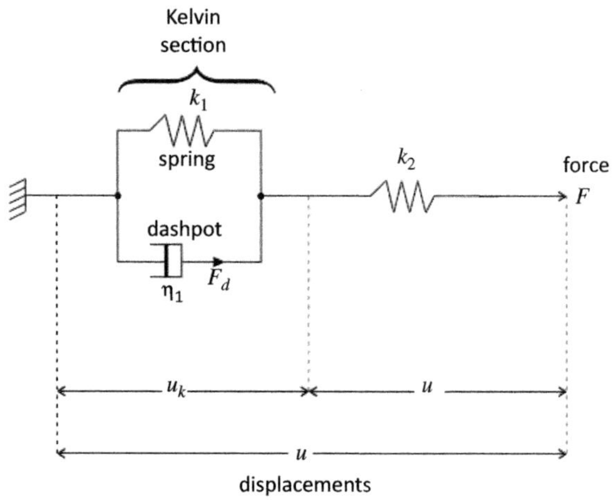

Figure 25 Degraded Burgers rheological model.

- The elastic–plastic constitutive model is replaced by the viscoelastic–plastic constitutive model to simulate the rheological deformation of the surrounding rock to 1.5 years.
- The lining is installed, and the calculation of the rheological parameters continues. At this point, the rheological deformation of the surrounding rock continues to occur over time, and the compressive stress inside the lining gradually increases.
- The calculation continues until 100 years after the tunnel excavation. In other words, the forces on the lining structure during its operation are examined over a time span of about 100 years.

According to the results in Figure 26, if the lag between the lining construction and the excavation is 0.5 years, the maximum compressive stresses at the spandrel, the arch waist and the arch foot should exceed the standard compressive strength of C30 concrete, which is 20.1 MPa. This means that the long-term safety of the structure is a problem. If the lag between the lining construction and the excavation exceeds 1 year, the maximum stress at the cross section of the lining after 100 years of long-term operation will not exceed the standard compressive strength of concrete.

7 SUMMARY AND OUTLOOK

The 21st century is the century when underground engineering progresses deeper than ever. For example, South Africa has started to assess underground extractions more than 4000 m deep. With the rapid development of China's economy, the center of national construction and development has gradually shifted to the west. A number of deep underground engineering projects related to water resource development and traffic construction need to be developed. Jinping II deep buried tunnel group, designed by HydroChina Huadong Engineering Corporation during the boom of construction development, has received widespread attention in engineering and academic fields both locally and abroad due to its unique characteristics.

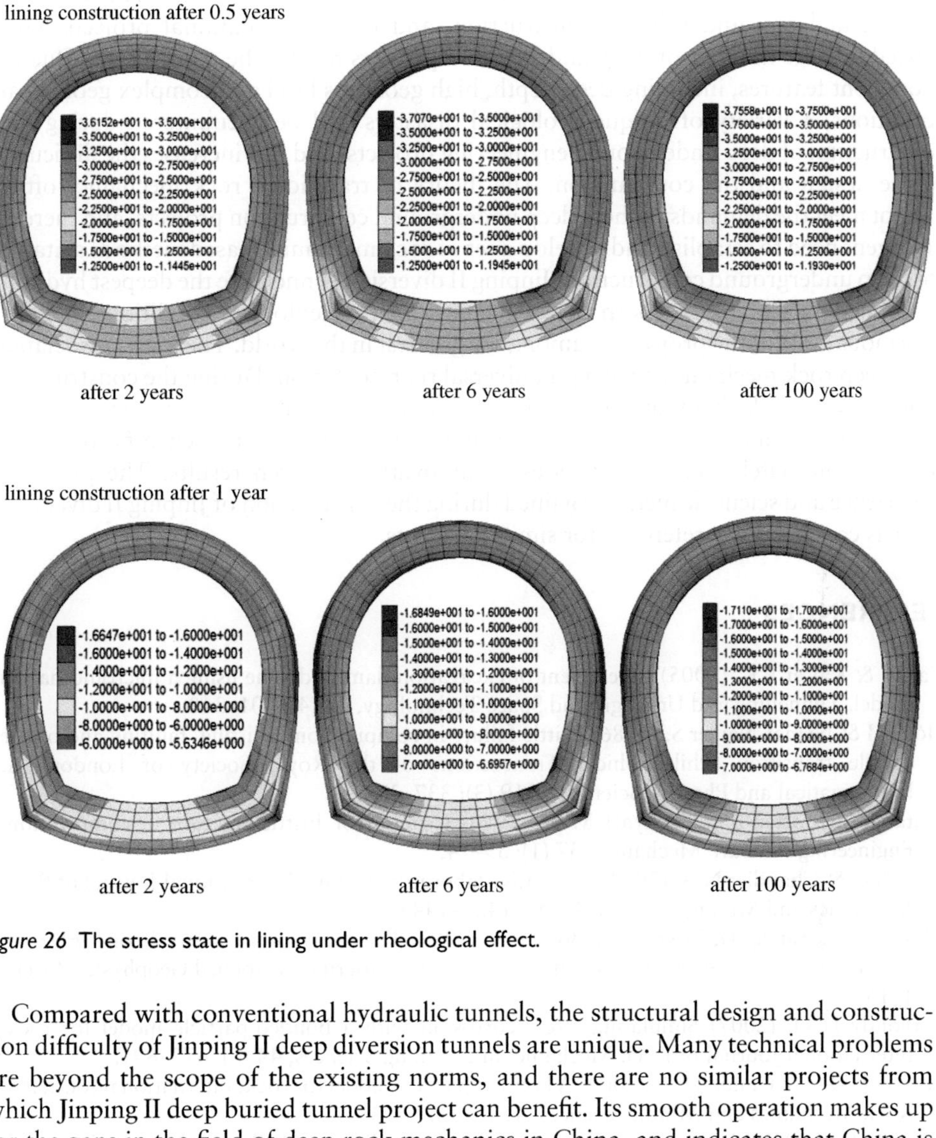

Figure 26 The stress state in lining under rheological effect.

Compared with conventional hydraulic tunnels, the structural design and construction difficulty of Jinping II deep diversion tunnels are unique. Many technical problems are beyond the scope of the existing norms, and there are no similar projects from which Jinping II deep buried tunnel project can benefit. Its smooth operation makes up for the gaps in the field of deep rock mechanics in China, and indicates that China is now at the international forefront in the design of deep buried tunnels. The latest technologies and research results employed in Jinping II deep diversion tunnels provide reliable technical support for the smooth operation of these diversion tunnels and the punctual power generation of Jinping II hydropower station. A number of previously unresolved technical difficulties, such as high geostress, high external water pressure, rock bursts, soft rock deformation and rupture lag, have been solved, and the results have been successfully applied to engineering practices to achieve satisfactory effects.

With the rapid development of China's economic construction, the number of infrastructure construction projects under complicated geological conditions has continued to increase. In recent years, especially, as the West-to-East Gas Transfer Project, the South-to-North Water Diversion Project, the West-to-East Power Transmission Project, highway and railway construction, and other key national projects have started, a large number of deep underground projects need to be constructed. Due to prominent features, including deep depth, high geostress level and complex geological conditions, a variety of unique geological disasters will be encountered during the construction of deep underground engineering projects, and this increases the difficulty of the design and the construction. The theory of traditional rock mechanics often cannot meet the demands of these deep underground construction projects, and there is an urgent need to establish and develop a set of rock mechanics-based methods suitable for deep underground construction. Jinping II diversion tunnels are the deepest hydraulic tunnels in China, and both the complexity of the geological conditions and the enormous scale of the project are among the greatest in the world. The issues associated with deep rock mechanics have high universal representation. During the construction stage, due to a lack of understanding on these particular issues, we encountered numerous detours. Through years of technical development and scientific research, we have ultimately achieved a series of innovative research results. The practical experience and scientific methods gained during the construction of Jinping II diversion tunnels can provide a reference for similar projects.

REFERENCES

Cai M & Kaiser P K. (2005) Assessment of excavation damaged zone using a micromechanics model. Tunnelling and Underground Space Technology, 20 (4): 301–310.

Horii H & Nemat-Nasser S. (1986) Brittle failure in compression: splitting, faulting and brittle ductile transition. Philosophical Transactions of the Royal Society of London (A, Mathematical and Physical Sciences), 319 (3): 337–374.

Lajtai E Z, Carter B J, & Ayari M L. (1990) Criteria for brittle fracture in compression. Engineering Fracture Mechanics, 37 (1): 59–74.

Lau J S O & Chandler N A. (2004) Innovative laboratory testing. International Journal of Rock Mechanics and Mining Sciences, 41 (8): 1427–1445.

Malmgren L, Saiang D, Toyra J, & Bodare A. (2007) The excavation disturbed zone (EDZ) at Kiirunavaara mine, Sweden-By seismic measurements. Journal of Applied Geophysics, 61 (1): 1–15.

Potyondy D O. (2007) Simulating stress corrosion with a bonded-particle model for rock. International Journal of Rock Mechanics and Mining Sciences, 44 (5): 677–691.

Potyondy D O & Cundall P A. (2004) A bonded-particle model for rock. International Journal of Rock Mechanics and Mining Sciences, 41 (8):1329–1364.

Read R S. (2004) 20 years of excavation response studies at AECL's underground research laboratory. International Journal of Rock Mechanics and Mining Sciences, 41 (8): 1251–1275.

Schmidtke R H & Lajtai E Z. (1985) The long-term strength of Lac du Bonnet granite. International Journal of Rock Mechanics and Mining Sciences and Geomechanics Abstracts, 22 (6): 461–465.

Zhang Chunsheng. (2007) Study on technical cruxes of diversion tunnels of Jingping hydropower Project on Yalong River. China Investigation and Design, 8: 41–44.

Zhang Chunsheng, Chen Xiangrong, & Hou Jing, *et al.* (2010) Study of mechanical behavior of deep-buried marble at Jinping II hydropower station. Chinese Journal of Rock Mechanics and Engineering, 29 (10): 1999–2009.

Chapter 10

Headrace tunnel of Jinping-II hydroelectric project

Wu Shiyong
Yalong River Hydropower Development Company, Ltd. Chengdu, China

1 PROJECT BACKGROUND

1.1 Comparison and selection of development proposals

Chengdu Engineering Corporation Limited completed the Report on Hydropower Planning for the Main Stream (from Kala to Jiangkou) of the Yalong River in Sichuan Province in October 1992. Sichuan Planning Commission and Water Resources and Hydropower Planning and Design Administration of the Ministry of Electric Power held an intermediate review meeting from February 29 to March 1, 1996, and experts and representatives from the following organizations attended the meeting: Department of Energy Business of State Development & Investment Corporation, Sichuan Construction Commission, State Grid Sichuan Electric Power Company and Sichuan Provincial Investment Group Co., Ltd., Ertan Hydropower Development Company Ltd. (now known as Yalong River Hydropower Development Company, Ltd.), and Chengdu Engineering Corporation Limited. The minutes of main review comments are as follows:

In terms of development proposal of the Jinping reaches, Chengdu Engineering Corporation Limited studied the following four proposals based on previous studies: the first proposal: Jinping one-cascade dam and diversion conduit type development; the second proposal: two-cascade development, *i.e.* Cascade-I Jinping high dam toe type development, and Cascade-II low-sluice diversion conduit type development; the third proposal: three-cascade development, *i.e.* Cascade-I Jinping high dam toe type development, Cascade-II Jiulong River Estuary dam and diversion conduit type development, and Cascade-III Dashuigou dam-toe development; the fourth proposal: four-cascade development along river, *i.e.* Jinping high dam, Jiulong River Estuary, Jiaozigou and Dashuigou dam-toe development. The report recommended two-cascade development, and considered the proposal of four-cascade development along river as a standby proposal. It can be seen from the review at the meeting that:

(1) As the one-cascade dam and diversion conduit type development proposal integrates such technical difficulties as high dam, deep and long tunnels, and high-head and high-capacity turbine units, it should not be adopted.
(2) Although the three-cascade development proposal achieves utilization of runoff of the Jiulong River and is featured with relatively high energy indexes, it has the worst economic indicators and does not avoid technical difficulties in deep and long tunnels, so it should not be adopted.

(3) The Jinping-I Hydroelectric Project is a cascade shared by the two-cascade development proposal and the proposal of four-cascade development along river. Since the Project enjoys a predominant geological location, relatively good reservoir regulation performance, significant energy storage, and outstanding cascade compensation benefits, it has an important strategic position in hydropower development of the Yalong River and Sichuan. According to review, it is approved that the Jinping-I Hydroelectric Project recommended in the report serves as the leading cascade in the cascade development of the main stream (from Kala to Jiangkou) of the Yalong River.

(4) During pre-feasibility study of the Jinping-II (Phase I) Hydroelectric Project, it needs to strive to make technological breakthrough in deep and long tunnels in details and detail the proposal of dam-toe development planning along river, to further confirm the development proposal for the river bend reaches downstream of the Jinping-I Hydroelectric Project, *i.e.* diversion conduit type development (low sluice + long tunnel) or dam-toe development along river.

1.2 Demonstration for feasibility of long headrace tunnel proposal

For the Jinping-II Hydroelectric Project, its preliminary survey and design was commenced in the 1960s. Shanghai Investigation, Design & Research Institute (now known as Shanghai Investigation, Design & Research Institute Co., Ltd.), through a large quantity of survey, tests and design, completed the *Report on Preliminary Design of Jinping Hydroelectric Project on the Yalong River of Sichuan Province* in 1969. Chengdu Engineering Corporation Limited of the former Ministry of Water Resources & Electric Power (now known as POWERCHINA Chengdu Engineering Corporation Limited) completed the *Report on Study of Development Proposal of Jinping Hydroelectric Project on the Yalong River* in 1979. In 1989, POWERCHINA Huadong Engineering Corporation Limited (hereinafter referred to as "POWERCHINA HUADONG") and Chengdu Engineering Corporation Limited commenced the pre-feasibility study of the Jinping-II (Phase I) Hydroelectric Project. In 1993, POWERCHINA HUADONG submitted the *Report on Pre-feasibility Study of Jinping-II (Phase I) Hydroelectric Project on Yalong River of Sichuan Province.* The former Ministry of Water Resources & Electric Power and competent authorities once approved "two-cascade development (high dam toe type power generation, and low sluice headrace and power generation)" for the Jinping reaches of the Yalong River. To carry out work in detail, POWERCHINA HUADONG had carried out survey and test for 5km long exploratory adits since October 1991. By May of 1995, the actual tunneling length of adits PD-1 and PD-2 was 3.948 km and 4.168 km respectively. Due to relatively large-scale concentrated gushing water in long exploratory adits, plugging of these exploratory adits was completed in 1996. During this period, part of high-pressure grouting tests and karst water tracer tests, and other tests had also been completed, and valuable data had been obtained.

Under the guidance of western development strategies implemented by the central government and the national development policy of "West-East Electricity Transmission Project", the hydropower construction in Sichuan faces further development opportunities. As a group of power supply points enjoying favorable development conditions, the Jinping-I and II Hydroelectric Projects on the Yalong River must be taken into overall consideration, to accelerate preliminary work. According to arrangement made by the State Development Planning Commission (now known as National Development and Reform Commission), under the support of project investors, and based on experience widely learned from domestic and foreign karst hydrogeology and long tunnel construction works, and experts' opinions, in June 2001, POWERCHINA HUADONG completed the following reports: *Study on Development Mode of the Large Jinping River Bend of the Yalong River — Report on Feasibility Study of Headrace Tunnel of Jinping-II Hydroelectric Project*, and corresponding special report 1 — *Special Report on Study of Construction Technology of Long Headrace Tunnel* (including Attachment: *Special Report on Design of Access Tunnel on the Jinping Mountain*), special report 2 — *Special Report on Study of Surrounding Rock Stability and Structure Design of Headrace Tunnel*, special report 3 — *Special Report on Engineering Geological Investigation of Headrace Route Area*, and special report 4 — *Special Report on Study of Karst Hydrogeology*.

From June 19 to 21, 2001, China Renewable Energy Engineering Institute of State Grid Corporation of China held a review meeting concerning special report on study of headrace tunnel of Jinping-II Hydroelectric Project on the Yalong River in Hangzhou. About 70 experts and representatives from the following organizations attended the meeting: China Development Bank, State Development & Investment Corporation, Sichuan Provincial Investment Group Co., Ltd., State Grid Sichuan Electric Power Company, Ertan Hydropower Development Company Ltd., China Academy of Railway Science, Sinohydro Foundation Engineering Co., Ltd., POWERCHINA Guiyang Engineering Corporation Limited, POWERCHINA Chengdu Engineering Corporation Limited and POWERCHINA Huadong Engineering Corporation Limited. Meanwhile, Pan Jiazheng, the Vice President of the Chinese Academy of Engineering and a consultant of State Power Corporation of China, Tan Jingyi and Lu Yaoru, academicians of the Chinese Academy of Engineering, and Ma Lin, President of Sichuan Provincial Investment Group Co., Ltd. attended the meeting and made important comments. At the meeting, an introduction to the abovementioned reports was made by POWERCHINA HUADONG, and reports on technical investigation of long tunnels in Norway and Germany conducted by teams from Ertan Hydropower Development Company Ltd. were heard. Through serious discussion and review of study reports and special reports by discipline teams, the major conclusions in the abovementioned reports were agreed, *i.e.* long headrace tunnels of the Jinping-II Hydroelectric Project were feasible, and corresponding design proposal was also feasible.

2 PROJECT OVERVIEW

2.1 Project overview

The Jinping-II Hydroelectric Project is huge in scale, and is developed for the purpose of power generation. The reaches developed are featured with deep-incised valleys, many shoals and rapids, and are unnavigable. Areas along the river are sparsely populated and distributed with decentralized cultivated land, without important towns, and

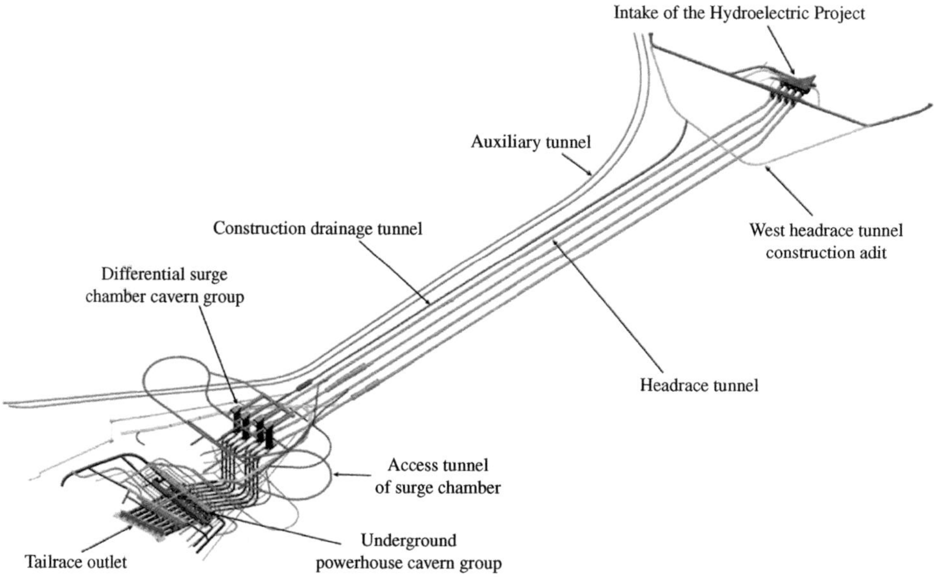

Figure 1 3D effect picture for headrace and power generation system of Jinping-II Hydroelectric Project.

industrial and mining enterprises. With the natural drop created by the 150km river bend at the downstream reaches (Kala to Jiangkou) of the Yalong River, the Jinping-II Hydroelectric Project is designed to cut the river bend through a headrace tunnel with a length of about 17 km, to form a water head of about 310 m. Compared to other hydroelectric projects on the Yalong River, the Project has the highest water head and the largest installed capacity. It has a total installed capacity of 4,800 MW, a unit capacity of 600 MW, a rated head of 288 m, an average annual energy output of 24.23×10^9 kW·h, a firm output of 1,972 MW, and annual operation hours of 5,048 h.

The Jinping-II Hydroelectric Project is of a low-sluice, long-tunnel and high-capacity conduit type, and its project mainly consists of low sluice for head structure, headrace system, underground powerhouse at the tail and other buildings. Refer to Figure 1 for details. The water retaining sluice dam for head structure is located in Maomaotan at the west end of the large Jinping River Bend of the Yalong River. It is 7.5 km away from the Jinping-I dam site at upper reaches, with the maximum dam height of 34 m. The catchment area upstream of the sluice site is about 103×10^3 km^2, with an annual average discharge of 1,230 m^3/s. The reservoir has a normal pool level of 1,646 m, a minimum pool level of 1,640 m and a daily regulating storage of 4.96×10^6 m^3. The intake of the Project is located at the Jingfeng Bridge 2.9 km upstream of the sluice site. In the layout plan of head structure, sluice dam and intake are separated from each other. The underground powerhouse is located in Dashuigou at the east end of the large Jinping River Bend of the Yalong River. The headrace tunnel routes from Jingfeng Bridge to Dashuigou. The arrangement mode of "four tunnels for eight units" is applied, and there are four headrace tunnels in total. The buried depth of overlying rock masses along these tunnels is 1,500–2,000 m in general, and the

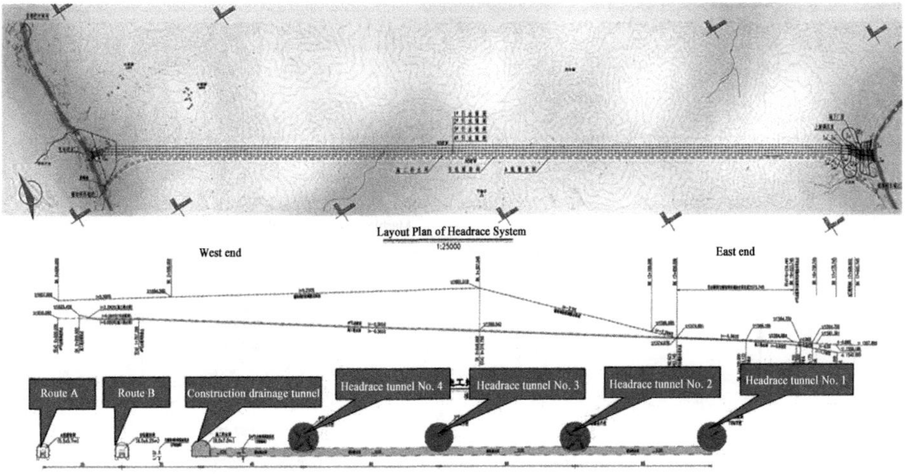

Figure 2 Project background of Jinping-II Hydroelectric Project and schematic diagram for arrangement of tunnels.

maximum buried depth is about 2,525 m. These headrace tunnels are buried deeply and long, and are large in tunnel diameter. They are hydraulic tunnel cavern group works with the largest scale in the world.

For the Jinping-II Hydroelectric Project, the construction of preparation works was carried out successively at the beginning of 2003. The feasibility study report of the Project passed review in December 2005, and the project application report was subjected to national approval on December 15, 2006. The construction of main works of the Project was formally commenced on January 30, 2007, and the first batch of two units was put into operation on December 31, 2012. The construction period before the operation of the first unit was 6 full years. All works were completed by the end of September 2015, with the total construction period of 8 years and 9 months. Refer to Figure 2 for specific project layout.

Figure 2 indicates that tunnels are in the axial direction of SE–NW. Tunneling from both ends is designed as an excavation mode, to make tunnels through at intermediate sections. For convenience, the two ends are called east and west ends for short in the project. In addition, engineering chainage is also quoted in this paper to represent specific locations. The intakes of deep headrace tunnels of the Jinping-II Hydroelectric Project are arranged at the west end, and corresponding chainage is calculated from openings. For example, Y(1)3+300 represents the location in the 1# headrace tunnel 3,300 m away from opening.

It can be seen from Figure 2 that 7 tunnels of parallel arrangement in total are designed for the Project, orderly including two ∩-shaped auxiliary tunnels, a circular construction drainage tunnel constructed by TBM tunneling, and four headrace tunnels from southwest to northeast. In addition to high stress caused by a large buried depth, preliminary investigation results also indicate that the Project also suffers from HP high-discharge groundwater seepage. As revealed during construction of the Project, the maximum water pressure is greater than 10 MPa, and the instantaneous flow rate at

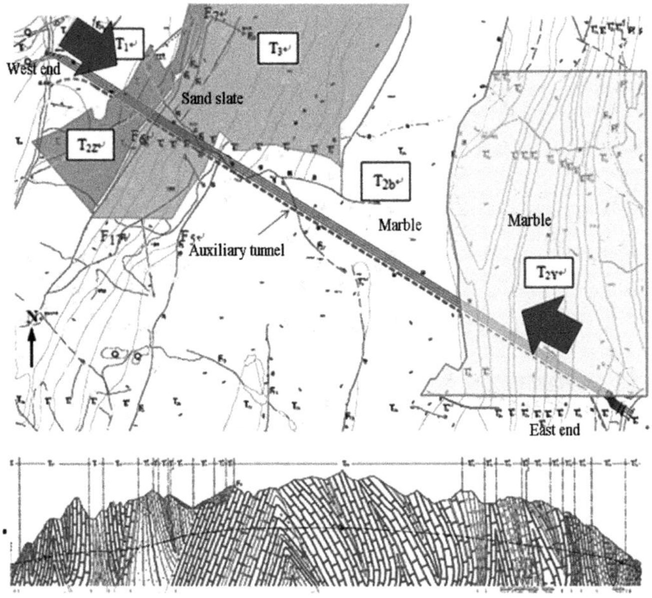

Figure 3 Geological background along deep tunnels of Jinping-II Hydroelectric Project.

single point is up to 7 m^3/s. For this reason, excavation of construction drainage tunnel is carried out prior to commencement of headrace tunnels, with one TBM having a diameter of 7.3 m employed for one-way tunneling from the east end to the west end. The TBM tunneling method of drainage tunnel has provided experience for the following TBM-based construction of headrace tunnels for direct reference.

As shown in Figure 3, the basic geological conditions along tunnels of the Jinping-II Hydroelectric Project are as follows: The strata are mainly of the Triassic marbles, supplemented by sand slates and pelitic schists distributed at hundreds of tunnel sections. Strata along these tunnels constitute a closed sharply angular structure, with a series of fold structures developed. The central section of the Jinping Mountain with the largest buried depth is the core of syncline. In addition to fold, fracture structures are also developed along these tunnels, mainly consisting of compressive structures basically in parallel with the strikes of rock strata, supplemented by tensional-shear structures in the strike of NWW. These tensional-shear structures are dominant water diversion structures in this area. Relevant studies and engineering practice indicate that the basic characteristics of initial stress fields along these tunnels are subjected to the control of lithology and structure conditions. In particular, there exists an intrinsic relation between structure conditions and stress surrounding rock failure occurring at site.

For the Jinping-II Hydroelectric Project, due to large buried depth of headrace tunnels, high gravity stress, strong high ground stress actions and tectonic actions, after excavation in high ground stress field, the stress environment where tunnel surrounding rocks are located will become much severer due to secondary concentrated stress. As a consequence, propagation of surrounding rock fractures and failures is unavoidable. Meanwhile, as tunnels are extremely long, with a length of about 16.67 km, they will run through various rock groups, and Class II, III and IV and other surrounding rocks. In addition, the project area is cut by various fault fracture zones, so the geological conditions are complicated.

2.2 Project characteristics

For the Jinping-II Hydroelectric Project, the total excavation quantities of underground works are greater than 13.20×10^6 m^3. Headrace tunnels, powerhouse caverns, access tunnels, construction adits, drainage tunnels and other caverns are crisscrossed and huge in scale. Headrace tunnels are famous for their scales and technological difficulties, and are world-class works recognized at home and abroad.

The axes of headrace tunnels of the Jinping-II Hydroelectric Project are almost orthogonal to the ridge line of the Jinping Mountain, with steep and strong mountains along the axes. The buried depth of headrace tunnel is greater than 2,000 m basically and 2,525 m at its maximum, and is greater than that (2,135 m) of the world-famous Simplon Tunnel, and close to that (2,619 m) of headrace tunnel of Shera Hydropower Project in France with the largest buried depth in the world at present. The arrangement of construction adits, inclined and vertical shafts are restricted by topographical conditions. Therefore, design and construction of headrace tunnel are major problems in the construction of the Jinping-II Hydroelectric Project. The maximum tunnel diameter of the Jinping tunnel is 13 m, far greater than the tunnel diameters of the Qinling Tunnel (D = 8.8 m) and Shera Tunnel (D = 5.8 m). Hydrogeological conditions along the Jinping tunnel are complicated. High external water pressure of greater than 1,000 m, long-term stable water source replenishment and ground stress of greater than 70 MPa bring a series of technical difficulties to construction and structure design of tunnels.

The deep tunnels of the Jinping-II Hydroelectric Project are featured with large buried depth, length and tunnel diameter, high ground stress level, complicated karst hydrogeological conditions and difficult construction layout, so the tunnel works, compared to other tunnel works completed and under construction in the world, is underground cavern group works with the largest overall scale and the highest comprehensive difficulty at present.

In August 1999, the Qinling Tunnel, the longest railway tunnel in China, was made through. The tunnel has a total length of 18.4 km, a diameter of 8.8 m and a maximum buried depth of 1,600 m. The construction of the Qinling Tunnel greatly accelerates design level and construction technology of tunnel in China.

The scale of hydraulic tunnel in China gradually increases along with hydropower construction. However, most of hydraulic tunnels are qualified for the arrangement of construction adits, and the length of tunneling from one end is mostly below 5 km. Table 1 shows the conditions of some long hydraulic tunnels completed in China.

Table 2 shows the examples of some foreign long hydraulic tunnels completed, and Table 3 shows the examples of some foreign deep hydraulic tunnels completed.

The hydraulic tunnels of the Jinping-II Hydroelectric Project are extremely complicated in geological conditions and huge in construction scale. Although the current tunnel technologies are changing rapidly, the hydraulic tunnel works, due to its own difficulties, is extremely challenging in the tunnel engineering sector.

Table 1 Some deep and long hydraulic tunnels completed in China.

Project Name	*Tunnel Length (km)*	*Section dimensions (m)*	*Maximum Buried Depth (m)*
Futang	19.3	9.0	700
Taipingyi	10.5	9.0	480
Tianshengqiao-II	9.8	8.7~9.8	760
Lubuge	9.4	8.0	/
Xi'er River-I	8.2	5.6	/
Yuzixi-II	7.6	6.5~7.4	800
Gutian-II	5.2	6.4~6.9	435
Yingxiuwan	3.8	8.0	570

Table 2 Some foreign long hydraulic tunnels completed.

Project Name	*Country*	*Tunnel Length (km)*	*Diameter (m) / Area (m^2)*	*Number of Construction Adit*
Helsinki	Finland	120	4.73m	24
Päijänne	Finland	120	15.5 m^2	21
Mahesi	Peru	95	/	/
Orange–Fish	South Africa	82.5	4.8m	7
Bolmen	Sweden	80	7.5 m^2	11
Headrace tunnel in California	USA	50	6m	4
Arpa-Sevan	Former Soviet Union	48.4	4.1m	4
Lesotho	South Africa	45	5m	/
Oukenbintumute	Australia	22.5	6.4m	1
Mohuienge	New Zealand	19.3	3.3m	None
Kelaixun. Dikesangsi	Switzerland	19.23	/	/
Arc-Isere	France	19	6.4m	None
Heizemu	Sweden	16.3	110 m^2	1
Kaimanuo	Canada	16.2	7.6m	1
Madingna	Switzerland	14.3	/	None
Tailamosen	Italy	14.5	5.3m	/
Stillwater	USA	13	3.1m	None
Makesijin	USA	10.7	7.2m	None

3 KEY TECHNICAL PROBLEMS IN ENGINEERING CONSTRUCTION

Hydrogeological conditions along the Jinping tunnel are complicated. High external water pressure of greater than 1,000 m, long-term stable water source replenishment and ground stress of greater than 70 MPa bring a series of technical difficulties to the construction of tunnels. The headrace tunnels of the Jinping-II Hydroelectric Project are featured with large buried depth, length and tunnel diameter, high ground stress level, complicated karst hydrogeological conditions and difficult construction layout, so the tunnel works, compared to other tunnel works completed and under

Table 3 Some foreign deep hydraulic tunnels completed.

Project Name	*Country*	*Buried Depth (m)*
Beileduona (Shera)	France	2619
Olmos	Peru	2000
Qikesiaoyi	Guatemala	1500
Yakanbu	Venezuela	1270
Arpa-Sevan	Former Soviet Union	1230
A'beitemake	Italy	1200
Delasinizi	Austria	1200
Wola	Austria	1000

construction in the world, is the largest in overall scale and the highest in comprehensive difficulty at present. The construction of headrace tunnel groups is confronted with a series of technical difficulties, including study and determination of construction proposal of cavern group, prevention and control of HP high-discharge groundwater, prevention and control of rock burst under high ground stress, coordinated ventilation of extremely long tunnel group, and organization and management of high-intensity transportation. Solutions to these technical difficulties restrict the completion and smooth operation of headrace tunnels.

4 PACKAGED TREATMENT TECHNOLOGY OF HP HIGH-DISCHARGE GROUNDWATER

4.1 Overall treatment principles of groundwater

According to engineering geological conditions and karst hydrogeological conditions, and in combination with the requirements of social environment in the project area, the following overall treatment principles of groundwater have been established: "exploration before tunneling; put plugging first and combine plugging with drainage; controlled drainage and plugging at the right time". The treatment mode for HP high-discharge concentrated gushing water points has taken into consideration the conformity of plugging effects with quality standard, conformity with the requirements of concrete lining and creation of conditions for construction of following works, as well as adverse impacts of water level rise after plugging of high-discharge water points on surrounding rock stability.

4.2 Identification of karst groundwater storage and risks in alpine and canyon regions

Carbonate rock strata are widely distributed in the river bend area of the Yalong River. Due to the particularities of natural geological environment and regional geological environment in this area, regional karsts are weakly developed on the whole, with a few typical karst morphologies. Through tracing tests of karst water and corrosion tests of marbles at high pressure and low temperature, and studies on basic geological conditions and karst development rules, karst water regime, karst hydrogeochemistry, karst water isotope and other systems in the project area at the river bend of the Yalong

River, the following works have been carried out to effectively identify karst groundwater storage and risks in alpine and canyon regions: zoning of karst development degree, division of karst hydrogeological units, analysis on relationship among groundwater recharge, runoff and discharge of each unit, balance calculation of karst water, and analysis on three-dimensional seepage field and karst development depth in the project area at the river bend, as well as prediction of maximum instantaneous water inflow and stable water inflow of headrace tunnel, and innovation of geological advanced prediction method.

(1) Development characteristics of karst groundwater
Through wide karst hydrogeological survey, long-term karst water dynamic observation, artificial and natural tracing tests of karst water and water balance method, and in coordination with boreholes and adits, complicated karst development rules and characteristics of hydrogeological system in the project area have been identified particularly by means of survey through 5km long exploratory adit and test technology with the longest tracing distance (14 km) in the world at present. Groundwater migration, discrete crack network statistic model and other theories and methods have been adopted for tracing test and analysis, to greatly improve extraction extent and interpretation precision of traced information.

(2) Karst water recharge area
According to characteristics of karst water from the large Jinping River Bend of the Yalong River such as high groundwater level and low water temperature, corrosion test of marbles is carried out at high pressure (10 MPa, 20 MPa) and low temperature (10°C). Accordingly, in combination with groundwater seepage field distribution rules, seepage gradient, and seepage velocity rules after tunnel excavation, karst water recharge areas are further determined by means of hydrochemistry and water isotope effects, as well as water balance analysis based on modified rainwater recharge formula.

(3) Gushing water prediction and external water pressure of headrace tunnel
Based on karst hydrogeological study and in combination with three prediction methods including hydrogeological analogy method, simple water balance method and three-dimensional seepage field analysis, without setting anti-seepage conditions, it is predicted that the total stable flow of two auxiliary tunnels will be about 10 ~ 13 m^3/s, and the total stable water inflow of seven tunnels will be about 27.43 ~ 29.93 m^3/s. However, during actual construction, the water prediction results will vary with prediction conditions due to grouting or plugging of some gushing water points. The maximum external water pressure of headrace tunnel may be up to 10 MPa, basically the same as the external water pressure acting on a long exploratory adit, with the maximum gushing water flow at single point being 5 ~ 7 m^3/s.

4.3 Comprehensive geological advanced prediction technology of groundwater

According to advanced prediction tests and practice of deep tunnels of the Jinping-II Hydroelectric Project, a multi-step prediction and early warning mechanism of

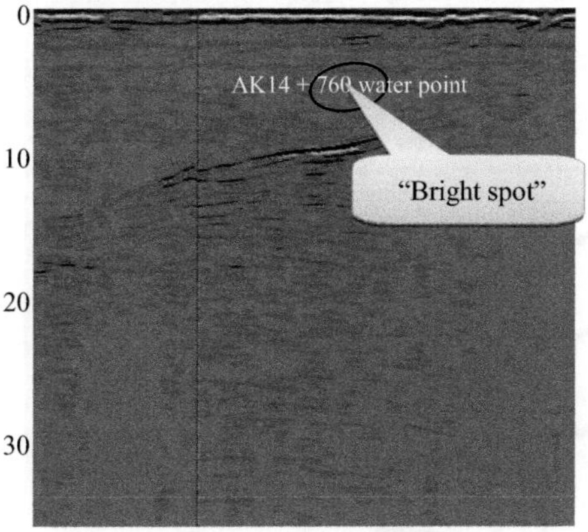

a) Bright spot in radar image

(b) Bursting and gushing water at opening revealed

Figure 4 Typical case for identification of "Bright spot" on radar image of gushing water point of auxiliary tunnel.

macroscopic geological advanced prediction (engineering geological method), long-term (long distance: 50 m ~ 200 m) geological advanced prediction (engineering geological method, TSP detection), short-term (short distance: 0 ~ 50 m) geological advanced prediction (geological radar, transient electromagnetism) is established, to constitute a comprehensive construction geological prediction system of karst tunnels for the purpose of advanced prediction of unfavorable geological bodies. Prediction results are obtained through comprehensive analysis based on engineering geological analysis and in combination with interpretation result test through instruments. During construction, field prediction information of each step is fed back to improve the interpretation precision of information, and corresponding geological advanced prediction reports are submitted through comprehensive judgment. Corresponding early warning scheme and treatment measures are formulated according to advanced prediction reports, to avoid occurrence of geological disasters and ensure construction safety of tunnels.

According to characteristics of tunnel structure, spatial distribution of geological bodies and surface radar detection method, the following methods have been proposed: determination of relative formation dielectric constant, "U"-type line layout, attitude detection of geological structural plane, initial wave phase method of geological radar, discrimination criteria of TSP-based prediction, and discrimination method of bursting and gushing water structures based on geological radar prediction. Refer to Figure 4 for typical cases.

4.4 Packaged comprehensive treatment technology of groundwater

- **Plugging technology of HP high-discharge underground bursting and gushing water points**

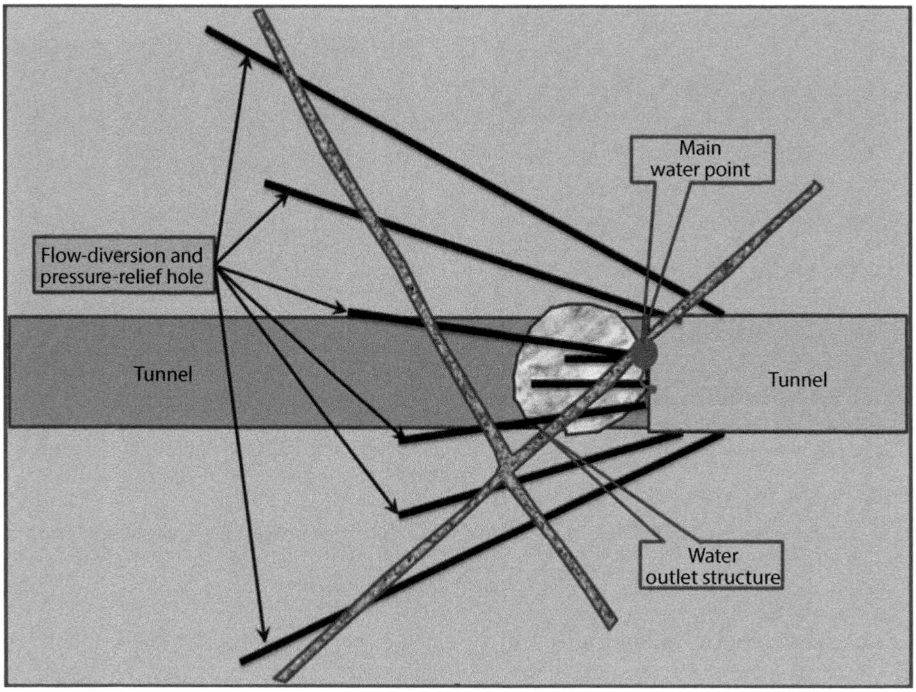

Figure 5 Layout profile of flow-diversion and pressure-relief hole.

(1) Plugging technology based on flow diversion and pressure relief
If plugging with grouting based on water sealing wall is adopted for HP bursting and gushing water points, a conventional method comprises the following procedures: diversion of HP bursting and gushing water, placing of water sealing wall, drilling of grouting holes when the strength of water sealing wall meets requirements, plugging with grouting, and excavation of water sealing wall. Such conventional method is complicated in treatment procedures, long in construction period, high in cost, difficulty and risk. For this reason, the **plugging technology based on "flow diversion and pressure relief"** (Figure 5) has been researched and developed, to directly plug HP bursting and gushing water without a water sealing wall.
Such plugging technology can be applied to simplify construction procedures and reduce construction risks. Relevant practice shows that such treatment technology can shorten treatment period by about 1–2 months, greatly shorten construction period and reduce treatment difficulty.

(2) Caisson-based plugging technology
"Caisson-based" plugging technology (Figures 6 and 7) has been researched and developed for high-discharge gushing water points on bottom plate, to solve problems existing in direct plugging of concentrated HP high-discharge groundwater. When the bottom plates of tunnels suffer from HP high-discharge bursting and gushing water, such technology is applied to directly plug gushing groundwater without any change in tunnel route.

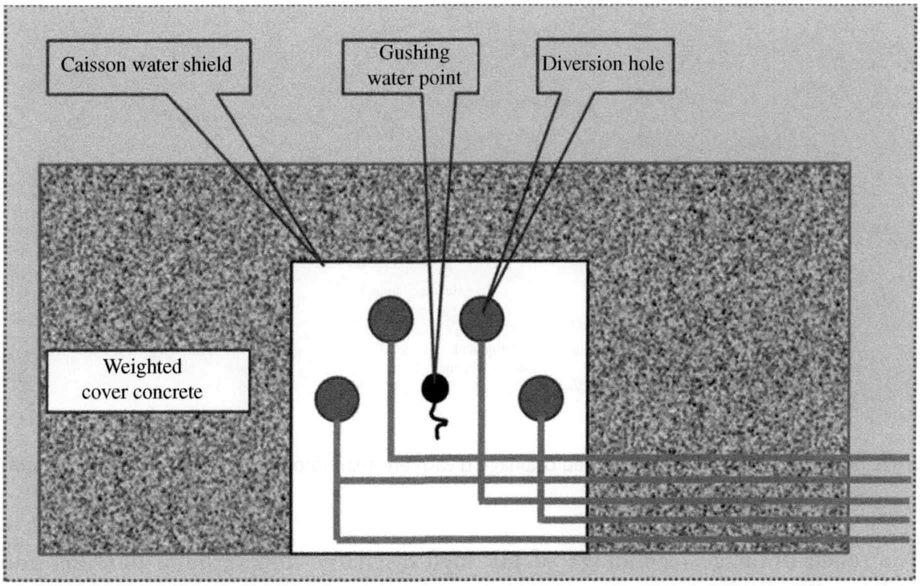

Figure 6 Layout plan for caisson structure of bottom plate.

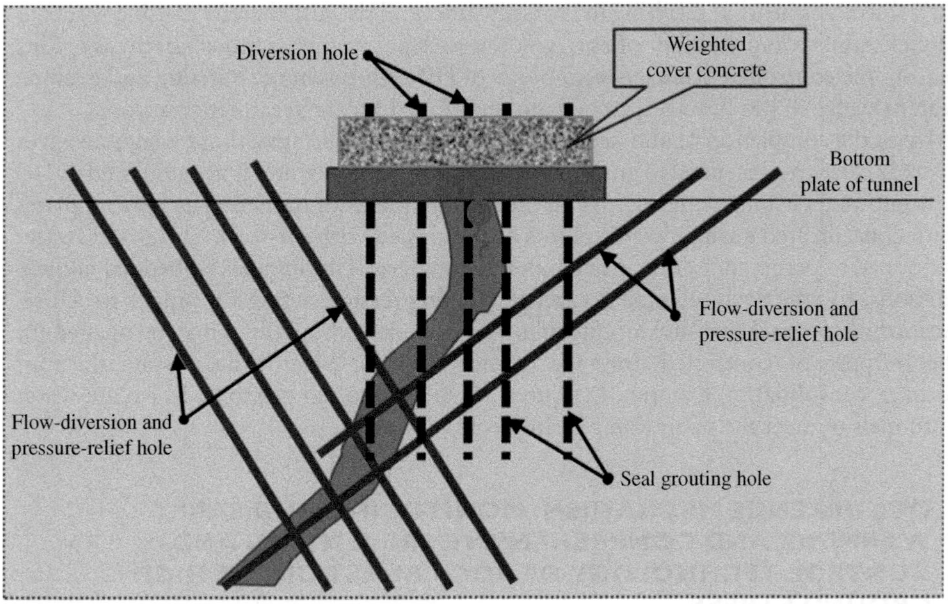

Figure 7 Layout profile for caisson structure of bottom plate.

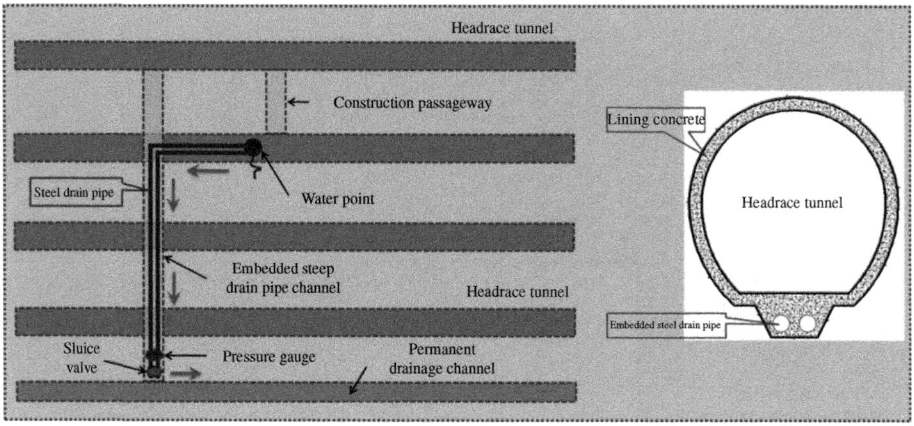

Figure 8 Schematic diagram for controlled drainage treatment technology of HP high-discharge bursting and gushing water.

- **Controlled drainage technology of HP high-discharge underground bursting and gushing water points**

With regard to treatment of HP high-discharge bursting and gushing water points, complete plugging of these water points are high in difficulty and long in construction period. Moreover, after plugging, HP high-discharge groundwater is accumulated at tunnel sections nearby, resulting in groundwater level rise and thereby causing threat to tunnel stability and stability of surrounding rocks at tunnel sections nearby. For this reason, **the controlled drainage technology of HP high-discharge bursting and gushing water** (Figure 8) has been researched and developed for the treatment purpose.

Upon the completion of anti-seepage consolidation grouting without weighted cover in areas close to concentrated gushing water points, groundwater is accumulated at 1–3 concentrated gushing water points on the bottom plates of tunnels. The bottom plates with concentrated gushing water points are equipped with caissons. The groundwater is drained to permanent drainage channels through steel drain pipes embedded in these caissons, to effectively eliminate external water pressure acting on tunnel structure. According to water pressure monitoring, valves on steel drain pipes may be opened for the purpose of controlled drainage outside tunnels. Meanwhile, systematic anti-seepage consolidation grouting measures are taken around caissons, to isolate water in tunnels to meet the operation requirements of tunnels.

5 OCCURRENCE MECHANISM, MONITORING AND EARLY WARNING, AND COMPREHENSIVE PREVENTION AND CONTROL TECHNOLOGY OF ROCK BURST UNDER HIGH GROUND STRESS

5.1 Study on ground stress in deep tunnel group works area

Ground stress is an important decisive factor of regional stability, and acting force for deformation and failure of rock and soil works during underground or ground excavation. With deep and long headrace tunnels of the Jinping-II Hydroelectric Project as the project background, a new study method for ground stress field in deep tunnel group works area has been proposed. The method consists of two important parts: macroscopic forward analysis method of regional ground stress field and inverse analysis method of local ground stress field. The former method starts with macroscopic factors such as regional tectonic movement history and regional landform control actions, to study the overall characteristics and the distribution rules of ground stress field in the works area by means of three-dimension numerical simulation technique of these macroscopic control factors. The latter method, based on the analysis of the former, integrates various analysis and judgment technologies of ground stress field according to engineering geological conditions and specific data of local deep tunnel sections, for mining and inversion of magnitude, direction and distribution rules of local ground stress field by means of ground stress test data, response of surrounding rocks during field excavation, rock monitoring information and numerical simulation technique. Moreover, the latter method also focuses on and systematically studies control actions and impacts of geological structure on local ground stress field. With the established forward and inverse fusion analysis technologies and methods of ground stress fields along deep and long tunnels, the characteristics and the distribution rules of ground stress fields in the headrace tunnel works area of the Jinping-II Hydroelectric Project have been comprehensively and systematically studied and cognized.

To reflect the characteristics of ground stress field in a more accurate manner, analysis and various test methods such as aperture method, hole wall method and acoustic emission method are carried out, but they fail to complete ground stress test directly or lead to distortion of test results. The test system based on conventional hydraulic fracturing method only ranges from 30 MPa to 50 MPa. For this reason, UHP pressure supply system, pipe system, packer system and impression system have been developed and improved, to enable the withstand pressure standard of the developed UHP ground stress test system to be 120 ~ 150 MPa, so the requirements of ground stress test for the Jinping deep tunnels are met. Meanwhile, by establishing new test methods, deformation caused by time effects and stress adjustment is minimized. Finally, the maximum ground stress of headrace tunnels of the Jinping-II Hydroelectric Project is tested successfully, *i.e.* 94.97 MPa. Refer to Figures 9 and 10 for test system based on UHP hydraulic fracturing method and corresponding test results.

5.2 Study on inoculation mechanism of marble rock burst

Indoor loading/unloading tests under different stress paths of the Jinping marbles, and field monitoring tests of typical tunnel sections are adopted for obtaining loading/unloading mechanical characteristics and excavation mechanical response characteristics of marbles in the Jinping deep tunnels. With these characteristics obtained, inoculation rules and mechanism of deep marble rock burst are revealed, and corresponding mechanical models for rock burst analysis under high stress are established, to provide theoretical basis for early warning and prediction of rock burst disasters.

(1) Strain-type rock burst: In terms of formation mechanism, high stress or induced secondary high stress field is an essential condition for the occurrence of such

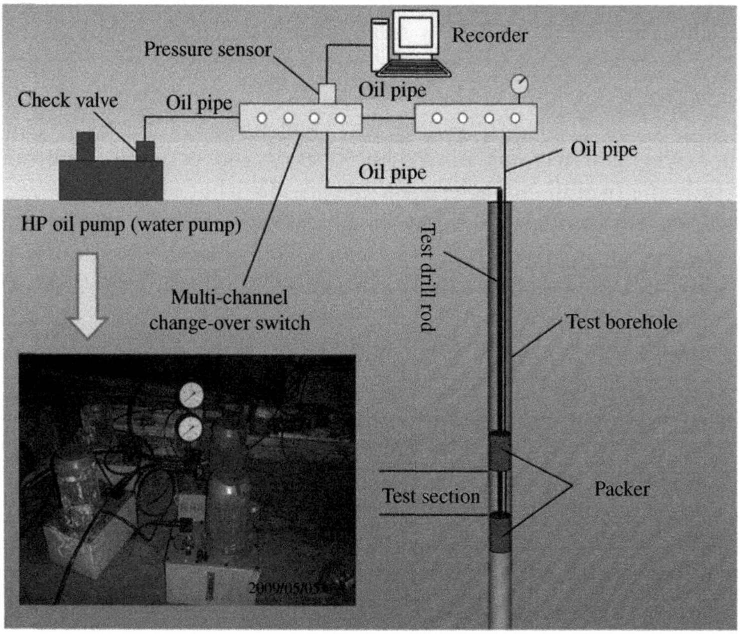

Figure 9 Ground stress test system based on UHP hydraulic fracturing method.

rock burst. After excavation of headrace tunnels, the original three-dimensional stress state close to heading face is out of balance and transits to a two-dimensional stress state, leading to redistribution of stress, loss of confining pressure and radial stress rise. The occurrence of surrounding rock burst will experience three stages including: cleavage, shearing and ejection.

(2) Structural rock burst: A weak structure in the NWW strike is the main cause of inducing structural rock burst of the Jinping-II Hydroelectric Project. In terms of internal mechanism, such rock burst is caused by local stress concentration areas at structure ends. Specifically, when heading faces are continuously approaching to stress concentration areas at structure ends, due to the impacts of local initial high stress, stress in secondary stress field may be extremely concentrated, and structures close to the field provide better conditions for failure, leading to a severe failure.

(3) Pillar-type rock burst: Due to superposition of stress concentration areas in front of two heading faces in opposite tunneling directions, rock burst failure is caused by deterioration of stress states of rock pillars. Such rock burst occurs, provided that there are stress concentration areas in front of heading faces.

Strength criterion of brittle rocks under deep high stress — generalized polyaxial strain energy criterion (*i.e.* GPSE criterion) has been established according to strength and failure characteristics of marble. The multi-axial test of the Jinping deep marbles shows that the GPSE criterion can properly describe the mechanical characteristics of hard rocks under high stress and interpret the occurrence mechanism of hard rock burst.

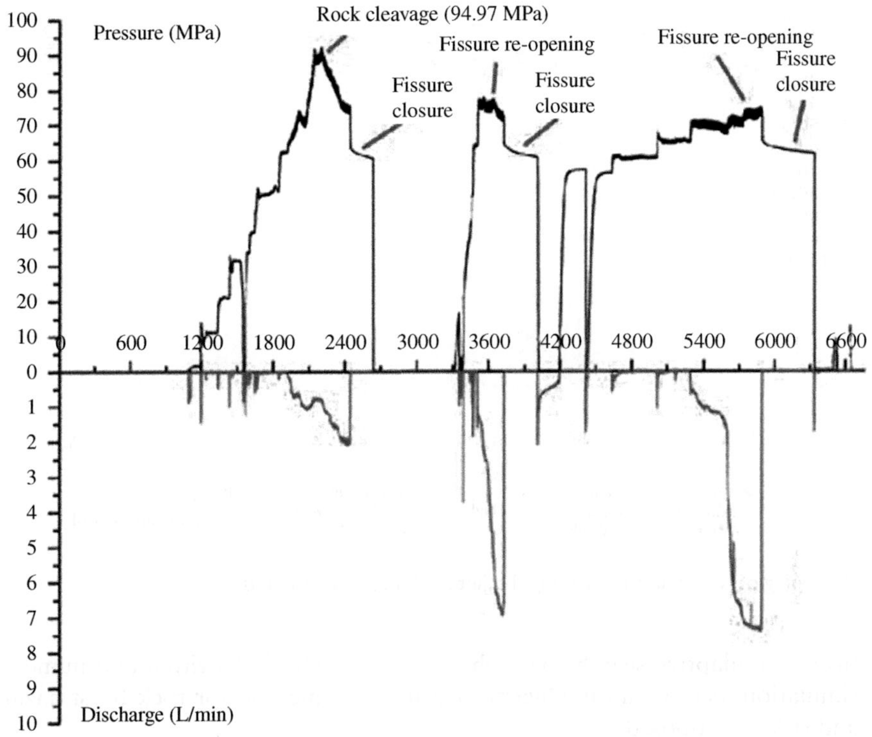

Figure 10 Test curve based on hydraulic fracturing method.

5.3 Identification of rock burst risk

(1) Under the guidance of "macroscopic comprehensive geological survey → survey of 5km ultralong exploratory adit → parallel survey of auxiliary tunnels", and through fusion of various technologies and methods such as ultrahigh ground stress survey technology and inverse analysis method, and acquisition technologies for physical and mechanical information of rock masses, study is carried out for the project area of the Jinping-II Hydroelectric Project with regard to acquisition methods and measures of such information as physical and mechanical information of rock masses, ground stress, geological structure, fault, and structural plane;

(2) Based on the multi-source information acquired, and by means of statistical analysis and other methods, study is carried out for the project area of the Jinping-II Hydroelectric Project with regard to physical and mechanical properties of rock masses, and ground stress distribution characteristics and rules. Meanwhile, a surrounding rock classification system capable of reflecting high ground stress and high external water pressure is established;

(3) By establishment of a stress analysis technology based on macroscopic analysis of stress field, field monitoring and comprehensive integration inverse analysis, and

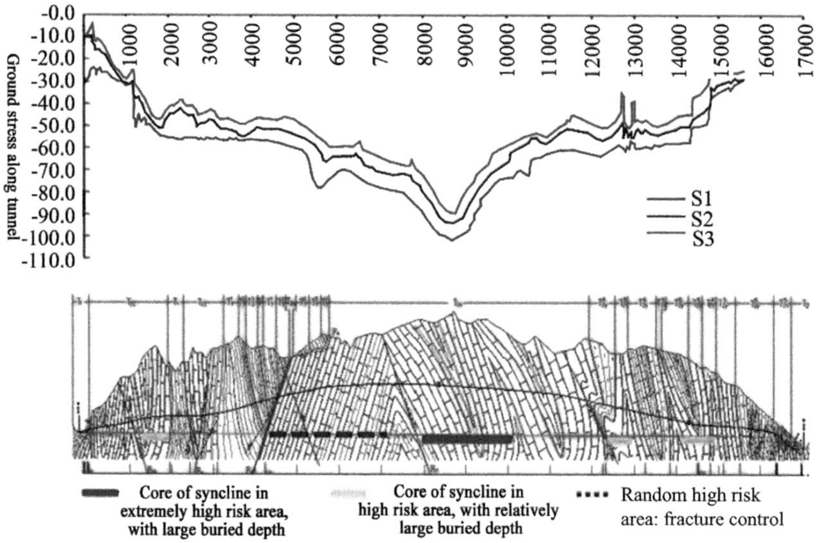

Figure 11 Zoning of rock burst risks along the Jinping-II headrace tunnels.

fusion of adaptive search space chaos-genetic hybrid algorithm and numerical simulation method, an intelligent identification method for rock burst hazards and risks is proposed;

(4) Through fusion of multi-information-based macroscopic intelligent identification methods for rock burst risks, macroscopic evaluation is carried out for rock burst risks along tunnels, and rock burst risk zoning along tunnels is obtained, as shown in Figure 11. Therefore, rock burst hazards and risks are displayed visually, to provide bases for prevention and control decisions of project hazards, avoid / reduce the occurrence of major project accidents, and provide powerful guarantee for the completion of the Jinping-II headrace tunnel group works.

5.4 Early warning of rock burst

A comprehensive rock burst prediction method has been established for the Project. Such method integrates geological prediction, numerical analysis, micro-seismic monitoring, geophysical detection and experience-based judgment, to implement real-time monitoring, analysis and early warning for rock burst risks by means of information acquired through different prediction methods. Details are as follows:

(1) Prior to tunnel excavation, zone and identify rock burst risks according to macroscopic geological conditions;

(2) Further evaluate possible rock burst risks according to damage and failure of surrounding rocks and geological conditions revealed at tunnel sections excavated and adjacent tunnel sections;

(3) According to geological conditions, ground stress field distribution and properties of rock masses, simulate the rock burst risks of tunnel sections to be

excavated by means of numerical simulation method, evaluate the occurrence risks of rock burst, and implement prediction during temporary excavation according to the prediction results obtained in the second step;

(4) Monitor and track the inoculation process of rock burst through real-time micro-seismic monitoring, to obtain precursor information and inoculation rules of rock burst, and implement real-time monitoring, analysis and early warning of rock burst, to continuously correct and improve prediction accuracy.

5.5 Prevention and control ideas and principles of rock burst

Based on the occurrence mechanism of rock burst, the comprehensive rock burst treatment idea combining active stress relief with passive control, and overall design principles for prevention and control of rock burst through support measures have been proposed:

(1) Always adhere to the principle that surrounding rocks are the main bearing structures of a tunnel;
(2) Take support measures (nano-shotcrete + water expansion anchors) close to heading faces, to maintain confining pressure where possible and limit propagation of surrounding rock fractures, and effectively utilize the ductility characteristics of marbles;
(3) By fully utilizing the arching effects of heading faces, take support measures (expansion shell prestressed anchors) behind heading faces in time, and control high stress failure of surrounding rocks;
(4) Take systematic shotcrete-anchorage support measures, to comprehensively reinforce areas suffering from damage and failure of surrounding rocks.

5.6 Comprehensive prevention and control method of rock burst hazard at D&B construction sections

Active stress relief and passive support control measures shall be taken for the prevention and control of rock burst at D&B excavation tunnel sections, to ensure the construction safety of tunnel sections suffering from rock burst. In other words, stress-relief blasting and timely and effective support shall be implemented, *i.e.* excavation by controlled blasting with a short advanced depth. For tunnel sections suffering from strong and extremely strong rock burst, the following measures shall be taken in a coordinated manner: excavation by stress-relief blasting → clearing of dangerous rocks on side and roof arches and HP water washing → timely shotcrete for covering heading faces → timely implementation of anchorage measures for prevention of rock burst (including rapid support with anchors, wire meshes and steel supports) → subsequent systematic support.

• Active stress relief method

The active stress relief method is as follows: During D&B tunneling, manual intervention can be implemented to intervene stress accumulation in rock masses ahead, to ensure subsequent construction safety. The following different rock burst control plans

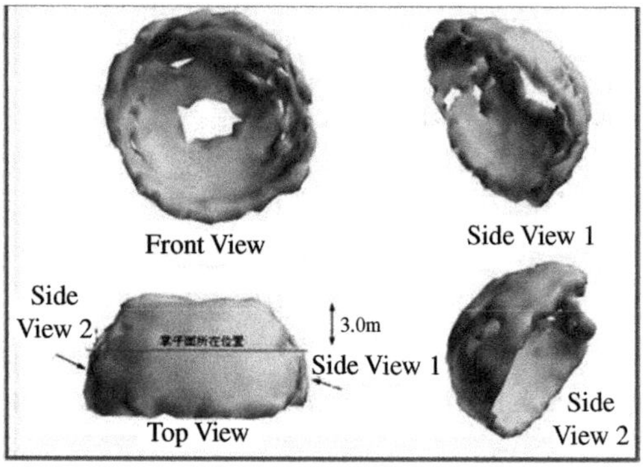

Figure 12 3D view for stress states of surrounding rocks during D&B excavation.

applicable to specific project characteristics and construction capabilities are mainly implemented at site:

(1) Shape modification of heading face. The plan is based on the distribution of high surrounding rock stress areas on heading faces (Figure 12), to enable the shapes of excavation faces to be adaptive to such stress distribution pattern. In this way, it is favorable for maintaining confining pressure level and strength of surrounding rocks, and achieving the aim of controlling rock burst by means of surrounding rock strength;
(2) Stress-relief blasting. The method (Figure 13) enables stress concentration areas in front of heading faces and around tunnels to be far away from excavation faces, to reduce the extent of stress concentration around tunnels and reduce the occurrence risk of strong rock burst. Moreover, stress-relief blasting can be implemented together with conventional blasting, with insignificant impacts on construction progress.

• Passive support control method
According to mechanical characteristics of deep marbles and mechanical properties of support units, the passive support control method for prevention and control of rock burst is summarized as follows:

(1) Anchors shall be long enough to penetrate areas suffering from damage and failure of surrounding rocks, to effectively prevent propagation of surrounding rock fractures and cracks, reinforce rock masses, improve shear strength of rock masses and structural planes, and improve stress distribution close to structural planes;
(2) Anchors shall have good rock burst impact resistance and supporting force. With comprehensive consideration given to durability, supporting force, impact resistance, economical efficiency and construction convenience, full resin anchors have relatively optimum comprehensive indexes;
(3) Support measures, as required, shall be taken in time, to play their roles timely and rapidly to ensure construction safety. The plan combining water expansion anchors and mechanical expansion shell prestressed anchors is employed at site. These anchors can easily and rapidly provide supporting force for surrounding

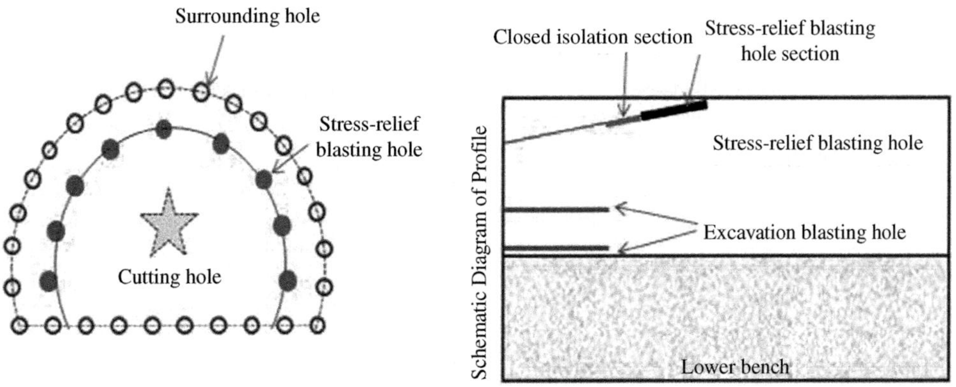

Figure 13 Stress-relief blasting plan.

rocks in the first time, and thereby makes up the defects that full resin anchors are slow in acting speed of supporting force due to relatively slow increment of mortar strength;

(4) Systematic support measures shall be taken as required. To adapt to unloading characteristics on the surfaces of surrounding rocks with high ground stress and impact failure characteristics of rock burst, all system anchors are equipped with external steel shims to improve the support effects of anchors, so that anchors, wire meshes and shotcrete layers can form a complete structure system;

(5) Nano-materials are mixed in shotcrete to improve early strength and thickness of shotcrete layer, and steel fibers or steel-wire-like organic fibers are mixed in shotcrete to improve the mechanical resistance to impact failure of shotcrete layers, to further improve the supporting force for the surfaces of surrounding rocks of tunnels, and thereby adapt to decentralized force transmission of anchor groups by impact force of rock burst.

5.7 Comprehensive prevention and control method of rock burst hazard at TBM tunneling sections

The prevention and control of rock burst at TBM tunneling sections still adheres to the idea of active stress relief and passive support control. Active prevention and control plan and measures are implemented, to avoid strong and extremely strong rock burst failure during TBM excavation where possible, and avoid major impacts on TBM construction safety and progress due to catastrophic surrounding rock collapse caused by rock burst.

• Prevention and control method of medium rock burst risk and below during TBM tunneling

(1) TBM equipment is relatively limited in active control measures of rock burst and corresponding effects. If stress-relief blasting cannot be carried out in front of heading faces, adjustment of tunneling parameters will be an active control measure of rock burst during TBM tunneling. Such measures as reduction of tunneling

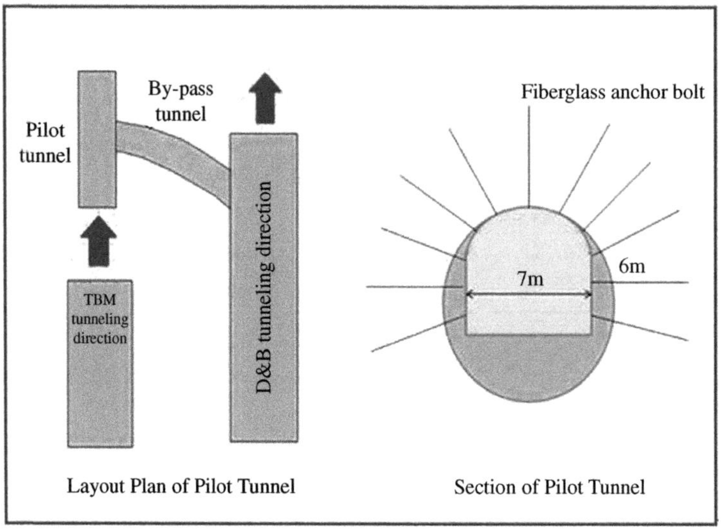

Figure 14 Excavation of pilot tunnel at tunnel section suffering from strong-extremely strong rock burst during TBM tunneling.

speed and adjustment of cutter pressure can be taken, to reduce disturbance of TBM tunneling on surrounding rocks and occurrence risks of rock burst;

(2) Specific support measures can be taken to rapidly form surrounding rock support force, such as the application of new materials and technologies including flexible wire mesh, nano-organic steel-wire-like fiber shotcrete, mechanical expansion shell hollow prestressed anchors or water expansion anchors.

• Prevention and control method of strong rock burst risk during TBM tunneling
During TBM tunneling at sections suffering from strong ~ extremely strong rock burst, to avoid damage to equipment by falling objects, rock burst control plan based on TBM tunneling safety has been proposed (Figure 14) as follows: Excavate a pilot tunnel by a D&B method at tunnel section with strong rock burst risks, to release high ground stress in advance through the pilot tunnel; take the pilot tunnel as an advanced geological exploratory adit and a working face for advanced pre-treatment and micro-seismic monitoring, to provide good practical conditions for advanced revelation, monitoring, analysis and treatment of strong and extremely strong rock burst. After excavation of pilot tunnel, secondary expanded excavation by TBM is carried out. A pilot tunnel can improve the passive situation that TBM equipment is limited in measures for strong rock burst.

6 SAFE AND RAPID CONSTRUCTION TECHNOLOGY OF HEADRACE TUNNEL

The Jinping-II Hydroelectric Project is huge in quantities of headrace tunnel group works, including about 10.00×10^6 m^3 of rock tunnel excavation, 3.60×10^6 m^3 of lining

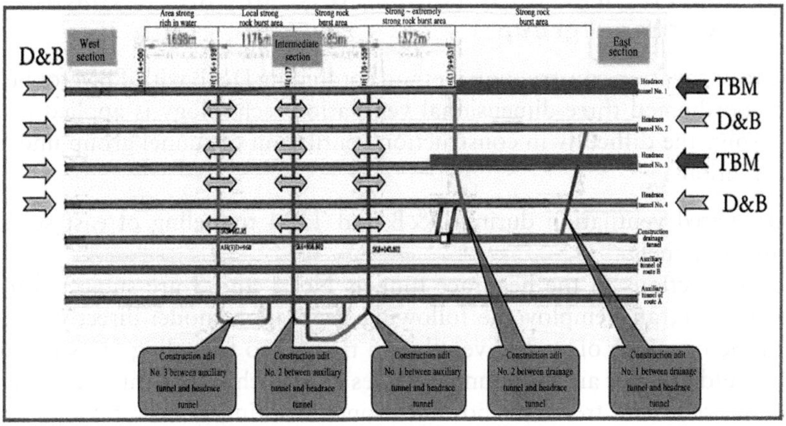

Figure 15 Schematic diagram for combined excavation method of headrace tunnel.

concrete and 3.00×10^6 m^3 of consolidation grouting holes. To put the Project into operation for power generation in time, study is required for safe and rapid construction technologies.

6.1 Construction technology combining D&B with TBM

The Jinping-II headrace tunnels are long in tunnel route, moderate in rock strength and good in drillability, so TBM construction is applicable, to exploit the advantage of TBM (*i.e.* rapid in tunneling) to the full. Intermediate deep sections of headrace tunnels are vulnerable to rock burst due to high ground stress, resulting in extremely large construction safety risk, so it is suitable to exploit the flexibility advantage of D&B method. Therefore, the excavation construction technology combining D&B with TBM has been proposed. Refer to Figure 15 for the schematic diagram of combined excavation method of headrace tunnel.

Through adits between auxiliary tunnels and headrace tunnels, and adits between drainage tunnels and headrace tunnels completed in July 2010, simultaneous tunneling of more than 30 working faces was achieved at intermediate tunnels sections with high ground stress of headrace tunnels. The excavation of 22km tunnel sections was completed in 2010. The maximum monthly advanced depth is 683 m during TBM tunneling, and is 302 m for single heading face during D&B upper half section excavation, and the total maximum monthly advanced depth is 3,612 m for excavation faces at west and east ends. Therefore, the guarantee rate and controllability of construction progress are greatly improved.

The headrace tunnel No. 1 of the Jinping-II Hydroelectric Project was made through completely on December 6, 2011, and all of the four headrace tunnels were made through in the same year. The construction period from commencement to completion is only 4 years.

6.2 Coordinated three-dimensional construction ventilation technology of tunnel group

According to excavation construction method combining D&B with TBM for headrace tunnels, a coordinated three-dimensional ventilation technology is applied to tunnel groups, to solve the difficulty in construction ventilation of tunnel group under complicated conditions.

(1) Coordinated ventilation during D&B and TBM tunneling of east sections of headrace tunnels
The east sections of the headrace tunnels No. 1 and 3 are subjected to TBM construction, and employ the following ventilation mode: directly induce air from the openings of special ventilation tunnels to tunneling faces through air hoses, and exhaust air from tunnel bodies after such treatment as dust removal. Since the return air from the abovementioned working faces is still relatively high in quality, it can serve as fresh air for the tunneling of other tunnels. Therefore, through theoretical calculation, it can be determined that the TBM return air from the headrace tunnels No. 1 and 3 is utilized by the D&B heading faces at the east sections of the headrace tunnels No. 2 and 4, to achieve coordinated ventilation. Refer to Figure 16 for details.

(2) Three-dimensional construction ventilation for various working faces at intermediate sections of headrace tunnels
By taking the headrace tunnel heading faces for the construction adit No. 2 between auxiliary tunnel and headrace tunnel as an example, fresh air is completely introduced to the heading faces from the construction adit No. 2

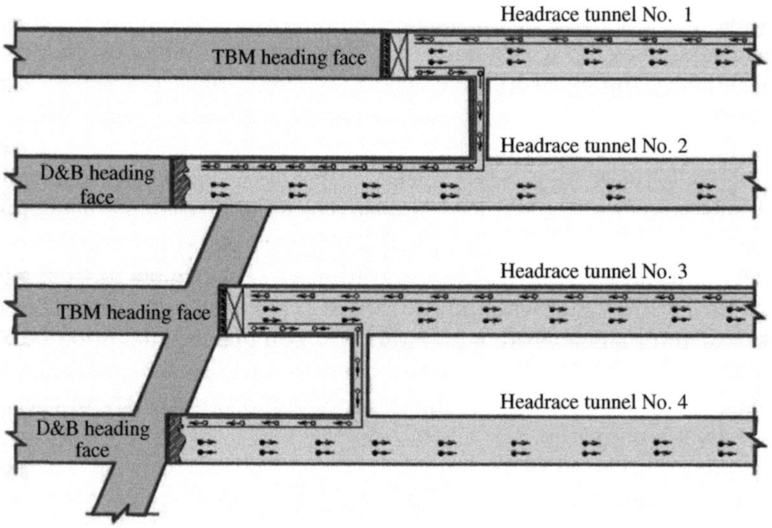

Figure 16 Schematic diagram for coordinated ventilation during D&B and TBM tunneling of east sections.

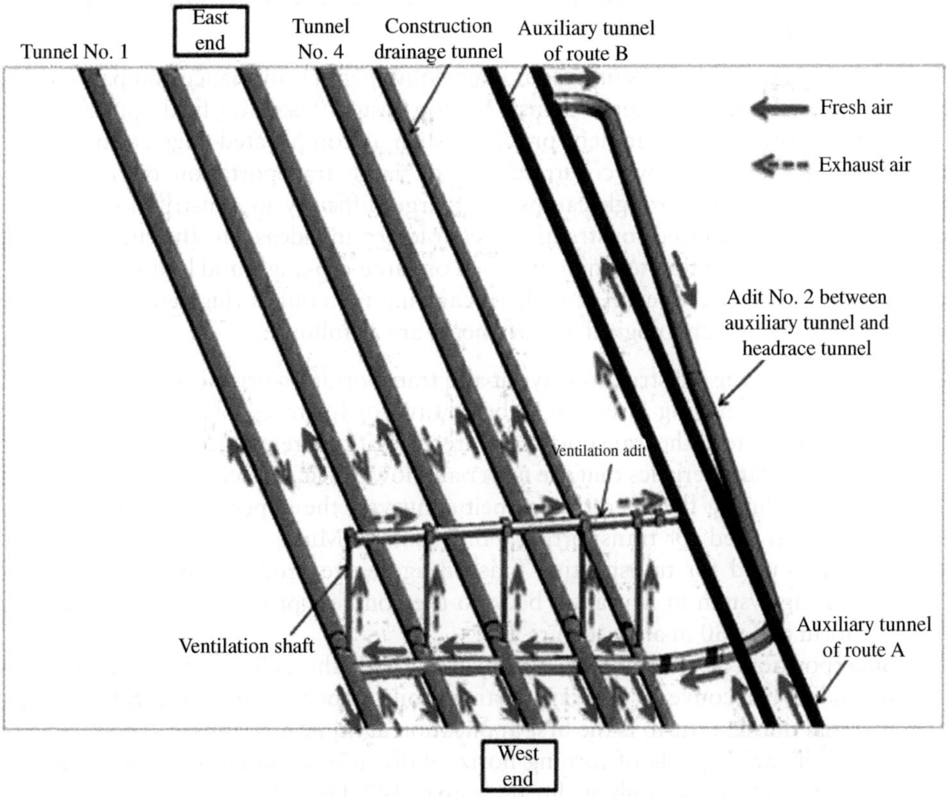

Figure 17 Schematic diagram for three-dimensional ventilation of intermediate section.

between auxiliary tunnel and headrace tunnel through the auxiliary tunnel of route A, and flows through ventilation shaft and ventilation adit from the heading faces along headrace tunnels, and finally exhausted through the auxiliary tunnel of route B (Figure 17). Theoretical calculation proves that during simultaneous construction of various working faces of headrace tunnels, the air velocity in the auxiliary tunnel of route A will not exceed standard requirements, and the auxiliary tunnel of route B can meet the requirements of operation and construction ventilation.

For design schemes of construction ventilation at all construction stages of headrace tunnels, detailed theoretical calculation has been carried out, and simulation and analysis have also been carried out by means of corresponding models established by large analysis system software FLUENT, to implement dynamic configuration for ventilation equipment according to simulation and analysis results. Meanwhile, field data detection and verification have been conducted by instruments, indicating that the ventilation effects are good, and investment and operation costs of ventilation equipment are saved.

6.3 Efficient construction material transportation technology for deep and long tunnels

Through comparative analysis on layout plan, equipment performance and parameter selection of construction material transportation system based on field construction conditions, according to prominent problems such as complicated engineering topographical conditions, narrow construction site, large transportation quantities of materials, need to pass through camps, and large difficulty in construction layout, and by fully use of advanced construction technologies and ideas, an efficient construction material transportation technology based on large-dip-angle and long-haul space-curve continuous belt conveyor capable of carrying material during return has been proposed. The main technological achievements are as follows:

(1) According to the construction layout and transportation organization characteristics (*i.e.* the moving direction of belt conveyor for transporting TBM spoil is rightly opposite to the transportation direction of concrete aggregate), and by full use of the characteristics that the upper and lower belt surfaces move in opposite directions during the operation of belt conveyor, the upper belt surface of belt conveyor is used for transporting TBM spoil to Musagou, and the lower belt surface is used for transporting finished aggregate produced by an aggregate processing system in Musagou back to the concrete production system on the platform at 1,560 m a.s.l. (Figure 18);

(2) Superposition of two large belt conveyors with the belt width of 1,200 mm (including belt conveyor for transporting spoil and belt conveyor for transporting material during return) is the first application at home and abroad. The two belt conveyors are capable of turning horizontally at a minimum turning radius of 1,200 m, and have a climbing dip angle up to 14°. The belt conveyors are of linear arrangement and can turn in a form of space curve at the largest dip angle at home and abroad, as shown in Figure 19. The application of such belt conveyors, to the largest extent, solves the difficulty in arrangement of belt conveyors due to complicated topographical conditions such as bends, reliefs and large-span ditches along the arrangement route of belt conveyors, simplifies the arrangement

Figure 18 Transportation of materials during return of belt conveyor.

Figure 19 Large-dip-angle continuous belt conveyor turning in aform of space curve.

of belt conveyors, and saves investment and cost of transportation equipment of finished concrete aggregate;

(3) Bridge with a steel structure and a span of 236 m will turn in a form of space curve. Due to particularities of topography and location, cranes or other conventional hoisting equipment is unavailable for erection. For this reason, "incremental launching erection for large-span curve steel structure truss" is applied for the first time at home and abroad;

(4) A new tensioning device of "tension trolley + tension tower + winch" is adopted, to reduce the height of heavy-hammer tension tower of long-haul belt conveyor. The application of such technology innovates and optimizes heavy-hammer tensioning devices commonly used at home and abroad at present, and makes up the defects that the tension towers for heavy-hammer tensioning devices of long-haul belt conveyors are relatively high (deep), and topography cannot meet corresponding requirements.

7 PROJECT OPERATION EVALUATION

7.1 Study on mechanical properties of marbles under high ground stress

To describe the brittleness – ductility – plasticity transition characteristics of marbles, a brittleness – ductility – plasticity constitutive model (BDP model for short) based on the Hoke-Brown strength criterion has been developed. Refer to Figure 20 for the results obtained by such numerical simulation test. When the confining pressure increases from 2 MPa to 6 MPa and 8 MPa, the model properly reflects the brittleness – ductility transition characteristics. When the confining pressure in the model increases to 30 MPa, the model properly simulates the plasticity characteristics based on presetting. Meanwhile, considering relatively wide application of the Mohr-Coulomb strength criterion in China, a GPSE constitutive model based on the Mohr-Coulomb strength criterion has also been developed for the Project, to describe the brittleness – ductility – plasticity transition characteristics of the Jinping marbles.

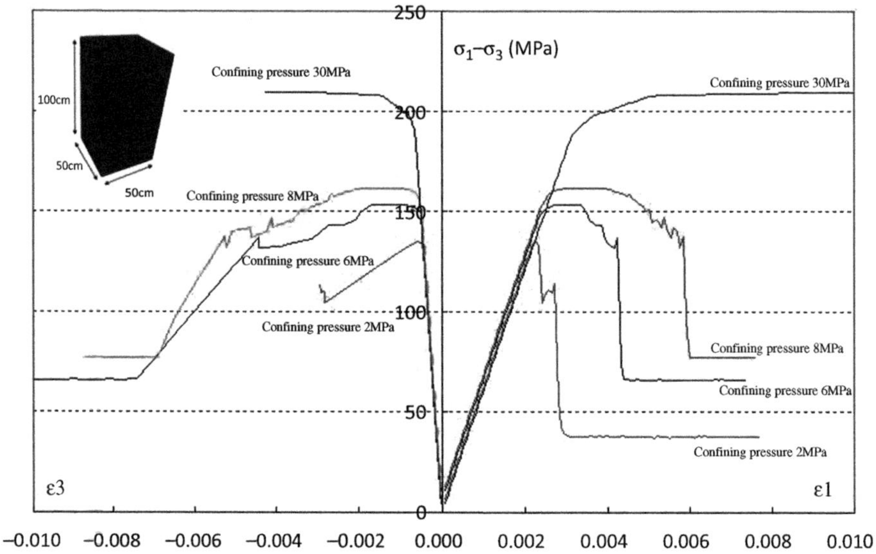

Figure 20 Numerical simulation test of brittleness – ductility – plasticity transition characteristics.

7.2 Stability analysis method for surrounding rocks of deep, long and large tunnels

Under a large buried depth, compression, opening or shearing of structural planes, and fresh fractures of rock masses will affect the stress concentration level of surrounding rocks and the depth of yield area. Therefore, it is necessary to introduce a discontinuous method, to reflect deformation of structural planes, and impacts of fractures on surrounding rock energy and stress distribution. It can be seen from Figure 21 that along with the increase of buried depth, discontinuous mechanical calculation results indicate the increase in depth of damage area, and more importantly, reveal changes in crack density (*i.e.* damage degree) of damage area. Compared to continuous mechanical calculation, discontinuous mechanical analysis method is more direct in problem description and has incomparable advantages. Therefore, during detailed and local analysis on complicated problems related to rock masses, particularly internal mechanism analysis, the discontinuous method is significantly more superior to the continuous one.

7.3 Surrounding rock stability control technology and bearing structure system for deep, long and large tunnels

During construction of deep, long and large tunnels, support principles of taking surrounding rocks as main bearing structures, and rapidly maintaining confining pressure by various combination forms at early stage have been proposed, and a combined bearing structure system considering external water pressure relief technology has been established (Figure 22), to ensure safety and stability of tunnels under large buried depth and high external water pressure.

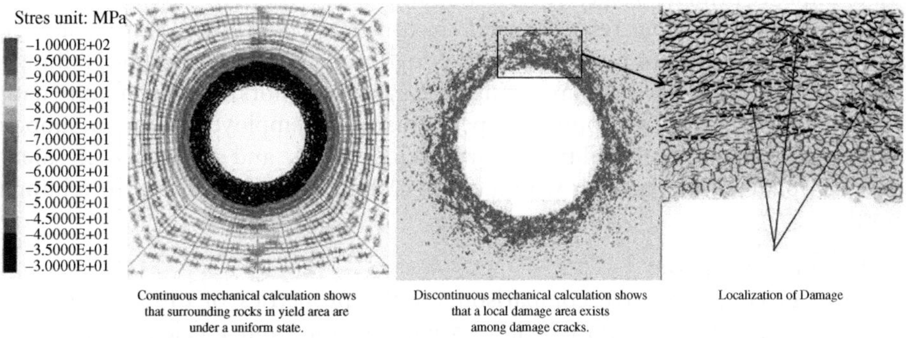

Figure 21 Differences between continuous and discontinuous mechanical methods with regard to description modes for surrounding rock damage areas.

The surrounding rocks of deep, long and large hydraulic structures are featured with a series of disadvantages such as prominent stability problems and high rock burst risks due to high ground stress. In combination with the results of new shotcrete tests conducted at site, the following types of shotcrete are mainly employed for the support purpose: double-mixture silica fume steel fiber concrete, nano steel fiber concrete, nano-organic steel-wire-like fiber concrete, and plain shotcrete with wire mesh. Generally, nano-materials are mixed in shotcrete for tunnel sections with high ground

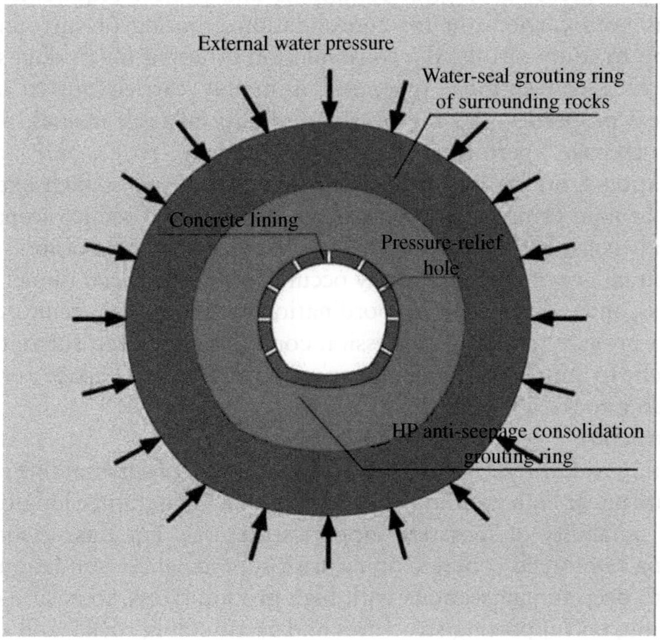

Figure 22 Bearing structure under high ground stress and high external water pressure.

stress, to improve primary shotcrete thickness and rapidly reach the strength required, to control deformation and relaxation failure of surrounding rocks. Strong support measures are required for tunnel sections suffering from rock burst, to control failure of rock masses. On the premise of supporting with system anchors, double-mixture nano steel fiber concrete with the optimum performance is employed as main support materials, to ensure the stability of surrounding rocks and construction safety. Meanwhile, plain shotcrete with wire mesh is carried out subsequently to reinforce support.

(1) Anchors for support
Through comprehensive comparison for the properties of different types of anchors, considering construction speed, difficulty and efficiency, and other factors at project site, and through continuous tests and comparison, the following main anchor types are finally selected in the design for use at site: water expansion anchor, ordinary mortar anchor and expansion shell prestressed anchor. The selection of water expansion anchor, ordinary mortar anchor and expansion shell prestressed anchor depends on mechanical properties of marbles of Baishan Formation and requirements for stability control of surrounding rocks at site, but is also affected by field application conditions. Through field application tests and comprehensive effect evaluation, it is finally determined that water expansion anchors are used as feature rock anchors and anchors for prevention and control of rock burst, local expansion shell prestressed anchors for timely and permanent support, and ordinary mortar anchors for lagging local permanent support.
(2) HP anti-seepage consolidation grouting
According to actual geological conditions and project requirements, detailed study has been carried out for consolidation grouting of surrounding rocks. According to study results, the consolidation grouting for broken surrounding rocks is of a conventional type, and is mainly carried out to improve the mechanical properties and the bearing capacity of rock masses, and improve the anti-seepage performance of surrounding rocks; HP anti-seepage consolidation grouting is a main measure taken to resist high external water pressure around tunnels, control seepage stability and reduce seepage volume; superficial consolidation grouting is designed for relaxation failure and expansion failure of marbles widely occurring on superficial tunnel walls under high stress, and is carried out in coordination with secondary reinforced concrete lining, to provide triaxial compression conditions for deep surrounding rocks, and thereby to improve the bearing capacity and the anti-seepage performance of surrounding rocks.
(3) Lining structure
Considering widespread occurrence and time effects of surrounding rock fracture and failure under high ground stress, it is difficult to guarantee long-term stability and safe reliability of shotcrete support structures. For this reason, the mode combining reinforced concrete lining with superficial consolidation grouting is applied to deep tunnel sections with high ground stress, to ensure coordinated stress acting on supporting structures and surrounding rocks, and to effectively transfer load of surrounding rocks and external and internal water pressure.

Meanwhile, surrounding rocks are enclosed, to provide strong confining pressure for surrounding rocks, avoid further development of relaxation of internal rock masses, and improve peak strength and residual strength of surrounding rocks under triaxial confining pressure, and thereby to ensure the stability of surrounding rocks of headrace tunnels.

(4) External water pressure relief technology
Deep hydraulic tunnels with large sections are fragile to bear external water pressure. To ensure long-term operation safety of tunnels, construction drainage tunnels are provided between headrace tunnels and auxiliary tunnels in parallel for long-term groundwater drainage, to release external water pressure along headrace tunnels as a whole. Such extreme conditions as rainstorm in rainy seasons will lead to sharp rise of external water pressure within a short period of time. For this reason, anti-seepage consolidation grouting is required for the full-length full sections of headrace tunnels, to form water-seal grouting rings of surrounding rocks. Meanwhile, system pressure relief holes are set on lining structures, to rapidly discharge external water in a balanced manner, and thereby to release the external water pressure at outer edges of lining structures and control the pressure always within the range of allowable design value. Two-way control check valves are provided to avoid seepage of internal water, release concentrated high external water pressure applied by corrosion pipe cracks, and protect lining structures against collapse and buckling failure due to extremely high external water pressure.

7.4 Time effects and control measures of propagation of deep marble fractures

All of monitoring data obtained at the site of the Jinping-II Hydroelectric Project, and indoor test results and detailed numerical analysis based on the monitoring data indicate that pressure hydraulic tunnels with large sections constructed at deep tunnel sections with high ground stress will lead to unavoidable surrounding rock fractures and damage. Meanwhile, fracture propagation and long-term strength of marbles have significant time effects. Therefore, reinforced concrete lining is necessary to reinforce surrounding rock bearing rings in combination with grouting, to give full play to the self-bearing capacity of surrounding rocks, and to ensure long-term stability of surrounding rocks by improving the anti-seepage performance of surrounding rocks. Meanwhile, external water pressure at outer edges of lining structures and force acting on lining structures are released by optimizing the arrangement of pressure relief holes, to ensure long-term safety of tunnels.

7.5 Long-term safety evaluation of deep hydraulic tunnel with high external water pressure

Along with gradual completion of construction period, the decreasing groundwater level will be gradually recovered along with implementation of plugging works and continuous operation of tunnels, so the distribution rules of external water pressure along tunnels will also change. As a consequence, external water pressure will certainly affect the long-term stability of surrounding rocks. Particularly for the headrace tunnels of the Jinping-II Hydroelectric Project, under combined effects of high ground stress, high external water pressure and internal water pressure, the long-term mechanical properties of surrounding rocks are completely different from shallow works, certainly leading to complicated long-term safety evaluation. For this reason, separate evaluation is required for seepage stability of surrounding rocks, stability of surrounding rocks, and safety of lining structure.

- Seepage stability of surrounding rocks: HP anti-seepage consolidation grouting is a main measure taken to resist high external water pressure around tunnels, control seepage stability and reduce seepage volume, so that the seepage gradient in surrounding rocks meets control standards.
- Stability of surrounding rocks: Under high ground stress and high external water pressure, the main discrimination criteria for the stability of surrounding rocks depend on three main principles: whether surrounding rock relaxation rings propagate continuously, whether surrounding rock deformation converges continuously, and whether anchor stress is controlled within the range of yield strength.
- Safety of lining structure: Construction of concrete lining is required to be carried out upon the basic completion of surrounding rock deformation, to reduce initial load acting on lining. Meanwhile, hydraulic tunnels are designed by limiting cracks. Crack width, concrete strength and reinforcement stress shall meet standard requirements simultaneously.

Inspection results of headrace tunnels No. 1, 2 and 3 obtained in construction period, and after water filling and emptying, and analysis on monitoring data in operation period indicate that anchor stress, deformation, reinforcement stress in concrete lining, and lining concrete strain are stable, and various monitoring data also stable (Figure 23). At present, the measured maximum external water pressure is only 1.1 MPa. No damage has been found along headrace tunnels in operation period, indicating that stability of surrounding rocks and structure safety are guaranteed.

8 CONCLUSIONS

Under the support of joint fund of Yalong River hydropower development research and the National Natural Science Foundation of China, and relying on headrace tunnel works of the Jinping-II Hydroelectric Project, Yalong Hydro, in cooperation with various organizations, has made a breakthrough in key construction technology of deep, long and large hydraulic tunnel group works through years of technical breakthrough activities. The main innovation achievements are as follows:

(1) Technical difficulties in deep, long and large hydraulic tunnels with regard to comprehensive prevention and control of rock burst under extremely high ground stress have been overcome. A study method for ground stress field in the area of deep tunnel group works has been proposed, and inoculation and generation mechanisms of deep marble rock burst have been revealed. Comprehensive integration technology for acquisition of information related to rock burst hazards and risks under extremely high ground stress, and a zoning, comprehensive identification and prediction system of hazards and risks have

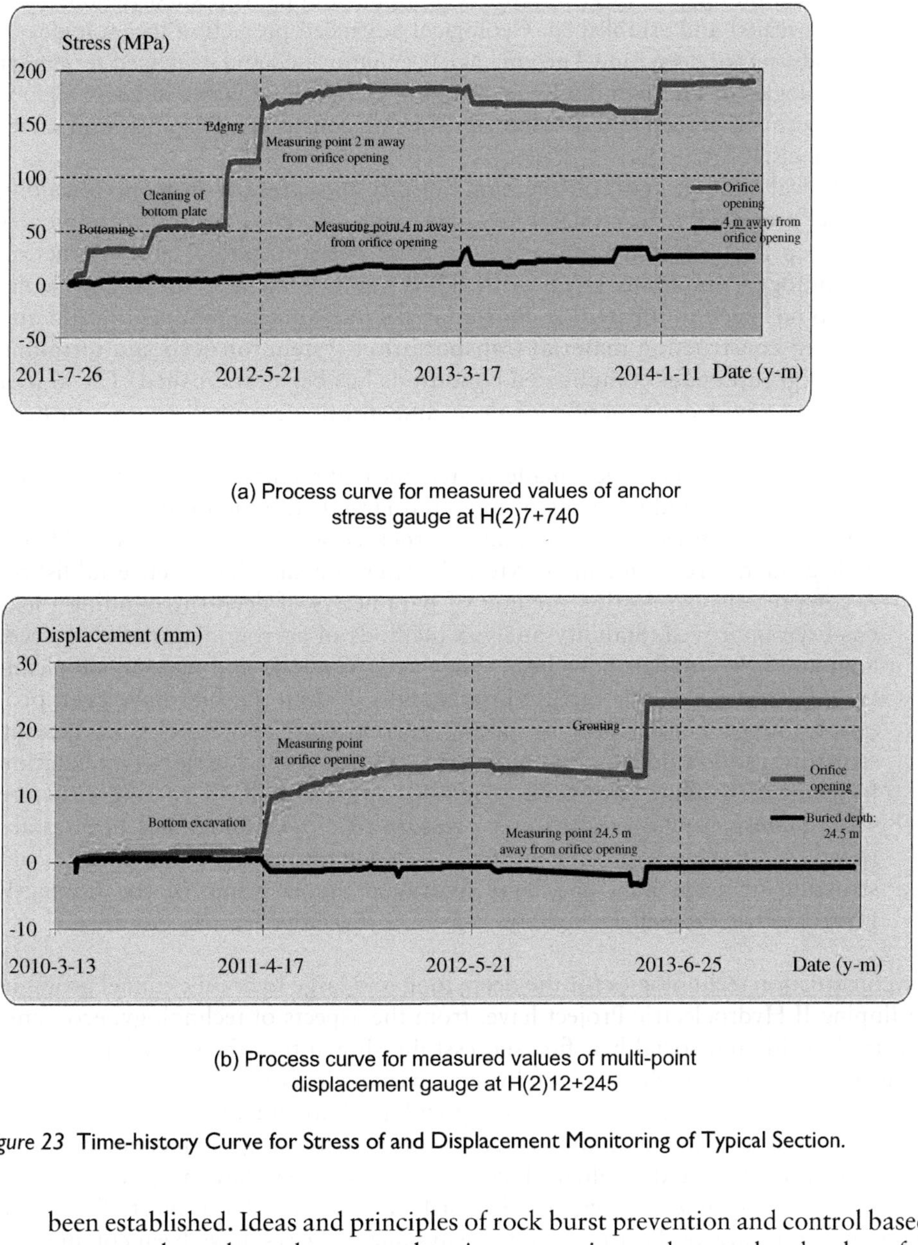

(a) Process curve for measured values of anchor stress gauge at H(2)7+740

(b) Process curve for measured values of multi-point displacement gauge at H(2)12+245

Figure 23 Time-history Curve for Stress of and Displacement Monitoring of Typical Section.

been established. Ideas and principles of rock burst prevention and control based on strength grade, and a comprehensive prevention and control technology for strong rock burst during D&B and TBM construction have been proposed, to ensure construction safety of tunnels.

(2) Packaged treatment technologies of HP high-discharge bursting and gushing water in deep, long and large hydraulic tunnels have been researched and developed. Groundwater treatment principles for deep hydraulic tunnel group in alpine and canyon karst regions have been proposed. Occurrence rules and risk

identification method of groundwater in alpine and canyon karst regions have been revealed and established. Geological advanced prediction technologies of groundwater of deep tunnel group have been innovated and packaged treatment technologies of HP high-discharge bursting and gushing water in karst regions have been researched and developed, to ensure construction rapidity and safety of tunnels.

(3) Technical difficulties in construction rapidity and safety of deep and ultralong tunnel group have been overcome. New packaged construction technologies for deep and ultralong tunnel group have been proposed. A construction technology combining D&B with TBM for a deep and ultralong tunnel group has been innovated. A coordinated three-dimensional ventilation and efficient construction material transportation system for deep and ultralong tunnel group under complicated conditions has been established. The aim of safe and rapid construction of deep and ultralong tunnel group has been achieved.

(4) A breakthrough has been made in technical difficulties related to stability control of surrounding rocks of long and large hydraulic tunnels under high external water pressure. A surrounding rock classification system suitable for high ground stress and high external water pressure has been established. A comprehensive selection method of parameters of deep surrounding rocks has been improved. Stability analysis methods of surrounding rocks of deep, long and large tunnels have been innovated. Methods and measures for controlling time effects of fracture propagation of deep marbles have been proposed. Support principles of taking surrounding rocks as main bearing structures, and rapidly maintaining confining pressure by various combination forms at early stage have been proposed, and a combined bearing structure system considering external water pressure relief technology has been established. All of these works have provided strong technical support for the construction of deep, long and large hydraulic tunnel group of the Jinping-II Hydroelectric Project.

Key construction technologies for the deep, long and large hydraulic tunnel group of the Jinping-II Hydroelectric Project have, from the aspects of technology, economy, industrial health and social benefits, successfully shown that the study has a broad application prospect in the field of hydropower engineering and other industries throughout the country. In particular, along with the booming of economic construction in China in recent years, infrastructure construction projects increase continuously under complicated geological conditions. National key projects, such as West-East Gas Pipeline Project, South-North Water Transfer Project, West-East Electricity Transmission Project, and highways and railways, have been commenced successively. Meanwhile, along with the development of the Yarlung Tsangpo River, there are more and more projects similar to the scale of the headrace tunnel works of the Jinping-II Hydroelectric Project, and construction of more deep, long and large tunnels is required. Moreover, these projects certainly will gradually become more difficult, and the requirements on quality will be much higher. The study achievements of the Project have important reference values and significances for other tunnel works.

REFERENCES

Cai M, Kaiser P K, Martin C D. Quantification of rock mass damage in underground excavations from microseismic event monitoring. International Journal of Rock Mechanics and Mining Sciences, 2001, 38 (8): 1135–1145.

Chen Wei-zhong, Wu Guo-jun, Dai Yong-hao, Chen Rong, Chen Li-bao. Stability analysis of diversion tunnel for Jinping hydropower station. Chinese Journal of Rock Mechanics and Engineering, 2008, 30 (8): 1184–1190 (in Chinese).

Cheng Yunhai, Jiang Fuxing, Zhang Xingmin, Mao Zhongyu, Ji Zhenwen. C-shaped strata spatial structure and stress field in longwall face monitored by microseismic monitoring. Chinese Journal of Rock Mechanics and Engineering, 2007, 26 (1): 102–107 (in Chinese).

Ge M C. Efficient mine microseismic monitoring. International Journal of Coal Geology, 2005, 64 (1/2): 44–56. Hirata A, Kameoka Y, Hirano T. Safety management based on detection of possible rockburst by AE monitoring during tunnel excavation. Rock Mechanics and Rock Engineering, 2007, 40 (6): 563–576.

Hou Jing, Zhang Chunsheng. Shan Zhigang. Rockburst characteristics and the control measures in the deep diversion tunnel of Jinping II Hydropower station. Chinese Journal of Rock Mechanics and Engineering, 2011, 7 (6): 1251–1257 (in Chinese).

Jiang Fuxing, Ye Genxi, Wang Cunwen, Zhang Dangyu, Guan Yongqiang. Application of high-precision microseismic monitoring technique to water inrush monitoring in coal mine. Chinese Journal of Rock Mechanics and Engineering, 2008, 27 (9): 1932–1938 (in Chinese).

Jimin Wang, Xionghui Zeng, Jifang Zhou. Practices on rockburst prevention and control in headrace tunnels of Jinping II Hydropower Station. Journal of Rock Mechanics and Engineering, 2014, 4 (3): 258–268.

Jing Feng, Bian Zhihua, Yang Huoping, Liu Yuankun. Geostress survey and rockburst prediction analysis in deep-buried long tunnel. Yangtze River, 2008, 39 (1): 80–83 (in Chinese).

Li Shulin, Yi Xiangang, Zheng Wenda, Cezar Trifu. Research on multi-channel microseismic monitoring system and its application to Fankoulead-zinc mine. Chinese Journal of Rock Mechanics and Engineering, 2005, 24 (12): 2048–2053 (in Chinese).

Liu Jianhong, Cui Qinyuan, Wang Yapeng, Li Xin. The application of micro-quake suspecting technology at Chang-6 layer of Nanniwan oil field. North Western Geology, 2004, 37 (1): 80–83 (in Chinese).

Lu Zhenhua, Zhang Liancheng. Evaluation of near-field monitoring efficiency of tremors in Mentougou mine. Earthquake, 1989, 10 (5): 32–39 (in Chinese).

Mercera R A, Bawden W F. A statistical approach for the integrated analysis of mine induced seismicity and numerical stress estimates, a case study—part II: evaluation of the relations. International Journal of Rock Mechanics and Mining Sciences, 2005, 42 (1): 73–94.

Tang Lizhong, Pan Changliang, Yang Chengxiang, Guo Ran. Micro-seismic monitoring system set-up and apply research in Donggua mountain copper mine. Mining Technology, 2006, 6 (3): 272–277 (in Chinese).

Wang H L, Ge M C. Acoustic emission/microseismic source location analysis for a limestone mine exhibiting high horizontal stresses. International Journal of Rock Mechanics and Mining Sciences, 2008, 45 (5): 720–728.

Wu Shiyong, Zhou Jifang, Chen Bingrui, Huang Manbin. Effect of excavation schemes of TBM on risk of rock burst of long tunnels at Jinping II hydropower station.Chinese Journal of Rock Mechanics and Engineering, 2013, 23 (4): 728–734 (in Chinese).

Wu Shiyong, Zhou Jifang. Research on safe and fast tunneling technology by open-type hard rock TBM under high geostress of long diversion tunnels of Jinping II hydropower station. Chinese Journal of Rock Mechanics and Engineering, 2012, 31 (8): 1657–1665 (in Chinese).

Wu Shi-yong, Ren Xu-hua, Chen Xiang-rong, Zhang Ji-xun. Stability analysis and supporting design of surrounding rocks of diversion tunnel for Jinping hydropower station. Chinese Journal of Rock Mechanics and Engineering, 2005, 24 (20): 3777–3782 (in Chinese).
Xu Ying, Zhang Jingyi, Shi Li. Rockburst prediction in underground engineering. Science and Technology Information, 2008, 25: 110 (in Chinese).
Zhou Chuiyi, Zhou Yong, Li Jun. Tunnel Construction Design for Jinping II hydropower station. Water Power, 2010, 36 (1): 14–16 and 56 (in Chinese).
Zhang Chunsheng, Chu Weijiang, Hou Jing, Chen Xiangrong, Wu Xumin, Liu Ning. In-situ test on diversion tunnel at Jinping II Hydropower Station I—test design. Chinese Journal of Rock Mechanics and Engineering, 2014, 33 (8): 1691–1701 (in Chinese).
Zhang Chunsheng, Chen Xiangrong, Hou Jing, Chu Weijiang. Study of mechanical behavior of deep-buried marble at Jinping II hydropower station. Chinese Journal of Rock Mechanics and Engineering, 2010, 29 (10): 1999–2009 (in Chinese).
Zhang Chunsheng, Chu Weijiang, Liu Ning, Zhu Yongsheng, Hou Jing. Laboratory tests and numerical simulations of brittle marble and squeezing schist at Jinping II hydropower station, China. Journal of Rock Mechanics and Geotechnical Engineering, 2011, 3 (1): 30–38.
Zhu Huanchun. Monographic study of surrounding rock stability, dynamic support design and rockburst in diversion tunnels of Jinping II hydropower station. Wuhan: Itasca China, 2009 (in Chinese).

Mining

Chapter 11

Mechanism of mining-associated seismic events recorded at Driefontein – Sibanye gold mine in South Africa

Jan Šílený[1] & Alexander Milev[2]

[1] *Institute of Geophysics, Academy of Sciences, Czech Republic*

[2] *Council of Scientific and Industrial Research, South Africa*

Abstract: Mode of fracturing of the rock mass is the key point in rock mechanics applied both to natural earthquakes foci and to sources of induced or triggered seismic events. The parameter providing the answer is the source mechanism. The moment tensor, currently used as a universal tool for descriptions of the mechanism, involves general balanced dipole sources. However, in case of small-scale seismic events, the moment tensor need not be always reliably determined. In an effort to fit the data, there may be notable non-shear components caused by the low quality of input data. It means that while the orientation of the fracture is usually estimated well, the mode of fracturing itself may be dubious. Constraining the source model to directly determine a simpler one is convenient for describing the physical phenomena expected for a particular focus. An opening of new fractures can be described, to a first approximation, by a tensile crack, optionally combined with a shear slip. Such an alternative model is called a shear-tensile crack (STC) source model. The combination is practical, and can be used to both identify events that reflect purely mode-I (tensile) failure and to determine the dilation angle of the fracture undergoing shear. The advantage of the STC is even enhanced in application to mining tremor foci, as implosion (*i.e.*, tensile process with a negative sign) can be reasonably expected there as the consequence of collapsing mined-out cavities. From the technical point of view, the STC inversion is more robust than the MT one thanks to smaller number of the model parameters (5 vs. 6), which is crucial in cases of a poor monitoring configuration.

We demonstrate the dominance of the STC model over the traditional MT in resolving the mode of fracturing having occurred in foci of several mining associated seismic events recorded at Driefontein now Sibanye gold mine in South Africa. Mechanisms of five events with magnitudes ranging from 2.5 to 2.8 are non-shear, comprising notable share of implosion in addition to a shear-slip, in accordance with the anticipation of collapsing into void – mined-out – cavities. The error analysis in terms of confidence regions for fracture orientation and the fracture mode parameter reveals better resolution of the STC compared to the MT, *i.e.* higher chance for fracture mode identification even in cases of sparse configurations of the monitoring.

1 INTRODUCTION

1.1 The moment tensor concept

Until recently, moment tensor (MT) has been used by default for the description of the source of both natural earthquakes and man-made seismic events. It replaced the historic era of the double-couple (DC) accepted as the equivalent of a shear slip, in accordance with the prevailing conviction about seismic events as phenomena originated by a shear displacement along a fault surface. Moment tensor, though originated during the 1970s in papers by Backus & Mulcahy (1976), became widespread in the 1980s, just after the Harvard initiative of routine determination of moderate to strong earthquake mechanisms in terms of the Centroid Moment Tensor solutions (CMT). During the decades, it became an indispensable tool to seismologists interpreting the mechanism of earthquakes having occurred in numerous zones with differing tectonic setup, and to geologists discussing the issues related to plate tectonics. It can be said that in the "early years" of the CMT business, it was the DC part of the solution which was in the focus, the rest having been omitted largely, so the generality of the MT concept has been used just from the advantage of yielding a linear inverse problem. Partly the expression of the preference of the DC among all the MT components is the fact that virtually all the agencies determining the CMT solutions routinely apply the deviatoric constraint, *i.e.* impose a zero trace of the MT excluding volumetric changes in the focus. The other reason is the increase of stability of the solution with respect to inconsistencies involved in the task of inversion of noisy data with inexact location of the hypocenter and improper velocity/attenuation modeling. In this way, several teams provide by sophisticated procedures fast solutions about the mechanism of earthquakes ranging from strong to moderate events both on world-wide and regional scales.

Solution of waveform inversion is obviously sensitive to the success in constructing an accurate response of the medium – the Green's function – in the frequencies which are to be processed. The records are low-pass filtered to avoid high frequencies sensitive to small-scale details of the structure, which are usually impossible to model. In Šílený & Milev (2006) waveforms of six events from a Kopanang gold mine in South Africa were inverted in this way. The simplest possible model of the rock mass – the isotropic homogeneous space – is acceptable only above the characteristic length of small-scale tectonic discontinuities, *i.e.* the wavelengths in the records to be inverted must be larger than this length. Then, the homogeneous model is an average of the properties of small-scale units, and its parameters should be considered with some uncertainties around the average values. Šílený & Milev (2006) assigned them ad hoc the Gaussian distribution the width of which they calibrated by using the simulated rockburst carried out underground in Kopanang gold mine (Milev *et al.*, 2001). In this way, they constructed estimates of the confidence regions of the resulting MT providing information about the reliability of the retrieved mechanism.

Contrary to tectonic earthquakes – at least the strong ones, where a shear slip along the fault plane is generally assumed, there is a good reason to expect a more complex process in foci of earthquakes occurring in volcanic regions (*e.g.*, Julian, 1994; Julian *et al.*, 1997), and of seismic events induced by industrial activities because of complex stress state in situ (in case of mining) or the technology applied (high-pressure fluid injection into oil/gas or geothermal wells). Comprehensive review of possible earthquake processes which may lead to non-DC mechanism was presented by Julian *et al.* (1998). The accompanying paper by Miller *et al.* (1998) reviews most important and convincing observations of non-DC earthquakes.

Events occurring in mines are phenomena with frequent non-DC components which can most probably be attributed to physical processes related to stress concentration around void spaces originated by mining. Predominant dilatational first polarities were frequently observed, which indicated incompatibility with a DC mechanism (*e.g.*, Wong & McGarr, 1990). More detailed studies inverting amplitudes or waveforms revealed non-DC source components for mining tremors from South Africa gold mines (*e.g.*, McGarr, 1992; Šílený & Milev, 2008) and Underground Research Laboratory in Canada (Baker & Young, 1997; Cai *et al.*, 1998).

Phenomena of particular importance are seismic events induced by hydro-fracture treatment of oil, gas or geothermal wells: contrary to tremors induced by mining, which are unwelcome phenomena devastating mine works, microearthquakes originated as a consequence of fracturing the rock mass in the vicinity of the borehole due to its pressurizing by injecting a high-pressure fluid are the expected output of this routine technological procedure in oil, gas and geothermal industry. Its aim is to increase the permeability of the rock mass to enhance the influx of the oil and gas from the hydrocarbon reservoir into the treatment well, or increase the connectivity of the system of boreholes penetrating the underground geothermal reservoir. For this purpose, it is of a prime importance to distinguish between fracture mode I and mode II – between tensile cracks opening a new void space, and shear fractures without a volume change. It is obvious that the desired fracturing in the hydro-fracture treatment of industrial boreholes is the mode I.

In exploiting the geothermal energy, in addition to occurrence of microearthquakes reflecting the penetration of injected fluid into the rock mass, sometimes a fairly strong seismicity is invoked which harms the local community. An example is the facility in Basel, where a M 3.4 earthquake (Dec., 2006) below the city with macroseismic effects brought the project into a standstill. Contrary to this case, the enhanced geothermal system (EGS) at Soultz-sous-Forets, Alsace, has already proceeded from the research stage into the operational status (Genter *et al.*, 2009). The seismicity associated to creation and treatment of geothermal reservoirs is routinely monitored and processed, among other parameters, into the MTs (*e.g.*, Julian *et al.* (2010) in the Coso, California geothermal field, Cuenot *et al.* (2006) and Horálek *et al.* (2010) for Soultz). In accordance with the expectation, in the Coso geothermal field Julian *et al.* (2010) found evidence of tensile fracturing combined with wings of shear displacement, while Horálek *et al.* concluded that mechanisms of events they treated were shear slips. This result probably indicates that the injected fluid escaped into the system of natural fractures and new cracks were not created, which means a failure of the injection.

In oil and gas industry, hydrofracturing is a routine operation for revitalizing the treatment wells. Seismic monitoring is a default method for observing the extent of penetration of the injected fluid into the rock mass in the vicinity of the well – how far the fluid penetrates out of the borehole and how quickly it migrates. Localization of the hypocentra yields the answer to the above questions. However, a question of equal importance is whether the rock mass has been fractured in the convenient way for increasing the permeability of the reservoir, *i.e.* whether tensile cracks have been created. The answer is contained in the mechanism of the microearthquakes, if determined in a more general description than a pure DC. During recent years several papers appeared dealing with the topic and presenting evidences of both DC (*e.g.*, Rutledge *et al.*, 2004) and non-DC mechanisms (Šílený *et al.*, 2009; Baig & Urbancic, 2010).

It is a well-known fact that among all the parameters influencing the resolution of the mechanism, the distribution of the observing points around the focus is of a prime importance. While in monitoring of mining tremors and events occurring during creation and operation of underground geothermal reservoirs seismometers – at least in the horizontal projection – usually surround the event hypocentra, monitoring of microearthquakes induced by hydrofracturing in oil and gas wells is mostly performed by linear strings of receivers situated in monitoring boreholes. Typical configuration is a single monitoring well, only rarely is there more than the single one, thus observation is available from the only azimuth then, which represent an extreme case of a sparse coverage of the focal sphere. It was pointed out by Nolen-Hoeksema & Ruff (2001) and revisited by Vavryčuk (2007) that the complete MT cannot be recovered from a linear array of receivers (if far-field P and S are used only). The usual assumption is the deviatoric constraint (Nolen-Hoeksema & Ruff, 2001), which means to give up the reconstruction of volume changes. The common belief is that the deviatoric solution obtained by inverting data from a non-deviatoric source should be close to the true orientation, which is however necessarily not always the case (Jechumtálová & Eisner, 2008). Another approach to overcome the problem is including more phases into the data set, *e.g.*, by inverting waveforms instead of mere amplitudes. This however demands for a detailed velocity/attenuation model, which is not available commonly.

1.2 Traditional MT decomposition: An unphysical CLVD

Moment tensor is a general dipole source allowing to describe mechanisms of seismic sources ranging from tectonic earthquakes to various types of non-shear foci, which include both natural events like volcanic phenomena and man-made tremors occurring in mines and during hydro-fracturing of oil and geothermal wells. It may be however a too general description, sometimes not needed in its full generality. Concept of the moment tensor is a convenient tool because it yields a linear inverse problem when retrieving the source mechanism from the data. As the moment tensor is a quantity not friendly enough to human perception, commonly it is decomposed into 'elementary mechanisms' mostly with direct physical interpretation. Immediate step is separation of the isotropic component (ISO) of the tensor. The decomposition of the remaining, deviatoric part of the MT is not unique and several ways have been proposed, some of them being even misleading (for an overview see Julian *et al.* (1998)). The most common and now widely accepted is the decomposition into a DC and a system of dipoles without moment, called compensated linear-vector dipole (CLVD).

The point is that while the ISO and DC are sources with a clear physical interpretation, namely an explosion/implosion and a shear-slip, respectively, the interpretation of the CLVD is a matter of debate. Obviously in the traditional decomposition, motivated by singling out the DC as the force equivalent to the anticipated earthquake shear slip, it is present as an appendix to complement the DC to full MT deviator. It is worth of noting that the traditional intuitive explanation of the CLVD, namely a crack opening/closing with simultaneous influx/discharge of a fluid, is not correct, confusing body forces with mass movement. In other words, the CLVD is largely an unphysical source.

1.3 Biased solution of the inverse problem

Contrary to events with 'genuine' components of the MT decomposition, which are originated by physics of the process in the focus, the individual components in the MT decomposition can also be spurious. It means they are originated not in the source process itself but due to, *e.g.*, improper modeling of the medium through which seismic waves propagate from the focus to the points of observation. The trouble may be illustrated by a symbolic formulation of the inverse problem for the source

$$\text{wavefield} = \text{source} \otimes \text{medium}. \tag{1}$$

The right-hand side of (1) represents the coupling of the source and propagation effects in the resulting wavefield. If the exact response of the medium were available, we could get unbiased estimate of the source with sufficient amount of data. However, inexact modeling of the medium introduces inconsistency into the inverse problem and the equality in (1) is no more valid

$$\text{observed wavefield} \approx \text{source} \otimes \text{simplified response of the medium}. \tag{2}$$

If we replace the sign of approximation by the equality sign during solving the inverse problem, the estimate of the source is biased by the inexact modeling of the medium:

$$\text{observed wavefield} = \text{distorted source model} \otimes \text{simplified response of the medium}. \tag{3}$$

The other factors yielding distortion of the recovered source is quality of the data, inexact location of the hypocenter, effects of a finite-extent source not accounted in the point approximation assumed in the MT.

2 SHEAR-TENSILE CRACK (STC) SOURCE MODEL

The MT is a general dipole source, but for particular practical situations it may be too general, its generality causing troubles during its reconstruction from noisy data in the inverse process, which may be additionally ill-conditioned due to inexact location of the hypocenter and/or availability of a rough velocity/attenuation model only. Then, the retrieved source may be biased, containing components which do not reflect the true source but which are artifacts of a low-quality data or the inconsistent inverse problem. It seems reasonable to avoid the trouble by assuming a simpler source model containing components which directly describe the mechanism anticipated in the particular focus, avoiding use of the unphysical CLVD.

In particular, in practice frequently the source is well approximated by a simple combination of a shear slip with a tensile crack. It may be a good model for both natural earthquakes, where the tensile crack simulates a fault opening due to, *e.g.*, its roughness or bending, and various types of seismic events induced by industrial activity. It is well suited for description of events involving an implosion of cavities in mines, and especially for phenomena associated with injection/migration of fluids, for crack opening during hydro-fracturing in oil or geothermal wells in particular. Examples of fluid-induced natural phenomena are volcanic earthquakes. Such a model was proposed

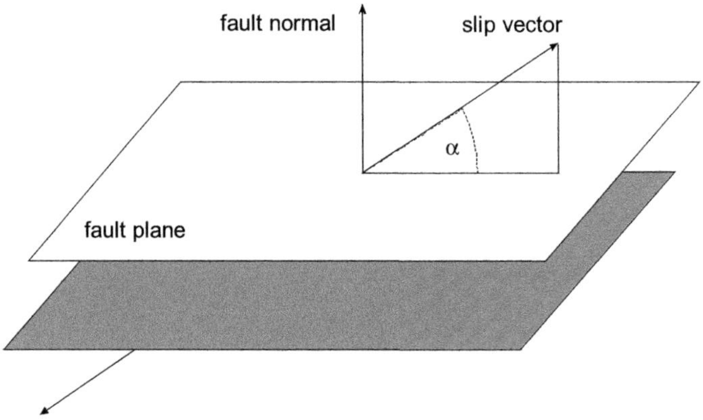

Figure 1 Shear-tensile/implosive source model: combination of a shear slip along the fault plane with an opening or closing the fault, represented by the slip vector inclined by angle α, the slope, from the fault plane. The slope value α = 0° describes the shear slip, α = 90° is a tensile crack, α = −90° an implosion.

early by Kozák & Šílený (1985) who were concerned with representing a tensile crack originating at the tip of a fault undergoing a shear slip. Slip along a fault creates a stress concentration at its tip that results in a tensile fracturing that radiates seismic energy almost synchronously with the shear slip. The reverse sequence of a slip on the fault, initiated by the jog collapse of a mined-out slot nearby, represents the shear-implosive source model (Teisseyre, 1980; Rudajev & Šílený, 1985). A simple shear-tensile source model composed of a shear-slip accompanied by a simultaneous opening or closing of the fault (Fig. 1) was introduced by Dufumier & Rivera (1997) and revisited by Vavryčuk (2001, 2011).

Independently, it was applied also by Minson *et al.* (2007). It is described by 5 parameters only, *i.e.* one parameter less than an unconstrained MT. In addition to the fact that this model consists from such components only, which represent simple physical phenomena, smaller number of parameters is an advantage in the inverse process, which should be more robust. The disadvantage of the formalism is that – contrary to the MT – the shear-tensile/implosion model is non-linear: it is described by 4 angles (for a couple of vectors – the normal to the fault plane and the slip vector) and a magnitude. In a slightly more general version of the model, assumption of the common plane for both shear-slip and the tensile/implosion movement is abandoned, and a couple of planes is assumed which form a non-zero angle (Julian *et al.*, 1998). It seems reasonable to suppose this angle fixed. This extension of the shear-tensile source could be a model for the coalescence of tensile and shear wings in the swarm model of Hill (1977). The drawback of non-linearity due to working with angles however offers the advantage of a convenient estimation of the quality of the solution "on the way", in the course of the iterations, see the next section.

Notable advantage of the STC is the fact that, contrary to the traditional MT, it is a physical model. Whereas components of the MT are body force couples acting in an unfractured medium, which radiate equally to the real fracture (Aki & Richards, 1980), the STC is formed from genuine physical phenomena occurring in the process of rock mass fracturing – tensile crack and shear-slip, namely modes I and II of the fracturing.

The simple scheme of a slip along a plane optionally with an off-plane component of the slip vector depicted in Fig. 1 offers to describe a remarkably wide class of mechanisms ranging from a pure shear to a pure tensile crack. The parameter controlling the transition is the slope angle α:$\alpha = 0$ describes a shear slip, when the slip vector is parallel to the fault plane, $\alpha = 90$ implies a tensile crack where the slip is just normal to the plane of opening. A series of mechanisms with varying slope angle is depicted in Fig. 2. Increasing the value α, compressional onsets of the P wave start to dominate the radiation pattern at the expense of the dilations, and the four-lobe pattern of the P radiation changes to ever compressions: in the traditional fault-plane solution plot the color zones flow over the nodal lines and fill the whole focal sphere – starting from about $\alpha = 30$ the dilations disappear completely. This plot is however flat, not presenting information on the shape of the radiation pattern. This is seen using the wireframe diagram of the radiation pattern in a 3-D view, where it is obvious that despite the same P-onsets for all the models with the slope angle exceeding roughly 30 degrees, the radiation is not isotropic but it has maxima along the slip vector (Fig. 2, 2nd row). In the pure shear-slip model, one of the nodal planes of the traditional fault-plane solution coincides with the fault, the other is perpendicular to the slip vector, *i.e.* they are mutually perpendicular. With a generally directed slip vector, they lose this clear physical interpretation. Therefore, Vavrycuk (2011) introduced the concept of source planes and source lines as their intersections with the focal sphere. One of the source planes coincides with the fault, the other is perpendicular to the slip vector again, now they however are no more perpendicular mutually: they tend to be closer and closer and merge into a single plane for the pure tensile crack (Fig. 2, 3rd row). The radiation of the S-wave is plotted using projections of the arrows on the focal sphere, as contrary to P, for the S wave also the direction of the displacement is changing. We can see that the four-lobe pattern of the S-radiation of a shear-slip gradually changes into a bi-polar pattern of the tensile crack (Fig. 2, 4th row). The radiation scheme of the shear-tensile source is depicted for two types of the mechanism in Fig. 2 – a vertical strike slip (upper half of the Figure), and a 45° inclined dip slip (lower half). Because of the capacity of the simple shear-tensile source model from Fig. 1 to absorb both the shear-slip and the tensile crack, we will call it a shear-tensile crack (STC) from now on. It is important to note, that in the terminology of fracture mechanics the tensile crack is so called Mode I and a shear-slip Mode II fracturing, thus the STC is flexible enough to comprise both the modes of fracturing.

2.1 Inverse scheme for the STC, confidence zones

To invert the data – amplitudes of P and S amplitudes, we apply a simple grid search in two steps: (1) a rough grid along whole model space, and (2) a fine grid in the vicinity of the solution of step 1. Thus, the procedure performs advantageously a mapping of the model space – with a reasonable sampling, the information is available on the goodness of fit in matching the synthetics to the data, which can be used to construct confidence regions of the individual parameters of the model: confidence regions for the T, N and P axes of the shear-slip part of the solution provide information on the certainty in retrieval of the source orientation, confidence region for the off-plane angle – the slope – expressing the deviation of slip vector out of the fault plane determine how

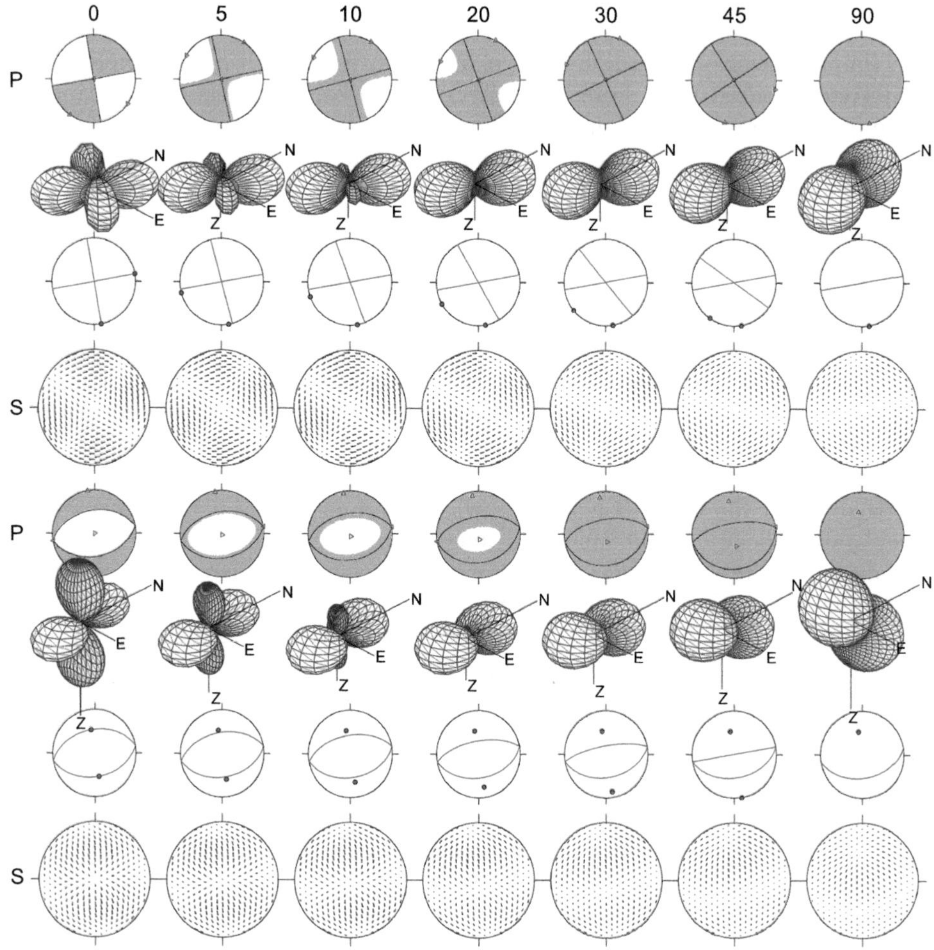

Figure 2 Plots of radiation pattern of P and S waves of the shear-tensile crack (STC) source in dependence of the slope angle acquiring, in turn, the values 0, 5, 10, 20, 30, 45 and 90 degrees. Upper half of the figure – vertical strike-slip mechanism, lower half – dip-slip inclined 45 degrees from the vertical. Row 1 – traditional fault-plane solution plot (nodal lines, zones of compression of P-waves in green, principal axes – red triangles (T-axis – triangle up, P-axis – triangle right, N-axis – triangle left)); Row 2 – wireframe diagram of P-radiation in 3-D (compressions – red, dilations – blue); Row 3 – source lines (red lines) and poles of the source planes (red circles); Row 4 – S-radiation pattern (projection of polarization vectors of S-waves); in rows 1,2,4 equal area projection of lower hemisphere used.

significant is non-shear slip component, in other words it would indicate whether the mechanism is tensile, shear or implosion. We considered the confidence zone of the model parameters, **m**, with the probability content p to be a set of points, **m**, satisfying the following:

$$\chi^2(\boldsymbol{m}) < \chi_p^2 \tag{4}$$

where $\chi^2(\boldsymbol{m})$ is the likelihood function constructed from the residual least squares of data and synthetics, and the dispersion, χ_p^2, is determined from the following condition:

$$\frac{\int\limits_{\chi^2(m)<\chi_p^2} \mathrm{PPD\,dm}}{\int\limits_{m} \mathrm{PPD\,dm}} = \mathrm{p} \tag{5}$$

that specifies the ratio of the cumulative probability in the region to the integral of the posterior probability density (PPD; *e.g.*, Tarantola, 2005) across the entire model space. The PPD is defined, as follows:

$$\mathrm{PPD} = \exp\left(-\frac{1}{2}\chi^2\right). \tag{6}$$

Using the function $\chi^2(\boldsymbol{m})$, we determined particular confidence zones as regions limited by the contour χ_p^2 and possessing the probability content requested (Šílený, 1998). Graphically, in a 1-D scheme it may be sketched as the interval around the minimum of χ^2 such that the area under the PPD above the value χ_p reaches, *e.g.* 90% of the total area under the whole PPD curve; then we have the 90% confidence interval (Fig. 3).

An illustrative interpretation of confidence zones is such that if these zones are small, the model parameters are retrieved safely; if they are large, the parameters obtained are uncertain. We construct separately confidence zones for the P, T, and N axes of the DC portion of the mechanism which provide us with information on the uncertainty for determining the orientation, and confidence zone for the slope angle α which yields

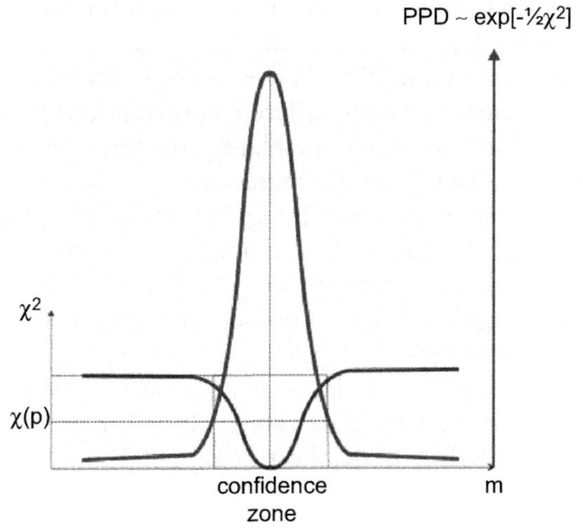

Figure 3 Cartoon illustrating construction of a 1-D confidence interval. χ^2 is a residual function minimized in the course of the inversion, PPD is the posterior probability density function, χ_p is the value satisfying (5).

information on the uncertainty of shear vs. non-shear content. A high content of a non-shear slip component is indicated by an increased uncertainty in the P and N axes in cases of the fault opening, and the T and N axes if the fault is closing. For a pure tensile crack, mode I of the fracturing, the P and N (or T, N) axes were undetermined and their confidence zones merged into a single belt within the plane perpendicular to the crack opening (or closing). For the case of the MT source model, we investigate the confidence zones of the P, T, and N axes again, as well as the MT decomposition in terms of the percentage of the double-couple (DC), the volumetric component (V), and the compensated linear vector dipole (CLVD).

3 SITE DESCRIPTION

The seismic events used in this study were recorded underground at Driefontein gold mine now Sibanye gold. The Mine, is located 70–80 kilometers west of Johannesburg, near Carletonville in the Gauteng province of South Africa, has produced more than 100 million ounces of gold during its 50-year life. Driefontein – Sibanye god is well-established deep to ultra-deep-level gold mine with its lowest working level some 3,400 meters below surface. In 2012 the unbundling of Sibanye gold from Gold Fields Limited was announced. Sibanye owns a 100% interest in the Driefontein Operations ("Driefontein"). The mining right is valid from 30 January 2007 to 29 January 2037 in respect of an area totaling 8,561 hectares. Figure 4 illustrates location of the Sibanye gold mine within Witwatersrand Basin.

The mine is part Witwatersrand super-group or Central Rand group located south of Johannesburg. The Central Rand group is predominantly comprised of quartzite conglomerate rocks with small percentage of Boysens formation and thin bed of lava (Visser, 1998). The most important faults in this area are strike faults which are parallel to the axis of deposition and ages of the basin (Pretorius, 1981). Three primary reefs are exploited; the Ventersdorp Contact Reef (VCR) located at the top of the Central Rand Group; the Carbon Leader Reef (CL) near the base; and the Middelvlei Reef (MR), which stratigraphically occurs some 50 to 75 meters above the CL. The mine comprises eight producing shaft systems, that mine different contributions from pillars and open ground, and three gold plants of which one plant processes mainly underground ore, with the remaining two processing surface material.

At the deepest section of the mine, the mining operations take place at depth of over 3400 m below the surface. At that depth the mining induced seismicity becomes an important issue. A relatively new mining method namely "Closely spaced dip-pillar mining layout" is used to minimize the amount of seismicity and especially the occurrence of damaging seismic events. The pillars are 40 m wide, the mined-out span between the pillars is 140 m. The pillar system is designed to reduce the overall seismicity by restricting the energy release rate to 30 MJ/m and average pillar stress up to 400 MPa at a depth of 4000 m (Klokow *et al.*, 2003).

4 SEISMIC DATA

The seismogenic area hosting the seismic events, used in this study is shown in Fig 4. The area comprises complex mixture of geological (faults and dikes) and mining

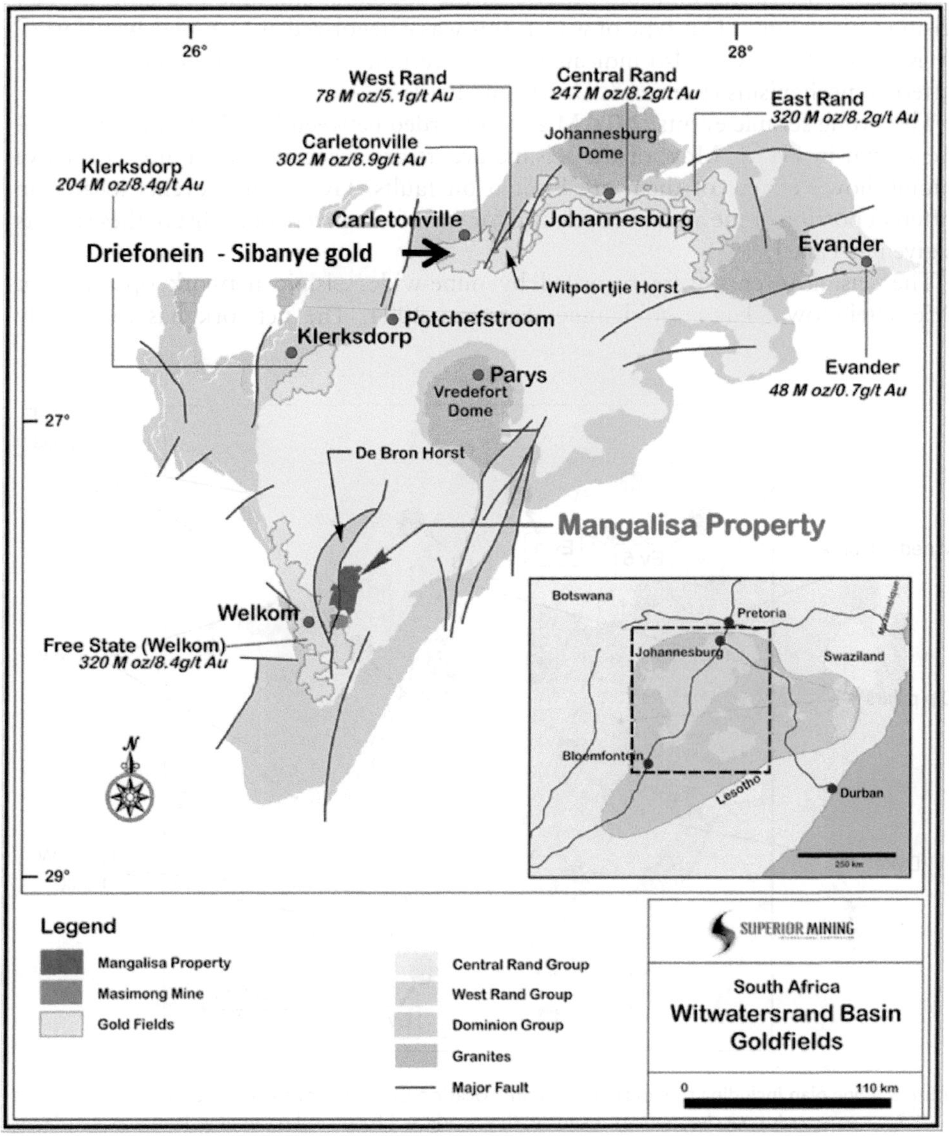

Figure 4 Location of Witwatersrand Basin including the position of Sibanye Gold Mines: Goldfields online images, South Africa, 2015.

(closely spaced dip-pillars, often intercepted by faults and dikes) structures. The fault system in this area is presented by strike-slip type of faults with large lateral component. The majority of the faults in this area are oriented NE/SW direction; however some faults in the upper part of the area have NW/SE orientation. The dikes crossing the area have almost N/S orientation (Riemer, 2006).

The waveforms of the seismic events recorded in this area have rather complex character. Some of the events show multiplication of the P- and S- arrivals on the seismogram indicating multiple rupture of the source or multiple events in a small seismogenic volume. This type of waveforms was categorized as 'complex waveforms' by Riemer (2005) who also indicated that in certain cases the sub-events may have different mechanisms (Riemer & Durrheim, 2012).

The strong seismic events, $2.0 \leq M \geq 3.0$ recorded between July 2003 and July 2004 are shown in Fig. 5. Most of the seismic events are associated with areas of active mining however few of them are located on faults. Five seismic events located in different portion of the area were selected for waveform inversion. A list of these events is given in Tab. 1.

The seismic events were recorded by mine-wide seismic network operating at Driefonteinnow Sibanye gold mine, Ferreira, 2004. The network has reasonably

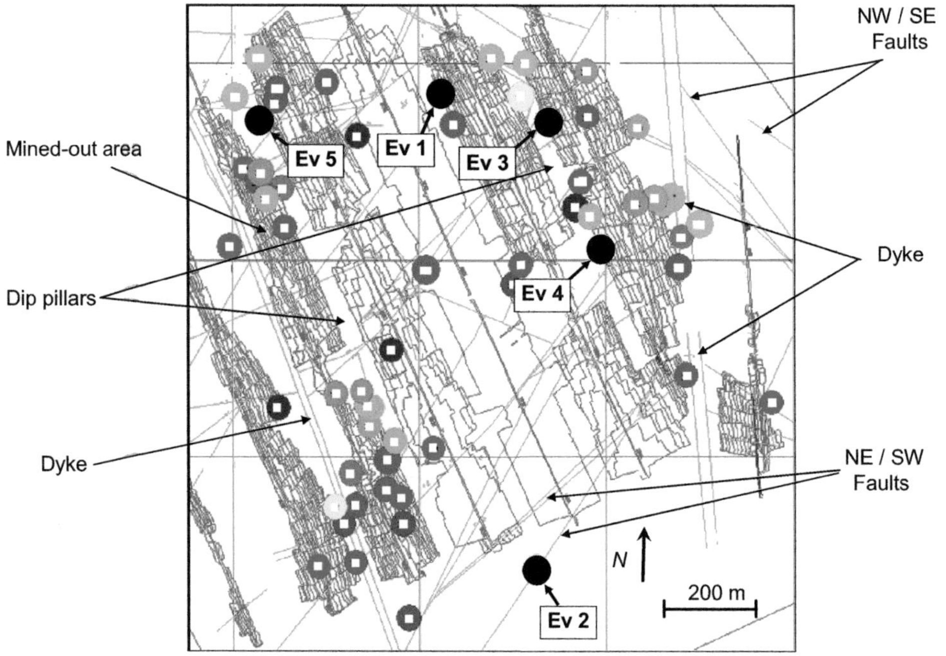

Figure 5 Mine plan including the local geology and location of the seismic events: events located in the area – color circles; events used in this study – black circles; mining layout – red lines, geological structures (faults and dikes) – green lines.

Table 1 Seismic events used in this study; the location of the seismic events is shown in Fig. 5.

No	*Date*	*Time*	*Mag*	*Mo [Nm]*	*Stress drop [MPa]*	*No Trig.*
1	18/09/03	16:01:51	2.5	12.67	1.24	13
2	13/12/03	14:14:49	2.8	12.94	1.68	18
3	20/01/04	20:00:24	2.8	12.91	3.31	24
4	25/01/04	08:05:57	2.6	12.82	1.08	22
5	03/06/04	15:24:25	2.7	12.96	1.12	18

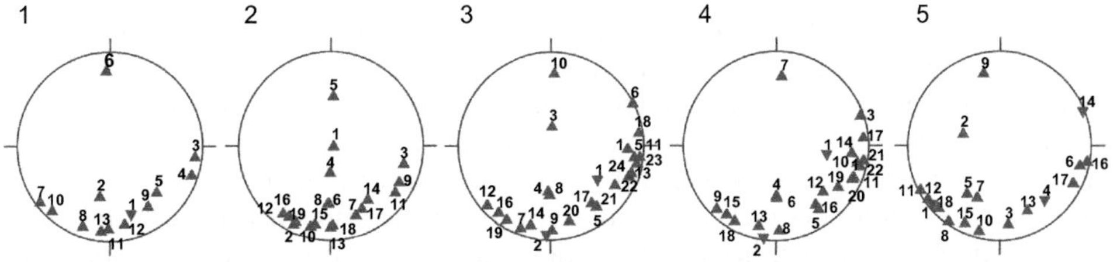

Figure 6 Station distribution: focal sphere coverage for the 5 events selected. Equal-area projection of lower hemisphere.

dense configuration and operates in high sampling rates, 0.2 kHz to 10 kHz, which enable location accuracy of several tens of meters (Riemer, 2005). The network is equipped with omnidirectional geophones with natural frequency 4.5 Hz, grouted in solid rocks. This type of geophones has flat frequency response between 3 Hz and 2000 Hz. The frequency range of recorded seismic events falls well within the flat response of the geophones. Thereafter, further correction of the signals was not necessary. The events were recorded by 13 to 22 stations with hypocenter distances between 0.3 km and 5.5 km with rather irregular distribution, see the station location in the mine map (Fig.5) and the focal sphere coverage for the events selected (Fig. 6). The P and S velocities were taken as V_P = 6200 m/s and V_S = 3650 m/s, and the value of density ρ = 2700 kg.m^{-3}.

5 STC ANALYSIS

Five events from the Driefontein now Sibanye gold mine were processed jointly in terms of the MT and STC source models and their confidences evaluated. We took advantage of our previous paper on the site, Šílený & Milev (2008), where we explored (1) inversion of seismic waveforms on one hand and (2) inversion of amplitudes of direct P/S phases picked from the waveforms on the other hand. The message of the paper was such that there is a little chance to model waveforms successfully despite they are recorded underground. The reason is a poor knowledge about the medium – only parameters of a homogeneous halfspace, *i.e.* the P and S velocities and the density, are known. The consequence is that the Green's functions are too simple – containing merely two pulses, namely direct P and direct S wave – to simulate successfully the observed seismograms. Then, the waveform inversion procedure, which compares observed records with the synthetic seismograms constructed from the Green's functions, the current mechanism and current source time function, moves all the observed record complexities not matched by the Green's function into the source time function. The source time function retrieved in this way is spurious and the mechanism is biased as well. With such a simple model, it is reasonable to rely on the robust part of the wavetrain, namely on direct waves only. In fact, modelling of the picked amplitudes (which equals parameterization of the seismograms) instead of the complete waveforms decreases the negative effect of imperfect knowledge of the medium. The benefit comes from skills of the interpreter who performs the picking. Although it brings certain level of subjectivity, it helps a lot in such situations where 'blind' waveform inversion may fail largely. And this is just the case of reflected and converted phases in the records not generated by the velocity model available. Then, the moment-time function is assumed to be the step function and the inversion problem is reduced to the estimation of the time-independent mechanism: six parameters of the MT or five parameters of the STC. In such a case the inversion is much more robust.

We take the amplitudes picked previously, which were also corrected carefully for the inclination from the predicted polarization and, in some cases, for the anisotropic splitting, see Šílený & Milev (2008), and invert them to the MT and STC. The new information gained from the MT inversion, which was performed already in the previous paper, is determination of its confidence. The STC inversion is a novelty as for the solution itself and its confidence as well.

Event 1 (Fig. 7). It yields a well constrained solution in terms of both MT and STC concerning the orientation: confidence regions of the MT principal axes, *i.e.* the T, P, N axes, are small. For the STC, we can assess the good resolution of its orientation both from small confidence zones of the source plane poles, and from small confidence zones of the T, P, N axes of the auxiliary moment tensor which we derive from the STC solution, see the top of Fig. 6. There is however another feature of the MT solution apart from the orientation of the mechanism, namely the decomposition of the MT. And this is not constrained too tightly, especially as for the DC and CLVD parts – the confidence zones expressed as the histograms are rather wide. The best-determined component is the ISO. Its histogram is well out of the value zero in all the probability levels, which we can interpret in such a way that there is significantly non-zero ISO component in the MT solution. Its sign is negative, *i.e.* it represents an implosion. Thus, regardless the ambiguity of the CLVD in the MT solution (its histogram is spread both in positive and negative values), the MT solution obtained is significantly non-DC. The STC solution is largely consistent with the MT solution. The orientation is the same, compare the T, P, N axes in both the options in Fig. 7. The STC orientation is better constrained a bit, as the confidence zones of the T, P, N axes are smaller than those for the MT, the improvement is however not large. Where the improvement is notable is the non-DC content: histograms of the ISO component are very similar but concerning DC and especially CLVD they are narrower in the STC solution. The improvement is significant: contrary to the MT where the CLVD histogram was spread around zero to both positive and negative values (*i.e.*, it remains undetermined in the confidence level chosen), in the STC solution it is fairly narrow and stays exclusively within negative values. It means that the STC yields a significant CLVD with a negative sign. It follows from the theory (Vavryčuk, 2001, 2011) that, unlike for the unconstrained MT, for the STC (or, strictly speaking, for the moment tensor evaluated for the particular STC model) the ISO and CLVD are not independent but related mutually by a factor depending on material parameters. Regardless the material, they are however always of equal sign, both positive or both negative. This is caused by the very design of the STC model as a shear fault which is at the same time either opening or closing. Here the ISO and CLVD are negative which means a closing. It parallel, it is expressed by the histogram of the slope, the off-plane angle of the slip vector: it remains all within negative values, which can be interpreted that the slope is significantly negative and, thus, the fault closes during the slip.

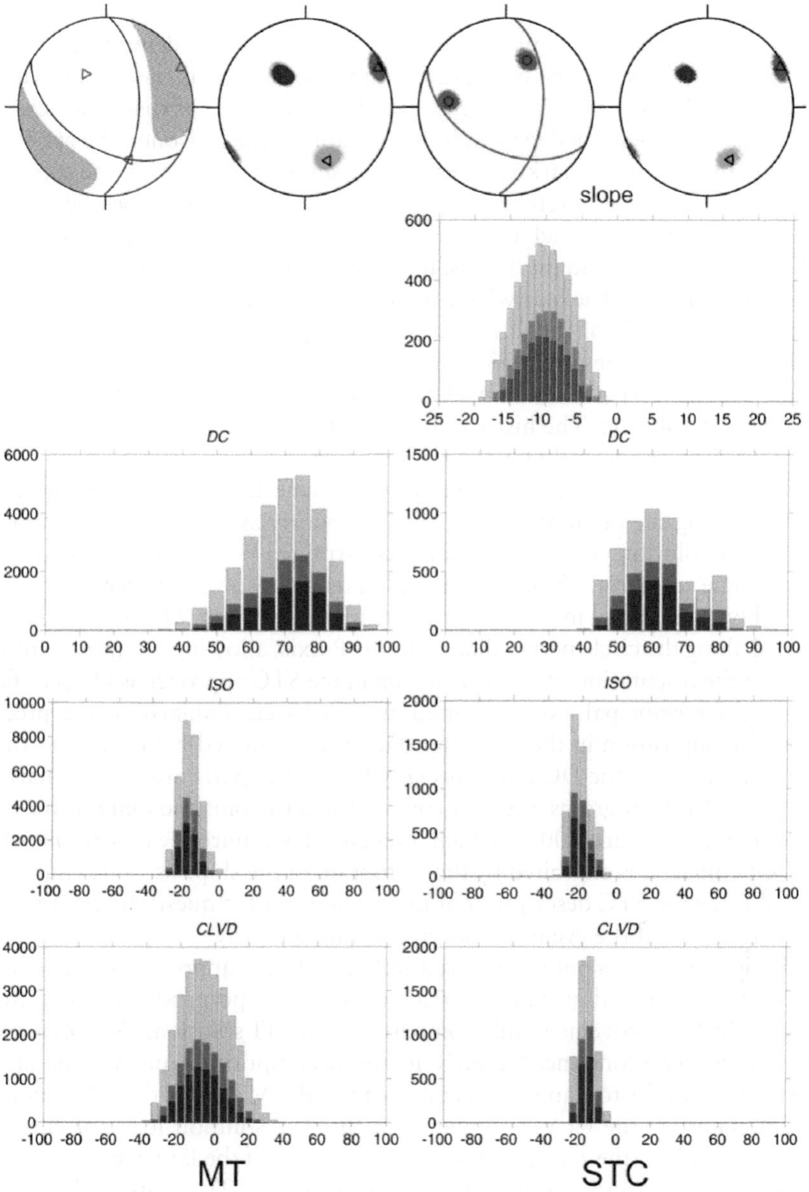

Figure 7 Event 1. Mechanism of the event in terms of the MT (left) and STC (right) source model. Top from left to right: fault-plane solution plot of the MT (nodal lines – black lines and areas of P-compressions – red-hatched areas, T,P,N axes – red triangles), T,P,N axes of the MT and their confidence zones (T – triangle pointing up, red color; P – triangle pointing right, blue color; N – triangle pointing left, green color); source lines of the STC (red lines) and poles of the source planes (black circles) with their confidence zones (red areas); T,P,N axes of the STC and their confidence zones (T – triangle pointing up, red color; P – triangle pointing right, blue color; N – triangle pointing left, green color). Equal-area projection of lower hemisphere of the focal sphere applied. Further from top to bottom: confidence zone of the slope angle of the STC expressed in the form of the histogram; in turn, confidence zones of the DC, ISO and CLVD of the MT (left) and STC (right). Different colors represent confidence zones with different probabilities. Dark, medium, and light colors correspond, in turn, to probabilities of 90, 95, and 99 %.

Event 2 (Fig. 8). This is example of an event, which is still determined well, but contrary to the previous one the orientation of the mechanism is less safe. Confidence zones of the T, P, N axes are bigger: only slightly in the STC approach (which means that the STC is still constrained very well) but notably in the MT one. The zone for the T-axis is still fairly compact marking a reasonable stability of this axis, the zones of the P and N-axes are however stretched along a circle in the plane perpendicular to the T-axis. The interpretation is such that this pattern of the uncertainty allows a rotation of the MT mechanism around the T-axis. It is due to a notable percentage of the CLVD in the MT mechanism retrieved, which possesses just the rotational symmetry with respect to the major axis of the CLVD system of dipoles. The STC solution is more constrained as for both the orientation of the mechanism and its decomposition – the histograms, especially of the ISO and CLVD components, are narrower that those related to the MT solution. The histogram of the slope angle is well in positive values, *i.e.*, the mechanism significantly involves an opening.

Event 3 (Fig. 9). This earthquake is an example of a mechanism with a still acceptable stability concerning the orientation, but resolved poorly as for its shear vs. non-shear nature. The MT solution yields a reasonably constrained P-axis, though its confidence zone allows some variation. The other principle axes are resolved less certainly: they are still acceptably constrained in 95% confidence level but not in 99% level, where they are stretched along the circle perpendicular to the P-axis, allowing a rotation around it. Resolution of the orientation of the mechanism in the STC approach is OK, confidence zones of all three principal axes are small and compact. Similarly to the preceding events, the decomposition in the STC approach is constrained better than in the MT one – the histograms of the DC and especially the CLVD parts are notable narrower. The ISO and CLVD histograms however contain the zero point and that one for the DC component extends toward 100%, which indicates that a pure shear is among eligible solutions. Of course, it is exhibited by the histogram of the slope as well. Thus, neither using the MT nor the STC description of the mechanism the question about the shear vs. non-shear nature of the event is answered unambiguously.

Event 4 (Fig. 10). T-axis only is resolved well in both the approaches, the remaining axes remain undetermined in wide intervals in the plane perpendicular to T (for the STC there is a little improvement only compared to the MT solution). The preference of the STC is however pronounced clearly in the decomposition plots – for all three components the STC histograms are narrower than the MT ones, the difference being notable especially for the CLVD. While from the MT solution it is not clear if the mechanism is non-DC (the zero point is just on the edge of the ISO histogram but well inside the CLVD one), this is significantly manifested in the STC solution – both the ISO and CLVD histograms (and the histogram of the slope as well) are markedly outside zero. The mechanism is significantly implosive.

Event 5 (Fig. 11). The earthquake is an example of a very poor resolution of the orientation but still possessing a significant evidence of its non-shear pattern. Unlike Event 4 where the only constrained axis was the T, here it is the P-axis the confidence region of which is still acceptably limited in its extent. The T and N axes are however totally uncertain – their confidence zones encircle all around the P axis "equator", mostly overlapping each other. This feature has no difference between the MT and STC. There is however a remarkable difference in resolution of the decomposition of the mechanism. It is rather poor in the MT approach – histograms of all three

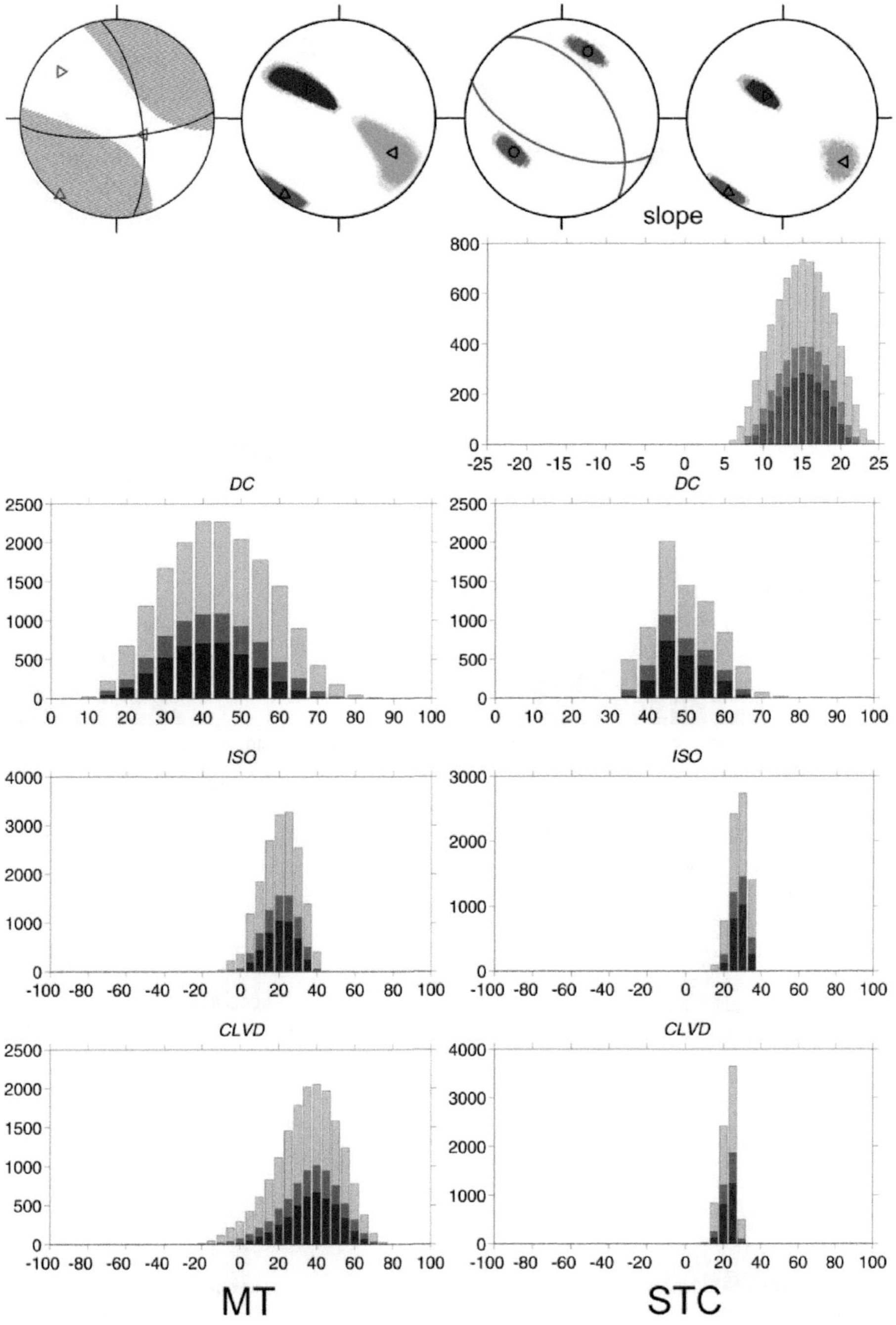

Figure 8 Event 2. For details see the caption of Fig. 7.

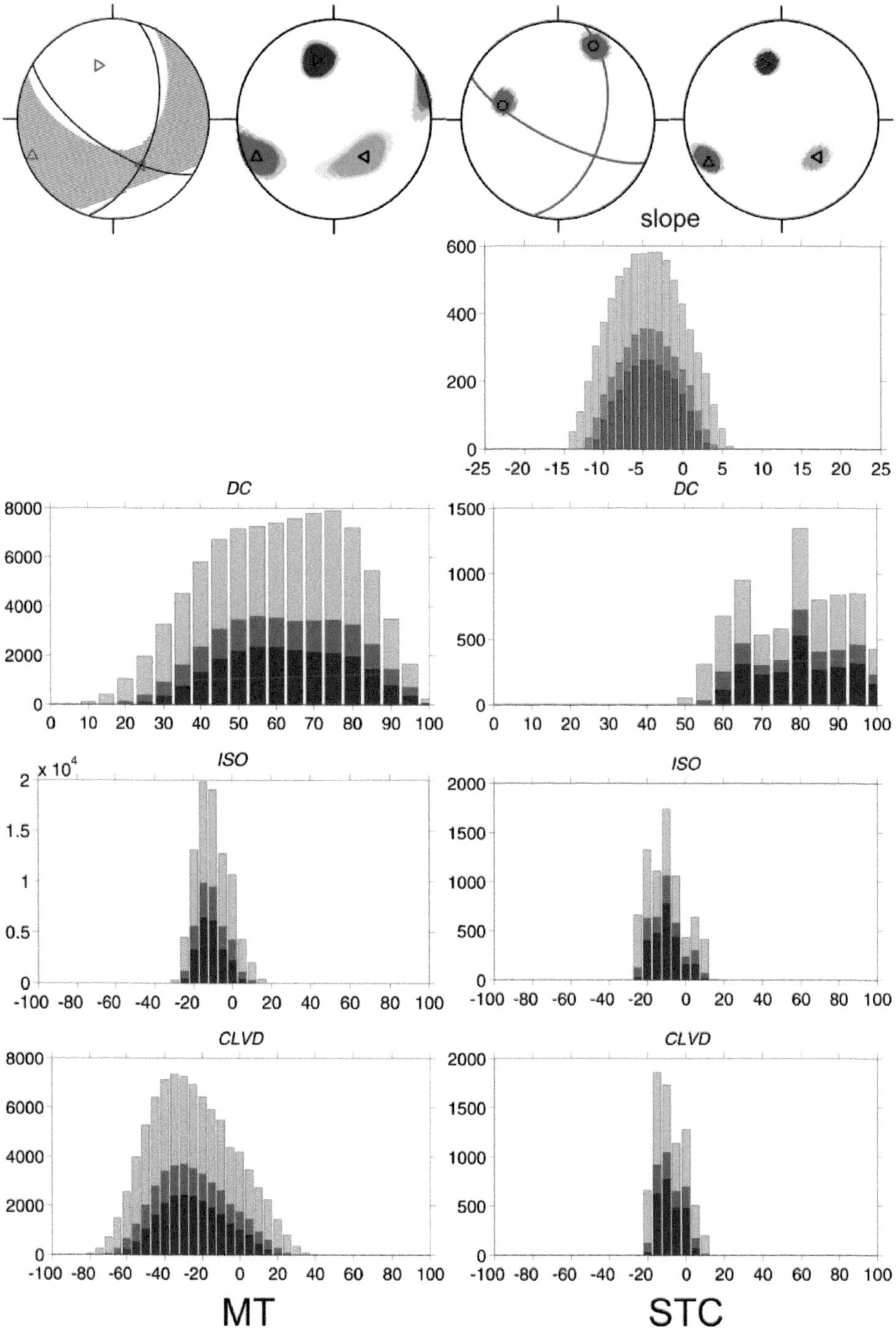

Figure 9 Event 3. For details see the caption of Fig. 7.

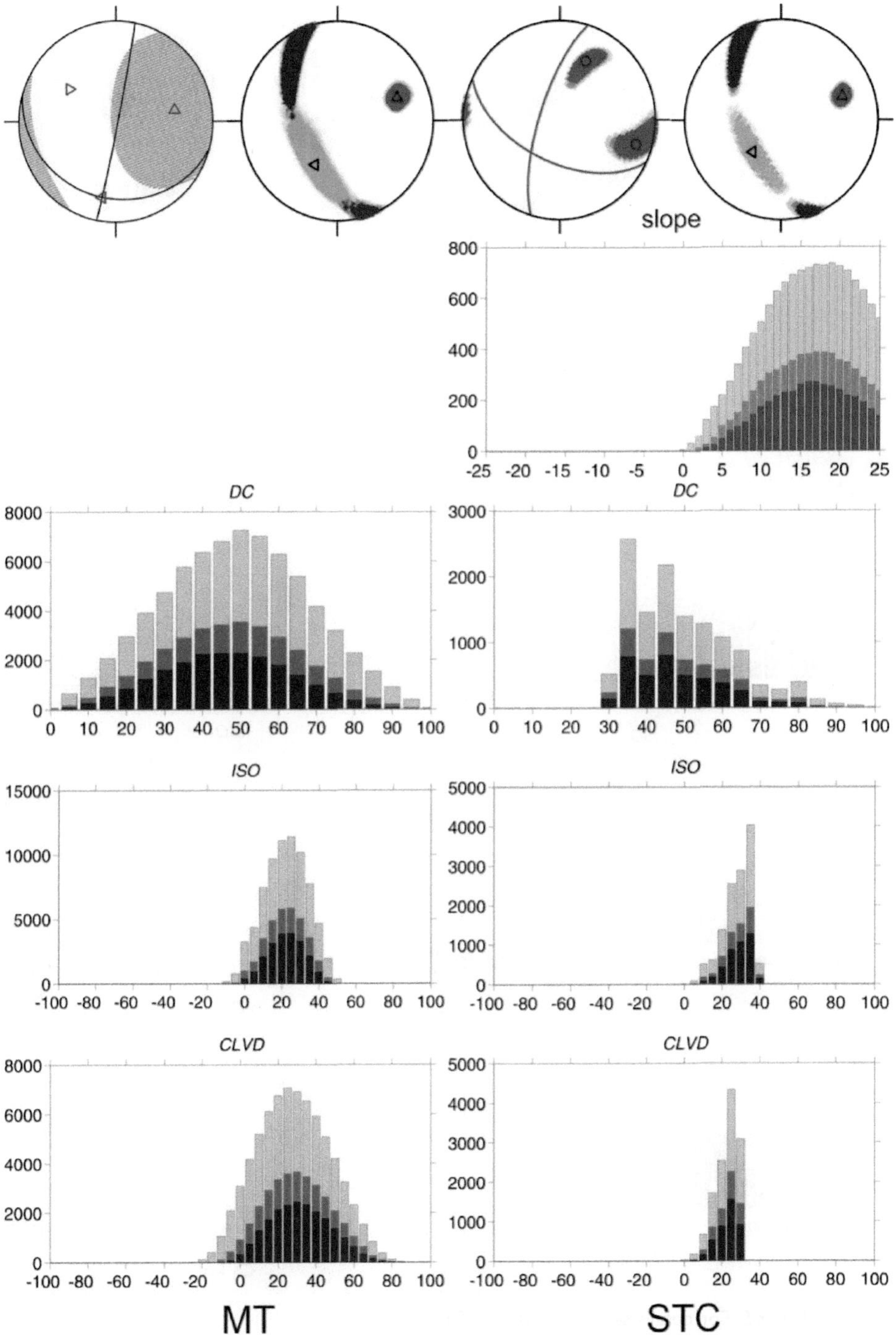

Figure 10 Event 4. For details see the caption of Fig. 7.

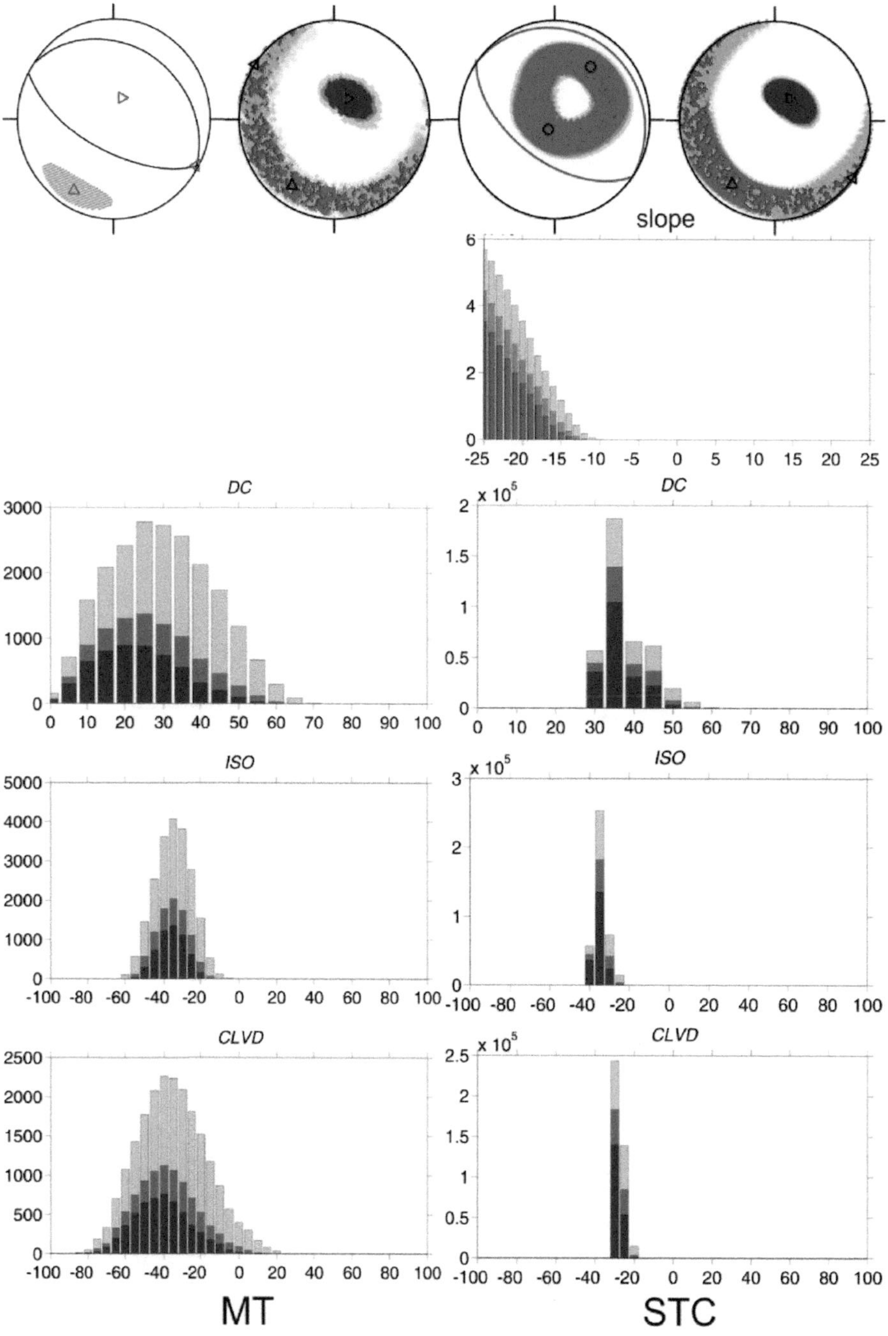

Figure 11 Event 5. For details see the caption of Fig. 7.

components are wide allowing large variation in their percentages, but markedly narrow in the STC, especially for ISO and CLVD components. In the MT solution, while the ISO is significantly determined, the significance of a non-zero CLVD is just on the edge (zero point is on the very margin of the CLVD histogram). Contrary to that, in the STC approach we obtain clearly a significant non-shear mechanism: the ISO and CLVD histograms (and that one for the slope as well, of course) are far apart from the zero point. These parameters acquire negative values, *i.e.* the mechanism is an implosion.

6 DISCUSSION

Despite the monitoring underground, which could evoke fairly simple seismograms and, thus a possibility to work with whole waveforms, the records are rather complex and definitely cannot be simulated in the frequencies observed by using simple velocity models (in fact, mostly a homogeneous half-space) available in the mine. Therefore, we had to step down from inverting waveforms and parameterized the records by picking amplitudes of direct phases P and S waves. This simplifies the task largely, as we neglected all the later parts of the wave-groups. However, even then the match of the synthetic amplitudes (more exactly, the 3-component polarization vectors) to the data is very often rather poor. It indicates that the minimum of the residual function may be not very sharp, and the question naturally arises which is the certainty of the mechanism determined. The tool for the search for the answer is the construction of the confidence regions of individual parameters of the mechanism – its orientation and the decomposition. Extending our previous study of the site by Šílený & Milev (2008), we determined the mechanisms of five earthquakes selected using two source models – moment tensor (MT) and shear-tensile crack (STC). The STC is a type of constrained MT, however still possessing the capacity to describe volume changes in the focus. The novelty of the study is a detailed assessment of the credibility of the parameters of the mechanisms retrieved – construction of their confidence regions at probability levels 90, 95 and 99%. The mechanisms in both descriptions are very close for all 5 events investigated, concerning both the orientation and their decomposition, *i.e.* the content of the individual source components DC, ISO and CLVD (with the STC mechanism, the corresponding moment tensor was evaluated using the particular STC parameters and decomposed subsequently into DC, ISO and CLVD again). While the difference in the parameters themselves of the MT and STC mechanisms for each particular event is not large, they differ – and sometimes enormously – in the reliability of their determination. The traditional approach, *i.e.* the MT description of the mechanism fails to answer the question about the existence of non-DC component unambiguously – in no single case the confidence zones of the individual MT components were small enough with respect to the percentages of these components themselves to justify the statement "the MT is significantly non-DC". Contrary to that, the STC provided the resolution high enough to be able to assign four out of five events as non-shear, either tensile or implosion, see Table 2.

The mechanisms of all five seismic events are mostly in agreement with the mining layouts and the geological settings shown in Fig 5. Event 1 and Event 5 are located at the pillar and their mechanism approaches a 1-D implosion with the axis inclined, but

Table 2 Qualitative assessment of resolution of the orientation of the mechanism and its decomposition by using the MT and STC source models for 5 earthquakes investigated.

Event			*1*	*2*	*3*	*4*	*5*
Resolution of orientation:	MT	single axis	good	fair	fair	excellent	fair
		all 3 axes	good	poor	poor	poor	very poor
	STC	single axis	excellent	good	good	excellent	fair
		all 3 axes	excellent	good	good	poor	very poor
Resolution of decomposition: MT			fair	poor	very poor	very poor	very poor
MT: DC vs. non-DC			none	none	none	none	none
Resolution of decomposition: STC			good	good	fair	fair	excellent
STC: shear vs. non-shear			**implosion**	**tensile**	**none**	**tensile**	**implosion**

closer to vertical direction expected in the case of a pillar failure. The view that pillar failure is normally associated with the implosion type of seismic events dominates within mine seismologist community and was supported by McGarr's view (McGarr, 2005) that out of the mining-induced seismic events with magnitudes well above zero, mechanisms with significant volume reductions, in addition to slip across the fault plane, are the most common.

An alternative point of view was also introduced by Riemer & Durrheim, (2012), that the pillar failure may occur due to failure of the hanging wall, in vicinity of the pillar, as a result of over strain induced by the tensional failure of the hanging wall. This type of mechanism is most probably the case with Event 4 which is located at the boundary of the wide pillar. The hypothesis of sub-vertical surface in the mine-out area which slips down could support this type of mechanism.

Event 3 is located in the intersection of pillar and transverse faulting area. From the geomechanical point of view, both a shear slip and implosion-type event could be feasible here. The quality of the solution of the inverse task both in terms of MT and STC source models however enables to determine satisfactorily only the orientation, while the decomposition – *i.e.* the type of fracturing – remains uncertain.

Event 2 is located on the fault away from the mining stopes. Its mechanism is constrained satisfactorily in the STC description only, where it results in a shear slip combined with an opening of the fault. Taking into account the location, this type of mining-induced seismic events should be the most similar to natural earthquakes. They are related to mine production but do not appear to interact with a particular nearby stope or other excavation (McGarr, 2005).

7 CONCLUSIONS

- Orientation of the mechanism is determined more robustly than its decomposition regardless the source model applied.
- Orientation of the mechanism is retrieved by using MT and STC equally.
- Decomposition is much more certain from the STC than from MT. It enables the STC to discern the type of fracturing even if the MT fails.
- Five Driefontein events from the viewpoint of geomechanics:

- None of them significantly non-DC if viewed by the MT, but four event significantly non-shear when studied by the STC
- Two implosion-type events may be associated with pillar failure, one of the two tensile events with a hanging wall failure.
- The mechanism of the event far from the stope and mining works is tensile, its interpretation is uncertain.

ACKNOWLEDGMENTS

The work was supported by the grants of the Czech Science Foundation, 'Constrained models of seismic source: in between a double-couple and moment tensor', Grant Agreement No. P210/12/2235, and 'Solid body fracturing mode by shear-tensile source model: acoustic emission laboratory study', Grant Agreements No. P108 16-03950S. The authors would like to express grateful acknowledgment to Mr R. Ferreira and Mr K. Riemer from Gold Fields South Africa (now Sibanye Gold), for providing the seismic data at stage of first processing as well as for the fruitful discussions.

REFERENCES

Aki, K. & Richards, P.G. (1980) Quantitative Seismology: Theory and Methods, Freeman & Co., 932 pp.

Backus, G.E. & Mulcahy, M. (1976a) Moment tensors and other phenomenological descriptions of seismic sources I – Continuous displacements, *Geophys. J. R. Astron. Soc.*, **46**, 341–362.

Backus, G.E. & Mulcahy, M. (1976b) Moment tensors and other phenomenological descriptions of seismic sources II – Discontinuous displacements, *Geophys. J. R. Astron. Soc.*, **47**, 301–330.

Baig, A. & Urbancic, T. (2010) Microseismic moment tensors: A path to understanding frac growth, *The Leading Edge*, **29**, 320–324.

Baker, C. & Young, R.P. (1997) Evidence for extensile crack initiation in point source time-dependent moment tensor solutions, *Bull. Seismol. Soc. Am.*, **87**, 1442–1453.

Cai, M, Kaiser, P. & Martin, D. (1998) A tensile model for the interpretation of microseismic events near underground openings, *Pure Appl. Geophys.*, **153**, 67–92.

Cuenot, N., Charléty, J., Dorbath, L. & Haessler, H. (2006) Faulting mechanisms and stress regime at the European HDR site of Soultz-sous-Forets, France, *Geothermics*, **35**, 561–575.

Dreger, D.S. & Helmberger, D.V. (1993) Determination of source parameters at regional distances with three-component sparse network data, *J. Geophys. Res.*, **98**, 8107–8125.

Dufumier, H. & Rivera, L. (1997) On the resolution of the isotropic component in moment tensor, *Geophys. J. Int.*, **131**, 595–606.

Dziewonski, A.M., Chou, T.A. & Woodhouse, J.H. (1981) Determination of earthquake source parameters from waveform data for studies of global and regional seismicity, *J. Geophys. Res.*, **86**, 2825–2852.

Ekstrom, G., Dziewonski, A.M., Maternovskaya, N.N. & Nettles, M. (2005) Global seismicity of 2003: Centroid-moment-tensor solutions for 1087 earthquakes, *Phys. Earth Planet. Int.*, **148**, 327–351.

Ferreira, R.I. Rock Mechanics Department, Driefontein Gold mine, Gold Fields, South Africa, (personal communication, 2004).

Genter, A., Fritsch, D., Cuenot, N., Baumgärtner, J. & Graff, J.J. (2009) Overview of the current activities of the European EGS Soultz Project: From exploration to electricity production,

Proceeding of the Thirty-Fourth Workshop on Geothermal Reservoir Engineering Stanford University, Stanford, California, February 9–11, 2009, SGP-TR-187.

Gold Fields South Africa Online Images, Witwatersrand Basin Gold Felds, Superior Mining International Corporation. (2015) https://www.google.co.za/search?q=map+goldfields&biw=1088&bih=527&tbm=isch&tbo=u&source=univ&sa=X&ei=bJRlVb7NJsG0Upz1gPgP&ved=0CBsQsAQ#imgrc=KF6s6sCsYVEkSM%253A%3BgrTzrrfiQyiE0M%3Bhttp%253A%252F%252Fsuperiormining.com%252F_resources%252Fmaps%252F20120213_witwatrand_basin.jpg%3Bhttp%253A%252F%252Fsuperiormining.com%252Fproperties%252Fsouth_africa%252F%3B2208%3B2578.

Hill, D.P. (1977) A model for earthquake swarms, *J. Geophys. Res.*, **82**, 1347–1352, doi:10.1029/JB082i008p01347.

Hirasawa, T. & Stauder, W. (1965) On the seismic body waves from a finite noviny source, *Bulletin of the Seismological Society of America*, **55**, 237–262.

Horálek, J. & Fischer, T. (2008) Role of crustal fluids in triggering the west Bohemia/Vogtland earthquake swarms: Just what we know (a review), *Stud. Geophys. Geod.*, **52**, 455–478.

Jakobsdóttir, S.S., Roberts, M.J., Gudmundsson, G.B., Geirsson, H. & Slunga, R. (2008) Earthquake swarms at Upptyppingar, north-east Iceland: A sign of magma intrusion? *Stud. Geophys.Geod.*, **52**, 513–528.

Jechumtálová, Z. & Eisner, L. (2008) Seismic source mechanism inversion from a linear array of receivers reveals non-double-couple seismic events induced by hydraulic fracturing in sedimentary formation, *Tectonophysics*, **460**, 124–133, doi:10.1016/j.tecto.2008.07.011.

Šílený, J. & Milev, A. (2006) Seismic moment tensor resolution in local scale: calibration blast in the Kopanang gold mine, South Africa, *Pure Appl. Geophys.*, **163**, 1495–1513, doi: 10.1007/s00024-006-0089-z.

Julian, B.R. (1994) Volcanic tremor: Nonlinear excitation by fluid flow, *J. Geophys. Res.*, **99**, 11859–11877.

Julian, B.R., Miller, A.D. & Foulger, G.R. (1997) Non-double-couple earthquake mechanisms at the Hengill-Grensdalur volcanic complex, southwest Iceland, *Geophys. Res. Lett.*, **24**, 743–746.

Julian, B.R., Miller, A.D. & Foulger, G.R. (1998) Non-double-couple earthquakes. 1. Theory, *Rev. of Geophys.*, **36**, 525–549.

Julian, B.R., Foulger, G.R., Monastero, F.C. & Bjornstad, S. (2010) Imaging hydraulic fractures in a geothermal reservoir, *Geophys. Res. Lett.*, 37, L07305, doi:10.1029/2009GL040933.

Kawakatsu, H. (1995) Automated near-realtime CMT inversion, *Geophys. Res. Lett.*, **22**, 2569–2572.

Kubo, A., Fukuyama, E., Kawai, H. & Nonomura, K. (2002) NIED seismic moment tensor catalog for regional earthquakes around Japan: Quality test and application, *Tectonophysics*, **356**, 23–48.

Kozák, J. & Šílený, J. (1985) Seismic events with non-shear component I. Shallow earthquakes with tensile source component, *Pure Appl. Geophys.*, **123**, 1–15.

Milev, A.M., Spottiswoode, S.M., Rorke, A. J. & Finnie, G.J. (2001) Seismic monitoring of a simulated rockburst on a wall of an underground tunnel, *J. S. Afr. Inst. Min. Metall.*, August 2001, 253–260.

McGarr, A. (1992) An implosive component in the seismic moment tensor of a mining-induced tremor, *Geophys. Res. Lett.*, **19**, 1579–1582.

McGarr, A. (2005) Observations concerning diverse mechanisms for mining-induced earthquakes. In: Proceedings of the 6th International Miller A.D., Foulger, G.R. & Julian, B.R., 1998. Non-double-couple earthquakes. 2. Observations, *Rev. Geophys.*, **36**, 551–568.

Milev, A.M., Spottiswoode, S.M., Rorke, A.J. & Finnie, G.J. (2001) Seismic monitoring of a simulated rockburst on a wall of an underground tunnel, *J. S. Afr. Inst. Min. Metall.*, August 2001, 253–260.

Minson, S.E., Dreger, D.S., Burgmann, R., Kanamori, H. & Larson, K.M. (2007) Seismically and geodetically determined non-double-couple source mechanisms from the 2000 Miyakejima volcanic earthquake swarm, *J. Geophys. Res.*, **112**, B10308, doi:10.1029/2006JB004847.

Nábělek, J. & Xia, G. (1995) Moment-tensor analysis using regional data: Application to the 25 March, 1993, Scotts Mills, Oregon, earthquake, *Geophys. Res. Lett.*, **22**, 13–16.

Nolen-Hoeksema, R.C. & Ruff, L.J. (2001) Moment tensor inversion of microseisms from the B-sand propped hydrofracture, M-site, Colorado, *Tectonophysics*, **336**, 163–181, doi: 10.1016/S0040–1951(01)00100–7.

Pondrelli, S., Morelli, A., Ekström, G., Mazza, S., Boschi, E. & Dziewonski, A.M. (2002) European-Mediterranean regional centroid-moment tensors: 1997–2000, *Phys. Earth Planet. Inert.*, **130**, 71–101.

Pretorius, D.A. (1981) Structural framework: The analysis of gravity data in Precambrian of the Southern Hemisphere. Hunter (Editor), Elsevier, Amsterdam, pp. 411–419.

Riemer, K.L. (2005) Interpreting complex waveforms from some mining induced seismic events. Proc. RaSiM 6. Potvin & Hudyma (Editors), Perth, Australia, pp. 247–257.

Riemer, K.L., Rock Mechanics Department, Driefontein Gold mine, Gold Fields, South Africa (personal communication, 2006).

Riemer, K.L. & Durrheim, R.J. (2012) Mining seismicity in the Witwatersrand Basin: Monitoring, mechanism and mitigation stratesies in perspective, *J. Rock Mech. Geotech. Eng.*, **4**(3), 228–249.

Ritsema, J. & Lay, T. (1993) Rapid source mechanism determination of large ($Mw \geq 4.5$) earthquakes in western United States, *Geophys. Res. Lett.*, 20, 1611–1614.

Rudajev, V. & Šílený, J. (1985) Seismic events with non-shear component II. Rock bursts with implosive source component. *Pure Appl. Geophys.*, **123**, 17–25.

Rutledge, J.T., Phillips, W.S. & Mayerhofer, M.J. (2004) Faulting induced by forced fluid injection and fluid flow forced by faulting: An interpretation of hydraulic fracture microseismicity, Carthage Cotton Valley gas field, Texas, *Bull. Seismol. Soc. Am.*, **94**(5), 1817–1830, doi: 10.1785/012003257.

Savage, J.C. (1965) The effect of rupture velocity upon seismic first motions, *Bull. Seismol. Soc. Am.*, **55**, 263–275.

Šílený, J. (1998) Earthquake source parameters and their confidence regions by a genetic algorithm with a "memory", *Geophys. J. Int.*, **134**, 228–242.

Šílený, J., Hill, D.P., Eisner, L. & Cornet, F.H. (2009) Non-double-couple mechanisms of microearthquakes induced by hydraulic fracturing, *J. Geophys. Res.*, **114**, B08307, doi: 10.1029/2008JB005987.

Šílený, J. & Milev, A. (2006). Seismic moment tensor resolution in local scale: calibration blast in the Kopanang gold mine, South Africa, *Pure Appl. Geophys.*, **163**, 1495–1513, doi: 10.1007/s00024–006–0089–z.

Šílený, J. & Milev, A. (2008) Seismic source mechanism – Source mechanism of mining Induced Seismic Events – Resolution of Double couple and non double couple models, *Tectonophysics*, **456**, 3–15, doi:10.1016/j.tecto.2006.09.021.

Sipkin, S.A. (1982) Estimation of earthquake source parameters by the inversion of waveform data: Synthetic waveforms, *Phys. Earth planet. Int.*, **30**, 242–259.

Sipkin, S.A. & Zirbes, M.D. (2004) Moment-tensor solutions estimated using optimal filter theory: global seismicity, 2002, *Phys. Earth Planet. Int.*, **145**, 203–217.

Stich, D., Ammon, C.J. & Morales, J. (2003) Moment tensor solutions for small and moderate earthquakes in the Ibero-Maghreb region, *J. Geophys. Res.*, **108**, Art. No. 2148.

Teisseyre, R. (1980). Some remarks on the source mechanism of rockhursts in mines and on the possible source extension. *Acta Montana CSAVPraha*, **58**, 7–13.

Vavryčuk, V. (2001) Inversion for parameters of tensile earthquakes: *J. Geophys. Res.*, **106**, 16339–16355, doi: 10.1029/2001JB000372.

Vavryčuk, V. (2007) On the retrievalof moment tensors from borehole data: *Geophys. Prospecting*, **55**, 381–391, doi: 10.1111/j.1365-2478.2007.00624.x.
Vavryčuk, V. (2011) Tensile earthquakes: Theory, modeling and inversion, *J. Geophys. Res.*, **116**, B12320, doi:10.1029/2011JB008770.
Visser, D.J.L. (1998) *The geotectonic evaluation of South Africa and offshore areas.* Council of geosciences, Pretoria, South Africa, p. 40.

Chapter 12

Review on rock mechanics in coal mining

M.C. He, G.L. Zhu & W.L. Gong
State Key Laboratory for Geomechanics & Deep Underground Engineering, China University of Mining & Technology (Beijing), Beijing, China

1 REVIEW OF LARGE-DEFORMATION SUPPORT UNDER ENGINEERING SOFT ROCK ENVIRONMENT WITH NPR BOLT/CABLE

With the growing demands for energy resources, the mining depth for underground coal mining in many districts has surpassed 600–800 m; some dozens of them operate at a depth below 1000 m, such as Dong Xing coal mine (1287 m) in Shangdong Province, Caitun colliery of Shenyang coal mine (1197 m), and Zhao Gezhuan colliery in Kailuan coal mine (1159 m).

It is expected that within the coming 20 years, many coal mines in China will be entering into mining depths ranging from 1000 m to 1500 m. With the increasing mining depths, the frequency of the various engineering disasters will be increasing, such as coal bumps, coal and gas outbursts, rockburst, as well as the large deformations of the country rocks (Salamon, 1970; Singh, 1987; Singh, 1989; Li & Ma, 2004; Pettitt & King, 2004; Cho *et al.*, 2005; Brady & Brown, 2007; He *et al.*, 2007). Therefore, control measures for these rock mechanics disasters, such as rock supports under large deformation risks, have been the focus of rock mechanics research across the world.

Under the condition of deep mining, support design should consider not only the strength but the large deformation of the country rockmass (Hoek *et al.*, 1995). Hence, the yieldable bolt or energy-absorbing bolt was developed which can provide constant working resistance over the yield phase of the rockmass and a large extent of the elongations (He, 2009a). The energy-absorbing bolt has been investigated overseas for around 20 years and domestically ever since the early 1980s, which is used mainly in the underground support system such roadway large deformation in coal mining.

This section presents the findings achieved by the author at the State Key Laboratory for Geomechanics and Deep Underground Engineering (LGDUE) in China, regarding the innovative work for the development of a new kind of the energy-absorbing bolt with negative Poisson's Ratio effect (NPR), which means the material is not necking under tensile condition, instead, it keeps large constant resistance and performs large deformation. The NPR bolt, so-called Constant-Resistance and Large-Deformation bolt (CRLD bolt), has been tested both in the laboratory and field, including the static pull-out test and weight-falling test (dynamical impact test), as well as field blasting test in deep coal mine roadway support systems. A case on large-deformation support under engineering soft rock environment with NPR bolt/cable in Shajihai coal mine in Xinjiang Province, China, is reported.

1.1 Background

Shajihai coal mine is located in the northwest coalfield in Xinjiang Province, China. The mining area is 68.26 square kilometers, with proved coal reserves of around 1.817 billion tons. The coal seam belongs to the Mesozoic group and the trend is NE-SW with the dip angle 7 degrees to 28 degrees to the southeast. The mine field development employed two inclined shafts and one vertical shaft, and the development was designed in three levels. The first level is undertaking and the excavation of the roadway is mainly in mudstone, sandy mudstone and fine mudstone layer, these rock layers are in poor diagenesis with low intensity, and the softening coefficient is from 0.03 to 0.44, belonging to soft rocks.

Take the +550 m level main roadway as the example. The roadway is a straight wall and a half arched shape, and tunneling section size is 5900 mm × 5200 mm. The surrounding rock layers from swallow to deep are the fine sandstone, medium-grained sandstone, silty mudstone and muddy sandstone. Sandstone layer is of low strength with fracture developed structure and calcium cementation. The strata inclination is 13, and the silty mudstone layer is rich in ground water with water dropping phenomenon during the excavation.

A striking mechanics feature of the rockmass in this area is that the strength of sandstone, mudstone, or sandy mudstone is lower than the coal, and the aquifer in the mining area is 110 meters thick and with the increase of engineering the water in-flow increase quickly. All these surrounding rock conditions enhance the occurrence of the roof caving, floor heave and other typical large deformation failures during the roadway excavating. The recorded large deformation in the field survey shows that the roadway faced serious floor heave when the mudstone and silty mudstone on the floor exposed to water; during the transport gateway excavation, roof caving and side failure occurred, making the existing U steel frame obviously deformed at the top beam and both sides. Besides, varying degrees of large deformation failures occurred in the process, seriously threaten the safety of underground construction and the coal production.

1.2 Large deformation support design

In view of nonlinear large deformation failure, including roof subsidence, sidewall bulge and floor heave, occurred in mesozoic compound soft-rock of Shajihai coal

Figure 1 A photograph of surrounding rock mass in the field.

mine in Xinjiang, China, large deformation supporting system was designed, on the basis of field geological survey, hydrological properties test of soft rock, deformation mechanism analysis, numerical simulating and parameter optimizing. An initiative supporting system, based on the NPR bolt with constant resistance and large deformation features, is proposed with the key technologies for coupling support design.

1.2.1 Deformation mechanism

By the means of field investigation, numerical modeling analysis and laboratory tests, the large deformation and failure mechanisms of field rockmass in Shajihai Mine are studied.

In the micro scale, the expanding of clay minerals, molecular combination between clay and water, is one of the reasons that cause the sandstone, mudstone and many other types of rock containing clay. Test results show that the mineral composition of rocks in the Mesozoic Jurassic coal-bearing strata in Shajihai Mine, has a large content of illite and smectite (I / S) mixed layer clay mineral, around 40% to 60%. And I / S mixed layer is composed of many parallel crystal unit cells with open cracks and low connection force, which can absorb amount of water and break the micro-structures. When the shale rock with water, the water molecules become more easily enter the cell room, so that cell spacing increases, and drives the rapid expansion of the mineral particles and expansion of rock volume in the macro performance [3].

Laboratory tests have been performed as can be seen from Figure 2, and the sampling was performed in the airway. The mudstone samples, numbered N-1, N-2, N-3, N-4 and N-5 water absorption characteristics at certain temperature and humidity environment were studied.

As can be seen from Figure 2, the initial rock samples in the gaseous water absorption, adsorption capacity growth is large, but the growing rate of adsorption has a sharp decline as time increases. The water absorption rate keeps reducing until it reaches zero and keep stable. The relationship between the water absorption and time can be expressed as a negative exponential function:

$$w(\mathrm{t}) = a(1 - e^{-bt}) \tag{1}$$

Where: w is the water absorption of rock sample at time t, %; a, b are fitting parameters; t is the time the sample water absorption, h.

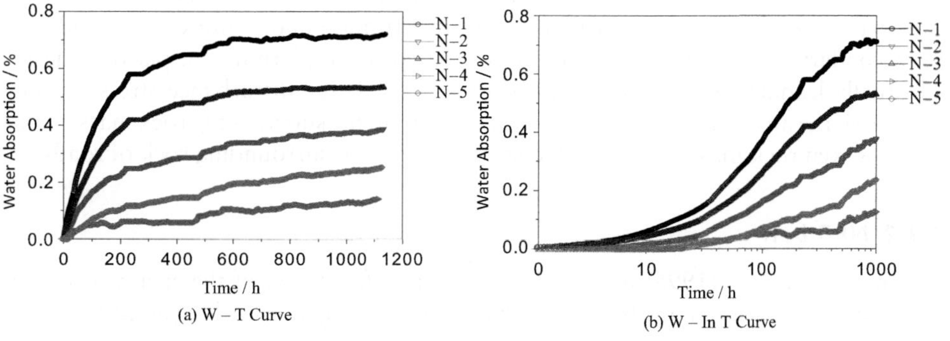

Figure 2 The characteristic curve of gaseous water adsorption.

Five absorption curves of rock samples for fitting analysis were performed and acquired each rock sample absorption process function. The mean value of fitting parameters are obtained that a = 0.564, b = 0.012. From Equation 1 can be seen, water adsorption of rock samples eventually reach equilibrium along with mutative growth over time, means a rock sample is saturated water absorption.

Thus, when the roadways excavated in the field, the roof, two sides and the floor rock masses are exposed to the air, and water molecules could cause rock swelling, weathering, softening, rock strength with increasing exposure time and reduced, thereby resulting in a weaker surrounding rock. In addition, this coal mine with sandy ground is rich in ground water, and measured data of ground inflow rate at the connection tunnel of auxiliary shaft can reach 60 m3 / h.

On the one hand, water causes a significant reduction of gateway rock intensity; on the other hand, gateway faces strong additional expansion stress caused by the rapid expansion of water-absorbing side rock. Non-uniform stress field, the additional expansion stress and ground stress, is superimposed, and plastic deformation is formed at the weakest part of the roadway with the formation of local loose rupture zone, while the zone loses all the support capacity. In the process of stress redistribution, high concentrate stress is applied to the destruction of the support body near the site, and it is difficult for traditional Poisson's Ratio material support system to adapt to the surrounding rock deformation, and thus induces the overall damage to the roadway.

For soft mudstone floor in Shajihai Coal Mine, no bottom support was applied for roadway support, and the open channel flow along the bottom cause intense rock swelling, plastic deformation of roadway and energy release. Severe floor heave, thereby, affect the stability of side rock and the roof. Thus, the rock softening and expansion is the main reason for the destruction of the roadway in the field undertakings.

In addition, the roadway deformation and failure are also related with the rock structure surface dislocation and excavation impact.

Geological exploration and test results show that the microstructure, joints and fractures in the rock masses easily become the weak interlayer under infiltration of groundwater that severely weakened the structure surface. Weak intercalations of low intensity can cause severe roof caving, floor heave and other large deformation failures.

Excavation impact is another important factor for the disturbance to soft rock roadways. During the first phase of construction in Shajihai mine, the connection of service shaft and roadway, the central pump house and many other substation chambers are facing large deformation problems during the construction. Excavation sequence had a great influence on the stress and stability of rock mass. In addition, the cavern excavation in the lateral side can produce the local concentration of deviatoric stress, and form the secondary stress redistribution and superposition effect, forming the plastic zone and eventually leading to local instability of rock mass. Large disturbance stress can also produce large amounts of induced joints and cracks in the surrounding rock mass, makes more broken rock mass, further reducing the stability of surrounding rock of roadway.

1.2.2 NPR bolt

McCreath & Kaiser (1995) proposed the principles for design of the energy-absorbing bolt and pointed out that the bolt should have at least elongations as long as 200–300 mm and have slip properties adapted to the deformation of the country rockmass.

Ansell (2005) performed the laboratory testing of a new type of energy absorbing rock bolt. In the early 1990s, Ortleep (1992) proposed the concept of energy absorbing supporting system (Ortlepp, 1992). Jaeger (1990) developed the energy absorbing bolt, *i.e.* cone bolt for the first time, and some of the developed cone bolt came out later (Simser, 2002). In recent years, with the increased demanding across the world for energy-absorbing bolts, various kinds of energy-absorbing bolts have been developed, such as Garford bolt, Durabar bolt, Yielding Secura bolt, and Roofex bolt (Varden *et al.*, 2007; Charette & Plouffe, 2007). These energy-absorbing bolts, however, realize elongations by improving the material property of the rod or using dot-frictional assembly in the bolt structures. Thus, their major deficiencies lie in the increasing or decreasing supporting resistance which cannot provide large deformation at constant resistance in the practical use.

For a continuous monitoring of the sliding force in a slope, the material with Negative Poisson's Ratio (NPR) effect is developed, which has been proved by large numbers of laboratory and field tests.

Figure 3 shows schematically the layout of CRLD bolt which is composed of the constant-resistance device, bolt rod, pallet, and nut. The constant-resistance device consists of a cone (a cone-like piston providing the constant resistance by its frictional force) and a sleeve pipe. The rod and the sleeve are threaded in order to lower their weight. The constant-resistance element has an outer diameter of 34 mm, sleeved at the end of the bolt rod, and the pallet and locking nut were screwed to the rod at the rear of the constant-resistant body. In the current design, the material strength of the sleeve is lower than that of the constant-resistant body, specially for protecting the constant-resistant body from destructed by the friction force with the slide-track sleeve due to its lower strength. The mechanical property of constant-resistance for the novel bolt will be deteriorated in the case of a damaged constant-resistant body.

The designed resistance of the CRLD bolt is around ninety percent of the yield strength of the bolt rock material to prevent the rod from undergoing yield deformation in the case of the external load exceeding the yield strength of the rod material. The end of the CRLD bolt was designed as a conical-shaped stirring ware which is suited for resin-anchored installation and cement-sand-pulp anchored installation schemes.

Excavations in the underground engineering will break the equilibrium of the original rockmass. Tensile stressed region (or plastic region) will exist in the surrounding rocks over the stress redistribution process. The degree of instability for the surrounding rocks will increase the looseness of the rock masses adjacent to the excavations and

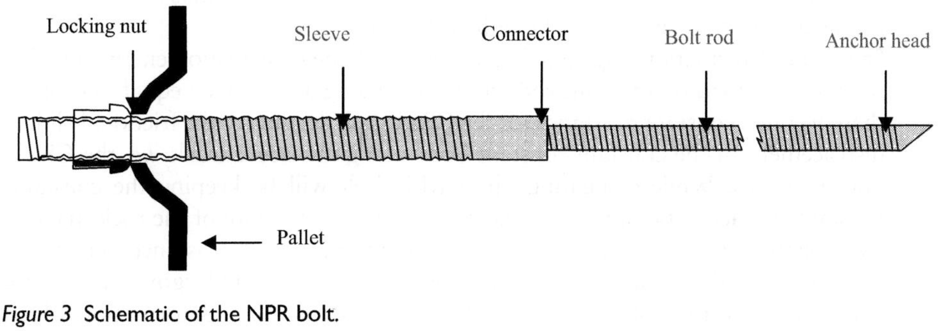

Figure 3 Schematic of the NPR bolt.

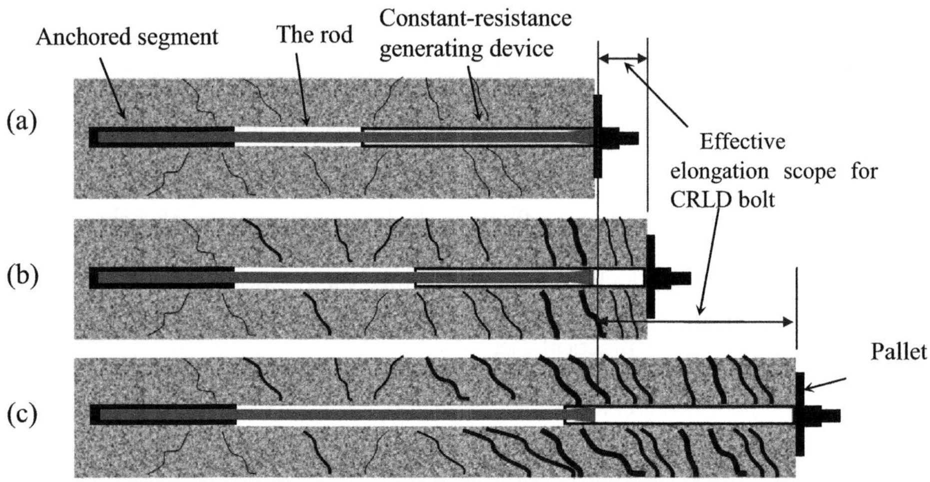

Figure 4 Working principle of the CRLD BOLT in supporting country rockmass with large deformation in deep ground. (a) elastic deformation of the working CRLD BOLT and country rockmass; (b) elongation (structural deformation) stage of the working CRLD BOLT, and (c) ultimate deformation stage of the CRLD BOLT.

the trend to the instability would be accelerated adding up the influence of the tectonic formations. Therefore, rocks masses adjacent to the excavations have to be supported with suitable supporting schemes so that the supporting coupled with rocks will form a sound and stable load-carrying circle, stabilizing the excavated roadway tunnels (He & Yuan, 2004).

Figure 4 shows schematically the working principle of the CRLD bolt in supporting the country rockmass with large deformation in deep ground. In general, the CRLD bolt support has three different work stages, that is:

(1) Elastic deformation stage (Figure 4a). The deformation energy of the surrounding rocks could be converted to the bolt rod in the bolt assembly through the pallet (the outer anchorage segment) and inner anchorage segment. In the case of a relative deformation for the surrounding rocks and the axial force loaded by the rock deformation less than the rated constant resistance for the CRLD bolt, the CRLD bolt will not elongate by the displacement of the constant-resistance element, but resist the deformation and failure of the rock mass solely relying on the elastic deformation of the bolt rod itself.
(2) Structural deformation stage (see Figure 4b). With the deformation energy building up, the axial load on the built rod will be increasing and may be equal to or large than the rated constant resistance for the CRLD bolt, leading to the frictional-sliding displacement of the constant-resistant body along the sleeve track, *i.e.* the CRLD bolt elongates. While elongating, the CRLD bolt will be keeping the constant-resistant characteristics and resist the deformation and failure of the rock mass by its elongation, *i.e.* the structural deformation of the constant-resistance element.
(3) The ultimate deformation state (see Figure 4c). After undergone the elastic deformation of the bolt rod itself at the first state and large deformation which

equals to the elongation of the constant-resistance element, the deformation energy for the rock mass in abutment to the roadway excavations has fully been released. In this case, the external axial load will be smaller than the rated constant resistance, and the constant-resistant body will stop sliding due to the fractional drags. Therefore, the surrounding rock mass for the roadway excavation has been stabilized once again.

Consequently, under the conditions that the surrounding rocks deformed with large extent gradually or instantaneously, the CRLD bolt could absorb the deformation energy of the rock mass, which will release the energy stored in the surrounding rocks. Over the structural deformation stage, this novel bolt is still able to elongate steadily while keeping its working resistance constant in responding to the external loading. Thus, the stabilization could be realized for the CRLD bolt supported rock masses adjacent to the excavations, mitigating the potential disasters such as roof fall, rock collapse, slabbing and splitting, and floor heaving.

1.2.3 Supporting system design

Based on the analysis of large roadway deformation mechanism in Shajihai mine, the support system is proposed to keep constant high resistance to the surrounding rock and to perform large deformation with it. With this idea, the principles of this rock mass stability control system are as follows:

Constant resistance and large deformation system is employed. The developed CRLD bolts take place of traditional bolts in the supporting system and experimental data suggest that constant resistance constant resistance from 120KN to 130KN and the largest deformation from 600 mm to 1000 mm. When the stress from the surrounding rock is smaller than bolt's constant resistance, the bolts support the roadway in stable state as normal bolts. If the stress continues to increase and reach the CR value, the device will increase the deformation with the expansion and plastic deformation off surrounding rock, where the energy and high stress has been fully released. At the same time, due to higher CRLD Bolt's resistance, effectively limiting the excessive damage of surrounding rock deformation. When the confining pressure reduced less than the CR value, the bolts becomes stable again and the surrounding rock – support system in equilibrium state, achieves the overall stability of the roadway.

Bolt and net coupling support is adapted to unite the single CRLD support unit. It is common that large plastic deformation, interlayer weak structure surface faulting and deformation caused by some weak positions between soft surrounding rock and support system are usually uncoordinated (difference deformation, asymmetric deformation, etc.), which easily cause the whole supporting system failed. According to net – anchor coupling support theory, by the first network anchor coupling active supporting shallow surrounding rock, to maximize the protection of their own carrying capacity of roadway surrounding rock through the second anchor coupling support fully mobilize the deep tunnel rock strength, so shallow to deep surrounding rock stress shift focus to expand, to stress uniform load – supporting the integration of surrounding rock and supporting structure coupled to the purpose, and to achieve a stable roadway by coupling support.

The corner support is another significant measurement in Shajihai coal mine roadways. As the failure mechanism analyzed, the weak layer and strong expansion of soft rock can easily form in the corner of roadways when exposed to water during the excavation, causing floor heave, softening, and large deformation failure of the roadway, thereby, affecting the stability of the two sides of roadway and the roof. Corner support can effectively cut off on both sides of the roadway plastic slip lines, weaken the floor heave, and keep the whole roadway stable. Thus, in the design and construction support, we should focus on strengthening the corner.

1.2.4 Numerical simulating and supporting design optimizing

Numerical modeling analysis of the roadway construction is performed for supporting system design optimizing.

For the large plastic deformation, load exerted influence on the process of rock deformation of underground roadways, same mechanics countermeasures but different construction sequences can reach different mechanical responses and roadway conditions, so it must emphasis on the comparison of different construction process analysis. FLAC3D is employed in the following study on the mechanical response during support system construction with three-dimensional finite difference numerical procedures. Comparative analyses are performed for the displacement and destruction of farm field distribution characteristics, to determine the best process of construction and a better supporting effect.

The +550 m level main roadway was taken as the example for the study. The roadway section size is tunneling section size 5900 mm × 5200 mm in a straight wall and a half arched shape. The surrounding rock layers from swallow to deep are fine sandstone, medium-grained sandstone, silty mudstone and muddy sandstone as introduced above. The entire range of three – dimensional numerical model for calculation, length × width × height = 20 m × 40 m × 40 m, was divided into 82,264 units and 88,335 nodes. The side horizontal movement of the model is restricted, the fixed bottom, and gravity stress of the overlying rock applying on the top surface boundary is 8.3MPa. In order to simulate the weak surrounding rock with water softening, large deformation and failure of plastic flow characteristics, in +550 level main roadway, the large deformation strain softening model and large deformation calculation method are used. The construction of the numerical simulation model is shown in Figure 5, the physical and mechanical parameters used in this calculation are shown in Table 1.

For the study on different control effect of supporting system in the construction process during roadway excavation, five options of excavation and support construction schemes are proposed in order to optimize the construction process, where excavation is up and down stairs excavation method, and the support in forms of anchor install, bolt and net, corner bolts, and shotcrete.

Respectively, the numerical simulation of each scheme were performed, in each case, the FISH language command stream was set extra 400 steps calculation after previous excavating or supporting step, and the deformation of roof, floor and two sides of roadway was monitored. The maximum deformation values were shown in Table 2. The horizontal and vertical displacement field of the model was shown in Figure 6 separately.

From table 2 and figure 6, it can be seen that the deformation of surrounding rock in scheme I without support reached a maximum of 376 mm, and scheme III

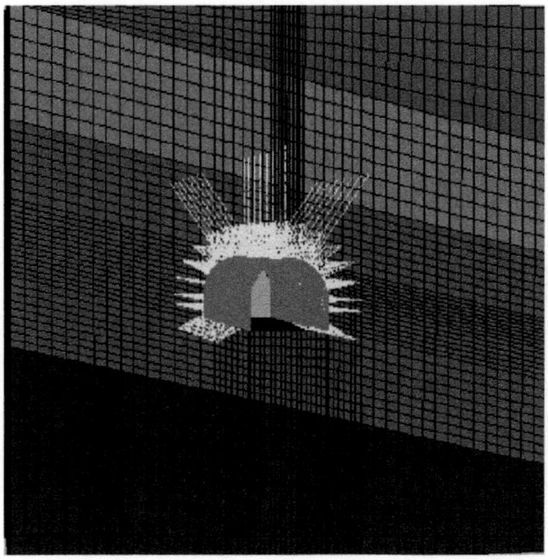

Figure 5 Numerical calculation model.

Table 1 Physical and mechanical parameters of rock.

Rock Type	*Symbol*	*E / GPa*	*Poisson's ratio*	*c / MPa*	*Φ / ( )*	*Tensile Strength /MPa*
Fine Sandstone	■	2.9	0.35	3.0	28	0.3
Medium-grained Sandstone	■	2.11	0.37	3.2	32	0.5
Silty Mudstone	■	1.7	0.22	1	24	0.1
Muddy Sandstone	■	1.74	0.31	1.5	30	0.2

Table 2 The deformation values of different schemes.

Scheme	*Maximum Displacement*			
	Roof /mm	*Floor/mm*	*Left Side/mm*	*Right Side/mm*
I	307	143	376	348
II	80	24	270	269
III	218	234	328	328
IV	105	37	185	180
V	31	22	146	142

in the roadway surrounding rock deformation of a maximum of 328 mm. Both of them have entered the nonlinear deformation failure stage. For scheme V and IV, the maximum roadway side shrank and roof subsidence reached 185 mm and 105 mm respectively, comparing to the scheme III, reduced by 43% and 52%, respectively, significantly improved the support effect. The reasons can be found that, compared with other schemes, plan IV and V performed the sprayed concrete support right after the excavation, which avoiding the surrounding rock exposing to air and water, greatly reducing the swelling large expansion deformation failure caused by clay mineral.

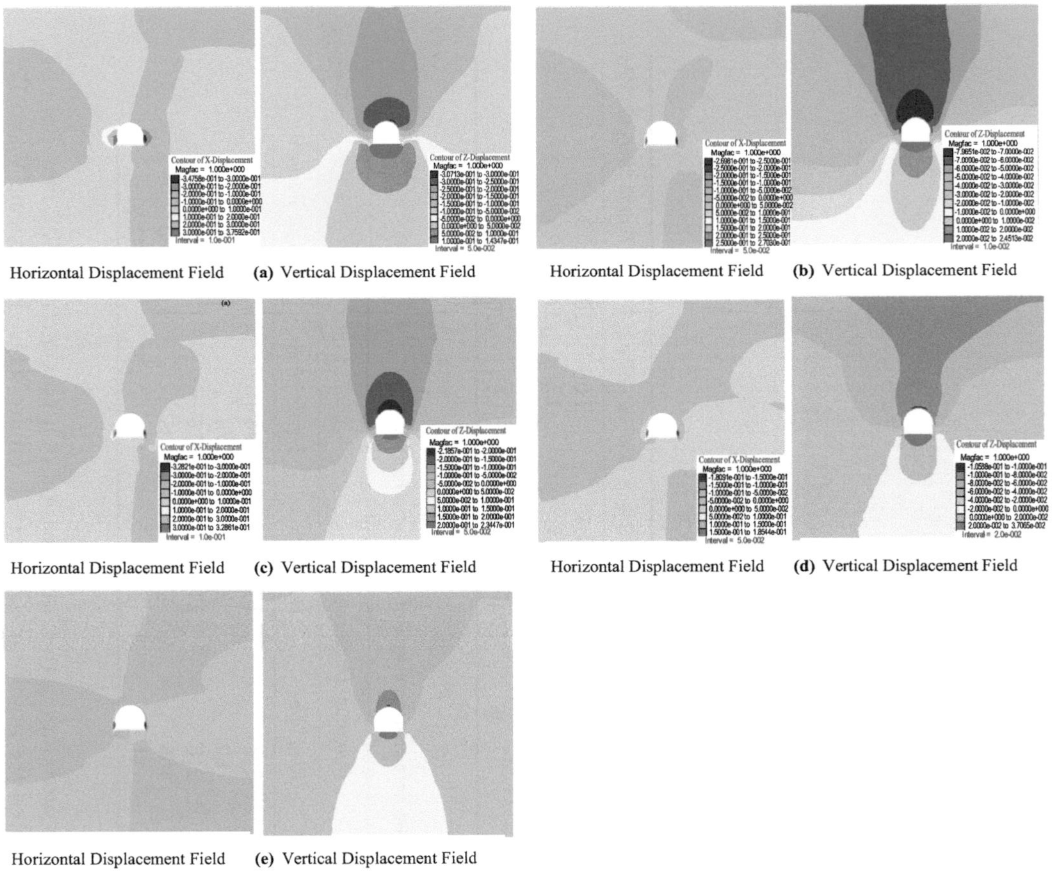

Figure 6 The displacement field distribution of different schemes; (a) Scheme I: Up and down stairs excavation without support; (b) Scheme II: Up stairs excavation → Bolt & net → Anchor → Shotcrete → Down stair excavation → Bolt & net → Shotcrete → Corner Bolt; (c) Scheme III: Up stairs excavation → Anchor → Bolt & net → Shotcrete → Down stair excavation → Bolt & net → Shotcrete → Corner Bolt; (d) Scheme IV: Up stairs excavation → Shotcrete → Anchor → Bolt & net → Down stair excavation → Shotcrete → Bolt & net → Corner Bolt; (e) Scheme V: Up stairs excavation → Shotcrete → Bolt & net→ Anchor→ Down stair excavation → Shotcrete → Bolt & net → Corner Bolt.

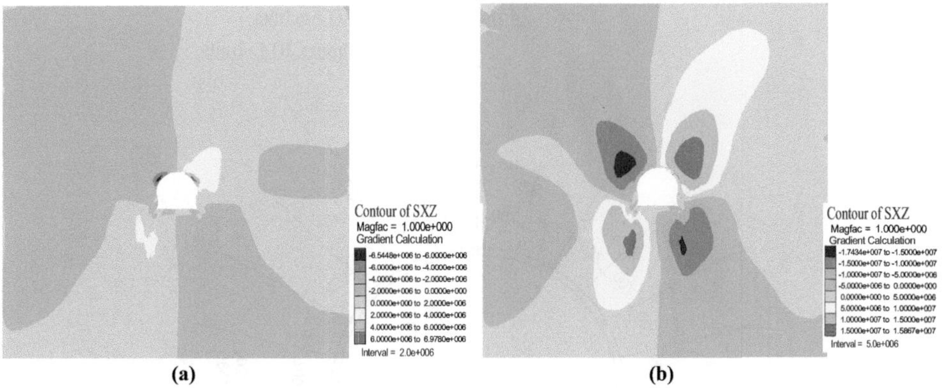

Figure 7 Distribution of shear stress field: (a) Scheme II: Up stairs excavation → Bolt & net → Anchor → Shotcrete → Down stair excavation → Bolt & net → Shotcrete → Corner Bolt; (b) Scheme III: Up stairs excavation → Anchor → Bolt & net → Shotcrete → Down stair excavation → Bolt & net → Shotcrete → Corner Bolt.

The distributions of shear stress field of Scheme II and III are showing separately in Figure 7. It can be seen that the prompt bolt-net support is effective in reducing the shear stress and deformation if it was preformed immediately after the excavation. Simulating results show that the maximum amount of shrinkage of roadway wall reduced from 656 mm to 539 mm by 18%, and the deformation between roof and floor reduced from 452 mm to 104 mm by 77%, the maximum stress concentration decreased from 1.67 (17.4 / 10.4) to 1.48 (6.97 / 4.69), by 11.4%, improving the stability of roadway.

According to the construction process of scheme II and III, a timely bolt and net support after the roadway excavation can effectively control the excessive deformation of shallow surrounding rock mass, and in the meantime, the CRLD Bolts allows the rock deformation in a certain extent, which benefits the whole supporting system and restrict the plastic zone of soft rock in a certain range. The following anchors in the support system can fully mobilize the deep rock mass strength, reducing the influence of stress concentration in overlying rock layers and promoting the stress field and strain field of surrounding rock to homogenization.

From the above analysis, Scheme V is the best construction and supporting process for the roadway excavation in Shajihai Mine.

1.3 Results

With the application of CRLD Bolts and numerical simulating optimizing, the roadway support parameters in the field are designed, showing in Figure 8.

The construction method and supporting system was applied in the under-construction cross-cut roadway, and the field monitoring station was set. The layout of displacement monitoring points was shown in Figure 9. Field monitoring has been performed for 5 months from September, 2011 and the typical monitoring curves for roadway deformation versus time were displayed in Figure 10.

As can be seen from Figure 10, roadway was in good condition and the rock deformation was effectively controlled, the maximum amount of two side-walls shrink

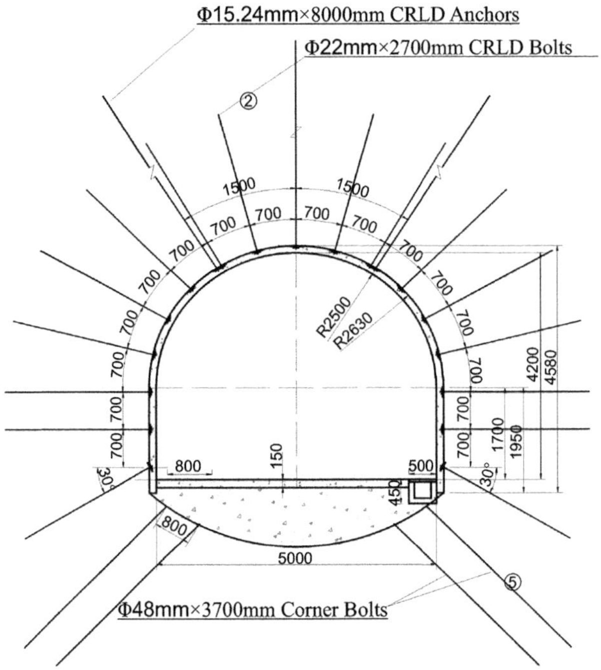

Figure 8 Roadway cross section of supporting design: (1) CRLD Anchor, spacing 1500 mm × 2100 mm; (2) CRLD Bolt, spacing 700 mm × 700 mm; (3) Wire netting, ϕ6.5 mm, meshing 100 × 100 mm, size 1700 × 900 mm for overlap joint; (4) Shotcrete, C20 concrete with water-proofing agent; (5) Corner Bolt, concrete filled steel tubular, ϕ48 mm, spacing 800 mm × 800 mm.

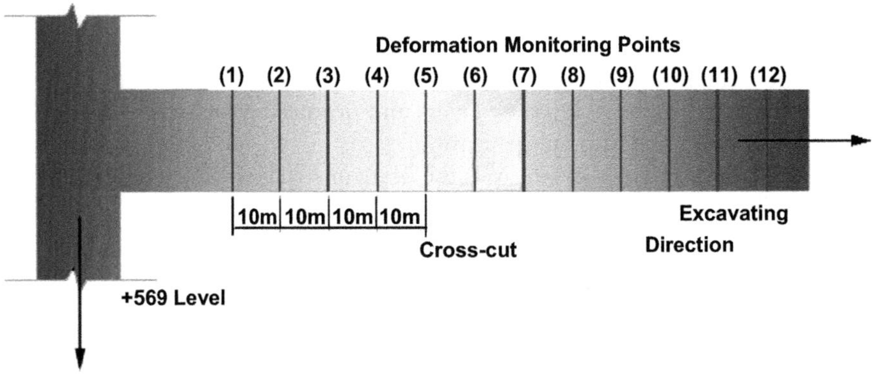

Figure 9 Layout of displacement monitoring station.

was 56 mm, the deformation between roof and floor was 48 mm. The early surrounding rock deformation was not obvious due to the surrounding pressure was lower than the constant resistance value of support materials; from 20 to 70 days during the excavation, the energy release and construction disturbance drove the deformation increase in both

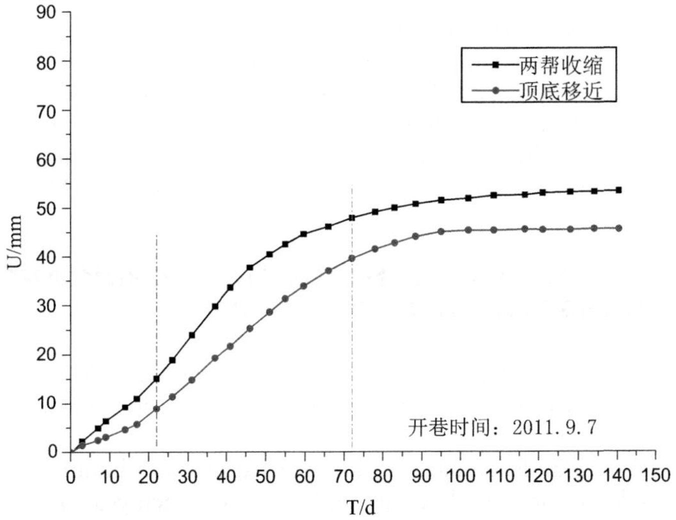

Figure 10 Typical monitoring curve on displacement-time of the roadway.

horizontal and vertical directions; finally, the deformation tended to be stable within the allowable range, and met the requires, indicating that this large deformation support system proved to be effective and reliable in Mesozoic Era soft rock roadway control.

1.4 Summary

Large deformation problems are of great significance with the increasing of underground projects and manual activities, and the control measures for these rock mechanics disasters, such as rock supports under large deformation have been the worldwide focus. In this chapter, an underground coal mine facing soft rock deformation problem in China is introduced, and the main conclusions are as following:

The deformation mechanism of soft rock roadway was studied by investigating and laboratory tests. The expanding of clay minerals when exposed to water, formation of weak interlayer under infiltration of groundwater, rock softening and excavating disturbance are the main reasons for Shajihai mine roadway deformation failure during construction. Thus, the water control and effective large deformation support system is urgently needed.

The new material, CRLD bolt with constant resistance and large deformation, was employed, and its unique mechanical properties achieved the energy absorbing ability, which will release the energy stored in the surrounding rocks. The CRLD Bolts allows the rock deformation in a certain extent and release the energy and the stabilization is realized for the CRLD bolt supported rock masses adjacent to the excavations, mitigating the potential disasters such as roof caving, slabbing and splitting, and floor heaving.

Coupled support measurements are employed, the bolt and net support, anchor support, shotcrete and corner bolt. And the construction and support process is optimized by numerical modeling method before the support system parameters finally determined.

The research results are applied in the +569 cross cut roadway construction in Shajihai coal mine. Deformation monitoring has been performed in the field, and the results showed that the maximum amount of two side shrink was 56 mm, and the maximum amount of roof and floor about 48 mm within the allowable range for normal working requirements. Besides, this large deformation support technology has been proved to be effective and reliable in the underground mining practice and Mesozoic soft rock large deformation control in Xinjiang Shajihai Mining.

2 REVIEW ON LANDSLIDE CONTROL, MONITORING AND FOREWARNING WITH NPR BOLT/CABLE

Landslides are a major geological disaster and the hazard is ranked only after earthquakes and volcanic eruptions. China is one of the countries in the world with the most widely distributed landslides and suffering the heaviest damage. According to the incomplete statistics, landslide hazards have caused death of hundreds of people and direct economic losses of several hundred million RMB each year. For coal mining in China, landslide is also one of the most serious disasters in open-pit coal mines. In the last century, open-pit coal mines were rapidly developed with the growing demands for energy resources, especially in Shanxi Province, Inner Mongolia and many other mining areas with good conditions at that era. But with the mining depth increase, landslides, especially landslide happened in the slopes under mining operation, has been the major threat to the normal production. Thus development of new techniques for forecasting and prediction of the landslides is imperative and of great importance.

Based on the analyses of the surface displacement landslide monitoring techniques, it was proved that displacement of the rock masses is the only necessary condition for a landslide event, and new landslide monitoring and forecasting technique should be developed for combating the increasing threats posed by the various landslides in natural and man-made slopes.

New landslide control idea incorporating the reinforcement, sliding force monitoring and forecasting was proposed by He (2009) based on the fact that the necessary and sufficient condition for a landslide could be derived from the balance of the forces acting on the slopes. The new idea was realized by the development of new material for landslide control, *i.e.* NPR cable with the capability to elongate as long as 2 m at high-constant resistance which is well suited for the slope reinforcement under large rock deformation.

A landslide control, monitoring and forewarning system with NPR cable has been developed and applied successfully in 245 landslide site, including open pit iron mine, open pit mine, underground gold mine, slope monitoring along the gas pipeline and highways, over 12 provinces in China. In this section, the integrated application of landslide control, monitoring and forewarning based on sliding force monitoring techniques in Pingzhuang Coal Mine, China, will be presented.

2.1 Background

Pingzhuang west open pit coal mine, affiliated to the Pingzhuang coal mining group, is located in Pingzhuang town southwest of Chifeng city in Inner Mongolia Autonomous

Region in Western China. The open pit mine was constructed in August, 1958 and put into production in 1965 with an annual output of 15,000 tons. At the beginning of this century, landslides became the major threat to its production. A total of 66 recorded landslides occurred in the mine, among which are 33 events of landslides that occurred in the slopes under mining.

The open-pit mine is located at the edge of a low mountainous area and is higher in the west and lower in the east with geology consisting of mudstone and sandstone, it intersects to the slope with a dip angle of 26 degrees into the slope, and a fault (F3) plane with a dip angle of 33 degrees parallel to the slope face. A bird's-eye view is shown in Figure 11a and the geological map showing the structural geology and

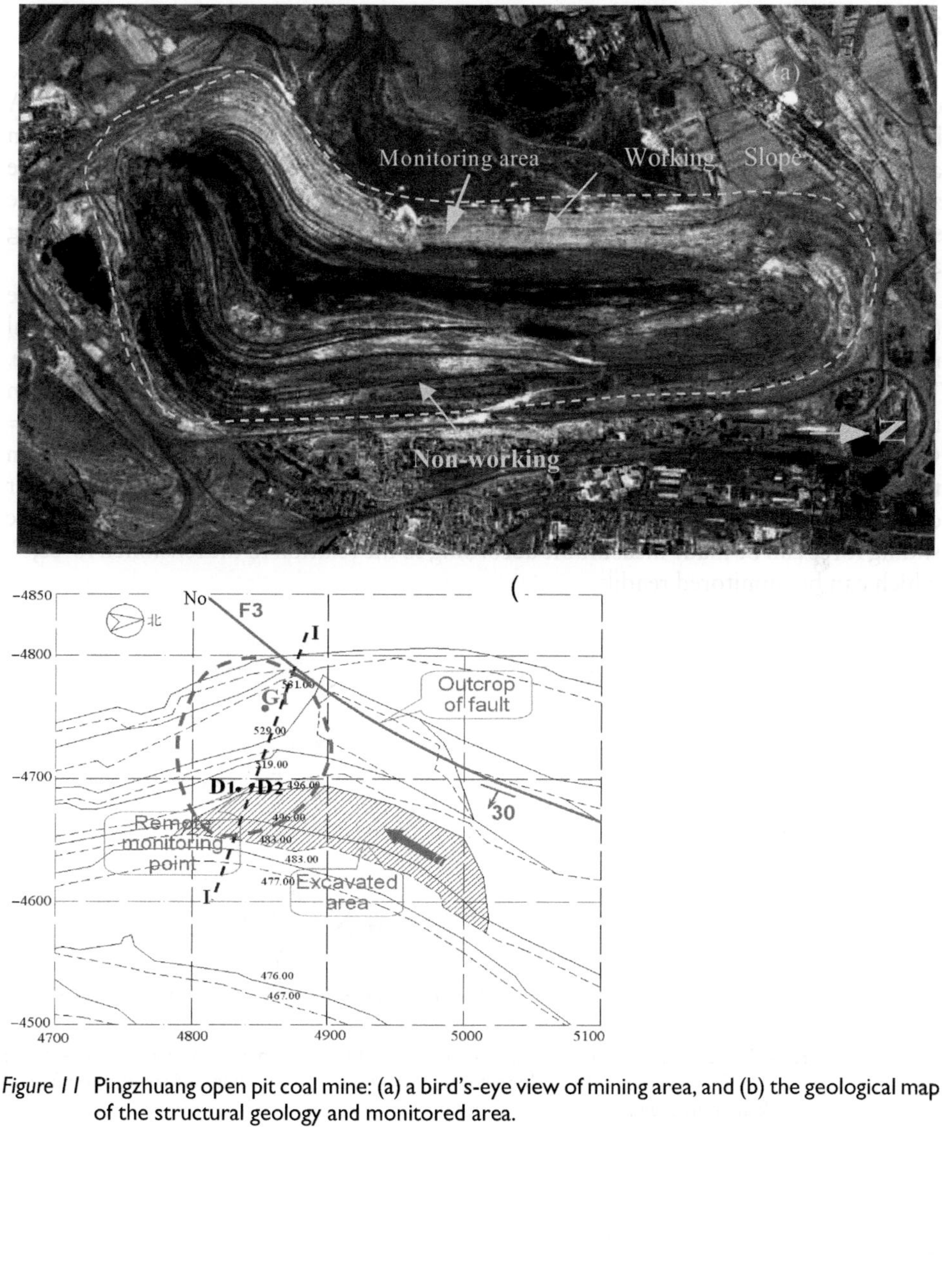

Figure 11 Pingzhuang open pit coal mine: (a) a bird's-eye view of mining area, and (b) the geological map of the structural geology and monitored area.

monitored area is Figure 11b, and D1 and D2, which is adjacent to the F3 fault had undergone landslides many times. The final height of the slope at present is about 358 m. The height of each mining bench is 35 m and the average angle of the slope on the western side is 65 degrees. The eastern side slope, the non-working slope, is a stable slope with a dip angle of around 25 degrees.

2.2 Landslide control, monitoring and forewarning system

To combat the landslide hazards, the new landslide control, monitoring and forewarning technology, developed by the State Key Laboratory for Geo-mechanics and Deep Underground Engineering, Beijing, China, is employed.

2.2.1 Working principles

The movement of the sliding mass, a mechanical system by nature, depends upon the changes of the balance state between the sliding force and sliding-resistant force. A landslide is a dynamical process including stable deformation, unstable de-formation and catastrophic failure. Monitoring the evolution features of the acting force in the sliding surface of a slope during a landslide will obtain the database for establishing the sufficient and necessary conditions for the evaluation of the slope stability and giving a high accuracy in forecasting the landslide disasters.

However, for the sliding forces, as the natural mechanics system is immeasurable, He (2009) proposed an idea of embedding an artificially measurable system into the natural mechanics system and the combined system makes the sliding force measurable. The approach for the proposed idea is schematically illustrated in Figure 12a. It is seen that the shearing forces along the sliding surface include the natural sliding force $T_1 = G \sin \alpha$ and man-made shear force ΔT which is the component force decomposed from the tensioned anchorage force P. The resistance, T_2, is also associated with the other component force of the tensioned force P. The natural forces, T_1 and T_2, therefore, the sliding force can be made measurable according to the relation to the tensioned force, P, which can be monitored readily.

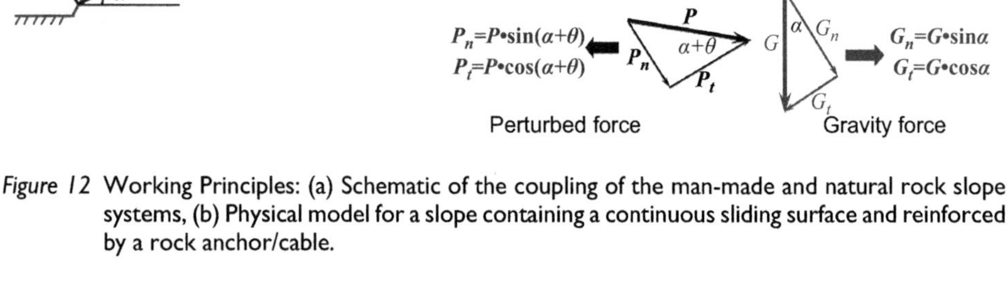

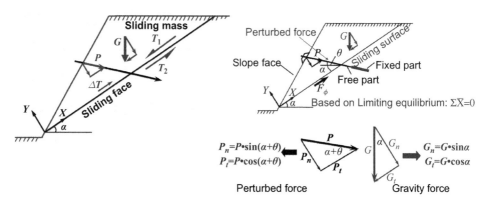

Figure 12 Working Principles: (a) Schematic of the coupling of the man-made and natural rock slope systems, (b) Physical model for a slope containing a continuous sliding surface and reinforced by a rock anchor/cable.

In fact, rock anchors/cables can be used as the artificial system embedded into the natural and excavated rock slopes to measure this important parameter, as shown in Figure 12b, thus a measurable artificial system was obtained. Based on this system, we can obtain the sliding force calculated by the measured perturbation force. The landslide monitoring principle shown in this figure is to use "puncture perturbation" technology, and make the rock anchor/cable through the landslide surface, meanwhile loading a small pre-stressing force on the cable, then this force has participated in the mechanical system of the landslide. Therefore, the functional relationship between the pre-stressing force and the sliding force can be established.

The mathematical model that was obtained according the model of the sliding system and the sliding force can be written as:

$$F_s = k_1 P + k_2 \tag{1}$$

And the parameters k_1 and k_2 are:

$$\begin{cases} k_1 = \cos(\alpha + \theta) + \sin(\alpha + \theta) \cdot \tan \overline{\phi} \\ k_2 = G \cdot \cos\alpha \cdot \tan \overline{\phi} + c \cdot l \end{cases} \tag{2}$$

Meanings of the parameters in Equation 1 are: the lithology influence coefficient including $\overline{\phi}$ and c, $\overline{\phi}$ is the internal friction angle, c is the cohesion. Geometry influence coefficient including l, α and l, l is the length of sliding surface, α is the angle between the sliding surface and the horizontal axis, θ is the anchor inclined angle below the horizontal. According to Equation 1, the immeasurability force G_t is a function of P, which is a measurable quantity. Thus the force G_t can be calculated according to Equation 1.

It was proved that the artificial mechanics system can be measured easily, but the problem is that the traditional Poisson's material of cables or anchors has limited elongation at break, and the tensioned force, P, would vanish when the landslide occurs. So a new material is urgently needed.

2.2.2 NPR bolt/anchor

For a continuous monitoring of the sliding force in a slope, the material with Negative Poisson's Ratio (NPR) effect is developed, which has been proved by large numbers of laboratory and field tests.

The unique structure allows the NPR bolt/cable to have a capacity of very large elongation and keep constant resistance if the appropriate materials for the sleeve and cone are chosen. Thus the NPR bolt/cable is also known as constant-resistance and large-deformation (CRLD) bolt/cable. NPR bolt has been applied widely in underground mining and civil engineering, *i.e.* as mentioned in the last section. In landslide engineering, NPR cable is generally applied. Figure 13 schematically shows the deformation mechanism of the NPR cable; Figure 13a depicts the original working state for the cable and Figure 13b illustrates the deformation state for the cable at which the cone body slides from the left side to the face plate side.

There are many specifications for the most widely used NPR cables in landslide control engineering projects, varying in constant force and the largest deformation value. In this project, the NPR cable was chosen based on the field investigation, with a constant resistance (CR) of 850kN and largest deformation (LD) of 2 m.

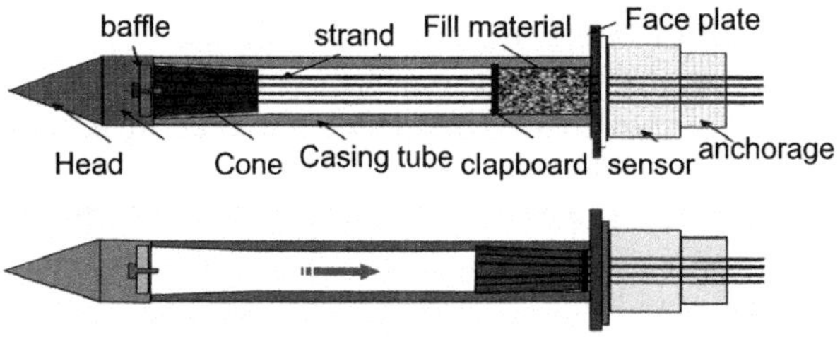

Figure 13 Schematic of the HE cable: (a) original state, and (b) sliding of the cone under external load.

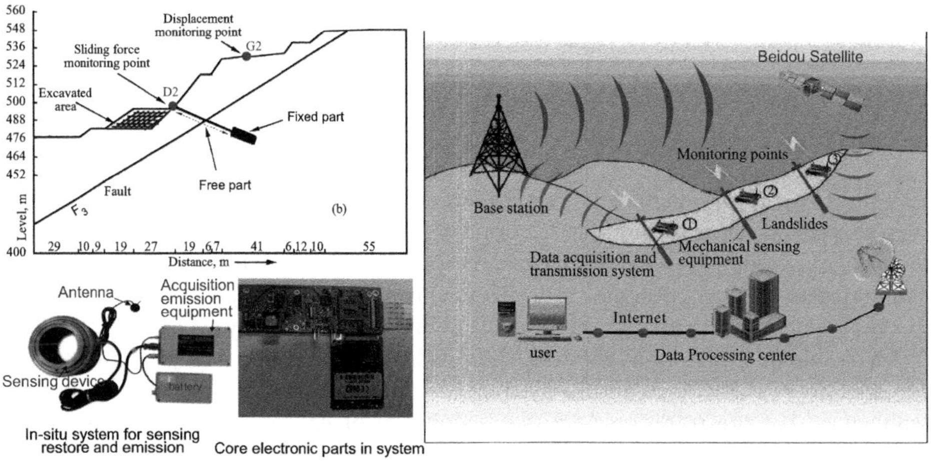

Figure 14 Remote-sensing based force measurement system for control and motoring of the landslides: (a) physical model of pre-stressed NPR cable sliding-resistant system; (b) Modern communication devices in the prevention of sliding body monitoring and forewarning system, and (c) Beidou satellite remote sensing system for monitoring the sliding force.

2.2.3 System design

With the application of NPR cable, the landslide control, monitoring and forewarning system is designed. Figure 14a shows the model of this system with NPR cable for stabilizing the slope and in-situ monitoring the sliding forces by the use of the sensor device installed between the face plate and fastening nut indicated at point D2. Modern communication technologies were employed for the force data collection, transformation and emission as shown in Figure 14b. The collected data transforms by the Beidou satellite based remote sensing system for the landslide monitoring, and a diagram of the overall process is shown in Figure 14c. The measured data from D2 are sensed by the receiver in the Beidou satellite and then transmitted to the indoor monitoring center. The acquired sliding force data are processed by using the computer system based on mathematical models described above, involving the resultant sliding force model in Equation 1 and the associated necessary and sufficient condition for the slope stability, to reveal the whole process for slope failure from stability, limit equilibrium, to the final sliding event.

Figure 15 Monitored area in Pingzhuan open pit coal mine: A photograph showing the monitored points on the slope.

Thus the landslide control, monitoring and forewarning system in Pingzhuang open-pit mine was contracted and undertaken from 2006. Figure 15 shows the monitored slope points, among them the monitoring points D1 and D2 are adjacent to the F3 fault mentioned in Figure 11b.

2.3 Results

The landslide control, monitoring and forewarning system with NPR cable has been put into use since 2006, and tens of thousands of precious data have been obtained from the in-situ monitoring points.

During the field monitoring, there were several success cases for landslide forewarning and control. One of them happened in the area around Point D2. Figure 16 shows evaluation of the sliding force (gray line) against time plotted based on the data remotely sensed by using the in-situ monitoring device installed on the NPR cable at Point D2 and for comparison, the vertical displacement (black line) against time plotted based on the data remotely sensed by other devices. The landslide monitoring began on November 8, 2006 and seven points, marked A – G, are the time instants at which the forewarning message were signaled. These seven points divided the curves into six parts with some early warning modes illustrated in Figure 16. Their characteristics could be summarized as follows.

It is seen from Figure 16 that the curve of the sliding force was almost constant at the initial segment around point A during which the slope is stable. In the second half of the AB segment, the sliding force gradually increased at first and increased rapidly when nearing point B which is the first forecasting point. During segment BC, the sliding force also increased slowly in the first half and rapidly neared point C which is the second forecasting point. However, at time instants B and C, no cracks were found on the crest of the slope. Figure 17a shows the photograph corresponding to time instant C. In segment CD, variation of the sliding force repeated the mode in BC. As the force level was relatively high, point D was taken as the first waning point indicating that partial slope failure occurred. Figure 17b is taken at time instant D showing the sparsely distributed cracks on the crest of the slope at the moment.

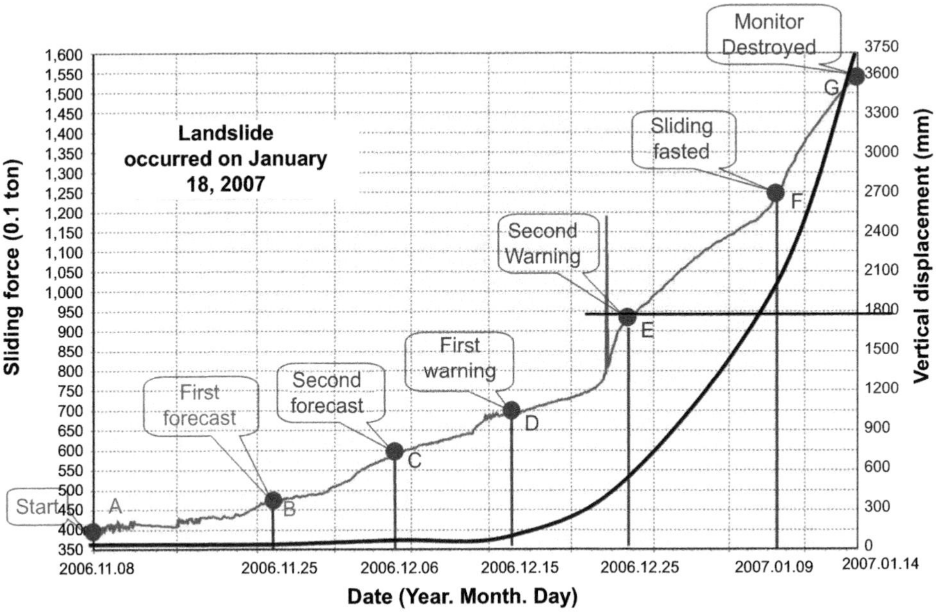

Figure 16 Variation of the sliding force (gray line) and vertical displacement (black line) of sliding block monitored in situ and remotely sensed by the Beidou satellite system.

In segment DE, the sliding force increased steadily over the greater part; and when approaching point E, the sliding force increased sharply with a single peak representing the coalescence of the cracks at time instant E as shown in Figure 17c. The opened cracks averaged 80 mm in width and the widest is as much as 500 mm. The monitored vertical displacement at point G2 at this moment was as much as 1700 mm as shown in the black line in Figure 16. At time instant E, the second warning message was signaled and the landslide occurred shortly after the warning was sent to the concerning parties. Because a timely warning was sent, severe injury and damage to the personnel and equipment was avoided. Figure 17d is a photograph taken on the crest of the slope showing the slope failure after point E. The average width of the crack was 300 mm and dropping distance was 330 mm between the two crack surfaces.

Figure 17e was taken on the crest of the slope corresponding to point G on the sliding force curve (Figure 16). It is seen from Figure 17e that the displacement of the sliding block was so large that the sensor device installed on the cable was destroyed. The width of the crack was as wide as 500 mm with a dropping variance of 860 mm between two side walls of the crack. Figure 17f shows the scene after the landslide. It is noted that (see Figure 16) the displacement (black line) is a monotonically increasing curve over which no abnormality was found; while the sliding force is more sensitive than the displacement to the stability state of the sliding block with the abnormal information such as the evolution modes and the singles peaks being the forewarning message. Sliding force based landslide forewarning is, therefore, more reliable and precise over the other currently used landslide prediction techniques.

Figure 17 Photographs taken showing fractures on the crest of the monitored slope: (a) corresponding to time instant C, and (b) corresponding to time instant D; (c) corresponding to time instant E, and (d) corresponding to time instant F; (e) corresponding to time instant G, and (f) scene after landslide.

2.4 Summary

Landslide hazards in the open pit mines in China are briefly reviewed and the landslide control, monitoring and forewarning system is introduced with a field application case of Pingzhuang open-pit mine in China. Main conclusions are as follows:

The principle of this technology monitoring the relative motion for sliding body and sliding bed was proposed. The advantage of this monitoring technique lies in force measurement, *i.e.* measuring the difference between the sliding force and the resistance force, which is the function of the many factors such as raining, underground water level, excavation and blasting etc. This novel force-monitoring technique, instead of dealing with many factors with complicated interactions, makes the prediction explicit and definite by using a single parameter, *i.e.* the sliding force measured during the sliding processes.

The new material, the NPR cable with constant resistance and large deformation (CRLD), was employed, and its unique mechanical properties achieved the reinforcement, controlling and prevention, monitoring, and forewarning in landslide control. It can provide enough deformation by the structural and material deformation to absorb the deformation energy triggered by rock mass sliding.

Based on the NPR cable, landslide control, monitoring forewarning system, as well as the software and hardware for the new landslide control technique, were developed incorporating multiple functions in one system including the reinforcement, sliding force monitoring, remote sensing and forecasting. It has been applied in open pit mines, an underground gold mine, slope monitoring along a gas pipeline, and so on, and has achieved remarkable economic benefits and social benefits. In the case of Pingzhuang open-pit mine, the system was applied and successfully forewarned the landslide disasters, giving advanced warning signals before the landslide according to the monitoring curve, which prevented casualties and property loss.

REFERENCES

Ansell, A. (2005) Laboratory testing of a new type of energy absorbing rock bolt. Tunnelling and Underground Space Technology, 20: 291–330.

Charette, F., Plouffe, M. (2007) Roofex-results of laboratory testing of a new concept of yieldable tendon. In: Potvin Y. (Ed.), Deep mining' 07. Perth: Australian Centre for Geomechanics, pp. 395–404.

Da Wa, Ming Debayier. (2008) Geological characteristics and processing for swelling soft rock of an engineering in Xinjiang. West-China Exploration Engineering, 11: 55–58. (In Chinese).

Dai Zhiyong, Yuan Yong, Liu Yongzhi. (2004) Research on monitoring system for landslides based on fiber optic strain sensing. Optics & Optoelectronic Technology, 2(3): 51–53.

He Jianguo, Liu Zhongfu, Gao Yin. (2006) Study on analysis of Dingshan soft rock tunnel in Xinjiang. Water Resources & Hydropower of Northeast, 24(2): 1–4. (In Chinese).

He Manchao. (1996) Supporting theory and practice of roadways in soft rocks in China coal mines. Xuzhou: China University of Mining and Technology Press. (In Chinese).

He Manchao. (2009) Real-time remote monitoring and forecasting system for geological disasters of landslides and its engineering application. Chinese Journal of Rock Mechanics and Engineering, 28(6): 1081–1090.

He Manchao, Qian Qihu. (2006) Summary of basic research on rock mechanics at great depth. Proceedings of the 9th Rock Mechanics and Engineering Conference. Beijing: Science Press, pp. 49–62. (In Chinese).

He Manchao, Jing Haihe, Sun Xiaoming. (2002) Soft rock engineering mechanics. Beijing: Science Press. (In Chinese).

He Manchao, Li Guofeng, Liu Zhe. (2007) Countermeasures aiming at the support for crossing roadway of deeply buried soft rocks in xing'an coal mine. Journal of Mining & Safety Engineering, 24(2): 127–131. (In Chinese).

He Man-chao, Tao Zhi-gang, Zhang Bin. (2009) Application of remote monitoring technology in landslides in the Luoshan mining area. Mining Science and Technology, 19(5): 609–614.

He Manchao, Xia Hongman, Jia Xuena, Gong Weili, Zhao Fei, Liang Kangyuan. (2012) Studies on classification, criteria, and control of rock bursts. Journal of Rock Mechanics and Geotechnical Engineering, 4(2): 97–192.

He Manchao, Xie Heping, Peng Suping, *et al.* (2005) Study on rock mechanics deep mining engineering. Chinese Journal of Rock Mechanics and Engineering, 24(16): 2803–2813. (In Chinese).

He Manchao, Zou Zhengsheng, Zou Youfeng. (1993) Introduction of soft rock tunnel engineering. Beijing: China University of Mining and Technology Press. (In Chinese).

Hou Chaojiong, He Ya'nan. (1995) Principle and applicationof extensive anchor bolt. Coal Technology of Northeast China, 5(1): 21–25. (In Chinese).

Hu Chuanting, Guo Aimin,Qiao Lixue. (2009) Application of deformation and yield bolt to deep mining gateway of Huxi Mine. Coal Science and Technology, 37(8): 18–23. (In Chinese).

Itasca Consulting Group Inc. (2003) $FLAC^{3D}$(Version 2.1) users manual. [S.l.]: Itasca Consulting Group Inc.

Jiang Binsong, Feng Qiang, Wang Tao, *et al.* (2011) Mechanical analysis of close type yieldable steel support. Rock and Soil Mechanics, 32(6): 1620–1624. (In Chinese).

Jun Han, Xiuli Ding, Jiebing Zhu. (2001) New progress on anchorage techniques in rock-soil. *Journal of Yangtze River Scientific Research Institute*, 5.

Kang Hongpu, Lin Jian, Wu Yongzheng. (2009) High pretensioned stress and intensive cable bolting technology set in full section and application in entry affected by dynamic pressure. Journal of China Coal Society, 34(9): 1153–1159. (In Chinese).

Leonardo Zan, Gilberto Latini, Evasio Piscina, Giovanni Polloni, Pieramelio Baldelli. (2002) Landslides early warning monitoring system. Geoscience and remote sensing symposium. IGARSS' 02. 2002 IEEE International, 1: 188–190.

Lian Chuanjie, Xu Weiya, Wang Zhihua. (2008) Analysis of deformation characteristic and supporting mechanism of a new-typed yielding anchor bolt. Journal of Disaster Prevention and Mitigation Engineering, 28(2): 242–247. (In Chinese).

Liu Bo, Han Yanhui. (2005) Guideline for principle, example and application of FLAC. Beijing: China Communications Press. 3–15. (In Chinese).

Manchao He, Weili Gong, Jiong Wang, Peng Qi, Zhigang Tao, Shuai Du, Yanyan Peng. (2014) Development of a novel energy-absorbing bolt with extraordinarily large elongation and constant resistance. International Journal of Rock Mechanics and Mining Sciences, 67: 29–42.

McCreath, D.R., Kaiser, P.K. (1995) Current support practices in burst-prone ground, Mining Research Directorate. In: Canadian Rock burst Research Project (1990–95), GRC, Laurentian University.

Ortlepp, W.D. (1992) Invited lecture: The design of support for the containment of rockburst damage in tunnels-An engineering approach. In Kaiser P. K. and McCreath D. R. (Eds.), Proceedings of the International Symposium on Rock Support, pp. 593–609, Sudbury, Canada, June 16–19. Rotterdam/Brookfield: A. A. Balkema.

Ortlepp, W.D. The design of support for the containment of rock burst damage in tunnels-an engineering approach. In: Kaiser P.K., McCreath D.R. (Eds.), Rock support in mining and underground construction. Rotterdam: Balkema, pp. 593–609.

Player, J.R., Villaescusa, E., Thompson, A.G. (2008) Dynamic testing of reinforcement system. In: Sixth international symposium on ground support in mining and civil construction. Cape Town, South Africa, SAIMM Jonannesburg, pp. 597–622.

Reevea, B.A., Stickley, G.F., Noon, D.A., Longstaff, I.D. (2000) Developments in monitoring mine slope stability using radar interferometry. Geoscience and remote sensing symposium. Proceedings. IGARSS 2000. IEEE International, 5: 2325–2327.

Simser, B. (2002) Modified cone bolt static and dynamic tests. Noranta Technology Centre Internal Report. Quebec, Canada.

Sun Xiaoming, He Manchao, Yang Xiaojie. (2006) Research on nonlinear mechanics design method of bolt-net-anchor coupling support for deep soft rock tunnel. Rock and Soil Mechanics, 27(7): 1061–1065. (In Chinese).

Thompson, A.G., Villaescusa, E., Windsor, C.R. (2012) Ground support terminology and classification: An update. Geotechnical and Geological Engineering, 30: 553–580.

Toshitaka Kamai. Monitoring the process of ground failure in repeated landslides and associated stability assessments. Engineering Geology, (1998)50: 71–84.

Varden, R., Lachenicht, R., Player, J., Thompson, A., Villaescusa, E. (2007) Development and implementation of the Garford dynamic bolt at the Kanowna Belle Mine. In: 10th underground operators' conference, Launceston, Australian Centre for Geomechanics, pp. 395–404.

Wang Dianming, Guo Qifeng, Feng Zhifeng. (2007) The development and application of QXY-5 memory borehole inclinometer. The Chinese Journal of Geological Hazard and Control, 12 (supp): 54–57.

Wang Lianguo, Li Mingyuan, Wang Xuezhi. (2005) Study on mechanisms and technology for bolting and grouting in special soft rock roadways under high stress. Chinese Journal of Rock Mechanics and Engineering, 24(16): 2890–2893. (In Chinese).

Wyllie, Duncan C., Mah, Christopher W. (2005) Rock slope engineering – civil and mining (4th ed.). London: Spon Press, Taylor & Francis Group.

Xu Likai, Li Shihai, Liu Xiaoyu, Feng Chun. (2007) Application of real-time telemetry technology to landslide in Tianchi Fengjie of Three Gorges Reservoir Region . Chinese Journal of Rock Mechanics and Engineering, 26(supp.2): 4477–4483. (In Chinese).

Yang Shengbin, He Manchao, Liu Wentao. (2008) Mechanics and application research on the floor anchor to control the floor heave of deep soft rock roadway. Chinese Journal of Rock Mechanics and Engineering, 7(Supp.1): 2913–2920. (In Chinese).

Zhang Guofeng, Yu Shibo, Li Guofeng. (2011) Research on complementary supporting system of constant resistance with load release for three-soft mining roadways in extremely thick coal seam. Chinese Journal of Rock Mechanics and Engineering, 30(8): 1619–1626. (In Chinese).

Zhou Jiawen, Xu Weiya, Li Mmingwei, *et al.* (2009) Application of rock strain softening model to numerical analysis of deep tunnel . Chinese Journal of Rock Mechanics and Engineering, 28(6): 1116–1127. (In Chinese).

Zhu Weiguo. (2011) Construction technology of Xiaband soft-rock tunnel in Xinjiang. Heilongjiang Science and Technology of Water Conservancy, 39(6): 59–62. (In Chinese).

Chapter 13

Status and prospects of underground thick coal seam mining methods

B.K. Hebblewhite
School of Mining Engineering, UNSW Australia, Sydney, NSW, Australia

Abstract: This chapter provides an overview of the international "state of the art" with respect to high performance underground coal mining methods used in seams of thickness greater than 4.5m, with particular emphasis placed on the countries which have significant, high performance thick seam operations – namely Australia and China. Consideration of some early developments with thick seam mining – both with bord and pillar and longwall systems – is provided, drawing particularly on Australian mining experience. The chapter then focuses on mining systems capable of delivering high production/productivity performance, in part through a review of a selection of high performing thick seam mining operations around the world – both Single Pass Longwall (SPL) and Longwall Top Coal Caving (LTCC). The chapter discusses both operational experiences, together with the ground control/geotechnical issues faced, and identifies some of the major, inherent geotechnical factors that can influence the mining method selection, design criteria and ultimate performance, from both an operational and safety perspective. The chapter also presents some geotechnical design approaches and research challenges relevant to thick seam mining operations.

1 INTRODUCTION

1.1 Terminology

The following terminology and definitions have been used in this chapter, based on the widely accepted terminologies used across the Australian coal mining industry, and elsewhere in the world. These terms and definitions are not unique but are adopted and recommended, to provide a consistent and clear understanding of the parameters and methods under discussion.

As the first point of definition, the term "thick seam" has commonly been applied historically to any minable seam thickness greater than the reach of development and longwall systems prevailing at the time. In the 1980s and 1990s, this was interpreted as 4.0m. However, with the advent of higher reach continuous miners and longwall systems, an arbitrary figure of 4.5m has subsequently been adopted in Australia for all recent studies into "thick seam mining". (Note: The Chinese mining industry uses a figure of +3.5m to define thick seam operations (Wang, 2011)).

Turning to the various major mining methods of interest, there are a number of mining methods adopted around the world for underground thick seam mining, and some simple acronyms and method names are widely used to describe these, as follows:

- SPL – refers to Single Pass Longwall. This is intended to describe a conventional longwall mining configuration and equipment (roof shields and shearer-based mining system), which has been extended in height to cut and support a higher face, within the definition of a thick seam, *i.e.* 4.5m or greater. It is also referred to as a high reach longwall (HRL). The key point to note is that the full mining height is extracted in a single pass, using extended height equipment.
- MSL – refers to Multi-Slice Longwall. This method involves the use of conventional longwall equipment operating at normal heights in various systems of extraction whereby multiple descending slices of the thick coal seam are mined separately and sequentially.
- Soutirage – This is a term used to describe the original European method of longwall mining in thick seams where the bottom section of the seam is mined conventionally, and the upper section of the seam is allowed to cave, and the coal is collected through some form of hatch or movable canopy within the longwall roof shields and is directed either onto the front armored face conveyor (AFC), or a supplementary AFC operating within the rear of the shields. The term soutirage derives from a French word meaning *bleeding-off*, or *decanting*, presumably referring to the drawing down of the caved coal onto the AFC.
- LTCC – refers to Longwall Top Coal Caving. This method is the Chinese derivation of the original Soutirage system, the difference being that a second AFC is towed behind the longwall shields to collect the caved coal from the upper sections of the thick seam. Both soutirage and LTCC could loosely be described as coal mining equivalents of sub-level caving, as used in a hard rock mining context.
- Hydraulic Mining – This is the method where the actual coal cutting is achieved by breakage of the coal as a result of impact from water pressure provided by a high pressure "water cannon" or hydraulic monitor. The water is also then typically used as the primary means of coal clearance from the production face or stope, with the assistance of gravity.

Further details are provided for each of these methods later in the chapter. One other term that should be defined at this stage is a universal underground coal mining term – *goaf* – the term used in Australia to describe the area of the mine from where the coal has been extracted, and where usually the unsupported roof has, or will fall in. Goaf is also described as *gob* in other parts of the world, with exactly the same meaning.

1.2 The thick seam mining challenge

Historically, the extraction of thick coal seams has been a challenge around the world, fraught with problems, but also opportunities, if the problems can be overcome. Countries such as Poland, France, Germany, Russia, Slovenia, India, China, South Africa, Australia, New Zealand and others have all faced the challenge of underground thick seam mining, with mixed degrees of success. Traditionally, the extraction of thick seams in many of these countries has been a matter of national or strategic necessity over and above the straight economic incentive. This can be due to various issues imposed either by regulatory bodies or the mines themselves (Bassier & Mez, 1999; Peng, 1998; Jian *et al.*, 1999). For instance, in China the government placed a requirement on all mines to achieve 75% – 80% overall recovery (by volume) of the coal resource within a lease. This type of strategic or political pressure has resulted in very significant engineering and technical developments and achievements within the field of thick seam mining methods, but not always to the extent required to satisfy the high sustained levels of safety standards, production, productivity and profitability that would be required, for example, in the modern Australian free market coal industry. There is much to gain from evaluating international thick seam mining experience, but even more to be done to then apply and upscale it to suit modern industry requirements and performance standards.

If one is to consider the challenges of mining seams of greater than 4.5m thickness, the challenges are many and varied. These include both operational and technical challenges, of which the geotechnical, or ground control challenges are foremost, accompanied by other technical issues such as ventilation, spontaneous combustion, gas management, dust, water and so on.

Many thick seam mining methods have been adopted in different parts of the world, but often with only limited success. In recent decades, conventional longwall faces have struggled to achieve optimum productivity as well as resource recovery when cutting at heights at or above 5m, and in some cases even less. Ground control problems associated with face stability are one of the major challenges faced by some of these high reach single pass longwalls. However, in more recent years, panels are being mined successfully with face heights up to 7m in parts of China – where conditions are ideally suited to the stability and extraction of such high faces (see later discussion).

Given the potential difficulties with face stability on some SPL faces, there are two alternatives available. The first is not very palatable, and that is to simply mine to a conventional <4.5m face height, and leave the additional coal in the ground – sterilized forever. The second alternative is to adopt one of the caving-based longwall systems such as LTCC. LTCC has the potential to provide a viable, high performance alternative to SPL where poor to average face conditions prevail. LTCC avoids the problem of high faces, but relies on good, predictable caving of top coal immediately behind the face in order to achieve optimum resource recovery while maintaining good productivity. In both mining systems – SPL and LTCC – apart from underground geotechnical challenges, surface subsidence is also a major geotechnical consideration which must be managed.

1.3 Australian thick seam resources

Australia has some of the largest black coal resources found anywhere in the world. Estimates put the value of measured black coal resources in this country in the order of 42 billion tonnes (Bt) (Simonis, 1997). Hebblewhite *et al.* (2002) reported that Australia has significant thick coal seam resources that require the application of alternative mining methods, if maximum resource recovery is to be achieved – beyond the conventional bord and pillar or longwall systems. Table 1 summarizes the available data regarding potential underground "thick seam coal" in Australia – both in terms of "measured" resources and "measured plus indicated" resources. These figures confirm that there are at least 6.4 billion tonnes of measured underground

Table 1 Australian thick seam resources (after Hebblewhite *et al.*, 2002).

Australia		*Number of Sites*	*Measured Resources (Mtonnes)*	*Measured Resources (%)*	*Ind. & Measured Resources (Mtonnes)*	*Ind. & Measured Resources (%)*
Seam Thickness	4.5m – 6.0m	29	2,249	34.8	7,425	42.2
	6.0m – 9.0m	33	3,310	51.2	8,397	47.8
	>9.0m	9	597	9.2	1,113	6.3
	No Info	2	310	4.8	650	3.7
	TOTAL	**73**	**6,466**	**100.0**	**17,585**	**100.0**
Seam Dip	<5 deg	43	3,173	49.1	11,684	66.4
	5 deg – 15 deg	21	2,256	34.9	3,698	21.0
	>15 deg	9	1,037	16.0	2,203	12.5
	TOTAL	**73**	**6,466**	**100.0**	**17,585**	**100.0**
Seam Depth	<150m	28	3,312	51.2	6,303	35.8
	150m – 300m	23	1,594	24.7	5,383	30.6
	>300m	19	1,508	23.3	5,698	32.4
	No Info	3	52	0.8	201	1.1
	TOTAL	**73**	**6,466**	**100.0**	**17,585**	**100.0**

thick seam resources, rising to 17.5 billion tonnes, once indicated resources are included.

Some significant features of these measured resources are:

- 86% are in seam thicknesses between 4.5m and 9m, with 51% of these tonnes in the 6m – 9m range (based on measured resources).
- 84% are in seams with a dip of less than 15°
- 76% are less than 300m depth.

On the basis of these extensive resources, the Australian coal mining industry has a very strong incentive, and a responsibility, to develop appropriate mining systems for the recovery of these thick seam resources in a safe, efficient and productive manner. The international incentive for identifying or developing new methods for underground thick seam mining is commonly for optimizing resource recovery. However, the Australian coal industry is an export-dominated free market industry where high productivity, sustainable financial viability and the highest safety standards are paramount. As such, the Australian requirement is for appropriate methods which first and foremost meet the Australian safety and productivity/financial performance criteria, or preferably improve on them, while at the same time still aiming to achieve maximized resource recovery, but coal recovery comes as a second tier priority. This is in contrast, for example, to countries like China, where resource recovery has been the number one priority, by government decree, possibly ahead of productivity performance – resulting in many of the mining system and equipment advances that have occurred in the last twenty-five years in that country.

2 HISTORICAL OVERVIEW OF MINING METHODS

2.1 Bord and pillar methods

A brief review of historical mining methods is provided here, based on Australian and international experience. Thick seam bord and pillar mining has taken place in Australian coalfields for over sixty years – in locations such as the Newcastle and Hunter Coalfields near Newcastle in New South Wales (NSW); the Ipswich Coalfield west of Brisbane in Queensland; and in Moura and Collinsville, in the Bowen Basin in Queensland. In most of these locations, the thicker seams were in the thickness range of 5m to 10m typically.

These early operations consisted of various different forms of bord and pillar mining, using continuous miners and shuttle cars (and some earlier hand-worked operations). The maximum reach of a typical continuous miner was between 3m and 4m, and so the normal method of mining was to develop panels of first working bord and pillar roadways in the top of the seam at these conventional mining heights. The panels were mined to their extremities, with normal roof support installed on the advance in this first pass. Then a process of second pass retreat mining commenced where the miner pulled back several hundred metres from the end of the panel, then commenced cutting the floor coal, in an advancing process, forming a ramp down to a lower horizon, possible 3m or so below the original floor horizon. All the floor coal was then removed from the existing roadways in this inbye section of the panel, forming

Figure 1 Early thick seam workings at Abermain Colliery, NSW.

increased height pillars, equivalent to the height of two passes. This incremental ramping and floor lifting was then repeated further outbye in the panel, until the full panel had been mined by a first workings layout, using two sequential passes in each roadway. Then, depending on seam thickness, the entire process of floor ramping and lifting could be repeated for a third pass.

Figure 1 shows early thick seam workings from the 1950s, in Abermain Colliery, near Cessnock, NSW. Figure 2 shows 9m high first workings pillars at Muswellbrook Colliery in the NSW Hunter Valley, mined during the 1980s. These workings have been mined in a three-pass operation.

There have also been a number of systems for achieving secondary extraction in thick seams using continuous miner-based mining. One such system was a trial conducted in the early 1980s of multi-slice pillar extraction in an experimental panel (27 Level) at the Bowen No. 2 Mine in Collinsville, Queensland. The trial was designed and managed by Australian Coal Industry Research Laboratories (ACIRL) Ltd (O'Beirne, 1982).

In this trial, the 5.4m thick seam was divided into three 1.8m thick slices. A conventional two-heading development panel was mined in the top slice, in preparation for extracting the adjacent panel of coal using the Wongawilli total extraction system with a continuous miner and two shuttle cars. The only difference to a normal development panel was that the middle slice of the seam (or septum), forming the floor of the top slice, was reinforced on the advance using fully-grouted 1.8m long wooden dowels, in

Figure 2 Pillars at Muswellbrook Colliery.

order to pre-support the septum, prior to it becoming the roof for the lower slice development. Figure 3 illustrates the sequence of mining in the three slices, while Figure 4 shows the bottom end of the wooden dowel supports intersected when the development of the lower third of the seam commenced. Figure 5 shows the end, or outbye view of the two development levels, with the reinforced septum in the middle of the seam.

Extraction of the top third of the adjacent panel was then carried out using conventional Wongawilli split and fender lifting. Extraction of the bottom third was done in the same manner, except that on the retreat, in each fender lifting sequence, the miner head was raised to cut the middle section of septum coal. The method was technically successful, but was never adopted on a wider scale.

One of the major geotechnical risk factors with many of these early bord and pillar thick seam mining systems was the issue of rib stability where high ribs were often present. In many instances no rib support was installed, resulting in the potential for rib buckling or toppling failure causing injuries to persons working or travelling beside such high ribs. Apart from this form of direct human safety concern, there were also the related overall mine stability problems brought about by high rib failures leading to local and regional pillar instability; potential

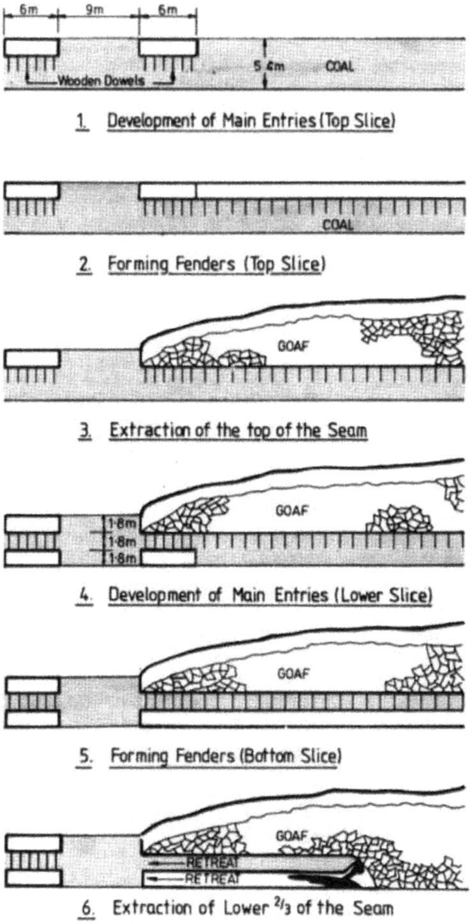

Figure 3 Mining sequence for thick seam pillar extraction at Collinsville in the 1980s.
Source: O'Beirne, 1982

Figure 4 Development of the lower third of the seam revealing the pre-installed wooden dowel reinforcement of the septum.
Source: O'Beirne, 1982

Figure 5 End view of the Collinsville multi-slice development panel.
Source: O'Beirne, 1982

roof falls due to increased spans; and potential for unwanted excessive surface subsidence.

2.2 Multi-slice longwall

In the context of identified extensive underground thick seam resources in Australia (see Table 1), studies were conducted by UNSW and others (Hebblewhite, 1999, 2001; Hebblewhite *et al.*, 2002) to review the Australian opportunities and available methods and technologies for safe and efficient underground thick seam mining. Arising from this review, four generic methods were identified as having potential. These were:

- extended height single pass longwall (SPL)
- multi-slice longwall (MSL)
- hydraulic mining (HM)
- caving longwall systems (CL), including longwall top coal caving (LTCC).

The option of extending the height of a conventional single pass longwall was considered at the time to have limited possibilities. It was apparent that technology was already increasing both shearer and support heights from 4m to 5m and above (Hamilton, 1999). However, limitations such as equipment size, weight and stability (representing both operational concerns and concerns for transporting around the mine), plus face conditions, were considered at the time to limit the application of this method to no more than 6m height, for many years to come.

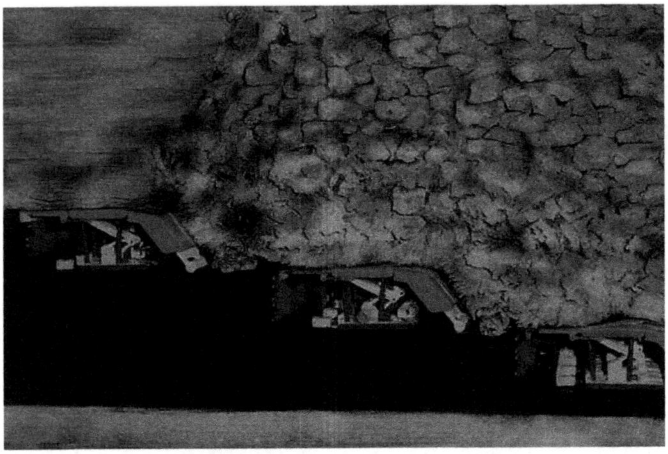

Figure 6 Schematic view of MSL operation.

The method was certainly considered to have great potential in the 4.5–6.0m mining height range.

Multi-slice longwall (MSL) has been used extensively over recent decades in both Europe and China. This method involved several variations of longwall faces operating in multiple, descending slices through a panel of thick seam coal. Figure 6 is a schematic diagram of three longwall faces operating together in an MSL panel, although use of only two faces was more common. The essential concept of the MSL system was to use conventional height longwall faces operating in multiple slices through a thick seam, in a descending sequence. This could be done in one of two main ways – either with a second longwall face immediately following the top face within close proximity, say 50–100m; or more commonly, completing the mining of the top slice panel in full, and then leaving the panel for a period of a year or more typically, to allow some level of goaf consolidation to occur in the top slice, prior to undermining with the second slice. This second scenario had two main benefits – firstly, that only one set of longwall equipment was needed; and secondly, that goaf consolidation would improve mining conditions in the second slice, especially where the presence of clay minerals may have assisted with goaf consolidation over time.

In the above scenarios, the most common means of additional support or stabilization between the descending slices has been the use of wire mesh. Mesh is laid out onto the floor behind the shield supports in the top slice. This is done using rolls of mesh, the same width as each shield, and can either be rolled out over the front of the roof canopy of each shield, or alternatively, rolled out directly behind the shield from beneath the rear canopy, as is illustrated in Figure 7.

The function of the mesh is then to form some level of constraint to the broken goaf material, prior to the second slice being mined directly beneath, in order to provide a stable roof for the longwall face on the subsequent descending slices. Figure 8 shows the mesh visible between two longwall shields mining in the second slice of a Chinese multi-

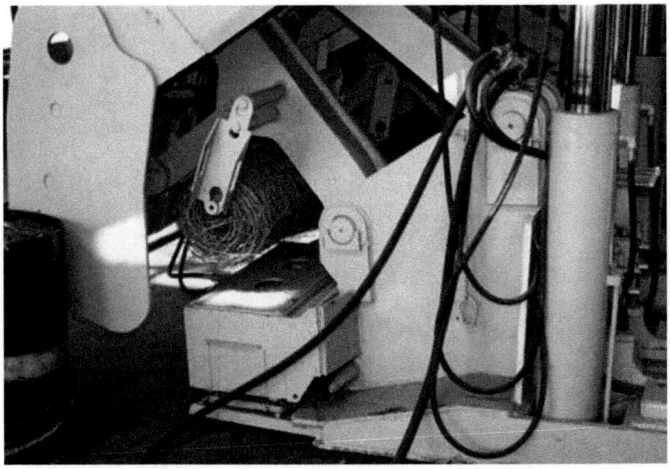

Figure 7 Wire mesh placement behind top slice of MSL face.

Figure 8 Mesh visible between shield supports.

slice longwall face. Figure 9 shows the conditions in the maingate/face-end area of the same longwall, where conditions can become extremely difficult to manage, as is evident. Even on the face itself, the degree of support offered by the lightweight steel mesh is only quite minimal, and the method is very dependent on the compaction and subsequent self-supporting nature of the broken overlying goaf material, in order to prevent roof falls ahead of the shields.

There are two other alternatives to use of mesh for goaf support above the lower slices of the MSL method. The first is to actually leave a layer of intact coal/rock to form a solid septum between the slices. This is ideal where stone bands might be present in the coal seam at a suitable horizon, but otherwise this approach can result in significant coal losses. A further alternative is to apply modern cemented paste fill technologies to

Figure 9 Mesh supporting the roof of the lower slice in the maingate/face-end area of a multi-slice longwall.

form an artificial septum in the goaf behind the top slice, which can become a more competent roof for the lower slice. A number of mines in Poland and elsewhere in Europe have trialled this concept, with mixed success, by laying out multiple paste placement pipes behind the face shields, and dragging them forward with the face, as the upper face retreats (Palarski, 1999; Bassier & Mez, 1999). This method is technically feasible, but can add significantly to the mining costs, and can potentially introduce delays to face productivity.

Apart from the above concerns regarding goaf stability for the lower slices, there are other major safety concerns with this method, when operating in a lower slice, beneath a previous goaf area. These include obvious issues of mining under potentially trapped and unplanned bodies of water in low points within the upper goaf; dangers associated with gas concentrations in the upper goaf; possible concerns with excessive dust on the lower faces; and a major concern with management of the ventilation circuit on the lower face after effectively holing into and re-ventilating the upper goaf area. This latter concern is one of the most serious ones, especially where propensity of the coal to spontaneous combustion presents a major hazard to the operations.

The other issue to consider with MSL operations is that for every slice (two or three), you are still required to mine a new set of development gate roads. So firstly, there is no economy of scale in terms of development metres-longwall production tonnes ratio, for the method. Secondly, there are serious geotechnical design issues associated with determining where the lower development roadways should be located – directly beneath the overlying roadways; beneath the edges of the overlying goaf (resulting in progressively wider chain pillars); or 100% offset beneath the middle of the goaf of the overlying panel slice, resulting in overlying chain pillars being located over the underlying longwall face, creating potential face weighting problems. All of these scenarios are technically feasible, but are complex from a design perspective, and can create difficult geotechnical and operational conditions.

On the basis of all of these concerns, MSL has never been pursued as a thick seam mining option in Australia.

2.3 Hydraulic mining

Hydraulic mining has been practiced in many parts of the world, including Germany, Russia, India, Japan and Canada. More recently it has been used at a number of mines on the South Island of New Zealand. It has been found to be a method with a significant potential in a limited range of applicable mining conditions. Under the right conditions it can offer significant financial benefits (low capital and operating costs), but with limited large scale production potential. It is therefore considered to be suitable as a "niche application" mining method, but not a universally applicable option.

The seams in which hydraulic mining techniques have previously been applied, were predominantly steeply dipping (> 15 degrees), and hence, mechanized methods had limited application. (However, in thick coal seams, it is possible to operate at apparent dips in excess of the actual seam dip, hence hydraulic mining can find broader application, even in flatter, thick coal seams).

Drawing on recent New Zealand experience, the Strongman No. 2 Mine (now closed) operated by Solid Energy, was producing in excess of 400,000 tonnes/yr using a single hydraulic monitor (plus conventional continuous miner development units). The technique and similar layout was then applied at the Spring Creek Mine, and was in the early stages of introduction at the Pike River Mine at the time of the tragic gas explosion in 2010 which claimed twenty-nine lives. Pike River had been planned to produce in excess of one million tonnes per annum, once fully operational, using multiple monitor panels.

Features of the method are relatively low capital cost and the flexibility of multiple operating faces – provided conditions are suitable. Figure 10 illustrates the concept of underground hydraulic coal mining, using conventional continuous miner or road-header systems for up-dip development, followed by retreating down-dip monitor extraction, in panels up to 30m wide (15m reach of the monitor, either side of center).

Hydraulic mining has several advantages over conventional mechanized mining methods. The advantages of hydraulic mining are:

- Mining layout is similar to that for conventional mechanized bord and pillar mining; however, the level of mechanical complexity is significantly reduced.

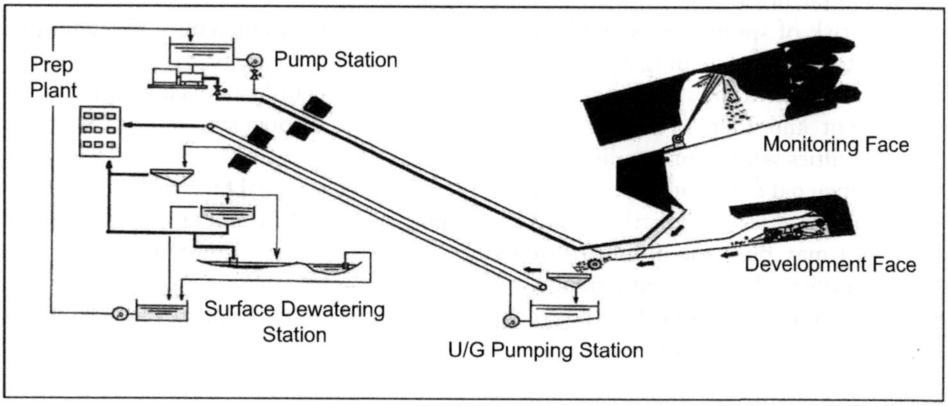

Figure 10 Concept of hydraulic mining (diagram courtesy of Solid Energy).

- Due to a reduction in the production of coal dust, the elimination of frictional ignition sources, the removal of personnel from the face area and the ability to automate equipment operation, hydraulic mining can be relatively safer than more traditional underground coal mining methods.
- Extraction of thin (0.3–1.5m), thick (> 4.5m) and steeply dipping seams has, in the past, challenged traditional mechanized techniques. However, hydraulic mining has had great success in such geologies.
- Typically the tonnages obtained from hydraulic mining operations are less than those obtained from mechanized methods. However, due to reduced manning requirements and lower capital costs, hydraulic mining is still highly productive.
- Hydraulic mining offers significantly lower capital and operating cost structures over conventional mines.
- Hydraulic mining is operationally flexible and can be used to extract areas of working mines where mechanized methods would otherwise encounter operational difficulties or would be economically unfeasible.
- Provided mine layout is designed appropriately, hydraulic mining can cope with large scale structural disturbances; and as a result the profitability of a hydraulic mining operation is less affected by these constraints.

Hydraulic mining also has some disadvantages – they are:

- The entire mine must be planned around the gravity-driven hydraulic transportation system; roadways must typically have an inclination of at least 5–10 degrees, even if this means coal is left in the floor.
- High influx of water can cause problems with acid mine drainage; this acidity increases with high sulfur coals.
- Hydraulic mining can require the consumption of larger volumes of water than conventional mechanized techniques – although the Strongman experience, where coal was hydraulically transported all the way to the surface, prior to dewatering, was typically achieving water losses of less than 10%.
- Coal is broken along its entire transportation route resulting in higher levels of fines in the run-of-mine product. This increases the capital and operating cost of dewatering facilities.
- The risk of spontaneous combustion is greater than for conventional techniques due to irregular goafing and difficulties in sealing-off old areas.
- Water reduces the strength of many rock types, especially where clay minerals are present and therefore there may be an increased propensity to roof falls, or difficulties with floor conditions.
- During coal extraction the operator is unable to see the coal face and as a result is unable to ascertain exactly what is happening. As a result, a sudden collapse of roof strata may bury the monitor/face equipment, resulting in production disruption and potential equipment damage.
- The method does not offer a high capacity mining solution – typically achieving less than 1MTPA production capacity.

In summary, hydraulic mining has limited potential, but in the right conditions, can be a highly productive, low capacity, thick seam mining method.

2.4 Caving methods

Underground hard rock, metalliferous mining has used caving mining methods for many years – particularly sub-level caving, and more recently block caving. The underlying principle of such methods is to make use of the natural inherent weakness of the in situ rock mass that can lead to rock failure and natural rock fragmentation under gravity. This geotechnical behavior – if able to be predicted and utilized in a continuous manner, can lead to very low cost mining operations, where in situ stress environments coupled with wide spans, inherently weak rock mass properties and gravity can result in very effective mining methods.

In underground coal mining, the same concept has been developed for use in certain thick seam environments, provided the coal can be encouraged to cave predictably and consistently. Coal mining traditionally relies on machine cutting electrical power to break up or fragment the coal mass. However in a caving-based mining system, the concept is to only conventionally cut the lower section of the coal seam, and then rely on natural caving and fragmentation of the upper sections of the seam – thereby significantly reducing the power requirements for coal breakage. The other major benefit of such a system is that the amount of conventional roadway and panel development that is required, per tonne of coal production, can be dramatically reduced, and this offers considerable savings in unit operating costs, since development is typically a much more expensive process than subsequent extraction, and can be a major constraint due to slower than ideal development rates.

European coal mining in thick seams developed what was known as the "soutirage" mining system, as illustrated in Figure 11. Figure 12 shows a typical soutirage longwall shield, in use to this day at some mines in Europe. This method found wide application across parts of Europe in the 1970s and 1980s.

The method could be described as a coal mining equivalent of sub-level caving. Essentially it consisted of developing the mine conventionally in the bottom section

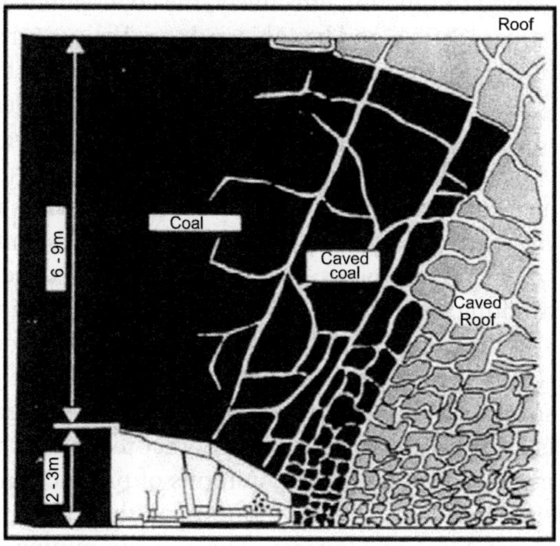

Figure 11 Soutirage mining concept.

Figure 12 Typical soutirage longwall shield (photograph courtesy of DBT).

of the coal seam at a normal working height (say 2–3m typically); installing a longwall face and shearer/AFC system at this lower horizon and operating the face with conventional cutting of the lower coal. However the upper sections of the coal seam were allowed to fracture above and directly behind the face shields, and then cave onto the rear of the shields where a hatch was located within the rear goaf canopy of the shields. Groups of these hatches would be progressively opened, moving along the face, allowing the broken coal to pass through the hatches onto a chute that would feed either onto the main front AFC (as shown in Figure 12), or alternatively would drop directly onto a small second AFC that was located behind the support legs, but within the shield.

In order for this method to function effectively, the coal seam was required to have a reasonably weak rock mass rating, and be subjected to sufficient vertical stress, such that it would cave immediately, into reasonably small lump sizes, so that it could be captured within the longwall system. Any delay in the caving process would lead to significant loss of top coal that would fall beyond the reach of the mining system. Other problems experienced with this method, apart from coal loss, included dilution (due to overlying strata caving with the caved coal); major restriction of walkways and ventilation flow along the face to the obstruction caused by the modified shields; significant dust problems on the face from the rear caving process; slow face retreat rates due to delays in the top coal caving process resulting in the onset of difficult mining conditions and face weighting and instability; and most importantly, spontaneous combustion susceptibility as a result of the large amount of fractured coal being left behind in the goaf.

So although the soutirage method offered significant improvements in thick coal seam recovery, it was also accompanied by many other technical challenges, and never became a high performance mining system, in terms of production levels.

A more recent and quite innovative variation of the caving methods has been developed at the Valenje Mine in Slovenia. The Valenje mining system grew out of the soutirage concept, and uses the same type of supports as illustrated in Figure 12.

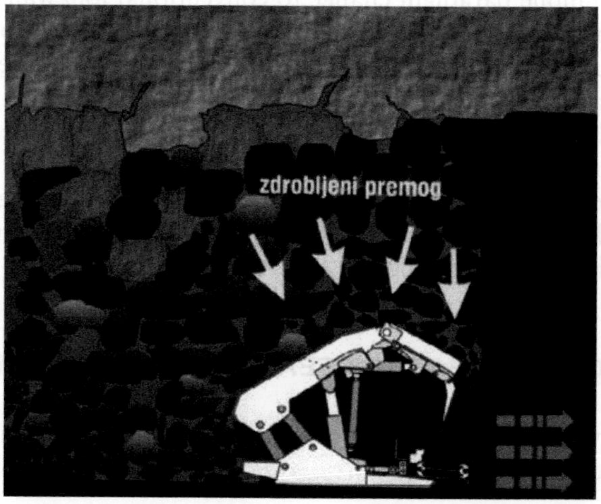

Figure 13 Valenje caving concept (diagram courtesy of Valenje Mine).

The mine is extracting an underground lignite deposit which is located in a seam or coalescence of seams that are up to 160m thick in places. Longwall panels are established at the bottom section of 20m thick sub-levels, working from the top of the seam and progressively descending in 20m high blocks. For each 20m block of coal, the lower conventional working height (3–4m) is extracted by a conventional longwall panel with shields, shearer and AFC. The overlying coal, up to 17m thick, is caved on each face slice, after the shearer has passed. Rather than caving this top coal through the hatch and chute fitted to the rear canopy, as in a soutirage face, this equipment is no longer used, and is kept "out of service". In contrast, the front or roof canopy of the shield is lowered across a small batch of shields, and the coal roof and the face itself is allowed to cave onto the AFC in front of the shields. Figure 13 illustrates this front caving concept, and Figure 14 is a photograph of the Valenje face, after completion of a caving and coal clearance sequence.

Figure 14 Longwall face at Valenje (photograph courtesy of Valenje Mine).

This is a quite unique paradigm change to conventional longwall wisdom which is totally focused on always maintaining the stability of the face and roof in the tip to face location. Valenje is successful for exactly the opposite reasons – it relies on failure of the roof and face ahead of the shields and upward propagation of the caving process within the 20m sub-level. Despite the different approach, the system works extremely successfully at Valenje. In situations where the coal is too strong to cave well, they sometimes mine a development roadway in the top of each 20m sub-level, along the length of the longwall panel, above the centerline, and undertake low energy blasting on retreat, just ahead or above the underlying longwall face, to facilitate the caving process.

3 HIGH PERFORMANCE THICK SEAM MINING SYSTEMS

There are now two dominant high performance thick seam mining methods being practised internationally:

- Single pass longwall (SPL)
- Longwall top coal caving (LTCC)

Both of these methods have broad application and both are capable of delivering "world's best" production performances, while achieving the highest possible levels of resource recovery, and doing so in a safe and productive manner. The following sections of the chapter summarize the key features of these methods and present a number of examples of each.

3.1 Single pass longwall

Single Pass Longwall, or High Reach Longwall, as defined earlier, simply refers to a conventional retreat longwall mining system operating at greater heights, *i.e.* with face heights of 4.5m or more, using larger, high reach shearers and face shield equipment, to cut the full face height in a single pass. In terms of high performance SPL operations around the world, the majority of these are located in either Australia or China, with a lesser number of operations in South Africa and parts of Europe.

It is significant to note that although a number of these SPL faces around the world have the capacity to operate at 5.5m and above, they are typically working at lower heights due to operational factors associated with ground control problems. These geotechnical factors, together with logistical issues associated with size, weight and stability concerns with the very large, high face equipment, are considered to be the major limitations to this method finding broad application beyond 6m height for the foreseeable future, although there are some exceptions to this limit, as described later.

Based on 2011 Australian statistics (Australian Longwall Magazine (2012)), using the definition of a ≥4.5m mining height, Table 2 lists Australia's SPL thick seam operations. (Note: Since this time, two of these faces – Broadmeadow and North Goonyella – have been converted from SPL to LTCC operations; and a number of other new SPL operations have commenced, including the 5m high SPL face at Whitehaven Coal's Narrabri North Mine, in the Gunnedah Basin of NSW, which produced over 8 MTPA in 2015).

Table 2 Australian thick seam longwall mines using SPL mining systems (2011/2012).

Mine	*Depth(m)*	*Seam Thickness (m)*	*Mining Height (m)*	*Shield Height (m)*	*Shield Capacity (tonnes)*
Broadmeadow	300	4.8	4.8	5.2	1,152
Carborough Downs	450	4.5	4.5	5.3	1,238
Mandalong	320	5.0	4.8	5.2	1,053
Moranbah North	320	4.5	4.5	5.0	1,750
Newlands Northern	240	5.0	4.5	5.0	1,040
North Goonyella	350	4.5	4.5	5.3	1,200
West Wallsend	205	4.6	4.6	5.3	900

In Table 2, Mining Height refers to the cutting height of the longwall face, whereas Shield Height refers to the maximum potential working range of the face shields. It is significant to note that even where seam thickness might permit, cutting height is usually less than the maximum shield range. This is a prudent precaution against roof falls on the face, to enable the shields to operate at higher than the cut height, when required, in order to maintain roof contact at all times, even with the presence of roof cavities. Figure 15 is a photo of the longwall face at West Wallsend Colliery when it was cutting up to 4.8m in height. It highlights one of the key geotechnical challenges for SPL faces – face spalling – which had been experienced regularly at West Wallsend. A further adverse consequence of face spalling is when it leads to an extended tip-to-face distance of unsupported roof, which can result in roof cavities forming ahead of the face, if not controlled quickly.

The problems of face spalling and instability, and potential roof cavities ahead of the face, are ones which have been experienced on a number of SPL faces around the world. Figure 16 shows the 6m high supports that were installed some years ago at Exxaro's Matla No. 2 Mine in South Africa – another high performance SPL operation.

Figure 15 4.8m high SPL face at West Wallsend Colliery (after Hamilton, 1999).

Figure 16 DBT 6m supports for Matla Mine, South Africa (photograph courtesy of DBT).

Matla No. 2 Mine has 6m high shields with a maximum operating height of 5.5m, but is typically cutting only 5m or less. Face height is reduced to about 4.3m for face recovery. Supports are two leg shields with a maximum capacity of 1,066 tonnes. The shields are only 1.75m in width, and the engineering/management personnel at the mine believe that 5m is an effective upper limit for these shields due to lack of rigidity in the supports leading to twisting of the lemniscates in each shield, resulting in face management problems and potential equipment damage when operating at greater heights. Matla staff believe that wider shields are necessary to operate at 5m or greater (Matla Mine, 2011).

Matla's 126m long face operates in the 2 Seam, at a relatively shallow depth (typically no more than 110m); with quite massive overburden strata and relatively strong coal. The overburden also includes old workings in the 4 Seam above, together with some massive sandstone units and dolerite sills. Even with such shallow depth, Matla has experienced major weighting and face-roof instability problems on occasions, when working in difficult ground conditions, and as a result has been forced to operate with a reduced cutting height at such times, as low as 4–4.5m under such conditions, and more commonly at less than 5m cutting height – well below the support height capacity.

When delayed caving does occur, this can lead to windblasts, especially associated with establishing the initial cave in each new panel. To overcome the initial caving problems, drilling and blasting of the immediate overburden has been practised at times, at the Matla No. 2 Mine. As a result of the relatively shallow depth, surface subsidence is also a major issue requiring a considerable amount of surface remediation and pre-mining preparation. In summary, the biggest challenges facing thick seam mining at Matla are the geological conditions (including relatively incompetent immediate roof and floor); the operational problems due to the low width:height ratio of the shields; and water management in a multi-seam mining operation, working beneath previous goaf areas.

Another example in the past decade has been the Sihe Mine in China. Sihe was operating a 225m long SPL face with 5.5m maximum support height, operating at 5.5m cutting height (*i.e.* with no excess height capacity in the supports to allow further extension to provide improved roof control). Sihe experienced ongoing difficulty holding the coal roof above the supports and ahead of the longwall face (between the tip of the roof supports and the coal face being cut by the shearer). As a result, after continuing ground control problems at the face, they proceeded to replace the face equipment with a 6.2m high set of shields – not necessarily to cut higher than 5.5m, but to ensure adequate height capacity on the face to deal with roof cavities.

In contrast to the above roof and face problems experienced by many SPL operations, there are also some extremely successful SPL faces, operating in very favorable ground conditions. There are an increasing number of SPL faces being installed in China, particularly where depths are low to moderate; overburden conditions are favorable; and most importantly, the coal is relatively strong, with minimal coal structure (cleat and bedding planes). These coal seam properties assist with face stability, but mitigate against the alternative method of an LTCC operation. A number of these newer Chinese faces have demonstrated world-class performance at face heights in excess of 7m.

The first such high performance, "super" mine, was the Bulianta Mine, located near Ordos, in Inner Mongolia, in northern China, owned by the Shenhua Group. Figure 17 shows the Bulianta face supports installed in a "mini-build" on the surface, prior to underground commissioning in 2011. Bulianta Mine is a relatively new mine within the Shenhua Group, together with its sister mines, Daliuta and Shang An – both of which

Figure 17 Bulianta Mine 7m high face supports (photograph courtesy of Shenhua).

were in the process of commencing or operating 7m SPL faces at the time of the Bulianta commissioning.

Bulianta operates a 7m high SPL face on a 286m long face using Chinese-built two leg shields, with 1,800 tonne capacity, 2.06m width, 76 tonne weight, with a 7.5m maximum height and an operating range of 3.2–7.5m. The full seam section (6.5–7.0m) is being mined. Web thickness is 865mm. A significant ground control feature of the shields is the three-stage flippers (visible in Figure 17 in a fully extended position) that can provide face support to almost 60% (4m) of the face height. Other equipment details included: a 3,000kW Eickhoff SL1000 shearer fitted with 3.5m diameter drums, capable of cutting at a speed of 14.5m/min; coal clearance was by way of a Caterpillar 6,000t/hr AFC powered by 3 x 1.6MW drives, feeding on to a 1.6m wide maingate conveyor belt.

The following production data from the Bulianta Mine was recorded during the first longwall panel of the 7m face at Bulianta, in 2011/2012 (Bulianta Mine, 2011):

- Total annualized mine production at the time was 30MTPA, coming from 3 SPL faces – one at 7m height and two each at 5.5m height;
- The 7m high face alone was producing at an annualized equivalent rate of 13MTPA, or an average of 43,400t/day;
- Production from each shear off the face was 2,170t/shear;
- Face retreat rate was averaging 430m/month.

On the basis of these production figures, it is believed that Bulianta could rightly claim at that time to be the highest producing longwall mine in the world, and certainly the 7m face could be said to be the highest performing single face in the world.

Conditions in the mine (operating at a depth of approximately 300m) were excellent – both from an operational and logistics perspective, as well as ground conditions. The access travel road, together with the full length of the maingate, was concrete-paved for high speed transport. The coal was an anthracitic coal with what appeared to be relatively high strength and very little sign of geological structure, such that the 7m high face stood with negligible sign of any spalling, apart from isolated small blocks, generally less than 0.5m in maximum dimension. Use of the extensive flipper face coverage was a critical component of face management. The mine had insignificant amounts of seam gas content. Gate roads were between 3.5m and 4m in height with roof and rib conditions standing "as cut", with no sign of any visible deformation or instability. Support density in the maingate was 6 bolts/m in the roof, plus some cables; plus 4 bolts/m in the ribs, which were also fully meshed.

3.2 Longwall top coal caving

Evaluation of different European and early Chinese experiences with the original "soutirage" mining concepts and equipment showed some promise, but performances were below the level required to be viable in a competitive market, not to mention health and safety concerns over issues such as dust and spontaneous combustion. However the development of the modern, double AFC, LTCC method in the 1990s, by the Chinese coal mining industry – largely replacing previous multi-slice longwalls and earlier soutirage caving systems – identified a viable and significant alternative to SPL, not only for the seam thicknesses between 4.5m and 6m, but particularly for the thicker 6m+ seams.

Xu (2001) provided an early insight into a modern LTCC face at the Chinese Dongtan Mine, owned by the Yankuang Group. Hebblewhite (2001) provided an assessment of the geotechnical challenges faced by the different thick seam mining options at that time, many of which are addressed in this chapter.

The major change developed by the Chinese industry for LTCC face equipment was the relocation of the top coal draw process to the rear of the longwall supports, rather than bringing coal through the roof/goaf canopy of the shield onto a conveyor within the shield structure, as in a soutirage face. The Chinese equipment has a pivoting supplementary goaf or tail canopy behind the support. Beneath this is a retractable second AFC, towed by chains behind each shield. With the rear AFC extended and the rear canopy lowered/retracted, caved top coal can fall vertically, directly onto the rear AFC, while production continues conventionally in front of the supports. The top coal caving process is managed by only caving a small bank of shields at any one time, usually immediately after the shearer has passed. In the retracted rear AFC position with the rear canopy raised, the supports and face operation can function conventionally, without top coaling. This is also an option if any problems are experienced with the rear caving system – it can effectively be shut down while the face continues to operate in conventional mode.

The Chinese industry has reported averages of 15,000 to 20,000 tonnes per day (t/day) from a number of LTCC faces across the country; and up to 75% recovery of 8m+ thick seams using a 3m operating height longwall; and 5–10 MTPA face production. There are now over 100 LTCC faces in China.

Figure 18 illustrates the geotechnical caving principles for LTCC operation, with upper or top coal commencing fracturing above the shield supports, and then fragmenting fully, behind the roof shield, and above the goaf shield, then falling onto the rear conveyor. Figures 19 and 20 are photographs showing a Chinese 4 leg LTCC shield in the factory, and the rear conveyor area (in the closed up configuration) in an operating LTCC face underground.

The major benefits of the LTCC method over other thick seam methods include:

- Operating Cost Reductions: The LTCC method enables potentially double (or greater) the longwall recoverable tonnes, per metre of gateroad development (compared to a conventional height face), thereby reducing the development cost/tonne significantly, and reducing the potential for development rate shortfalls leading to longwall production disruption.
- Resource Recovery and Mine Financial Performance: The LTCC method offers a viable means of extracting up to 75% to 80% of seams in the 5–9m thickness range, and above. (Single pass longwall is considered to be currently limited to an upper height of approximately 6m in typical mining conditions).
- Mine Safety: Lower face heights (relative to high reach single pass longwall) result in improved face control, smaller and less expensive equipment and improved spontaneous combustion control in thick seams, through removal of the majority of top coal from the goaf.

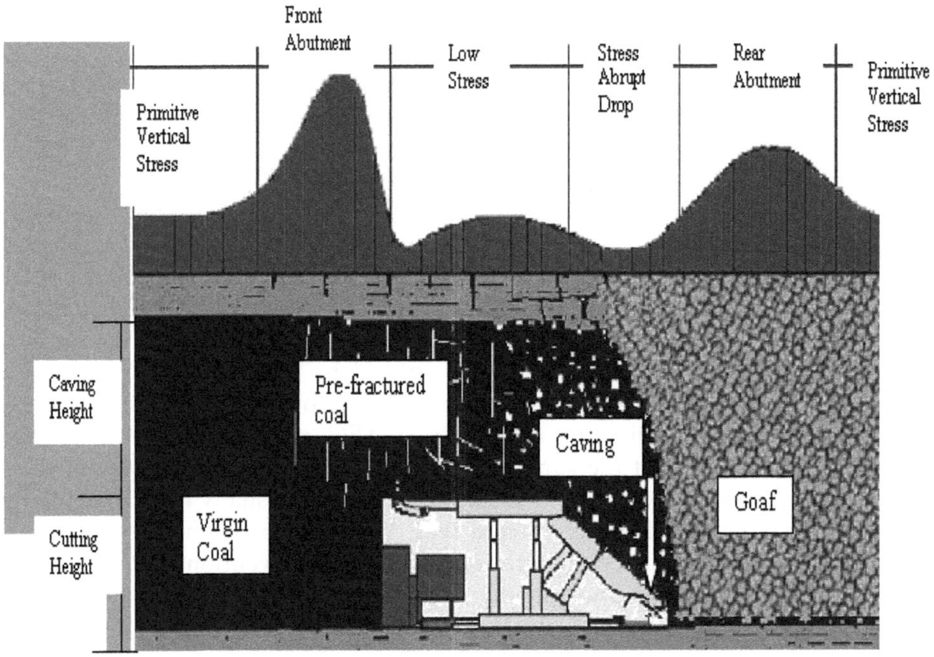

Figure 18 Conceptual model of LTCC System (after Xu, 2001).

Figure 19 A Chinese 4 leg LTCC shield.

- Production Capacity: LTCC is capable of achieving at least equivalent long-term annual production tonnage, compared to conventional height longwalls, if not better, subject to coal clearance capacities.
- Flexibility: LTCC offers very flexible face equipment for negotiating structural disturbances at the face such as faulting, by being able to continue to operate in a conventional cutting mode at reduced heights through fault zones.

Figure 20 Rear AFC area of an operating LTCC face.

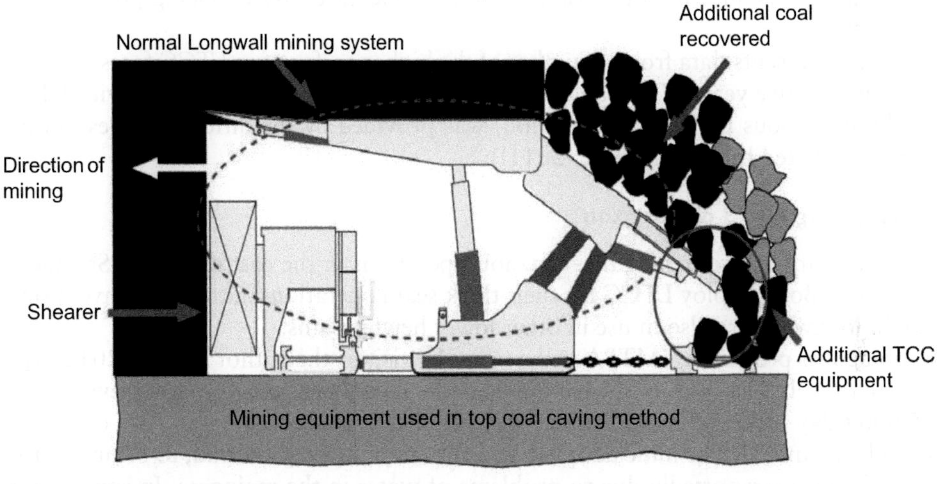

Figure 21 Schematic diagram of LTCC face equipment and caving process. (after Moodie & Anderson, 2011).

The first LTCC faces to be operated outside of China have been in Australia. Austar Colliery was the first to operate an LTCC face in Australia. Austar is owned by the Chinese company, Yancoal Australia Ltd, whose Chinese parent company is Yanzhou, part of the Yankuang Group, which is also one of the leading LTCC performers in China. Austar was the first mine in the world to introduce two leg LTCC shields (see Figure 21), through a collaborative development project between Yankuang and DBT.

Previously all Chinese LTCC faces used 4 leg shields. Subsequently the 2 leg shield design has been adopted in some of the newer Chinese LTCC operations. Moodie &

Table 3 A selection of Chinese LTCC faces.

Mine	*Owner*	*Depth*	*Mining Height*	*Cave: Cut Ratio*	*Face Length*
Jining No. 3	Yanzhou	700m	Cut 3m in 7m seam	1.3	290m
Dongtan	Yanzhou	710m	Cut 3–3.5m in 6m seam	0.7 – 1.0	223m
Gaohe	Shanxi Gaohe Energy	470m	Cut 2.5–2.8m in 6.5m seam	1.3 – 1.6	236m
Tashan	Datang Tashan Coal Mine Co.	400m	Cut 3.5m in 12–20m thick seam	2.4 – 4.7	235m

Anderson (2011) provided an assessment of the successful implementation of LTCC at Austar, including the geotechnical challenges of caving management. Austar mines the 6.5m thick Greta Seam at depths of 500–600m, with a cutting height of 2.9m, hence a cave to cut ratio of approximately 1.2.

It should also be noted that two further LTCC faces have since been introduced in Australia. These are at the Broadmeadow Mine, operated by BMA; and also the North Goonyella Mine, owned by Peabody. Both of these mines are located in the Bowen Basin in Queensland. In both cases, the new LTCC faces are replacing previous SPL faces.

Table 3 presents data from a number of the higher performing LTCC faces in China over the last five years. (Data and reports of performance and experience with LTCC (and the previous Bulianta information) was provided by the different mines, during visits conducted by Hebblewhite (2011)).

3.2.1 Jining No. 3 & Dongtan

These two mines are part of the Yanzhou operations in the east of China (Shandong Province). Both employ LTCC for their thick seam operations, although conventional height longwalls are also in use in other lower height seams.

Jining was producing 7MTPA at the time of a visit by the author in 2011/2012, with approximately 85% of this production from the LTCC face which was achieving up to 15,000 t/day production, and an average volumetric coal recovery of 70% of the 7m seam being mined. The mine adopted an unusual mid-face gate road, to bring services onto the face – reportedly due to problems of water in the maingate. In spite of this complication, the mine was achieving a face retreat rate of approximately 6m/day.

Dongtan Mine has been using LTCC since 1992, having previously used multi-slice longwall. In 2006, following the development of the DBT 2 leg LTCC shield, a new 2 leg, 850 tonne capacity shield face was introduced to Dongtan with a production capacity of 6MTPA. This produces 15,000 – 16,000 t/day. The cutting height at Dongtan is dictated by a mid-seam stone band at approximately 3.5m above the floor.

Conditions on both the Jining and Dongtan faces were very good, with no evidence of any significant face or roof stability problems, and good coal caving (and good fragmentation) immediately behind the face onto the rear AFC. Dongtan only caves one shield at a time, immediately behind the shearer, and claimed a volumetric coal recovery of up to 86%.

Discussion with engineers from Yanzhou revealed that all mines in the group were now using LTCC rather than the previous multi-slice longwall system. This had resulted in a significant improvement in mining conditions and stability, and also marked improvement in production (daily average face production up from 10,000 t/day to 15,000 t/day), and significant reduction in development requirements. Depending on the cave:cut ratio for any given seam, they were also considering higher capacity rear AFCs to handle increased caving ratios. (They believe that the optimum cave:cut ratio is approximately 3:1 for current LTCC technology).

On the subject of deciding between SPL and LTCC, the Yankuang view expressed was that it was very much related to geological conditions (especially coal seam conditions). In their eastern Chinese conditions, SPL was used for any seams up to 5m; but LTCC was adopted for any seams above 5m where face and possible coal roof stability became a problem with SPL at such heights, due to face instability caused by coal seam structure and strength. By comparison with the new SPL high faces in northern and western China (see earlier Bulianta discussion), Yankuang believe that the geotechnical differences that allow for high SPL faces are stronger coal, shallower depth and lower horizontal stresses allowing for improved roof conditions above and ahead of the shields.

3.2.2 *Gaohe*

The Gaohe Mine is a relatively new mine operation located near Changzhi, in Shanxi Province, south of Taiyuan city. The mine is owned by a joint venture – 55% to the local Lu'an Coal Bureau, and 45% to Asian-American Coal which is now wholly owned by Banpu from Thailand.

Gaohe had initially been designed in the early years of the 21st century with a unique shield concept that would allow either SPL or LTCC mining options – in other words, a shield with a 6m height range, which also incorporated an LTCC rear configuration. While initial design work was done for such a shield, it never eventuated and the mine subsequently was developed as an LTCC mine – particularly due to concerns about lower than expected coal strength and the prevalence of structure within the coal. Gaohe engineers refer to a Chinese empirical coal cavability index referred to as an F Factor. Inquiries made among Chinese researchers/geotechnical experts into the F Factor have revealed that it is simply an indicator of strength, determined as UCS (in MPa), divided by 10. According to them, Gaohe coal has an F Factor of 0.2 (representing extremely weak coal), compared to 0.6 for other mines in the same area, but this contrasts with a value of 6.0 for Bulianta – a very significant difference. Clearly, based on these figures, SPL would certainly not have been an option for Gaohe.

Gaohe was designed for a production capacity of 6MTPA. Shields were 800 tonne capacity, cutting at between 2.5m and 2.8m height. The face conditions included a mid-seam stone band plus regular cleating and jointing, and a number of low angled faults (all of which mitigated against SPL as an option). Other problems for Gaohe included high gas levels (18 m^3/tonne CH_4); and higher than normal subsidence due to the top 200m of overburden being unconsolidated soils and clays.

3.2.3 *Tashan*

The Tashan Mine, located near Datong, in Shaanxi Province, was at the time the highest performing LTCC mine in China (and hence the world). The mine was operating two LTCC faces and one conventional height longwall, with a total annual mine production of 23MTPA – 10MTPA of which was coming from each of the LTCC faces. These were averaging 35,000 t/day, with a maximum of 50,000 t/day achieved. Seam thickness varied from 12m to 20m, and the mine claimed to be recovering between 70% and 80% of the seam in each panel, by volume – with a cutting height of only 3.5m (giving a cave:cut ratio as high as 4.7), and a web thickness of 800mm. Overall face retreat rate was between 3m and 5m per day (face length of 235m at a depth of 400m).

The Tashan LTCC faces consisted of Chinese built four leg, 1,300 tonne capacity shields. Coal clearance was achieved by AFC capacities of 2,000 t/hr (front) and 3,000 t/hr (rear), plus a maingate belt capacity of 3,500 t/hr, using a 1.4m wide belt. Underground conditions on the face were excellent in all respects. Typical coal fragmentation onto the rear AFC was in the range of 100mm – 200mm particle size, with only isolated blocks requiring any additional breakage. Face shields were rarely yielding, other than some groups of rear legs in yield – often associated with top coal fracturing initiating just above/ahead of the start of the goaf shield.

3.3 Geotechnical considerations and future research challenges

It is clear from the above discussions that there are a number of geotechnical issues and challenges associated with both LTCC and SPL thick seam mining. In terms of LTCC, one of the major considerations is the assessment, and then predictability of effective, consistent and controlled top coal caving from above the goaf shield and immediately behind the face. Both cavability, and also fragmentation, are critical factors in the success of LTCC. If the coal hangs up even for as little as 2m behind the shields, prior to failing, it will cave beyond the reach of the rear conveyor and not be recoverable.

Chinese experience suggests that cavability and fragmentation are primarily controlled by a combination of in situ stress, coal strength, and coal structure (cleat and bedding planes). The Chinese industry, with the benefit of a large number of LTCC operations, relies heavily on empirical relationships derived from experience. For example, in terms of depth, there is a general view that for typical Chinese coal properties, depths of at least 150m – 180m are required, as a minimum, in order to generate sufficient vertical loading in the coal to initiate the coal fracture process above the shields. Coal strengths are generally required to be at or below about 20MPa, if good caving is to be expected – but this is very much a function of the coal structure density also. Higher structural density in stronger coals may also cave well. In contrast, for SPL to be successful from a geotechnical perspective, coal strengths must be moderate to high, with minimum structure present in the coal, in order to be able to maintain stable high face conditions.

Outside of the extensive empirical approaches adopted in China, some further advances have been made internationally to address this challenge of assessing coal seams for cavability or otherwise. Vakili & Hebblewhite (2010) developed an empirical Top Coal Caving Rating (TCCR) based on a combination of various seam geometric factors; stress conditions; coal properties and coal seam structure. This was calibrated using a limited amount of available Australian and Chinese caving data. This rating system also made use of numerical modeling using discrete element modeling, in order to provide appropriate parametric inputs. There is certainly scope for far more refinement and development in this area, together with improved numerical modeling of top coal cavability.

The impacts on cavability of both high horizontal stress (especially parallel to the longwall face orientation); and also the presence of massive overburden strata units above the coal are also areas where further research effort is required. Chinese experience does not include high or extreme conditions relating to these two parameters, so there is very little experience to draw on, in order to understand how horizontal stress and/or massive overburden can impact the cavability of the top coal.

A more recent and obvious area where further geotechnical research studies are required is to provide a means (empirical or numerical) for assessing SPL face stability, in order to make a geotechnical judgment on the choice between LTCC and SPL. This is clearly a function of depth (hence face loading); coal properties and strength; and coal seam structure, and may well be closely linked to any predictive capability for top coal caving also.

The above two factors – cavability/fragmentation assessment, and high face stability for SPL – are the most pressing issues for consideration at the mining method selection/design stage of a thick seam project. The issue of face shield stiffness is a further one warranting ongoing consideration. The concerns expressed by Matla about high shield aspect ratios are one part of this issue. But overall stiffness of the structure and the hydraulic system in these very tall supports is another factor worthy of consideration.

Another operational performance parameter which is closely linked to geotechnical considerations is the rate of retreat of a longwall face in thick seam conditions. Clearly, for the same face length as a conventional height face, there is more recoverable coal per metre of face retreat in a thick seam operation, which, everything else being equal, will result in a slower retreat rate to achieve maximum recovery. While this is a good thing in terms of relieving pressure on development rates, and operational aspects of services retraction, it can also create problems. It is a well-known fact that slower retreat rates can lead to face stability and shield closure problems, when mining in difficult ground conditions. Maintaining a faster and consistent retreat rate is one of the most effective remedies when mining through difficult ground. Therefore in such conditions, if the overall mining, caving and coal clearance rates cannot be increased on a thick seam face, it may be necessary to design the thick seam longwall panels (LTCC or SPL) with a shorter face length, in order to achieve an acceptable minimum face retreat rate at all times. This same issue of slow retreat rate can also have adverse implications for spontaneous combustion management.

Finally, for any form of thick seam mining, surface subsidence is an inevitable consequence which must be taken into account. A 7m thick coal seam may result in surface subsidence of between 2m and 4m or more, depending on depth and face width. This can require modifications to underground panel designs (*e.g.* narrow panels), or considerable surface preparation and rehabilitation. Thick and even conventional seam height operations in some Polish mines place sand backfill behind their longwall faces in order to reduce the maximum potential subsidence from a typical value of 60% or more of extraction height, to a more acceptable level of below 30%. However this comes at a great cost in terms of underground operational cost and performance.

ACKNOWLEDGMENTS

The author wishes to thank all management and operations personnel at a number of international thick seam mining operations; together with longwall equipment manufacturers; for their willingness to provide information and share their experiences, and assist with logistical arrangements for mine visits that have informed the content of this chapter.

REFERENCES

Australian Longwall Magazine (2012). Australian Longwall Statistics, prepared by Ken Cram, Coal Services Pty Ltd, March 2012, pp48–56.

Bassier R & Mez W (1999). Application of paste fill in active longwalls and for stowage. 2nd Intl Underground Coal Conf., UNSW, Sydney, Australia, 15–18 June 1999, pp47–54.

Bulianta Mine staff (2011). Personal communication.

Hamilton N (1999). Single pass thick seam longwall experience at West Wallsend Colliery. 2nd Intl Underground Coal Conf., UNSW, Sydney, Australia, 15–18 June 1999, pp55–61.

Hebblewhite B (1999). Overview of Australian thick seam mining prospects. 2nd Intl Underground Coal Conf., UNSW, Sydney, Australia, 15–18 June 1999, pp29–36.

Hebblewhite B (2001). Risk assessment of geotechnical factors associated with underground thick seam mining methods. 20th Intl Conf. on Ground Control in Coal Mining, Morgantown, USA, August 2001, pp26–33.

Hebblewhite B, Simonis A & Cai Y (2002). Technology and feasibility of potential underground thick seam mining methods. School of Mining Engineering, UNSW/CMTE ACARP Project C8009 Final Report UMRC 2/02.

Hebblewhite B (2011). Mine data provided by various Chinese mines (personal communication).

Jian W, Xianrui M & Yaodong J (1999). Development of Longwall Top Coal Caving technology in China, in Proceedings of the 1999 International Workshop on Underground Thick Seam Mining (Ed: Jian, W and Jiachen, W), p105, The National Nature Science Foundation of China, Beijing.

Matla Mine staff (2011). Personal communication.

Moodie A & Anderson J (2011). Geotechnical considerations for Longwall Top Coal Caving at Austar Coal Mine. 30th Intl Conf. on Ground Control in Coal Mining, Morgantown, USA, July 2011, pp238–247.

O'Beirne T (1982). Design and trial of a multi-lift Wongawilli mining method for thick coal seams. ACIRL Published Report PR82-4, Australian Coal Industry Research Laboratories Ltd, 102pp.

Palarski J (1999). Multi-slice longwalling with backfill. 2nd Intl Underground Coal Conf., UNSW, Sydney, Australia, 15–18 June 1999, pp37–46.

Peng S (1998). Recent developments and some problems in the application of fully mechanised longwall mining, using sub-level caving, in China. Coal International, November, p229.

Simonis AC (1997). The underground mining of thick coal seams. Internal Report for the Centre for Mining Technology and Equipment (CMTE), Brisbane. Unpublished.

Vakili A & Hebblewhite B (2010). A new cavability assessment criterion for Longwall Top Coal Caving. IJRMMS Vol 47, No. 8, pp1317–1329.

Wang J (2011). The technical progress and problems to be solved of thick coal seam mining in China. 30th Intl Conf. on Ground Control in Coal Mining, Morgantown, USA, July 2011, pp363–368.

Xu B (2001). The longwall top coal caving method for maximizing recovery at Dongtan Mine. 3rd Intl Underground Coal Conf., UNSW, Sydney, Australia, 12–15 June 2001, unpag.

Chapter 14

Pillar design issues in coal mines

Ebrahim Ghasemi
Department of Mining Engineering, Isfahan University of Technology, Isfahan, Iran

Abstract: Coal mine pillar design has been the subject of sustained and intensive researches in the major coal producing countries of the world. Pillar design and stability are two of the most complicated problems in mining related to rock mechanics and ground control subjects. Without pillars to support the great weight of overburden, underground coal mining would be practically impossible. Coal pillars are employed in a wide variety of mining operations, from shallow room and pillar mines to deep longwall mines. In this chapter, first, different types of coal pillars are introduced and then the design method for each type is described. It should be mentioned that all presented equations are in their original forms. Therefore, the units are not the same, *i.e.* some are in metric (SI) system and some in English engineering system.

1 INTRODUCTION

Coal pillars can be defined as the in situ coal between two or more underground openings (Bieniawski, 1987; Peng, 2008). Their main function is to support the weight of overburden and to protect the integrity of adjacent entries and crosscuts, thereby allowing miners to extract the coal between them and to travel safely. Coal pillars are employed in a wide variety of mining operations, from shallow room-and-pillar mines to deep longwall mines. The various pillars used in coal mines can be classified into several categories based on objective and mechanism, as can be seen in Table 1 and 2 (Peng, 2008).

Pillar stability is one of the prerequisites for safe working conditions in underground coal mines. When the pillar design is improper, three types of failures can occur (Mark *et al.*, 2003): (1) pillar squeeze: where the pillars slowly fail in a controlled manner and the coal mine roof and floor undergo excessive convergence, (2) massive pillar collapse: where a large area of undersized pillars suddenly fails, and (3) coal pillar bumps: where the pillar fails dynamically and violently ejects coal into the surrounding entries.

Many researchers have devoted a part of their studies to prevention of pillar squeezes, massive pillar collapses and pillar bumps (Khair & Peng, 1985; Campoli *et al.*, 1989; Maleki, 1992; Mucho *et al.*, 1993; Chase *et al.*, 1994; Zipf, 1996; Mark *et al.*, 1997; Zipf & Mark, 1997; Zipf, 1999; Maleki *et al.*, 1999; Chen *et al.*, 1999; Zingano *et al.*, 2004; Iannacchione & Tadolini, 2008; Esterhuizen *et al.*, 2010; Poeck *et al.*, 2015; etc.). The occurrence of pillar failure in underground coal mines entails detrimental effects on miners in the form of injury, disability or fatality as well as

Table 1 Different types of pillars based on objective.

Name	*Definition*
Panel pillars	Rows of in situ block of coal aligned in the main and submain panels of room and pillar coal mines.
Chain pillars	Rows of in situ block of coal aligned in the gateroads and bleeder systems of longwall coal mines.
Barrier pillars	a) Internal barrier pillars: solid coal blocks designed to isolate the effect of mining on one side of the block from the other side. b) Outcrop barrier pillars: blocks of coal left around the outcrop perimeter of an above drainage mine.
Web pillars	The blocks of coal left between adjacent cuts in highwall mining.

Table 2 Different types of pillars based on mechanism.

Name	*Definition*
Stiff (abutment) pillars	Pillars that are designed to support the expected load that the pillars experience throughout their service life.
Yield pillars	Pillars that are designed in deep mines to yield at the proper time and rate.

mining company due to downtimes, interruptions in the mining operations, equipment breakdowns, and etc. Proper pillar design is the key to prevention of pillar failure and reduction of related accidents. Coal mine pillar design has been the subject of sustained and intensive researches in the major coal producing countries of the world. Pillar design and stability are two of the most complicated and extensive problems in mining related to rock mechanics and ground control subjects. The main aim of this chapter is to describe the pillar design issues in coal mines. This chapter is organized as follows: panel pillars issues are explained in section 2. In section 3, the design procedures of chain pillars are described. In section 4, design formulas for barrier pillars are presented. The design considerations of web pillars are explained in Section 5 and finally, the yield pillar design issues are presented in Section 6.

2 PANEL PILLARS IN ROOM AND PILLAR COAL MINES

In underground coal mining, room and pillar is the method of working preferable for flat and tabular deposits in thin seams, where rooms or entries are driven in the solid coal to form pillars in the development panels (Hustrulid, 1982; Hartman, 1987). Pillars of coal are left behind to support the roof and prevent its collapse. In some cases, the pillars are extracted partly or fully in a later operation, known as retreat mining (also known as secondary mining or pillar recovery operation). Pillar design is one of the most fundamental principles of room and pillar mining. Primarily, the dimensions of pillar were largely determined by experience based on trial and error, intuition or established rules of thumb. But nowadays, various pillar design formulas are developed, based upon laboratory testing, full-scale pillar testing, and back-analysis of failed and successful case histories.

The pillar design methods can be basically divided into two approaches of classic and novel. The main difference of these two approaches is in estimation of load applied on pillar. The classic methods estimate pillar load through the calculation of overburden weight immediately above the pillar, which is called development load. The classic approach of pillar design is suitable when the pillar recovery operation is not intended to be carried out. Nowadays in most of the room and pillar mines in order to increase the recovery and productivity, remnant pillars in panels are extracted by retreat mining. Since the classic approach neglect the abutment loads due to retreat mining and creation of a mined out gob, they are not appropriate for pillar design in retreat mining. Regarding this shortcoming, novel methods of pillar design have been presented recently, which consider both development loads and the abutment loads induced by retreat mining. These approaches are explained in details as follows.

2.1 Classic approach

In 1980, field studies conducted by the US Bureau of Mines developed the classic pillar design methodology. It consists of three steps (Mark, 2006): (1) Estimating the pillar load, (2) Estimating the pillar strength, and (3) Calculating the pillar safety factor.

A number of approaches are available for estimating the pillar load during development phase (development load). The simplest approach to determine the pillar load is by the tributary area theory. Based on this theory, the area supported by a pillar covers the area above it and neighboring areas tributary to it. In the other words, a pillar uniformly supports the weight of rock overlying the pillar and one-half the width of rooms or entries on each side of the pillar (Peng, 2008). The tributary area concept for rectangular pillars is shown graphically in Figure 1. Based on this theory, the load on rectangular pillars can be calculated by Equation 1:

$$\sigma_p = \sigma_z \frac{A_t}{A_p} = \gamma H \frac{(w+B)(l+B)}{wl} \tag{1}$$

where σ_p is average pillar load (stress), σ_z is vertical stress, A_t and A_p are tributary area and pillar area respectively, w and l are width and length of pillar respectively, B is entry width, H is depth of cover, and γ is unit weight of overburden rock.

A room and pillar panel is often described by the extraction ratio (e), or recovery ratio, which refers to the ratio of the mined-out area to total area. For rectangular pillars the extraction ratio can be written as:

$$e = 1 - \frac{A_p}{A_t} = 1 - \frac{wl}{(w+B)(l+B)} \tag{2}$$

Therefore, the pillar development load can be determined as a function of the extraction ratio:

$$\sigma_p = \frac{\gamma H}{1-e} \tag{3}$$

Pillar strength can be defined as the maximum resistance of a pillar to axial compression (Brady & Brown, 1993). Empirical evidence suggests that pillar strength is related to both its volume and its geometric shape (Salamon & Munro, 1967; Brady & Brown,

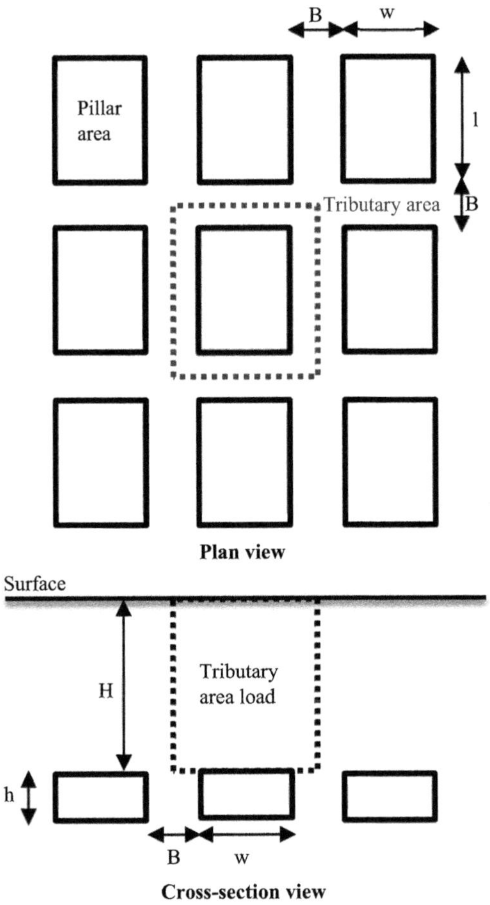

Figure 1 Schematic of tributary area theory for rectangular pillars.

1993). Numerous formulas have been developed for estimation of pillar strength in coal mines, most of which have been presented in Table 3 (Peng, 2008; Du *et al.*, 2008; Arioglu & Tokgoz, 2011). Each of these formulas estimates the pillar strength in terms of two variables; width to height ratio and in situ coal strength.

As can be seen, most of pillar strength formulas have been developed for square pillars and for estimating rectangular pillar strength, the effective width concept should be used instead of pillar width in these formulas. This concept converts the cross sectional area of a rectangular pillar into a square pillar of equivalent strength. There are three equations for calculating the effective width of rectangular pillars (Peng, 2008):

- Salamon & Oravecz (1976): $w_{eff} = \sqrt{wl}$
- Wagner (1980): $w_{eff} = 4A/P$; where A is the area of the pillar and P is the pillar circumference.
- Hsiung & Peng (1985): $w_{eff} = w^{0.85}l^{0.15}$

Table 3 Some empirical formulas for estimating coal pillar strength.

Pillar strength formula	*Presenter (year)*	*Pillar shape*	*Remarks*	
$S_p = 6.9\left(0.7 + 0.3\frac{w}{h}\right)$	Bunting (1911)	square	–	
$S_p = 19.2\frac{\sqrt{w}}{\sqrt[6]{h^5}}$	Greenwald (1939)	square	–	
$S_p = k\frac{\sqrt{w}}{h}$	Holland & Gaddy (1957)	Square	–	
$S_p = S_1\left(0.778 + 0.222\frac{w}{h}\right)$	Obert & Duvall (1967)	Square	–	
$S_p = 7.2\frac{w^{0.46}}{h^{0.66}}$	Salamon & Munro (1967)	Square	–	
$S_p = S_1\left(0.64 + 0.36\frac{w}{h}\right)$	Bieniawski (1968)	Square	–	
$S_p = S_1\sqrt{\frac{w}{h}}$	Holland (1973)	Square	–	
$S_p = 7.2\frac{R_o^{0.5933}}{V^{0.0667}}\left\{\frac{0.5933}{\varepsilon}\left[\left(\frac{R}{R_o}\right)^{\varepsilon} - 1\right] + 1\right\}$	Wagner & Madden (1984)	Square	(R≥5)	
$S_p = 0.27\sigma_c h^{-0.36} + \left(\frac{H}{250} + 1\right)\left(\frac{w}{h} - 1\right)$	Sheorey *et al.* (1987)	Square	–	
$S_p = S_1\left(0.64 + 0.54\frac{w}{h} - 0.18\frac{w^2}{lh}\right)$	Mark & Bieniawski (1988)	Rectangular	–	
$S_p = 5.24\frac{w^{0.63}}{h^{0.78}}$	Madden (1991)	Square	–	
$S_p = 6.88\frac{\sqrt{w_e\Theta}}{h^{0.7}}$	Galvin *et al.* (1999)	Parallel-piped	w/h<5	$\Theta = \frac{2l}{w+l}$
$S_p = \frac{19.05\sqrt{\Theta}}{w_e^{0.133}h^{0.066}}\left\{0.253\left[\left(\frac{w_e}{5h}\right)^{2.5} - 1\right] + 1\right\}$			w/h≥5	$w_e = w\sin\theta$
$S_p = 3.5\frac{w}{h}$	Van der Merwe (2002)	square	Normal coal	
$S_p = 1.5\frac{w}{h}$			Weak coal	

S_p: pillar strength (MPa), *w*: pillar width (m), *h*: pillar height (m), *l*: pillar length (m), S_l: strength of critical cubical coal pillar (MPa), *k*: Gaddy factor (MPa), σ_c: strength of a 25 mm cube coal sample (MPa), *R*: width to height ratio, R_o: the critical width to height ratio, *V*: pillar volume (m^3), ϵ: the rate of strength increase, w_e: effective width of parallel-piped pillars, θ: internal angle for parallel-piped pillars, and Θ: a dimensionless factor.

The safety factor is defined as the ratio of pillar strength to pillar load. When designing a system of pillars, the factor of safety must be selected carefully, because it must compensate for the variability and uncertainty related to pillar strength and stress and mining inconsistencies. An appropriate safety factor can be selected based on a subjective assessment of pillar performance or statistical analysis of failed and stable cases (Salamon & Munro, 1967). As the factor of safety decreases, the probability of pillar failure can be expected to increase. Theoretically, the safety factor value greater than one means that the pillar is stable, while the values lower than one means unstable. A safety factor of one implies only a 50% probability of stability i.e. there is an even chance of the system failure or stability. Therefore, a safety factor significantly greater than one should be selected in the design stage of room and pillar panels. It is recommended that the safety factor should be 1.5 for short term pillars and 2 for long term pillars (Bieniawski, 1992).

Based on classic approach, Bieniawski (1981) presented a logical, step-by-step approach for sizing coal pillars in room and pillar mines. These steps can be summarized as follows (Bieniawski, 1987):

Step 1: from geological data, borehole logs and rock and coal specimen testing (54 mm core or cubes), tabulate the following: uniaxial compressive strength of roof rock and coal, spacing, condition and orientation of geological discontinuities and ground water conditions.

Step 2: based on uniaxial compressive strength of coal (σ_c), determine the value of Gaddy factor (k) for the pillar locality:

$$k = \sigma_c \sqrt{D} \tag{4}$$

where D is specimen diameter or cube size.

Step 3: select a pillar strength formula to estimate the pillar width (w) for a known seam height (h):

$$S_p = S_1 \left(0.64 + 0.36 \frac{w}{h}\right) \tag{5}$$

S_1 for pillars with height more than 36 inches (0.9 m) can be calculated by the following equation:

$$S_1 = \frac{k}{\sqrt{36}} \tag{6}$$

and for pillars with height less than 36 inches (0.9 m), S_1 can be determined by below equation:

$$S_1 = \frac{k}{\sqrt{h}} \tag{7}$$

Step 4: determine the roof span (entry width; B) based on roof rock quality.

Step 5: calculate the pillar load based on the tributary area theory.

Step 6: select a safety factor (SF) and equate $S_p = \sigma_p / SF$. Solve for pillar width (w).

Step 7: for economic considerations, check whether the extraction ratio (e) is acceptable for profitable mining. Extraction ratio for rectangular pillar plans can be calculated using the Equation 2, as mentioned before and for square pillar plan the Equation 8, can be used:

$$r = 1 - \left(\frac{w}{w+B}\right)^2 \tag{8}$$

Step 8: if the extraction ratio is not acceptable and needs to be increased by decreasing the pillar width, select the pillar width from step 7 (and pillar length if required) which would give the required coal extraction and determine whether this is acceptable for mine stability. This requires calculation of the factor of safety:

$$SF = \frac{S_p}{\sigma_p} \tag{9}$$

The safety factor should be between 1.5 and 2.

Step 8: Exercise engineering judgment, by considering a range of mining and geologic parameters, to assess the various options for mine planning. Consider the effect of the floor condition.

2.2 Novel approach

As mentioned before, since the classic approaches do not consider the abutment loads, they are not suitable for mines where retreat mining is going to be done. In retreat mining, the creation of a mined-out gob area generates abutment loads. Abutment loads affect the pillars adjacent to the pillar line and a load more than the one estimated by tributary area theory applies on pillar. Pillar design without the abutment loads leads to failure of pillars during retreat mining. So, proper pillar design is the key to prevent the pillar failure and reduce accidents related to retreat mining. In the US, the LaModel and the Analysis of Retreat Mining Pillar Stability (ARMPS) programs are used successfully for designing safe retreat mining (Tulu *et al.*, 2010). LaModel is a PC-based program for calculating the stresses and displacements in coal mines or other thin seam or vein type deposits (Heasley & Barton, 1999; Heasley, 2009). It is primarily designed to be utilized by mining engineers for investigating and optimizing pillar dimensions and layouts in relation to overburden, abutment and multiple seam stresses. The program was developed based on displacement-discontinuity variation of the boundary element method. The ARMPS computer program was developed by National Institute for Occupational Safety and Health (NIOSH) to aid the design of pillar recovery operations (Mark & Chase, 1997). ARMPS is an empirical method in which the statistical analysis is used to derive design guidelines that separate the "successful" case histories (those where the entire panel was mined without pillar failure) from those that are "unsuccessful". The goal is to minimize the risk of the most hazardous types of pillar failures (collapses and bumps). Like classic pillar design methodology, ARMPS consists of three basic steps: (1) estimate the applied loads, including development and abutment loads; (2) estimate the load bearing capacity of the coal pillars; and (3) compare the load to the capacity, and employ engineering criteria to determine whether the design is adequate.

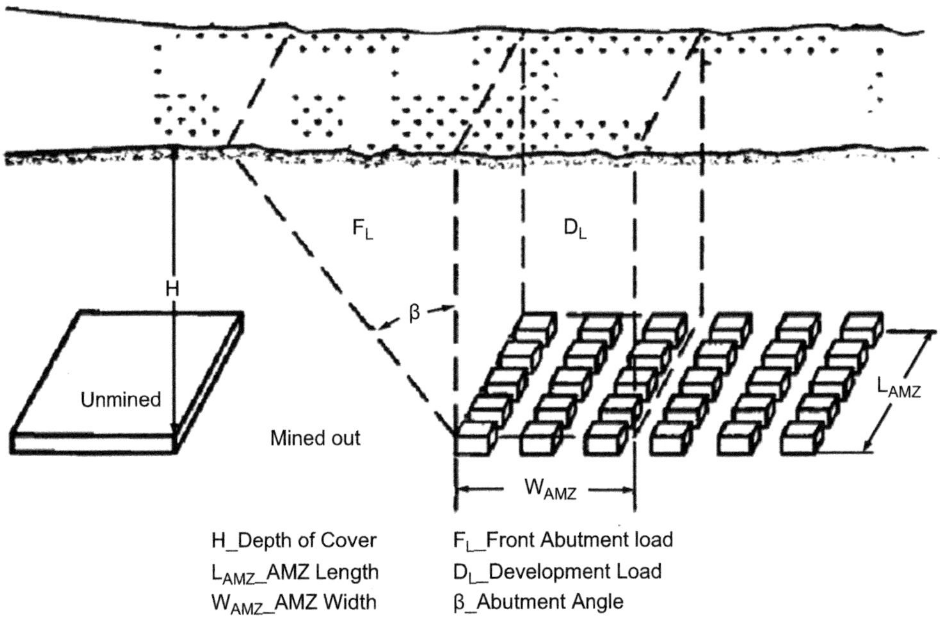

Figure 2 Schematic show of the AMZ (Mark & Chase, 1997).

For pillar design, ARMPS program considers the pillars in the active mining zone (AMZ). As shown in Figure 2, AMZ includes all of pillars on the extraction front (or pillar line), and extends out by the pillar line a distance of 5 times the square root of the depth of cover. This width of AMZ was selected because measurements of abutment load distributions show that 90% of front abutment loads fall within its boundaries (Mark & Chase, 1997). To estimate the development loads, ARMPS starts with the tributary area theory. The abutment angle concept is used to estimate the abutment loads transferred to the pillars during the various stages of pillar extraction process. To calculate the strength of the pillars within the AMZ, ARMPS uses the Mark-Bieniawski formula. Each pillar's load bearing capacity is simply its strength multiplied by its load bearing area. A stability factor (SF) is then calculated by dividing the total load bearing capacity of all pillars within the AMZ by the total load. ARMPS also calculates a SF for each barrier pillar that is part of the design, and if the barrier pillars are too small, transfers additional loads to the AMZ. The power of ARMPS is not derived from the accuracy of its calculations, but rather from the large data base of retreat mining case histories that it has been calibrated against. Statistical analysis has been used to help derive guidelines for selecting an appropriate ARMPS SF for design. After the Crandall Canyon mine disaster, the NIOSH revisited the issue of pillar design for deep cover retreat mining. The analysis focused on the development of a "pressure arch" loading model for ARMPS. The revised loading model has been implemented in an updated version of the ARMPS computer program, called ARMPS Version 6 (2010), in which the size of ARMPS database has increased from 150 case histories to almost 650 (Mark, 2010; Mark *et al.*, 2011). Ghasemi & Shahriar (2012) presented a new method for designing coal pillars in room and pillar mines in order to enhance the safety of retreat mining and increase the recovery of coal reserves. In this method, the abutment loads due to retreat mining are estimated using empirical equations. This method is similar to ARMPS program in structure, but the main difference is that optimum pillar width is calculated using this method in order to decrease the pillar failure risk during retreat mining. Whereas, in ARMPS program by inputting parameters such as pillar width, panel width, depth of cover and etc., stability of pillars in preliminary and secondary stages are evaluated. It means that in proposed method the pillar width is an unknown parameter (*i.e.* pillar width is the main output) whereas in ARMPS program the pillar width is one of inputs.

Similar to the ARMPS program, the proposed method by Ghasemi & Shahriar (2012) considers the pillars in the AMZ because these pillars are exposed to maximum load throughout mining process therefore the pillar dimensions obtained by this method are more satisfactory. The Ghasemi & Shahriar method is made up of twelve steps which are described below. Figure 3 also illustrates different steps of this method in a flowchart plot. The symbols used here are provided in Table 4.

Step 1: Gathering essential data

Essential data to determine the optimum pillar dimensions in this method are as following:

1. Depth of cover: average overburden thickness over the pillar system.
2. Pillar height (Mining height): note that the value of pillar height is not necessarily equal to the seam thickness.
3. Entry width: entry width is usually determined based on roof rock quality, production rate and operational width of equipment. In this method, crosscuts are assumed to have the same width as the entries.
4. In situ coal strength
5. Mean unit weight of the overburden
6. Abutment angle: the abutment angle determines how much load is carried by gob. Measurements of abutment loads in US coal mines indicated that an abutment angle of 21 degrees is appropriate for normal caving conditions (Mark & Chase, 1997).
7. Panel width: panel width is usually determined based on geotechnical conditions, stress state in the region, economic criteria, and environmental conditions. Panel width affects stress distribution, loading conditions and caving mechanism. An increase in panel width results in an increase of abutment loads applied on the pillars adjacent to the gob area. The tension zone height developed in the roof of gob area also increases as the panel width increases and may lead to a large failure in overburden (Bieniawski, 1987). Based on width to depth ratio (P/H), panels are divided into two categories: (1) Sub-critical panels ($P/H<2\tan\beta$), and (2) Super-critical panels ($P/H\geq 2\tan\beta$).
8. Coal Mine Roof Rating (CMRR): this index is used to evaluate roof rock quality. In 1994 the CMRR was developed to fill the gap between geologic characterization and engineering design (Mark & Molinda, 2005). This classification system considers geotechnical factors such as roof rock strength, bedding and other discontinuities, moisture sensitivity of the roof rock, groundwater, etc. CMRR varies between zero and 100. Based on this index, roof rocks in coal mines are put in three categories (Chase *et al.*, 2002): Weak (CMRR<45), Intermediate (45<CMRR<65), and Strong (CMRR>65).

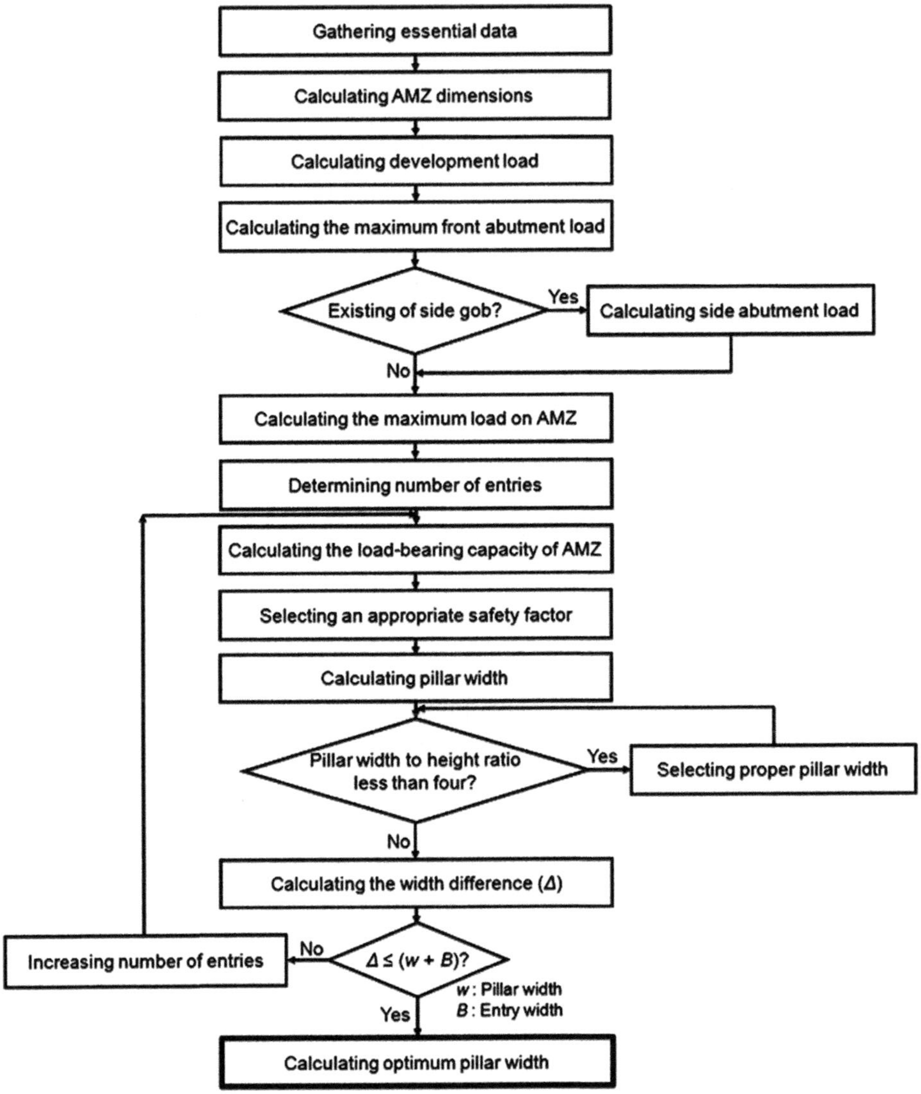

Figure 3 Flowchart for proposed coal pillar design method.

Step 2: Calculating AMZ dimensions

AMZ length and width are determined from Equations 10 and 11 respectively:

$$L_{AMZ} = P - B \tag{10}$$

$$W_{AMZ} = 5\sqrt{H} \tag{11}$$

Step 3: Calculating development load

Development loads are resulted from the overburden weight over active mining zone. Based on tributary area theory, development loads are obtained from Equation 12:

Table 4 Used symbols in proposed coal pillar design method.

Symbol	*Description (unit)*
AMZ	Active mining zone
H	Depth of cover (m)
P	Panel width (m)
h	Pillar height (m)
B	Entry width (m)
γ	Mean unit weight of the overburden (KN/m^3)
β	Abutment angle (deg.)
L_{AMZ}	AMZ length (m)
W_{AMZ}	AMZ width (m)
D_L	Development load (KN)
F_L	Maximum front abutment load (KN)
S_L	Side abutment load (KN)
W_{SG}	Side gob width (m)
W_B	Barrier pillar width (m)
R	Transfer rate (%)
M_L	Maximum load applied on AMZ (KN)
S_P	Pillar strength (MPa)
N_E	Number of entries
N_P	Number of pillars in AMZ
T_{LC}	Overall load-bearing capacity of AMZ (KN)
SF	Safety factor
w	Pillar width (m)
Δ	Width difference (m)
w_P	Optimum pillar width (m)

$$D_L = \gamma H L_{AMZ} W_{AMZ} \tag{12}$$

Step 4: Calculating the maximum front abutment load

Retreat mining starts with the extraction of panel pillars. When a sufficient number of pillars has been extracted, the overburden strata above the extracted pillars start to cave. As a result of this roof caving, the active gob is formed. Some portion of the overburden load above the gob is carried by the gob, but a considerable amount of the original overburden load over the gob is transferred to the pillars in AMZ and barrier pillars as a front abutment load (see Figure 2). Front abutment load is calculated based on abutment angle concept (Mark, 1992; Tulu *et al.*, 2010) and its distribution is different in sub-critical and super-critical panels (see Figure 4). Depending on whether the panel is sub-critical or super-critical, the maximum front abutment load is given by Equations 13 and 14, respectively:

$$F_L = 0.9\,\gamma\left[-\frac{L_{AMZ}^3}{24(\tan\beta)} + \frac{H\,L_{AMZ}^2}{4}\right] \tag{13}$$

$$F_L = 0.9\,\gamma\left[\frac{H^2\,L_{AMZ}}{2}(\tan\beta) - \frac{H^3}{3}(\tan\beta)^2\right] \tag{14}$$

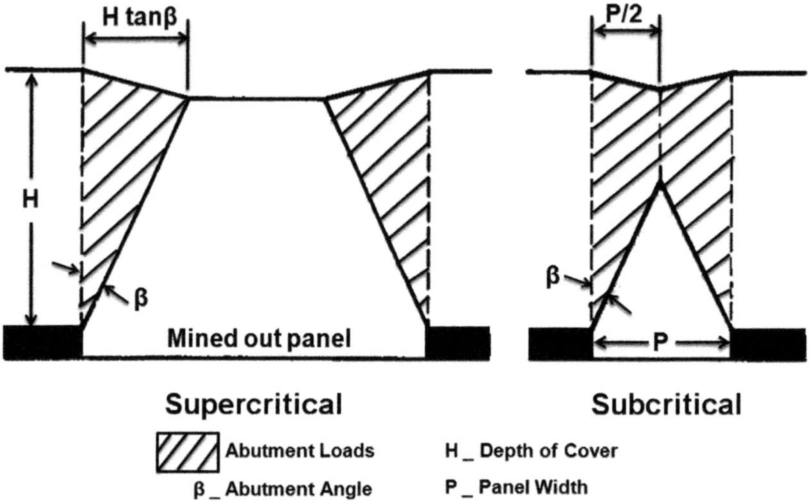

Figure 4 Abutment angle concept in sub-critical and super-critical panels (Mark, 1992).

Step 5: Calculating side abutment load

The gob area beside the mining panel is the source of side abutment load. Two gob areas may exist beside each mining panel. The side abutment load is shared between the barrier pillar and the AMZ. This load the same as front abutment load is calculated by abutment angle concept. Gob area width and barrier pillar width are required to calculate side abutment load applied on AMZ. Depending on whether the side gob is sub-critical or super-critical, side abutment load is given by Equations 15 and 16, respectively:

$$S_L = \left(\frac{H\,W_{SG}}{2} - \frac{W_{SG}^2}{8(\tan\beta)}\right)\gamma \times W_{AMZ} \times R \tag{15}$$

$$S_L = \frac{H^2}{2}(\tan\beta)\,\gamma \times W_{AMZ} \times R \tag{16}$$

In both of them, regarding Equation 17, R is:

$$R = \left(\frac{9.3\sqrt{H} - (W_B + (B/2))}{9.3\sqrt{H}}\right)^3 \tag{17}$$

Factor *R* is transfer rate that shows the percentage of total side abutment load applied to AMZ.

Step 6: Calculating the maximum load on AMZ

The maximum load applied on the pillars in AMZ is calculated by summation of development load, maximum front abutment load, and side abutment load according to Equation 18:

$$M_L = D_L + F_L + S_L \tag{18}$$

Step 7: Determining number of entries

The number of existing entries is usually determined based on panel width, rock mechanics conditions, operational equipment, and production rate. At least four entries are needed; one for accommodating the conveyor, one for fresh air, and two others in two sides of panel to take the air out (Stefanko, 1983). Economically and operationally, this number of entries is not adequate in continuous (mechanized) mining method and at least five entries should be planned in panel and this number increases up to seven entries in mines with high production rate (Hartman, 1987).

Step 8: Calculating the load-bearing capacity of AMZ

The load-bearing capacity of the pillars in AMZ is calculated by summing the load-bearing capacities of all pillars within its boundaries. The load-bearing capacity of each pillar is determined by multiplying their strength by their load-bearing area (Mark & Chase, 1997). In this method, pillar strength is estimated using the Bieniawski's strength formula. The number of existing pillars in AMZ, according to Equation 19 is:

$$N_P = \frac{W_{AMZ}}{(w+B)} \times (N_E - 1) \tag{19}$$

Hence, the overall load-bearing capacity of pillars in AMZ is given by Equation 20:

$$T_{LC} = (N_P \times S_P \times w^2) \times 10^3 \tag{20}$$

Step 9: Selecting an appropriate safety factor

The studies of Chase *et al.* (2002) provides suggested safety factors for stability of the pillars in AMZ, as shown in Table 5. These values are obtained from 250 analyses of panel design in US and as can be seen from the table, safety factor depends on CMRR as well as depth of cover.

Step 10: Calculating pillar width

In this step, putting the safety factor in Equation 21 and solving it, pillar width is obtained:

$$T_{LC} = SF \times M_L \tag{21}$$

Step 11: Correcting pillar width to decrease the pillar failure risk

One of the ways to decrease the pillar failure risk, especially large pillar collapses, is to choose a pillar width to height ratio larger than four. In this step, if the ratio of the

Table 5 Suggested safety factors for stability of the pillars in AMZ.

Depth of cover (m)	*Weak and intermediate roof (CMRR≤65)*	*Strong roof (CMRR>65)*
$H \leq 200$	≥1.5	≥1.4
200 $H \leq 400$	1.5–[(*H*-200)/333]	1.4–[(*H*-200)/333]
400 $H \leq 600$	0.9	0.8

obtained width from the previous step to pillar height is smaller than four, pillar width is increased so a pillar width to height ratio larger than four is reached. Of course, in order to control and avoid excessive increase of pillar width, the recovery rate is taken into consider. According to experiments and considering economic purposes in preliminary mining stage, the most suitable recovery rate varies from 40% to 60%. It should be noticed that 0.5 meters is added to the pillar width each time in this step.

Step 12: Determining the optimum pillar width

In this step, the width obtained from previous step is corrected so that the optimum pillar width is determined based on the number of pillars in each row and the panel width. To achieve this purpose, at first Δ should be calculated using Equation 22. If Δ is less than or equal to the sum of pillar width and entry width, the optimum pillar width is obtained from Equation 23. Otherwise, the number of entries is added depending on Δ value and calculations are repeated from step 8:

$$\Delta = L_{AMZ} - [(w + B) \times (N_E - 1)] \tag{22}$$

$$w_P = w + \frac{\Delta}{(N_E - 1)} \tag{23}$$

In addition to the mentioned methods, the probabilistic methods (Griffiths *et al.*, 2002; Cauvin *et al.*, 2009; Ghasemi *et al.*, 2010; Recio-Gordo & Jimenez, 2012; Wattimena *et al.*, 2013), numerical modeling (Su & Hasenfus, 1996; Gale, 1999; Mohan *et al.*, 2001; Heasley, 2009; Jaiswal & Shrivastva, 2009; Esterhuizen & Mark, 2009; Heasley *et al.*, 2010; Esterhuizen *et al.*, 2010) and artificial intelligence based methods (Jian *et al.*, 2011; Ghasemi *et al.*, 2014a, b) have been successfully used for designing coal pillars in room and pillar mines.

3 CHAIN PILLARS IN LONGWALL MINES

Longwall mining is a highly mechanized production technique that was developed in Europe in the early twentieth century and gained popularity very fast in all coal producing countries (Bise, 2013). This mining method similar to room and pillar mining is applied for extracting flat and nearly flat seams. Figure 4 shows a typical layout of a longwall mine. As can be seen, each panel is surrounded by three entries, namely headgate, tailgate and bleeder entries, which are also called panel entries. Panel entries affect the safety and productivity of longwall mining remarkably. Thus, in order to increase the safety, the panel entries are made as multi-entries. The panel entries consist of two, three, four, or sometimes five entries. The most common is the three-entry system. By law, this is the minimum number of entries required for developing the panel entries, because any entry development requires at least one separate entry for intake and return air escapeway, and a separate belt entry in neutral air (Peng, 2006).

In multi-entry system, each entry is separated from adjacent entry by a row of pillars which are called chain pillars. Chain pillars are left to support the overlying strata. Proper design of chain pillars in longwall mines is very vital in order to provide safe access roads for travel to and from the longwall face. The selection of proper

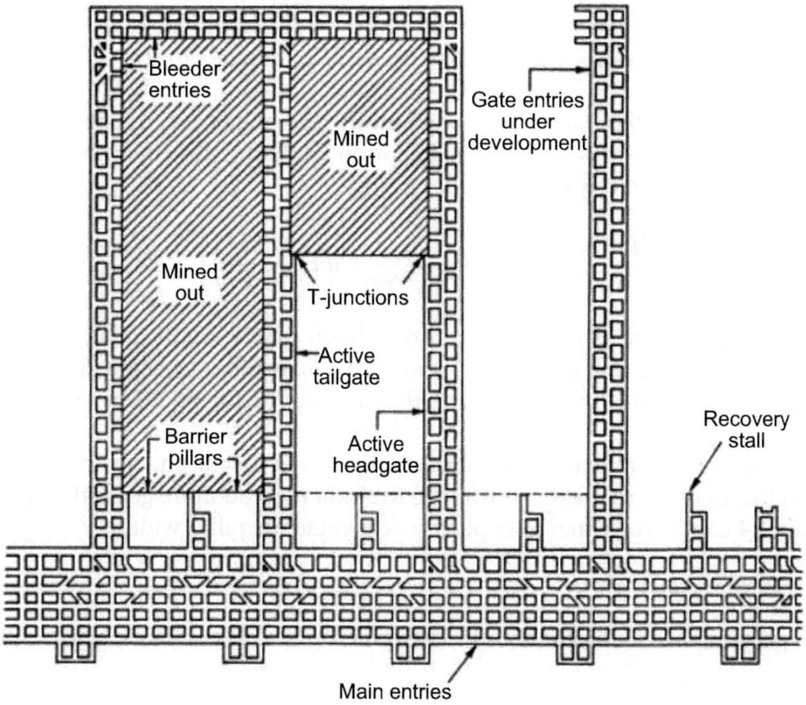

Figure 5 Typical longwall panel layout (Mark, 1990).

pillar size is particularly important in the longwall retreating phase since the high abutment loads can damage entries ahead of the longwall face and the tailgate for the next panel.

Several methods have been presented for designing chain pillars including Choi & McCain, Carr & Wilson, Hsuing & Peng, and analysis of longwall pillar stability (ALPS) methods which are explained in the following (Mark, 1990; Badr, 2004; Peng, 2008). It should be mentioned that the Carr & Wilson method is used for designing stiff and yield chain pillars but this section concentrates on design of stiff chain pillars and the design methodology for yield pillars will be described in section 6 in detail.

3.1 Choi & McCain method

Choi & McCain (1980) presented the first method for chain pillar design based on field studies, numerical modeling and practical experience on Pittsburgh coal seam, US. The Choi & McCain method was developed for three-entry system with a large stiff pillar next to a 9.75 m (32 ft) width yield pillar.

Longwall chain pillars experience various loads throughout mining process which can be divided into two types: (1) development loads: which are present before longwall mining, (2) abutment loads (front and side): which occur during longwall panel extraction. In the proposed method by Choi and McCain, development loads, front abutment loads and side abutment loads are estimated using tributary area theory, empirical equations and two-dimensional simplified subsidence model, respectively. Furthermore, the pillar strength is calculated by the Holland-Gaddy equation.

Considering the safety factor equation and its rearrangement, the final form of the Choi & McCain method is obtained as Equation 24. Using this equation, the width of stiff pillar can be indirectly determined.

$$P = 0.6H - 1.2\left[\frac{H^2}{4} - \frac{5}{3}\left(\frac{wl}{l+B} \times \frac{S_p}{24.9SF} - wH - \frac{BH}{2}\right)\right]^{0.5} \tag{24}$$

where P is panel width (ft), H is overburden depth (ft), w is stiff pillar width (ft), l is pillar length (ft), B is entry (crosscut) width (ft), S_p is pillar strength (psi), and SF is safety factor. Choi and McCain suggest that a safety factor of 1.3 be employed with their method.

As can be seen, the above equation is actually solved for the panel width rather than pillar width. In order to make the Choi & McCain method more generally applicable, Equation 24 can be rewritten to explicitly determine the pillar width:

$$0 = c_1 w^{1.5} - c_2 w + c_3 \tag{25}$$

where c_1, c_2, and c_3 are constant coefficients which are calculated using below equations:

$$c_1 = \frac{lk}{24.9SF(l+B)h} \tag{26}$$

$$c_2 = H \tag{27}$$

$$c_3 = \frac{3}{5}\left[\left(\frac{0.6H - P}{1.2}\right)^2 - \frac{H^2}{4}\right] - \frac{BH}{2} \tag{28}$$

where h is pillar height (ft) and k is Gaddy factor (psi).

The Equation 25 can be applied for rectangular pillars when w is smaller than l. When the required pillar width is greater than the desired pillar length, it is usually necessary to use square pillars to maintain pillar strength. If square pillars are used, then $w=l$ and Equation 25 must be rewritten as:

$$0 = c_0 \cdot w^{2.5} + c_1 \cdot w^{1.5} - c_2 \cdot w + c_3 \tag{29}$$

where c_0 is constant coefficient and calculated as follows:

$$c_0 = \frac{k}{24.9 \cdot SF \cdot h} \tag{30}$$

Finally, the Choi & McCain method can be used for back analysis of case histories by rearranging Equation 25 to be solved for the safety factor:

$$SF = \frac{c_4 \cdot w^{1.5}}{c_2 \cdot w - c_3} \tag{31}$$

where c_4 is a constant coefficient which can be calculated using below equation:

$$c_4 = \frac{lk}{24.9 \cdot h \cdot (l + B)} \tag{32}$$

It is worth mentioning that when the panel width is supercritical meaning that $P > 0.6H$, then a value of $P = 0.6H$ must be used in Equations 25 through 32.

3.2 Hsuing & Peng method

Hsuing & Peng (1985) presented a method for designing three-entry longwall systems using equal-sized pillars. Their method was developed based on three dimensional finite-element modeling. The three dimensional modeling allows the evaluation of pillar stability at the critical headgate and tailgate T-junctions. During modeling, Hsuing & Peng varied important design and rock mechanics parameters in order to assess chain pillar stability. These parameters included overburden depth, pillar width, compressive strength of coal specimens, panel dimensions, stiffness of the roof and floor, and thickness of the main and immediate roofs. Finally, Hsuing & Peng analyzed the model results using multivariate regression analysis and developed a simple equation that predicts the required chain pillar width for use in three-entry longwall systems:

$$\begin{aligned} w = & -4.676 \times 10^{-3}(E_i/E_c) - 4.04 \times 10^{-3}(E_m/E_c) - 3.33 \times 10^{-2}\log(E_f/E_c) \\ & -7.89 \times 10^{-2}\log S_c + 0.5144\log H + 4.94 \times 10^{-2}\log(L/P) + 0.1941\log P \end{aligned} \tag{33}$$

where w is chain pillar width (ft), E_c is elastic modulus of coal (psi), E_i is elastic modulus of immediate roof (psi), E_m is elastic modulus of main roof (psi), E_f is elastic modulus of floor (psi), S_c is compressive strength of laboratory coal specimen (psi), H is overburden depth (ft), L is longwall panel length (ft), and P is panel width (ft).

Hsuing & Peng suggest that for preliminary design purposes the lowest possible modulus ratios be used, or $E_i/E_c = 0$, $E_m/E_c = 0$, and $E_f/E_c = 1$.

3.3 Carr & Wilson method

This method was proposed originally for designing four-entry longwall systems (Carr & Wilson, 1982). In this system, three rows of chain pillars are made of two yield pillars on the sides and a stiff pillar in the center, commonly called the yield-stiff-yield system. The method attempts to optimize the pillar design in terms of speed of advance, reserves of coal, ventilation, and flexibility in operation in multi-entry systems. Generally, the Carr & Wilson method includes three main steps: (1) estimation of pillar strength, (2) estimation of pillar load, and (3) assessment of stability.

3.3.1 Estimation of pillar strength

Carr & Wilson used Wilson confined core model to estimate the pillar strength. Wilson confined core model may be summarized as follows. When a pillar is initially developed, it consists of two zones: an outer yield zone and an elastic inner core. The yield zone has failed and can take no more loads, but it provides confinement to the core, which usually provides most of the load-bearing capacity of the pillar. Immediately after development, the greatest stresses in the pillar are found at the boundary between the yield zone and the core. As the longwall progresses and additional loading are applied, the average stress in the pillar core increases until it equals the peak stress at the yield zone boundary. Up until this point, which Wilson calls the limit of roadway stability (LRS), both the pillar and entries adjacent to it are expected to be stable. Further loading on the pillar causes the yield zone to expand, resulting in increased horizontal stresses that can damage the nearby roadways. Finally, the ultimate limit (UL) is reached when the entire core has yielded. Any additional loads will now be transferred to adjacent pillars.

Wilson provided equations for determining the stress distributions in the pillar at both the LRS and the UL for two different boundary conditions, one in which the surrounding rock is rigid (rigid roof and floor, or RRF, conditions) and the other in which yielding takes place all around the entry (yielding roof and floor, or YRF, conditions). The strength (load-bearing capacity) of pillar can be determined by integrating these stress distributions over the area of pillar. When RRF conditions are assumed, the UL is unrealistically large, so the more conservative YRF conditions are almost always employed for coal mine applications. In YRF conditions, pillar strength for LRS and UL stages is calculated using Equations 34 and 35, respectively:

$$LRS = 8Y_1 + 2Y_2 + 3Y_3 + Y_4 \tag{34}$$

$$UL = 8Y_1 + 2Y_2 \tag{35}$$

Where Y_i is a constant coefficient and can be determined by:

$$Y_1 = \frac{hp^*}{2}\left[\frac{\left(\frac{2x_b}{h}+1\right)^{(K+1)} - 1}{(K+1)\frac{2}{h}} - x_b\right] \tag{36}$$

$$Y_2 = (l - 2x_b)\left(\frac{hp^*}{2}\right)\left[\left(\frac{2x_b}{h}+1\right)^{K} - 1\right] \tag{37}$$

$$Y_3 = (w - 2x_b)\left(\frac{hp^*}{2}\right)\left[\left(\frac{2x_b}{h}+1\right)^{K} - 1\right] \tag{38}$$

$$Y_4 = (l - 2x_b)(w - 2x_b)(Kq + S_1) \tag{39}$$

where h is pillar height (ft), $K=[(1+\sin\phi)/(1-\sin\phi)]$ is triaxial stress factor, ϕ is angle of internal friction (deg), l is pillar length (ft), w is pillar width (ft), p^* is uniaxial strength of fractured coal (pounds per square foot), $q=\gamma H$ is overburden pressure (pounds per square foot), H is overburden depth (ft), γ is unit weight of the overburden (pounds per cubic foot), S_1 is intact coal strength (pounds per square foot), x_b is width of yield zone (ft) for YRF conditions which is calculated by below equation:

$$x_b = \frac{h}{2}\left[\left(\frac{\gamma H}{p + p^*}\right)^{\frac{1}{K-1}} - 1\right] \tag{40}$$

where p is the resistance offered by the support system (pounds per square foot).

3.3.2 Estimation of pillar load

Carr & Wilson similarly to Choi & McCain for estimation of load imposing on chain pillars used tributary area theory and simplified subsidence model. They presented a step-by-step procedure in order to estimate the loads applied on each pillar. The first step is to calculate the total side abutment load. Side abutment load for subcritical ($P<0.6H$) and supercritical ($P>0.6H$) panels is calculated using Equations 41 and 42, respectively:

$$L_{ss} = 0.5P\gamma[H - (P/1.2)] \tag{41}$$

$$L_s = 0.15\gamma H^2 \tag{42}$$

where L_{ss} and L_s are side abutment load (pounds per foot of entry), P is panel width (ft).

In the next step, the peak abutment stress and the shape constant are calculated:

$$\sigma_{max} = Kq + S_1 \tag{43}$$

$$C = \frac{L_s or L_{ss}}{\sigma_{max} - q} \tag{44}$$

where σ_{max} is peak abutment stress (pounds per square foot), C is shape constant (feet).

Now the average abutment stress before any load transfer may be calculated for each pillar. If the pillar is bounded at roadway centers of x_1 and x_2 (expressed in feet from the extracted panel), then the average abutment stress may be calculated as:

$$\sigma = \frac{\sigma_{max} - q}{x_2 - x_1}\left[C\left(e^{\frac{-x_1}{C}} - e^{\frac{-x_2}{C}}\right)\right] \tag{45}$$

where σ is average abutment stress applied on each pillar (pounds per square foot).

The total initial average pillar stress is equal to average abutment stress plus the cover stress:

$$\sigma_p = \sigma + q \tag{46}$$

3.3.3 Assessment of stability

In this step, to assess the pillar stability, the stress applied on each pillar is compared to its strength. Comparison of stress and strength begins with the nearest located pillar to the mined-out panel. Three cases are possible. If the applied stress is less than the LRS, then no load transfer occurs and the adjacent entry furthest from the gob side of the pillar is presumed stable. If the applied stress exceeds the LRS but is still less than the UL, then the entry may be damaged but still no load transfer occurs. Finally, if the applied stress is greater than the UL, the additional load is transferred to the pillar in the next row. Because a failed pillar is assumed to maintain all of its peak load-bearing capacity, only the additional load, called the transferred remnant load (TRL), is carried over the adjacent pillar. The analysis is then repeated for the pillar in the next row, except that any TRL must be added to the total initial pillar load already calculated. The process continues until the last pillar is reached. In longwall pillar design, the stability of the future tailgate entry is generally of greatest concern. Therefore, if the tailgate pillar load, including TRL, exceeds the LRS of the tailgate pillar, then the design is assumed to be acceptable. Other design criterion is that the TRL from the tailgate row of pillars should be less than 10000 ton/ft along the length of the tailgate in order to avoid damage to the tailgate. When TRL is more than 10000 ton/ft, damage to the tailgate will be severe and supplementary supports will be required.

3.4 Analysis of Longwall Pillar Stability (ALPS) method

The ALPS method was developed by Mark & Bieniawski (1986) in order to design abutment chain pillars based on field data collected from US longwall mines. As a conventional pillar design method, ALPS consists of three basic components (Mark, 1990, 1992): estimation of the load applied to the pillar system, estimation of the strength of the pillar system, and determination of a stability factor and comparison with a design criterion.

3.4.1 Estimation of longwall pillar loads

As mentioned before chain pillars in longwall mines experience two types of loads during mining process; development loads and abutments loads (side and front). The tributary area theory is used by ALPS to estimate the longwall development load as:

$$L_d = HW_t\gamma \tag{47}$$

where L_d is development load (pounds per foot of gate entry), and W_t is width of the pillar system (ft).

In ALPS, side abutment load is calculated similarly to the Carr & Wilson method based on over simplified subsidence model proposed by King & Whittaker (1971; see Figure 3). Based on Figure 3, side abutment load is the wedge of overburden defined by the abutment angle (β). Depending on whether the panel is subcritical ($P/H < 2\tan\beta$) or supercritical ($P/H \geq 2\tan\beta$), the side abutment load is given by Equations 48 and 49, respectively:

$$L_{ss} = \left[\frac{HP}{2} - \frac{P^2}{8\tan\beta}\right]\gamma \tag{48}$$

$$L_s = H^2(\tan\beta)\left(\frac{\gamma}{2}\right) \tag{49}$$

Another type of abutment loads which chain pillars experience at the face ends or T-junctions is front abutment load. The magnitude of the front abutment (L_f) is more difficult to determine analytically, because the load transfer at the T-junctions is a complicated three dimensional problem. The empirical approach used in ALPS begins

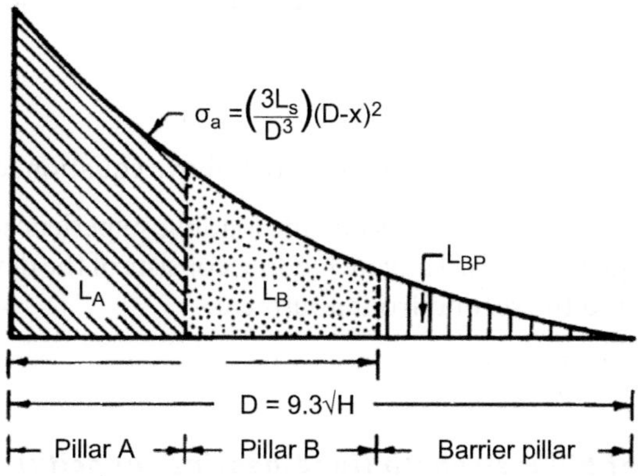

Figure 6 Distribution of side abutment load, σ_a is abutment stress distribution function, L_A, L_B, L_{BP} is abutment load on pillar A, B, and barrier pillar (Mark, 1992).

with the concept that the front abutment load is friction of the side abutment load and can be represented as:

$$L_f = F(L_{ss}\, or\, L_s) \tag{50}$$

where F is a front abutment factor with a value of less than 1. Two front abutment factors are needed, one for a headgate (or first panel) front abutment (F_h) and the other for the tailgate front abutment (F_t). The actual magnitudes of F_h and F_t are affected by local geology and based on US mining conditions the appropriate values for them have been determined as 0.5 and 0.7, respectively.

When the first panel adjacent to a gate entry system is mined, the pillars may not carry the entire side abutment load because some portion may be transferred to the nearby barrier pillar or unmined panel (Figure 5). Thus, the final step of load estimation problem is determination of abutment fraction. The abutment fraction indicates the percentage of side abutment load applied on chain pillars and can be calculated by the below equation.

$$R = 1 - \left[\frac{D - W_t}{D}\right]^3 \tag{51}$$

where D is the extent of the side abutment influence zone, which is equal to $9.3(H)^{0.5}$. Where W_t is greater than D, or where there is no adjacent unmined panel or barrier pillar, then R=1.

Now the maximum loading to which the pillar system is subjected (L) can be determined. The maximum loading depends on the services for which the pillar system will be used. Three possible loading conditions may be defined. (1): The loading experienced by pillars at the T-junctions in the headgate, or in the tailgate during first panel mining, is called headgate loading. Headgate loading consists of the development load plus the first front abutment load:

$$L_H = [L_d + L_s F_h R] \tag{52}$$

where L_H is headgate loading (pounds per foot of gate entry).

(2): Pillars that are expected to protect bleeder entries will be subjected to the development load and the first full side abutment:

$$L_B = [L_d + L_s R] \tag{53}$$

Barrier pillar loads may also be determined from above equation, expect that $R = 1$.

(3): The most severe longwall loading is tailgate loading, experienced during the mining of second and subsequent panels. Tailgate loading consists of the development load, the first side abutment, and the second front abutment:

$$L_T = [L_d + (1 + F_t)L_s] \tag{54}$$

3.4.2 Estimation of pillar strength (load-bearing capacity)

The second component of ALPS is the estimation of the load-bearing capacity of the longwall pillar system. For multi-entry gates, it is first necessary to determine the strength of the individual pillars. ALPS uses the Bieniawski formula for calculation of pillar strength. Then the load bearing capacity of a longwall pillar system is calculated as the sum of the individual pillar resistances:

$$B_c = \sum_{i=1}^{n} \frac{144 S_p A_p}{C} \tag{55}$$

where B_c is load bearing capacity of pillar system (pounds per foot of gate entry), A_p is load bearing area of pillar (square foot), S_p is pillar strength based on Bieniawski formula (pounds per square inch, psi), and C is spacing between crosscuts (ft).

3.4.3 Determination of stability factor

The stability factor (SF) is defined as the maximum load bearing capacity of the longwall pillar system divided by the maximum applied load:

$$SF = \frac{B_c}{L} \tag{56}$$

where L is maximum applied load determined by one of Equations 52, 53, or 54.

The final step in the analysis is the comparison of the stability factor determined in Equation 56 with a design criterion. If no previous longwall experience is available, a stability factor in the range of 1 to 1.3 should be used to size gate entry pillars. For barrier pillars that are expected to protect the main entries for a long period of time, higher stability factor (in the range of 1.5 to 2) should be used. Based on field studies conducted by Mark *et al.* (1994), there is a meaningful relationship between CMRR and chain pillar stability factor. They suggested the below equation to calculate the stability factor for chain pillar design.

$$SF = 1.76 - 0.014CMRR \tag{57}$$

Colwell *et al.* (1999), surveying longwall mining operations in Australia, presented a design method for chain pillars. The starting point or basis of their method was the

ALPS methodology. The final output of their research project is called Analysis of Longwall Tailgate Serviceability (ALTS). This method has been updated in order to accommodate with Australian mining conditions (ALTS II: Colwell *et al.*, 2003).

Artificial intelligence based methods (Zhang *et al.*, 1995), probabilistic analysis (Najafi *et al.*, 2011; Recio-Gordo & Jimenez, 2012), analytical methods (Scovazzo, 2010; Hosseini *et al.*, 2010) and numerical modeling (Torano *et al.*, 2000; Su & Haesenfus, 2010; Shabanimashcool & Li, 2012) have also been applied for chain pillar design.

4 BARRIER PILLARS

As mentioned in Table 1, there are two types of barrier pillars: internal and outcrop barriers (Peng, 2008). Internal barrier pillars are intact blocks of coal left to support and separate main entries from the mined panels (longwall mining), or to separate mined panels (room and pillar mining and highwall mining). Outcrop barriers are those left around the perimeter of the outcrop in the above drainage mines. The internal barrier pillars are primarily designed to support the abutment load transferred over to them, whereas the outcrop barrier pillars, and in some cases internal barrier pillars, are not only designed to support the abutment load, but also to prevent excessive water inflow from the flooded portion of the mine. There are many empirical design formulas for both types of barrier pillars, several of which are introduced in the following (Majdi, 1991; Koehler & Tadolinin, 1995; Rangasamy *et al.*, 2001; Beck, 2004; Kendorski & Bunnell, 2007; Peng, 2008).

4.1 Internal barrier pillar

a) Dunn's rule:

$$W_B = \frac{H - 180}{20} + 15 \tag{58}$$

where W_B is barrier pillar width (ft) and H is depth of mining cover (ft).

b) Pennsylvania mine inspector's formula:

$$W_B = 20 + 4h + 0.1H \tag{59}$$

where h is mining height (ft).

c) British coal rule of thumb:

$$W_B = \frac{H}{10} + 45 \tag{60}$$

d) North American formula:

$$W_B = \frac{H \cdot P_{adj}}{7000 - H} \tag{61}$$

where P_{adj} is width of adjacent panel (ft).

e) Peng formula:

$$W_B = 9.3\sqrt{h} \tag{62}$$

Pressure arch formula:

$$W_B = \frac{P + W_{pa}}{2} \tag{63}$$

where P is actual panel width (ft) and W_{pa} is maximum width of pressure arch (ft) which is calculated using Equation 64:

$$W_{pa} = 3\left(\frac{H}{20} + 20\right) \tag{64}$$

g) Holland convergence formula:

$$W_B = \frac{5\left(\log(50.8C)\right)}{E \cdot \log e} \tag{65}$$

where C is the estimated convergence on the high-stress side of the barrier pillar (in), E is the coefficient for the degree of extraction adjacent to the barrier, and e is base of the natural system of logarithm. Value $E = 0.07$ should be used if adjacent panels are hydraulically backfilled, value $E = 0.08$ should be used if strip pack-walls are built next to the barrier pillar, value $E = 0.085$ should be used if partial extraction is practiced, and value $E = 0.09$ should be used if complete caving will occur in adjacent panel. C is calculated from the relationship:

$$C = 0.001333 \times H \times \frac{h}{7} \times \frac{3000}{\sigma_c} \tag{66}$$

where σ_c is unconfined compressive strength of a coal sample (pound per square inch).

h) Wilson formula:

$$W_B = 2(c + x_b) \tag{67}$$

where, x_b is width of yield zone (ft; Equation 40), and c is exponential decay factor (ft) which is calculated using the following equation.

$$c = \frac{(L_s or L_{ss}) - \frac{h}{2}}{(K - 1) + \frac{S_1}{\gamma H}} \tag{68}$$

where L_{ss} and L_s are side abutment loads (pounds per foot of entry) for sub-critical and super-critical panels, respectively (Equations 41 and 42), K is triaxial stress factor, and S_1 is intact coal strength (pounds per square foot).

In addition to the above equations, the ALPS method can be applied for barrier pillar design in longwall mining (see section 3.4). ARMPS program can be also used for barrier pillar design in room and pillar mining. Furthermore, the barrier pillar design method for highwall mining will be described in section 5.

4.2 Outcrop barrier pillar

a) Old English barrier pillar law:

$$W_B = \frac{h \cdot H_{hy}}{100} + 5h \tag{69}$$

where H_{hy} is the hydraulic head, or depth below drainage level (ft).

b) Pennsylvania mine inspector's formula:

$$W_B = 20 + 4h + 0.1H \tag{70}$$

where H is depth of mining cover, or height of hydraulic head if it is greater than the thickness of overburden (ft).

c) Ash & Eaton impoundment formula:

$$W_B = 50 + 0.426H \tag{71}$$

d) Kentucky formula:

$$W_B = 50 + H_{hy} \tag{72}$$

e) Kohli & Block formula:

$$W_B = S(0.385h + 0.48H_{hy}) \tag{73}$$

where S is surface slope expressed by the horizontal over vertical distance.

In addition to the explained empirical formulas, Luo *et al.* (2001), Shabanimashcool & Li (2013) and Tadolini *et al.* (2013) have successfully applied numerical methods for barrier pillar design.

5 WEB PILLARS IN HIGHWALL MINING

The highwall mining method is a relatively new semi-surface and semi-underground coal mining method that evolved from auger mining (Zipf & Mark, 2005; Luo, 2013). Highwall mining is a technique utilized after the open cut portion of a reserve has been mined, sometimes prior to the introduction of underground mining. Generally, in highwall mining, the coal seam is divided into several webs or panels for safe production. In each panel, coal is extracted in several parallel entries (production entries) by a remotely controlled continuous miner (CM). The width of entries depends on the type of excavation machine and varies between 2.7 and 3.6 m. Once a production entry, is completed, the machine will move to another pre-determined location to start another production entry. A continuous solid coal pillar with a pre-determined width, called web pillar, will be left between the two adjacent production entries. Also, a solid coal pillar called barrier pillar, is left between two panels, which separates the effects of

Figure 7 Highwall mining, showing a barrier pillar (left) and a web pillar (right).

mining on each panel from the adjacent panel. The schematic of web pillar and barrier pillar can be seen in Figure 6. The duty of these pillars (web and barrier) is to support the overburden during and after highwall mining. Stable pillars are essential for a successful highwall mining operation. The failure of the pillars during active mining time will endanger the underground mining machines and the highwall stability. Pillars failing in a large contiguous area could also cause chimney or trough subsidence of the ground surface. Thus, for a successful highwall mining operation, adequately sized web and barrier pillars are essential. In the following, the method of designing web and barrier pillars is described.

Zipf (1999) suggested a three step method for designing web pillars. These steps are:

1. Estimation of pillar strength: since the web pillars are very long, narrow (slender) and rectangular, the best equation for estimating strength is Mark-Bieniawski formula. This formula for long web pillars, whose length is much greater than their width, takes the following form:

$$S_p = S_1\left(0.64 + 0.54\frac{W_{web}}{h}\right) \tag{74}$$

   where W_{web} is web pillar width.

2. Estimation of pillar stress: tributary area theory is used to estimate the pillar stress. Average stress on a web pillar is:

$$\sigma_p = \gamma H\frac{(W_{web} + W_o)}{W_{web}} \tag{75}$$

   where W_o is width of highwall miner hole.

3. Calculation of safety factor: the safety factor for web pillars can be calculated using below equation:

$$SF_{web} = \frac{S_p}{\sigma_p} = \frac{S_1\left(0.64 + 0.54\frac{W_{web}}{h}\right)}{\gamma H\frac{(W_{web}+W_o)}{W_{web}}} \tag{76}$$

For design purposes, the stability factor for web pillars typically ranges from 1.3 to 1.6. It is recommended that the web pillar width to height ratio be maintained at 1.0 or higher, to help maintain web pillar integrity.

A barrier pillar is commonly used to separate adjacent panels and prevent ground control problems from cascading along the entire length of highwall. Zipf (1999) also suggested a three step method for designing barrier pillars in highwall mining. These steps are:

1. Estimation of pillar strength (Mark-Bieniawski formula):

$$S_p = S_1\left(0.64 + 0.54\frac{W_B}{h}\right) \tag{77}$$

where W_B is the barrier pillar width.

2. Estimation of pillar stress (tributary area theory):

$$\sigma_p = \gamma H\frac{(W_B + P)}{W_B} \tag{78}$$

where P is highwall panel width which can be calculated using Equation 79:

$$P = N(W_{web} + W_o) + W_e \tag{79}$$

where N is number of web pillar in a highwall panel.

3. Calculation of safety factor:

$$SF = \frac{S_p}{\sigma_p} = \frac{S_1\left(0.64 + 0.54\frac{W_B}{h}\right)}{\gamma H\frac{(W_B+P)}{W_B}} \tag{80}$$

Since the stress carried by web pillars within a panel is neglected, the stability factor for barrier pillars can be as low as 1. Barrier pillars with a width to height ratio greater than 3 are superior for sound geomechanics reasons.

Zipf (2005) developed design charts for web and barrier pillars based on the methodologies described above. Two examples of these charts are shown in Figure 8 and 9 for web and barrier pillars, respectively. To use these figures, the user begins with the design depth on the x axis, moves up vertically to the applicable mining height and then moves left horizontally to the y axis where the suggested web (or barrier) pillar width is read.

Lue (2013) developed a spreadsheet program (called highwall mining design program) for the highwall mine pillar design based on Zipf's design concept and methodology with some modifications.

NIOSH has developed the Analysis of Retreat Mining Pillar Stability-Highwall Mining (ARMPS–HWM) computer program to assist mine planners with pillar design

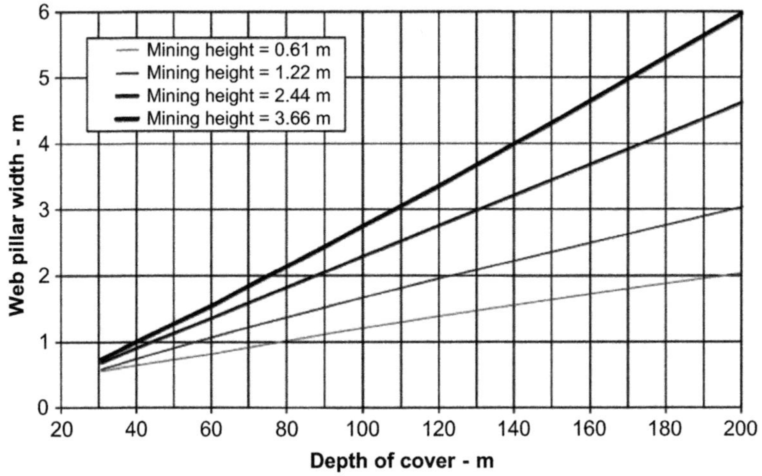

Figure 8 Suggested web pillar width with stability factor of 1.3, coal strength of 6.2 MPa and 2.75 m wide hole (Zipf, 2005).

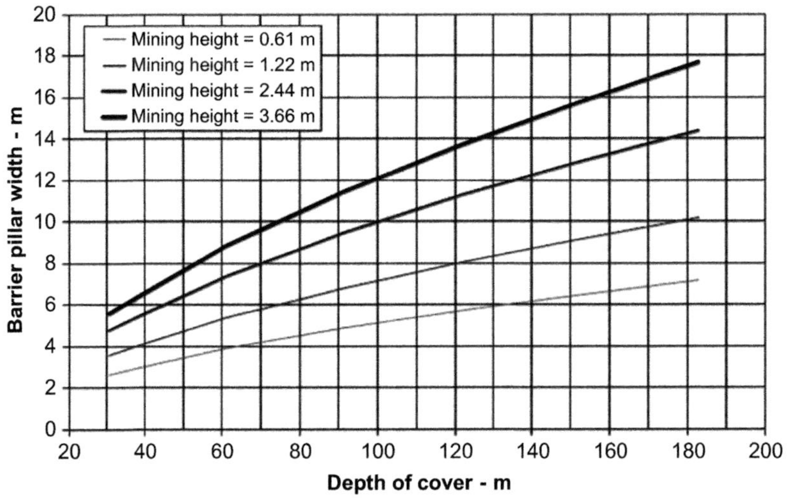

Figure 9 Suggested barrier pillar width for 61 m wide panel assuming coal strength of 6.2 MPa and stability factor of 1 (Zipf, 2005).

(Mark, 2006). ARMPS-HWM is a modification of ARMPS software for designing web and barrier pillars in Highwall Mining. ARMPS-HWM calculates stability factors based on estimation of loads applied to, and the load-bearing capacities of, web and barrier pillars. ARMPS-HWM uses the Mark-Bieniawski formula to estimate the strength of web and barrier pillars. It uses tributary area theory for estimation of load applied to web pillars. But, estimation of load applied to barrier pillars was

Table 6 ARMPS-HWM design guidelines (Mark, 2006).

Web pillar stability factor	1.6	When the panel width (excluding the barrier) exceeds approximately 200 ft (60 m).
	1.3	When the panel width (excluding the barrier) is less than approximately 200 ft (60 m).
Barrier pillar stability factor	2	When the barrier's width to height ratio is less than 4.
	1.5	When the barrier's width to height ratio is equal to or more than 4.
Overall stability factor	2	Applicable to all conditions.

a little more complicated. ARMPS-HWM assumed a worst case where the webs on either side of the barrier might have failed. Thus, it uses abutment angle concept for estimation of load applied to barrier pillars. Table 6 summarizes the design criteria used in ARMPS-HWM. In addition to the stability factors for the webs and barriers, ARMPS-HWM also calculates an overall stability factor for the pillar system consisting of one barrier and one panel of webs. The suggested minimum stability factor for the system is 2.0.

It is worth mentioning that numerical modeling has been also used widely for web pillar design (Duncan Fama *et al.*, 1995; Adhikary *et al.*, 2002; Hamanaka *et al.*, 2010; Sasaoka *et al.*, 2010; Porathur *et al.*, 2013; Verma *et al.*, 2014; Perry *et al.*, 2015).

6 YIELD PILLARS

Yield pillars were originality designed for deep mines (Mark, 1990; Morsy, 2003; Peng, 2008). They are much smaller than stiff pillars, therefore, the recovery of coal increases drastically in comparison to the coal produced by stiff pillars. Yield pillars are designed to yield as loads are applied to them, whereas stiff pillars resist and support full loads. Ideally a yield pillar will yield at the proper time and rate depending on the characteristics of the surrounding strata and stages of mining. Many definitions have been proposed for yield pillars. Tadolini & Haramy (1992) defined the yield pillar as a pillar that yields or fails upon isolation from the coal seam during the longwall development stage. Mark (1990) recommended that the yield pillars should be designed to yield at the development load. Pen (1994) defined the yielding pillar as a pillar that yields completely before the maximum load arrives. Gauna (1985) defined a yield pillar as one which yields upon isolation from coal seam; *i.e.* during development.

Yield pillars have been proposed for several purposes, including reducing floor heave, improving tailgate stability, eliminating pillar bumps, and reducing stress related roof falls during development. DeMarco (1994) lists the following requirements for a successful yielding pillar gate road:

1. That sufficient depth exists to initiate pillar yielding;
2. That the mine roof be of sufficient quality to withstand the deformation resulting from pillar yielding;
3. That sufficient floor quality exist to withstand excessive floor heave and/or pillar punching; and
4. That existing mining depth and/or unique seam conditions do not prevent pillar yielding.

The idea of yield pillars for ground control is not new; it was originally popularized in the US by Holland as part of the pressure arch concept (Mark, 1990). There are two types of methods for yield pillar sizing (Morsy, 2003): (1) historical and (2) theoretical methods.

6.1 Historical methods

In these methods the size of yield pillar is determined based on successful case histories and experience gained by trial and error. Mark (1990) based on field observations suggested that upper limit of stability factor for yield pillar design might be equal to 0.5. For calculating the stability factor of yield pillars, the pillar strength is estimated using Bieniawski formula and pillar load is estimated by tributary area theory. Mark (1990) also suggested that the minimum yield pillar size might be the seam height plus 4 ft.

DeMarco (1994) examined western US longwall experiences with yield pillar design and concluded the following findings: (1) No successful yield pillar designs were achieved in mines where the CMRR was less than 50, and (2) No successful yield pillar designs employed width to height ratios greater than 5.

Kneisley (1995) developed an empirical method for sizing gate road yield pillars based on US Bureau of Mines field studies in several Western US coal mines. For each case, pillar width to height ratios, development-induced extraction ratios, and the pillar development factors (the ratio of the pillar strength, based on the Bieniawski pillar formula, to the tributary load resulting from development) were calculated. Discriminant analysis was used to generate expressions for each of the above parameters as functions of the CMRR, and to separate the case histories into either satisfactory or unsatisfactory populations. The discriminant analysis was performed for the 51 data sets having CMRR values exceeding 50 (no satisfactorily performing entry systems were documented for CMRR values less than 50). The obtained equations are as follows:

$$\frac{w}{h} = 13.337 - 0.087CMRR \tag{81}$$

$$ER = 0.355 + 0.002CMRR \tag{82}$$

$$DF = 2.06 - 0.02CMRR \tag{83}$$

where $\frac{w}{h}$, ER, and DF are yield pillar width to height ratio, extraction ratio resulting from pillar development, and development factor, respectively.

Successfully performing pillar systems, for CMRR greater than 50, are assumed to consist of those configurations that meet all of the following criteria:

1. Pillar width to height ratio must not exceed the maximum value as determined by Equation 81;
2. The development-induced extraction ratio must not be less than the minimum allowable value by Equation 82; and
3. The pillar development factor must not exceed the maximum value allowed by Equation 83.

Schissler (2002) collected and reviewed the case histories of yield pillar in US coal mines and found that the width to height ratio for successful yield pillars ranges from 2.8 to 6.0 and development load stability factor ranges from 0.4 to 0.6.

6.2 Theoretical methods

In these methods the size of yield pillar is determined based upon analytical or numerical techniques. The most popular theoretical methods for yield pillar design are the Carr & Wilson method, the Chen method, the Tsang method, the JS Chen method, and the Morsy method (Peng, 2008).

6.2.1 Carr and Wilson method

This method is the first analytical method for yield pillar design, which is based on Wilson's confined core approach (Carr & Wilson, 1982). Based on this method, a yield pillar should not be greater than twice the yield zone width. Therefore, the maximum yield pillar width can be calculated for both rigid roof and floor (RRF) and for yielding roof and floor (YRF) conditions (Equation 84 and 85, respectively) as follows (Haramy & Kneisley, 1990; Kneisley & Haramy, 1992):

$$w_y = 2\frac{h}{F}\ln\left(\frac{\gamma H}{p + p^*}\right) \tag{84}$$

$$w_y = h\left[\left(\frac{\gamma H}{p + p^*}\right)^{\frac{1}{K-1}} - 1\right] \tag{85}$$

where w_y is yield pillar width (ft) and F is a function of triaxial stress factor (K) which is calculated using Equation 86:

$$F = \left(\frac{K-1}{\sqrt{K}}\right) + \left(\frac{K-1}{\sqrt{K}}\right)^2 \tan^{-1}\sqrt{K} \tag{86}$$

where $\tan^{-1}$ is in radians.

6.2.2 Chen method

Chen (1989) developed a method for yield pillar design. The method is a combination of numerical modeling and the Wilson's confined core concept (Chen & Karmis, 1988). Chen used an elastic-plastic model to describe the pillar yielding. The basis of this model is Drucker-Prager yield criterion. Chen conducted a parametric study using 2D finite element models to define the stress distribution in the pillar yield zone. Based on the statistical analysis for the results of finite element simulations, he developed a mathematical model concerning the stress distribution in the yield zone. Furthermore, he established the formulae to determine the pillar bearing capacity and the width of yield zone. Finally, based on the method of estimating pillar loading, Chen proposed three possible yield pillar sizes; namely the maximum, the recommended, and the minimum.

The maximum (critical) yield pillar width (w_{max}) is considered to be twice the width of the yield zone predicted by Equation 87.

$$w_{max} = 2x_b$$
$$= 2h\left(9.61\cos\left[\frac{1}{3}cos^{-1}\left(\frac{\gamma H \times 10^{-5}}{K^{1.7}(0.17v^2 + 0.057v - 0.028)} - 1\right)\right] - 4.8\right) \tag{87}$$

where x_b is yield zone width, h is pillar height, K is the triaxial stress factor, H is overburden depth, γ is unit weight of the overburden, and v is Poisson's ratio.

The recommended yield pillar (w_r) is defined as the pillar width in which the peak vertical stress at the center of a completely yielded pillar equals the average tributary stress:

$$\frac{\gamma H(w_r + B)^2}{w_r^2} = K^{2.7}(0.17v^2 + 0.057v - 0.028)\left[454\left(\frac{w_r}{2h}\right)^3 + 6545\left(\frac{w_r}{2h}\right)^2\right] \tag{88}$$

where B is the entry width.

The minimum pillar width (w_{min}) is the width of yielded pillar that support the roof strata below the pressure arch:

$$\frac{2}{3}\gamma D W_T(w_{\min} + B) = K^{2.7}(0.17v^2 + 0.057v - 0.028)\left(273\frac{w_{\min}^4}{h^2} + 5.68\frac{w_{\min}^5}{h^3}\right) \tag{89}$$

where $W_T = w_{min} + 2B$ is the width of pressure arch, and $D = 2W_T$ is the height of pressure arch.

6.2.3 Tsang method

Tsang (1992) developed a method for yield pillar design in three-entry gateroad systems. Tsang developed regression equations to present the relationships between the stability conditions of entry system; *i.e.* roof, pillar and floor, and the studied variable for different roof and floor conditions (Table 7). To develop the regression equations, Tsang used two-dimensions finite element models combined with fractional factorial experiment design. The used finite element models consider the plastic failure properties and time-dependent behaviors of rock. To better organize the finite element model simulation, the orthogonal experiment design is used to arrange the finite element models and proved to be very effective (Tsang & Peng, 1993). Based on the recommended pillar safety factor (Table 8), Tsang method can be used to design different types of three-entry systems; *i.e.* Stiff-Stiff (S-S), Yield-Stiff (Y-S) and Yield-Yield (Y-Y). Tsang assumed that the pillar widths estimated by the regression equations in Table 7, are initial pillar widths. These initial pillar widths should be increased in the followings cases: (1) If the roof tensile strength was less than the estimated maximum tensile stress, and (2) If the floor and/or roof safety factors, were less than 1.3.

Table 7 Regression equations for different roof/floor conditions.

	Regression equations
Strong roof and strong floor	$w_1 = \left(\frac{C}{H}\right)^{-0.4048}(52.7884SF_1 - 15.4181SF_2 + 0.0005hP)$
	$w_2 = \left(\frac{C}{H}\right)^{-0.5581}(44.9214SF_2 - 23.1242SF_1 + 0.0005hP)$
	$T_r = H^{0.7454}\left(\frac{2.9877}{\sqrt{w_1}} + \frac{1.0168}{\sqrt{w_2}} + 2.8 \times 10^{-9}BE_r\right)$
Strong roof and weak floor	$w_1 = \left(\frac{C}{H}\right)^{-0.4162}\left(57.1280SF_1 - 24.9814SF_2 + 1.03 \times 10^3 \frac{hP}{E_f}\right)$
	$w_2 = \left(\frac{C}{H}\right)^{-0.4931}\left(52.9712SF_2 - 30.3162SF_1 + 1.03 \times 10^3 \frac{hP}{E_f}\right)$
	$T_r = H^{0.6775}\left(\frac{3.7869}{\sqrt{w_1}} + \frac{1.2029}{\sqrt{w_2}} + 9.7 \times 10^{-9}BE_r\right)$
	$SF_f = 2.162 \times 10^{-3}\left(\frac{E_f}{H}\right)^{0.4442}\left(0.2548w_1 + 0.1403w_2 + 6.4\frac{\sigma_h}{P}\right)$
Weak roof and strong floor	$w_1 = \left(\frac{C}{H}\right)^{-0.4135}\left(62.7666SF_1 - 22.8459SF_2 + 8.15\frac{hP}{E_r}\right)$
	$w_2 = \left(\frac{C}{H}\right)^{-0.4616}\left(62.8519SF_2 - 29.7170SF_1 + 1.8 \times 10^2 \frac{hP}{E_r}\right)$
	$T_f = H^{0.8759}\left(\frac{1.3043}{\sqrt{w_1}} + \frac{0.4907}{\sqrt{w_2}} + 3.4 \times 10^{-4}BE_r + \frac{0.9236}{\sigma_h}\right)$
	$SF_r = 2.398 \times 10^{-3}\left(\frac{E_r}{H}\right)^{0.4367}\left(0.2384w_1 + 0.1217w_2 + 4.5\frac{\sigma_h}{P}\right)$
Weak roof and weak floor	$w_1 = \left(\frac{C}{H}\right)^{-0.5235}\left(52.7862SF_1 - 15.2109SF_2 + 1.17 \times 10^7 \frac{hP}{E_rE_f}\right)$
	$w_2 = \left(\frac{C}{H}\right)^{-0.5341}\left(53.7009SF_2 - 23.4415SF_1 + 2.02 \times 10^7 \frac{hP}{E_rE_f}\right)$
	$SF_r = 4.521 \times 10^{-3}\left(\frac{E_r}{H}\right)^{0.3908}\left(0.1339w_1 + 0.1532w_2 + 6.5989\frac{\sigma_h}{P}\right)$
	$SF_f = 2.4 \times 10^{-4}\left(\frac{E_f}{H}\right)^{0.6032}\left(0.5568w_1 + 0.8454w_2 + 53.8824\frac{\sigma_h}{P}\right)$

w_1 and w_2 are the pillar width for pillar 1 and 2, respectively, (ft); SF_1 and SF_2 are the safety factor for pillar 1 and 2 , respectively; SF_r and SF_f are the minimum safety factors of the roof and floor, respectively; C is the cohesion strength of the coal (psi); H is the depth of cover (ft); h is the pillar height (ft); B is the entry width (ft); P is panel width (ft); E_r and E_f are the Young's module of the immediate roof and floor, respectively, (psi); σ_h is the virgin horizontal stress (psi); and T_r and T_f are the maximum tensile stress in the roof and floor, respectively, (psi).

Table 8 Recommended pillar safety factors for Tsang's method.

Design layout	SF_1	SF_2
Stiff-Stiff (S-S)	1.3	1.3
Yield-Yield (Y-Y)	0.9	0.9
Stiff-Yield (S-Y)	1.5	0.9

6.2.4 JS Chen method

JS Chen *et al.* (1999) introduced three criteria for yield pillar design in bump-prone longwall mines: tailgate stability factor (SF_t), width to yield zone of the yield pillar (R_y), and width to height ratio of the yield pillar (R_H). The tailgate stability factor is defined as the ratio of yield pillar strength and stress under the tailgate load as follows:

$$SF_t = \frac{S_p}{L_T} \tag{90}$$

where S_p is the pillar strength determined by the Mark-Beiniawski pillar strength formula, and L_T is the tailgate pillar stress which can be calculated by Equation 54. The width to yield zone of the yield pillar is defined as:

$$R_y = \frac{w_y}{2x_b} \tag{91}$$

where w_y is width of yield pillar and x_b is width of yield zone for yielding roof and floor conditions and can be determined by Equation 40.

Analysis of the field observations indicate that tailgate pillar bumps can be eliminated for a purely yield pillar system if all the following conditions can be met: $SF_t \leq 0.18$, $R_y \leq 1.5$, and $R_H \leq 5$. If all three criteria are greater than their threshold (critical) values, the intended yield pillar will yield in a violent manner. If only one of the three criteria is greater than its corresponding threshold value, a tailgate bump of light significance may be possible.

6.2.5 Morsy method

Morsy (2003) and Morsy & Peng (2003) introduced a new method for evaluating the stability of yield pillars. This method was conducted in three steps:

6.2.5.1 Estimation of pillar loading

Finite element technique (ABAQUS program) was used to estimate the state of stress, strain, elastic strain energy, plastic energy, and etc. at every point within the pillar. Using the adapted Drucker-Prager model and the rock/coal frictional model, the non-uniform stress distribution in the yield pillar was simulated.

6.2.5.2 Defining the yield pillar loading zones

Once the yield pillar is formed, three zones of different amount of confinement could be defined namely; core, transition and rib zones (Figure 10). More detailed explanation for these zones is presented below:

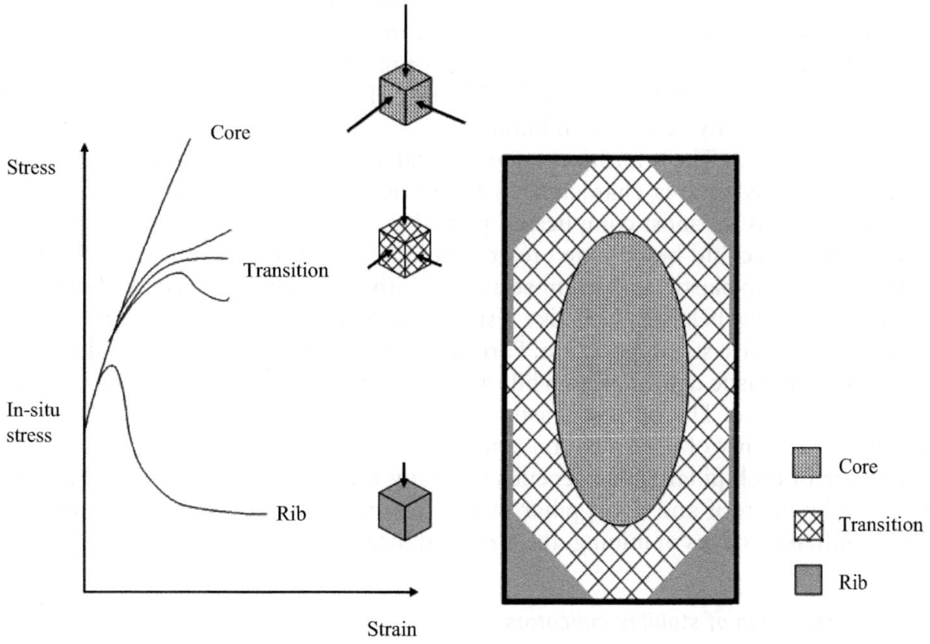

Figure 10 View of yield pillar loading.

Rib zone: The rib zone occupies the pillar corners (Figure 10). It is the weakest part of the pillar because it is bounded by free faces on two sides; *i.e.*, low confining pressures. The pillar yielding starts from this zone toward the pillar core. The rib zone is defined as the instable parts of the pillar which have post-peak stiffness larger than or equal to the local mining stiffness:

$$|K_p| \geq |K_{LMS}| \tag{92}$$

where $|K_{LMS}|$ is the absolute value of the local mine stiffness, and $|K_p|$ is the absolute value of pillar post-peak stiffness. The numerical methods for estimation of $|K_p|$ and $|K_{LMS}|$ have been published by Morsy & Peng (2002).

The rib zone behaves as a brittle material under uniaxial compression. Numerically, the rib zone is capable of carrying relatively little stress but it can't store large amount of elastic strain energy. Most of the loading energy of the rib zone is dissipated in the form of plastic deformation. The instability of the rib zone can be manifested in the form of coal bounces.

Core zone: The core zone occupies the center portion of the pillar (Figure 10). This zone is defined as the part of the pillar that does not experience any plastic deformations. The elastic behavior of the core zone is a result of high horizontal stresses in this zone. The core zone is capable to withstand extremely high stresses even when the pillar has been compressed beyond its maximum resistance, which is traditionally regarded as the strength of the pillar. Therefore, the core stores a significant amount of elastic strain energy. Since the yield pillars have relatively small widths and operate under high overburden depths, the state of stress in the core zone is usually very close to the yielding condition. Hence the stability of the core zone is crucial when the longwall face approaches the yield pillar. At this stage of mining, the core zone starts to yield. The more the elastic strain energy retained in the yielded parts, the more likely the pillar experiences instability such as coal bumps.

Transition zone: The transition zone is located between the rib and core zones (Figure 10). This zone is characterized by a wide range of confining stresses. The transition zone is the most complicated part of the pillar where three types of stress-strain behaviors could be observed for the points located in this zone; namely strain hardening, elastic-plastic and strain-softening with high residual strength. Generally, the transition zone behaves more like a strain-hardening material because of the high triaxial compressive stress condition. Part of the energy of transition zone is dissipated in the form of plastic deformation while a significant amount of elastic strain energy is stored in this zone. The ratio of elastic strain energy to the dissipated plastic energy governs the stability of the transition zone.

The size of each pillar zones depends on many factors, such as; the end constraint provided by the roof and floor, pillar width, overburden depth, stage of mining, etc. During different stages of mining, the sizes and locations of pillar zones change.

6.2.5.3 Estimation of stability indicators

As illustrated before, the yield pillar is composed of different zones with different load-deformation behaviors. Therefore, more than one stability criterion should be used to evaluate the stability of yield pillar. The pillar stability is checked by three stability indices, such as; Core Stability Factor (*CSF*) Pillar Bump Index (*PBI*) and Rib Instability Index (*RIF*). Throughout the following discussion, the yield pillar is composed of a number of elements. The stability/instability of the pillar zone will be evaluated in two steps: first; the stability factors of the elements inside the pillar zone are determined, second; the stability of the pillar zone is determined as an average for the stability/instability factors of the elements located inside the studied zone.

Core stability factor (*CSF*): *CSF* is applied to evaluate the stability of the core zone. The strength criterion evaluates the state of stress in the core with respect to the yield strength. The element stability factor (*ESF*) is determined by the linear Drucker-Prager yield criterion (Equation 93) as follows:

$$ESF = \frac{k + J_1 \tan\alpha}{J_{2D}} \tag{93}$$

where α and k are the material property constants related to cohesion and angle of internal friction, J_1 is the first invariant of stress tensor, and J_{2D} is the second invariant of the deviator stress tensor.

Numerically, it is not possible to have ESF less than 1. The possible values for ESF are either 1 for those elements experiencing plastic deformation or greater than 1 for elastic elements. The core stability factor *CSF* is determined by averaging the *ESF*s of all elements located within the core zone.

As mentioned earlier, the core zone has the ability to store a significant amount of elastic strain energy. As long as the core stability factor CSF is greater than 1, the elastic strain energy stored in this zone will be in stable condition. This condition can be strongly obvious in the stiff pillar design, where the core zone characterizes most of the pillar and the stress level inside the stiff pillar is far from its yielding strength. On the other hand, the stability of the core is crucial for the yield pillar, especially when the longwall face approaches the yield pillar. At that stage of mining, some of the elements inside the core zone start to yield and join the transition zone. Therefore the stored elastic strain energy could be released to the surrounding in a violent fashion.

Pillar bump index (PBI): Figure 11 shows a typical stress-strain curve for an element located in the transition zone. At any stress level (σ_c) the shaded triangle CED represents the amount of stored strain energy (W_e) while the area of polygon OABCE represents the dissipated energy (W_p). Finite element technique was used to estimate the state of elastic strain energy and plastic dissipated energy at every element within the pillar. Hence; using Equation 94, the element bump index can be calculated as follows:

$$PBI = \frac{W_e}{W_p} \tag{94}$$

The pillar bump index for the transition zone is determined by averaging the local strain energy storage indices. The larger the bump index the more elastic energy stored in the pillar and the more tendency for serious coal bumps.

Rib instability factor (*RIF*): The local mine stiffness criterion is used to evaluate the instability of the rib zone. The element instability factor (*EIF*) for the rib zone elements, is determined by:

$$EIF = \frac{|K_P|}{|K_{LMS}|} \tag{95}$$

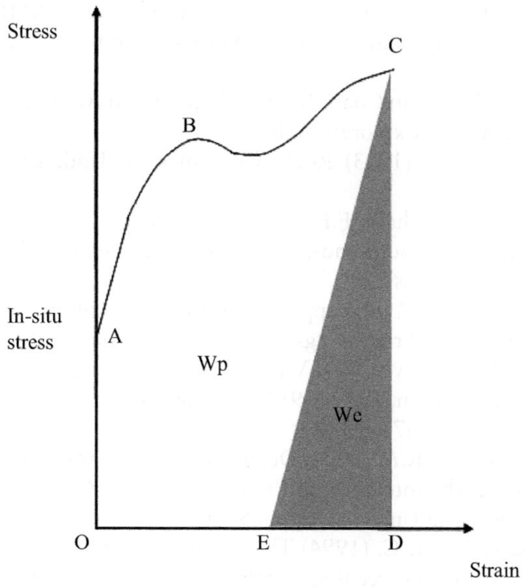

Figure 11 Strain energy storage index (PBI) for transition zone.

According to the definition of rib zone (Equation 92), the *EIF* is usually greater than 1. The RIF is determined by averaging *EIF*s of all the elements located in the rib zone. The *RIF* covers a wide range of instabilities, from unrecognizable bounces to serious ones. The larger the rib instability factors, the larger the energy that can be released to the surrounding.

In addition to the explained theoretical methods, numerical modeling has been widely used for design and stability assessment of yield pillars (Smith *et al.*, 1995; Yavuz & Fowell, 2001; Badr *et al.*, 2003; Badr, 2004; Ozbay & Badr, 2010; Li *et al.*, 2015).

REFERENCES

Adhikary, D.P., Shen, B. & Duncan Fama M.E. (2002) A study of highwall mining panel stability. International Journal of Rock Mechanics and Mining Sciences, 39, 643–659.

Arioglu, E. & Tokgoz N. (2011) Design Essential of Hard Rock Mass and Coal Strength: With Practical Solved problems. Istanbul, Evrim Publisher.

Badr, S. (2004) Numerical analysis of coal yield pillars at deep longwall mines. Colorado School of Mines, PhD Dissertation.

Badr, S., Ozbay, U., Kieffer, S. & Salamon, M. (2003) Three-dimensional strain softening modeling of deep longwall coal mine layouts. In: Proceedings of the third International FLAC Symposium. Sudbury, Ontario, Canada, pp. 233–240.

Beck, K.D., Mishra, M., Chen, J., Morsy, K., Gadde, M. & Peng, S. (2004) Effect of the approaching longwall faces on barrier and entry stability. In: Proceedings of the 23rd International Conference on Ground Control in Mining. Morgantown: West Virginia University, pp. 20–26.

Bieniawski, Z.T. (1981) Improved design of coal pillars for US mining conditions. In: Proceedings of the 1st International Conference on Ground Control in Mining. Morgantown: West Virginia University, pp. 13–22.

Bieniawski, Z.T. (1987) Strata Control in Mineral Engineering. Rotterdam, AA Balkema.

Bieniawski, Z.T. (1992) Ground control. In: Hartman, H.L. (eds.) SME Mining Engineering Handbook. 2nd edition. Colorado, Society for Mining, Metallurgy, and Exploration, Inc. pp. 897–937.

Bise, C.J. (2013) Modern American Coal Mining: Methods and Applications. Colorado, Society for Mining, Metallurgy, and Exploration, Inc.

Brady, B.H.G. & Brown, E.T. (1993) Rock Mechanics for Underground Mining. London, Chapman & Hall.

Campoli, A.A., Oyler, D.C. & Chase, F.E. (1989) Performance of a novel bump control pillar extracting technique during room-and-pillar retreat coal mining. Pittsburgh: US Bureau of Mines, Report of Investigations 9240.

Carr, F. & Wilson, A.H. (1982) A new approach to the design of multi-entry development for retreat longwall mining. In: Proceedings of the 2nd International Conference on Ground Control in Mining. Morgantown: West Virginia University, pp. 1–21.

Cauvin, M., Verdel, T. & Salmon, R. (2009) Modeling uncertainties in mining pillar stability analysis. Risk Analysis, 29, 1371–1380.

Chase, F., Mark, C. & Heasley, K.A. (2002) Deep cover pillar extraction in the US coalfields. In: Proceedings of the 21th International Conference on Ground Control in Mining. Morgantown: West Virginia University, pp. 69–80.

Chase, F.E., Zipf, R.K. & Mark, C. (1994) The massive collapse of coal pillars: case histories from the United States. In: Proceedings of the 13th International Conference on Ground Control in Mining. Morgantown: West Virginia University, pp. 69–80.

Chen, G. & Karmis M. (1988) Computer modeling of yield pillar behavior using post-failure criteria. In: Proceedings of the 7th International Conference on Ground Control in Mining. Morgantown: West Virginia University, pp. 116–125.

Chen, G. (1989) Investigation into yield pillar behavior and design consideration. Virginia Polytechnic Institute and State University, PhD Dissertation.

Chen, J., Mishra, M. & DeMichiei J. (1999) Design considerations for bump-prone longwall mines. In: Proceedings of the 18th International Conference on Ground Control in Mining. Morgantown: West Virginia University, pp. 124–135.

Chen, J.S., Mishra, M. & DeMichiei, J. (1999) Design consideration for bump-prone longwall mines. In: Proceedings of the 18th International Conference on Ground Control in Mining. Morgantown: West Virginia University, pp. 124–135.

Choi, D. & McCain, D.L. (1980) Design of longwall systems. SME Transactions, 268, 1761–1764.

Colwell, M., Frith R. & Mark C. (1999) Analysis of longwall tailgate serviceability (ALTS): A chain pillar design methodology for Australian conditions. In: Proceedings of the second International Workshop on Coal Pillar Mechanics and Design. US National Institute for Occupational Safety and Health (NIOSH), Information Circular 9448, pp. 33–48.

Colwell, M., Hill, D. & Frith, R. (2003) ALTS II- a longwall gateroad design methodology for Australian collieries. In: Proceedings of the 1st Australasian Ground Control in Mining Conference. Scientia: University of New South Wales, pp. 123–135.

DeMarco, M.J. (1994) Yielding pillar gate road design considerations for longwall mining. In: Proceedings of New Technology for Longwall Ground Control. US Bureau of Mines Technology Transfer Seminar, pp. 19–36.

Du, X., Lu, J., Morsy, K. & Peng, S. (2008) Coal pillar design formulae review and analysis. In: Proceedings of the 27th International Conference on Ground Control in Mining. Morgantown: West Virginia University, pp. 254–261.

Duncan Fama, M.E., Trueman, R. & Craig, M.S. (1995) Two-and three-dimensional elasto-plastic analysis for coal pillar design and its application to highwall mining. International Journal of Rock Mechanics and Mining Sciences and Geomechanics Abstracts, 32, 215–225.

Esterhuizen, E., Mark, C. & Murphy M.M. (2010) The ground response curve, pillar loading and pillar failure in coal mines. In: Proceedings of the 29th International Conference on Ground Control in Mining. Morgantown: West Virginia University, p. 9.

Esterhuizen, G.S. & Mark, C. (2009) Three-dimensional modeling of large arrays of pillars for coal mine design. In: Proceedings of the International Workshop on Numerical Modeling for Underground Mine Excavation Design. US National Institute for Occupational Safety and Health, NIOSH publication no. 9512, pp. 37–46.

Esterhuizen, G.S., Mark, C. & Murphy, M.M. (2010) Numerical model calibration for simulating pillars, gob and overburden response in coal mines. In: Proceedings of the 29th International Conference on Ground Control in Mining. Morgantown: West Virginia University, p. 12.

Gale, W.J. (1999) Experience of field measurement and computer simulation methods for pillar design. In: Proceedings of the Second International Workshop on Coal Pillar Mechanics and Design. US National Institute for Occupational Safety and Health, NIOSH publication no. 9448, pp. 49–61.

Gauna, M., Price, K.R. & Martin, E. (1985) Yield Pillar Usage in Longwall Mining at deep, No. 4 Mine, Brookwood, Alabama. In: Proceedings of the 26th US Symposium on Rock Mechanics. Rapid City: South Dakota School of Mines and Technology, pp. 695–702.

Ghasemi, E. & Shahriar, K. (2012) A new coal pillars design method in order to enhance safety of the retreat mining in room and pillar mines. Safety Science, 50, 579–585.

Ghasemi, E., Ataei, M. & Shahriar, K. (2014a) A intelligent approach to predict pillar sizing in designing room and pillar coal mines. International Journal of Rock Mechanics and Mining Sciences, 65, 86–95.

Ghasemi, E., Ataei, M. & Shahriar, K. (2014b) Prediction of global stability in room and pillar coal mines. Natural Hazards, 72, 405–422.

Ghasemi, E., Shahriar, K., Sharifzadeh, M. & Hashemolhosseini, H. (2010) Quantifying the uncertainty of pillar safety factor by Monte Carlo simulation—a case study. Archives of Mining Sciences, 55, 623–635.

Griffiths, D.V., Fenton, G.A. & Lemons C.B (2002) Probabilistic analysis of underground pillar stability. International Journal of Numerical and Analytical Methods in Geomechanics, 26, 775–791.

Hamanaka, A., Sasaoka, T., Shimada, H., Matsui, K., Meechumna, P., Laowattanabandit, P. & Takamoto, H. (2010) Application of highwall mining system at surface coal mine in Thailand. In: Proceedings of the 29th International Conference on Ground Control in Mining. Morgantown: West Virginia University, p. 8.

Haramy, K.Y. & Kneisley, R.O. (1990) Yield pillars for stress control in longwall mines-case study. International Journal of Mining and Geological Engineering, 8, 287–304.

Hartman, H.L. (1987) Introductory Mining Engineering. New York, John Wiley & Sons.

Heasley, K.A. & Barton, T.M. (1999) Coal mine subsidence prediction using a boundary-element program. Preprint for SME 1999 Annual meeting. Littleton, Colorado, pp. 6.

Heasley, K.A. (2009) An overview of calibrating and using the LaModel program for coal mine design. In: Proceedings of the International Workshop on Numerical Modeling for Underground Mine Excavation Design. US National Institute for Occupational Safety and Health (NIOSH), Information Circular 9512, pp. 63–74.

Heasley, K.A., Sears, M.M., Tulu, I.B., Calderon-Arteaga, C.H. & Jimison, L.W. (2010) Calibrating the LaModel program for deep cover pillar retreat coal mining. In: Proceedings of the 3rd International Workshop on Coal Pillar Mechanics and Design. Morgantown: West Virginia University, pp. 47–57.

Hosseini, N., Oraee, K. & Gholinejad, M. (2010) Optimization of chain pillars design in longwall mining method. In: Proceedings of the 29th International Conference on Ground Control in Mining. Morgantown: West Virginia University, p. 4.

Hsiung, S.M. & Peng, S.S. (1985) Chain pillar design for U.S. longwall panels. Mining Science and Technology, 2, 279–305.

Hustrulid, W.A. (1982) Underground Mining Methods Handbook. New York, SME-AIME.

Iannacchione, A.T. & Tadolini, S.C. (2008) Coal mine burst prevention controls. In: Proceedings of the 27th International Conference on Ground Control in Mining. Morgantown: West Virginia University, pp. 20–28.

Jaiswal, A. & Shrivastva, B.K. (2009) Numerical simulation of coal pillar strength. International Journal of Rock Mechanics and Mining Sciences, 46, 779–788.

Jian, Z., Xi-bing, L., Xiu-zhi, S., Wei, W. & Bang-biao, W. (2011) Predicting pillar stability for underground mine using Fisher discriminant analysis and SVM methods. Transactions of Nonferrous Metals Society of China, 21, 2734–2743.

Kendorski, F.S. & Bunnell M.D. (2007) Design and performance of a longwall coal mine water-barrier pillar. In: Proceedings of the 26th International Conference on Ground Control in Mining. Morgantown: West Virginia University, pp. 97–103.

Khair, A.W. & Peng, S.S. (1985) Causes mechanisms of massive pillar failure in a southern West Virginia coal mine. Mining Engineering, 37, 323–328.

King, H.J. & Whittaker B.N. (1971) A review of current knowledge on roadway behavior. In: Proceedings of the Symposium on Roadway Strata Control. Institute of Mining and Metallurgy, pp. 73–87.

Kneisley, R.O. & Haramy, K.Y. (1992) Large-scale strata response to longwall mining: a case study. US Bureau of Mines, Report of Investigation 9427.

Kneisley, R.O. (1995) Yield pillar sizing: an empirical approach. US Bureau of Mines, Report of Investigation 9593.

Koehler, J.R. & Tadolini S.C. (1995). Practical Design Methods for Barrier Pillars. US Bureau of Mines, Information Circular 9427.
Li, W., Bai, J., Peng, S., Wang, X. & Xu, Y. (2015) Numerical modeling for yield pillar design: a case study. Rock Mechanics and Rock Engineering, 48, 305–318.
Luo, Y. (2013) Highwall mining: design methodology, safety and suitability. Coal and Energy Research Bureau, West Virginia University, Report No. 2013–004.
Luo, Y., Peng, S.S. & Zhang, Y.Q. (2001) Simulation of water seepage through and stability of coal mine barrier pillars. SME, Preprint No. 01–131, 6 pp.
Majdi, A., Hassani, F.P. & Cain, P. (1991) Two new barrier pillar design methods for longwall mining. Mining Science and Technology, 13, 323–336.
Maleki, H. (1992) In situ pillar strength and failure mechanisms for US coal seams. In: Proceedings of the workshop on coal pillar mechanics and design. Pittsburgh: US Bureau of Mines, Information Circular 9315, pp. 73–77.
Maleki, H., Zahl, E.G. & Dunford, J.P. (1999) A hybrid statistical–analytical method for assessing violent failure in US coal mines. In: Proceedings of the Second International Workshop on Coal Pillar Mechanics and Design. US National Institute for Occupational Safety and Health, NIOSH publication no. 9448, pp. 139–144.
Mark, C. & Bieniawski, Z.T. (1986) An empirical method for the design of chain pillars in longwall mining. In: Proceedings of the 27th US Symposium on Rock Mechanics. Tuscaloosa: SME-AIME, pp. 415–422.
Mark, C. (1990) Pillar design methods for longwall mining. US Bureau of Mines, Information Circular 9247.
Mark, C. (1992) Analysis of Longwall Pillar Stability (ALPS): an update. In: Proceedings of the Workshop on Coal Pillar Mechanics and Design. US Bureau of Mines, Information Circular 9315, pp. 238–249.
Mark, C. & Chase, F. (1997) Analysis of Retreat Mining Pillar Stability (ARMPS). In: Proceedings of New Technology for Ground Control in Retreat Mining. US National Institute for Occupational Safety and Health (NIOSH), Information Circular 9446, pp. 17–34.
Mark, C. & Molinda, G.M. (2005) The Coal Mine Roof Rating (CMRR)-a decade of experience. International Journal of Coal Geology, 64, 85–103.
Mark, C. (2006) The evolution of intelligent coal pillar design: 1981–2006. In: Proceedings of the 25th International Conference on Ground Control in Mining. Morgantown: West Virginia University, pp. 325–334.
Mark, C. (2010) Pillar design for deep cover retreat mining: ARMPS version 6. In: Proceedings of the Third International Workshop on Coal Pillar Mechanics and Design. Morgantown: West Virginia University, pp. 106–121.
Mark, C., Chase, F.E. & Molinda, G.M. (1994) Design of longwall gate entry systems using roof classification. In: Proceedings of the New Technology for Longwall Ground Control. Pittsburgh: US Bureau of Mines, pp. 5–17.
Mark, C., Chase, F.E. & Zipf, R.K. (1997) Preventing massive pillar collapses in coal mines. In: Proceedings of New Technology for Ground Control in Retreat Mining. US National Institute for Occupational Safety and Health, NIOSH publication No. 9446, pp. 35–48.
Mark, C., Chase, F.E. & Pappas, D.M. (2003) Reducing the risk of ground falls during pillar recovery. Transactions of the Society of Mining Engineers, 314, 153–160.
Mark, C., Gauna, M., Cybulski, J. & Karabin, G. (2011) Applications of ARMPS (Version 6) to practical pillar design problems. In: Proceedings of the 30th International Conference on Ground Control in Mining. Morgantown: West Virginia University, p. 9.
Mohan, G.M., Sheorey, P.R. & Kushwaha, A. (2001) Numerical estimation of pillar strength in coal mines. International Journal of Rock Mechanics and Mining Sciences, 38, 1185–1192.
Morsy, K. (2003) Design consideration for longwall yield pillar stability. West Virginia University, Morgantown, PhD Dissertation.

Morsy, K. & Peng, S.S. (2002) Evaluation of a mine panel failure using the local mine stiffness criterion-a case study. Transactions of SME, 312, 8–19.

Morsy, K. & Peng, S. (2003) New approach to evaluate the stability of yield pillars. In: Proceedings of the 22nd International Conference on Ground Control in Mining. Morgantown: West Virginia University, pp. 371–377.

Mucho, T.P., Barton, T.M. & Compton, C.S. (1993) Room-and-pillar mining in bump-prone conditions and thin pillar mining as a bump mitigation technique. Pittsburgh: US Bureau of Mines, Report of Investigations 9489.

Najafi, M., Jalali, S.E., Yarahmadi Bafghi, A.R. & Sereshki, F. (2011) Prediction of the confidence interval for stability analysis of chain pillars in coal mines. Safety Science, 49, 651–657.

Ozbay, U. & Badr S. (2010) Numerical modeling of yielding chain pillars in deep longwall coal mines. In: Proceedings of the Third International Workshop on Coal Pillar Mechanics and Design. Morgantown: West Virginia University, p. 8.

Pen, Y. (1994) Chain pillar design in longwall mining for bump-prone strata. University of Alberta, Edmonton, Canada, PhD Dissertation.

Peng, S.S. (2006) Longwall Mining. Morgantown: West Virginia University, Department of Mining Engineering.

Peng, S.S. (2008) Coal Mine Ground Control. Morgantown: West Virginia University, Department of Mining Engineering.

Perry, K.A., Raffaldi, M.J. & Harris, K.W. (2015) Influence of highwall mining progression on web and barrier pillar stability. Mining Engineering, 67, 59–67.

Poeck, E.C., Zhang, K., Garvey, R. & Ozbay, U. (2015) Energy concepts in the analysis of unstable coal pillar failures. In: Proceedings of the 34th International Conference on Ground Control in Mining. Morgantown: West Virginia University, p. 6.

Porathur, J.L., Karekal, S. & Palroy, P. (2013) Web pillar design approach for Highwall Mining extraction. International Journal of Rock Mechanics and Mining Sciences, 64, 73–83.

Rangasamy, T., Leach, A.R., van Vuuren, J.J., Cook, A.P. & Brummer R. (2001) Current practice and guidelines for the safe design of water barrier pillars. Braamfontein, South Africa: The Safety in Mines Research Advisory Committee (SIMRAC), COL 702.

Recio-Gordo, D. & Jimenez, R. (2012) A probabilistic extension to the empirical ALPS and ARMPS systems for coal pillar design. International Journal of Rock Mechanics and Mining Sciences, 52, 181–187.

Salamon, M.D.G. & Munro, A.H. (1967) A study of the strength of coal pillars. Journal of the South African Institute of Mining and Metallurgy, 68, 56–67.

Sasaoka, T., Shimada, H., Matsui, K. & Takamoto, H. (2010) Geotechnical considerations in highwall mining applications in Indonesia. In: Proceedings of the 29th International Conference on Ground Control in Mining. Morgantown: West Virginia University, p. 6.

Schissler, A. (2002) Yield pillar design for coal mines based on critical review of case histories. Colorado School of Mines, Golden Colorado, PhD Dissertation.

Scovazzo, V.A. (2010) Analytical design procedure using the Wilson equation. In: Proceedings of the Third International Workshop on Coal Pillar Mechanics and Design. Morgantown: West Virginia University, p. 11.

Shabanimashcool, M. & Li, C.C. (2012) Numerical modeling of longwall mining and stability analysis of the gates in a coal mine. International Journal of Rock Mechanics and Mining Sciences, 51, 24–34.

Shabanimashcool, M. & Li, C.C. (2013) A numerical study of stress changes in barrie pillars and a border area in a longwall coal mine. International Journal of Coal Geology, 106, 39–47.

Smith, W., Dolinar, D. & Haramy, K. (1995) Nonlinear approach for determining design criteria for yield pillar performance. In: Proceedings of the 14th International Conference on Ground Control in Mining. Morgantown: West Virginia University, pp. 124–133.

Stefanko, R. (1983) Coal Mining Technology: Theory and Practice. New York, SME-AIME.

Su, D.W. & Hasenfus G.J. (2010) A retrospective assessment of coal pillar design methods. In: Proceedings of the Third International Workshop on Coal Pillar Mechanics and Design. Morgantown: West Virginia University, p. 8.

Su, D.W.H. & Hasenfus, G.J. (1996) Practical coal pillar design considerations based on numerical modeling. In: Proceedings of 15th International Conference on Ground Control in Mining. Golden, Colorado: Colorado School of Mines, pp. 768–789.

Tadolini, S.C. & Haramy, K.Y. (1992) Gateroads with yield pillars for stress control. In: Proceedings of the 4th Conference on Ground Control for Midwestern US Coal Mines. Illinois: Southern Illinois University, pp. 179–194.

Tadolini, S.C., Porter, C. & Jarvis, J. (2013) Barrier pillar design for safe and productive longwall mining. In: Proceedings of the 32nd International Conference on Ground Control in Mining. Morgantown: West Virginia University, p. 8.

Torano, J., Rodriguez, R. & Cuesta, A. (2000) Determination and optimization of pillar size in underground mining through numerical methods. In: Proceedings of the European Congress on Computational Methods in Applied Sciences and Engineering. Barcelona: Spain, p. 18.

Tsang, P. (1992) Yield pillar design for US longwall mining. West Virginia University, Morgantown, PhD Dissertation.

Tsang, P. & Peng, S.S. (1993) A new method for longwall pillar design. In: Proceedings of the 12th International Conference on Ground Control in Mining. Morgantown: West Virginia University, pp. 261–273.

Tulu, I.B., Heasley, K.A., & Mark, C. (2010) A comparison of the overburden loading in ARMPS and LaModel. In: Proceedings of the 29th International Conference on Ground Control in Mining. Morgantown: West Virginia University, pp. 28–37.

Verma, C.P., Porathur, J.L., Thote, N.R., Pal Roy, P. & Karekal, S. (2014) Empirical approaches for design of web pillars in highwall mining: review and analysis. Geotechnical and Geological Engineering, 32, 587–599.

Wattimena, R.K., Kramadibrata, S., Sidi, I.D. & Azizi, M.A. (2013) Developing coal pillar stability chart using logistic regression. International Journal of Rock Mechanics and Mining Sciences, 58, 55–60.

Yavuz, H. & Fowell, R.J. (2001) FDM prediction of a yield pillar performance in conjunction with a field trial. In: Proceedings of the 20th International Conference on Ground Control in Mining. Morgantown: West Virginia University, pp. 78–85.

Zhang, Y., Chugh, Y.P. & Yang, G. (1995) Fuzzy neural network for chain pillar design in longwall coal mining. In: Proceedings of the 35th US Symposium on Rock Mechanics. Reno: University of Nevada, p. 6

Zingano, A., Koppe, J. & Costa, J. (2004) Violent coal pillar collapse-a case study. In: Proceedings of the 23th International Conference on Ground Control in Mining. Morgantown: West Virginia University, pp. 60–67.

Zipf, K. (1999) Catastrophic collapse of highwall web pillars and preventative design measures. In: Proceedings of the 18th International Conference on Ground Control in Mining. Morgantown: West Virginia University, pp. 18–28.

Zipf, R.K. (1996) Simulation of cascading pillar failure in room-and-pillar mines using boundary-element method. In: Proceedings of the Second North American Rock Mechanics Symposium. Rotterdam, AA Balkema, pp. 1887–1892.

Zipf, R.K. (2005) Ground control design for highwall mining. SME, Preprint No. 05–82, p. 12.

Zipf, R.K. & Mark, C. (1997) Design methods to control violent pillar failures in room-and-pillar mines. Transactions of the Institution of Mining and Metallurgy, 106, 124–132.

Zipf, R.K. & Mark, C. (2005) Ground control for highwall mining in the United States. International Journal of Surface Mining, Reclamation and Environment, 19, 188–217.

Petroleum Engineering

Chapter 15

Rock mechanical property testing for petroleum geomechanical engineering applications

M.A. Addis
Rockfield Global Technologies, Swansea, UK

Abstract: Acquiring sub-surface geomechanical properties to support the development of oil and gas fields draws on best practices from rock and mining engineering as well as from civil engineering. The determination of these rock mechanical properties through laboratory tests is discussed based on best practices, industry standards as well as commonly used and accepted methods. Geomechanical characterization of the sub-surface has to overcome several challenges including the low sampling density and that representative and undisturbed samples are obtained from the recovered cores, sometimes from depths in excess of 5km. Once the cores are retrieved the laboratory testing techniques must be appropriate for the wide range of lithologies and samples encountered in oil and gas fields, taking into consideration any sample heterogeneity.

The use of rock mechanical properties in different aspects of petroleum engineering design is discussed and the need to upscale laboratory-measured geomechanical properties for well-based analysis and field-wide numerical modeling. The motivation to geomechanically characterize the sub-surface is to minimize risk and uncertainty and to ensure successful field developments.

1 GEOMECHANICAL TESTING IN THE PETROLEUM INDUSTRY

Drilling oil and gas wells subjects rock formations to an open excavation analogous to a small tunnel, held open and supported only by the hydrostatic column of drilling fluid in the wellbore. This commonly leads to extreme stress differences developing around the wellbore walls. On a larger scale, the production of hydrocarbons from a field typically involves pore pressure changes and deformations in the reservoir rocks and in the surrounding formations. This interaction between human activity during the development of hydrocarbon resources and the response of the rock formations to excavation and stress relief, or to effective stress changes resulting from pore pressure depletion or injection, is addressed by Petroleum Geomechanical Engineers. The geomechanical engineer may address design issues based on an empirical approach, or use analytical approaches based on elasticity, pressure-dependent elasticity or even elasto-plasticity for idealized geometries of wellbores, perforations, or simplified field geometries. With improved computing and software capabilities, engineers increasingly turn to numerical analyses to address more complex and realistic rock behavior, as well as irregular and unique field geometries.

Throughout the analyses, the incremental changes in stress and strain are compared with the elastic properties of the rock and the yield and rock failure criteria, to determine the onset of non-linear behavior and the development of plasticity. This occurs through either homogeneous plastic deformation and creep, or more commonly through the development of shear bands and brittle failure, which transforms the rock mass from a continuum to one with new discontinuities, along which strain is concentrated.

The formations drilled during any field development have almost infinite variability, ranging from weak unconsolidated sediments at the surface of the well, or below the mudline in offshore locations, to more competent rock formations at greater depths with unconfined strengths which can exceed 200 MPa for 'tight' low porosity gas-bearing sands. These reservoir formations exist at depths that can exceed 5 km, though typical reservoir depths are in the region of 1,500m to 3,000 km true vertical depth (TVD).

The development of oil fields before the 1980s relied on drilling vertical or sub-vertical wells. The advent of more difficult offshore drilling from a limited number of platforms, was accompanied by the development of inclined and horizontal drilling through relatively thin reservoirs. Currently, the most extreme high angle wells are over 12km long drilled at inclinations of > 80°, where the end, or toe, of the wells are located 11.7km horizontally away from the surface well location. Horizontal wells also predominate the development of shale resources, as a way to increase the contact between the well and reservoir.

Since reservoirs occur over a large range of depths, the wells used to drain these oil and gas fields are often drilled through a large number of formations with contrasting mechanical properties, and subjected to large variations in both the stress magnitudes and stress regimes, which must be considered in any analysis. The stress regimes vary from a near-surface, uniaxial-strain (K_o) conditions representative of normal burial, to extensional environments comprising normal fault stress regimes, or to compressional environments characterized by strike-slip regimes and thrust fault regimes (Anderson, 1905, 1951).

The formation pore pressures also vary considerably. Near the surface, these pore pressures may be hydrostatic resulting only from the weight of the pore fluid column. However, with increasing burial and at greater depths, the rocks compress, the permeability reduces, and due to either continued 'undrained' or 'partially drained' burial, chemical changes or tectonic compression, the pore pressures increase above the hydrostatic pressure, become trapped and unable to drain even over geological time. These over-pressures can reach magnitudes where the pore pressure is almost equal to the vertical (overburden) total stress.

The initial stress regime and pore pressure magnitudes, as well as their evolution during field development are key considerations in defining any laboratory test program. The mechanical properties and characteristics obtained from the laboratory tests are then used in the geomechanical analysis to predict the formation response to pressure changes and excavation. The aim of which is to optimize the design and operation of the field development during hydrocarbon production and recovery, by minimizing risks, reducing costs and increasing production and hydrocarbon recovery from the field development.

2 SAMPLING DENSITY

Most reservoir and field characterization is performed early in the exploration and appraisal (E&A) stages of a development. For a typical exploration well, core samples are taken across the length of the hydrocarbon-bearing reservoir interval. This depth range may vary from 20m to several hundred meters in the larger oil fields and geomechanical testing of 15 – 20 core sample intervals from the reservoir formation would comprise a typical sampling density for an E&A well. This may be repeated in up to 10 wells across the field, but geomechanical testing in more than 2–3 exploration and appraisal wells per field is uncommon.

For any one well drilled through a 100m thick reservoir, this corresponds to a sampling density of 0.011 – 0.015 (1–1.5%) of the reservoir interval, based on 7.5 cm (3") long samples used for mechanical property determination. While this sampling rate for the reservoir interval in a particular well may be reasonable, the sampling rate along the entire well length, which typically is 3km in total depth (TVD) decreases to $3*10^{-4}$ to $5*10^{-4}$. For a 3D reservoir volume, such a sampling rate for one well drilled in a reservoir with a radius of 3km, corresponds to a sampling density of $0.4*10^{-12}$ to $0.6*10^{-12}$; less than a part per trillion. This very low sampling rate illustrates why petrophysical well logs and 3D seismic data, calibrated to the laboratory mechanical property measurements, are used to extrapolate the measured rock properties across the entire reservoir, and overburden, critical for any 3D geomechanical characterization and analysis of the field (Figure 1).

Well logs record data every 15cm along the wellbore as a standard, and comprise a range of geophysical or petrophysical measurements including electrical and acoustic measurements of the formations intersected by the well. The logs typically have a shallow depth of investigation of the near wellbore region, ranging from centimeters to a meter inside the wellbore wall: Newer log measurements enable investigation out to 20-60m for deep resistivity and acoustic measurements, but these currently focus on identifying discontinuities and bedding planes between different formations. Well logs are used for geomechanical analysis as they are run along the entire reservoir section for characterization purposes and allow the rock properties measured in the laboratory, for discrete well depths, to be extrapolated along the length of the wellbore, using suitable correlations between the well logs and the laboratory measured rock properties. A sufficient number of mechanical tests are required to satisfactorily characterize

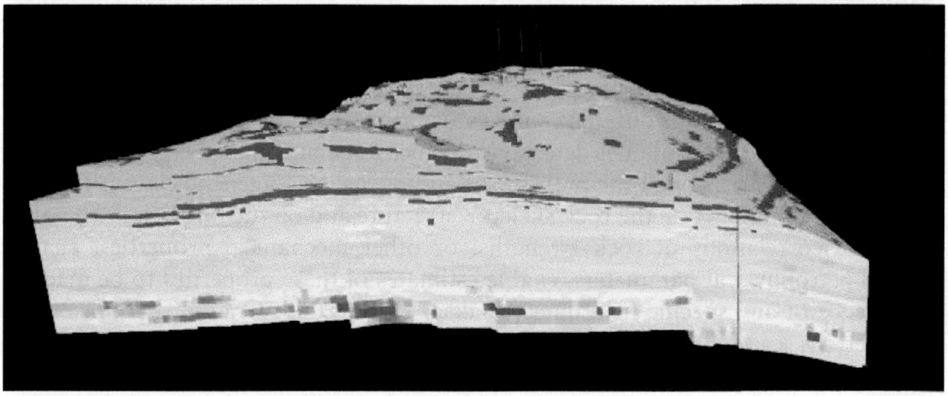

Figure 1 Distribution of mechanical rock properties across a field from core-based laboratory measurements, well logs and 3D geophysical seismic surveys.

the reservoir and overburden, based on the size of the reservoir, the number of formations and facies (geological depositional environments), and the potential lateral variations. This enables formation-specific and facies-specific mechanical property correlations to be used to adequately characterize the field.

The petrophysical well logs are compared to the properties from the seismic surveys, whose depth resolution is generally in the range of 10-20m, to ensure compatibility at different well locations over the structure of the oil field. The geophysical responses and attributes obtained from surface seismic analysis enable average mechanical properties along the wellbore to be extrapolated away from the wellbores into the entire oil field reservoir and in some case into the overburden above the reservoir.

This methodology is routinely implemented by the industry (Herwanger & Koutsabeloulis, 2011), but as with the other geological reservoir properties and subsurface characterization, the low sampling density and the use of correlations to estimate values across the field introduces considerable uncertainties in the geomechanical characterization of the reservoir.

2.1 Influence of geological processes on rock strength

The variation of mechanical properties within a reservoir, resulting from diagenetic/facies variations is illustrated in Figure 2, which shows a cross section of the Lunskoye field containing 10 different layers. The strength vs porosity correlations for the Daghinsky sandstone reservoir in the Lunskoye Field, are described by two trends; a low strength correlation for the upper layers (I-VI) of the reservoir, the higher strength correlation for the deeper reservoir layers (VII – X) (Addis *et al.*, 2008).

These strength vs porosity correlations were established from laboratory tests measurements of both unconfined compressive strengths and hollow cylinder tests (see section 7.2).

The higher strength trend shows values 30% higher than the lower trend for the higher porosity samples in the different layers of the Daghinsky reservoir. This difference in sandstone strengths within the one reservoir, was attributed to variations in the geochemistry of the cementation in the reservoir layers, even though this was not reflected in the porosity measurements.

A similar impact of sedimentary and diagenetic factors is observed in a collation of sandstone strengths from different sources described by Plumb (1994), where for a given porosity, the clay volumes determine the measured unconfined compressive rock strength (Figure 3). Though the presence of authogenic or diagenetic clays is likely to have a different impact on the rock skeleton structure and on the formation strengths.

Generic correlations of rock strengths, or other mechanical properties, against selected petrophysical parameters, enable estimates of these properties to be made in the absence of any specific field data (Khaksar *et al.*, 2009; Crawford *et al.*, 2010). However, the scatter around these correlations is generally large, introducing considerable uncertainty into any analysis. Consequently, mechanical strength data measured on core samples from the reservoir should be used to calibrate global correlations, in order to generate field- and reservoir-specific correlations and reduce the uncertainty for any subsequent analysis.

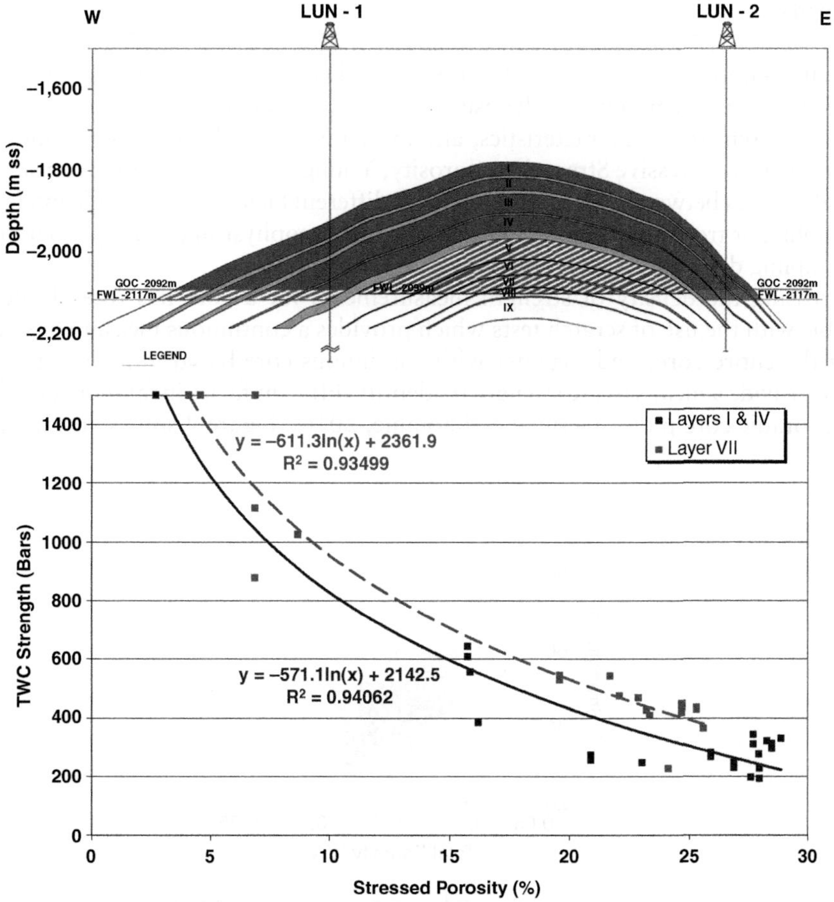

Figure 2 Daghinsky formation reservoir strength trends.

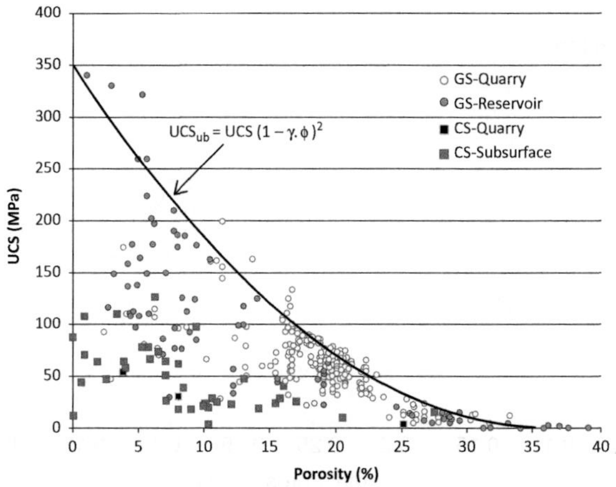

Figure 3 Variation of rock strength for sandstones with differing porosity and clay contents (From Plumb, 1994).

Greater sampling density provides more confidence when defining any lithological trends. Different formations, or depositional facies, comprising a reservoir are likely to have a different set of characteristics, and mechanical–geophysical correlations (*e.g.*, Unconfined Compressive Strength vs porosity, Young's modulus vs sonic transit time). The differences between these correlations for different facies can be significant, leading to doubling or trebling of the strength for the same geophysical property, even though the sampling depths may be juxtaposed within the core.

Increasing the density of strength measurements for a core has recently become possible with the use of scratch tests which provides a continuous measure of strength along the entire core, and together with continuous core-based ultrasonic measurements, provide a means to use strength to identify different facies in far more detail than with sporadic core plugs (Germay & Lhomme, 2016). Figure 4 shows an example of

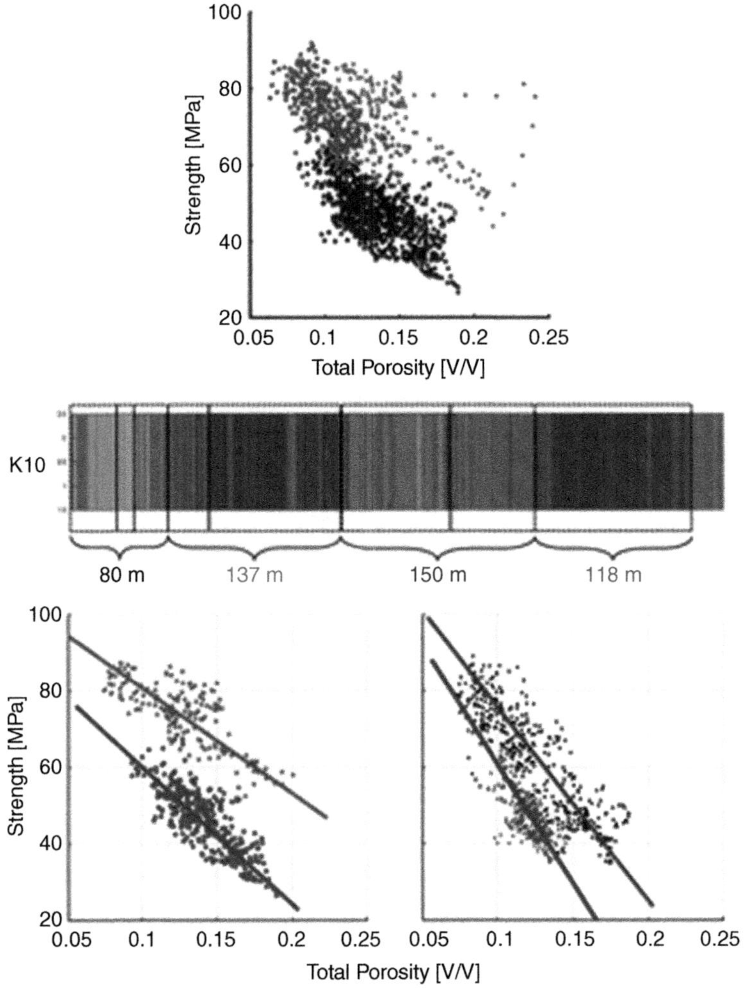

Figure 4 Identification of different facies using strength vs log cross plots, based on continuous strength and p-wave velocity measurements on cores (after Noufal *et al.*, 2015).

scratch test-based strength estimates over a 400m long core section, showing a large scatter of data when plotted against the corresponding log porosity. When re-analyzed using a clustering scheme, 4 different facies are identified enabling unique strength correlations to be established for each facies.

Both standard measurements on core plugs and the more detailed measurements of strength from the scratch tests provide independent evidence to support the use of facies-based strength correlations for detailed analysis. These scratch tests also provide the data to identify core intervals with uniform properties to sample for standard laboratory mechanical property tests, especially for triaxial tests where several identical samples are required to define failure envelopes.

The variations in sandstone formations, described above, may be predictable across the reservoir formations, with relatively little lateral variation. However, considerably more geological variation may exist within carbonate reservoirs with consequently more uncertainty in both the geomechanical characterization and in any analysis. Consequently, fewer strength correlations exist for carbonate lithologies compared to sand-shale correlations (Ameen *et al.*, 2009).

3 ROCK TESTING OUTLINE FOR PETROLEUM GEOMECHANICS APPLICATIONS

Various analyses are performed to address both well-based and reservoir-based geomechanical responses and potential risks during the life of a well or field. The expected geomechanical responses will depend upon the lithologies and field structure, but also upon the type of development and the type of wells employed to extract the oil and gas from the reservoir.

The geomechanical considerations that are important for the development of a field depend on whether the field is produced by continually depleting the reservoir pressure possibly leading to reservoir compaction and surface subsidence, or whether injection is used to maintain or raise the reservoir formation (pore) pressure and potentially cool the reservoir. Thermal stimulation, using steam to reduce the viscosities of heavy oil, also leads to geomechanical changes, including formation expansion in the reservoir and surface heave.

At specific locations over the field, wells are drilled through numerous rock formations each with potentially different pore pressures and stress magnitudes. The wells may have different trajectories (vertical, inclined or horizontal) going through these formations and experience wellbore instability during drilling. Formation pore pressure reductions that accompany hydrocarbon production may lead to time dependent, or delayed, well failure, leading to sand or solids production with the hydrocarbons. These potential well-based geomechanical issues as well as field–based issues resulting from field development strategies and recovery methods, are illustrated in the geomechanical screening table (Figure 5). Such tables vary for different geographical, geological and operating conditions (*e.g.*, onshore vs offshore locations, good land access vs limited platform locations) and operational practices, but provide a high-level guide to the potential geomechanical risks and concerns for specific field development strategies, using a traffic light color scheme.

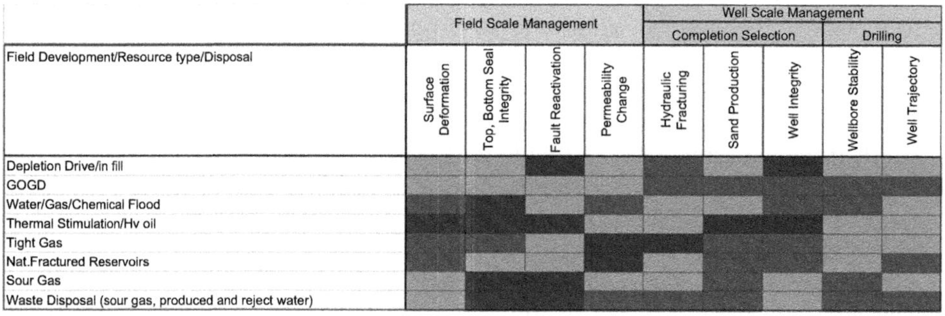
Field Scale Management
Well Scale Management
Completion Selection
Drilling
Field Development/Resource type/Disposal
Surface Deformation
Top, Bottom Seal Integrity
Fault Reactivation
Permeability Change
Hydraulic Fracturing
Sand Production
Well Integrity
Wellbore Stability
Well Trajectory
Depletion Drive/in fill
GOGD
Water/Gas/Chemical Flood
Thermal Stimulation/Hv oil
Tight Gas
Nat.Fractured Reservoirs
Sour Gas
Waste Disposal (sour gas, produced and reject water)

Figure 5 Example of petroleum geomechanical engineering issues associated with different field development strategies

The following geomechanical analyses are generally included at some point in a field development:

- Stress profile and fracture gradient assessment,
- Wellbore stability,
- Sand production,
- Compaction and subsidence,
- Hydraulic fracturing,
- Produced water re-injection
- Cuttings disposal and re-injection
- 3D Field modeling,
- IOR/EOR & Thermal recovery,
- Unconventional resources (shale gas and shale/tight oil),
- Natural fracture conductivity.

These may be addressed using analytical or numerical tools, however, as the analyses become more rigorous, and detailed, more complex material behavior models are used for these analyses requiring more intensive laboratory test programs. The petroleum industry generally lacks, not only geomechanical data at depth, but also recognized standards, guidelines, or a consensus, with which to analyze these design issues, or to assess risks. Consequently, the type of analyses performed varies from company to company, and along with this the rock mechanical tests required to support these analyses.

This section therefore provides an overview of the more common analyses performed in the industry and the testing required for typical well types and field developments, as illustrated in the decision tree in Figure 6. When the design and operational consequences of these challenges becomes large, exceeding $100s million or even billions of dollars, more detailed analysis is often undertaken with research analysis tools. The inputs to these more specialized tools are specific to the numerical model and will not be discussed in detail in this chapter.

The mechanical property tests used for standard analyses and design issues are separated into three types;

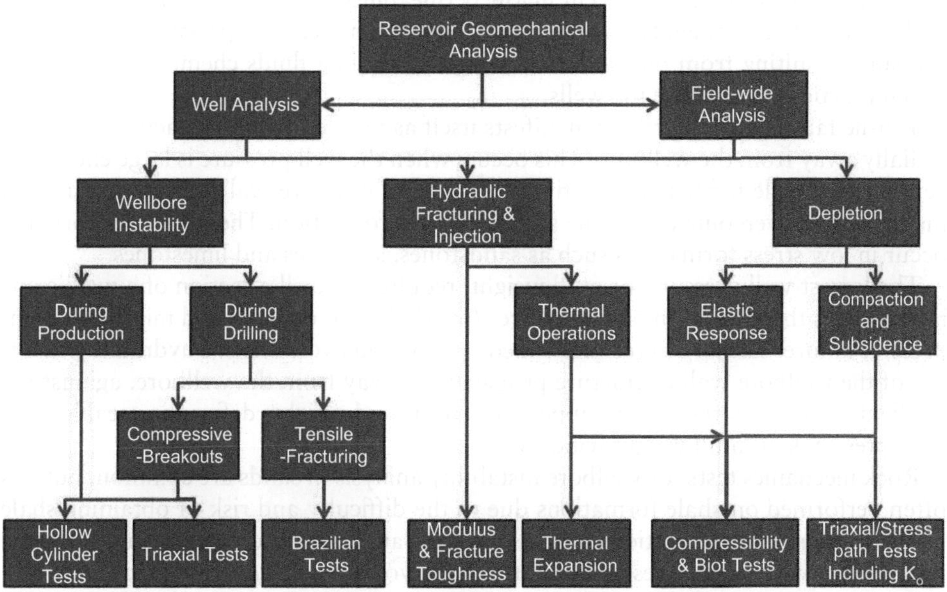

Figure 6 Decision tree illustrating the required laboratory tests used for standard analyses.

a) Wellbore failures: wellbore instability during drilling, or sand and solids production during production.
b) Fracturing: comprising hydraulic fracturing for increased production (including 'unconventional' shale formations), produced water re-injection, for reservoir pressure maintenance during development, and fracturing for cuttings and waste disposal.
c) Field-wide management: Including reservoir depletion, or injection EOR/IOR, compaction and subsidence, and natural fracture conductivity.

3.1 Rock mechanical property determination for wellbore failure analysis

Failure of wellbores occurs whilst drilling the well or during hydrocarbon production. During drilling it is referred to as wellbore (or borehole) instability, during hydrocarbon production in the reservoir section it is referred to as sand (or solids) production.

3.1.1 Wellbore instability during drilling

Wellbore instability occurs in two main forms; compressive failure and tensile failure. The compressive failure is analogous to tunnel spalling, with either excavation-parallel tensile cracks or shear planes developed at the wellbore wall. The compressive failure generally focuses on shear failure of the wellbore wall forming breakouts (Bradley, 1979; Bell & Gough, 1979; Hottman, Smith & Purcell, 1979), resulting from insufficient wellbore pressure to stabilize the wellbore.

Breakouts most commonly occur in low permeability shale or mudstone formations, with remedial recommendations relying on optimum wellbore pressure (hydrostatic pressures resulting from different mud weights), drilling fluids chemistry and operational practices to stabilize the wells.

Tensile failure during drilling manifests itself as tensile hydraulic fractures oriented radially away from the wellbore. This occurs when the well pressure is large enough to generate a tensile tangential effective stress in the wellbore wall that is sufficient in magnitude to overcome the tensile strength of the formation. These failures normally occur in low stress formations such as sandstones, siltstones and limestones.

The lowest well pressure, or mud weight, required to drill a section of a well is one that exceeds the formation pore pressures (overbalanced drilling) and minimizes compressive failure. The largest recommended mud weight avoids tensile hydraulic fracturing of the wellbore wall or fracture propagation away from the wellbore, against the far-field minimum stress. These upper and lower mud weights define the useable wellbore pressures or 'mud weight window'.

Rock mechanics tests for wellbore instability analysis in sands are common, but less often performed on shale formations due to the difficulty and risk of obtaining shale cores from unstable formations. However, shale and mudstone formations which are interbedded with sandstones and limestone reservoir intervals are often recovered as a part of a coring program, but only suitable for testing if well preserved and handled in line with best practice (Ewy, 2016).

When it is possible to sample plugs for testing, the following suites of tests are performed;

a) Triaxial tests: Drained tests in sandstones (and permeable formations) and undrained test conditions for shales, with confining stress magnitudes chosen to reflect the low radial stress magnitudes acting in and near the wellbore wall. The confining pressures selected for this test suite should enable an estimate of unconfined compressive strength to be made.

   The triaxial test sample plugs are normally cored perpendicular to the bedding planes in the core to maximize the number of samples available for any testing program. The orientation of the sample plugs, relative to bedding planes, will affect the relative success rate of the sampling, with sampling plugging parallel to the bedding often being more successful.

b) Many wellbore instability problems in shales occur in inclined wellbores, where the wellbore stresses leading to rock failure are inclined to the bedding planes. Consequently, when possible, plugs of shale are taken from the core orientated parallel and inclined to the bedding planes, to assess the weakening effect of any bedding planes on the failure analysis (Aadnøy & Chenevert, 1987; Ambrose *et al.*, 2014).

In addition to static mechanical triaxial tests, core samples can be characterized during testing by measuring acoustic velocities across the samples at different stress magnitudes (Holt *et al.*, 2005; Sarker & Batzle, 2010; Sharf-Aldin *et al.*, 2013) to determine both isotropic and anisotropic dynamic properties.

The tensile failure strength of the sandstones, siltstones or limestones, can be assessed using Brazilian tests (Section 6.9), or through the use of the Unconfined Compressive Strength (UCS) tests (Section 6.7.1) obtained from the triaxial test suite, along with general correlations linking the tensile strength and UCS.

3.1.2 Sand and solids production: wellbore instability during hydrocarbon production

Wellbores in reservoir formations can also experience rock failure during production of hydrocarbons, or sand (solids) production, which result from effective stress increases due to either lower well or reservoir pressures. Sand failure occurs in a range of well configurations, but is analyzed using rock mechanics principles and testing in only two types of completion: The first is where no sand screen, or filter, is deployed across the reservoir section; the openhole completion. The second is a cased and perforated completion; where the wellbore is sealed with steel casing and cemented in place. In this completion, contact between the reservoir formation and the wellbore is achieved by shooting perforations (shaped explosive charges) through the steel casing into the formation. These perforations form conical shaped cavities directed radially out from the wellbore. These perforations are approximately 5–10 mm (0.2" to 0.4") in diameter at the wellbore wall and extend between 30–90 cm (12 to 36") into the formation.

Sand production can also occur in completions with sand screens and gravel packs placed inside openholes or cased and perforated completions. This sand production occurs when formation failure leads to screen erosion by the sand.

These two completion types have different geometries but are analyzed using the two most common methods of sand production prediction: The hollow cylinder-based analytical approach, or a numerical modeling, finite element approach (Figure 7).

The analytical approach uses hollow cylinders as index tests that represent the perforation geometry (see section 7.2). This approach was developed by Shell in the 1970s and 1980s (Antheunis *et al.*, 1976; Veeken *et al.*, 1991) and has since become widely accepted as a pragmatic approach for sand production prediction (Willson *et al.*, 2002). In this approach, approximately 20 sample locations are selected across each reservoir formation, with each sample location comprising at least 15cm long of intact rock. Ten locations are selected from high porosity intervals, 5 sample locations each from medium and low porosity intervals. From the 15 cm sampling locations, plugs are taken for thick-walled hollow cylinder (TWC) and Unconfined Compressive Strength (UCS) tests, as well as plugs for formation characterization, such as porosity, density and dynamic velocities (V_p and V_s), in order to establish correlations between the strengths and the core derived petrophysical properties. The hollow cylinder samples

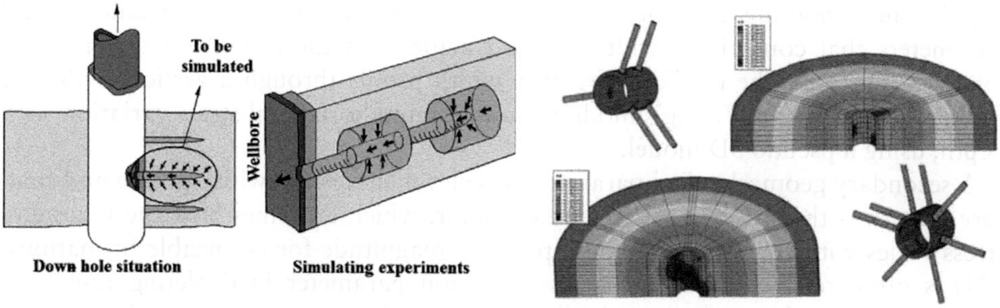

Figure 7 Cased and perforated completions and the corresponding use of hollow cylinder tests and finite element modeling for analysis (Courtesy of Shell and GMI/Baker Hughes).

are cut with the long axis parallel to bedding, and if sufficient core material is available an additional sample is taken perpendicular to the bedding.

The Finite Element approach for analyzing sand production has been championed by Morita and co-workers (Morita *et al.*, 1987; Morita, 1994), and again has been adopted by the industry (Vaziri *et al.*, 2002). The methodology relies on a series of triaxial tests to characterize the reservoir. At least three samples from each sample (and porosity) location are used for the triaxial tests, to establish stress-strain curves and a failure envelope for each interval.

The stress-strain curve for the test is reproduced, either digitally, or with an appropriate constitutive model, up to failure, where pure plasticity is normally assumed in the analysis. The onset of sand failure is predicted using a plastic strain criterion, often selected through calibration with field data (Morita, 1994), or by comparison with laboratory-based Advanced Hollow Cylinder Tests. Samples are normally selected for this FEM analysis from the weaker parts of the reservoir formation, which are still intact after coring.

3.2 Rock mechanical property determination for hydraulic fracture design

Hydraulic fractures are used in field developments to increase the connectivity and contact area between the wellbore and the reservoir formation. Hydraulic fracture stimulation increases production from a reservoir (Economides & Martin, 2007; Ghalambor, *et al.*, 2009) but is also used to increase injectivity, either for; injecting produced water to maintain reservoir pressures and to dispose of drilling waste (Nagel & McLennan, 2010).

These applications require slightly different input into the design, however the main rock property used to model the growth of hydraulic fractures is the plane strain Young's modulus, or plane strain confined modulus when measured at in-situ stress conditions.

The plane strain Young's modulus (E') is calculated from the Young's modulus (E) and the Poisson's ratio (v) (Obtaining these will be discussed in Section 6.1):

$$E' = \frac{E}{(1 - v^2)}$$

The Young's modulus and the minimum in-situ stress are two key geomechanical parameters that control the width of the fracture and the growth of the fracture. Figure 8 illustrates the predicted fracture width profile through a vertical hydraulic fracture, based on the Young's modulus or minimum horizontal stress variation with depth, using a pseudo 3D model.

A secondary geomechanical parameter which influences the width profile and fracture growth is the stress-pore pressure constant, which describes how the minimum stress varies with changes in the pore pressure magnitude for permeable formations. This is often referred to as the stress-depletion parameter in depleting reservoirs (Referred to as either γ_h or A in the literature). During the formation and growth of hydraulic fractures, the injected fluid leaks from the hydraulic fracture into the formation through the walls of the newly formed hydraulic fracture. The pore pressure in the

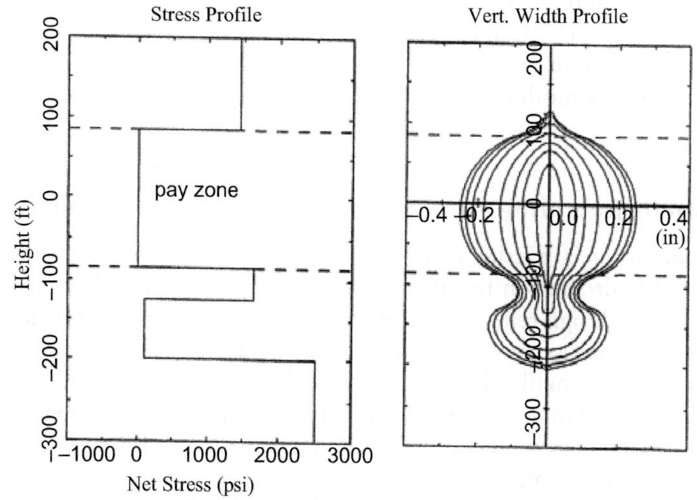

Figure 8 Prediction of hydraulic fracture width based on the Stress/Young's modulus profile (after Warpinski *et al.*, 1993).

formation increases and results in an increase in the near-fracture minimum stress magnitude, often called a 'back-stress'.

If the injected fluid is produced water or seawater and cooler than the reservoir formation, this area of fluid leak-off is also accompanied by a thermal change and contraction. Whilst if the fluid injected is steam, as used in the development of heavy oil fields, and a fracture forms during the injection, then thermal expansion of the rock adjacent to the fracture occurs.

As a result the three geomechanical properties normally used for hydraulic fracture analysis includes, the plane strain Young's modulus, or the confined modulus, the stress – depletion parameter (commonly estimated from the Poisson's ratio) and the thermal expansion coefficient. The first two are measured in the laboratory, whilst the third, the thermal expansion coefficient is calculated or generic values taken. The thermal expansion coefficient can also be measured for specific formations, when required.

Fracture initiation of hydraulic fractures is normally addressed in simulators using a constant or nominal value for the fracture toughness (K_I), though again this can be measured and introduced to the analysis. Generally hydraulic fracture simulators are designed to obtain the optimum hydraulic fracture geometry and conductivity (Economides & Martin, 2007) and enabling history matching with pressure and rate data from stimulation treatments, with less emphasis on the initiation. Simulators for produced water or cuttings injection do not consider the fracture initiation mechanism, focusing predominantly on the growth of any hydraulic fractures and the design parameters.

The Young's, or confined, modulus should be measured on cored samples at several points along the reservoir, and preferably in the bounding formations that are expected to contain any fracture height growth. Poorly constrained fracture height growth, leads to poor fracture stimulation, therefore these high stress and high moduli formations are key depth intervals to characterize.

The testing procedure for the Young's modulus is a standard within rock mechanics (ASTM D2938-71a, 1979; ISRM, 1983). During the test the elastic portion of the stress-strain response can be used to determine the Poisson's ratio from which the Plane Strain Young's modulus and the stress-depletion parameter can be estimated (Section 6.1).

3.2.1 Hydraulic fracturing in unconventional resource developments

Hydraulic fracturing is a critical technology that has enabled the development of unconventional hydrocarbon resources such as tight gas and shale gas and oil plays. Hydraulic fracture design relies on the parameters and measurements described above, as well as operational experience. The hydraulic fractures used to access unconventional resources are normally placed at specified depths along horizontal wellbores, and rely on the fractures remaining conductive and held open with proppant or sands. As such, best practice relies on placing the hydraulic fractures in layers of higher stiffness, or 'brittle' formations, which are more likely to remain conductive by reducing proppant embedment and creep (Economides & Martin, 2007).

Shale gas and oil developments have characterized the hydrocarbon-bearing formations with some measure of stiffness, including; Young's modulus, hardness or brittleness. There is no standard definition of 'brittleness' used for this shale characterization (Holt *et al.*, 2011), but most of the definitions rely on both Young's modulus and Poisson's ratio. These values are normally dynamically derived either from well logs and surface seismic attribute analysis but may be correlated with static measurements from the laboratory, as described above. This use of the term 'brittleness' contrasts with the common geomechanics usage, which describes the post-failure softening behavior of samples: The percentage drop in strength from the peak to the residual strength (Bishop, 1974).

3.3 Field-wide operations: Reservoir depletion, compaction and subsidence, or injection EOR/IOR, and natural fracture responses

Field management of oil and gas fields involves a number of geomechanical reservoir responses, which in turn depend on the selected field development strategy; depletion, improved oil recovery (IOR) such as water or polymer flooding, or Enhanced Oil Recovery through thermal injection and recovery.

Depletion is a standard primary recovery strategy, relying on the reservoir pressures to drive production. This gradual reduction in the formation pore pressures, and increased effective stresses, leads to reservoir compaction in high porosity reservoirs and surface subsidence. The stress evolution and porosity decrease result initially from an elastic response of the formations to the depletion, but become far more significant when the formations experience effective stress magnitudes sufficient to develop plastic strains (Jones *et al.*, 1987). The changes in the stress and strain accompanying depletion can be monitored using 4D (repeat 3D) surface seismic surveys, which indicate travel time changes in both the reservoir and overburden rocks.

Depletion can also influence the production in stress-sensitive naturally fractured reservoirs, as the changes in the effective stresses accompanying depletion can affect the mechanical aperture and the apparent conductivity of the natural fractures, if not cemented or propped open.

Field wide injection using water, polymer or steam are used to maintain the reservoir pressure or increase the reservoir pressure, during the displacement of oil in the reservoir from the injector wells to the production wells. These operations lead to pore pressure and stress changes in the reservoir as well as any associated thermal-stress changes.

3.3.1 Reservoir depletion, compaction and subsidence

Depletion leading to reservoir compaction and surface subsidence is modeled at a well scale for initial screening of the compaction risks, and at a field scale using finite element modeling, and less commonly finite difference modeling. These numerical modeling efforts aim to assess both compaction and subsidence risks, as well as any potential strain concentrations associated with bedding plane slip or fault reactivation which can take place in either the reservoir or overburden. The compaction estimates rely on elastic compressibility tests and Uniaxial strain (zero-lateral strain, or K_o) tests (Section 7.1), which mimics the compaction of a formation in a laterally extensive formation. For more complex, or realistic, field geometries and non-uniform drainage and depletion profiles the entire constitutive behavior of the reservoir and overburden formations are required as input into the numerical models. The laboratory test program required to build these constitutive model includes measuring; the Uniaxial strain response; the triaxial and stress path test responses, to adequately define the yield and failure envelopes as well as the plastic constitutive behavior and; Isotropic (or hydrostatic) tests, that are used to define the end of the yield cap for conditions of zero shear stress.

3.3.2 Injection and Ior/Eor field developments

Fluid and gas injection into reservoirs involved in Improved Oil Recovery (IOR) and Enhanced Oil Recovery (EOR) reservoir development strategies, may be initiated at the start of a field development. However, they are more commonly considered secondary recovery mechanisms, following the primary depletion drive recovery. As a result, the stresses at the start of IOR and EOR may have evolved during the depletion stage and differ significantly from the initial stress conditions in the field.

During the secondary injection phase of IOR or EOR, if the pressure increases, the stresses will again change (Santarelli *et al.*, 1998; Santarelli *et al.*, 2008). The pressure changes will be unequal over the field, and 3D field modeling is used to estimate the stress changes and associated strains, requiring comprehensive constitutive models, for both the reservoir and the overburden, in order to estimate the potential onset of hydraulic fracturing, the direction of any fracturing, potential risks to well integrity, fault movement and cap rock integrity. The constitutive models are derived using triaxial tests and potentially stress path tests, which mimic the stress changes during production.

During injection, two strategies can be adopted: injecting fluids at downhole pressures below the fracturing pressures in the reservoir, where the fluid is injected into the rock matrix through the pore system: matrix injection. This is the most commonly adopted injection strategy, as reservoir simulators commonly consider matrix injection. The second injection strategy involves injecting at downhole pressures above the fracturing pressures, where a hydraulic fracture is generated and maintained open by high injection pressures. Again the thermal conditions should be taken into account, but the test measurements for these IOR/EOR operations above fracturing pressure are similar to those discussed in the hydraulic fracture section above.

Matrix injection into sandstone and carbonate formations is accepted to have a limited duration, especially for injected Produced Water Re-Injection (PWRI) or seawater, which contains solids even after filtering. These solids during injection gradually lead to plugging of the pores, to lower permeabilities and higher injection pressures: The injection pressures gradually increase, over a period of months, until the fracturing pressures in the reservoir are reached whereupon a hydraulic fracture is generated, which improves injectivity. For both sandstone and carbonate formations, even when planning to inject under matrix conditions, assessing the likelihood and extent of an induced hydraulic fracture is prudent. PWRI injection models are based on this approach, but use a fracture criteria modified from standard hydraulic fracture models.

Having discussed different application of geomechanics to field developments, the analysis requires field-specific mechanical property data, the only source for which is core-based laboratory measurements. This data also acts as a basis for correlations of material properties and petrophysical and geophysical measurements for 3D characterization. The first stage of this workflow involves core recovery and ensuring representative samples are obtained.

4 CORE RECOVERY AND CORE DAMAGE

Rock Samples used for mechanical testing are obtained from rotary drilling cores that are typically 100–125 mm (4–5") in diameter. Normal coring procedures involve drilling cores with core barrels 10m long, which result in 'trips' to surface to recover the cores before cutting the next 10m. This is a time consuming process and to get around this longer core barrels of up to 55m may be employed. The cores are taken from reservoir depths ranging between 0.5 and 6 km true vertical depth. The deeper formations may contain significant pore pressures, large stresses and high temperatures, which are relieved during coring and become equal to the ambient conditions upon retrieval of the core to the surface at the rig site. This equilibration involves the release of the vertical stress, reduction of the pore pressures and horizontal stresses, with the accompanying expansion of the cores and cooling of the cores leading to contraction.

Cores typically continue to expand for up to two days after reaching the surface as observed from the Anelastic Strain Recovery measurements of Teufel and co-workers in the 1980s (Teufel, 1982, 1983; Blanton & Teufel, 1986; Warpinski & Teufel, 1989).

Teufel showed that the cores expand anisotropically, the magnitude of the expansion, upon retrieval from the wellbore, is proportional to the principal in situ stresses at depth, under which the cores were cut (Figure 9).

The strains measured upon retrieval are typically around 200 microstrain (0.02%), but the total expansion from coring may be 3–4 times this magnitude.

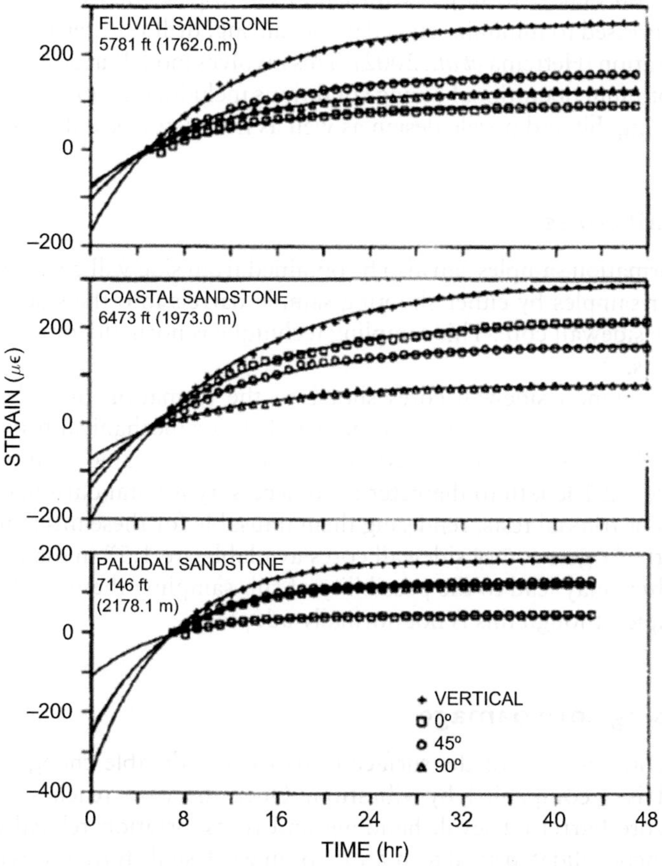

Figure 9 Examples of anelastic strain recovery on three sandstones (Warpinski & Teufel, 1989).

The pore pressure response during core retrieval is dependent upon the permeability of the sample. In highly permeable formations the pore pressure in the formation will equilibrate with the drilling fluid pressure during coring, and gradually reduce as the core is pulled out of the wellbore, in equilibrium with the hydrostatic head changes of the drilling fluid in the wellbore at different depths. Formations with low permeability will respond to coring in an undrained manner, with core retrieval resulting in an undrained pore pressure response associated with the expansion of the core due to reducing the confining and vertical stresses (Dyke, 1989, Santarelli & Dusseault, 1991; Ewy, 2015).

The impact of coring and stress retrieval on weak formations has been assessed experimentally by Holt, Kenter and Brignoli and co-workers in a series of papers (Holt *et al.*, 1994, 1998; Brignoli *et al.*, 1998), the focus of their work has in part been to assess the impact of core damage on compaction estimates of the Groningen field in northern Netherlands.

In addition to the stress, pressure and thermal equilibration accompanying core retrieval, the impact of vibrations on induced core damage and core handling is significant in cores taken from weak formations, and particular operational practices have been addressed to minimize such damage during, drill-out, retrieval, rig handling and transportation (Hettema *et al.*, 2002). This involves modeling downhole vibrations to identify optimum operational coring conditions including, controlling the weight on bit during coring, bit and nozzle design as well as flow rates used during coring.

4.1 Sidewall cores

Downhole formation samples can also be obtained from sidewall coring which take 1–1.5" diameter samples by either shooting sample barrels into the side of the well, or using a rotary sidewall core. This sampling technique is normally only used in competent formations.

The more common sidewall cores shot into the formation often lead to sample fragmentation and are not used, or recommended, for mechanical testing due to the core damage. The rotary sidewall cores may be of acceptable quality, but do not generally have a 2:1 length to diameter ratio necessary for standard unconfined compression tests or triaxial tests, rendering them unusable for these mechanical tests.

Currently the largest rotary sidewall cores available are 1.5" in diameter and up to 2.5" long, which may lead to the possibility of sub-sampling these for 1" diameter 2" long test samples, though this is not normally adopted.

4.2 Assessing core damage

Coring and core retrieval to the surface leads to considerable changes in stress and pressure and is accompanied by vibration. Once the core reaches the surface, it experiences core barrel retrieval, handling and transportation related damage. It is common in unconsolidated sand formations or in weak shale formations (*e.g.*, between sand formations) to be unable to recover sections of core, and for the core recovery from each barrel to be less than 100%. The success of coring is recorded as a percentage recovery from each core run (Santarelli & Dusseault, 1991).

Once retrieved cores may be stored at atmospheric ambient conditions, or preserved, where the core is cut into sections, typically about 30cm long, wrapped in plastic film, aluminum and sealed with hot wax. The purpose of this preservation is to maintain the core fluids. A common, and effective, method of both preservation and mechanical stabilization is to fill the annulus between the core and the inner core barrel with a resin, or a hardening foam, and then place rubber caps on the ends sealed and secured with hose clamps (Ewy, 2015).

Weak samples of unconsolidated sands can be frozen upon core retrieval (Torsaeter & Beldring, 1987) using either solid $CO2$ or the vapor from liquid nitrogen, in order to maintain their integrity, and kept frozen during sampling until they are placed in a triaxial cell in preparation for testing.

Accurately measuring the rock properties relies on testing representative and undamaged samples, which reflect in-situ conditions. Sections of cores are commonly evaluated while still in core barrels using CT scans to assess the presence of fractures, bedding planes and heterogeneities, to select suitable locations for sampling plugs to be taken.

Once the core sections are selected, the degree of damage can be estimated using sample plugs, or offcuts, for measurements for porosity, density or sonic velocities which are compared with well-log derived values. The well-log measurements correspond to the properties in the near wellbore region, and are considered less disturbed than the cores, and treated as a baseline for comparison for any potential core damage. The porosity is most commonly used parameter for these comparisons and any differences between the core-measured porosities and the log-derived porosities are attributed to core damage.

Dusseault and van Domselaar (1982) evaluated core damage on heavy oil sand formations in Canada and Venezuela, which are prone to core damage, based on a comparison of core- and log-based porosity. They established thresholds and guidelines for quality controlling samples for petrophysical testing, and for mechanical testing by defining an Index of Disturbance (I_D) as illustrated in Table 1.

$$I_D = 100\,(\phi - \phi_i)/\phi_i$$

Where, ϕ = laboratory measured porosity, and ϕ_i = well-log measured porosity. Dusseault and van Domselaar recommend that any core with an Index of Disturbance (I_D) greater than 10% is unsuitable for rock mechanical testing.

Methods to improve core quality and reduce the impact of coring on weak rock samples, includes the use of small tolerance core barrels to limit the radial expansion of cores upon retrieval, or injection of foam between the core barrels and the core, again to limit radial expansion (P. Collins, pers. comm.). These methods have predominantly used for coring uncon-solidated or weakly consolidated sand formations containing gaseous heavy oil. Specialist coring barrels have also been proposed as well as freeze coring but have not become common practice in the industry (Dusseault & van Domselaar, 1982).

Coring and retrieving shale formations can be troublesome, not only due to the stress, pore pressure and thermal changes but also due to the interaction of drilling fluids with the clay minerals. Sanatrelli and Carminatti (1995) along with the comprehensive review of shale handling recommendations by Ewy (2015) provide insight into the physical mechanisms involved in retrieving, handling and preserving argillaceous shale cores and preparing shale samples for testing. Ewy (2015) recommends only retrieving shale cores using oil-based drilling fluids to avoid interaction with water-based systems, and during preparation "do not allow the native water content to increase, or to decrease, unless you are doing so on purpose and under controlled conditions". His recommended approach is to "1) minimize exposure to air (unless in a

Table 1 Guidelines for petrophysical and geomechanical tests on core, based on the measured index of disturbance (After Dusseault & van Domselaar, 1982).

Index of Disturbance (I_D)	*Description*
$I_D < 10\%$	Intact or slightly disturbed
$10\% < I_D < 20\%$	Intermediate disturbance
$20\% < I_D < 40\%$	Highly disturbed
$40\% < I_D$	Disrupted generally

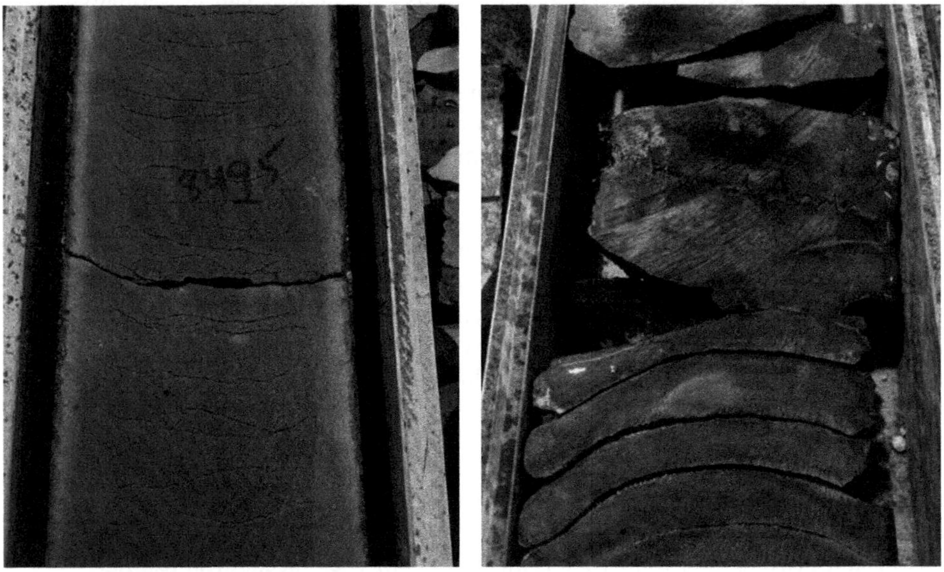

Figure 10 Examples of incipient core disking and well developed disking in a core of lower cretaceous formation, Onshore Abu Dhabi.

controlled relative humidity environment) and 2) do not contact with water or brine unless the shale is under stress". Ewy (2015) evaluated the saturation of well-preserved shales, by assessing their degree of saturation following exposure to different relative humidity atmospheres in a vacuum desiccator.

Core damage also occurs in shales and high strength formations, through core disking, which can pervade the entire core and render the core unusable for mechanical testing. This core disking is normally related to well defined bedding planes in shale formations, particularly in poorly cemented shale formations. In high strength formations, the core disking in cores taken from vertical wells has been related to large horizontal stress magnitudes and anisotropy, and potentially to the drilling operations (Dyke, 1988, 1989; Kutter, 1991; Venet *et al.*, 1989).

Figure 10, shows two examples of slabbed core showing disking. The cores were retrieved from carbonate formations from a field located onshore Abu Dhabi. The core on the left shows clear signs of incipient disking, but the core is still intact, whereas the core on the right is clay-rich and shows pervasive disking, making these core unusable for standard mechanical laboratory testing.

5 SAMPLE PREPARATION CONSIDERATIONS

Cutting and retrieving cores face several challenges as discussed in the previous section, especially with weak formations and laminated argillaceous shale samples. The shales and clay/mudstones considered here are clay-rich, clay-supported rocks and differ from the 'shales' comprising unconventional resources, which are fine grained sedimentary rocks with generally low clay contents. Once cores have been retrieved, scanned, and found suitable for sampling in the laboratory, damage can be induced when samples plugs are drilled from the core for testing. This damage results from the vibration of the sampling barrel that can render sample plugs unusable for mechanical testing.

The challenges of sampling cores is very common for weak or poorly cemented argillaceous shales due to their laminated structure, resulting in low success rates for plugging samples in poorly preserved cores. Argillaceous shale and mudstone cores dry out very quickly, and plugged samples dry out often in a matter of minutes and need have their moisture content maintained as close to the natural state as possible. Once plugged from the cores, samples should maintain the sample saturation by either storage in controlled high humidity atmospheres or under hydrocarbon in a sealed container (Ewy, 2015). Shale cores that have dried out often have unknown saturations, and can develop desiccation cracks and permanent damage, and are not considered suitable for testing. The success rate for sampling well preserved shale plugs can be increased by using Sintef's sample barrel that contains a pressurized piston. The piston sits inside the sample core barrel and applies a small axial stress on the core during the plugging, maintaining a normal force across the laminations and bedding planes, to maintain the sample integrity (Fjaer *et al.*, 2008). This is currently considered best practice to increase the success rate when sampling plugs from shale cores.

For weak unconsolidated sand formations, freezing cores upon retrieval from the well is used to keep the core intact during transportation. Cutting the samples plugs from the cores also require the use of liquid nitrogen to maintain plug integrity. The selection of fluids for plugging samples from cores is also a concern in shales, as these formations are not only highly laminated but can be smectite-rich and chemically reactive with aqueous fluids, resulting in swelling when exposed to the fluids used to lubricate the core barrels used for plugging cores. To avoid any samples swelling during plugging, the use of non-aqueous fluids such as decane or mineral oil is recommended (Ewy, 2015; Milton-Tayler, *pers.com*). However, modified procedures may be required for shales containing high percentages of organic matter.

The rock mechanical property measurements used for geomechanical characterization of the sub-surface oil fields are based predominantly on tests performed on right cylindrical samples, in line with standard rock mechanics and civil engineering practice. However, specialist tests described in a later section (section 10) also consider cubic samples.

The cylindrical plugs cut from the cores should, to the extent possible, comply with ISRM and ASTM recommended practices (ASTM D4543-08, 2008; ISRM, 1983) consistent with the standard testing procedures (ASTM D2938-71a, 1979; ISRM, 1983; Dudley *et al.*, 2016).

5.1 Plug sample sizes and dimensions

Rock sample plugs taken for the cores for material property measurements comprise right circular cylinders with height to diameter ratios equal to between 2.0 – 3.0 in line with standard rock mechanics and civil engineering practice. Two common sizes of samples are taken from cores for petroleum geomechanical sampling, with diameters of 25.4 mm (1.0") and 38 mm (1.5"), but larger samples *e.g.*, 101.6 cm (4") diameters are occasionally used for specialist tests such as Advanced Hollow Cylinder, or Advanced Thick Walled Cylinder tests. The ISRM recommendation is for sample diameters of no less than 54mm, and ASTM recommendations of samples with diameters great than 47mm are generally not practical for sampling from the downhole cores obtained.

The plug sample size used is dictated by the amount of available core material: Even though 4" or 5" diameter cores are drilled through the reservoir, these are not always entirely available for testing. Regulatory bodies around the world specify that when cores are taken, ½ or ⅓ of the core cut axially along the core, is reserved and not used for sampling to ensure a complete geological record and profile through the core is preserved. Also geomechanical testing is normally performed after geological and petrophysical sampling of the core, further limiting core availability. This reduces the amount of core available for testing, and the possible length of any samples when taking samples perpendicular to the core axis. It should be noted that the reuse of samples used for petrophysical characterization tests, such as poro-perm measurements, is not recommended, are these are generally not of the 2 x 1 (L/D) minimum ratio required for mechanical testing and have been cleaned and already subjected to stresses These constraints on the core availability also dictate the diameter and number of samples that can be taken at the same sample depth, taken parallel to the core axis (Figure 11).

Plugs taken from core also need to satisfy the need for sample homogeneity, which can present challenges in formations such as mélanges and conglomerates where grain diameters may be greater than 1cm, or even larger than the sample plug sizes. Reefal and platform carbonates that contain vuggy porosity also presents challenges for obtaining homogeneous representative samples that fit in a standard testing machine. In such circumstances, the merits of performing a testing program on such formations should be questioned, especially when numerous identical samples are required, as in the case of a triaxial test suite. As a guideline, it is generally accepted that the minimum sample plug diameter for well-sorted clastics should exceed 10 times the largest grain diameter (ISRM, 1983), a similar guideline should also be used for carbonate formations, for example in the case of vuggy limestones, 10 grains or grain assemblages bounded by 'vuggy' pores: this obviously excludes cores with large pores from testing programs in standard test equipment and standard sample sizes.

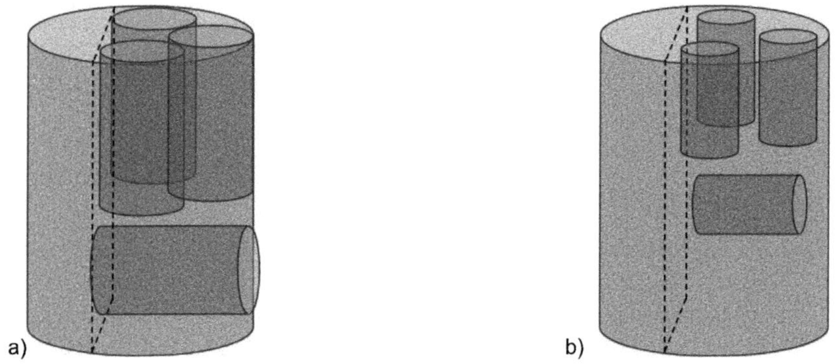

Figure 11 Illustration of sampling possibilities from a 4" core with 1/3 reserved, with a) 1.5", and b) 1" diameter sample plugs.

5.2 Core plug orientation

Rock mechanical tests for petroleum geomechanical applications are typically performed on samples plugged either perpendicular or parallel to sedimentary bedding, depending upon the application (Section 3). The bedding plane orientation in the cores may be determined visually or using X-ray computed tomography (CT) scans. The impact of core textures and laminations on the selection of the sample plugs is illustrated in Figure 12 (Dudley *et al.*, 2016).

- A shows cylindrically drilled plugs orientated in a core for measuring the permeability (k_H and k_V) or mechanical properties parallel and perpendicular to the core axis.
- B and C show a laminated rock with test plugs drilled normal to bedding and parallel to the core axis.
- D indicates that the test plug has been obtained in a natural or induced fracture, not representative of larger core or formation.
- E and F show heterogeneous core or vuggy porosity for which smaller plugs are also likely to be unrepresentative of bulk behavior.

5.3 Sample plug preparation

The preparation of the right cylindrical samples required for the mechanical tests should be in line with industry best practices (*e.g.*, ASTM D4543-08, ISRM, 1983). The specimen ends should be parallel to within 0.05 mm per 50 mm of sample length,

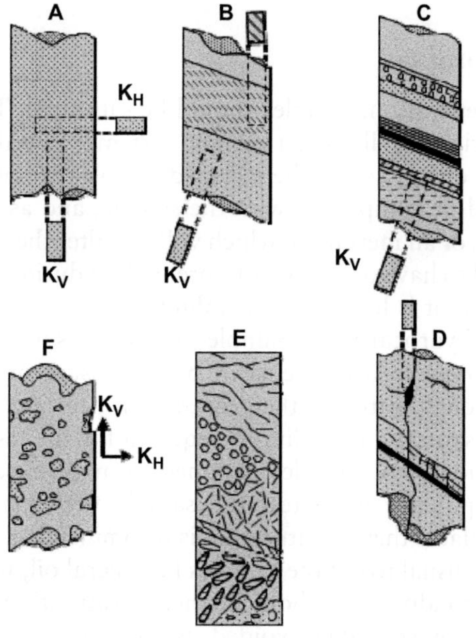

Figure 12 Variations of textures and laminations for 2-3' core, illustrating sample plug location and orientation options (courtesy of Ove Wilson, from Dudley *et al.*, 2016).

and at right angles to the long axis of the samples to within 0.25° for spherical seats and 0.13° for fixed end platens (ASTM D4534-08), though ISRM (1983) recommends 0.05°. The axis of the sample plug needs to be a cylinder within a tolerance of between 0.5 (0.02 in) and 0.3 mm (0.01 in) along the sample length. The differences here reflect the recommendations from the ISRM and ASTM guidelines, the ISRM generally having lower tolerances.

The ends of the cylindrical samples should ideally be smooth and free from irregularities, and flat to 0.025–0.02 mm. However in practice, especially with very weak and large grained samples, plucking of the grains occurs during the milling of the ends of the samples. Flatness of the ends of the sample plug should be as free of irregularities as possible, to avoid non-linearities and 'bedding-in effects' being observed in the stress-strain behavior. Filter stones to aid drainage of the sample pore fluids are commonly fitted to both ends of the sample in the triaxial cell and the sample surrounded by a thin flexible rubber membrane.

5.4 Triaxial cell requirements

Triaxial cells are available in a range of sizes and specifications, however for Petroleum applications, certain aspects of the testing set up are preferred, and recommended due to the weak nature of many of the formations evaluated in a test program.

Preferred practice is to use triaxial cells for any material testing using thin rubber membranes, typically 0.25 mm (10/1000") in thickness, that are flexible enough to avoid providing any resistance to sample strains. Thick and stiff membranes should be avoided.

5.4.1 Pore fluid saturation

During the sample preparation, samples should be saturated. The saturation may be achieved outside the triaxial cell using a vacuum system, or inside the cell, once a small confining stress has been applied to the jacketed sample. In both cases the saturating pore fluid used should be compatible with the sample, and as similar to the original pore fluids as possible, or an inert fluid, which will not alter the mechanical response of the samples. It is usual to have one phase of pore of fluid during testing, however when using preserved samples it is likely that two fluid phases may be present in the pores (formation water and hydrocarbons). Sample saturation is normally required prior to testing to ensure accurate measurements of pore pressure or pore volume changes during the tests. One qualification is the procedure for testing core samples from gas reservoirs, where the saturation conditions required for the test ideally reproduce *in situ* saturation conditions. For samples obtained from gas reservoirs in unpreserved core, then common practice is to test the sample with 'as received' conditions; if connate brine is available then re-saturation is recommended. However, if preserved cores are available it is usual to cut the samples in mineral oil, wrap in Saranwrap and test in the preserved conditions without further re-saturation (Milton-Tayler, pers. com.). Testing dry samples should be avoided, as aged and dried samples of sandstones and limestones are known to be significantly (100%) stronger than saturated samples, and shales samples weaken due to desiccation cracking and induced damage.

In low permeability and porosity formations the pore pressure and pore volume measuring systems need to have a small volume and compressibility and calibrated for pressure and temperature effects to ensure accurate pore pressure measurements (Wissa, 1969; Bishop, 1976). The axial and radial strains should be measured either internally in the triaxial cell or externally with any appropriate corrections applied to obtain the final measurements used in the property determinations. The ISRM (1983) recommends that measurements should have a strain sensitivity of the order of 5×10^{-6}. The confining stresses and axial loads applied to the samples should also be measured internally where possible, or externally, with any appropriate corrections for the triaxial testing system, which may include axial friction or cell pressure dependencies on load cells or deformation devices.

6 STANDARD TESTS AND PARAMETER DETERMINATIONS FROM TESTS

Standard Rock Mechanical tests are used to characterize the geomechanical properties and failure characteristics of the sub-surface formations, as in other branches of soil and rock mechanics. As a first stage of many investigations, pure elastic properties are measured and used in the analysis and design of different aspects of field developments.

6.1 Linear elastic parameters

Elastic properties are used for standard characterization of formations, and for calibration and comparison with elastic properties measured using 'dynamic' acoustic measurements, either from seismic surveys or petrophysical well logs. Elastic compressibilities, including bulk compressibilities, and pore volume compressibilities, are used in field-wide models to assess volume changes and oil or gas volumes in place, which are recoverable. These compressibilities are generally an estimate, as the stress state commonly used in the tests is not representative of the stress state and stress changes that occur in-situ when the reservoir depletes. A more representative compressibility reflecting the depletion conditions in a reservoir is obtained from the Uniaxial 'zero-lateral' strain (K_o) test (Section 7).

The Young's modulus, or the confined moduli where non-zero confining pressures are used, is a commonly used elastic parameter in many analyses and is measured for the design of hydraulic fractures in order to estimate the potential width of the hydraulic fracture.

The Young's modulus is estimated from the change in stress resulting from a change in strain, and is determined from Unconfined Compressive Strength test. Three estimates of Young's modulus, are commonly quoted (Figure 13):

- at an axial stress equivalent to 50% of the UCS (E_{50});
- measured over the entire linear portion of the stress strain curve (E_{Av});
- Secant modulus (E_s), the ratio of the stress to the strain at any point, relative to zero loading.

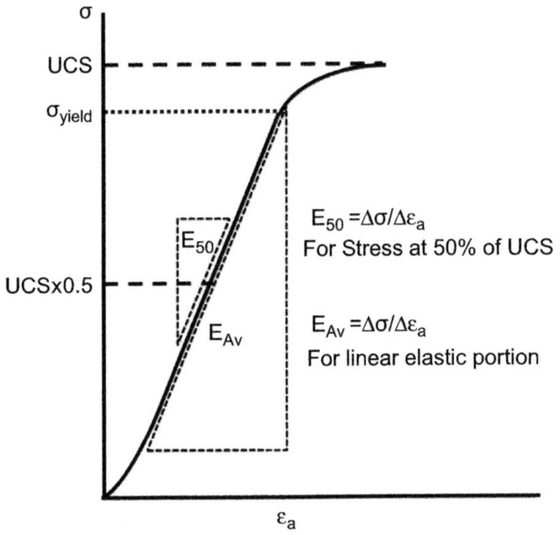

Figure 13 Estimation of Young's modulus from unconfined compressive strength tests (modified from ASTM D2938-71a, 1979.)

The first two definitions are most commonly used by petroleum geomechanical engineers, the second is more common where the linear elastic region is small in extent.

The Poisson's ratio determined for the elastic section of any stress-strain curve should be obtained in a consistent manner to the Young's modulus. Typically this requires that the Poisson's ratio is determined by taking the slope of the radial strain vs axial strain curve, either at the point where the E_{50} is taken, or over the range of stresses that the E_{Av} is determined.

Even when formations appear linear elastic in such tests, the moduli obtained from loading often differ from the moduli obtained during unloading cycles. These stiffer unloading moduli generally agree more closely with the dynamic, small-strain, stiffness moduli derived from ultrasonic velocity measurements on core samples.

6.2 Non-linear elasticity

Linear elasticity is rarely observed in petroleum formations though it is unclear whether this is an inherent property resulting from the granular nature of many of the formations, or whether this is a result of core expansion and damage, as discussed in Section 4. However, it may still be representative of downhole near-wellbore conditions, since as a result of drilling, the near-wellbore region undergoes some expansion into the wellbore with a resulting stress re-distribution around the wellbore, and potentially some mechanical property changes, including microcrack development, leading to stress dependent elasticity (Ewy & Cook, 1990; Wu & Hudson, 1991).

The Young's modulus measured on core plugs at ambient conditions with zero confining pressure, differs from the moduli or stiffnesses measured at elevated confining pressures, these confined moduli increase with the confining stress asymptotically up to a maximum stiffness value (Figure 14). This pressure dependent elasticity is

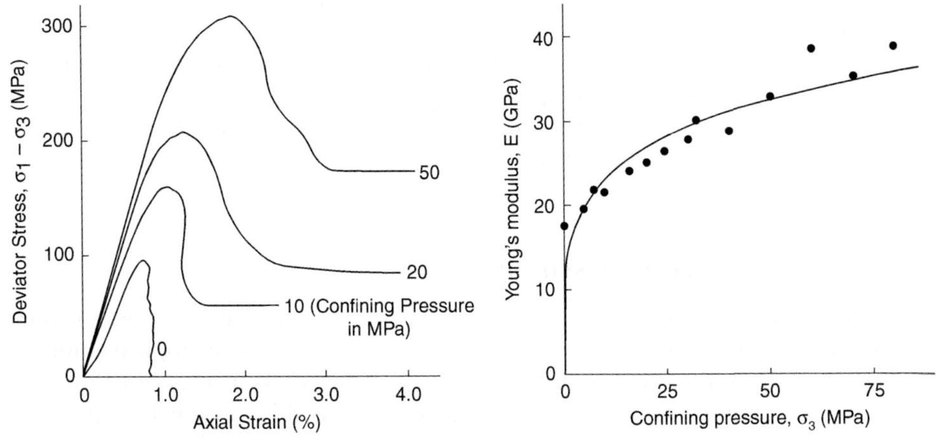

Figure 14 Example of stress strain response of sandstones under different confining pressures illustrating the confining stress-dependent elastic moduli (After Santarelli, 1987).

described by Kulhawy (1975) and the impact on the analysis of wellbore stability by Santarelli (1987) and Santarelli and Brown (1987) for uniform horizontal stresses, and by McLean (1987) and Duncan Fama and Brown (1989) for anisotropic horizontal stresses using numerical analysis.

Teeuw (1971) also addresses the non-linear elastic response of weakly consolidated sand and the impact of stress conditions on the apparent stiffness of the sand during uniaxial loading for reservoir compaction estimates.

Addis and Wu (1993) demonstrate the divergence between linear elastic predictions up to the point of yield and observed behavior in thick walled hollow cylinder tests on sandstone samples. This pre-yield response reflects the broader discrepancy between linear elastic predictions and material responses in these tests for the onset of breakouts. A formations response to stress changes at depth is often simplified to a linear-elastic assumption prior to yield, for ease of analysis and due to a lack of detailed rock mechanics data. However, this may not be a true reflection of the elastic material behavior, due the pressure dependency described above.

6.3 Bulk compressibility

The bulk compressibility can be measured for both drained and undrained conditions using samples subjected to increments of isotropic ($\sigma_1 = \sigma_2 = \sigma_3$), or hydrostatic, stress increase. For saturated samples under drained conditions with constant pore pressure, the bulk compressibility (C_{bc}) will be calculated from the bulk volume strains during the isotropic confining pressure increase. An alternative method for measuring bulk compressibility is to vary the pore pressure at constant isotropic confining pressures (C_{bp}). For identical samples at the same effective stress levels the comparison of the drained and undrained compressibilities provide an estimate of the grain compressibility (C_g) resulting from pore pressure changes.

For undrained conditions, the isotropic confining pressure increase will result in a pore pressure increase, defined by Skempton's B pore pressure parameter (Skempton,

1954) for unconsolidated materials and for low compressibility materials by Bishop (1973). This pore pressure increase will result in grain deformation, in addition to any skeleton deformation. The undrained bulk compressibility will be less than the drained compressibility, due to the low compressibility of the pore fluid.

The measurement of poroelastic parameters is addressed in Detournay and Cheng (1993) and Charlez (1991), the later also addresses measurements of thermo-poro-elastic parameters.

6.4 Pore volume compressibility

The Pore Volume Compressibility (PVC) is a measure of the changes in pore volume resulting from changes in either pore pressures (V_{pp}), or externally applied confining stresses (V_{pc}). The estimation of the PVC using confining stress increase is common in standard core analysis for petroleum engineers, who use this to estimate how much incremental oil or gas production can be expected as the reservoir depletes. However, this does not reproduce the conditions during the life of a depleting reservoir, consequently, the most important and representative measurement is to assess this change in pore volume resulting from pore pressure changes.

The pore pressure compressibility is normally used by petroleum engineers and measured under isotropic stress conditions. This is again unrepresentative of depleting conditions, as the vertical effective stress magnitudes change more than the horizontal effective stresses. As such, for use in reservoir simulators, a pore volume compressibility measured with Uniaxial strain compression (K_o, zero-lateral strain conditions) is more representative of field conditions during depletion ($V_{pp\ uniax}$), (Chertov & Suarez-Rivera, 2014). The relationships between these different reservoir compressibility parameters are summarized by Dudley *et al.* (2016).

6.5 Biot's effective stress, or pore pressure, coefficient

Cemented rock materials behave similarly to uncemented materials, but the cementation and resultant stiffness leads to differences in how pore pressure changes affect the bulk response of the formation for isotropic stress increases. This is reflected in Biot's effective stress coefficient (α) which ranges in value from 0 to 1;

$$\alpha = 1 - \left(C_g/C_{bc}\right)$$
$$\sigma' = \sigma - \alpha Pp$$

The Biot re-formulation of the effective stress equation, is a more general form of Terzaghi's relationship, in which the Biot effective stress coefficient is taken as 1 for the unconsolidated materials which were the subject of Terzaghi's work (Terzaghi, 1943). The Biot formulation of effective stress shows that for cemented materials where the sample bulk compressibility is small and approaches the grain compressibility, the Biot coefficient is less than unity, but greater than the fractional porosity (Fjaer *et al.*, 2008). Detournay *et al.* (1989) and Detournay and Cheng (1993) present data showing the Biot coefficients varying from 0.19 to 0.85 for a range of competent rocks with porosities ranging from 1% to 26%. This results in higher effective stresses acting throughout the sample compared to similar stress and pore pressure conditions applied to unconsolidated materials.

The Biot coefficient can be measured using a number of approaches, but is often estimated using the values of bulk compressibility (section 6.2) and the grain compressibility.

The grain compressibility (C_g) can be estimated from a comparison of (C_{bc}) and (C_{bp}), or from the compressibility measurements of unjacketed samples where the change in the confining isotropic stress equals the pore pressure for high permeability samples (Charlez, 1991; Detournay & Cheng, 1993). The bulk compressibility varies with confining pressure, even at stress levels below the isotropic yield point, resulting in a stress dependent Biot coefficient (Azeemuddin *et al.*, 2002; Alam & Fabricius, 2012).

Other methods to determine the Biot effective stress (or pore pressure) coefficient are discussed by Zimmerman (2000), Wu (2001), and Al-Tahini *et al.* (2005). In general, the different techniques used to estimate the Biot coefficient result in significantly different estimates of the coefficient.

Non-unity Biot coefficients clearly have a very significant influence of the effective stresses acting in reservoirs: The stiffer and tighter the formations, the more significant its influence, for example in tight gas and oil reservoirs. However, the challenges involved in accurately estimating the Biot pore pressure coefficient have limited its application and the industry's use of more complex formulations such as Biot coefficients for anisotropic formations (Azeemuddin *et al.*, 2001; Al-Tahini & Abousleiman, 2008), naturally fractured reservoirs (Abbas *et al.*, 2009) and scaling of the Biot coefficient to larger rock volumes for use in field-wide modeling.

6.5.1 Poroelastic constants

The determination of poroelastic coefficients focuses on the Biot coefficient, described above. However, other poroelastic mechanical parameters may be required for more advanced models which include poroelasticity (Biot, 1956; Geertsma, 1957, 1966; Rice & Cleary, 1976; Detournay & Cheng, 1993). This section will focus on the measurement of the main isotropic poroelastic parameters, assessing the anisotropic poroelastic parameters still remains an activity for research groups.

The tests used to define the poroelastic coefficients comprise three types, for different incremental confining effective stress and pore pressure changes (ΔP_c, ΔP_p):

- The Drained test (ΔP_c, ΔP_p = O); where the confining pressure is varied, while the pore pressure is held constant with drained conditions, which may be at atmospheric pressure;
- The Undrained test (ΔP_c, ΔP_p = dependent variable); where the confining pressure is varied but no pore fluid is allowed to leave or enter the sample and the resultant pore pressure change is measured;
- The Unjacketed test ($\Delta P_c = \Delta P_p$); where the applied confining pressure and pore pressures are changed equally.

The confining stress and pore pressure increments applied in these determinations are generally of a few megapascals (MPa). As the reservoir materials are generally non-linear elastic, the stresses under which these tests are performed should reflect field conditions, and tangential material responses should be deduced from these incremental changes. The measurements and parameters derived from these tests are summarized by Detournay and Cheng (1993) in Table 2.

Table 2 Summary of poroelastic parameters and the test methods for determining the values.

Test	*Boundary Conditions*	*Measurements*	*Poroelastic Parameters*
Drained	$P_c = P_c^\circ + \Delta P_c$	$\Delta V/V$	Drained Compressibility (K)
	$P_p = P_p^\circ$	$\Delta V_f/V$	Biot coefficient (α)
Undrained	$P_c = P_c^\circ + \Delta P_c$	$\Delta V/V$	Undrained Compressibility (K_u)
	$\zeta = \Delta\phi = 0$	ΔP_p	Skempton's coefficient (B)
Unjacketed	$P_c = P_c^\circ + \Delta P_p'$	$\Delta V/V$	(K_s')
	$P_p = P_p^\circ + \Delta P_p'$	$\Delta V_f/V$	(γ)

Evaluating these poroelastic parameters relies on three basic measurements; the bulk volume change (ΔV), the fluid volume drained form the sample (ΔV_f) and the induced pore pressure change (ΔP_p), which involve small measurements at elevated pressures for the different combinations of applied confining pressure and pore pressure increments. Consequently, equipment calibration is paramount in order to remove any artifacts resulting from pressure sensitivity of the strain measuring devices. For undrained measurements, the volume and compressibility of the pore volume system should be minimized, with a recommended ratio of dead fluid volume to the pore fluid volume of less than 0.003, and any corrections applied to the measured pore pressures (Wissa, 1969; Bishop, 1976; Detournay & Cheng, 1993).

6.6 Plastic parameters and rock failure

The mechanical properties of formations encountered during drilling and production vary tremendously, from the weakest cohesionless sands and muds with strengths of less than 100 kPa, to competent formations with uniaxial compressive strengths exceeding 200 MPa.

Reservoir formations containing oil and gas commonly have high porosities and straddle the boundary between competent rock and weakly consolidated and cemented soils where a considerable portion of their mechanical behavior falls into the plastic region for the stresses which these formations are subjected to during the life of a well, or field. The plasticity is commonly expressed in shear failure where significant deviatoric or shear stresses act on the formation, *e.g.*, around the wellbore during drilling and production, or in the re-activation of pre-existing faults. In the laboratory, this shear deformation is observed and defined by the triaxial test results (Figure 15).

Plasticity is also observed during the depletion of fields comprising high porosity reservoirs such as weakly cemented sands which commonly occur in Tertiary deltaic sequences, such as the Gulf of Mexico, Offshore Nigeria, and offshore S.E. Asia, as well as in older Cretaceous high porosity (30–50%) chalk formations in the Central North Sea, and in the Central Luconia region of Malaysia. These formations, even though they are at depths of a few kilometers, often exist under low effective stress and low anisotropic stress conditions. During reservoir depletion they exhibit elastic behavior initially until yield after which the plastic compaction response predominates; they transition from a classic rock response to a plastic soil over the effective stress ranges induced during production.

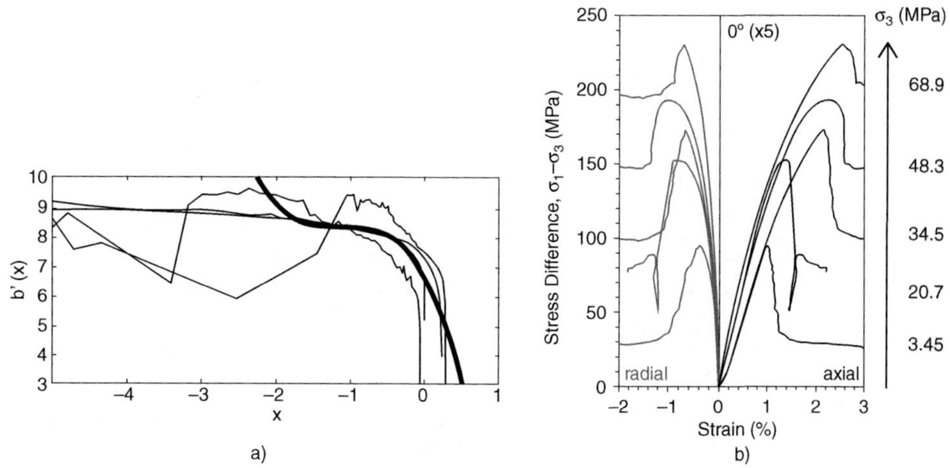

Figure 15 Triaxial test behavior and interpretation from soft and competent formations; a) Sandstone from Offshore Norway, measured behavior are dashed lines, solid line are the modeled fit (Raaen, 1998), and b) Outcrop shale showing brittle post peak mechanical behavior (Crawford *et al.*, 2012).

The failure styles of these high porosity formations differ in the low and high confining stress regions. At the higher stresses the cementation is broken and sample failure during triaxial tests can be defined by classical shear failure envelopes described by Mohr-Coulomb, Drucker-Prager, or Lade-type failure envelopes. These formations also undergo plastic deformation during isotropic loading, in the absence of shear stresses, where the pseudo-elastic cementation fails leaving behind an uncemented granular pack, with its own failure characteristics. This transition from elastic behavior to plastic granular soil-like behavior is delineated by a yield envelope, or sub-loading surface, which may be separate and independent of the shear failure surface. This is commonly referred to as end cap behavior in models, as it has been likened to the soils response, similar to the Rendulic or Roscoe surface, or the modified Cam-Clay, Delft Egg, yield surfaces. This end-cap behavior is observed in high porosity limestones and unconsolidated reservoir sands (Elliot & Brown, 1986; Crawford *et al.*, 2004), and the behavior of these high porosity formations is illustrated in Figure 16, for a high porosity chalk.

This end-cap behavior is often referred to in the petroleum industry as 'pore-collapse', but differs from the classical yield surface behavior in soils models by a very significant porosity reduction accompanying the breaking of cementation bonds defining the elastic surface. This pore volume reduction can occur under constant stress conditions satisfying yield, in the absence of significant shear stress, until the pore volume is sufficiently reduced to form a compact cohesionless frictional granular material with a porosity in equilibrium with the stress conditions at yield (Addis, 1987). This behavior is typical of weak high porosity rocks, cemented and structured soils (Vaughan & Leroueil, 1990; Burland, 1990).

This behavior results from early cementation preserving the porosity during the burial history, and where trapped undrained pore pressures have preserved the porosity

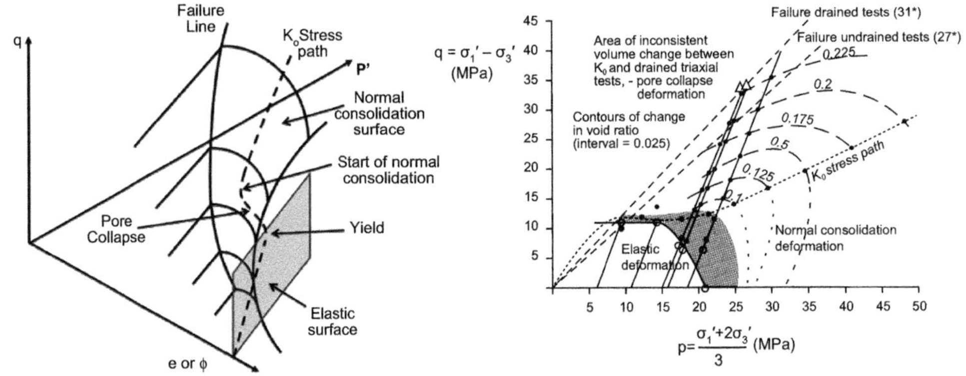

Figure 16 Illustration of the mechanical behavior of high porosity formations, with both shear failure and plastic yield envelopes (After Addis, 1989)

and resisted normal grain to grain compaction, common in more permeable formations such as sands. The occurrence of unusually high porosity formations for the depth of burial also occurs as a result of secondary porosity formation, resulting from the dissolution of the formation grains, and enhancement of the porosity at depth.

Another form of plasticity observed during field developments is experienced in evaporitic lithologies (salts and Anhydrite), which are encountered in the industry typically as flat layered formations comprising cap rocks and seals above reservoirs (*e.g.*, Gotnia or Hith formations in the Arabian plate) or as discrete sheets and massive diapirs and intrusions into shallower formations (Zechstein in the North Sea, or the Louann salt in the Gulf of Mexico, and the salts of the South Atlantic, offshore Brazil and West Africa). The plasticity experienced in these formations comprises plastic grain-grain deformation, and is akin to plasticity in metals. The deformation of these evaporitic formations is generally evaluated using creep tests (Section 6.11, (Dusseault, 1985; Fossum & Fredrich, 2002), though constant strain rate tests are also used to describe the behavior under different conditions (Fokker & Kenter, 1994).

6.7 Triaxial tests

Triaxial tests comprise testing a suite of jacketed cylindrical samples that are subjected to pore pressures, confining pressures (from the cell pressures acting on the cylindrical outer surface of the sample), and an axial load applied to the end of the samples. The axial load may equal the confining pressures at the start of the tests, and in the standard compressional tests is increased until the sample fails through the development of plastic deformation bands, or through the development of a single shear band, where the sample changes from a continuum to a discontinuum.

Compressional triaxial tests are the most common tests to determine the elastic and plastic mechanical properties, (section 3.1.1) and in most cases result in a compressive shear failure of the sample, which in more brittle and higher strength formations can result from the coalescence of extensional, axis-parallel cracks, in the post-yield to pre-peak section of the stress-strain curve. This leads to the start of bifurcation (transition from a continuum to a discontinuum) just before the peak strength is attained for consolidated samples. In weakly consolidated and unconsolidated samples, the shear failure is accompanied by grain translation, rotation, grain cracking and barreling of samples with multiple deformation bands.

Extension tests where the confining stresses acting around the cylinder surface of the samples exceed the stresses acting axially, are rarely performed in triaxial cells, but provide an indication of the impact of the intermediate principal stress on failure and the appropriate failure envelope to use in an analysis. The extension test involves a loading cycle where the axial stress is held constant, and the confining stress in increased until the sample fails.

Soil Mechanics investigations developed 'stress-path' testing, a variation of standard triaxial testing outlined above, which has been adopted for many studies on weak rocks. These stress path tests involve applying a small isotropic stress to stabilize and 'bed-in' the sample, the stresses are then increased to represent in-situ stress conditions. Once these sample come into equilibrium with these stresses, the loading cycle is applied which comprises changing the axial and confining (cell) pressures in pre-defined ratios (Figure 17, dotted stress path lines), to define the behavior of the samples and the yield and failure envelopes (Bishop & Wesley, 1975). These tests can be performed in either drained or undrained conditions (Figures 17b & c).

6.7.1 Unconfined compressive strength test

The Unconfined Compressive Strength (UCS) test is the most common test used in the petroleum industry to describe the mechanical properties of a formation. The UCS test is a triaxial test performed at zero confining stress. The UCS is used as an index test to distinguish between samples from different formations and porosities, while the results from confined triaxial tests are required for many analysis as the elastic (Young's modulus and Poisson's ratio), plastic and failure parameters of the samples are obtained for a range of confining stresses.

The standards for performing and interpreting UCS tests are given in ASTM D2938-71a (1979) and Bieniawski and Bernede (1979). Due to the sample variability, and the potential for core damage during retrieval, considerable variability is observed in the Unconfined Compressive Strength for any one formation and porosity, even in plugs taken adjacent to one another in a core. For this reason the ISRM suggested method (Bieniawski & Bernede, 1979) proposes that when possible the test is performed on 5 duplicate samples, though this is rarely possible using plug samples from core. Less variability of the failure stresses is observed for different samples if a small confining stress is applied to the samples, as any microcracks in the sample become closed and contribute to the frictional resistance inside the sample.

To minimize the impact of sample damage and the presence of microcracks on the results of Unconfined Compressive Strength tests, an alternative Constant Mean Stress (CMS) test has been proposed as an index for comparison between different samples (Kenter *et al.*, 1997). This modified test involves applying a small confining pressure of 50% of the expected UCS, then a stress path is applied where axial loading of the sample is accompanied by reducing the cell pressure to zero: a form of the constant mean stress, stress path. Alternative approaches to minimize the impact of sample damage on the UCS, as a rock index measure, include; applying a small confining

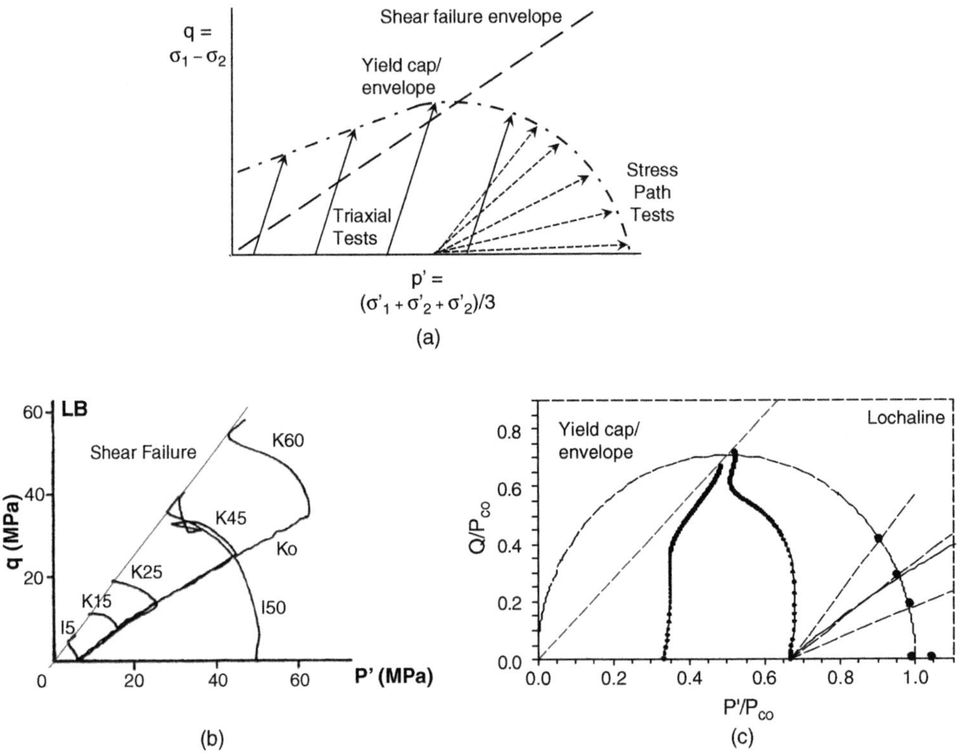

Figure 17 a) Plot of triaxial tests loading and Mohr-Coulomb failure envelopes; b) Example of stress path tests from a Ko loading followed by undrained loading of Lagon Bouffe clay (Yassir, 1989); c) Example of stress paths using Isotropic loading followed by undrained or drained loading of Lochaline sands (Crawford *et al.*, 2004).

pressure of 0.5 MPa, and use the resultant failure strengths as a basis of comparison between different samples and formations; or loading samples to *in situ* conditions then unloading the confining pressure, until failure occurs. None of these approaches has been formally adopted by the industry to date.

6.7.2 Confined triaxial tests

The elastic, work hardening plastic parameters and failure stresses for a formation are obtained from the standard triaxial tests. These tests comprise loading identical samples to failure for different magnitudes of confining stress, while recording the stresses and strains during loading. The tests are performed in line with industry standards: ASTM D2938-71a (1979); Bieniawski and Bernede (1979); ISRM (1981); Kovari *et al.* (1983); ISRM (2007).

The samples used in the triaxial tests are obtained from cores taken across the reservoir formations and the overburden, when available. The cores are cut across those formations of most interest to mechanically characterize the reservoir. The heterogeneity of the formation has been discussed in section 2.1 and often results in a limited number of samples being available for testing. Three adjacent and identical samples are the minimum required to establish a linear failure envelope from these tests. At least five samples at different confining pressures, in addition to a UCS test, comprise an acceptable triaxial test suite suggested by the ISRM (Bieniawski & Bernede, 1979), however this is not always possible given the limitations cores as a source of the samples.

Accurate measurement of the pore pressures during testing is important, to enable the effective stresses acting in the samples to be calculated. Best practice generally requires that the samples are saturated, and fluid pore pressures are uniform throughout the samples during testing. Inadequate sampling, handling and storage may require re-saturation of the samples with a compatible or inert pore fluid, and the application of a pore pressure to ensure saturation prior to loading. Samples with small pore throats, if unsaturated, may be subject to capillary pressures (suctions) resulting in inaccurate or unknown effective stresses being applied to the sample. This is particularly the case with fine-grained formations such as argillaceous, mudstone or shale formations and fine-grained carbonates, such as chalk, but is also observed at low moisture contents in sandstones.

The confining pressures recommended for triaxial tests depend upon the geomechanical application being addressed. Near wellbore issues, such as wellbore instability and sand production, where low effective confining pressures are expected, require a program of triaxial tests that includes unconfined tests and low confining stress suite of tests, relative to the magnitude of far field in-situ minimum principal stresses. Field-wide and field development applications such as, compaction and subsidence, hydraulic fracturing, or analysis of fault stability and caprock integrity, requires a program of triaxial tests where the confining stresses applied during the tests reflect the confining stresses expected over the life of the reservoir. This should include confining pressures equal to the initial minimum (normally horizontal) *in situ* stress, and stress changes resulting from depletion or injection expected in the reservoir during the development, which enables the elastic parameters, compressibility and failure envelope to be defined for the whole of the field life.

Once samples have equilibrated to the confining pressures, the recommended axial loading rates for triaxial tests on rocks specify that failure will occur in 5–15 minutes, or using stress loading rates of between 0.5 and 1.0 MPa/s (Brown, 1978, ASTM D7012-14). For low permeability saturated samples, unreasonably slow loading rates during a test may be required to ensure uniform pore pressure distribution and adequate pore fluid drainage from the sample, and to avoid excessive pore pressure generation resulting from the loading. For such samples, triaxial tests should be performed in undrained conditions, in line with best civil engineering practice (Bishop & Henkel, 1982; Head, 1998). Loading rates used in these undrained tests should ensure uniform pore pressure generation across the sample, and shale samples that are isotropically consolidated at a particular confining pressure may take two weeks to reach failure during loading to undrained shear. For such undrained loading, the pore pressures at both ends of the sample should monitored during the loading to ensure the pore pressures and effective stresses acting across the samples are known for any test interpretation. Best practice in civil engineering also includes pore pressures measured at the mid-point of weak argillaceous samples in triaxial cells, to avoid any the impact of end effects on the pore pressure measurements, though this is not currently adopted for sub-surface geomechanical testing discussed here.

Tests performed on low permeability clay rich samples with rapid loading and inadequate pore pressure measurement systems, or tests open to atmospheric conditions, commonly result in very low friction angles indicating that high undrained pore pressures have been generated during the pseudo-undrained loading.

Triaxial tests are typically performed using monotonic loading, however, inclusion of a partial unloading cycle during the elastic portion of the stress-strain response is not uncommon. These cycles are included to compare the unloading vs loading moduli. The unloading modulus shows a greater stiffness and may be closer to the dynamic modulus, measured using geophysical methods.

6.7.3 Rock failure criteria

It is not uncommon, even with three adjacent samples plugged from the core, to have test samples with different porosities or degrees of damage, and as a result different strength characteristics. This core heterogeneity leads to triaxial test samples tested at different confining stresses showing inconsistent friction angles, which do not define a straight-line or regular failure envelope. Consequently, more samples may be required when the core is sufficiently uniform to enable a reliable failure envelope to be defined. Additional indicators such as scratch tests (Noufal *et al.*, 2015) performed along the core, or petrophysical parameters such as density, porosity or acoustic velocities may be used to identify comparable samples suitable for triaxial tests.

The stresses and pore pressures defining the failure of the samples can be plotted in different stress space associated with different failure criteria. The choice of the preferred failure criteria is subject to the analysis and workflow, and depending upon the criterion may include the contribution of the intermediate stress to the sample compressive strength. For any set of compressional triaxial tests, the impact of the intermediate stress is generally unknown, as insufficient samples are available from the core to perform both the triaxial compression and extension tests that are required to assess the influence of the intermediate principal stress on the failure characteristics of the formation.

The most common failure criteria used for interpreting triaxial tests include; the Mohr-Coulomb, Drucker-Prager, Modified Lade and Critical State failure envelopes, though the use of variations of these is not uncommon. The role of the intermediate stress on the strength of the formations and the application to field problems, has been primarily considered for wellbore instability analysis. The interpretation of standard triaxial tests data using different failure criteria result in very different field recommendations for drilling fluid pressures required to stabilize the wellbore (McLean & Addis, 1990a, 1990b; Ewy, 1999).

6.8 Multistage loading tests

Stress-strain behavior which help to define the elastic, work hardening, and post-peak deformation of the samples, as well as the failure envelope determination are key data for many applications of petroleum geomechanical engineering to field problems. As discussed above, these mechanical properties obtained from triaxial tests require preferably six or more identical samples; five triaxial and one UCS plugs. However, because of the need to measure the rock properties on reservoir and overburden

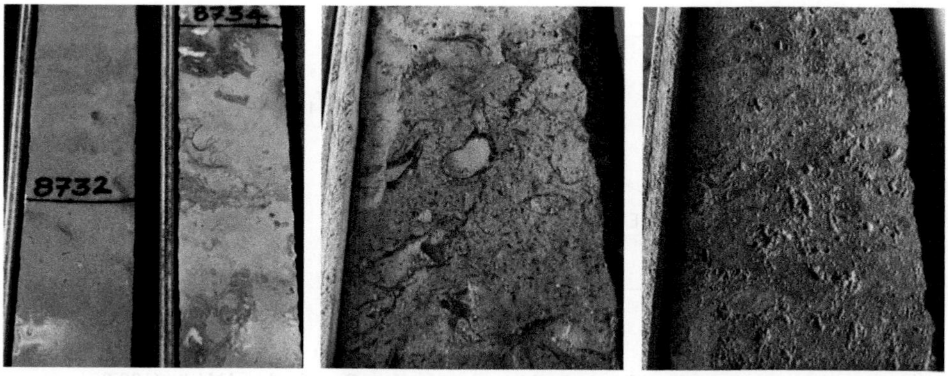

Figure 18 Cores cut from lower cretaceous carbonate reservoirs from onshore Abu Dhabi in the United Arab Emirates, illustrating the heterogeneity within formations.

formations obtained from 4" – 5" diameter cores at considerable depths, this can present challenges. Even short core intervals vary in porosity, texture, and strengths, with many lithological variations occurring on a meter-by-meter if not a centimeter-by-centimeter basis (Figure 18). This makes plugging six identical triaxial test samples, required for failure envelope definition, almost impossible in many formations.

To minimize the impact of the core heterogeneity on the determination of the failure envelope, Multistage tests can be used. These Multi-Stage Loading (MSL) test only use one sample to define the entire failure envelope, over the stress range of interest. Multistage testing is described by Kim and Ko (1979), and subsequently by a number of workers including: Kovari *et al.* (1983), Crawford and Wylie (1987), Cain *et al.* (1987) and Blumel (2010). These tests involves axially loading one sample at an initial low confining pressure until some non-linearity and work hardening is observed, then changing the testing conditions to avoid sample failure. This change can include unloading the axial stress, subsequently applying an incrementally higher cell pressure, then re-loading to a new higher axial stress. This sequential loading at different confining pressures, avoids sample failure, and is repeated until the required maximum confining pressure is reached, whereupon the sample is loaded to failure.

The second approach is to stop the loading at the axial stress where the sample is close to failure, then increase the cell pressure before continuing the loading, this again is repeated until the required range of confining pressures is reached to define a failure envelope. Upon reaching the final cell pressure the sample is taken to failure.

A third approach involves hydrostatic loading followed by axial loading at a constant rate until failure is imminent, whereupon the confining pressure is continually increased to prevent failure. This continues to a maximum confining pressure, when the axial stress increases to failure of the sample.

The first and third loading sequences comprise the ISRM suggested methods outlined by Kovari *et al.* (1983). By plotting the axial stresses at the points close to failure for each cell pressure an approximation of the failure envelope can be obtained. The first two multistage stress paths described above are illustrated in Figure 19.

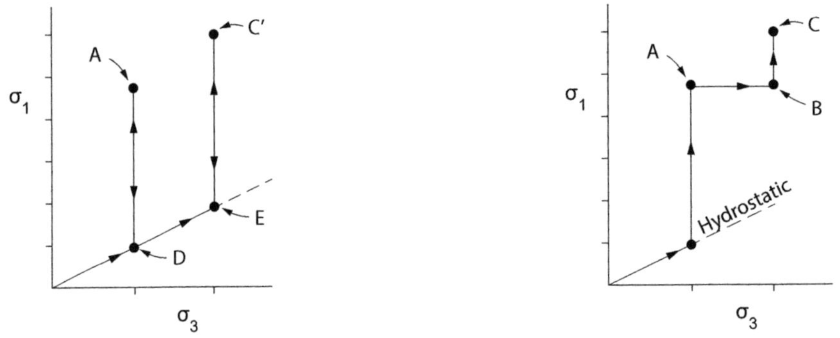

Figure 19 Stress conditions for different multistage tests, after Crawford and Wylie (1987).

Other stress paths have been proposed for Multistage loading test, however these two stress paths are most commonly adopted.

One of the challenges with these tests is determining the point during the loading cycle at which the loading is stopped, and the cell pressure increased, or the axial stress unloaded, prior to the subsequent axial loading. The point at which a loading cycle is terminated depends on the measurements that are used to show the sample is approaching peak stress, for the selected confining pressure. These measurements commonly include; the axial strain rate, or rate of deviatoric stress increase using tangent or secant moduli, radial strain rate, or the rate of volume change.

These tests are useful to determine an approximate failure envelope for very variable formations, but control of the tests, by limiting the onset of failure as peak stress is approached, has proved challenging for very brittle formations (Taheri & Chanda, 2013).

Failure envelopes obtained from multistage loading tests and standard triaxial tests have been compared for a number of lithologies by different workers, in some cases these show close agreement, in others a 5–10% lower friction angle is observed with the MSL test (Kim & Ko, 1979; Cain, 1987; Youn & Tonon, 2010; Sharma *et al.*, 2011, Taheri & Chanda, 2013). These discrepancies however may result from lithology-related behavior or from the test procedures used in the comparisons. Taheri and Tan (2008) performed comparisons using undrained tests on a mudstone and show most comparable results were obtained by using the secant modulus as the point to reverse the axial load.

6.9 Tensile failure: Brazil tests

Two methods for measuring the tensile strength of a rock are included in the ISRM suggested methods; a direct method involving application of tensile (extensional) load to the ends of a cylindrical sample, and an indirect method which takes advantage of an induced principal tensile stress generated in a direction perpendicular to an applied compressive stress. The preferred method for estimating the tensile strength of formations in the petroleum industry is using the indirect 'Brazil' or 'Brazilian' test. The test procedure, guidelines and recommendations are detailed in ISRM Suggested Methods for Determining Tensile Strength of Rock Materials (Bieniawski & Hawkes, 1978).

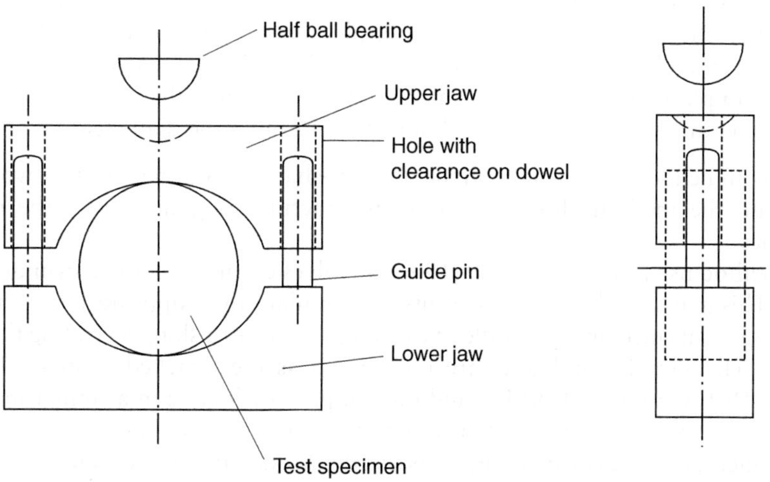

Figure 20 Apparatus for Brazil test for tensile strength measurements.

This test involves compressing disks or short cylinders of rock along one diametric axis of the disk. This compression results in tensile stresses being generated inside the disk, resulting in a tensile failure mode.

The cylindrical sample size recommended for this test procedure has a diameter greater than 54mm, with the length of the cylinder approximately equal to the sample radius (Figure 20). The sample is loaded at a continuous rate such that failure occurs in weak samples within 15–30 seconds, with a recommended loading rate of 200 N/s.

The load at failure is defined by the first or primary failure, but the load may increase past this primary failure in some cases. The tensile failure is related to the flaws and microcracks in a sample, and the tensile strengths measured on a number of 'identical' samples result in a large range of values. As such, it is recommended to obtain sufficient number of tests to assess the statistical range of tensile strengths for any one formation. The ISRM recommends 10 samples, which should be considered a minimum.

6.10 Fracture toughness tests: K_I

Fracture toughness is required as input into fracturing simulators. The ISRM suggested methods for measuring fracture toughness is detailed in Ouchterlony *et al.* (1988). An alternative approach is presented by Fowell *et al.* (1995). The 1988 suggested methods involve stressing a sample of rock containing a V-shaped saw drilled slot, formed by two intersecting saw cuts, to form a V-shaped notch in the sample. Two methods are described by Ouchterlony *et al.* (op cit);

– A V-shaped notch cut perpendicular to the axis of the core, where the initiation and extension of the fracture from the tip of the saw cut is achieve by applying a bending load to the sample, which expands the aperture of the saw cut slot, and initiates the fracture. This Chevron Bend test uses a long sample, which is approximately 4 times as long as the diameter, the diameter chosen should exceed 10 grain diameters.

- The Short Rod test involves a V-shaped notch cut parallel to the axis of the core sample. The sample is subjected to a tensile stress applied perpendicular to the V-shaped cut, which initiate the fracture at the tip of the intact V-shaped intact rock. The sample diameter should again be at least greater than 10 times the grain diameter and the sample length should be 1.45 times the diameter.

In the recommended methods the shape of the V-shaped cut differs for the two test set-ups, as do the methods for calculating the fracture toughness from the two test geometries.

The method used by Fowell *et al.* (1995), involves cutting a slot in a cylindrical disk sample. This is achieved by using two cuts of a circular saw at opposite ends of the disk: The two saw cuts meet in the middle of the sample to form a slot penetrating the axis of the disk. The samples and tests are referred to as the Cracked Chevron Notched Brazilian Disk samples (CCNBD), and the samples are loaded in a similar manner to the Brazil test, with the cut slot aligned with the loading direction.

Other methods are also available to assess the Mode I fracture toughness (Kuruppu *et al.*, 2014).

6.11 Creep tests

Creep tests in the petroleum industry are used to assess the time-dependent rate of deformation of evaporitic formations, specifically for modeling of the movement of salt layers and diapirs to assess loadings on casing (Cheatham & McEver, 1964; Willson *et al.*, 2003) and in dynamics 3d numerical models, or for closure rate estimations for boreholes drilled through salt formations (Fuenkajorn & Daeman, 1988; Dusseault *et al.*, 2004; Willson & Fredrich, 2005), and for closure estimated in salt caverns used for gas storage (Dusseault *et al.*, 1985; Fokker & Kenter, 1994). The tests are not normally performed as part of a standard testing program, but may be commissioned for specific field issues. ISRM Standards for creep testing are detailed in Aydan *et al.* (2015).

Time-dependent deformation at constant load, creep, also occurs during reservoir compaction during the life of a field development. As such, it is common to assess the time dependent deformation during compaction tests (K_o), whilst maintaining zero lateral strain boundary conditions (Dudley *et al.*, 1998). This differs from the classical 'creep' test, as outlined in Aydan *et al.* which considers creep measurement for uniaxial compression, triaxial compression and Brazilian test configurations. Other variations of test procedures have also been used to assess the time-dependent deformation of rock formations at a constant load, and the impact of this creep on ultimate reservoir compaction and deformation, by performing constant loading rate tests for uniaxial strain (K_o) conditions – The RTCM Rate Type Compaction Modeling (de Waal, 1986; de Waal & Smits, 1988). The RTCM approach imposes different loading rates on the sample during a monotonic loading cycle, which enables the time-dependent creep deformation to be assessed in the slower loading rate tests. These tests are addressed below in Section 7.1, Uniaxial-strain testing.

The standard creep tests utilize samples prepared by standard procedures for uniaxial compression or triaxial tests, with Length to Diameter ratios between 2–3. The tests are performed in line with standard uniaxial and triaxial tests, and the creep tests performed at pre-defined stress levels where the load is held constant and the deformation of the sample under the constant load is carefully monitored.

The ISRM standards, recommend that if elastoviscoplastic behavior is being assessed, the irrecoverable strain associated with the creep tests should be evaluated by unloading the sample to a load level of 1% of the applied load for the creep test, then re-loaded using the same rate as used during the initial loading step.

The results from these standard creep tests are presented as: Strain vs time, for the different loading step; and as time to failure for different loads or stress ratios.

To assess the implications of salt behavior on petroleum applications, the creep tests should be performed at reservoir temperature, as salts and anhydrites are recognized to have significant temperature dependence on the creep rate (Fossum & Fredrich, 2002).

7 SPECIALIZED TESTING SPECIFIC FOR PETROLEUM GEOMECHANICAL ANALYSIS

The previous sections discussed test protocols used to determine the mechanical properties of rock and weakly consolidated samples involved in petroleum field developments, along with the challenges encountered with obtaining representative samples from cores taken through the reservoir rocks. These tests are standard rock mechanics tests, used across different industries and as much as possible rely on the international standards from the ISRM and ASTM to guide and define the test protocols and standards.

This section will discuss mechanical property tests specifically used and originating from the petroleum industry. The two tests which fall into this category include; Uniaxial-Strain testing, often referred to as K_o testing, from soil mechanics literature, zero-lateral strain tests, or uniaxial strain tests, and secondly Hollow Cylinder Tests (HCT) or Thick-Walled Cylinder tests (TWC).

A Suggested Method for Uniaxial-Strain testing performed on reservoir formations has recently been issued by the ISRM, and this section draws on this guideline. The set-up, procedures and interpretation of this test for reservoir compaction estimates is discussed by Jones *et al.* (1987) and Fjaer *et al.* (2008).

7.1 Uniaxial-strain testing (zero lateral strain, or K_0 testing)

Uniaxial-Strain testing is commonly used in civil engineering to evaluate the compaction of soils and granular materials. This testing involves compressing a right cylinder of soil or rock either in a stepwise manner with incremental loads, or continually increasing the stress applied to the ends of the sample, while maintaining the diameter of the sample to the original dimension. Zero radial, or diametrical, strain is maintained across the diameter, perpendicular to the loading direction. This test is an approximation to what happens during compression, during burial under passive basin conditions, or underneath the center point a structure with a large area, or in an idealized cylindrical oil reservoir during uniform pore pressure depletion, accompanying the production of oil and gas.

Axially loading a sample will tend to result in radial strain as a result of Poisson's ratio of the sample in the elastic region of the deformation, and due to dilatancy in the plastic region of the deformation. The aim of the Uniaxial-strain test is to prevent this radial or lateral deformation. This uniaxial-strain test is described in the literature using several names; Uniaxial-strain test, the zero-lateral strain test, K_o test (from civil engineering), and the Oedometer test.

7.1.1 Oedometer tests

The Oedometer test is a particular and standard form of this test in civil engineering, which uses a sample constrained diametrically by a steel ring. The sample is carefully cut to ensure the diameter of the sample is exactly the same as the steel ring, the steel ring ensures that as the sample is vertically compressed no lateral strain is possible. Special modifications of this test allow the lateral stress magnitude to be measured during compaction (Brooker & Ireland, 1965).

This test has been adopted in some specialized testing in the oil industry (Teeuw, 1971; de Waal & Smits, 1988) to evaluate the vertical compressibility of sands and sandstones, and the rate dependency of the deformation.

The steel ring constraining the samples restricts the tests to remolded or completely uncemented samples. Intact cemented samples require careful preparation to ensure that the gap between the sample diameter and the steel ring is minimal. If this is not achieved, axial loading will result in shearing at the outer edges of the sample, and a true uniaxial compaction will not been achieved.

To eliminate the gap between the oedometer wall and the sample, some procedures include filling the gap with remolded or disaggregated material, but this is not recommended practice. To avoid these testing problems with cemented or consolidated intact samples, the K_o test performed in a triaxial test cell has been adopted and is the preferred approach for Uniaxial strain tests. This will be discussed in the following section 7.1.2.

The loading stages of the Oedometer test comprise either incremental loading or continual rate of loading (either constant rate of stress or constant rate of deformation). The vertical compaction resulting from this loading results in shear stresses generated between the sample and the oedometer wall, as the sample compacts. Minimizing these wall shear stresses is addressed by lubricating the steel walls of the oedometer, recommended lubricants include silicon grease, molybdenum disulfide or polytetrafluoroethylene. These shear stresses impact the compaction deformation of the sample, for this reason the sample dimensions used in an oedometer have been chosen to minimize the impact of the shear stresses on the overall deformation of the sample. ASTM D2435-04 standard recommends the sample diameter is 2.5 times the height of the sample, D/H = 2.5, though samples with a ratio of D/H = 4 is sometimes used. This contrasts with the sample dimensions for a standard triaxial test, discussed in Section 5, due to the different test set-up, and the shear stresses generated during these different tests.

The sample measurements recorded at the start of the test include the porosity, or voids ratio, and the sample height in the oedometer test cell. During the test the measurements recorded in the oedometer test include the axial load applied to the sample vs time, the pore pressures developed at the top and base of the sample and the axial deformation.

Performing the test should be in accordance with industry standards, as specified in ASTM D2435-04, bearing in mind these standards are defined for low stress tests and civil engineering applications compared to the high stress tests required for the oil and gas industry.

7.1.2 K_o, Zero lateral strain tests, or uniaxial-strain tests

The K_o or Zero Lateral strain test is a uniaxial-strain test performed in a standard triaxial test cell. The sample is compressed along the axis of a cylindrical sample in a controlled manner where the cell pressure is continually adjusted to ensure that the sample undergoes no lateral strain (Bishop, 1958; Teeuw, 1971; Jones *et al.*, 1987; Fjaer *et al.*, 2008). As the axial stress increases the lateral stress, or cell pressure, required to maintain zero lateral strain typically increases by 0.5 to 0.9 times the axial stress increase, depending upon the material properties of the sample tested.

This test uses a standard triaxial cylindrical sample with a height to diameter ratio of 2 (H/D = 2). The sample is isolated from the cell pressure using a thin rubber jacket that transmits the cell pressure to the sample, to control the sample diameter. The rubber jacket stiffness and thickness can affect the control of the test and deformation of the sample, as such the rubber jacket should be thin and flexible enough to avoid affecting the test results.

Recent ISRM guidelines have been published outlining the requirements of this test (Dudley *et al.*, 2016)

The motivation for performing these tests is to evaluate reservoir compaction, typically in sandstone or limestone reservoirs with high porosities and permeabilities. However, the test has broader application as it provides a more accurate estimate of the Pore Volume Compressibilities required in reservoir simulators. The loading rates imposed during the tests have been reasonably fast (0.5 to 10 MPa/hour), as drainage of the samples and excess pore pressure build-up resulting in partial drainage of the samples has not been a concern. However in the 1980's these tests were applied to chalks with lower permeabilities, where pore pressures in the samples were observed to increase during the loading, and as a result the loading rate now is calculated based on standard practice (Bishop & Henkel, 1982).

It is usual to perform these tests with pore pressures measured at both ends of the cylindrical sample, but one end drained. The pore pressure measurement at the undrained end of the sample is used to observe the pore pressure dissipation and minimize any excess pore pressure build-up.

The mechanical responses of granular and indurated rocks are known to be strain rate dependent. This is also the case with the strains developed in the K_o test. As such, for field compaction estimates, tests can be performed at different strain rates or, more commonly in the petroleum industry, at different stress loading rates to account for the differences between the deformation measurements recorded at standard laboratory loading rates and the loading rates experienced in a depleting reservoir (de Waal, 1986; de Waal & Smits, 1988).

One of the primary differences between K_o, or zero-lateral strain loading tests in civil engineering and in the petroleum industry is the load path. In civil engineering the sample deformation is investigated to assess the impact of a man-made structure on the deformation of the soil. In Petroleum engineering, and in aquifer drainage in civil engineering, the aim is to assess the impact of lowering the pore pressure, or formation pressure, on the deformation of the formation.

This has led to two test procedures used in the petroleum industry. The first procedure starts at low confining and axial stresses and a constant pore pressure (CPP) reflecting the initial *in situ* effective stresses, which is followed by gradually increasing the axial stress to mimic the effective vertical stress changes anticipated in the reservoir, during depletion. The second method involves gradually applying the *in situ* magnitudes of the vertical and horizontal total stresses along with the reservoir pressure to the sample. The reservoir depletion in this test procedure is reproduced by maintaining the axial stress on the sample while reducing the pore pressure (PPD) in line with the expected range of reservoir pressures in the field.

The tests are performed with either constant deformation rate or a constant stress-loading rate, which may be preferred when the test is performed to reproduce reservoir stress loading rates. A variation on these loading schemes is the step-hold method, which maintains a constant stress at specified stress levels during the loading cycle.

The boundary condition of zero radial deformation for these tests is achieved by using a servo-control feedback loop: As the sample compresses vertically, a radial deformation develops and is measured whereupon the cell pressure is increased to bring the radial strain back to zero. This raises the question of what is an allowable radial strain before it is returned to zero, in order to conform to uniaxial-strain conditions. If excessive radial strain is developed in the sample before the cell pressure is adjusted, the sample may become permanently deformed and may no longer correspond to a uniaxial compression, with zero lateral strain. Radial deformations of 20 microns have been observed to irreversibly affect the stress-strain response of fine grained (2–10 micron grain sized) chalk samples during uniaxial-strain tests. It is recommended to maintain the radial strain to ±25 microstrain (approximately 1 µm for a 38 mm diameter sample) during the test in order to maintain zero-lateral strain conditions (Dudley *et al.*, 2016).

Uniaxial Strain tests are run at ambient and reservoir temperatures, upto 150°C in most cases, to assess the impact of in-situ reservoir temperatures on the deformation and compressibility of the formation. Uniaxial tests where the types of pore fluids are changed during the tests should be undertaken to assess the impact of waterflooding of the reservoir, the displacement of oil by water, on the deformation properties and compaction of the formation (Brignoli *et al.*, 1995; Maury *et al.*, 1996).

Uniaxial test measurements: The measurements commonly acquired during a Uniaxial-Strain (K_o) test in the triaxial cell include; the axial stress, axial deformation, cell pressure, radial, diametrical or circumferential deformation, and pore pressures at the top and bottom of the samples. During the PPD test, the stress-depletion ratio ($d\sigma_h/dP_p$) is also obtained for uniaxial strain conditions. In specialized testing equipment additional measurements may include axial and/or radial ultrasonic transducers to monitor the evolution of the acoustic travel times accompanying compaction of the samples (Azeemuddin *et al.,* 1994).

The degree to which accurate tests can be performed is related to the calibration of the test set-up: Unlike most rock mechanics testing, the Uniaxial Strain test requires the cell pressure to continually change to maintain zero lateral strain. Consequently the test apparatus and especially the radial strain deformation gauges should be carefully calibrated against changes in the cell pressure. LVDTs and strain gauges used in these radial & circumferential deformation gauges are both sensitive to the cell pressure, which if left uncalibrated can lead to erroneous radial strain measurements and inaccurate test control.

Uniaxial test procedures: Uniaxial Strain (K_o) tests can be performed using two different procedures (Dudley *et al.*, 2016);

- Constant Pore Pressure (CPP); this relies on simulating the axial effective stress changes on a sample of the reservoir formation, by increasing the axial load on the sample, whilst maintaining a constant pore pressure and sample diameter. To ensure the sample diameter remains constant, simulating an ideal reservoir compaction, the cell pressure has to continually increase to counteract the lateral expansion and Poisson's effect as the axial stress and strain increases.
- Pore Pressure Depletion (PPD); this is a refinement of the CPP procedure, whereby the samples are taken to the approximate reservoir total stress condition with the pore pressure also at the initial reservoir pressures. To simulate reservoir compaction, the total axial stress is maintained constant, and the magnitude of the pore pressure in the sample is lowered to simulate production from the reservoir. The resulting increase in the effective axial stress results in shortening of the sample and the cell pressure has to be reduced to maintain a constant sample diameter to maintain zero lateral strain (K_o) conditions.

The compressibilities obtained from the two procedures differ and are outlined in the ISRM Suggested Method for this Uniaxial Strain, K_o test (Dudley *et al.*, 2016). The Pore Pressure Depletion methodology, includes the impact of grain compressibility resulting from pore pressure changes on the overall sample deformation, and allows the Biot pore pressure effective stress coefficient to be evaluated for the range of stresses experienced during reservoir depletion.

The Uniaxial strain, zero-lateral strain (K_o) test has become a standard test in the industry to assess reservoir compaction. It is also used in more comprehensive testing programs designed to characterize the yield end-cap models of weak rocks, (Charlez, 1991; Crawford *et al.*, 2004) which are included in standard FEM packages.

These tests are performed on permeable reservoir formations. However, K_o tests have also been performed on low permeability shale formations which overlay the reservoir and which may drain in-situ, in response to reservoir drainage (Addis, 1987). These tests on shale caprock sample can be used to assess whether there is any contribution of compaction from the overburden over the production lifetime of the reservoir.

7.2 Hollow Cylinder Tests (HCTs)

Hollow cylinder tests have been used in the rock mechanics community for decades, since the start of the 20th Century (see References in Geertsma, 1985). These tests are performed on cylindrical samples that contain an inner cylindrical cavity. These cylindrical cavities are normally drilled through the entire length of the sample, however some test set-ups contain a cavity that is drilled only part way through the length of the sample. The samples used are relatively large, typically 3 – 4" outer diameter with an internal hole diameter 1/3rd to 1/5th the size of external diameter.

The jacketed samples are tested in a triaxial cell by increasing the stresses until failure is recorded on the inner diameter of the samples (Figure 21). Commonly the cell pressure is applied to the top and sample circumference until the sample deforms or

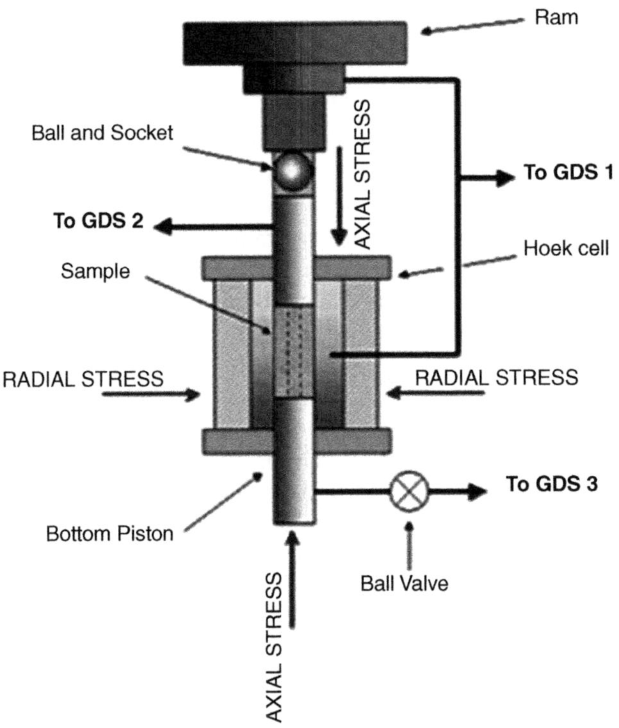

Figure 21 Illustration of a hollow cylinder test set-up (Courtesy of FracTech Ltd.)

fails. The axial stress applied to the ends of the samples and the cell pressures applied to the outer circumference of the sample can also be independently controlled, with the sample is loaded using a constant axial/confining pressure ratio.

The petroleum industry has adopted and modified these tests as a method to assess the onset of sand production. Sand production is the disaggregation of sandstone reservoir formations at the wellbore wall accompanying production, and the subsequent transport of the sand through the well and to the surface facilities. This is a form of wellbore stability across the reservoir interval, but also occurs across limestone and coals reservoir sections, in which it is referred to as Solids Production.

The use of Hollow Cylinder Tests, commonly referred to as Thick Walled Cylinder (TWC) tests for sand production, was developed by Shell over a number of years since the 1970s (Antheunis *et al.*, 1976; Geertmsa, 1985; Veeken *et al.*, 1991; Kooijman, 1991) and is an index test, rather than a mechanical property test which has been adopted by a number of companies (Willson *et al.*, 2002; Rahman *et al.*, 2010). The TWC samples originally used by Shell are small sample plugs obtained from 4" diameter cores retrieved from the reservoir. The TWC samples are 1" outer diameter, 2" length and 1/3" inner diameter. The deformation of the inner cavity is not monitored as in the larger hollow cylinder samples. The samples are jacketed and tested in a triaxial cell where the isotropic cell pressure is increased at a rate of (1 – 2.5 MPa/min) until the sample fails, this is referred to as the TWC strength, the 'thick walled cylinder' strength, observed as a collapse of part of the outer circumference of the sample, and identified by a drop in the cell pressure during the test.

This measured strength is a structural failure of the sample and it involves several stages: elastic response at low stress levels; a yield and initial failure of the inner diameter, which occurs at 0 to 30% below the TWC strength (Veeken *et al.*, 1991); the development of shear bands at the inner diameter of the sample; and growth of these shear bands (breakouts) until the strength of the remaining intact elastic sample and the growing plastic shear zones is insufficient to withstand the cell pressure. Typically the shear zones grow to approximately ½ the sample wall thickness when this sample collapse occurs.

These tests became a standard as Shell evaluated the sand production tendencies of the Groningen reservoir in Northern Holland. These wells were completed with deep penetrating perforations: Consequently the use of 1/3" for the Inner Diameter of the Thick Walled Cylinder tests is designed to reflect the geometry of the perforation diameter (Veeken *et al.*, 1991).

The more widespread adoption of Thick Walled Cylinder tests to assess sand production has led to different hollow cylinder dimensions being used by different companies and testing laboratories. The most common variation from the Shell sample size is the use of a triaxial sample size, of 1.5" outer diameter, 3 inch length, and 0.5" inner diameter (Willson *et al.*, 2002).

Shell's development work showed that these hollow cylinder samples were scale dependent and empirical plots were developed of TWC strength for different inner diameters for samples with constant OD/ID ratios of 3 (Kooijman *et al.*, 1991; van den Hoek *et al.*, 1994) (Figure 22a). For moderate rock strengths of reservoir sandstones, the 1" x 2" x 0.33" (Shell) TWC samples are approximately 20% stronger than the 1.5" x 3" x 0.5" samples, *i.e.* the Shell samples due to their smaller size can withstand 20% higher cell pressures and stresses before sample collapse compared to the larger samples.

Similar size dependency of cylindrical cavities strengths have been observed in true triaxial (polyaxial) tests by Herrick and Haimson (1994) using Alabama limestone

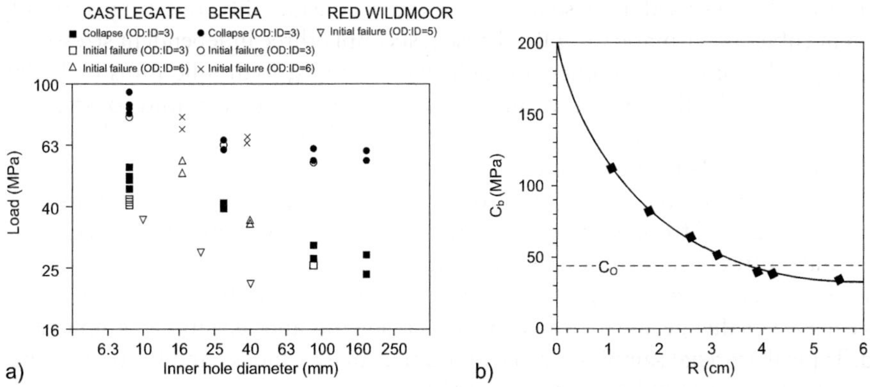

Figure 22 Scale dependency of a) Thick Walled Cylinder samples for different sandstones (Kooijman *et al.*, 1992; van den Hoek *et al.*, 1994), and b) true triaxial test on Alabama Limestone (Herrick & Haimson, 1994)

samples, for different hole diameters drilled into constant sized prismatic true triaxial samples (Figure 22b). These are not directly comparable with the TWC size dependency, but illustrate similar scale dependent behavior.

The measurements taken during these Thick Walled Cylinder tests typically only include cell pressure, due to the simple nature of the tests. However, additional parameters can be measured depending upon the sample sizes and the test cell capabilities. Commonly, these include, sample pore pressure, axial stress, if controlled independently from the cell pressure, inner diameter deformation (using a deformation gauge or strain gauge). If an internal rubber sleeve is used to isolate the inner cavity, the inner cylindrical cavity pressure and volume can also be monitored (Addis & Wu, 1993).

At the end of production gas reservoirs can be used for gas storage or as peak shaver facilities, drawing on the gas at periods of high demand, and storing gas in the reservoir at periods of low demand. This peak shaver demand can be seasonal, storing gas in the summer for increased winter demand, or diurnal, storing gas at night for use during the daily demand. The impact of these pressure cycles, on the likelihood of sanding has been assessed by the cyclical loading of hollow cylinder tests. The amplitude and the number of cycles performed on the thick walled cylinder samples reflects the annual or diurnal nature of the gas storage.

8 MODEL TESTS – TRUE TRIAXIAL TESTS

True triaxial tests are used by the petroleum industry to perform scaled model tests to investigate specific practical problems, these include: wellbore stability, breakout formation and salt closure (Haimson & Herrick, 1986, Fuenkajorn & Daemen, 1988; Addis *et al.*, 1990); hydraulic fracture geometry and connectivity to wellbore (Papadopoulos *et al.*, 1983; Hallam & Last, 1991; Weijers & De Pater, 1992, 1994; Meng & de Pater, 2010); hydraulic fracture and leak off initiation (Black *et al.*, reported in, Morita *et al.,* 1990); perforation charge performance (Halleck & Behrmann, 1990; Kooijman *et al.*, 1992); Anelastic Strain Recovery (Yassir *et al.*, 1998) and cavity completions (Wold *et al.*, 1995).

The sample sizes used in these model tests vary considerably, but range typically up to 0.5 m cubic or prismatic samples. These tests apply three independent stresses to the opposing sides of cubic samples through two methods; hydraulic rams or hydraulic flatjacks. Generally the larger the sample sizes, the lower the magnitude of stresses that is practical to apply, due to the large forces involved, however these test cells commonly apply up to 70 MPa to the samples.

The size of the samples required for these test cells means that the materials used for testing are obtained from outcrops. The petroleum industry commonly uses Castlegate, Berea and Saltwash sandstones, Pierre shales (I & II), Indiana and Alabama limestone samples, and cubic coal samples. Obtaining identical repeatable samples for these test cells can be a challenge, as such, artificial homogeneous and repeatable samples are also used. Typical artificial samples include gypsum or plaster blocks and Portland cemented high porosity sand samples (Minaeian, 2014).

True Triaxial Tests are also performed to assess the impact of the intermediate principal stress (σ_2) on the properties and failure of materials, and to evaluate the most appropriate failure criteria to use in analysis of field problems. An overview of this aspect of testing and application of True Triaxial Tests can be found in Minaeian (2014) and an overview of the dependency of failure criteria to the intermediate stress is addressed by a number of authors in the ISRM Suggested Methods (2015).

The use of true triaxial (polyaxial) testing applied to Petroleum Geomechanical Engineering issues remains focused on research and development activities, which aim to provide insight into the general issues, faced by the industry. This section briefly discusses some of the testing related aspects of these tests, and some of the procedures used to minimize any testing artifacts. Standard sample preparation and test procedures for these tests are not available. The preparation and procedures are developed within each laboratory and defined for each specific test program, depending upon the issue being addressed, wellbore stability, sanding, fracturing etc.

Some common themes and testing procedure issues are however recognized and addressed in most testing schedules.

8.1 Sample saturation

In most cases, especially when using outcrop samples, the saturation of the samples cannot be guaranteed; sample cutting and preparation of these large samples may also affect the saturation. Re-saturating samples is generally not possible, partly due to the large blocks sizes which range up to 0.5m x 0.5m x 0.5m, and because most true triaxial test facilities are not capable of sealing the samples with rubber membranes, or similar, or of providing back pressures during true triaxial (polyaxial) loading of the blocks.

8.2 Stress application and friction effects

Most of the samples used in the True Triaxial Tests are cubic, or prismatic. In standard triaxial tests the sample is maintained with a height to diameter ratio of 2:1 to minimize the impact of the shear stresses on the end platens on the failure of the sample. The use of cubic or prismatic samples with all six sides stressed through solid platens, face the same challenge, but for all three axes of the sample.

To ensure that the stresses generated by the hydraulic rams or flatjacks are those applied to the sample sides, considerable effort is taken to minimize the friction on the sides of the cubic samples, and to avoid shear stresses on the sides of the samples, for anisotropic stresses used in the tests. The friction reduction commonly utilizes combinations of Teflon sheets or rubber sheets, with lubricants. These methods do not eliminate the friction from the test set-up, as such it is important to assess the magnitude of the friction acting during the tests at different stress levels.

8.3 Test procedures and loading sequence

The benefit of these tests is the ability to independently control the three stresses applied to the six sides of the samples. Commonly the test loading involves application of small initial isotropic stresses, with all three stresses being equal, to ensure the sample is bedded in, at the center of the cell, with sufficient space or travel to enable the sample to deform.

The loading can be applied stepwise, or in a continual ramp, with the stresses increased in a pre-defined ratio. Once failure is observed, or if the maximum stresses for the test is reached, the stress unloading normally follows the reverse of the loading cycle, bringing the sample to an isotropic stress state before complete unloading. This may be varied if the unloading cycle will influence the sample or sample failure and if the sample is required after the test for analysis of failure modes.

These tests are generally Research and Development focused tests, so these procedures vary according to the laboratory and focus of the investigation.

9 FORMATION SPECIFIC TESTING REQUIREMENTS

The rock formations encountered in the petroleum industry typically comprise sedimentary lithologies: sandstones, siltstones, conglomerates, carbonate limestones and dolomites, argillaceous mudstones and shales, and evaporitic formations such as salts and anhydrites. Less frequently, igneous formations such as granites and basalts are also encountered, along with weathered products from these more competent formations.

Determining the mechanical properties of the most common reservoir and overburden lithologies, sandstones, carbonates and shales, are assessed using similar testing requirements, however some differences in their behavior justify modifications to the standards. This section discusses some of the issues that arise for these different lithologies during the design of a mechanical property test program.

9.1 Sands and sandstones

The mechanical properties of sands and sandstones cover the extreme bounds from elastic brittle behavior to fully plastic behavior. The more competent formations can be tested using international standards, but adequate permeability and drainage from these samples should be ensured as well as the degree of saturation, to ensure that the effective stresses acting across the sample are known.

Weaker, high porosity samples are often difficult to core and recover intact from the core barrel. When core recovery from sandstones reservoirs is less than 100% the missing core is generally from depth intervals containing the weaker formations, or from areas where faults and fractures are intercepted during coring. These areas of poor core recovery signify weak formations and have been used as an indicator of potential problems for reservoir compaction (Willson, pers. comm.) and sand production. Often unsuccessful recovery is exacerbated by the presence of thin shale layers, which may be reactive and swell as they come into contact with incompatible drilling and coring fluids. Sands that are unconsolidated and difficult to handle without causing damage can be frozen at the rig-site to keep them intact, during core handling and transportation prior to testing. However, this freezing can result in a disturbance of the core and weakening (Torsaeter & Beldring, 1987). Sands from heavy oil field require special coring and handling procedures (Collins, 2007) to minimize core disturbance at atmospheric conditions, especially to minimize the effects of gas expansion.

Sandstone samples can be tested with the mixture of hydrocarbon fluids and connate, residual formation water. Alternatively, the samples can be cleaned of the hydrocarbon fluids using different solvents, using Soxlet equipment or flow-through cleaning, and re-saturated with inert fluids or simulated reservoir brine. This process may lead to sample alteration, and minimizing the impact of any core plug cleaning on the mechanical properties is paramount. The saturation of the sample is key to accurate mechanical property determination. West (1994), shows that the higher strengths measured in dry samples of sandstones and limestones are related to the suction present at low moisture contents, accounting which he estimated to account for up to 30% strength increase compared to saturated samples, in a similar manner to the increase of strengths observed in partially saturated soils. However, dry samples exhibit significantly larger strengths than saturated samples (Rao *et al.*, 1987; Dyke & Dobreiner, 1991) with the elastic modulus, the onset of dilatancy and the strength increasing significantly when moisture contents fall below 1%. Nunes (1989) reported in Dyke and Dobreiner (1991), shows that the unconfined compressive strength of dry sandstones is between 1 and 2.5 times the saturated strengths, with the weaker sandstones being more significantly affected by moisture content variations. Similarly, Barefield and Shakoor (2006) found dry sandstones strengths of up to 3.5 times the saturated strengths, though in contrast with Dyke and Dobreiner (1991) they found the effect more pronounced in stronger samples.

Figure 23 shows repeat results from scratch tests and ultrasonic measurements performed along the lengths of 2 cores (Germay & Lhomme, 2016). The scratch & ultrasonic data were measured whilst the cores remained preserved, showing a reasonably constant strength over a period of 250 days. The cores were subsequently opened to atmospheric conditions and allowed to dry out, which was accompanied by a significant increase in the measured strengths over the same section of cores. These continuous core measurements, reflect what is observed in standard UCS tests, that competent sandstone cores experience a significant increase in strength accompanying drying.

9.2 Carbonates

Carbonate formations are typically heterogeneous as shown in cores and plugs of chalks, where despite continuous deposition and apparently homogeneous cores, different depositional textures and cementation occur. Also in ramp (Figure 18) and reefal carbonates, large variations in depositional textures occur, making the results of test programs difficult to interpret. The presence of large primary porosity vugs or

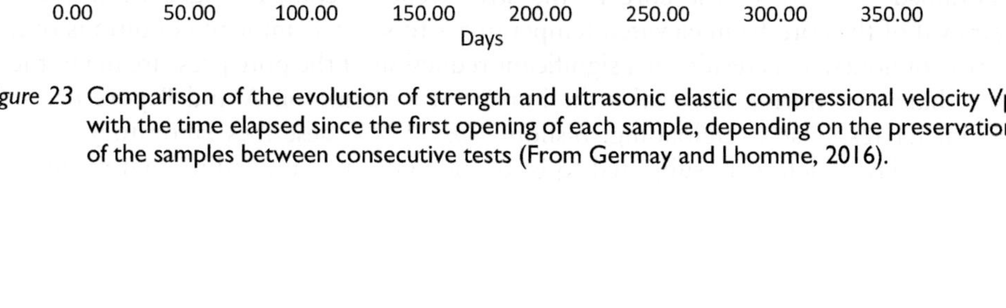

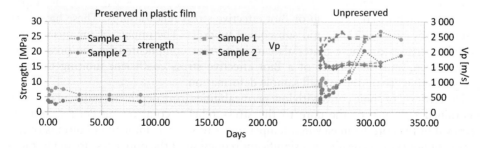

Figure 23 Comparison of the evolution of strength and ultrasonic elastic compressional velocity Vp with the time elapsed since the first opening of each sample, depending on the preservation of the samples between consecutive tests (From Germay and Lhomme, 2016).

secondary porosity vugs resulting from dissolution, may also lead to textural heterogeneity in the core and comprise large percentages of any sample plugs taken. This is problematic for the geomechanical engineer, as even if intact test plugs can be sampled from these sections, the measured strength from any intact material may not be representative of the larger rock mass containing the vuggy porosity.

Cementation of carbonate sediments can occur at very shallow burial depths (less than 10 meters burial) compared to sandstones, locking in high porosities in these sediments. When subjected to elevated magnitudes of stress during testing, the cementation can fail before stresses are large enough to shear through grains resulting in pore collapse behavior where the elastic rock transforms into a disaggregated granular plastic material. This pore collapse is accompanied by a large pore volume reduction in drained test conditions (Addis, 1989) and by a large pore pressure increase in undrained conditions (Jones & Leddra, 1989), for relatively small stress increases during the tests. This pore collapse can occur over a small stress range as defined by the yield surface for the formation. For some tests, *e.g.,* Uniaxial-strain (K_o or zero lateral strain tests) this can lead to challenges in controlling the zero-lateral strain boundary condition.

9.3 Shales and mudstones

Argillaceous lithologies such as mudstones and shales characteristically have very low permeabilities, down to nano-Darcies (10^{-21} m^2), requiring undrained tests to determine their mechanical behavior. The need for saturation in these materials is important in order to know the effective stresses acting in the sample during testing.

9.3.1 Sample saturation

Due to the small pore sizes and pore throat dimensions in these fine-grained formations, any air or vapor in the sample will result in unsaturated conditions and affect the magnitude of effective stresses acting in the sample. The unsaturated effective stresses are a result of the very large capillary pressures which are generated by the liquid menisci around the grains, providing a fluid tension (essentially a confining pressure) which holds the grains together, which is seen most notably in uncemented muds and clays, which when dry can behave as competent strong materials.

These capillary pressures (or suction) lead to effective stresses which are not represented by the normal Terzhagi or Biot effective stress relationships. These have been discussed by Bishop (1959) and subsequent workers. The impact of the unsaturated effective stresses on these materials can be further complicated by the presence of cementation, and the impact of a Biot effective stress coefficient limiting the pore pressure influence on the overall sample deformation.

The retrieval of shale cores with low permeabilities from depth inevitably involves an undrained response of the core to the stress relief accompanying coring, and the retrieval of the core from elevated temperatures to surface ambient conditions over a period of hours. This results in a significant reduction of the pore pressure magnitudes in a shale core during retrieval – the impact of a Skempton A and B pore pressure parameter response to unloading. Bishop *et al.* (1975) describes laboratory tests on London clay which were subjected to isotropic undrained unloading that resulted in

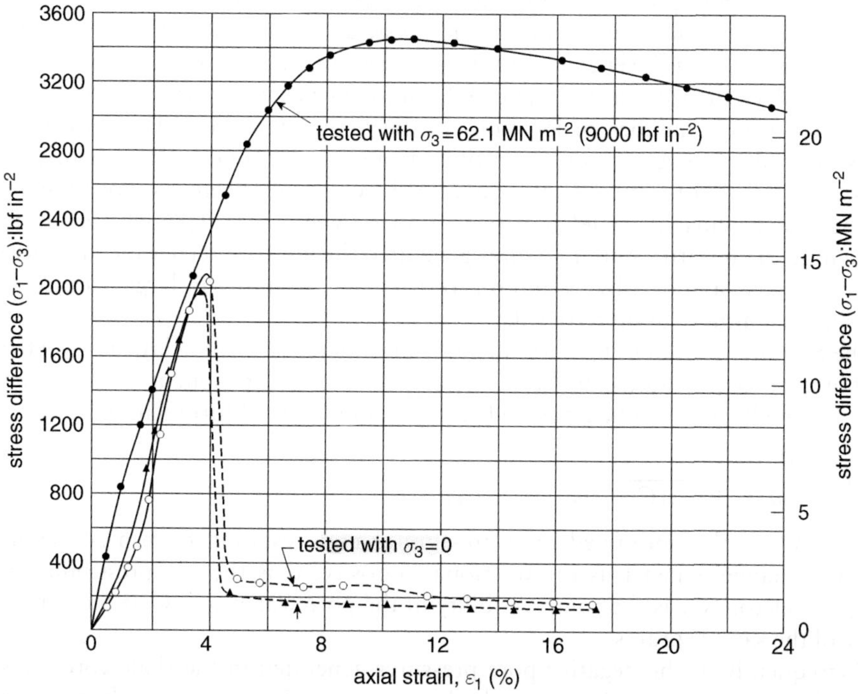

Figure 24 The impact of stress release on London Clay samples and the change from plastic to brittle behavior resulting from reducing the confining pressures on the undrained sample from 62.1 MPa to 0 MPa (Bishop *et al.*, 1975).

negative pore pressures sufficient in magnitude to exceed the tensile strength of the water. This results in water cavitation in the pores of the clay and ultimately unsaturated samples filled with water vapor. This response was accompanied by a transition from typical plastic London clay material behavior for saturated conditions to an uncharacteristic elastic-brittle response of the London clay samples following the generation of the negative pore pressures upon unloading (Figure 24).

It is likely that shale cores recovered from the depths of typical of oil fields will experience the same undrained response, with low or negative pore pressures developing during core retrieval and potential cavitation of the pore fluids. Ewy (2016) presents an equation to estimate the possible magnitude of the negative pore pressures based on Skempton's pore pressure coefficients (1954), for an elastic undrained shale sample during retrieval from depth with in a wellbore:

$$P_w = (1 - B).\sigma_m - \sigma'_m$$

Where; P_w is the pore water pressure after unloading or core retrieval (negative pore water pressures values indicate suction), B is Skempton's pore pressure coefficient for hydrostatic confining pressure changes, σ_m is the *in situ* mean total stress at the depth where the core is cut and σ'_m is the corresponding *in situ* mean effective stress.

Skempton's A pore pressure coefficient relates undrained pore pressure changes in a triaxial test to the change in the deviatoric stress, $\Delta q = \Delta (\sigma_1 - \sigma_3)$. In Ewy's derivation, the Skempton A value is set equal to 1/3, a value representative of elastic materials, and using B values of between 0.75 and 0.9, estimates negative pore pressures in cores recovered from a number of locations in the range -5 to -15 MPa. Yassir (1989) measured values of A ≥ 1 for reconstituted samples consolidated to mean effective stresses of 60 MPa for mud samples obtained from mud volcanoes. For these plastic normally consolidated shale/mudstone samples where A = 1, Ewy's equation for estimating the negative pore pressures generated during core retrieval can be used by substituting σ_m and σ'_m, by σ_1 and σ'_1, respectively, resulting in larger negative pore pressure magnitudes upon retrieval.

The Skempton B values for saturated unconsolidated soils and clays at low effective confining stresses is taken to be 1. However, in more consolidated and cemented samples, the B value can be estimated using Bishop's (1973) formulation:

$$B = \frac{\Delta P_p}{\Delta \sigma} = \frac{1}{1 + n(C_w - C_s)/(C - C_s)}$$

Where: ΔP_p is the undrained pore pressure change resulting from a $\Delta\sigma$, confining pressure change, when n is the fractional porosity, C is the compressibility of the sample skeleton, C_w is the compressibility of the pore water, and C_s is the compressibility of the sample grains.

Consequently, if the negative pore pressures generated in the shale core are sufficiently larger, cavitation of the pore fluid can occur and the cores may be unsaturated upon retrieval to the rig floor (Santarelli & Carminati, 1995). Such cores may exhibit a rim inside the circumference of the core, to a depth of up to a centimeter, as the negative pore pressures result in imbibition of the drilling fluids used for coring into the shale core, though this is less likely with oil based muds due to the high capillary entry pressures of oil into water saturated shales.

Re-saturating the shale samples prior to testing and demonstrating saturation is the key to accurately understanding the stress state in the samples during the mechanical tests, and to an interpretation of the mechanical behavior from the laboratory tests (Ewy, 2016). Re-saturation and exposure of shale and clay samples to fluids during testing must take into account the mineralogy and the potential for swelling clay minerals in the samples. Consequently, Ewy (2016) recommends ensuring saturation of preserved shale samples through vapor adsorption, by exposing samples to different relative humidity environments in a vacuum desiccator, following Steiger and Leung (1991), though this can take 2–3 weeks to obtain a constant sample weight with no additional water adsorption. Any fluids used during these tests on shale and mudstone samples needs to be chemically inert with these samples, and even then re-saturation through direct brine contact should be achieved under stresses to minimize swelling. Thereafter samples should be tested in the triaxial apparatus with an applied back pressure on the samples to ensure saturation is maintained during the entire test, which may preclude the use of standard UCS tests. Saturation can be checked in the triaxial cell during the consolidation stages of the tests using undrained incremental cell pressure increases to monitor the Skempton B value, though this is likely to show values less than unity for well consolidated and cemented samples (Bishop, 1973, 1976).

9.3.2 Anisotropy

An important characteristic of argillaceous shales is the alignment of the clay minerals during burial and the development of an anisotropic texture, dominated by fissility and splitting along the bedding planes or laminations which defines the texture. Shales are often considered transversely isotropic with material properties in the directions parallel to the bedding being equal, but differing from the material properties measured perpendicular to the bedding and laminations. True 3D material anisotropy is rarely considered as a part of any constitutive behavior.

The consequence of this anisotropy for laboratory testing, is that the measured mechanical properties will depend upon the direction in which the sample is loaded. There are two aspects of the mechanical properties that need to be considered for these transversely anisotropic shales:

- Elastic anisotropy;
- Strength anisotropy.

The elastic anisotropy is usually investigated using dynamic velocity measurements on samples. Many studies have been performed illustrating this elastic anisotropy (Dewhurst & Siggins, 2004), demonstrating both sedimentary and stress induced velocity anisotropy.

Strength anisotropy is investigated through standard laboratory mechanical tests, typically UCS and triaxial tests, using samples plugged from the core at different orientations to the bedding, β (Figure 25). Samples with significant strength anisotropy show a significant reduction in strength when the applied maximum stress is orientated approximately 45° to the bedding plane. The use of this anisotropic strength variation

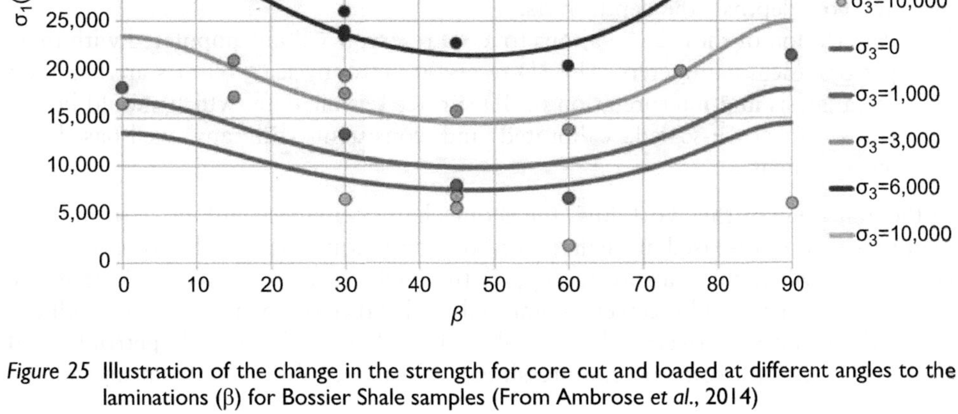

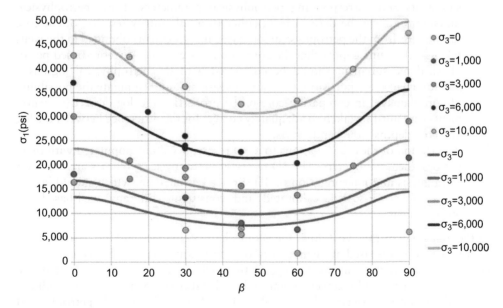

Figure 25 Illustration of the change in the strength for core cut and loaded at different angles to the laminations (β) for Bossier Shale samples (From Ambrose *et al.*, 2014)

is particularly important to model the failure around wellbores drilled through fissile shales.

Shale formations may be challenging to core and retrieve in an undisturbed state, unless the coring, retrieval, handling, storage and preservation and testing are not correctly controlled, and best practices adopted (Ewy, 2016). The core and samples need to be tested undrained to obtain strength parameters in a reasonable timeframe, but this is only of value if the effective stress state in the samples is known, requiring an understanding and control of the saturation, with representative pore pressure measurements during the tests.

10 APPLICATION OF TEST RESULTS FOR DESIGN

The mechanical characterization of formations ranging from unconsolidated, weakly consolidated through to well-cemented strong rocks has been described based on standard and specialized laboratory mechanical rock property tests. In addition, the challenges of obtaining undamaged, representative core samples from great depth, and under highly anisotropic horizontal stress conditions have been discussed. Despite the many challenges, the rock mechanics tests provide the only measurements of the static mechanical properties that are required for geomechanical analysis.

The limited volume of samples obtained from the reservoir and overburden upon which mechanical tests are performed, requires a workflow to apply these core-based mechanical property data to characterize the entire reservoir and overburden. This is achieved through the use of;

a) Correlations between the core-based laboratory mechanical rock property measurements and corresponding petrophysical parameters. These petrophysical measurements can be derived from either from laboratory tests on offcut core samples, or from the petrophysical log-based measurements obtained in the wells taken at the same depths that the core samples were cut. Any core-based petrophysical measurement should be compared with the petrophysical log-based measurement (*e.g.*, porosity), to ensure consistency, and eliminate the possibility of core damage. In comparing core-based properties and log-derived measurements care must be taken to adequately correct for the difference between the log and core depths – the depth shifts.
b) Application of these correlations to a 3D reservoir volume populated with these petrophysical parameters. The 3D reservoir petrophysical volumes are based on the analysis and interpretation of 3D surface seismic data, which should be up-scaled or down-scaled, calibrated and consistent with any well-based log measurements.

In the first stage of this workflow, the geomechanics engineer will move from using depth-specific core-based measurements to continuous log-based estimates. This conversion will include an averaging, as the well-log averages the rock response between the source and receivers, commonly a distance of 15cm along the wellbore wall. Differences between the core-based and well log-based petrophysical parameter, *e.g.*, porosity or compressional velocity, could result from any of the following:

- Averaging of the well-log measurement, particularly in finely laminated and thinly-bedded variable formations;
- Core damage during the coring, retrieval, rig handling and transportation;
- Wellbore effects such as overgauge hole, though these can be corrected for. But less known is damage and incipient breakouts resulting from microfractures in the wellbore wall, or fluid invasion;
- Fluid saturation differences in the core and the wellbore wall, which can affect the compressional velocity measurements (Gassman, 1951), and any fluid substitution 'Gassman' corrections applied to the well logs during normal processing and interpretation.

The upscaling from core data to log scale, may be subject to sample bias, and the use of limited samples. Recent application of scratch tests, that provide a continuous strength estimate at the cm scale along the core, show that integration of the continuous data to a well-log data scale (15 cm) can provide a reliable means to replicate the well-log response, and provide more reliable strength correlations (Figure 26).

The correlations between the laboratory strength measurements and the petrophysical parameter can then be applied to the field-wide 3D reservoir volume, following the appropriate upscaling of the log-based mechanical properties to the seismic scale.

Upscaling from the petrophysical scale of 15 cm depth resolution along the wellbore to surface seismic depth resolution of 10-20m depth is a standard practice for 3D reservoir characterization of petrophysical parameters, a but relatively recent development for geomechanical 3D characterization. The petrophysical upscaling of well-log data commonly 2 or 3 steps, depending upon the geological sub-divisions (or Zones in Figure 27), to the seismic scale of 10–20 m.

Upscaling of mechanical properties measured on core of 5 cm to 7.5 cm long to a 10m vertical depth interval, consistent with the seismic resolution, is not routine, and requires different averaging algorithms, for example for average strengths which may require an arithmetic average, whilst the Young's modulus calculated from average velocity measurements may require a harmonic average, or the use of Effective Media methods. Upscaling methodologies applied to 3D numerical modeling of reservoirs has recently been addressed by Qiu *et al.* (2005) and DeDontney *et al.* (2013) (Figure 28).

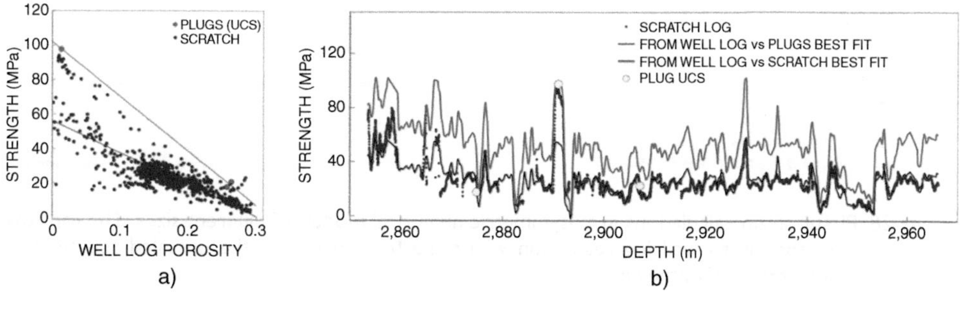

Figure 26 Upscaling from core measurements to log measurements, using scratch tests (from Germay & Lhomme, 2016)

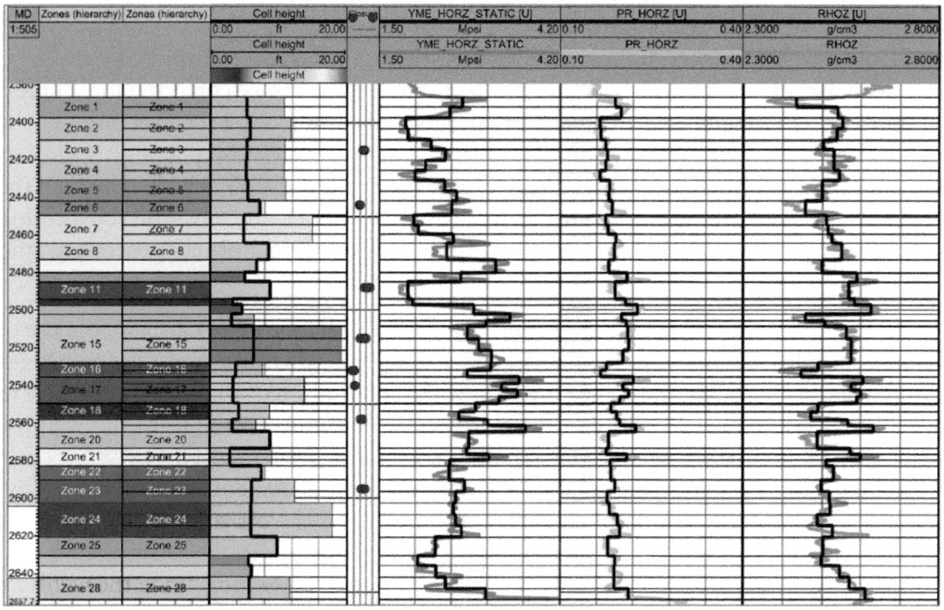

Figure 27 Example of petrophysical to seismic upscaling (Berard *et al.*, 2015)

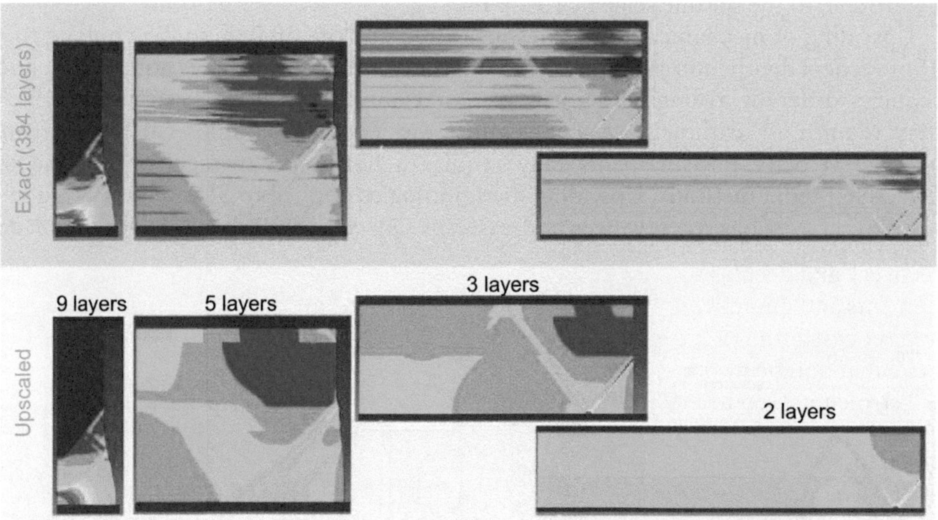

Figure 28 Plastic strain calculations in axisymmetric numerical models of different sizes: The top row corresponds to well log resolution whilst the bottom row is the upscaled seismic-type resolution (DeDontney *et al.*, 2013).

Upscaling in rock mechanics is well known, and has been evaluated for uniform or fractured formations (Goodman, 1989; Kulatilake, 1985), where decreased stiffness and strengths are observed and predicted for larger samples sizes. This results from the increased likelihood of cracks and flaws being present in samples with larger volumes.

Upscaling to obtain a representative 3D geomechanical reservoir volume based on the 3D seismic volume, which averages the velocity response over a 10 – 20 m depth interval, contains numerous lithologies with different velocities and mineralogies in the depth interval. As such, any field-wide upscaling should consider:

- The standard rock mechanics reduction of measured mechanical properties in the reservoir for large formation volumes, as well as;
- The impact of mechanical properties of the different formations and lithology which will comprise the seismic velocity interval measured by the surface seismic survey.

For analysis of well-based problems, this field-wide upscaling is less relevant, but prediction of wellbore failure or sand production still requires accurate correlations between the core-based strength and the petrophysical log-based measurements along the wellbore.

11 FUTURE TRENDS

The adoption of Petroleum Geomechanics Engineering as a component of field development of oil and gas industry since the late 1970s, has accompanied the increasing complexity of these developments. The changes in the field have seen a move from drilling vertical wells to inclined or extended reach and horizontal wells, with accompanying increasing completion complexity, most recently with multi-stage hydraulic fractures. As existing fields mature, the reservoir pressures commonly reduce leading to compacting fields and subsidence, and changes in the conductivity of naturally fractured reservoirs. Efforts to maintain the pressures require water, polymer and gas injection, Improved and Enhanced Oil Recovery, involving field-wide injection all of which have driven this uptake in geomechanics.

In line with the momentum for developing more complex reservoirs, increasing recovery and decreasing costs, the industry is looking towards more sustainable developments and minimizing HSE risks. In the future this focus may involve more injection and sequestration of greenhouse gasses, as well as re-developments of older fields, and field abandonment.

Petroleum Geomechanical Engineering issues which are now addressed at the research and early deployment stage include the use of more elaborate field dynamic simulators, combining traditional reservoir matrix depletion and injection pressure changes with the dynamic response of natural fractures, where natural fractures are modeled using representative fracture networks rather than through simplified dual porosity systems. The dynamic response of stress-sensitive natural fractures, result from pore pressures changes within an anisotropic in-situ stress regime, leading to 3D deformation maps and changes in natural fracture conductivities in response to both normal opening and shear-related fracture aperture changes (Barton, 1978).

The tendency to move to lower permeability or 'tight' formations for production, reflecting the developments of tight gas/shale gas and shale oil deposits in North

America, will increasingly address key technical aspects of the mechanical characteristics of these formations, notably:

- Effective stresses for partially saturated formations in tight low permeability formations;
- Improved reservoir characterization, modulus anisotropy and scale dependency;
- Uncertainty and probabilistic analysis of field-wide Petroleum Geomechanics modeling and analysis.

Given the large challenges of characterizing the sub-surface for geological and geomechanical properties, further integration between Geoscience, Well Engineering, Petroleum Engineering disciplines and Petroleum Geomechanical Engineering provides a means to increase the value derived from applying 'Geomechanics' to future field developments and re-developments.

12 CONCLUDING REMARKS ON LABORATORY TESTING FOR ROCK MECHANICAL PROPERTY DETERMINATION IN THE PETROLEUM INDUSTRY

This chapter has discussed the current practices for determining rock mechanical properties through laboratory testing for Petroleum Geomechanical Engineering applications to oil and gas field developments.

The chapter provides an overview of the current practices of mechanical testing of formations and lithologies encountered in the Petroleum industry, drawing on both rock mechanics best practice, captured in the ASTM standards and the ISRM suggested methods, as well as the best practices from soil mechanics in civil engineering. Given the large range of formations and mechanical properties encountered in the sub-surface by the oil and gas industry, the Petroleum Geomechanical Engineer is in a unique position to draw on the best practices from the different industry sectors.

Characterization of oil and gas reservoirs, the overburden and the surrounding formations, through the use of accurate testing provides the only 'ground truth' for any mechanical analysis. However, several obstacles have to be overcome, including the possible damage and de-saturation during core cutting and recovery, as well as the re-saturation with inert fluids. The impact of heterogeneity on the ability to derive mechanical properties from core materials has also been discussed along with the recommended practices.

Having derived the rock properties for 1 – 2" long cores, the application of these results to well-specific or field-wide analysis requires both translation from mechanical to dynamic properties, as well as upscaling to the appropriate length-scale for analysis and numerical modeling.

Petroleum Geomechanical Engineering began in earnest in the 1950's and has gradually become an integral part of the industry. The most recent adoption of Petroleum Geomechanics into field development has included identifying 'sweetspots' to drill lateral horizontal wells in 'brittle' formations as well as understanding the interaction of hydraulic fractures and natural fracture networks, in order to create an artificial permeability network to overcome the naturally very low permeability:

Petroleum Geomechanics has become a central discipline in understanding how to effectively drain these very tight formations.

As the complexity of field developments increases, the application of Petroleum Geomechanical Engineering to mechanically characterize the sub-surface will be a key step to ensure safe, economic oil and gas extraction.

ACKNOWLEDGEMENTS

I would like to acknowledge the support of Rockfield Software Limited in preparing this paper, and to thank David Milton-Tayler of FracTech and both Christophe Germay and Tanguy Lhomme of Epslog for their contributions. Finally, I would like to thank Russ Ewy of Chevron and Najwa Yassir for their valuable suggestions while reviewing and editing this chapter.

REFERENCES

Aadnøy, B.S. and Chenevert, M.E., 1987. Stability of highly inclined boreholes. SPE/IADC 16052. SPE/IADC Drilling Conference, New Orleans, La. USA. pp.25–41.

Abass, H.H., Tahini, A.M., Abousleiman, Y.N. and Khan, M., 2009. New Technique to Determine Biot Coefficient for Stress Sensitive Dual Porosity Reservoirs. SPE 124484. SPE Annual Technical Conference and Exhibition held in New Orleans, Louisiana, USA, 4–7 October 2009.

Addis, M.A., 1987. Mechanisms of sediment compaction responsible for oil field subsidence. Ph. D Thesis (Unpubl.), University of London.

Addis, M.A., 1989. The behavior and modelling of weak rocks. Proc. Int. Rock Mechanics Symp.: Rock at Great Depth, Pau, September 1989, V. Maury and D. Fourmaintraux (eds.), Vol. 2, pp.899–905.

Addis, M.A., Barton, N.R., Bandis, S.C. and Henri, J.P., 1990. Laboratory studies on the stability of vertical and deviated wellbores. SPE Paper 20406, 65th Annual SPE Conference, New Orleans, September 1990.

Addis, M.A., Gunningham, M.C., Brassart, Ph., Webers, J. Subhi, H., and Hother, J.A., 2008. Sand Quantification: The Impact on Sandface Completion Selection and Design, Facilities Design and Risk Evaluation. SPE 116713. 2008 SPE Annual Technical Conference and Exhibition, Denver, Colorado, USA, 21–24 September 2008.

Addis, M.A. and Wu, B., 1993. The Role of the Intermediate Principal Stress on Wellbore Stability Studies: Evidence from Hollow Cylinder Tests. *International Journal of Rock Mechanics and Mining Sciences and Geomechanics Abstracts*, 30, (7), pp.1027–1030.

Alam, M. M. and Fabricius, I.L., 2012. Effective stress coefficient for uniaxial strain condition. ARMA 12–302. 46th US Rock Mechanics / Geomechanics Symposium held in Chicago, IL, USA, 24–27 June 2012.

Al-Tahini, A.M., Abousleiman, Y.N., and Brumley, J.L., 2005. Acoustic and Quasistatic Laboratory Measurement and Calibration of the Pore Pressure Prediction Coefficient in the Poroelastic Theory. SPE 95825. SPE Annual Technical Conference and Exhibition held in Dallas, Texas, U.S.A., 9–12 October 2005.

Al-Tahini, A. and Abousleiman, Y., 2008. Pore Pressure Coefficient Anisotropy Measurements for Intrinsic and Induced Anisotropy in Sandstone. SPE 116129. SPE Annual Technical Conference and Exhibition, Denver, Colorado, USA. 21–24th September 2008.

Ambrose, J., Zimmerman, R.W. and Suarez-Rivera, R., 2014, Failure of Shales under Triaxial Compressive Stress. ARMA 14–7506. 48th US Rock Mechanics / Geomechanics Symposium held in Minneapolis, MN, USA, 1–4 June 2014.

Ameen, M.S., Smart, B.G.D., Somerville, Hammilton, S. and Naji, N.A., 2009. Predicting rock mechanical properties of carbonates from wireline logs (A case study: Arab-D reservoir, Ghawar field, Saudi Arabia). *Marine and Petroleum Geology, 26 (2009)*, pp.430–444.

American Petroleum Institute, 1998: *Recommended Practices for Core Analysis*, Recommended Practice 40, Second Edition, February 1998. API Publications and Distribution, API, 1220 L Street, N.W., Washington, D.C. 20005.

Anderson, E, M., 1905. The Dynamics of Faulting. *Transactions of the Edinburgh Geological Society*, 8, (1905), pp.387–402.

Anderson, E.M., 1951. *The Dynamics of Faulting and dyke formations with applications to Britain.* Publ. Oliver and Boyd.

Antheunis, D., Vriezen, P.B., Schipper, B.A. and van der Vlis, A.C., 1976. *Perforation Collapse: Failure of perforated friable sandstones.* SPE 5750. In: Proceedings SPE-European Meeting, Amsterdam, April 1976.

ASTM D2435/2435M, 2004. *Standard Test Methods for One-Dimensional Consolidation Properties of Soils Using Incremental Loading.* p. 15.

ASTM D2938-71a, 1979, *ASTM Book of Standards*, Standard Test Method for Unconfined Compressive Strength of Intact Rock Core Specimens, pp. 440–442.

ASTM D4543-08, 2008, *Standard Practices for Preparing Rock Core as Cylindrical Test Specimens and Verifying Conformance to Dimensional and Shape Tolerances.* p.9

ASTM D7012-14, 2014. *Standard Test Methods for Compressive Strength and Elastic Moduli of Intact Rock Core Specimens under Varying States of Stress and Temperatures.* p.9.

Aydan, O., Ito, T., Ozbay, U., Kwasniewski, M., Shariar, K., Okuno, T., Ozgenoglu, A, Malan, D.F. and Okada, T., 2015. ISRM Suggested Methods for Determining the Creep Characteristics of Rock. Publ., in R Ulusay (ed.), *The ISRM Suggested Methods for Rock Characterisation Testing and Monitoring: 2007–2014*, pp.115–130.

Barefield, E. and Shakoor, A. 2006. The Effect of Degree of Saturation on the Unconfined Compressive Strength of Selected Sandstones. IAEG 2006. 10th IAEG International Congress, Nottingham, United Kingdom, 6–10 September 2006. p.11.

Barton, N., *et al.*, 1978. Suggested Methods for the Quantitative Description of Discontinuities in Rock Masses. International Journal of Rock Mechanics and Mining Sciences and Geomechanics Abstracts, 15(6),pp.319–368.

Bauer, A., Christian Lehr, Frans Korndorffer, Arjan van der Linden, John Dudley, Tony Addis, Keith Love, and Michael Myers, M., 2008. *Stress and pore-pressure dependence of sound velocities in shales*: Poroelastic effects in time- lapse seismic, SEG Annual Meeting, Las vegas, pp.1630–1634.

Bell, J.S. and Gough, I., 1979. Northeast-southwest compressive stress in Alberta: evidence from oil wells. *Earth and Planetary Science Letters*, 45, pp.475–482.

Berard, T., Desroches, J., Yang, Y., Weng, X. and Olson, K., 2015. High-Resolution 3D Structural Geomechanics Modeling for Hydraulic Fracturing. SPE 173362. SPE Hydraulic Fracturing Conference, The Woodlands, Texas, 3–6th February 2015.

Bieniawski, Z.T. and Beneke, M.J., 1979, Suggested methods for determining the uniaxial compressive strength and deformability of rock materials. *International Journal of Rock Mechanics and Mining Sciences and Geomechanics Abstracts, 16 (2),April 1979*, pp.138–140.

Bieniawski, Z.T. and Hawkes, I., 1978. Suggested methods for determining tensile strength of rock materials. International Journal of Rock Mechanics and Mining Sciences and Geomechanics Abstracts, 15(3), pp.99–103.

Biot, M.A. 1956. General solutions of the equations of elasticity and consolidation for a porous material. *Journal of Applied Mechanics*, 78, pp.91–96.

Bishop, A.W. 1958. *Test requirements for measuring the coefficient of the earth pressure at rest.* In: Proceedings Brussels Conf. on Earth Pressure Problems, pp.2–14.

Bishop, A.W., 1959. The principle of effective stress. *Teknisk Ukeblad (October 1959)*, 106, (39), pp.859–863.

Bishop, A.W., 1973. The influence of an undrained change in stress on the pore pressure in porous media of low compressibility. *Geotechnique*, 23, pp. 435–442.

Bishop, A.W., 1974. The strength of crustal materials. *Engineering Geology*, 8, pp.139–153.

Bishop, A.W., 1976. The influence of system compressibility on the observed pore-pressure response to an undrained change in stress in saturated rock. *Geotechnique*, 26, pp. 371–375.

Bishop, A.W. and Henkel, D.J. 1982. *The measurement of soil properties in the triaxial test.* 2nd Edition, Edward Arnold, ISBN: 0 7131 3004 0.

Bishop, A.W., Kumapley, N.K. and El-Ruwayih, A., 1975. The influence of pore-water tension on the strength of clays. Philosophical Transactions of the Royal Society of London A: Mathematical, *Physical and Engineering Science*, 278, pp.511–544.

Bishop, A.W. and Wesley, L.D., 1975. A hydraulic triaxial apparatus for controlled stress path testing. Geotechnique, 25(4),pp.657–670.

Blanton, T.L. and Teufel, L.W., 1986. A Critical Evaluation of Recovery Deformation Measurements as a Means of In-Situ Stress Determination. SPE 15217. Unconventional Gas Technology Symposium of the Society of Petroleum Engineers held m Louisville, KY. 18–21 May 1986.

Blumel, M., 2010. Comparison of single and multiple failure triaxial tests. ISRM Eurock 2009-035. Rock Engineering in Difficult Ground Conditions – Soft Rocks and Karst, VRKljan (ed.), pp.239–242.

Bradley, W.B., 1979. Mathematical stress cloud – stress cloud can predict borehole failure. Oil & Gas Journal, 77(8), February., pp.92–102.

Brignoli, M, Papamichos,. and Santarelli, F.J., 1995. Capillary effects in sedimentary rocks: application to reservoir water-flooding. In: Proceedings of the 35th U.S. Symposium, 1995, pp.619–625.

Brignoli, M., Fanual, P., Holt, R.M. and Kenter, C.J., 1998. Effects on Core Quality of a Bias Stress Applied During Coring. SPE/ISRM 47262, SPE/ISRM Eurock '98, Conference, Trondheim, Norway, 8–10 July 1998.

Brooker, E.W. and Ireland, H.O., 1965. Earth pressure at rest related to stress history. *Canadian Geotechnical Journal,2(1), February 1965*, pp.1–15.

Brown, E.T., 1981, *Rock Characterization*, Testing and Monitoring: ISRM Suggested Methods. Pergammon Press.

Burland, J.B., 1990. On the compressibility and shear strength of natural clays. Geotechnique, 40, No. 3, pp. 329–378.

Cain, P., Yuen, C. M. K., Le Bel, G. R., Crawford, A. M., and Lau, D. H. C., 1987. "Triaxlal Testing of Brittle Sandstone Using a Multiple Failure State Method," Geotechnical Testing Journal, GTJODJ, 10(4), December 1987, pp.213–217.

Charlez, Ph. A., 1991. *Rock Mechanics: Theoretical fundamentals*, Volume 1. Editions TECHNIP, p.333.

Cheatham, J.B. and McEver, J.W., 1964. Behavior of Casing Subjected to Salt Loading. SPE 828. *Journal of Petroleum Technology, September* 1964, pp.1069–1075.

Chertov, M.A. and Suarez-Rivera, R., 2014. Practical Laboratory Methods for Pore Volume Compressibility Characterization in Different Rock Types. ARMA 14-7532. 48th US Rock Mechanics / Geomechanics Symposium held in Minneapolis, MN, USA, 1–4 June 2014.

Collins, P.M., 2007. Geomechanical Effects on the SAGD Process. SPE 97905. *SPE Reservoir Evaluation & Engineering, August* 2007. pp.367–375.

Crawford, A.M. and Wylie, D.A., 1987. A modified multiple failure state triaxial testing method. 28th US Symposium on Rock Mechanics, Tucson, 9 June–1 July 1987. pp.133–140.

Crawford, B.R., Gooch, M.J. and Webb, D.W., 2004. Textural Controls on Constitutive Behavior in Unconsolidated Sands: Micromechanics and Cap Plasticity. ARMA/NARMS 04–611. Gulf Rocks 2004. 6th North American Rock Mechanics Symposium (NARMS): Rock Mechanics Across Boders and Disciplines, Houston, Texas, 5–9th June 2004.

Crawford, B.R., Gaillot, P.J. and Alramahi, B., 2010. Petrophysical Methodology for Predicting Compressive Strength in Siliciclastic "sandstone-to-shale" Rocks. ARMA 10–196. 44 US Rock Mechanics Symposium and 5 U.S.-Canada Rock Mechanics Symposium, Salt Lake City, UT 27–30 June 2010.

Crawford, B.R., DeDontney, N.L., Alramahi, B. and Ottesen, S., 2012. Shear Strength Anisotropy in Fine-grained Rocks. ARMA 12–290. 46th US Rock Mechanics / Geomechanics Symposium held in Chicago, IL, USA, 24–27 June 2012

DeDontney, N., Crawford, B. and Alramahi, B., 2013. Generating Mechanical Stratigraphy in Layered Rock Masses Using Numerical Averaging of Cohesive-frictional Strength. ARMA 13–332. 47th US Rock Mechanics / Geomechanics Symposium held in San Francisco, CA, USA, 23–26 June 2013.

Detournay, E. and A.H.-D. Cheng. 1993. *"Fundamentals of poroelasticity", in Comprehensive Rock Engineering: Principles*, Practice and Projects, Vol. II, Analysis and Design Method, ed. C. Fairhurst, Pergamon Press, pp.113–171.

Detournay, E., Cheng, A.H.-D., Roegiers, J.-C. and McLennan, J.D., 1989. Poroelasticity considerations in In Situ stress determination by hydraulic fracturing. *International Journal of* Rock Mechanics and Mining Sciences and Geomechanics Abstracts. 26(6), pp.507–513, 1989.

De Waal, J.A., 1986. On the rate type compaction behavior of sandstone reservoir rock. Ph.D. Thesis Delft Technical University.

De Waal, J.A., Smits, R.M.M., 1988. *Prediction of Reservoir Compaction and Surface Subsidence: Field Application of a New Model.* SPE 14214. SPE Formation Evaluation, June.

Dewhurst, D.N. and Siggins, A.F., 2004. Impact of stress and sedimentary anisotropies on velocity anisotropy in shale. SEG-2004–1650. SEG Intnl. Exposition and 74th Annual Meeting, Denver, Colorado, 10–15 October, 2004.

Dudley, J.W., Brignoli, M., Crawford, B. R., Ewy, R. T., Love, D. K., McLennan, J. D., Ramos, G. G. Shafer, J. L., Sharf-Aldin, M. H., Siebrits, E., Boyer, J., and Chertov, M. A., 2016. Suggested methods for uniaxial-strain compressibility testing for reservoir geomechanics. *Rock Mechanics and Rock Engineering.* 49(10), pp. 4153–4178.

Dudley, J. W., Myers, M. T., Shew, R. D., and Arasteh, M. M., 1998, "Measuring Compaction and Compressibilities in Unconsolidated Reservoir Materials via Time-Scaling Creep", SPE 51324, SPE Reservoir Evaluation & Engineering, October.

Duncan-Fama, M.E. and Brown, E.T., 1989. *Influence of stress dependent elastic moduli on plane strain solutions for boreholes.* SPE/ISRM Rock at Great Depth, Maury & Fourmaintraux (eds.). pp. 819–826.

Dusseault, M.B. and van Domselaar, H.R. 1982. Unconsolidated sand sampling in Canadian and Venzuelan oil sands. In: Proceedings 2nd UNITAR Conf., pp. 336–348.

Dusseault, M.B., Maury, V., Sanfilippo, F. and Santarelli, F.J., 2004. Drilling Through Salt: Constitutive Behavior and Drilling Strategies. ARMA/NARMS 04–608. Gulf Rocks 2004. 6th North American Rock Mechanics Symposium (NARMS): Rock Mechanics across Borders and Disciplines, Houston, Texas, 5–9 June 2004.

Dusseault M.B., Mraz, D., Unrau, J. and Fordham, C., 1985. Test Procedures for Salt Rock. ARMA-85–0313. 26th US Symposium on Rock Mechanics, Rapid City, SD, 26–28 June 1985. pp. 313–319.

Dyke C G, 1988. In situ stress indicators for rock at great depth. Ph.D Thesis, Univ. of London.

Dyke, C.G., 1989. Core Discing: Its Potential as an Indicator of Principal In Situ Stress Directions. SPE/ISRM Rock at Great Depth, Maury & Fourmaintraux (eds), pp. 1057–1064.

Dyke, C.G. and Dobreiner, L., 1991. Evaluating the strength and deformability of sandstones. *Quarterly Journal of Engineering Geology and Hydrogeology*, 24, pp.123–134.

Economides, M. J. and Martin, T., 2007. *Modern Fracturing, Enhancing Natural Gas Production*, (hardbound) Energy Tribune Publishing, Houston, 2007.

Elliot, G.M. and Brown, E.T., 1986. Further Development of a Plasticity Approach to Yield in Porous Rock. International Journal of Rock Mechanics and Mining Sciences and Geomechanics Abstracts, 23(2), pp. 151–156, 1986.

Ewy, R.T., 1999. Wellbore-Stability Predictions by Use of a Modified Lade Criterion. *SPE Drilling & Completions J.*, 14(2), June 1999, pp.85–91.

Ewy, R.T., 2015. Shale/claystone response to air and liquid exposure, *and implications for handling, sampling and testing. International Journal of Rock Mechanics and Mining Sciences*, 80 (2015), pp. 388–401.

Ewy, R. T. and Cook, N. G. W., 1990. Deformation and fracture around cylindrical openings in rock – I. *Observations and analysis of deformations. International Journal of Rock Mechanics and Mining Sciences and Geomechanics Abstracts*, 27:387–408.

Fjaer, E., Holt, R.M., Horsrud, P., Raaen, A.M., and Risnes, R. 2008. *Petroleum Related Rock Mechanics*. 2nd Edition, Elsevier. ISBN: 978–0-444–50260-5 (0376–7361)

Fokker, P.A. and Kenter, C.J., 1994. The micro mechanical description of rocksalt plasticity. SPE 28117. Eurock SPE/ISRM Rock Mechanics in Petroleum Engineering Conference, Delft, The Netherlands, 29–31 August 1994. pp. 705–713.

Fossum, A. F. and Fredrich, J. T., 2002. *Salt Mechanics Primer for Near-Salt and Sub-Salt Deepwater Gulf of Mexico Field Developments*. Sandia Report, SAND2002–2063 Unlimited Release, July 2002, p. 67.

Fowell, R.J., Hudson, J.A., Xu, C., Chen, J.F. and Zhao, X. 1995. Suggested Method for Determining Mode I Fracture Toughness Using Cracked Chevron Notched Brazilian Disc (CCNBD) Specimens. International Journal of Rock Mechanics and Mining Sciences and Geomechanics Abstracts, 22(1), pp. 57–64.

Fuenkajorn, K. and Daemen, J.J.K., 1988. Borehole Closure in Salt. Key Questions in Rock Mechanics, Cundall *et al.* (eds.). In Proceedings 29th U.S. Symposium on Rock Mechanics (USRMS), 13–15 June, Minneapolis, Minnesota. pp. 191–198.

Gassmann, F., 1951, Uber die Elastizitat poroser Medien: *Veirteljahrsschrift der Naturforschenden Gesellschaft in Zurich*, 96, 1–23.

Geertsma, J., 1966. Problems of Rock Mechanics in Petroleum Production Engineering. *ISRM Congress, 1966–099*, pp.585–594.

Geertsma, J., 1957. The Effect of Fluid Pressure Decline on Volumetric Changes of Porous Rocks. SPE 728. Petroleum Transactions J, AIME, pp.331–340.

Geertsma, J., 1985. Some Rock-Mechanics Aspects of Oil and Gas Completions. SPE Journal, *December* 1985, pp.848–856.

Germay, C. and Lhomme, T., 2016. Upscaling of Rock Mechanical Properties Measured on Plug Samples. *Epslog Technical Note*, pp.17.

Ghalambor, A., Ali, S., and Norman, W.D., (Eds.), 2009. Frac Packing Handbook. Society of Petroleum Engineers, ISBN:978–1-55563–137-6.

Goodman, R.E., 1989. *Introduction to Rock Mechanics*, 2nd ed., John Wiley & sons.

Haimson, B.C. and Herrick, C.G., 1986. Borehole breakouts – a new tool for estimating in situ stress? Proc. Int. Symp. Rock Stress Meas., Stockholm, 1-3rd September, 1986, pp.271–280.

Hallam, S.D. and Last, N.C., 1991. Geometry of Hydraulic Fractures From Modestly Deviated Wellbores. SPE 20656. *Journal of Petroleum Technology, June* 1991, pp.742–748.

Halleck, P.M. and Behrmann, L.A., 1990. Penetration of shaped charges in stressed rock. Rock Mechanics Contributions and Challenges, Hustrulid & Johnson (eds.), pp.629–636.

Head, K.H., 1998. *Manual of soil laboratory testing. Effective Stress Tests. Vol. 3*, Chichester, UK: John Wiley & Sons, 1998.

Herrick, C.G. and Haimson, B.C. 1994. Modelling of episodic failure leading to borehole breakouts in Alabama limestone. ARMA 1994–0217, Rock Mechanics, Nelson and Lauback (eds.) 1994, pp.217–224.

Herwanger, J. and Koutsabeloulis, N., 2011. Seismic Geomechanics: How to Build and Calibrate Geomechanical Models Using 3D and 4D Seismic Data. EAGE Publications. ISBN 978–90-73834–10-1.

Hettema, M.H.H., Hanssen, T.H. and B.L. Jones, B.L., 2002. Minimizing Coring-Induced Damage in Consolidated Rock, SPE/ISRM 78156, SPE/ISRM Rock Mechanics Conference held in Irving, Texas, 20–23 October 2002.

Holt, R.M., Brignoli, M., Fjaer, E., Unander, T.E. and Kenter, C.J., 1994. Core Damage effects on compaction behavior. SPE 28027. Eurock SPE/ISRM Rock mechanics in Petroleum Engineering Conference, Delft, The Netherlands, 29–31st August 1994. pp.55–62.

Holt, R.M., Brignoli, M., Kenter, C.J., Meij, R. and Schutjens, P.M.T.M., 1998. From core compaction to reservoir compaction: Correction for core damage effects. SPE/ISRM Eurock '98, Trondheim, Norway. 8-10th July 1998. pp.311–320.

Holt, R.M., Fjaer, E., Nes, O.-M. and Alassi, H.T., 2001. A Shaly look at Brittleness. ARMA 11–366. 45th US Rock mechanics / Geomechanics Symposium, San Francisco, CA, 26–29th June 2011.

Holt, R.M., Bakk, A., Fjaer, E. and Stenebraten, J.F., 2005. Stress Sensitivity of Wave Velocities in Shale. SEG Houston 2005 Annual Meeting. pp.1593–1598.

Hottman, C.E., Smith, J.H. and Purcell, W.R., 1979. Relationship Among Earth Stresses, Pore Pressure, and Drilling Problems Offshore Gulf of Alaska. SPE 7501. Journal of Petroleum Technology, 31, Issue 11, November 1979. pp. 1477–1484.

ISRM, 1981. *Rock Characterization, Testing and Monitoring: ISRM Suggested Methods*, Ed. E. T. Brown, IRSM, Pergammon Press.

ISRM, 1983, Suggested Methods for Determining the Strength of Rock materials in Triaxial Compression. Coordinator, J. Franklin. *International Journal of Rock Mechanics and Mining Sciences & Geomechanics Abstracts*, 20(6), pp. 285–290. (Reprinted in ISRM, 2007, The Complete ISRM Suggested Methods for Rock Characterization, Testing and Monitoring: 1974–2006, Eds. R. Ulusay and J.A. Hudson, IRSM.)

ISRM, 2007. *The Complete ISRM Suggested Methods for Rock Characterization, Testing and Monitoring: 1974–2006, Ed. R. Ulusay, and Hudson*, J.A., IRSM.

ISRM, 2015. The Complete ISRM Suggested Methods for Rock Characterization, Testing and Monitoring: 2007–2014, Ed. R. Ulusay, *IRSM*.

Jones, M.E., Leddra, M.J. and Addis, M.A., 1987. *Reservoir compaction and surface subsidence due to hydrocarbon extraction*. Offshore Technology Report, OTH 87 276. Department of Energy, HMSO, 1987.

Jones, M.E. and Leddra, M.J., 1989. *Compaction and flow of porous rocks at depth*. In: Proceedings Int. Rock Mechanics Symp.: Rock at Great Depth, Pau, September 1989, V. Maury and D. Fourmaintraux (eds.), Vol. 2, pp. 891–898.

Kenter, C.J., Brignoli, M. and Holt, R.M. 1997. CMS (constant mean stress) VS. UCS (unconfined strength) tests: A tool to reduce core damage effects. International Journal of Rock Mechanics and Mining Sciences, 34(3–4), April–June 1997, paper No. 129. p.11.

Khaksar, A., Taylor, P.G., Fang, Z., Kayes, T., Salazar, A., Rahman, K., 2009. Rock Strength from Core and Logs: Where We Stand and Ways to Go. SPE 121972. SPE Europe/

EAGE Annual; Conference and Exhibition, Amsterdam, The Netherlands, 8-11th June 2009.

Kim, M. M. and Ko, H. Y., 1979. "Multistage Triaxial Testing of Rocks," Geotechnical Testing Journal GTJODJ, 2 (2), June 1979, pp. 98–105.

Kooijman, A.P., van den Elzen, M.G.A, and Veeken, C.A.M, 1991. Hollow-cylinder collapse: Measurement of deformation and failure in an X-ray CT scanner, observation of size effect. US Rock Mechanics Symposium: Rock Mechanics as a Multidisciplinary Science (Roegiers (ed.)), pp. 657–666.

Kooijman, A.P., Halleck, P.M., de Bree, Ph., Veeken, C.A.M. and Kenter, C.J. 1992. Large-scale Laboratory Sand Produciton Test. SPE 24798. 67th Annual Technical Conference & Exhibition, Washington, DC, 4th October 1992.

Kovari, K., Tisa, A., Einstein, H. H., and Franklin, J. A., 1983. "'*Suggested Methods for Determining the Strength of Rock Materials in Triaxial Compression, International Journal of Rock Mechanics, Mineral Science* and Geomechanics Abstracts. 20(6), 1983, pp. 283–290 (revised version).

Kulatilake, P.H.S.W., 1985. Estimating elastic constants and strength of discontinuous rock. *Journal of geotechnical engineering, ASCE, 111 (7),July 1985*, pp.847–863.

Kulhawy, F.H., 1975. Stress deformation properties rock and rock discontinuities. *Engng. Geol.*, 9, pp.327–350.

Kutter, H.K., 1991. *Infuence of drilling method on borehole breakouts and core disking.* 7th ISRM Congress, 16–20 September, Aachen, Germany, pp. 1659–1664.

Kuruppu, M.D., Obara, Y., Ayatollahi, M.R., Chong, K.P. and Funatsu, T, 2015. *ISRM-Suggested Method for Determining the Mode I Static Fracture Toughness Using Semi-Circular Bend Specimen.* Publ., in R Ulusay (ed.), The ISRM Suggested Methods for Rock Characterisation Testing and Monitoring: 2007–2014, pp. 107–114.

Leroueil, S. and Vaughan, P.R., 1990. The general and congruent effects of structure in natural soils and weak rocks. Geotechnique, 40, No. 3, pp. 467–488.

Maury, V.M., Santarelli, F.J. and Henry, J.P., 1988. *Core Discing: A Review. In: Proceedings 1st African Conf. on Rock Mech.*, Swaziland.

Maury, V., Piau, J.-M. and Halle G., 1996. Subsidence Induced by Water Injection in Water Sensitive Reservoir Rocks : The Example of Ekofisk. SPE 36890. 996 SPE European Petroleum Conference, Milan Italy 22–24 October 1996, pp. 153–169.

McLean, M.R., 1987. Wellbore stability analysis. Ph.D. Thesis (Unpublished), Univ. of London.

McLean, M.R. and Addis, M.A. 1990a. Wellbore stability: A review of current methods of analysis and their field application. IADC/SPE Paper 19941; IADC/SPE Drilling Conference, Houston, February 1990.

McLean, M.R. and Addis, M.A. 1990b. Wellbore stability: the effect of strength criteria on mud weight recommendations. SPE Paper 20405, 65th Annual SPE Conference, New Orleans, September 1990.

Meng, C. and de Pater, C.J., 2010. Hydraulic Fracture propagation in pre-fractured natural rocks. ARMA 10–318. 44th US Rock Mechanics Symposium and 5th Canadian Rock Mechanics Symposium, Salt Lake City, 27–30th June 2010.

Minaeian, V., 2014. True Triaxial Testing of Sandstones and Shales. Ph.D Thesis (unpublished), Curtin University.

Morita, N., 1994. Field and Laboratory Verification of Sand-Production Prediction Models. SPE Drilling and Completion J., *December* 1994. pp.227–235.

Morita, N., Black, A.D. and Fuh, G-F., 1990. Theory of Lost Circulation Pressure. SPE 20409. 65th SPE Annual Technical Conference and Exhibition, New Orleans, LA, September 23–28th, 1990.

Morita, N., Whitfill, D.L., Massie, I. and Knudsen, T.W., 1987. Realistic Sand Production Prediction: Numerical Approach. SPE 16989. 62nd SPE Annual Technical Conference and Exhibition, Dallas, TX, September 27–30th, 1987.

Nagel, N.B. and McLennan, J.D., 2010. Solids Injection (SPE Monograph Series, 24). *Society of Petroleum Engineers. ISBN-10*: 1555632564

Noufal, A., Germay, C., Lhomme, T., Hegazy, G., and Richard, T., 2015. Enhanced Core Analysis Workflow for the Geomechanical Characterization of Reservoirs in a Giant Offshore Field, Abu Dhabi. SPE-177520. Abu Dhabi International Petroleum Exhibition and Conference held in Abu Dhabi, UAE, 9–12 November 2015. p. 23.

Nunes, A. L. L., 1989. Um Estodo Sobre as Caracteristicas de Resistencia e Deformabilidade de Arenitos. MSc Thesis, Pontificia Universidade Catrlica do Rio de Janeiro.

Ouchterlony, F., Franklin, J.A., Zongqi, S., Rummel, F., Muller, W., Nishimatsu, H., Atkinson, B.K., Meredith, P.G., Costin, L.R., Ingraffea, A.R. and Bobrov, G.F., 1988. Suggested Methods for Determining the Fracture Toughness of Rock. International Journal of Rock Mechanics and Mining Sciences and Geomechanics Abstracts, 25(2), pp. 71–96.

Papadopoulos, J.M., Narendran, V.M. and Cleary, M.P., 1983. Laboratory simulations of Hydraulic Fracturing. SPE/DOE 11618. SPE/DOE Symposium on Low Permeability, Denver, Colorado, 14–16th March, 1983, pp.161–174.

Plumb, R.A., 1994. Influence of Composition and Texture on the Failure Properties of Clastic Rocks. SPE 28022. 1994 Eurock SPE/ISRM Rock Mechanics in Petroleum Engineering Conference, Delft, The Netherlands, 29–31 August 1994.

Qiu, K., Marsden, J.R., Solovyov, Y., Safdar, M. and Chardac, O., 2005. Downscaling Geomechanics Data for Thin Beds Using Petrophysical Techniques. SPE 93605. 14th SPE Middle East Oil & Gas Show and Conference, Bahrain, 12–15th March 2005.

Raaen, A. M., 1996. Efficient Determination of the Parameters of an Elastoplastic Model. SPE/ISRM 47362. SPE/ISRM 47238. SPE/ISRM Eurock '98, Conference, Trondheim, Norway, 8–10 July 1998. pp. 277–283.

Rahman, K., Khaksar, A. and Kayes, T., 2010. An Integrated Geomechanical and Passive Sand-Control Approach to Minimizing Sanding Risk From Openhole and Cased-and-Perforated Wells. SPE Drilling and Completion J., *June* 2010, pp.155–167.

Rao, K.S., Venkatappa Rao, G. and Ramamurthy, T., 1987. Strength of sandstones in saturated and partially saturated conditions. Geotechnical Engineering, 18(1987), pp.99–127.

Rice, J.R. and Cleary, M.P., 1976. Some Basic Stress Diffusion Solutions for Fluid-Saturated Elastic Porous Media with Compressible Constituents. Rev. Geophys. Space Physics, 14(2), May 1976, pp. 227–241.

Ringstad, C., Addis, M.A., Brevik, I., and Santarelli, F.J. 1993. *Scale effects in hollow cylinder tests*. Scale Effects in Rock Masses, ed. Pinto da Cunha, Balkema.

Santarelli, F.J., 1987. Theoretical and experimental investigation of the stability of the axisymmetric wellbore. Ph.D. Thesis, Univ. of London.

Santarelli, F.J. and Brown, E.T. 1987. Performance of deep wellbores in rock with a confining pressure-dependent elastic modulus. ISRM Congress, paper 225. pp. 1217–1222.

Santarelli, F.J. and Carminati, S., 1995. Do Shales Swell? A Critical Review of available Evidence. SPE/IADC 29421. SPE/IADC Drilling Conference, Amsterdam, 26 February–2 March 1995.

Santarelli, F.J. and Dusseault, M.B., 1991. Core Quality Control in Petroleum Engineering, US Rock Mechanics Symposium. ARMA-91-111. Rock Mechanics as a Multidisciplinary Science, Roegiers (ed.), pp. 111–120.

Sharf-Aldin, M., Rosen, R., Narasimhan, S. and PaiAngle, M., 2013. Experience Using a Novel 45 Degree Transducer to Develop a General Unconventional Shale Geomechanical Model.

ARMA 13-317. 47th US Rock Mechanics / Geomechanics Symposium, San Francisco, CA, USA. 23–26th June 2013.

Sharma, M.S.R., Baxter, C.D.P, Moran, K., Vaziri, H., and Narayanasamy, R., 2011. Strength of Weakly Cemented Sands from Drained Multistage Triaxial Tests. Journal of Geotechnical and Geoenvironmental Engineering, *December* 2011, pp. 1202–1210.

Sarker, R. and Batzle, M., 2010. *Anisotropic elastic moduli of the Mancos B Shale – An experimental study*. SEG Denver 2010 Annual Meeting. pp. 2600–2605.

Skempton, A. W., 1954, *The pore-pressure coefficients A and B: Geotechnique* 4, 143–147

Taheri, A. and Tani, K., 2008. Proposal of a New Multiple-Step Loading Triaxial Compression Testing Method. ISRM International Symposium 2008, 5th Asian Rock Mechanics Symposium (ARMS5), 24–26 November 2008 Tehran, Iran. pp. 517–524.

Taheri, A. & Chanda, E., 2013. A new multiple-step loading triaxial test method for brittle rocks. Proc. 19th NZGS Geotechnical Symposium. Chin (ed.), Queenstown, NZ. pp.1–8.

Teeuw, D., 1971. Prediction of Formation Compaction from Laboratory Compressibility Data. SPE 2973. *Journal of Petroleum Technology, September* 1971. pp.263–271.

Terzaghi, K., 1943. *Theoretical Soil Mechanics*, John Wiley and Sons, New York. ISBN 0-471-85305-4.

Teufel, L. W., 1982. Prediction of Hydraulic Fracture Azimuth from Anelastic Strain Recovery Measurements of Oriented Core. 23rd U.S Symposium on Rock Mechanics (USRMS). 25–27 August 1982, Berkeley, California, ARMA-82-238, pp. 238–245.

Teufel, L. W., 1983. Determination of In-Situ Stress From Anelastic Strain Recovery Measurements of Oriented Core. SPE/DOE 11649. 1983 SPE/DOE Symposium on Low Permeability held in Denver, Colorado, March 14–16, 1983,

Torsaeter, O., and Beldring, B., 1987. The effect of Freezing of Slightly Consolidated Cores. SPE Formation Evaluation J., *September* 1987. pp.357–360.

Van den Hoek, P.J., Smit, D.-J., Kooijman, A.P., de Bree, Ph., and Kenter, C.J., 1994. Size Dependency of hollow-cylinder stability. SPE 28051. SPE/ISRM Rock Mechanics in Petroleum Engineering Conference, Delft, The Netherlands. 29–31st August 1994. pp.191–198.

Vaziri, H., Xiao, Y., and Palmer, I., 2002. Assessment of several sand prediction models with particular reference to HPHT wells. SPE/ISRM 78235. SPE/ISRM Rock Mechanics Conference, Irving, Tx, 20–23rd October 2002.

Veeken, C.A.M., Davies, D.R., Kenter, C.J. and Kooijman, A.P., 1991. Sand Production Prediction Review: Developing an Integrated Approach. SPE 22792. 68th SPE Annual Technical Conference and Exhibition, Dallas, TX., 5–9th October 1991.

Venet, V., Henry, J.P., Santarelli, F., Maury, V., 1989. Estimation of stresses at depth by core discing analysis (in French). SPE/ISRM Rock at Great Depth, Maury & Fourmaintraux (eds.). pp.1551–1557.

Warpinskl, N.R. and Teufel, L.W., 1989. In-Situ Stresses in Low-Permeability, *Nonmarine Rocks. Journal of Petroleum Technology, April* 1989. pp.405–414.

Warpinski, N.R, Abou-Sayed, I. S., Moschovidis, Z. and Parker, C., 1993. Hydraulic Fracture Model Comparison Study: Complete Results. GRI Report GRI-93-0109, Sandia Report 93-7042, February 1993, pp. 180.

Weijers, L. and de Pater, C.J., 1992. Interaction and link-up of hydraulic fractures close to a perforated wellbore. SPE 28077. SPE/ISRM Rock Mechanics in Petroleum Engineering Conference, Delft, The Netherlands, 29–31st August 1994, pp.409–416.

Weijers, L. and de Pater, C.J., 1994. *Fracture Reorientation in Model Tests*. SPE 23790. SPE Intl. Symp. Formation Damage, Lafayette, Louisiana, 26–27th February 1992, pp.259–266.

West, G, 1994. Effect of Suction on the Strength of Rock. *Quarterly Journal of Engineering Geology and Hydrogeology*, 27, pp. 51–56.

Willson, S.M., Moschovidiv, Z.A., Cameron, J.R. and Palmer, I.D., 2002. New Model for Predicting the Rate of Sand Production. SPE/ISRM 78168. SPE/ISRM Rock Mechanics Conference, Irving, Texas, 20–23rd October 2002.

Willson, S.M., Fossum, A.F. and Fredrich, J.T., 2003. Assessment of Salt Loading on Well Casings. SPE 81820. *SPE Drilling and Completion, March* 2003. pp. 13–21.

Willson, S.M. and Fredrich, J.T., 2005. Geomechanics Considerations for Through- and Near-Salt Well Design. SPE 95621. SPE Annual Technical Conference and Exhibition held in Dallas, Texas, U.S.A., 9–12 October 2005.

Wissa, A.E.Z., 1969. Pore pressure measurement in saturated stiff soils. J.Soils Mech., *Found. Div., In: Proceedings ASCE*, Vol. 95, SM4, pp. 1063–1073.

Wold, M.B., Davidson, S.C., Wu., B., Choi, S.K. and Koenig, R.A., 1995. Cavity Completion for Coalbed Methane Stimulation – An Integrated Investigation and Trial in The Bowen Basin, Queensland. SPE 30733. SPE Annual Technical Conference and Exhibition, Dallas, TX., 22–25th October 1995, pp.323–337.

Wu, B., 2001. Biot's effective stress coefficient evaluation: static and dynamic approaches. ARMA Conference 2001, Paper 082, pp. 369–372.

Wu, B. and J.A. Hudson, J.A., 1991. Stress-induced anisotropy in rock and its influence on wellbore stability. Rock Mechanics as a Multi-Disciplinary Science, Roegiers (ed.), In: Proceedings 32nd US Rock Mechanics Symp., Norman, OK., 10–12 July 1991. pp. 941–950.

Yassir, N.A., 1989. Undrained shear characteristics of clay at high total stresses. In: Proceedings. Int. Rock Mechanics Symp.: Rock at Great Depth, Pau, September 1989, V. Maury and D. Fourmaintraux (eds.), Vol. 2, pp. 907–913.

Yassir, N., Wang, D.-F., Enever J.R., and Davies, P.J., 1998. An Experimental Analysis of Anelastic Strain Recovery of Synthetic Sandstone Subjected to Polyaxial Stress. SPE/ISRM 47238. SPE/ISRM Eurock '98, Conference, Trondheim, Norway, 8–10 July 1998.

Youn, H. and Tonon, F., *Multi-stage triaxial test on brittle rock. International Journal of Rock Mechanics and Mining Sciences*, 47 (2010), pp. 678–684.

Zimmerman, R.W., 2000. Implications of static Poroelasticity for reservoir compaction. Pacific Rocks Conference, Girard, Breeds and Doe (eds.), pp.169–172.

Chapter 16

Four critical issues for successful hydraulic fracturing applications

A.P. Bunger[1,2] *& B. Lecampion*[3]
[1]*Department of Civil and Environmental Engineering, University of Pittsburgh, Pittsburgh, PA, USA*
[2]*Department of Chemical and Petroleum Engineering, University of Pittsburgh, Pittsburgh, PA, USA*
[3]*Geo-Energy Laboratory – Gaznat Chair on Geo-Energy, EPFL-ENAC-IIC-GEL, École Polytechnique Fédérale de Lausanne, Lausanne, Switzerland*

Abstract: This paper reviews four critical mechanisms for successful applications of hydraulic fracturing that have emerged or grown in importance over the past 2 decades. These critical issues are managing height growth, decreasing near-wellbore tortuosity, predicting and engineering network versus localized growth geometry, and promoting simultaneous growth of multiple hydraulic fractures. Building on the foundation of decades of research relevant to each area but with an emphasis on advances within the past 20 years, the review presents available field evidence, laboratory data, and modeling insights that comprise the current state of knowledge. Fluid viscosity, injection rate, and in situ stresses are shown to appear as controlling parameters across all of these issues. The past contributions and limitations on future developments point to a future in which advances are enabled by a combination of fully coupled 3D simulations, carefully designed laboratory experiments, vastly expanded characterization capabilities, and order of magnitude improvements in monitoring resolution.

1 INTRODUCTION

Hydraulic fracturing for well stimulation initially entailed injection of relatively small volumes of fluid (gelled hydrocarbons or aqueous gel mixed with sand proppant) in order to bypass near wellbore damage in conventional oil reservoirs (Montgomery & Smith, 2010; Economides & Nolte, 2000, Chap. 5A). The technology evolved through its first 4 decades, the 1950s through the 1980s, bringing several changes. The fluid compositions transitioned from predominantly hydrocarbon-based to predominately water-based fluids, as summarized in the first review paper on hydraulic fracturing by Hassebroek and Waters (1964). The fluid volumes grew from the early, hundreds or thousands of gallons treatments to 250,000+ gallon treatments, with these larger stimulations carrying $\sim 1,000,000$ pounds of sand proppant, eventually targeting low permeability (unconventional) reservoirs (Montgomery & Smith, 2010).

Still, the evolution of the technology did not result in fundamental changes in the models used to simulate hydraulic fracture (HF) growth and guide engineering decisions. The importance of modeling fully 3D fracture propagation was recognized early on. Still, throughout this period the predominant directions of modeling innovation followed the basic elasto-hydrodynamic model set out in the early, seminal papers (Khristianovic & Zheltov, 1955; Perkins & Kern, 1961; Geertsma & de Klerk, 1969;

Nordgren, 1972) and progressively generalized to obtain approximate but rapid simulation of HF growth via the so-called Pseudo 3D (P3D) method (Settari & Cleary, 1984). These advances are well documented in past, model-oriented review articles (Medelsohn, 1984b; Medelsohn, 1984a; Adachi *et al.*, 2007).

More recently, treatments have switched from exclusive application to vertical wells to widespread application for stimulation of horizontal wells. By the early 1990s the concept of generating multiple HFs from horizontal wells was being developed for targeting tight sandstone and shale reservoirs (Soliman *et al.*, 1990). By the late 1990s success has been achieved in the Cotton Valley, Texas, using water with relatively small proppant concentration (Mayerhofer *et al.*, 1997), bringing slickwater fracturing, developed in the 1950s (Harmon, 1957), to the mainstream. At this point, concepts of complexity of HF geometry and stress interaction between multiple branches of HFs were transitioning from infancy to childhood (see the review of Mahrer, 1999). Since then there has been a focus on topics including growth of HF networks (for a review of natural fractures in US shale plays (see Li, 2014)), stress shadow interactions among multiple HFs distributed along horizontal wells, managing near wellbore tortuosity associated with initiation of HFs from horizontal wells, and fracture containment in shale reservoirs. While there exists some new terminology and the emphasis on these topics is unprecedented, all of these issues are appropriately understood in terms not only of the relevant contributions in the last decade but also with respect to the roots each of these have in the classical hydraulic fracturing literature.

The goal of this review article is therefore to approach four critical issues for successful applications: 1) Managing hydraulic fracture height growth, 2) Decreasing near-wellbore tortuosity, 3) Predicting and engineering network versus localized growth geometry, and 4) Promoting simultaneous growth of multiple hydraulic fractures. While not purporting to be an exhaustive list of critical issues for successful applications, all of these have received unprecedented attention in recent years. Furthermore, all 4 issues are best viewed with a synthesis of model predictions, laboratory experiments, and field evidences, all the while setting each topic in its context from previous decades and highlighting some of the most relevant contributions and corresponding insights from the past 1–2 decades. Hence, this article is organized so as to highlight the importance and background for each topic followed by insights from modelling, field data, and laboratory experiments. Taken together these considerations give a snapshot of the state of knowledge for each of these issues, leading naturally to proposed considerations for ongoing investigations.

2 MANAGING HYDRAULIC FRACTURE HEIGHT GROWTH

2.1 Importance and background

Height growth refers to the growth of HFs in the vertical direction and therefore it relates to the propensity of HFs to cut through multiple strata. While sufficient height growth is necessary in order to contact the entirety of the targeted reservoir, excessive height growth is detrimental to the effectiveness of the treatments because it results in delivery of fluid and proppant to unproductive zones or, in an even worse scenario, in the stimulation of water bearing strata resulting in increased water inflow to the well (*e.g.* Economides & Nolte, 2000). Height growth, which also is referred to as "fracture containment", has also been a topic of discussion in the context of preventing unwanted migration of fluids to strata that contain potable groundwater. This concern is potentially warranted in shallow wells that are separated by less than a hundred meters or so from groundwater resources; some coal seam methane wells are examples (*e.g.* EPA, 2004) as well as approximately 100 hydraulic fractures each year in the U.S. that are high volume ($> 10^6$ gallons) and relatively shallow (< 3000 ft depth) Jackson *et al.* (2015).

A seminal contribution in height growth prediction is Simonson *et al.* (1978). Their model is based on the equilibrium height of a plane strain cross section of a blade-shaped hydraulic fracture. That is, they look for the height at which the computed stress intensity factor, K_I, is exactly balanced by the stress intensity factor, K_{Ic}, for each cross section of a hydraulic fracture that is long relative to its height ($\ell \gg H$). In the limiting case of $K_{Ic} = 0$ they give an upper bound on the fracture height (H) relative to the payzone height (H_{pay}) that is implicit in the relationship

$$\frac{p_{net}}{\Delta\sigma} = -\frac{2}{\pi} ArcSin\left(\frac{H_{pay}}{H}\right) + 1 \tag{1}$$

where p_{net} is the difference between the fluid pressure and the stress opposing HF opening and $\Delta\sigma$ is the increase in compressive stress encountered between the payzone and the barrier. Equation 1 therefore represents an exact balancing between the stress intensity factor generated by the pressurization of the hydraulic fracture and the clamping stress generated by the elevated stress in the barrier region. This approach has been recently extended to account for equilibrium planar growth through up to 6 layers by Liu and Valko (2015).

While it has become generally accepted that stress contrast between the reservoir and bounding layers is the most important factor determining height growth (Warpinski & Teufel, 1987; Nolte & Smith, 1981), the role of a variety of other factors were explored in the years that followed the work of Simonson *et al.* (1978). These include:

- High permeability bounding layers, which reduce the rate of height growth due to leakoff of fluid into the bounding layer (Quinn, 1994). Drawing on laboratory experiments, these barriers are thought to be effective when the permeability of the barrier layer is 3–4 times the payzone permeability (de Pater & Dong, 2009).
- Contrasting stiffness of layers, which impacts not only through the fact that stiffer layers tend to carry larger in situ stresses (*e.g.* Prats & Maraven, 1981), but also because they can impede crack growth (*e.g.* Simonson *et al.* 1978) and restrict fluid flow (van Eekelen, 1982; Smith *et al.*, 2001). Nonetheless, it is generally considered that stiffness contrasts do not have substantial impact on height growth except through the impact on in situ stress (Abou-Sayed *et al.*, 1984; Ben, 1990; Gu & Siebrits, 2008). Weak bedding planes between the payzone and the bounding layer(s) can also serve to blunt or deflect fracture height growth (*e.g.* Daneshy, 1978b; Daneshy, 2009).

The following reviews the field, lab, and numerical evidences for the various conditions that limit height growth, highlighting the critical issue of current and ongoing research of the role of the weak bedding plane that often comprises the interface between the reservoir and the barrier layers above and/or below.

2.2 Field evidence

Height growth is most often ascertained in the field indirectly – in contrast to direct observation – from microseismic data. Most of this data is not published, and, of the published data, there is a tendency to present plan views of event clouds that give no information about height growth. Furthermore, there may be a tendency to under-represent problematic height growth in the data that is actually published. Nonetheless, there are some useful examples in the literature that include:

- *Barnett Shale*: Several horizontal wells completed in the Lower Barnett formation exhibit hydraulic fracture growth that is contained within the Lower Barnett with a fracture height of 350 ft (Fisher *et al.*, 2004). In a compendium of otherwise mostly unpublished data from wells ranging in depth between about 4000–8500 ft, Fisher and Warpinski (2012) show height growth that is generally increasing with depth and that appears particularly prone to downward bias, especially at shallow depths. Davies *et al.* (2012) analyzed the data of Fisher and Warpinski (2012) and found that the median of all Barnett hydraulic fractures, without accounting for depth, is about 410 ft with 10% of fractures exceeding 500 ft in height above the injection point. Similarly, the median downward height growth is about 260 ft with 10% of fractures exceeding 820 ft.
- *Marcellus Shale*: Two horizontal wells completed in the Lower Marcellus formation exhibit hydraulic fracture growth penetrating the 200 ft thick combined Upper/Lower Marcellus, extending 150 ft downward into the Onandaga formation and 250 ft upward into the Skaneateles shale formation for a total fracture height of about 600 ft (Mayerhofer *et al.* 2011). This example lies within the upper half of the range of heights in the data set of Fisher and Warpinski (2012) as analyzed by Davies *et al.* (2012), wherein the median upward height growth is about 410 ft with 10% of fractures exceeding 820 ft. The median downward growth is found to be about 100 ft with 10% of fractures exceeding about 160 ft.
- *Eagle Ford Shale and Woodford Shale*: Based on the data of Fisher and Warpinski (2012), Davies *et al.* (2012) find the median upward growth in both formations to be around 60 m with 10% of fractures exceeding 115 m. The downward growth is less in the Eagle Ford, with a median value of 160 ft and 10% exceeding 490 ft compared to the Woodford with a median value of 230 ft with 10% exceeding 650 ft.
- *Bakken Shale*: Dohmen *et al.* (2013) present data indicating typical 400–500 ft fracture height for 2 wells completed in the Middle Bakken formation. An additional $\sim$200 ft of downward growth is observed in some cases and many cases are accompanied by clusters of events about 800 ft above the top of the Middle Bakken that are attributed to activation of initially fluid-filled faults. Dohmen *et al.* (2014) use a similar monitoring method and argue that the data evidence the stress shadow from previous stages is able to increase height growth. (See further discussion of stress shadow in Section 5.)
- *Haynesville Shale*: Warpinski (2014) presents a case of a stimulation that traverses the 100 ft thick Haynesville formation, growing to at least 330 ft above and 160 ft below for a total fracture height of about 590 ft.

Besides inferring height growth from microseismicity, it is also possible to obtain measurements using tracers that either tag the fluid or the proppant. For example,

Saldungaray *et al.* (2012) present several case studies where propped fracture height in vertical wells is measured at the wellbore using either Compensated Neutron (CNT) or Pulsed Neutron Capture (PNC) tools that detect proppant in which a high thermal neutron capture compound is incorporated, leading to a reduced neutron count rate over the propped portion of the hydraulic fracture. As a further example, Hammack *et al.* (2013) present results wherein Perfluorocarbon tracers were injected along with hydraulic fracturing fluids in a well completed in the Marcellus formation. Monitoring of overlying Upper Devonian wells ranging from 3000–4000 ft above the Marcellus well confirmed that there was no measurable fluid migration during hydraulic fracturing and in the 8-month monitoring period that followed.

Direct observations of hydraulic fracture height growth are limited to shallow formations, namely coal seams, in which hydraulic fractures are placed and subsequently exposed as mining proceeds. While it is important at the outset to realize that shallow coal seam reservoirs typically have lower magnitude vertical stress relative to the horizontal principal stresses than in most deeper formations, there are nonetheless some important insights that can be gained. Most strikingly, these minethrough experiments not only show containment – *i.e.* arrest of growth in the vertical direction at the contact between the coal and the roof/floor rock, Figure 1a – but also they show deflection along the horizontal contacts especially when the bounding formation is stiffer and/or stronger than the coal, Figure 1b (Elder, 1977; Lambert *et al.*, 1980; Diamond & Oyler, 1987; Jeffrey *et al.*, 1992). Furthermore, these minethrough experiments enforce an appreciation of the complications associated with predicting hydraulic fracture growth in the presence of multiple layers. For example, hydraulic fractures are observed to cross thin layers even if they are stiff/strong and they are observed to grow on top of the contact above a thin stiff layer that is separated from a thick stiff layer by a thin soft layer (Figure 1c) (Lambert *et al.*, 1980; Diamond & Oyler, 1987; Jeffrey *et al.*, 1992; Jeffrey *et al.*, 1995).

Additionally, naturally-occurring dikes that form as fluid-driven cracks and which are mechanical analog to hydraulic fractures, have been observed to cross, arrest, or deflect when they encounter potential barriers (Figure 2) (Gudmundsson *et al.*, 1999; Gudmundsson & Loetveit, 2005; Gudmundsson, 2011), producing similar geometric observations as for hydraulic fracture minethrough experiments in coal seams.

2.3 Laboratory investigations

Experiments in gelatin have demonstrated dependence of hydraulic fracture growth through a contact between two layers on the relative material properties, especially the stiffness (see review in Rivalta *et al.*, 2015; Rivalta *et al.*, 2005; Kavanagh *et al.*, 2006). The persistent challenge, though, has been to obtain analogue experiments for the case of stress contrasts. This analogue is important because simulations suggest stress contrasts are the most important factors determining height growth (Warpinski & Teufel, 1987; Nolte & Smith, 1981; Warpinski *et al.*, 1982).

El-Rabaa (1987) presents a comprehensive laboratory parametric study using sandstone, limestone, marble, and hydrostone specimens. These experiments demonstrate the tendency of hydraulic fractures to favor growth in lower stress zone as well as the apparent suppression of height growth by weak interfaces (after the prediction of Daneshy, 1978a). However, they do not entail sharp stress jumps as expected at many

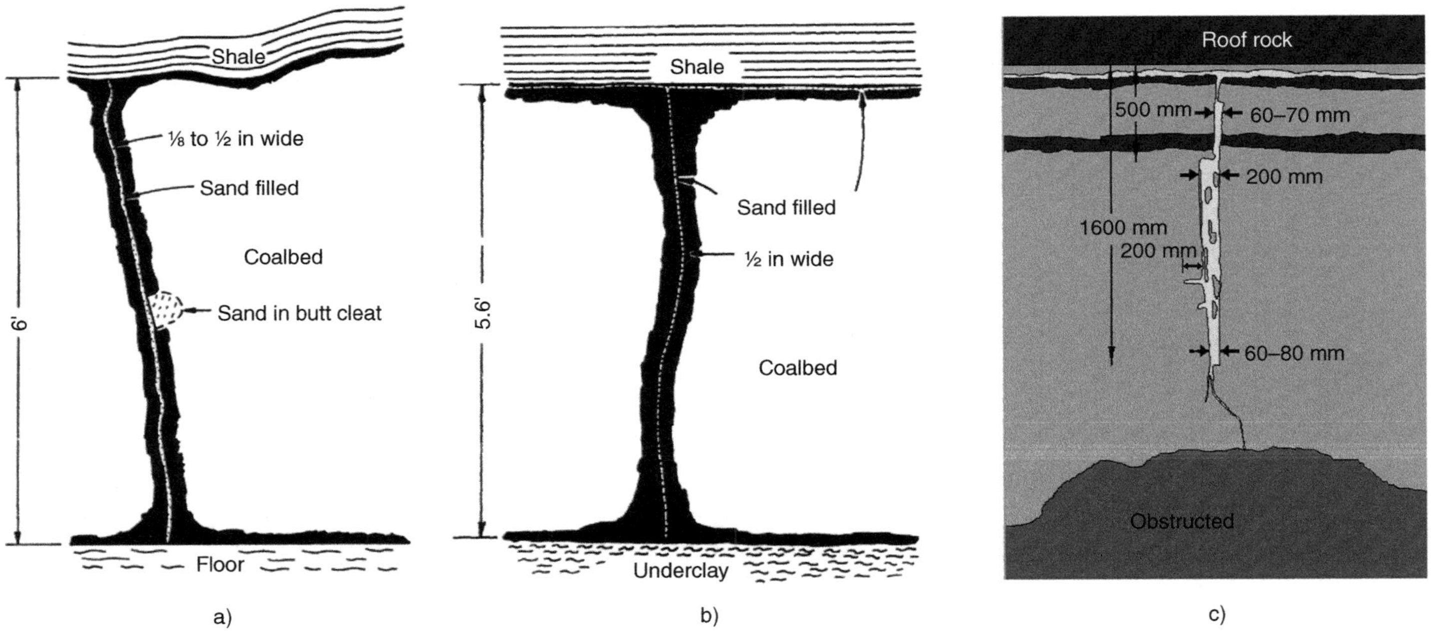

Figure 1 Mapped hydraulic fractures in coal showing: a) Arrest at relatively stiff/strong roof and floor layers, from Diamond and Oyler (1987), b) Formation of a horizontal component along the contact between the coal seam and a relatively stiff/strong roof rock, from Diamond and Oyler (1987), c) Direct crossing of the stiff/strong but thin layer between two coal layers and growth of a horizontal component above a relatively stiff/strong layer that is separated from the stiff/strong roof rock by a relatively thin layer of soft/weak coal, redrawn after Jeffrey *et al.* (1992) (all figures after Rivalta *et al.*, 2015, with permission).

a) b) c)

Figure 2 Dike interactions with layered geological media. a) A vertical, 6-m-thick dike that penetrates a sequence of lava flows in North Iceland. b) A vertical, 0.3-m-thick dike that arrests at the contact between a pyroclastic layer and a basaltic lava flow in Tenerife, Canary Islands. C) The dike that deflects along the contact to form a sill between a basaltic sheet and a lava flow, Southwest Iceland. From Gudmundsson (2011) (after Rivalta *et al.*, 2015, with permission).

lithological boundaries, but instead they obtain a smoothly-varying applied stress across the eventual plane of HF growth. In fact, obtaining a closer analog to geomechanical layering is challenging to achieve because simply constructing a system with layers of contrasting stiffness and loading the specimen on its boundaries generates spurious shear stresses along the layer contacts. In one early investigation, Johnson and Cleary (1991) portray contrasting cases of radial growth and fixed height, with height growth completely restricted, but this study does not investigate finite height growth.

More recent experiments make use of PMMA specimens that are machined with surface profiles such that a prescribed stress profile is generated when the blocks are pressed together (Jeffrey & Bunger, 2009, Figure 3a). These experiments show the contrast between moderate height growth when a low stress reservoir zone is bounded by higher stress barriers (Figure 3b) and preferential growth into a low stress layer that bounds the reservoir layer (Figure 3c). These experiments have provided benchmarks that have been used in the development of numerical simulators (*e.g.* Wu *et al.*, 2008; Dontsov & Peirce, 2015). Extending these experiments to include a weak horizontal interface (analogue bedding plane), Xing *et al.* (2016) present laboratory experiments demonstrating the impact of the applied stresses on height growth in layered, transparent specimens.

2.4 Modeling insights

The shortcomings of Pseudo 3D models in modeling height growth were recognized early on (*e.g.* Medelsohn, 1984a) and motivated the developments of Planar 3D models

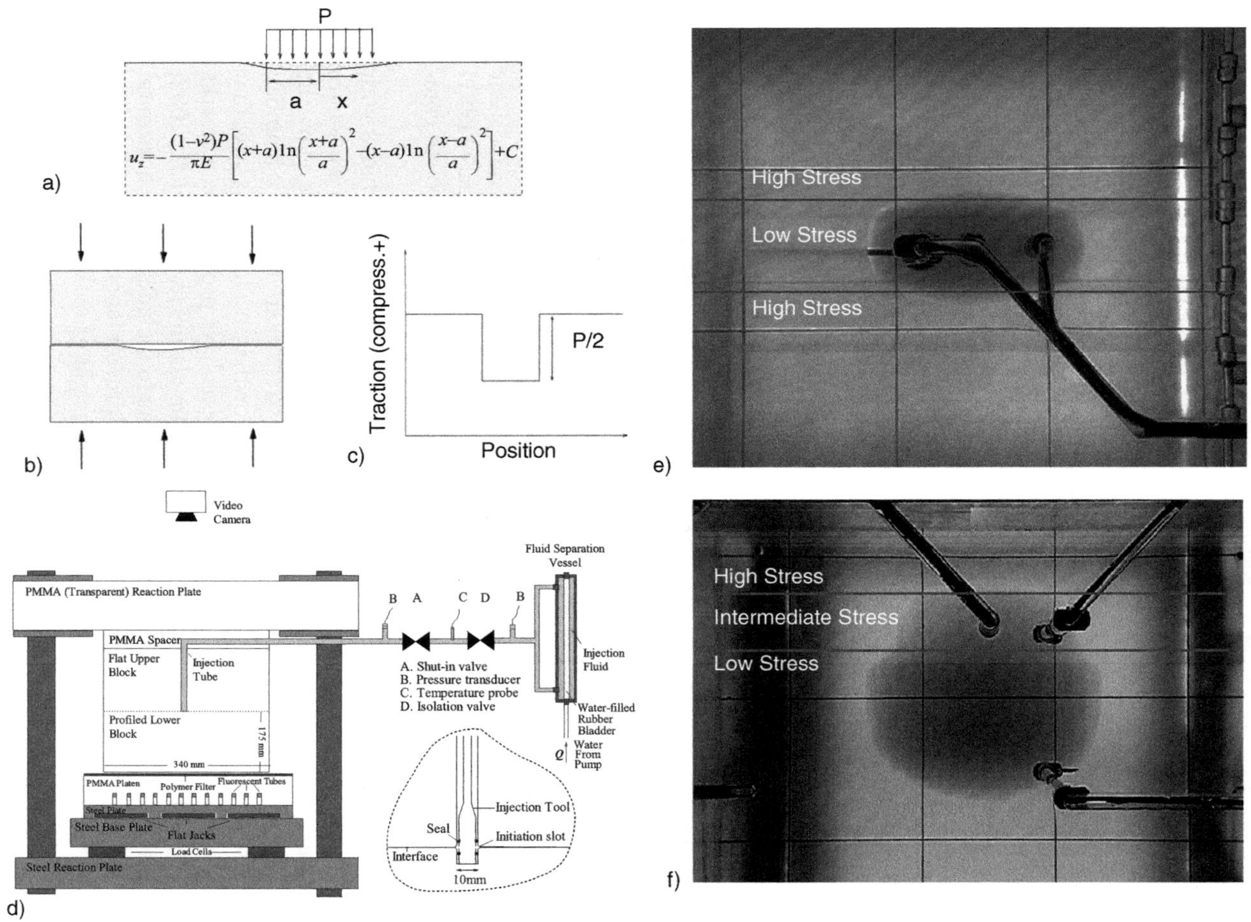

Figure 3 Laboratory experiments on height growth (after Jeffrey & Bunger, 2007; Wu *et al.*, 2008; Jeffrey & Bunger, 2009). a-c) Illustration of the method used to generate the stress jump by machining block surface(s) to the solution for the displacement due to a uniform strip loading, d) Laboratory apparatus used for the experiments, e) Moderate height growth into higher stress bounding layers, f) Preferential growth into a low stress layer.

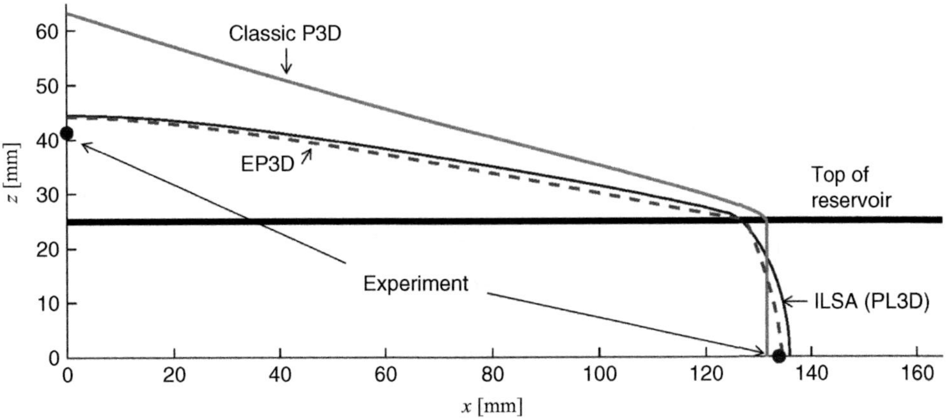

Figure 4 Comparison of fracture shape for pseudo 3D (P3D) and Planar 3D (ILSA PL3D) predictions with the experiments from figure 3b at time=604 seconds (modified from and courtesy of E. Dontsov Dontsov & Peirce, 2015).

for HF design in the early 1990s (*e.g.* Ben Naceur & Touboul, 1990; Advani *et al.*, 1990). The most obvious area for improvement of Pseudo 3D is the handling of the role of the layered geology ubiquitous to the sedimentary basins in which oil and gas deposits occur. The challenges are illustrated by comparison of the experiments of (Figure 3b) with Pseudo 3D ("P3D", *e.g.* Settari & Cleary, 1984) predictions (Wu *et al.*, 2008; Dontsov & Peirce, 2015). The comparison also includes predictions of the simulator ILSA (Implicit Level Set Algorithm, Peirce & Detournay, 2008), which is a planar 3D simulator accounting for full 3D elasticity and generalized growth (relative to P3D) of a planar hydraulic fracture. From this comparison it is apparent that P3D severely overestimates height growth in the viscosity dominated regime and, conversely, underestimates height growth in the toughness dominated regime. Dontsov and Peirce (2015) use this comparison to argue the necessity for enhancements to the modeling approach that seek to maintain the rapid computation of P3D while better approximating the 3D elasticity relationships and the equations governing the advance of the hydraulic fracture front.

Even in the case of a fully coupled planar 3D hydraulic fracture simulation, the restriction of fracture growth to a single plane may often miss the role of the layering in determining height growth. Indeed there is ample evidence from direction observation, such as Figures 1 and 2, for arrest, offset, and/or deflection of hydraulic fractures at the margins between layers. The most obvious reason for the deflection along these boundaries is relative weakness of the contact (*e.g* Daneshy, 1978b; Daneshy, 2009; Abbas *et al.*, 2014). However, weak contacts are almost certainly not the only cause of deviation to T-shaped growth. Very thin layers of relatively soft rock (Young's modulus 1–2 orders of magnitude smaller than surrounding rock) have been shown through modeling to dissipate the near crack-tip tensile stresses, thus leading to arrest of height growth and subsequent transition to T-shaped growth (Figure 5). While these authors were concerned with the related problem of arrest of ascending dikes, a similar mechanism could be the cause of the horizontal component of the hydraulic fracture forming

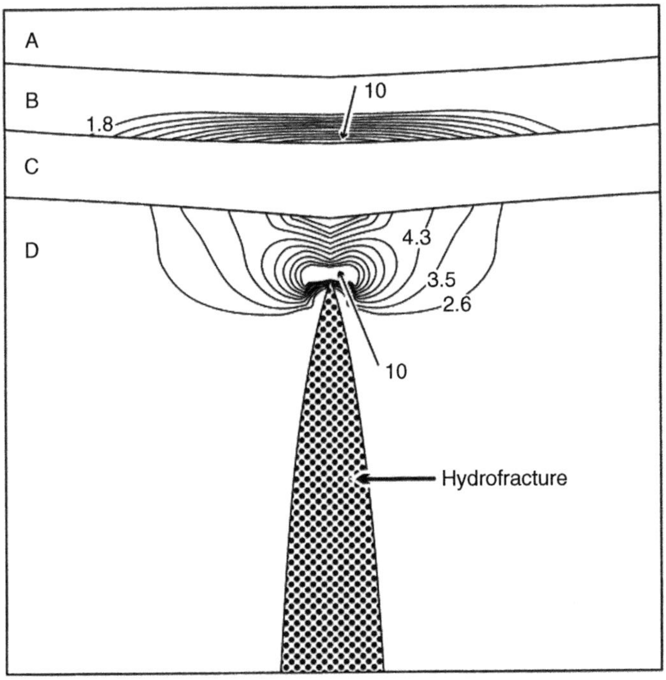

Figure 5 Reduction of fracture induced stress by a thin, soft layer, illustrated by contours of maximum principal stress. Layers B and D (100 GPa) are much stiffer than Layers A and C (E=1 GPa) (From Gudmundsson & Brenner, 2001, with permission).

below a thin layer of soft coal separating two layers of stiffer and stronger rock portrayed in Figure 1c.

While the impact of thin, soft layers should arguably receive more attention than it has to date, many consider the weakness of the interface between the reservoir and a potential barrier layer to be its defining characteristic. Indeed the ability of weak interfaces to suppress hydraulic fracture height growth has been long-recognized (*e.g.* Daneshy, 1978b; Teufel & Clark, 1984). Implementation of mechanisms associated with interaction with a weak interface into hydraulic fracture simulators has accelerated substantially in the past 1–2 decades. The common thread of predictions is that a weak enough interface subjected to a low enough vertical stress (*i.e.* normal stress across the horizontal interface) will cause it to suppress height growth. In essence this principle is accounted for in the early works on hydraulic fracture interaction with natural fractures (Blanton, 1982; Warpinski & Teufel, 1987; Renshaw & Pollard, 1995). However a range of approaches have been developed that place varying degrees of emphasis on the strength of the interface or the coupled fluid flow problem that arises in hydraulic fracturing. The classical Renshaw and Pollard (1995) criterion provides the interface strength-focused end member as it essentially compares the ability of a frictional interface to transmit a fracture-induced tensile stress that is sufficient to nucleate a crack on the opposite side of the interface. This criterion was adapted for prediction of fracture containment by Gu *et al.* (2008). On the other hand, Chuprakov *et al.* (2013) and Chuprakov *et al.*

(2014) present a criterion for HF crossing of natural fractures that depends on fluid flow and is therefore sensitive also to the permeability of the nature fracture, and this has been adapted for prediction of height growth by Chuprakov and Prioul (2015).

Even when the hydraulic fracture is able to cross a weak interface to grow in height into an overlying or underlying layer, the growth through the weak interface can lead to offsets. Offsets have been shown to also suppress height growth. This result is anticipated through the reduction of stress intensity at the fracture tips (Sheibani & Olson, 2013) and can be shown to reduce height growth by around 10% for offset angles of 60–70 degrees and by 20% or more for offset angles of 80–90 degrees (Abbas *et al.*, 2014).

2.5 State of knowledge

Indirect observations of height growth are valuable but also possessing important caveats. Microseismic events that locate out of zone could be the result of location error and, even if located appropriately, the connection to fluid flow is not direct and there is no indication of whether proppant reached that location. On the other hand, fluid tracer studies provide information for only one point and cannot comment on whether fluid reached other out-of-zone locations. Furthermore, tagged proppant is limited in that it only indicates the propped portion of the hydraulic fracture and only at the wellbore itself; it is possible that the height indicated by tagged proppant logs at the wellbore greatly overestimates the height at a short distance from the wellbore.

The observations of Dohmen *et al.* (2013) regarding activation of events by perturbing initially fluid-filled faults in the Bakken Shale raises another important caveat. While events are generated remotely, the timing suggests they are not associated with hydraulic fracture growth but rather pressure disturbance of the fluid already within a fault. A similar observation is given by Hammack *et al.* (2013) for a well completed in the Marcellus Shale. These can lead to reported fracture heights in data sets based on furthest extent of events, such as Fisher and Warpinski (2012), that are not attributable to height growth *per se* or, for that matter, to hydraulic fracture growth along a fault. These can be essentially remotely triggered events through stress and/or pore pressure perturbation. Hence the reported ranges of fracture heights based on microseismicity are potentially misleading in these cases of fault activation.

Ambiguities and uncertainties motivate direct observation and modeling. Laboratory experiments experiments show P3D can overestimate height growth, thus motivating alternatives that preserve the rapid computation of P3D but that improves accuracy. Field experiments, albeit in relatively shallow formations such as coal seams, nonetheless demonstrate the importance of layers and the bedding planes between layers. Hence, these direct observations point toward the importance of ongoing efforts to appropriately account for the impact of layering and discontinuities on height growth.

3 DECREASING NEAR-WELLBORE TORTUOSITY

3.1 Importance and background

In all applications, HFs are initiated from a wellbore drilled down to the target reservoir formation. Most of the time, the well is cased and cemented but so-called open-hole completions are also used. The location of the initiation point of the hydraulic fracture is controlled in most cases by detonating a set of shaped charges. It results in a number of perforation tunnels penetrating in the formation. The length of the perforation gun tool dictates the length of the well interval perforated (typically from 5 to 30 feet). In the case of a cemented wellbore, the perforation tunnels go through casing, cement and rocks and are the only possible "entry" points for the injected fluid into the formation. Abrasive jetting is also sometimes used in order to create an initiation flaw in the rock. Even with all the possible care, it is difficult in practice to optimally create a flaw such that the hydraulic fracture initiates and propagates directly from the wellbore in the direction of the preferred far-field in-situ stress orientation.

Another option is not to create any flaws, therefore relying on defects in the rock, but to limit the interval where a fracture can initiate by ensuring that the fluid can pressurized (and penetrate) the formation only over a finite interval of the well. For the case of an open-hole wellbore, this can be achieved by using a fracturing completion string, inflating packers to isolate a given zone of the well and pumping the fluid through a tubing. In cased and cemented wellbore, cemented sleeves can be used in conjunction with packers / plug systems on coil-tubing for example. The HF then initiates in the region controlled by near-wellbore stress concentration, which depends primarily on the orientation of the well with respect to the in-situ stresses but also the geometry of the created flaws if any (perforations or notch). The presence of cement (and its quality) can also play a significant role: interface debonding can occur thus providing more open-hole like conditions even for a cased and cemented wellbore (Weijers & de Pater, 1992; Behrmann & Nolte, 1999).

After initiating, the HF rotates from the local near-wellbore stress concentration towards its far-field preferred plane of propagation governed by the in-situ stress field. The extent of the region over which such a re-orientation takes place has been found experimentally to be about two to five times the wellbore radius. It is important to bear in mind that the fracture re-orientation toward the preferred far-field stress direction is not always smooth, as illustrated by the laboratory example shown in Figure 6. The reason is that mode II and III stress intensity factors (shear and tearing modes in addition to the opening mode I) may develop during re-orientation leading to curving and possibly segmentation of the fracture front. Depending on the wellbore and perforations orientation, several distinct fractures may initiate with different orientations (*e.g.* along the axis of the well and transverse to the well, or nearly parallel in the case of a deviated well) with a dominant one further propagating in the far-field with some merging and others not (Behrmann & Elbel, 1991; Weijers *et al.*, 1994). Of course, in the case where the far-field stresses are close to one another (*e.g.* similar horizontal stresses), the orientation of the fracture in the far-field is not constrained. Multiple fractures initiating from the wellbore have been observed in the laboratory as a result in such case (Doe & Boyce, 1989).

The directly-measurable consequence of near-wellbore fracture tortuosity is often an increase in the pumping pressure. Such a larger pressure – often referred to as near-wellbore entry friction or near-wellbore pressure loss – can be simply understood. If an HF is not oriented in the direction favored by the in-situ stress in the near-wellbore region, during its re-orientation the local normal stress acting on the fracture faces can be locally larger than the minimum in-situ stress. This larger local normal stress directly increases the fluid pressure required to open and propagate the HF. Depending on the details of the fracture tortuosity (for example smooth versus highly irregular with

Figure 6 Complex HF geometry in the near-wellbore region exhibiting completion interface debonding (micro-annulus), re-orientation toward the far-field preferred fracture plane with fracture front segmentation due to mixed mode fracturing (Photo courtesy of Rob Jeffrey & Jean Desroches). Experiments performed in PMMA with a Newtonian fracturing fluid.

several fracture branches), pinch-points can develop with possible detrimental consequences on the placement of proppant. The proppant can bridge at such pinch-points, then possibly packing back to the wellbore and causing a near-wellbore screen-out. When this happens slurry can no longer enter the fracture, the injection pressure exponentially increases, and the injection has to be stopped. However, a high treating pressure does not necessarily imply a higher risk of near-wellbore screen-out. The degree of the pressure increase and the associated width restriction is related in a non-trivial manner to the degree of complexity of the HF(s) geometry in the near-wellbore as well as to treatment parameters. For example, higher viscosity and injection rate directly increase the near-wellbore pressure loss, but appear to also play a role in the development of the near-wellbore fracture geometry.

Several mitigation techniques to reduce near wellbore tortuosity have been put forward in practice (sometimes combined). These include: 1) oriented perforating (or notching) to try to reduce re-orientation and complex fracture geometry in the near-wellbore region, 2) injection of acid/sand-slugs/gel slugs after formation breakdown to erode the near-wellbore tortuosity and/or 3) the use of highly viscous fluids to generate sufficient fracture width to avoid near-wellbore screen-out. The following reviews field evidence, laboratory experiments, and modeling insights for the presence and mitigation of near wellbore tortuosity.

3.2 Field evidence

The presence of near-wellbore fracture tortuosity is typically inferred indirectly from pressure records and diagnostic tests characterizing entry friction. Direct observations are limited to the trace of the fracture on the wellbore wall as observed by image logs performed after fracturing, of which little examples have been published (see Jeffrey *et al.*, 2014, for one such example).

The near-wellbore entry friction can be quantitatively assessed by performing a so-called step-down test (during a single-entry treatment). Such a test consists in lowering the injection rate by steps during fracturing from the initial constant injection rate to zero while recording the treating pressure (ideally downhole at the level of the perforations). At least four to five steps are required in order to decipher between a classical perforation friction term associated with the choking effect of the holes in the casing and another non-linear term associated with near-wellbore fracture tortuosity (see Economides and Nolte, 2000, chapter 9, pages 9–32 / 9–33 for details). Large near-wellbore friction associated with fracture tortuosity have been reported for both vertical (Roberts *et al.*, 2000) and horizontal wells (among others Kogsbø, 1993; Desroches *et al.*, 2014; Pandya & Jaripatke, 2014).

Another indirect line of evidence to detect near-wellbore tortuosity is the reduction in treating pressure often observed after injection of acid, sand/gel slugs, or even just proppant. Acid, sand or gel slugs appear to erode the near-wellbore tortuosity therefore reducing the near-wellbore pressure loss. Examples of reduction of 1000 psi or more are typical (Cleary *et al.*, 1993; Kogsbø, 1993; Chipperfield *et al.*, 2000; McDaniel *et al.*, 2001).

Analysis of treatments in wells drilled in the same formation but with different deviation (from vertical) have shown a correlation between larger fracturing pressure and well deviation (Veeken *et al.*, 1989). Correlation between horizontal well azimuth and fracturing pressure have also been reported (Owens *et al.*, 1992) with larger fracturing pressure observed for wells in the direction of the minimum stress (with axial HFs re-orienting to transverse in the far-field). However, without step-down tests, it is difficult to conclude that higher than expected treating pressure necessarily indicates larger near-wellbore tortuosity. Interestingly, in vertical wells in the Cooper basin Australia, large near-wellbore pressure losses associated with fracture tortuosity do not correlate at all with either the treating pressure (fracture gradient) or the measured stress magnitude (Mc Gowen *et al.*, 2007). It appears in that particular case that natural micro-cracks (*i.e.* rock fabric) are mostly responsible for large near-wellbore pressure losses. Neither sand slugs, acid nor oriented perforating were actually 100% successful in reducing near wellbore pressure losses in that particular case.

Jeffrey *et al.* (2015) reports a field study for open-hole vertical wells (in a stress regime favoring horizontal fractures) notched via abrasive jetting to promote the initiation of transverse fractures from the well directly. Multiple single entry treatments were performed and image logs were run post-frac. Axial (vertical in that case) fractures were found on image logs for a number of treatments despite the transverse notch. However, the breakdown pressures did not significantly differ between cases where an axial or a transverse fracture initiated, and higher near-wellbore friction was consistently observed when axial fractures were detected on image logs. This is a typical example of the effect of re-orientation from an axial fracture in the near wellbore to a transverse fracture in the far-field.

Large difference in near-wellbore pressure losses between single-entry fracturing jobs performed in the same horizontal well (completed with cemented sleeves and drilled in the direction of the minimum stress) have been reported in Desroches *et al.* (2014). Although the formation was found to be very homogeneous along the horizontal section, variations in near-wellbore tortuosity-related pressure drops of more than 1,000 psi (at the same injection rate) between fracture treatments were reported. Such large variation may be due to the cemented sleeve system (*i.e.* whether or not debonding occurs away from the sleeve) or just small scale rock fabric heterogeneities that have not been characterized.

There are a number of examples showing the benefit of oriented 180 degree perforating (*i.e.* the technique of orienting a line of perforations in the preferred far-field fracture plane) in reducing fracturing pressure in the case of vertical wells (Manrique & Venkitaraman, 2001; Behrmann & Nolte, 1999). An accuracy of 20 to 30 degrees of the orientation of the 180 degree perforating guns has been found to be sufficient to avoid large friction (Behrmann & Nolte, 1999). Uncertainty on stress direction is such that 60 degree phasing is often found to be operationally more robust while still providing reasonable results for proppant placement (*e.g.* Ceccarelli *et al.*, 2010). Oriented perforating is more complicated for deviated wells, and even more so for horizontal wells due to the intrinsic limitation of the perforation tool geometry. Nonetheless, some benefits have been reported in wells up to 65 degree deviation (Pearson *et al.*, 1992; Pospisil *et al.*, 1995). For horizontal wells drilled in the direction of the minimum stress, the accepted best practice is to cluster perforations over a rather short interval (*i.e.* the perforations cannot lie within a single plane in that case); multiple fractures that may or may not all link up away from the wellbore are to be expected (Behrmann & Nolte, 1999). The use of oriented abrasive jetting has been found beneficial in a number of cases. For both vertical and a horizontal well, abrasive jetting has allowed the placement of proppant in difficult situations (*e.g.* highly stress formation etc.) as discussed in McDaniel and Surjaatmadja (2009), Ceccarelli *et al.* (2010). It is not always successful, though, depending on rock fabric effects (Strain, 1962) or cleaning issues (Jeffrey *et al.*, 2015).

Finally, the practice of using highly viscous fluid has been found beneficial (Aud *et al.*, 1994; Hainey *et al.*, 1995). Higher viscosities create more width for proppant transport and are also though to help promoting a dominant single fluid path in the near-wellbore.

3.3 Laboratory investigations

Near–wellbore tortuosity can be observed in detail in the laboratory, although the scaling of the effect of the perforations, casing and cement is far-from trivial especially knowing the quasi-brittle behavior of rocks and the inherent size effects associated with failure of those materials. Nevertheless, a number of laboratory experiments (Daneshy, 1973a; Veeken *et al.*, 1989; Doe & Boyce, 1989; Weijers *et al.*, 1994; Behrmann & Elbel, 1991; Abass *et al.*, 1996; Fallahzadeh *et al.*, 2014) have documented the effect of wellbore, in-situ stress misalignment and magnitudes as well as the presence of perforations (see Figure 7 for a summary). The deviation of the HF from simple axial and planar geometry was first observed by Daneshy (1973b) in the laboratory for slightly deviated wells.

One important aspect (often overlooked in early contributions) relates to the completion sequence of block test investigating the effect of cemented completion. It is particularly important to cement the casing while the block is already loaded. Doing otherwise greatly enhances the bonding stress between the casing/cement and the block. Initiation at the tip of the perforations is always favored as a result, whereas fluid debonding of these interfaces might occur in reality when the bonding stress is lower (see for examples and discussion Weijers and de Pater 1992).

A large experimental effort was performed at TU Delft in the 1990s during the early stage of highly deviated and horizontal wellbore developments (Weijers & de Pater, 1992; Weijers *et al.*, 1994; van de Ketterij & de Pater, 1997; van de Ketterij, 2001). Besides the effect of stress and wellbore misalignment (Figure 7b), injection rate and viscosity were found to play an important role. The re-orientation of axial fractures into transverse fractures appear more gradual and smooth in the case of larger viscosity/rate (see Weijers *et al.*, 1994 for discussion). It is also worthwhile to remember that maximum pressure (breakdown) is increasing substantially with viscosity and rate for similar conditions (*e.g.* Weijers *et al.*, 1994; Detournay & Carbonell, 1997; Lecampion *et al.*, 2015a). Care must therefore be taken not to link higher treating pressures necessarily to higher fracture tortuosity. In that respect, the example of two strictly similar experiments (denoted as #13 and #14 in van de Ketterij and de Pater, 1999) for a fully deviated well is particularly striking (see Figure 7c). The resulting fracture surface for experiment #14 is quite complex exhibiting multiple fracture fronts splitting during the re-orientation from a more axial fracture to the well to the far-field preferred plane of propagation. In comparison, the fracture of experiment #13 is smoother during its reorientation and more symmetric. Interestingly, experiment #13 exhibits a maximum pressure about 8 MPa larger than experiment #14. The main difference between the two experiments is due to debonding of the casing/cement/block interfaces in experiment #14 (see van de Ketterij and de Pater, 1999, for more details and discussion). This example highlights the complex interactions between stress-orientation, fluid infiltration and mixed mode fracture propagation in the vicinity of the wellbore.

It is important to note that higher injection rate and viscosity tends to promote a dominant fluid pathway compared to low viscosity and rate which allows a more uniform pressurization of the different defects. It ultimately results in the creation of different fracture geometries in the near-wellbore region: *e.g.* higher injection rate and viscosity have also been found to decrease fracture tortuosity in experiments performed in a pre-fractured medium (Beugelsdijk *et al.*, 2000).

It is worth mentioning that laboratory experiments in homogeneous rock with slotted/notched wellbores in the direction of the preferred far-field fracture plane typically did not report any fracture tortuosity, which makes slotting an attractive technique (Surjaatmadja *et al.*, 1994; Chang *et al.*, 2014). The impact is also shown to be more pronounced for longer slots. In a recent study (Burghardt *et al.*, 2015), the effect of rock fabric (mineralized natural fractures and weak bedding planes) was shown to be significant in some cases, especially natural fractures in the near-wellbore region can arrest HF initiating from some slots and create significant tortuosity, a result consistent with some of the field observations described in Mc Gowen *et al.* (2007).

3.4 Modeling insights

Most modeling efforts entail only a static stress-analysis in order to discern the highest stress location and fracture initiation pressure depending on the geometrical configuration of the well, perforation and far-field in-situ stress (*e.g.* Alekseenko *et al.*, 2012; Li *et al.*, 2015, and many others). This unfortunately does not provide information on the actual degree of fracture tortuosity to be expected, and the associated pressure loss during fluid injection. In an analytical study, Weng (1993) has combined stress analysis and fracture mechanics arguments for the link up of fractures initiating at different perforations (but neglecting fluid flow) and provided some rationale for engineering perforation design. Other authors pre-suppose the fracture geometry to compute the corresponding pressure drop (for an example of 2D plane-strain simulations see Cherny *et al.*, 2009). Such computations, however, do not provide any information on the parameters affecting the development of HF near-wellbore tortuosity.

Coupled simulations of HF propagation and re-orientation from the wellbore accounting for elastic deformation, fluid flow in the newly propagating fractures are rare. Under plane-strain conditions (2D), Zhang *et al.* (2011) report an extensive numerical study applicable to the case of a vertical well. Their results show that a HF initiating from a defect misaligned with the minimum stress curves toward the preferred fracture plane over one a distance of about two wellbore radius (consistent with experimental observations). They have also introduced and verified dimensionless numbers that control the re-orientation in either toughness or viscosity dominated propagation regimes. These dimensionless numbers combine the minimum and maximum horizontal stress, σ_h and σ_H, respectively, with the fracture toughness of the rock (K_{Ic}), the radius of the wellbore (R), the injection rate (Q_o), the fluid viscosity (μ), and the plane strain elastic modulus of the rock ($E^{'}$). This dimensionless number is given by $(\sigma_H - \sigma_h)\sqrt{R}/K_{Ic}$ in the toughness dominated regime (negligible viscosity), and by $(\sigma_H - \sigma_h)\sqrt{R}/(12\mu Q_o E^{'3})^{1/4}$ in the viscous dominated regime of propagation (which is typically of more practical relevance). Fracture re-orientation occurs over shorter distance for larger value of such dimensionless parameter, with one practical consequence being that initiating an HF with higher viscosity and injection rate leads to a more gradual re-orientation. This result is consistent with the experimental observations of Weijers *et al.* (1994). Their simulations also display significant increase of wellbore pressure (up to 25 MPa) during further propagation for sharper re-orientation. The authors have also investigated the effect of multiple possible initiation flaws. In those cases, multiple HFs initiate then re-orient and compete, with one main HF finally dominating in the far-field.

Three dimensional, fully coupled near-wellbore HF simulations are very limited. Carter *et al.* (1999); Carter *et al.* (2000) were able to partly compare their results to some of the experiments of Weijers *et al.* (1994) but were not able to reproduce crack front segmentation as observed experimentally (see Figures 7b and c). This was recognized to be associated with the lack of a robust propagation criteria accounting for combined tensile and anti-plane shear propagation (mode I and III). Properly capturing such a mixed mode of propagation in 3D coupled with fluid flow remains a challenge today. Recent numerical modeling reported by Sherman *et al.* (2015) provide a step in that direction.

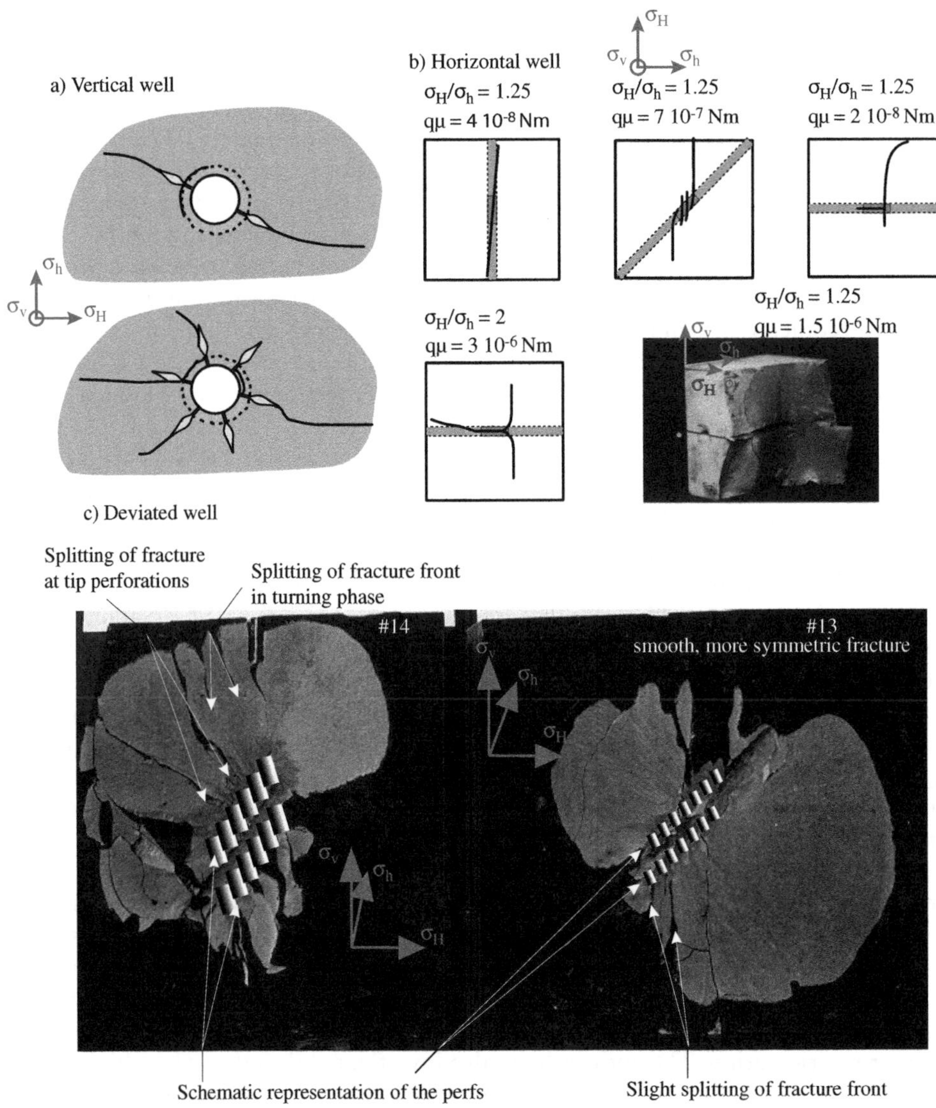

Figure 7 Schematic examples of near-wellbore fracture tortuosity. a) fracture(s) re-orientation due to misalignment of 180 or 60 degree phasing perforating in a vertical well. b) effects of horizontal well azimuth, horizontal stress ratio and fluid effect (injection rate q times viscosity μ) on the initiation of axial or/and transverse fracture(s) – Open-hole experimental results of Weijers (1995) (horizontal trace of the created HFs 5 cm above the plane containing the horizontal wellbore). c) Fractures surfaces for two strictly similar experiments: initiation from a cemented deviated well (49 degree deviation, 60 degree azimuth from the preferred far-field fracture plane) with top/bottom lines of perforations (van de Ketterij & de Pater, 1999; van de Ketterij, 2001), casing debonding occurred during experiment #14 (which also exhibited lower pressure) but not during experiment #13. Complete mixed mode (I, II and III) fracture propagation with fracture front splitting is more severe for experiment #14 (adapted from van de Ketterij and de Pater, 1999, reproduced with permission).

3.5 State of knowledge

Although, the main causes for HF tortuosity in the near-wellbore region are now well recognized (well/perforations/in-situ stress misalignment, fluid viscosity and injection rate), detailed quantitative predictions of the complex 3D HF geometry in the near-wellbore and its impact on fracturing pressures and screen-out risk remains a challenge. The main fundamental difficulties lie: i) in a good prediction of 3D mixed mode propagation – especially mixed mode I, II and III which leads to fracture front break-up, and ii) proper fluid-solid coupling in complex 3D fracture geometry. Both are needed in order to better understand the impact of stress, well orientation and treatment parameters on near-wellbore fracture geometry. However, one has also to keep in mind experiments #13 and #14 of van de Ketterij and de Pater (1999), which, although strictly similar, yielded widely different fracture re-orientation due to casing de-bonding. Small heterogeneities (difficult to characterize) at the scale of the wellbore may possibly play a large role in the development of the fracture geometry in the near-wellbore.

Practically today, the best technique to quantify near-wellbore tortuosity problem is via diagnostic tests (*e.g.* step-downs). Estimation of near-wellbore entry friction can then be used in order to subsequently refine the stimulation design (*e.g.* Gulrajani & Romero, 1996). Depending on the situation, some mitigation measures (for example sand-slugs, oriented perforating, and high viscosity fluid) may or may not work. Careful trial of these different techniques combined with properly-performed diagnostic tests remains today the most pragmatic approach to address near-wellbore tortuosity problems and lower the risk of early screen-out.

4 PREDICTING AND ENGINEERING NETWORK VERSUS LOCALIZED GROWTH GEOMETRY

4.1 Importance and background

Generation of complex networks of hydraulic fractures is a relatively new concept. Its current form arose out of field trials in the Barnett Shale, Texas, in the late 1990s and early 2000s wherein multiply-branched HFs were inferred from lineaments of microseisms (Maxwell *et al.*, 2002; Urbancic & Maxwell, 2002; Warpinski *et al.*, 2005). An example of evidence of network-like growth is given by Figure 8a, where the contrast to microseismic data taken to imply simpler, bi-wing geometry, Figure 8b, is clear.

These observations became profoundly influential. For starters, successful production of gas from the Barnett Shale – the first major success in hydrocarbon production from shale reservoir — was presumed to be tied to the large contact surface generated by this "complexity" (*e.g.* review of King, 2010). Hence, the practice began for low permeability gas, and later oil, wells of quantifying the so-called "Stimulated Reservoir Volume" (SRV), which is essentially the volume of reservoir within a box that contains most of the microseisms generated by a stimulation (from Warpinski *et al.*, 2005, although the concept and terminology was apparently developed previously for Enhanced Geothermal Systems, *e.g.* Kruger and Yamaguchi, 1993). SRV, or more specifically the width of the box defining what eventually because known as SRV,

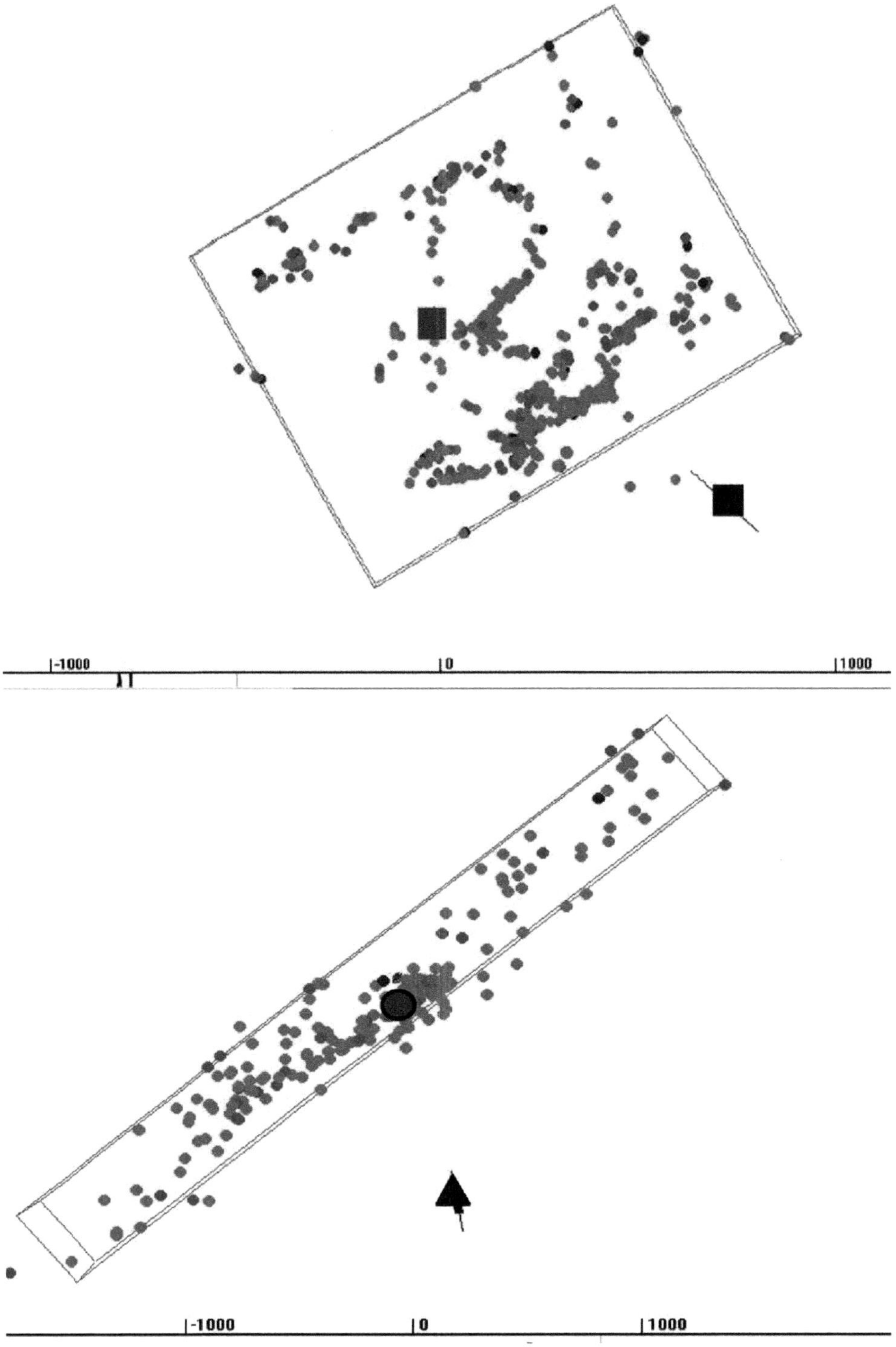

Figure 8 Microseismic data for vertical wells in the Barnett Shale. a) Example of network growth (from with permission Maxwell *et al.*, 2002, with permission). b) Example of simple geometry (from Urbancic & Maxwell, 2002, with permission). In both cases the scale is shown in feet, the light gray square/circle indicates the well and the dark gray square/triangle the monitoring well.

was found to correlate with production for Barnett Shale wells by Fisher *et al.* (2002) and therefore its size became a metric of success of a stimulation.

As the concept of "complexity" and the measurement of SRV took hold in the industry, significant effort was aimed at development of models to predict HF growth in reservoirs with a preponderance of natural fractures (NFs). Concomitantly, previous studies of fracture intersection with natural fractures were brought to the forefront. Not only were new models for predicting HF growth in NF reservoirs considered vital for optimizing stimulation in the Barnett Shale, but also the expansion of shale gas and oil development to other plays was driving a need to understand basic conditions for network versus localized growth. This section is aimed at summarizing the basic conditions for network growth that have been identified through these efforts.

4.2 Field evidence

Network growth is perhaps most strikingly shown in microseismic data from vertical well stimulation in the Barnett Shale, Figure 8a (Maxwell *et al.*, 2002). This data set shows distinct, parallel lineaments connected by oblique lineaments, thus forming the inferred network of HFs. There are some important features common to these early field studies. The first is that all involve injection of water rather than higher viscosity gel-type fluids. Indeed, in complimentary work, Warpinski *et al.* (2005) contrast water and gel stimulation for a horizontal well, again in the Barnett Shale. They show a larger microseismic cloud for the treatment that used water (Figure 9a-b). This result, along with several other studies showing superior production from water fractures (*e.g.* Mayerhofer *et al.*, 1997, and see review in King, 2010), became cornerstones for the argument in favor of water fractures in situation where a network of fractures is desired. It is, however, important and often overlooked that use of water seems to be necessary but not sufficient for network growth, even in the Barnett; two of the six examples presented by (Urbancic & Maxwell, 2002) indicate relatively simple, bi-wing geometry.

A second observation of these early Barnett studies is that the lineaments tend to be distinct and separated by several hundred feet. The spacing between the lineaments is therefore similar to the fracture height. Furthermore, the distinction of the lineaments is warranted via microseismic events because their separation is far greater than the typical, tens-of-feet location uncertainty.

A third observation is that evidence for network growth comes almost exclusively from the Barnett Shale. One reason is that nearly all shale stimulations in other formations have entailed injection to entry points separated by tens of feet distributed along horizontal wells. Hence, the nominal fracture spacing and the location uncertainty are approximately the same so that a complex fracture and single-stranded fracture are indiscernible from the available data. Another possible explanation lie in the fact that the Barnett Shale exhibit nearly equal in-situ horizontal stresses compared to other US shale basin; larger in-situ deviatoric stresses constraining fracture growth to a dominant plane.

In spite of a growing trend toward predictions of complex growth from models and assumption of complex growth underlying the generation of microseismic clouds, direct observation of hydraulic fractures shows arguably simpler tendency. Core-through data from the Multiwell/M-Site Project, Piceance Basin, Colorado (Northrop

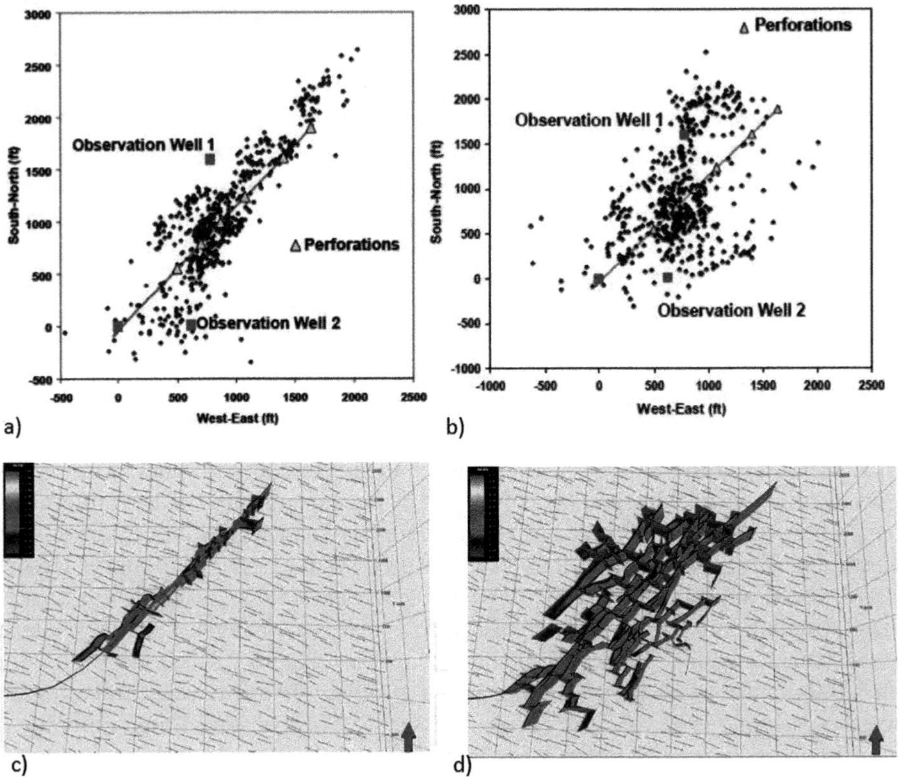

Figure 9 Images of microseismic data (a-b, from Warpinski *et al.* (2005), with permission) and numerical simulations (c-d, from Kresse *et al.*, 2013b) for a horizontal well in the Barnett Shale. Contrasting cases of presumably-localized gel fractures (left, a and c) and presumably network-like water fracture (right, b and d).

& Frohne, 1990), shows a cluster of 30 fracture strands covering an interval of about 4 feet. Of these, about 10 strands contain a white substance presumed to be dehydrated gel and 4 dominant strands contain clumps of 6–10 sand grains presumed to be residual proppant (Warpinski *et al.*, 1993). On the one hand, this finding comprised one of the first and most influential evidences of HF complexity, and similar multi-stranded morphology has been observed in core-through HF mapping in other basins (Fast *et al.*, 1994). Nonetheless, it is important to realize that the complexity only covered a section a few feet in width and therefore comprises an observation at a completely different scale than the hundreds-of-feet scale of complexity that comprises the focus of discussions since the Barnett Shale studies (*e.g.* Urbancic & Maxwell, 2002). Furthermore, even this very localized complexity has been argued to comprise a worst case scenario due to the specific conditions in the Piceance Basin (Nolte, 1993). Also note that microseismic clouds generated by growth of similar (but subsequent) HF experiments in this basin are on the order of 100 feet in width Warpinski *et al.* (1996), which is on the order of the location uncertainty and, based on comparison with the several feet of width of the "complexity" observed for previous treatments, is not likely to be an indicator of with width of the stimulated zone.

Direct observation of HFs is also available from experiments in which hydraulic fractures are placed in orebodies or coal so that the HFs are exposed by mining activities. For example, Jeffrey *et al.* (2009) find that propped HFs in a heavily-fractured and faulted orebody are typically found in single, quasi-planar strands. Occasionally two, and at one point three, strands were observed. But secondary strands are always of limited persistence. Furthermore, the observed fracture orientation is consistent overall with the orientation of the minimum in situ stress, with limited offsets perturbing otherwise quasi-planar HFs to form a stair-stepping geometry. In general, then, the observed HFs are not network like in this minethrough experiment, and this is in contrast to the predicted complexity from a DFN-type HF simulator (Rogers *et al.*, 2011). Moreover, persistence of network-like propped fractures are yet to be observed in spite of the experience of other mine through experiments in both hard rocks (*e.g.* van As, 2000; Jeffrey *et al.*, 2009; Bunger *et al.*, 2011) and coal (*e.g.* Elder, 1977; Lambert *et al.*, 1980; Diamond & Oyler, 1987; Jeffrey *et al.*, 1992; Jeffrey *et al.*, 1995).

4.3 Modeling insights

Numerical explorations provide one means of discerning the conditions favoring network growth. Three main conclusions can be drawn from available results:

(a) Network growth is promoted when the horizontal stress components are equal and is increasingly suppressed as the relative difference is increased (Olson & Dahi-Taleghani, 2009; Kresse *et al.*, 2013a).

(b) Low viscosity fluids will be the most amenable to network growth while high viscosity fluids will be the most amenable to forming a single dominant hydraulic fracture (*e.g.* Kresse *et al.* 2013b).

(c) The continuous joints should be oblique to the minimum stress direction (Olson & Dahi-Taleghani, 2009; Cipolla *et al.*, 2011).

4.4 Laboratory investigations

Besides these basic conditions, one of the critical observations from simulators is that the predicted HF geometry strongly depends on the criterion used to determine whether an HF will cross or be deflected when it encounters an NF, especially for simulating injection of high viscosity fluids (Kresse *et al.*, 2013b). This observation is a strong motivation for laboratory experiments aimed at understanding the mechanisms governing HF-NF interaction. In these experiments, a common theme is that HF crossing NFs is promoted by larger differential stresses (Warpinski & Teufel, 1987; Llanos *et al.*, 2006; Gu *et al.*, 2011; Bunger *et al.*, 2015). In this regard, HF lab experiments are similar to the "dry" crack experiments of Renshaw and Pollard (1995) (R&P). However, there are important distinctions.

Firstly, experiments with fluid driven cracks do not show dependence on friction coefficient as in the theory and experiments of R&P (Bunger *et al.*, 2015). Furthermore, use of high viscosity fluids promote localized growth relative to low viscosity fluids in fractured block experiments (de Pater & Beugelsdijk, 2005). Hence, attributing the role

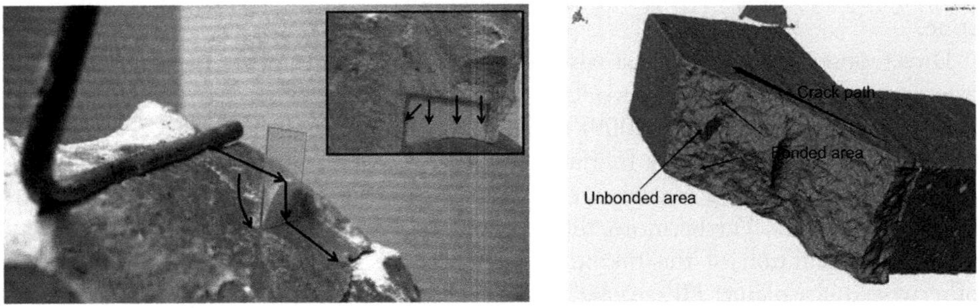

a) b)

Figure 10 Figure illustrating HF engulfing smaller discontinuities. a) HF path diverting around a strong inclusions (glass inclusion in hydrostone), from Olson *et al.* (2012), with permission. b) Digitized HF path showing offset as it engulfed a debonded region of an otherwise bonded natural fracture in a mortar specimen, after Fu *et al.* (2015).

of stresses would seem to not be related to the impact on the frictional strength of the interface. Rather, as discussed by Bunger *et al.* (2015), the stresses impact on the permeability of the interface, which is probably more relevant for fluid-driven cracks (and which is accounted for, for example, via the model of Chuprakov *et al.* (2014)). In support of the impact of stress on network growth, note also that under conditions of 300 psi vertical stress and zero horizontal stresses, Abass *et al.* (1990) observe extreme complexity at laboratory scale in blocks of coal, while Suarez-Rivera *et al.* (2013) demonstrate that HF paths are more directed and ostensibly less "complex" when shale laboratory specimens are subjected to increasing horizontal stress differences.

In summary, one theme is that lab experiments indicate that fluid flow plays a vital role in determining whether HFs will cross or deflect to NFs. A second, emerging theme is that HF-NF interaction often entails mechanisms and geometries that are not readily captured by 2D models. For example, ultrasound monitoring indicates NF-parallel (*i.e.* out-of-plane for a 2D model) growth of the HF prior to crossing when initially circular HFs approach discontinuities Bunger *et al.* (2015). Growth that is initially halted by a discontinuity is also inferred from laboratory experiments with HFs interacting with relatively strong and high permeability (compared to gypsum matrix) sandstone lenses (Blair *et al.*, 1990). But perhaps the most striking is the ability of an HF to engulf hard inclusions (Olson *et al.*, 2012) or weak discontinuities (Bahorich *et al.*, 2012; Fu *et al.*, 2015), as illustrated in Figure 10. This ability to bypass a discontinuity is undoubtedly a contributor to persistence of HFs through fractured and otherwise heterogeneous rocks. Furthermore, it is incompatible with the typical assumption of models, particularly 2D and pseudo 3D models, that NFs persist with uniform properties through the full height of the reservoir with uniform properties.

4.5 State of knowledge

While it is broadly agreed that isotropic stress, low viscosity fluid, and natural fractures that are oblique to the principal stress direction are basic ingredients for network growth, there are many unresolved issues. These include the role of leakoff and proppant transport. But these issues can be addressed by models that include more coupled physical processes and or make fewer geometric restrictions. In essence these are solvable problems with a series of extensions to existing approaches. There are, however, at least two issues that pose deeper challenges. The first stems from limitations on existing monitoring. Microseismic monitoring cannot distinguish between network and localized growth when entry points are spaced similarly to the location uncertainty (tens of feet). But besides direct observation, such as core through or mine through mapping, microseismic monitoring is the most direct and by far the most commonly applied method for inferring hydraulic fracturing growth geometry in the field. Problematically though, its resolution limits can leave models predicting simple versus complex growth geometry with similar degrees of model-data agreement.

Secondly, a drive toward including more physical processes and or more geometric complexity in models is inevitably a drive toward more required characterization. But some critical parameters are nearly impossible to characterize. Perhaps most notable of problematic parameters is the size (*i.e.* extent) of natural fractures and other heterogeneities in the reservoir. NF characterization is currently performed from well logs and it is not possible to discern the extent of a fracture from its intersection with the wellbore.

Taken together, these challenges make network versus localized growth prediction one of the most challenging issues in hydraulic fracturing. Overcoming these challenges will require 3-D models that are informed by characterization that includes natural fracture(s) strength, permeability, orientation, and size. Moreover, monitoring must be improved so as to definitively benchmark models using direct observation from offset wells and/or from microseismic monitoring with an order of magnitude better resolution than is currently available.

5 PROMOTING SIMULTANEOUS GROWTH OF MULTIPLE HYDRAULIC FRACTURES

5.1 Importance and background

Multiple hydraulic fracture growth from a horizontal well was first inferred from microseismic monitoring by Fisher *et al.* (2004), making it one of the most recent major developments in hydraulic fracturing. They observe "multiple linear features at roughly 500 foot spacing," compared to roughly 300 foot fracture height, "regardless of perforation cluster location." Their observations show that injection from a horizontal well can lead to multiple HF growth, which is distinct from network growth in that fractures are fed directly by the wellbore rather than by forming a network with other fractures. Hence, in its simplest form, multiple HF, as referenced here, consists of an array of planar and parallel HFs.

Following early observations, simultaneously generating multiple HFs has now become the essential goal of multistage HF treatments, particularly those using the so-called plug and perforate method. In this approach, the wellbore is divided into intervals on the order of hundreds of feet in length. Clusters of perforations are placed within these intervals. Typically there are 3–6, 2 foot long clusters of perforations holes distributed with roughly 30–100 feet of separation between them. The desired result is uniform stimulation resulting from uniform distribution of the fluid and proppant among the entry points (perforation clusters). This section describes data and modeling efforts aimed at discerning the degree to which uniform distribution is achieved and, in turn, devising approaches to improve the uniformity of stimulation associated with multiple simultaneous HF growth from horizontal wells.

5.2 Modeling insights

The emergence of multiple HFs growing with a spacing that is ~1.2–2.5 times the fracture height, as observed by Fisher *et al.* (2004), corresponds to an energy-minimizing geometry for an array of simultaneously-growing HFs (Bunger, 2013). In essence, this energetically-preferred spacing emerges from a competition. On the one hand, the HFs avoid growing very close to each other (relative to the fracture height) because more energy is required in order to overcome the elevated stresses caused by the fractures' neighbors (as previously pointed out by among other Fisher *et al.*, 2004). But, opposing this tendency to avoid growing near one another, minimization of energy dissipation associated with viscous fluid flow drives the system to split the fluid among all possible growing HFs (Bunger, 2013). The result of this competition is an energetically preferred spacing of 1.2–2.5 times the fracture height, where closer spacing in within this range corresponds to cases with larger pressure losses across the perforations (Bunger *et al.*, 2014).

In this context, current practice of multistage HF invariably places perforation clusters close to each other relative to the fracture height. The mechanical impact of the close proximity is typically referred to as "stress shadow, and has been associated with:

1. Tendency of the growth of central fractures to be suppressed due to the higher stress in the central part of the array compared with the ends (*e.g.* Germanovich *et al.* 1997; Fisher *et al.* 2004; Olson, 2008; Abass *et al.*, 2009; Meyer & Bazan, 2011; Kresse *et al.*, 2013a; Wu & Olson, 2013; Lecampion *et al.*, 2015b). See an example in Figure 11a.
2. Deflection of the HF path toward the regions of lower stress (*e.g.* Roussel & Sharma, 2010; Roussel & Sharma, 2011; Bunger *et al.*, 2012; Sesetty & Ghassemi, 2013; Wu & Olson, 2013; Daneshy, 2015) or, in some cases, complete re-orientation due to fracture-induced changes in the principal stress directions (Roussel *et al.*, 2012).

Note that there is a coupling between these behaviors. Daneshy (2015) shows that the non-uniformity of the growing hydraulic fractures can impact the deflection of their paths. The deflection/curving of simultaneously growing HFs also appears to be suppressed when the difference between the two horizontal stresses increase (*e.g.* Xu and Wong, 2013; Lecampion *et al.*, 2015b).

Besides predicting the challenges associated with stress shadow effects, models have been used in conjunction with field experience to suggest approaches that can promote multiple HF growth from horizontal wells. Some promising techniques include:

1. *Limited Entry*: This approach for horizontal wells draws inspiration from decades of application for multi-zonal stimulation from vertical wells (*e.g.* Howard & Fast, 1970). It entails the use of fewer or smaller perforation holes in order to

increase pressure across the perforations, especially for the outer and/or heel-ward clusters of the stage. In this way the technique promotes fluid/proppant injection to the central/toe-ward clusters. Its advantage is that it is operationally simple to execute. Model predictions demonstrate its potential to promote uniform fluid/ proppant distributions (Lecampion & Desroches, 2015a,b; Daneshy, 2015). However, Lecampion *et al.* (2015b) show that success is sensitive to the details of the design. Hence, it is unlikely that a rough guess at perforation cluster design for limited entry will be successful and so model-driven design is required. But, more problematically, this sensitivity to details shows a possible lack of robustness to perturbations from the designed entry losses, for example due to unexpected near wellbore fracture tortuosity as well as plugging and/or erosion of perforation holes as the stimulation progresses.

2. *Diverting Agents*: Diverting Agents are additives that are intended to collect at the entry of the fracture that are taking the most fluid, eventually blocking them so that fluid is diverted to other clusters. They are designed to naturally degrade and/ or dislodge and flow out of the well during flow back and/or production and, like limited entry methods, their origins like in multi-zonal stimulation from vertical wells (Gallus & Pye, 1972). Field trials show that diversion is response to pumping a diverter is incomplete, yet often deemed sufficient to be practically effective. For example, Potapenko *et al.* (2009) show that the spatial distribution of microseismicity shifts along a horizontal well in the Barnett Shale after a diverting agent is pumped. Microseismicity continues to be generated from the region that was apparently the focus of stimulation before the diverter was pumped so diversion is apparently not complete although it is certainly demonstrable. Similar conclusions can be drawn from a case study of 3 wells in the Eagle Ford Shale, where the monitoring included not only microseismicity but also unique radioactive tracers that were pumped before and after diverting agents were introduced (Viswanathan *et al.*, 2014). Again these results show that not every diverter pill results in measureable diversion and the diversion, when effective, is not complete. In the absence of advanced monitoring, diversion is usually said to have been successful when the well head pressure rises after injection of the diversion pill, and another pressure peak/drop ("breakdown") is observed, although obviously this does not ensure uniformity of the treatment.
3. *Log-Driven Perforation Cluster Placement*: This approach employs (typically) multipole sonic (wavespeed) data, interpreted in order to estimate the distribution of stress along the wellbore (Slocombe *et al.*, 2013). Perforation cluster locations are chosen so as to minimize variation of stress among the clusters within each stage. A major field trial in the Eagle Ford Shale demonstrated that this approach decreases the variation of stress among the clusters within each stage from upwards of 1000 psi to a couple hundred psi. In so doing, the percentage of non-producing clusters is approximately cut in half relative to the control group of wells wherein perforation clusters are uniformly-spaced without regard for stress or other reservoir properties. Slocombe *et al.* (2013) do not, however, report the impact on productivity of the wells. On the other hand, Lim *et al.* (2014) report production impacts relative to a control group for 4 wells in the Marcellus Shale. The log-driven placement cases produced over 20% more gas over their first 400 days than the control group. However, when production is normalized for

differences in number of perforation clusters among the wells, impact on production is only in the range of a 5–10% increase. Nonetheless, the potential is clear. However, a major drawback of the approach is cost; execution requires logging and interpretation that is almost never carried out otherwise for onshore, low permeability gas/oil wells.

4. *"Interference Fracturing"*: This so-far model-based approach essentially seeks to balance the impact of stress shadow on each fracture by varying the spacing between HFs. It is therefore a corollary of log-based selective of cluster location, but instead of focusing on pre-existing stress it focuses on HF induced stress. For a five cluster per stage example, Peirce and Bunger (2015) show that stress shadow can be balanced by moving the second and fourth clusters closer to the first and fifth clusters, respectively (Figure 11b). In this way, stress shadow is increased on clusters 1 and 5 and decreased on cluster 3, leading to more balanced distribution of fluid relative to uniform spacing. Interestingly, as some point during injection clusters 2 and 5 begin taking fluid preferentially to 1, 3, and 5 so that a relatively uniform final distribution of fractures is achieved (Figure 11c).
5. *Promoting Growth in Viscosity Dominated Regime*: Analytical and numerical models agree that growth of multiple HFs (planar as opposed to network-like) is more stable for viscosity- dominated HFs (Bunger, 2013; Bunger & Peirce, 2014; Lecampion, 2015). While still untested at lab or field scale, one implication is that water-driven HFs may be more evenly distributed if the water injection is preceded by a high viscosity gel pad to promote initiation and growth from all perforation clusters.

5.3 Laboratory investigations

Simultaneous growth of multiple hydraulic fractures poses a significant challenge for laboratory experimentation. The problem is mainly associated with initiation of multiple hydraulic fractures at laboratory length and time scales. Whether in the laboratory or the field, the first to initiate among several possible entry points will correspond to lowest stress or the weakest rock. Subsequent initiation is hypothesized to rely upon delayed rock failure (Bunger and Lu, 2015) or the continued increase in pressure that can occur following HF initiation (Bunger *et al.*, 2010; Abbas & Lecampion 2013). At laboratory scale, the experimental duration is often too short for the former mechanism and the fluid viscosity too small for the latter. Furthermore, even in laboratory experiments where there is evidence of multiple HF initiation, all but one of the initiated HFs stops growing so that most of the growth is concentrated in just a single HF (El-Rabaa, 1989).

Nonetheless, after many documented challenges, Crosby (1999) (see also Crosby *et al.*, 2002) was able to obtain two simultaneously growing hydraulic fractures in concrete blocks. One of these initially-parallel HFs curved toward and eventually intersected the other, a type of attractive curving that is predicted by models (*e.g.* prediction 1, above Bunger *et al.*, 2012; Sesetty & Ghassemi, 2013; Daneshy, 2015). This attractive curving is also observed in laboratory experiments for sequential HF growth in crystalline rock by Bunger *et al.* (2011), who show that the curving is suppressed by increasing the minimum stress, as predicted by the simulations of

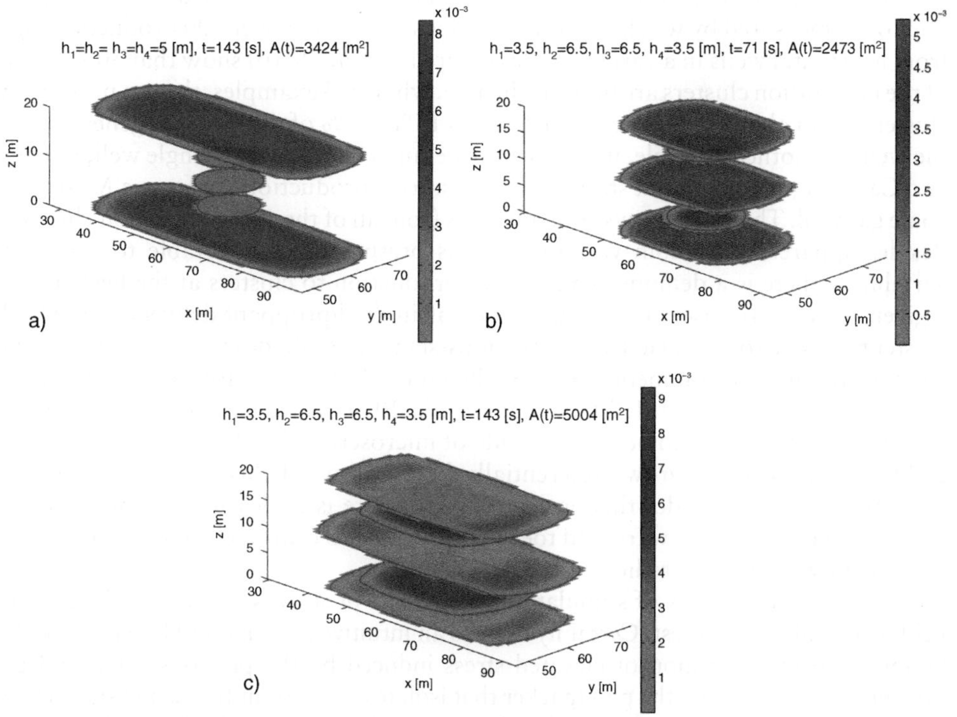

Figure 11 Planar 3D model results illustrating stress shadow and "interference" fracturing (from Peirce and Bunger, 2015, with permission). a) Stress shadow suppression of central fracture for 5 uniformly-spaced HFs, b-c) Improved uniformity obtained by moving entry point for HFs 2 and 4 to be closer to HFs 1 and 5. Here (b) illustrates how this change encourages growth of HF 3 (central HF) early in the growth, after which (c) illustrates the transition to injection preferentially going to HF 2 and 4. Note the z axis in these plots is the direction of the horizontal well.

Bunger *et al.* (2012) (As an aside, the simulations indicate the curving can also be *reduced* but not suppressed by increasing the deviatoric stress).

5.4 Field evidence

In spite of the goal of uniform stimulation, field evidence indicates that stimulation is most often far from uniform. For example, distributed acoustic sensing (DAS) using fiber optic sensors along a horizontal well in a shale reservoir shows good initiation from all 9–10 clusters in each stage of this study. However, as the treatment goes on, one cluster dominates, taking in total 30%–70% of the fluid and proppant that is injected (Sookprasong *et al.*, 2014). Similarly, growth from multiple perforation clusters but with dominance of one cluster can be inferred from microseismic data from the Barnett Shale (Daniels *et al.*, 2007).

Concurring evidence of non-uniform stimulation is provided by distribution production as measured by wireline production logs. Analyzing over 100 production logs from horizontal wells in a variety of plays, Cipolla *et al.* (2010) show that 30%–40% of the perforation clusters are non-productive, while in 4 examples, the top producing cluster (out of about 30 in all cases) produces 15%–45% of the total gas generated by the well. In another example, which is more detailed but for only a single well, Bunger and Cardella (2015) present statistical analysis of a production log from a Marcellus Shale gas well. This data shows contributions from all of the 105 perforation clusters. Besides apparently random variation this is presumably attributable to reservoir variability, there is a demonstrable bias of production to clusters at the heel end of a given stage. This observation suggests a bias in fluid/proppant delivery to the heel cluster relative to others. Hence, the previous-stage stress shadow, in the past inferred from increasing Instantaneous Shut in Pressure (ISIP) from toe to heel in wells (Vermylen & Zoback, 2011; Manchanda *et al.*, 2014) as well as changes in correlation between ISIP and moment magnitude of microseismicity (Vermylen & Zoback, 2011), also makes HFs grow preferentially from heel-ward clusters where the stress shadow is smaller. Besides this heel-ward bias, there is a measureable bias also to outer clusters (both heel-ward and toe-ward) relative to central clusters in each stage (consistent with model prediction 1, above).

Besides non-uniformity of stimulation, the impact of stress shadow on HF paths at field scale is also of interest. Certainly there is an intuitive notion that HF paths should be impacted by the region of elevated stress induced by the previous stage. Indeed pressure increases below the plug/packer that isolates the current fracturing stage from previous stages have been presented and interpreted as evidence for HF deflection/re-orientation that resulted in stress interaction and/or intersection between the growing HF(s) and HFs from previous stage(s) (Daneshy, 2014). Hence the deflection/re-orientation hypothesis has some support from field data. However, as valuable as the data is, the evidence is indirect, without a sense of the proportion of wells that exhibit this below plug/packer pressurization, and the interpretation is dependent on an underlying assumption that there is no leakage along the wellbore through the cement nor is there any fluid bypass of the plug/packer.

Moving beyond indirect evidence, some direct observations also support the hypothesis of HF curving while also pointing to a caveat that HF curving or re-orientation is not *always* obtained, even for very closely spaced HFs. Some relevant evidence comes from a multiple sequential fracturing experiment from a vertical well carried out by Oak Ridge National Lab (de Laguna *et al.*, 1968). In this investigation, HF paths are inferred from intersections with offset wells and, as pointed out in the interpretation from Bunger *et al.* (2012), two of the five fractures in one component of this trial appear to have been deflected away from previous fractures. This behavior is consistent with simulations (Bunger *et al.*, 2012). Counter-intuitively, the HFs that curved had a larger spacing between them than those that did not; this, too, is consistent with the same, counter-intuitive model prediction associated with the fact that principal stress orientations are nearly fracture-parallel both very close to and very far from a previously-placed HF (Bunger *et al.*, 2012). A lack of curving, again consistent with model predictions, is also demonstrated in mine through data (Bunger *et al.*, 2011). Unfortunately, definitive evidence of HF deflection, or lack thereof, has proven challenging to obtain in shale reservoirs themselves that are subjected to very closely spaced hydraulic fracturing. Hence, the importance of HF deflection at that scale and under actual reservoir conditions remains unknown.

5.5 State of knowledge

In summary, the field-tested and model-suggested approaches to promote multiple fracture growth either increase energy dissipation associated with fluid flow (see discussion of Bunger *et al.*, 2014), balance stresses resisting HF growth, or divert the fluid at the wellbore. But, in spite of the lack of conflict among the approaches, to date there is no known experience combining them in a given treatment. For example, log-based locations that minimize stress differences, injection of high viscosity pads, and limited entry are probably most important for initiation and early stages of growth. And they certainly can in principle be used together to increase their collective robustness. As the treatment continues, the induced stress shadow increases and hence interference fracturing, which balances the induced stress shadow among the fractures, could be important to ensure continued growth. Simulations show, however, that eventually influx to some fractures diminishes as interference fracturing growth. At this point, introduction of a diverter would promote influx to these otherwise suppressed fractures. Taken together, hybrid approaches that leverage multiple methods for promoting simultaneous HF growth have significant potential to improve the effectiveness of multistage, multi-cluster horizontal well completions.

There is one important footnote to the subject of generating multiple, simultaneously-growing hydraulic fractures. An obvious alternative is to avoid the need to simultaneously generate hydraulic fractures by individually isolating the zones that are intended to be stimulated. These "pinpoint" or "single-entry" methods exist; indeed, like limited entry methods, they have their roots in stimulation of multiple zones from vertical wells (*e.g.* Houser & Hernandez, 2009). These single entry methods, when applied to multi-stage stimulation from horizontal wells, have distinct advantages and, in a recent case study carried out in the Granite Wash, Anadarko Basin, Texas, have been shown to drastically reduce production decline and approximately double the wells Estimated Ultimate Recovery (EUR) (Maxwell *et al.*, 2013). The main drawback is the cost of these systems. Hence, concurrent with the efforts to improve simultaneous generation of hydraulic fractures from multiple entry points are concerted efforts to develop increasingly-economical single-entry stimulation systems.

6 CONCLUSIONS

This review has focused on height growth, near wellbore tortuosity, network HF growth in naturally fractured reservoirs, and simultaneous growth of multiple HFs. When examining all of these together, we see the importance of certain issues cross-cutting multiple topics. For example, high viscosity and high injection rate promote height growth through increasing the net pressure, diminish near wellbore tortuosity, suppress complex network-like growth geometry, and promote simultaneous growth of HFs from multiple entry points. Hence, the fluid choice and pumping schedule impacts in a multiplicity of ways.

Similarly, the in situ stresses impact multiple aspects of HF growth. For starters, height growth is mainly controlled by the difference between the minimum stress in the reservoir and neighboring formations. At the same time, increasing the difference between the two horizontal principal stresses promotes simpler fracture geometry, both from the perspective of diminishing fracture tortuosity associated with initiation and promoting localized over network-like growth.

Of the monitoring methods, microseismic monitoring is shown to give insights that crosscut multiple areas of interest. It is used to estimate the extent of height growth, to ascertain the effectiveness of diverting agents for promoting multiple HF growth, and, in some cases, to discern growth of multiple HFs and/or network versus localized growth. However, the location uncertainty (order of tens of feet) limits resolution so that microseismicity is unable to distinguish network from localized growth, or to discern growth from multiple entry points, when the entry points are separated by tens of feet.

The future advances in each of these areas also possess some common themes. Firstly, modeling insights in all areas are limited by the open challenges associated with 3D modeling that maintains the appropriate coupling among fluid flow, rock deformation, and rock breakage. Fully coupled, fully 3D HF simulation would enable unprecedented insight into the role of layering and weak interfaces in HF height growth, the controlling factors in near wellbore fracture tortuosity, and the impact of nature fractures and hard/soft or high/low permeability inclusions on HF paths and possible formation of fracture networks.

Development of coupled 3D simulators, however, not only generates insights but also highlights a second area for future advances. As modeling pushes forward, the critical limitation increasingly becomes characterization. Height growth will be controlled by potentially poorly constrained in situ stresses and mechanical details of the lithological boundaries that are unresolvable with current measurements. Near wellbore tortuosity is almost certainly controlled by heterogeneity at a currently unresolvable scale and/or at a depth that, while still in the near-wellbore region from the perspective of the HF, is out of the reach of current well logging technology. Network growth is dependent not only on the density, orientation, and mechanical properties of natural fracture sets, but also on the extent of the natural fractures and other heterogeneities. These details are all beyond the reach of the state of the art in characterization and highlight the need for concerted research efforts. In the absence of these advances, the problem will remain intractable due to limited data even with unbounded growth in simulation capability.

Finally, all areas are limited by the resolution of the available monitoring methods. Profound gains would be enabled with, for example, the ability to resolve millimeter-scale detail in the near wellbore region and reducing the location uncertainty associated with microseismic monitoring by one order of magnitude.

ACKNOWLEDGMENTS

The authors wish to thank Egor Dontsov and Anthony Peirce for providing unpublished data and Robert Hurt for pointing us to several relevant sets of published field data regarding stress shadow and fracture re-orientation. Numerous discussions with

Jean Desroches on near-wellbore effects and HF in general are also gratefully acknowledged.

REFERENCES

Abass, H., M. van Domelen, and W. El-Rabaa (1990). Experimental observations of hydraulic fracture propagation through coal blocks. In *SPE Eastern Regional Meeting*, Columbus, Ohio. SPE 21289.

Abass, H. H., S. Hedayati, and D. L. Meadows (1996). Nonplanar fracture propagation from a horizontal wellbore: Experimental study. *SPE Production & Facilities 11*(3), 133–137.

Abass, H. H., M. Y. Soliman, A. M. Tahini, J. Surjaatmadja, D. L. Meadows, and L. Sierra (2009, October 4–7). Oriented fracturing: A new technique to hydraulically fracture an openhole horizontal well. In *Proceedings SPE Annual Technical Conference and Exhibition*, New Orleans, LA, USA. SPE 124483.

Abbas, S., E. Gordeliy, A. P. Peirce, B. Lecampion, D. Chuprakov, and R. Prioul (2014). Limited height growth and reduced opening of hydraulic fractures due to fractureoffsets: An XFEM application. In *SPE Hydraulic Fracturing Technology Conference*, The Woodlands, Texas, USA. SPE 168622.

Abbas, S. and B. Lecampion (2013). Initiation and breakdown of an axisymmetric hydraulic fracture transverse to a horizontal wellbore. In A. P. Bunger, J. McLennan, and R. G. Jeffrey (Eds.), *Effective and Sustainable Hydraulic Fracturing*, Chapter 19. Rijeka, Croatia: Intech.

Abou-Sayed, A., R. Clifton, R. Dougherty, and R. Morales (1984, May 13–15). Evaluation of the influence of in-situ reservoir conditions on the geometry of hydraulic fractures using a 3-d simulator: Part 2-case studies. In *SPE Unconventional Gas Recovery Symposium*, Pittsburgh, PA, USA. SPE 12878.

Adachi, J., E. Siebrits, A. Peirce, and J. Desroches (2007). Computer simulation of hydraulic fractures. *International Journal of Rock Mechanics and Mining Sciences*, *44*, 739–757.

Advani, S. H., T. S. Lee, and J. Lee (1990). Three-dimensional modeling of hydraulic fractures in layered media: Part ifinite element formulations. *Journal of Energy Resources Technology 112*(1), 1–9.

Alekseenko, O. P., D. I. Potapenko, S. G. Cherny, D. V. Esipov, D. S. Kuranakov, and V. N. Lapin (2012, February 6–8). 3-d modeling of fracture initiation from perforated non-cemented wellbore. In *SPE Hydraulic Fracturing Technology Conference*, The Woodlands, Texas, USA. SPE 151585.

Aud, W., T. Wright, C. Cipolla, and J. Harkrider (1994, September 25–28). The effect of viscosity on near-wellbore tortuosity and premature screenouts. In *SPE Annual Technical Conference and Exhibition*, New Orleans, Louisiana, USA. SPE 28492.

Bahorich, B., J. E. Olson, and J. Holder (2012). Examining the effect of cemented natural fractures on hydraulic fracture propagation in hydrostone block experiments. In *SPE Annual Technical Conference and Exhibition*, San Antonio, Texas, USA. SPE 160197.

Behrmann, L. and K. Nolte (1999). Perforating requirements for fracture stimulations. *SPE Drilling & Completion 14*(4). SPE 59480.

Behrmann, L. A. and J. L. Elbel (1991). Effect of perforations on fracture initiation. *Journal of Petroleum Technology 43*(5), 608–615. SPE 20661.

Ben Naceur, K. and E. Touboul (1990). Mechanisms controlling fracture-height growth in layered media. *SPE Production Engineering 5*(2), 142–150.

Beugelsdijk, L., C. J. de Pater, and K. Sato (2000, April 25–26). Experimental hydraulic fracture propagation in a multi-fractured medium. In *SPE Asia Pacific Conference on Integrated Modelling for Asset Management*, Yokohama, Japan. SPE 59419.

Blair, S. C., R. K. Thorpe, and F. E. Heuze (1990). Propagation of fluid-driven fractures in jointed rock: Part 2 physical tests on blocks with an interface or lens. *International Journal of Rock Mechanics and Mining Sciences*, *27*(4), 255–268.

Blanton, T. L. (1982). An experimental study of interaction between hydraulic fractures and pre-existing fractures. In *Proceeding SPE/DOE Unconventional Gas Recovery Symposium*, Pittsburgh, PA, USA, pp. 613–627. SPE 10847.

Bunger, A. P. (2013). Analysis of the power input needed to propagate multiple hydraulic fractures. *International Journal of Solids and Structures 50*, 1538–1549.

Bunger, A. P. and D. J. Cardella (2015). Spatial distribution of production in a marcellus shale well: Evidence for hydraulic fracture stress interaction. *Journal of Petroleum Science and Engineering 113*, 162–166.

Bunger, A. P., R. G. Jeffrey, and X. Zhang (2011, June 26–29). Experimental investigation of the interaction among closely spaced hydraulic fractures. In *Proceedings 45th U.S. Rock Mechanics Symposium*, San Francisco, CA, USA. Paper No. 11-318.

Bunger, A. P., R. G. Jeffrey, and X. Zhang (2014). Constraints on simultaneous growth of hydraulic fractures from multiple perforation clusters in horizontal wells. *Soc. Pet. Eng. J. 19* (04), 608–620.

Bunger, A. P., J. Kear, R. G. Jeffrey, R. Prioul, and D. Chuprakov (2015, May 10–13). Laboratory investigation of hydraulic fracture growth through weak discontinuities with active ultrasound monitoring. In *Proceedings 13th International Society for Rock Mechanics (ISRM) Congress*, Montreal, Quebec, Canada.

Bunger, A. P., A. Lakirouhani, and E. Detournay (2010, August 25–27). Modelling the effect of injection system compressibility and viscous fluid flow on hydraulic fracture breakdown pressure. In *Proceedings 5th International Symposium on In-situ Rock Stress*, Beijing, P.R. China. ISRSV-06-010, pp. 59–67.

Bunger, A. P. and G. Lu (2015). Time-Dependent Initiation of Multiple Hydraulic Fractures in a Formation with Varying Stresses. *SPE Journal*, *20*(6), 1317–1325.

Bunger, A. P. and A. P. Peirce (2014, June 21–23). Numerical simulation of simultaneous growth of multiple interacting hydraulic fractures from horizontal wells. In *Proceedings ASCE Shale Energy Engineering Conference*, Pittsburgh, PA, USA, pp. 201–210.

Bunger, A. P., X. Zhang, and R. G. Jeffrey (2012). Parameters effecting the interaction among closely spaced hydraulic fractures. *Soc. Pet. Eng. J. 17*(1), 292–306.

Burghardt, J., J. Desroches, B. Lecampion, S. Stanchits, A. Surdi, N. Whitney, and M. Houston (2015, May 10–13). Laboratory study of the effect of well orientation, completion design and rock fabric on near-wellbore hydraulic fracture geometry in shales. In *13th ISRM International Symposium on Rock Mechanics*, Montreal, Quebec, Canada.

Carter, B., J. Desroches, A. Ingraffea, and P. Wawrzynek (2000). *Modeling in geomechanics*, Volume 200, Chapter Simulating fully 3D hydraulic fracturing, pp. 525–557. Wiley Chichester.

Carter, B. J., X. Weng, J. Desroches, and A. Ingraffea (1999). Hydraulic fracture reorientation: Influence of 3d geometry. In *Hydraulic Fracturing Workshop, 37th US Rock Mechanics Symposium*, Vail, Co.

Ceccarelli, R., G. Pace, A. Casero, A. Ciuca, and M. Tambini (2010). Perforating for fracturing: Theory vs. field experiences. In *SPE International Symposium and Exhibition on Formation Damage Control*. SPE 128270.

Chang, F., K. Bartko, S. Dyer, G. Aidagulov, R. Suarez-Rivera, and J. Lund (2014, February 4–6). Multiple fracture initiation in openhole without mechanical isolation: First step to fulfill an ambition. In *SPE Hydraulic Fracturing Technology Conference*, The Woodlands, Texas, USA. SPE 168638.

Cherny, S., D. Chirkov, V. Lapin, A. Muranov, D. Bannikov, M. Miller, D. Willberg, O. Medvedev, and O. Alekseenko (2009). Two-dimensional modeling of the near-wellbore fracture tortuosity effect. *International Journal of Rock Mechanics and Mining Sciences 46*(6), 992–1000.

Chipperfield, S. T., G. Roberts, W. K. Miller II, and R. Vandersypen (2000, April 3–5). Gel slugs: A near-wellbore pressure-loss remediation technique for propped fracturing. In *SPE/CERI Gas Technology Symposium*, Calgary, Alberta, Canada. SPE 59777.

Chuprakov, D., O. Melchaeva, and R. Prioul (2013). Hydraulic fracture propagation across a weak discontinuity controlled by fluid injection. In A. P. Bunger, J. McLennan, and R. G. Jeffrey (Eds.), *Effective and Sustainable Hydraulic Fracturing*, Chapter 8. Rijeka, Croatia: Intech.

Chuprakov, D., O. Melchaeva, and R. Prioul (2014). Injection-sensitive mechanics of hydraulic fracture interaction with discontinuities. *Rock Mechanics and Rock Engineering, 47*(5), 1625–1640.

Chuprakov, D. A. and R. Prioul (2015). Hydraulic fracture height containment by weak horizontal interfaces. In *SPE Hydraulic Fracturing Technology Conference*, The Woodlands, Texas, USA. SPE 173337.

Cipolla, C., X. Weng, H. Onda, T. Nadaraja, U. Ganguly, and R. Malpani (2011, October 30–November 2). New algorithms and integrated workflow for tight gas and shale completions. In *Proceedings SPE Annual Technology Conference and Exhibition*, Denver, Colorado, USA. SPE 146872.

Cipolla, C. L., M. G. Mack, and S. C. Maxwell (2010). Reducing exploration and appraisal risk in low permeability reservoirs using microseismic fracture mapping – Part 2. In *SPE Latin American and Caribbean Petroleum Engineering Conference*, Lima, Peru. SPE 138103.

Cleary, M., D. Johnson, H. Kogsbøll, K. Owens, K. Perry, C. De Pater, A. Stachel, H. Schmidt, and M. Tambini (1993, April 26–28). Field implementation of proppant slugs to avoid premature screen-out of hydraulic fractures with adequate proppant concentration. In *Low permeability reservoirs symposium*, Denver, Colorado, USA.. SPE 25892.

Crosby, D. G. (1999). *The Initiation and Propagation of, and Interaction Between, Hydraulic Fractures from Horizontal Wellbores*. Ph. D. thesis, University of New South Wales, Australia.

Crosby, D. G., M. M. Rahman, M. K. Rahman, and S. S. Rahman (2002). Single and multiple transverse fracture initiation from horizontal wells. *Journal of Petroleum Science and Engineering 35*(3–4), 191–204.

Daneshy, A. (1973a, October). Experimental investigation of hydraulic fracturing through perforations. *Journal of Petroleum Technology 25*, SPE 4333.

Daneshy, A. (1973b, April). A study of inclined hydraulic fractures. *SPE Journal, 13*(02), 61–68.

Daneshy, A. (1978a, January). Numerical solution of sand transport in hydraulic fracturing. *Journal of Petroleum Technology*, 132–140. SPE 5636.

Daneshy, A. (2009, January 19–21). Factors controlling the vertical growth of hydraulic fractures. In *SPE Hydraulic Fracturing Technology Conference*, The Woodlands, Texas, USA. SPE 118789.

Daneshy, A. A. (1978b). Hydraulic fracture propagation in layered formations. *Soc. Pet. Eng. J. 18*(01), 33–41. SPE 6088.

Daneshy, A. A. (2014, October 27–29). Fracture shadowing: Theory, applications and implications. In *Proceedings SPE Annual Technical Conference and Exhibition*, Amsterdam, The Netherlands. SPE 170611.

Daneshy, A. A. (2015). Dynamic interaction within multiple limited entry fractures in horizontal wells: Theory, implications, and field verification. In *SPE Hydraulic Fracturing Technology Conference*, The Woodlands, Texas, USA. SPE 173344.

Daniels, J., G. Waters, J. LeCalvez, J. Lassek, and D. Bentley (2007, October 12–14). Contacting more of the Barnett Shale through and integration of real-time microseismic monitoring, petrophysics, and hydraulic fracture design. In *Proceedings SPE Annual Technical Conference and Exhibition*, Anaheim, California, USA. SPE 110562.

Davies, R. J., S. A. Mathias, J. Moss, S. Hustoft, and L. Newport (2012). Hydraulic fractures: How far can they go? *Marine and Petroleum Geology 37*(1), 1–6.

de Laguna, W., E. G. Struxness, T. Tamara, W. C. McClain, H. O. Weeren, and R. C. Sexton (1968). Engineering development of hydraulic fracturing as a method for permanent disposal of radioactive wastes. Technical Report ORNL-4259, Oak Ridge National Laboratory, Oak Ridge, Tennessee, USA. www.ornl.gov/info/reports/1968/3445605101608.pdf.

de Pater, C. J. and L. J. L. Beugelsdijk (2005). Experiments and numerical simulation of hydraulic fracturing in naturally fractured rock. In *Proceedings 40th US Symposium on Rock Mechanics*, Anchorage, Alaska, June 25–29.

de Pater, C. J. and Y. Dong (2009, January 19–21). Fracture containment in soft sands by permeability or strength contrasts. In *Proceedings SPE Hydraulic Fracturing Technology Conference and Exhibition*, The Woodlands, Texas, USA. SPE 119634.

Desroches, J., B. Lecampion, H. Ramakrishnan, R. Prioul, and E. Brown (2014, September 30–October 2). Benefits of controlled hydraulic fracture placement: Theory and field experiment. In *SPE/CSUR Unconventional Resources Conference*, Calgary, Alberta, Canada. SPE 171667.

Detournay, E. and E. Carbonell (1997, August). Fracture mechanics analysis of the breakdown process in minifracture or leakoff test. *SPE Production & Facilities*, 195–199.

Diamond, W. P. and D. C. Oyler (1987). Effects of stimulation treatments on coalbeds and surrounding strata. Technical Report Report of Investigations 9083, United States Bureau of Mines.

Doe, T. and G. Boyce (1989). Orientation of hydraulic fractures in salt under hydrostatic and non-hydrostatic stresses. *International Journal of Rock Mechanics and Mining Sciences 26*(6), 605–611.

Dohmen, T., J. Zhang, and J. P. Blangy (2014, October 27–29). Measurement and analysis of 3D stress shadowing related to the spacing of hydraulic fracturing in unconventional reservoirs. In *Proceedings SPE Annual Technical Conference and Exhibition*, Amsterdam, The Netherlands. SPE 170924.

Dohmen, T., J. Zhang, C. Li, J. P. Blangy, K. M. Simon, D. N. Valleau, J. D. Eules, S. Morton, and S. Checkles (2013, September 30–October 2). A new surveillance method for delineation of depletion using microseismic, and its application to development of unconventional reservoirs. In *Proceedings SPE Annual Technical Conference and Exhibition*, New Orleans, LA, USA. SPE 166274.

Dontsov, E. V. and A. P. Peirce (2015). An enhanced pseudo-3d model for hydraulic fracturing accounting for viscous height growth, non-local elasticity, and lateral toughness. *Engineering Fracture Mechanics 142*, 116–139.

Economides, M. J. and K. G. Nolte (2000). Reservoir stimulation. Schlumberger: John Wiley & Sons.

El-Rabaa, W. (1987). Hydraulic fracture propagation in the presence of stress variation. In *SPE Annual Technical Conference and Exhibition*, Dallas, Texas, USA. SPE 16898.

El-Rabaa, W. (1989, October 8–11). Experimental study of hydraulic fracture geometry initiated from horizontal wells. In *Proceedings SPE Annual Technical Conference and Exhibition*, San Antonio, Texas, USA. SPE 19720.

Elder, C. H. (1977). Effects of hydraulic stimulation on coalbeds and assciated strata. Technical Report Report of Investigations 8260, United States Bureau of Mines.

EPA (2004). Evaluation of impacts to underground sources of drinking water by hydraulic fracturing of coalbed methane reservoirs. Technical Report EPA 816-R-04-003, United States Environmental Protection Agency, Washington, DC.

Fallahzadeh, S. H., V. Rasouli, and M. Sarmadivaleh (2014). An investigation of hydraulic fracturing initiation and near-wellbore propagation from perforated boreholes in tight formations. *Rock Mechanics and Rock Engineering 48*(2), 573–584.

Fast, R., A. Murer, and R. Timmer (1994). Description and analysis of cored hydraulic fractures, Lost Hills Field, Kern County, California. *SPE Production & Facilities 9*(2), 107–114. SPE 24853.

Fisher, K. and N. Warpinski (2012). Hydraulic fracture height growth: Real data. *SPE Production and Operations 27*(1), 8–19. SPE 145949.

Fisher, M., C. Wright, B. Davidson, A. Goodwin, E. Fielder, W. Buckler, and N. Steinsberger (2002). Integrating fracture mapping technologies to optimize stimulations in the Barnett Shale. In *SPE Annual Technical Conference and Exhibition*, San Antonio, Texas, USA. SPE 77441.

Fisher, M. K., J. R. Heinze, C. D. Harris, B. M. Davidson, C. A. Wright, and K. P. Dunn (2004, September 29). Optimizing horizontal completion techniques in the barnett shale using microseismic fracture mapping. In *Proceedings SPE Annual Technology Conference and Exhibition*, Houston, Texas, USA. SPE 90051.

Fu, W., B. C. Ames, A. P. Bunger, and A. A. Savitski (2015, June 28–July 1). An experimental study on interaction between hydraulic fractures and partially-cemented natural fractures. In *Proceedings 49th U.S. Rock Mechanics Symposium*, San Francisco, CA, USA. Paper No. 15-132.

Gallus, J. P. and D. S. Pye (1972). Fluid diversion to improve well stimulation. In *Joint AIME-MMIJ Meeting*, Tokyo, Japan. SPE 3811.

Geertsma, J. and F. de Klerk (1969). A rapid method of predicting width and extent of hydraulic induced fractures. *Journal of Petroleum Technology 246*, 1571–1581. (SPE 2458).

Germanovich, L. N., L. M. Ring, D. K. Astakhov, J. Shlyopobersky, and M. J. Mayerhofer (1997). Hydraulic fracture with multiple segments II: Modeling. *International Journal of Rock Mechanics and Mining Sciences 34*(3–4), 472.

Gu, H. and E. Siebrits (2008). Effect of formation modulus contrast on hydraulic fracture height containment. *SPE Production & Operations 23*(2), 170–176.

Gu, H., E. Siebrits, and A. Sabourov (2008). Hydraulic fracture modeling with bedding plane interfacial slip. In *SPE Eastern Regional/AAPG Eastern Section Joint Meeting*, Pittsburgh, PA, USA. SPE 117445.

Gu, H., X. Weng, J. Lund, M. Mack, U. Ganguly, and R. Suarez-Rivera (2011, January 24–26). Hydraulic fracture crossing natural fracture at non-orthogonal angles, A criterion, its validation and applications. In *Proceedings SPE Hydraulic Fracturing Technology Conference and Exhibition*, The Woodlands, Texas, USA. SPE 139984.

Gudmundsson, A. (2011). Deflection of dykes into sills at discontinuities and magma-chamber formation. *Tectonophysics 500*(1–4), 50–64.

Gudmundsson, A. and S. L. Brenner (2001). How hydrofractures become arrested. *Terra Nova 13*, 456–462.

Gudmundsson, A. and I. F. Loetveit (2005). Dyke emplacement in a layered and faulted rift zone. *Journal of Volcanology and Geothermal Research 144*(1–4), 311–327.

Gudmundsson, A., L. B. Marinoni, and J. Marti (1999). Injection and arrest of dykes: Implications for volcanic hazards. *Journal of Volcanology and Geothermal Research 88* (1–2), 1–13.

Gulrajani, S. N. and J. Romero (1996, October 22–24). Evaluation and modification of fracture treatments showing near-wellbore effects. In *European Petroleum Conference*, Milan, Italy. SPE 36901.

Hainey, B., X. Weng, and R. Stoisits (1995, October 22–25). Mitigation of multiple fractures from deviated wellbores. In *SPE Annual Technical Conference and Exhibition*, Dallas, Texas, USA. SPE 30482.

Hammack, R., S. Sharma, R. Capo, E. Zorn, H. Siriwardane, and W. Harbert (2013, August 20–22). An evaluation of zonal isolation after hydraulic fracturing; results from horizontal Marcellus Shale gas wells at NETL's Greene County test site in Southwestern Pennsylvania. In *SPE Eastern Regional Meeting*, Pittsburgh, Pennsylvania, USA. SPE 165720.

Harmon, J. A. (1957, January 1). The chemistry of fresh-water fracturing. In *Drilling and Production Practice*, New York, NY, USA. American Petroleum Institute. API-57-050.

Hassebroek, W. and A. Waters (1964). Advancements through 15 years of fracturing. *Journal of Petroleum Technology 16*(7), 760–764.

Houser, J. A. and R. A. Hernandez (2009). Pinpoint fracturing using a multiple-cutting process. In *SPE Rocky Mountain Petroleum Technology Conference*, Denver, Colorado. SPE 122949.

Howard, G. and C. Fast (Eds.) (1970). Volume 2. New York: Henry L. Doherty Fund, SPE.

Jackson, R. B., E. R. Lowry, A. Pickle, M. Kang, D. DiGiulio, and K. Zhao (2015). The depths of hydraulic fracturing and accompanying water use across the united states. *Environmental Science and Technology 49*(15), 8969–8976.

Jeffrey, R. and A. P. Bunger (2007, Jan 29–31). A detailed comparison of experimental and numerical data on hydraulic fracture height growth through stress contrasts. In *Proceedings SPE Hydraulic Fracturing Technology Conference*, College Station, Texas, USA. SPE 106030.

Jeffrey, R. and A. P. Bunger (2009). A detailed comparison of experimental and numerical data on hydraulic fracture height growth through stress contrasts. *Soc. Pet. Eng. J. 14*(3), 413–422.

Jeffrey, RG, ZR. Chen, X. Zhang, AP. Bunger, and KW. Mills (2015). Measurement and analysis of full-scale hydraulic fracture initiation and orientation. *Rock Mechanics and Rock Engineering, 48*(6), 2497–2512.

Jeffrey, R. G., R. P. Brynes, and D. J. Ling (1992, May 18–21). An analysis of hydraulic fracture and mineback data for a treatment in the german creek coal seam. In *Proceedings SPE Rocky Mountain Regional Meeting*, Casper, Wyoming, USA. SPE 24362.

Jeffrey, R. G., A. P. Bunger, B. Lecampion, X. Zhang, Z. R. Chen, A. van As, D. Allison, W. D. Beer, J. W. Dudley, E. Siebrits, M. Thiercelin, and M. Mainguy (2009, October 4–7). Measuring hydraulic fracture growth in naturally fractured rock. In *Proceedings SPE Annual Technical Conference and Exhibition*, New Orleans, Louisiana, USA. SPE 124919.

Jeffrey, R. G., A. Settari, and N. P. Smith (1995, October 22–25). A comparison of hydraulic fracture field experiments, including mineback geometry data, with numerical fracture model simulations. In *Proceedings SPE Annual Technical Conference and Exhibition*, Dallas, Texas, USA. SPE 30508.

Johnson, E. and M. Cleary (1991). Implications of recent laboratory experimental results for hydraulic fractures. In *Proceeding Rock Mountains Regional Meeting and Low-Permeability Reservoirs Symposium*, Denver, CO, pp. 413–428. SPE 21846.

Kavanagh, J. L., T. Menand, and R. S. J. Sparks (2006). An experimental investigation of sill formation and propagation in layered elastic media. *Earth and Planetary Science Letters 245* (34), 799–813.

Khristianovic, S. and Y. Zheltov (1955, June 6–15). Formation of vertical fractures by means of highly viscous fluids. In *Proceeding 4th World Petroleum Congress*, Rome, pp. 579–586. Carlo Colombo, Rome.

King, G. E. (2010, September 19–22). Thirty years of gas shale fracturing: What have we learned? In *Proceedings SPE Annual Technical Conference and Exhibition*, Florence, Italy. SPE 133256.

Kogsbøll, H., M. Pitts, and K. Owens (1993, September 7–10). Effects of tortuosity in fracture stimulation of horizontal wells- a case study of the Dan Field. In *Offshore Europe*, Aberdeen, United Kingdom. SPE 26796.

Kresse, O., X. Weng, H. Gu, and R. Wu (2013a). Numerical modeling of hydraulic fractures interaction in complex naturally fractured formations. *Rock Mechanics and Rock Engineering 46*(3), 555–568.

Kresse, O., X. Weng, D. Chuprakov, R. Prioul, and C. Cohen (2013b). Effect of flow rate and viscosity on complex fracture development in UFM model. In A. P. Bunger, J. McLennan, and R. G. Jeffrey (Eds.), *Effective and Sustainable Hydraulic Fracturing*, Chapter 9. Rijeka, Croatia: Intech.

Kruger, P. and T. Yamaguchi (1993, January 26–28). Thermal drawdown analysis of the Hijiori HDR 90-day circulation test. In *Proceedings 18th Stanford Workshop on Geothermal Reservoir Engineering*, Stanford, CA, USA.

Lambert, S. W., M. A. Trevits, and P. F. Steidl (1980). Vertical borehole design and completion practices to remove methane gas from mineable coalbeds. Technical Report DOE/CMTC/TR-80/2, United States Department of Energy.

Lecampion, B. and J. Desroches (2015a). Simultaneous initiation and growth of multiple radial hydraulic fractures from a horizontal wellbore. *Journal of the Mechanics and Physics of Solids. 82*, 235–258.

Lecampion, B. and J. Desroches (2015b). Robustness to formation geological heterogeneities of the limited entry technique for multi-stage fracturing of horizontal wells. *Rock Mechanics and Rock Engineering, 48*(6), 2637–2644.

Lecampion, B., J. Desroches, R. Jeffrey, A. Bunger, and J. Burghardt (2015a, May 10–13). Initiation versus breakdown pressure of transverse hydraulic fracture: theory and experiments. In *13th ISRM International Symposium on Rock Mechanics*, Montreal, Quebec, Canada.

Lecampion, B., J. Desroches, X. Weng, J. Burghardt, and J. E. Brown (2015b). Can we engineer better multistage horizontal completions? Evidence of the importance of near-wellbore fracture geometry from theory, lab and field experiments. In *SPE Hydraulic Fracturing Technology Conference*, The Woodlands, Texas, USA. SPE 173363.

Li, B. (2014). Natural fractures in unconventional shale reservoirs in US and their roles in well completion design and improving hydraulic fracturing stimulation efficiency and production. In *SPE Annual Technical Conference and Exhibition*, Amsterdam, The Netherlands. SPE 170934.

Li, Y., G. Liu, J. Li, L. Yu, T. Zhang, and J. Lu (2015). Improving fracture initiation predictions of a horizontal wellbore in laminated anisotropy shales. *Journal of Natural Gas Science and Engineering 24*, 390–399.

Lim, P. V., P. Goddard, J. Sink, and I. S. Abou-sayed (2014). Hydraulic fracturing: A Marcellus case study of an engineered staging completion based on rock properties. In *SPE/CSUR Unconventional Resources Conference*, Calgary, Alberta, Canada. SPE 171618.

Liu, S. and P. Valk'o (2015, February 3–5). An improved equilibrium-height model for predicting hydraulic fracture height migration in multi-layer formations. In *SPE Hydraulic Fracturing Technology Conference*, The Woodlands, Texas, USA. SPE 173335.

Llanos, E. M., R. G. Jeffrey, R. R. Hillis, and X. Zhang (2006). Study of the interaction between hydraulic fractures and geological discontinuities. In *Rock mechanics in underground construction: ISRM International Symposium 2006 and 4th Asian Rock Mechanics Symposium*, Singapore, p. 378. World Scientific Pub Co Inc.

Mahrer, K. D. (1999). A review and perspective on far-field hydraulic fracture geometry studies. *Journal of Petroleum Science and Engineering 24*(1), 13–28.

Manchanda, R., M. M. Sharma, and S. Holzhauser (2014). Time-dependent fracture-interference effects in pad wells. *SPE Production & Operations 29*(04), 274–287.

Manrique, J. and A. Venkitaraman (2001, September 30–October 3). Oriented fracturing – A practical technique for production optimization. In *SPE Annual Technical Conference and Exhibition*, New Orleans, Louisiana, USA. SPE 71652.

Maxwell, S., A. Pirogov, C. Bass, and L. Castro (2013). A comparison of proppant placement, well performance, and estimated ultimate recovery between horizontal wells completed with multi-cluster plug and perf and hydraulically activated frac ports in a tight gas reservoir. In *SPE Hydraulic Fracturing Technology Conference*, The Woodlands, Texas, USA. SPE 163820.

Maxwell, S., T. Urbancic, N. Steinsberger, and R. Zinno (2002, September 29– October 2). Microseismic imaging of hydraulic fracture complexity in the Barnett shale. In *Proceedings SPE Annual Technical Conference and Exhibition*, San Antonio, Texas, USA. SPE 77440.

Mayerhofer, M., M. Richardson, R. Walker Jr., D. Meehan, M. Oehler, and R. Browning Jr. (1997, October 5–8). Proppants? we don't need no proppants. In *Proceedings SPE Annual Technical Conference and Exhibition*, San Antonio, Texas, USA. SPE 38611.

Mayerhofer, M. J., N. A. Stegent, J. O. Barth, and K. M. Ryan (2011, October 30–November 2). Integrating fracture diagnostics and engineering data in the Marcellus Shale. In *Proceedings SPE Annual Technical Conference and Exhibition*, Denver, CO, USA. SPE 145463.

Mc Gowen, J., J. Gilbert, and E. Samari (2007, January 29–31). Hydraulic fracturing down under. In *SPE Hydraulic Fracturing Technology Conference*, College Station, Texas, USA. SPE 106051.

McDaniel, B., D. McMechan, and N. Stegent (2001). Proper use of proppant slugs and viscous gel slugs can improve proppant placement during hydraulic fracturing applications. In *SPE Annual Technical Conference and Exhibition*. SPE 71661.

McDaniel, B. and J. Surjaatmadja (2009, December 7–9). Using hydrajetting applications in horizontal completions to improve hydraulic fracturing stimulations and lower costs. In *International Petroleum Technology Conference*, Doha, Qatar. IPTC 13775.

Medelsohn, D. A. (1984a). A review of hydraulic fracture modeling II: 3D modeling and vertical growth in layered rock. *ASME Journal of Energy Resources Technology*, *106*(4), 543–553.

Medelsohn, D. A. (1984b). A review of hydraulic fracture modeling part I: General concepts, 2D models, motivation for 3D modeling. *ASME Journal of Energy Resources Technology*, *106* (3), 369–376.

Meyer, B. and L. Bazan (2011, January 24–26). A discrete fracture network model for hydraulically induced fractures-theory, parametric and case studies. In *Proceedings SPE Hydraulic Fracturing Technology Conference and Exhibition*, The Woodlands, Texas, USA. SPE 140514.

Montgomery, C. T. and M. B. Smith (2010). Hydraulic fracturing: History of an enduring technology. *Journal of Petroleum Technology* *62*(12), 26–40.

Nolte, K. G. (1993). Discussion of examination of a cored hydraulic fracture in a deep gas well. *SPE Production & Facilities* *8*(03), 159–164. SPE 26302.

Nolte, K. G. and M. B. Smith (1981). Interpretation of fracturing pressures. *Journal of Petroleum Technology* *33*(9), 1767–1775.

Nordgren, R. (1972). Propagation of vertical hydraulic fractures. *Journal of Petroleum Technology* *253*, 306–314. SPE 3009.

Northrop, D. A. and K.-H. Frohne (1990). The Multiwell Experiment – A field laboratory in tight gas sandstone reservoirs. *Journal of Petroleum Technology* *42*(6), 772–779. SPE 18286.

Olson, J. E. (2008, June 29–July 2). Multi-fracture propagation modeling: Applications to hydraulic fracturing in shales and tight gas sands. In *Proceedings 42nd US Rock Mechanics Symposium*, San Francisco, CA, USA. ARMA 08-327.

Olson, J. E., B. Bahorich, and J. Holder (2012). Examining hydraulic fracture: Natural fracture interaction in hydrostone block experiments. In *SPE Hydraulic Fracturing Technology Conference*, The Woodlands, Texas, USA. SPE 152618.

Olson, J. E. and A. Dahi-Taleghani (2009, January 19–21). Modeling simultaneous growth of multiple hydraulic fractures and their interaction with natural fractures. In *Proceedings SPE Hydraulic Fracturing Technology Conference and Exhibition*, The Woodlands, Texas, USA. SPE 119739.

Owens, K., S. Andersen, and M. Economides (1992, October 4–7). Fracturing pressures for horizontal wells. In *SPE Annual Technical Conference and Exhibition*, Washington, DC, USA. SPE 24822.

Pandya, N. and O. Jaripatke (2014). Rate step-down analysis improves placement efficiency of stimulation treatments in unconventional resource play. In *Unconventional Resources Technology Conference (URTeC)*, Denver, Co. URTeC 1943637.

Pearson, C., A. Bond, M. Eck, and J. Schmidt (1992, January). Results of stress-oriented and aligned perforating in fracturing deviated wells. *Journal of Petroleum Technology* *44*(1), 10–18. SPE 22836.

Peirce, A. and E. Detournay (2008). An implicit level set method for modeling hydraulically driven fractures. *Computer Methods in Applied Mechanics and Engineering 197*, 2858–2885.

Peirce, A. P. and A. P. Bunger (2015). Interference Fracturing: Non-Uniform Distributions of Perforation Clusters that Promote Simultaneous Growth of Multiple Hydraulic Fractures. *SPE Journal*, *20*(2), 384–395.

Perkins, T. and L. Kern (1961). Widths of hydraulic fractures. *Journal of Petroleum Technology, Trans. AIME 222*, 937–949.

Pospisil, G., C. Carpenter, and C. Pearson (1995, March 8–10). Impacts of oriented perforating on fracture stimulation treatments: Kuparuk river field, alaska. In *SPE Western Regional Meeting*, Bakersfield, CA,. SPE 29645.

Potapenko, D. I., S. K. Tinkham, B. Lecerf, C. N. Fredd, M. L. Samuelson, M. R. Gillard, J. H. L. Calvez, and J. L. Daniels (2009). Barnett Shale refracture stimulations using a novel diversion technique. In *SPE Hydraulic Fracturing Technology Conference*, The Woodlands, Texas, USA. SPE 119636.

Prats, M. and S. A. Maraven (1981). Effect of burial history on the subsurface horizontal stresses of formations having different material properties. *Soc. Pet. Eng. J. 21*(06), 658–662. SPE 9017.

Quinn, T. S. (1994). *Experimental Analysis of Permeability Barriers to Hydraulic Fracture Propagation*. Ph. D. thesis, Massachusetts Institute of Technology, Cambride, MA, USA.

Renshaw, C. E. and D. D. Pollard (1995). An experimentally verified criterion for propagation across unbounded frictional interfaces in brittle, linear elastic materials. *International Journal of Rock Mechanics and Mining Sciences*, *32*(3), 237–249.

Rivalta, E., M. Böttinger, and D. T (2005, jun). Buoyancy-driven fracture ascent: Experiments in layered gelatin. *Journal of Volcanology and Geothermal Research*, *144*(1–4), 273–285.

Rivalta, E., B. Taisne, A. P. Bunger, and R. Katz (2015). A review of mechanical models of dike propagation: schools of thought, results and future directions. *Tectonophysics 638*, 1–42.

Roberts, G. A., S. T. Chipperfield, and W. K. Miller II (2000). The evolution of a high near-wellbore pressure loss treatment strategy for the australian cooper basin. In *SPE Annual Technical Conference and Exhibition*, Dallas, Texas, USA. SPE 63029.

Rogers, S., D. Elmo, and W. Dershowitz (2011, June 26–29). Understanding hydraulic fracture geometry and interactions in pre-conditioning through DFN and numerical modeling. San Francisco, CA, USA. Paper No. 11–439.

Roussel, N. P., R. Manchanda, and M. M. Sharma (2012, February 6–8). Implications of fracturing pressure data recorded during a horizontal completion on stage spacing design. In *Proceedings SPE Hydraulic Fracturing Technology Conference*, The Woodlands, Texas, USA. SPE 152631.

Roussel, N. P. and M. M. Sharma (2010, September 19–22). Role of stress reorientation in the success of refracture treatments in tight gas sands. In *Proceedings SPE Annual Technical Conference and Exhibition*, Florence, Italy. SPE 134491.

Roussel, N. P. and M. M. Sharma (2011, May). Optimizing fracture spacing and sequencing in horizontal-well fracturing. *SPE Production & Operations 26*(2), 173–184.

Saldungaray, P. M., T. T. Palisch, and R. Duenckel (2012, March 20–22). Novel traceable proppant enables propped frac height measurement while reducing the environmental impact. In *SPE/EAGE European Unconventional Resources Conference and Exhibition*, Vienna, Austria. SPE 151696.

Sesetty, V. and A. Ghassemi (2013). Numerical simulation of sequential and simultaneous hydraulic fracturing. In A. P. Bunger, J. McLennan, and R. G. Jeffrey (Eds.), *Effective and Sustainable Hydraulic Fracturing*, Chapter 33. Rijeka, Croatia: Intech.

Settari, A. and M. P. Cleary (1984). Three-dimensional simulation of hydraulic fracturing. *Journal of Petroleum Technology 36*(8), 1177–1190.

Sheibani, F. and J. Olson (2013). Stress intensity factor determination for three-dimensional crack using the displacement discontinuity method with applications to hydraulic fracture height growth and non-planar propagation paths. In A. P. Bunger, J. McLennan, and R. G. Jeffrey (Eds.), *Effective and Sustainable Hydraulic Fracturing*, Chapter 37. Rijeka, Croatia: Intech.

Sherman, C., L. Aarons, J. Morris, S. Johnson, A. Savitski, and M. Geilikman (2015). Finite element modeling of curving hydraulic fractures and near- wellbore hydraulic fracture complexity. In *49th US Rock Mechanics Symposium / Geomechanics Symposium*. ARMA 15-0530.

Simonson, E. R., A. S. Abou-Sayed, and R. J. Clifton (1978). Containment of massive hydraulic fractures. *SPE Journal 18*(1), 27–32.

Slocombe, R., A. Acock, K. Fisher, A. Viswanathan, C. Chadwick, R. Reischman, and E. Wigger (2013, September 30–2 October). Eagle Ford completion optimization using horizontal log data. In *Proceedings SPE Annual Technology Conference and Exhibition*, New Orleans, Lousiana, USA. SPE 166242.

Smith, M., A. Bale, L. Britt, H. Klein, E. Siebrits, and X. Dang (2001, September 30–October 3). Layered modulus effects on fracture propagation, proppant placement, and fracture modeling. In *SPE Annual Technical Conference and Exhibition*, New Orleans, Louisiana, USA. SPE 71654.

Soliman, M. Y., J. L. Hunt, and A. M. El-Raaba (1990). Fracturing aspects of horizontal wells. *Journal of Petroleum Technology 42*(8), 966–973.

Sookprasong, P. A., R. S. Hurt, and R. F. LaFollette (2014, October 27–29). Fiber optic DAS and DTS in multicluster, multistage horizontal well fracturing: Interpreting hydraulic fracture initiation and propagation through diagnostics. In *Proceedings SPE Annual Technology Conference and Exhibition*, Amsterdam, The Netherlands. SPE 170723.

Strain, H. (1962, December). Well-bore notching and hydraulic fracturing. *Journal of Canadian Petroleum Technology 1*. PETSOC-62-04-01.

Suarez-Rivera, R., J. Burghardt, S. Stanchits, E. Edelman, and A. Surdi (2013). Understanding the effect of rock fabric on fracture complexity for improving completion design and well performance. In *International Petroleum Technology Conference*, Beijing, China. IPTC 17018.

Surjaatmadja, J., H. H. Abass, and J. Brumley (1994, 7–10 November). Elimination of near-wellbore tortuosities by means of hydrojetting. In *Asia Pacific Oil and Gas Conference*, Melbourne, Australia. SPE 28761.

Teufel, L. W. and J. A. Clark (1984). Hydraulic fracture propagation in layered rock: Experimental studies of fracture containment. *SPE Journal 24*(1), 19–32.

Urbancic, T. and S. Maxwell (2002). Microseismic imaging of fracture behavior in naturally fractured reservoirs. In *SPE/ISRM Rock Mechanics Conference*, Irving, Texas, USA. SPE 78229.

van As, A. and R. Jeffrey (2000). Caving induced by hydraulic fracturing at Northparkes Mines. In J. Girard, M. Liebman, C. Breeds, and T. Doe (Eds.), *Pacific Rocks 2000 – Proceeding 4th North American Rock Mechanics Symposium*, Seatle, WA, pp. 353–360. Balkema.

van de Ketterij, R. G. (2001). *Optimisation of the Near-wellbore geometry of hydraulic fractures propagating from cased perforated completions*. Ph. D. thesis, TU Delft.

van de Ketterij, R. G. and C. J. de Pater (1997, June 2–3). Experimental study on the impact of perforations on hydraulic fracture tortuosity. In *SPE European Formation Damage Conference*, The Hague, Netherlands. SPE 38149.

van de Ketterij, R. G. and C. J. de Pater (1999). Impact of perforations on hydraulic fracture tortuosity. *SPE Production & Facilities 14*(2), 131–138.

van Eekelen, H. A. M. (1982). Hydraulic fracture geometry: Fracture containment in layered formations. *Soc. Pet. Eng. J. 22*(3), 341–349.

Veeken, C., D. Davies, and J. Walters (1989, March 6–8). Limited communication between hydraulic fracture and (deviated) wellbore. In *Low Permeability Reservoirs Symposium*, Denver, Co, USA. SPE 18982.

Vermylen, J. P. and M. D. Zoback (2011, January 24–26). Hydraulic fracturing, microseismic magnitudes, and stress evolution in the Barnett Shale, In *Proceedings SPE Hydraulic Fracturing Technology Conference and Exhibition*, The Woodlands, Texas, USA. SPE 140507.

Viswanathan, A., H. Watkins, J. Reese, A. Corman, and B. V. Sinosic (2014). Sequenced fracture treatment diversion enhances horizontal well completions in the Eagle Ford Shale. In *SPE/CSUR Unconventional Resources Conference*, Calgary, Alberta, Canada. SPE 171660.

Warpinski, N., R. C. Kramm, J. R. Heinze, and C. K. Waltman (2005). Comparison of single-and dual-array microseismic mapping techniques in the Barnett Shale. In *SPE Annual Technical Conference and Exhibition*, Dallas, Texas, USA. SPE 95568.

Warpinski, N., J. Lorenz, P. Branagan, F. Myal, and B. Gall (1993). Examination of a cored hydraulic fracture in a deep gas well. *SPE Production & Facilities 8*(03), 150–158. SPE 22876.

Warpinski, N., R. Schmidt, and D. Northrop (1982). In-situ stresses: The predominant influence of hydraulic fracture containment. *Journal of Petroleum Technology 34*(3), 653–664. SPE 8932.

Warpinski, N., T. Wright, J. Uhl, B. Engler, P. Drozda, R. Peterson, and P. Branagan (1996). Microseismic monitoring of the B-Sand hydraulic fracture experiment at the DOE/GRI Multi-Site Project. In *SPE Annual Technical Conference and Exhibition*, Denver, Colorado. SPE 36450.

Warpinski, N. R. (2014, June 1–4)). A review of hydraulic-fracture induced microseismicity. In *Proceedings 48th US Rock Mechanics Symposium*, Minneapolis, MN, USA. ARMA 14-7774.

Warpinski, N. R. and L. W. Teufel (1987). Influence of geologic discontinuities on hydraulic fracture propagation. *Soc. Pet. Eng. J. 39*(2), 209–220.

Weijers, L. (1995). *The near-wellbore geometry of hydraulic fractures initiated from horizontal and deviated wells*. Ph. D. thesis, Delft University of Technology.

Weijers, L. and C. J. de Pater (1992, February 26–27). Fracture reorientation in model tests. In *SPE International Symposium on formation damage control*, Lafayette, Louisiana, USA. SPE 23790.

Weijers, L., C. J. de Pater, K. Owens, and H. Kogsbøll (1994). Geometry of hydraulic fractures induced from horizontal wellbores. *SPE Production & Facilities 9*(2), 87–92. SPE 25049.

Weng, X. (1993, October 3–6). Fracture initiation and propagation from deviated wellbores. In *SPE Annual Technical Conference and Exhibition*, Houston, Texas, USA. SPE 26597.

Wu, K. and J. E. Olson (2013). Investigation of critical in situ and injection factors in multi-frac treatments: Guidelines for controlling fracture complexity. In *SPE Hydraulic Fracturing Technology Conference*, The Woodlands, Texas, USA. SPE 163821.

Wu, R., A. Bunger, R. Jeffrey, and E. Siebrits (2008, June 29–July 2). A comparison of numerical and experimental results of hydraulic fracture growth into a zone of lower confining stress. In *Proceedings 2nd U.S.-Canada Rock Mechanics Symposium*, San Francisco, USA. Paper No. 08–267.

Xing, P., AP. Bunger, K. Yoshioka, J. Adachi, and A. El-Fayoumi (2016). Experimental study of hydraulic fracture containment in layered reservoirs. *Proceedings 50th U.S. Rock Mechanics Symposium*, Houston, Texas, USA, 26–29 June 2016. ARMA-16-49.

Xu, G. and S. Wong (2013, March 26–28). Interaction of multiple non-planar hydraulic fractures in horizontal wells. In *IPTC 2013: International Petroleum Technology Conference*, Beijing, China. IPTC 17043-MS.

Zhang, Z., R. G. Jeffrey, A. P. Bunger, and M. Thiercelin (2011). Initiation and growth of a hydraulic fracture from a circular wellbore. *International Journal of Rock Mechanics and Mining Sciences*, *48*, 984–995.

Chapter 17

Hydromechanical behavior of fault zones in petroleum reservoirs

S.A.B. da Fontoura, N. Inoue, G.L. Righetto & C.E.R. Lautenschläger
Department of Civil Engineering, Pontifical Catholic University of Rio de Janeiro, PUC-Rio, Brazil

Abstract: The presence of faults is expected to affect the flow of fluids through areas of the reservoir and to have an effect on the mechanical behavior of the reservoir itself. It is a known fact that hydrocarbon production and fluid injection into the earth crust may induce seismicity and undesired reservoir fluid leakage due to localized movements along faults. The behavior of a fault zone depends on the stresses acting upon it and the stresses do change during reservoir production. This chapter discusses the mechanical and hydraulic properties of fault zones within petroleum reservoirs. We offer suggestions on how to couple the hydromechanical processes in order to evaluate the possibility of fault reactivation.

Keywords: Fault zones, hydromechanical behavior, petroleum reservoir, coupled problem

1 INTRODUCTION

1.1 Generalities

The presence of faults is expected to affect the flow of fluids through areas of the reservoir and to have an effect on the mechanical behavior of the reservoir itself. During hydrocarbon production, fluid withdrawal and/or injection may cause displacements and strains along the faults present in the reservoir. These displacements may be responsible for the loss on reservoir sealing and also may generate some minor earthquakes.

Seismicity associated with strain energy release due to high pore pressures resulting from fluid injection at depth has been registered in several cases (Healy *et al.*, 1968; Raleigh *et al.*, 1976; Zoback & Harjes, 1997). Donnelly (2009) reports on the mechanics of shallow depth fault reactivation associated with mining exploration and on-shore fluid extraction activities. Large areas at surface may be affected, presenting evidences of subsidence and development of scarps. Suckale (2010) presents data associated with induced seismicity around producing hydrocarbon fields and CO_2 injection locations, associated with fault reactivation. Other examples do exist and certainly new cases will occur. The Groningen gas field (Sanz *et al.*, 2015) is a good example of induced seismicity, caused by very large volume of hydrocarbon extraction that brought about the State Government decision of reducing production in order to control the seismicity.

Reservoir fluid leakages during field development are, luckily, less common. Wiprut & Zoback (2000) describe a well-documented case of gas leakage in the North Sea. These authors claim that the fluid leakage is associated with fault reactivation. This hypothesis was verified using a simplified stress analysis to determine the stresses acting upon sections of the fault, combined with the verification of failure making use of Mohr-Coulomb criterion along the fault. Reservoir natural sealing systems seem to be both efficient and mechanically strong and have resisted throughout geological times. However, this system may be destroyed by poor engineering practice during reservoir development. Cases of reservoir leakage are less frequently reported in the literature due both to the difficulty in observing the leakage but also to legal issues normally associated with leakage.

In order to capture the behavior of a petroleum reservoir and to extract the fluids in an orderly manner, it is necessary to understand the role of the fault zones present in the surrounding rocks. Fault zones are 3D features with a peculiar architecture, which conditions both its mechanical and flow properties. The presence of faults may impose limits in injection pressure or pressure drawdown in reservoirs during hydrocarbon production on the accounts of its strength and mechanical properties. For the analysis of the long term safety of the reservoir sealing system, some elements are needed, in particular the properties that should be assigned to a given fault and the methodology that should be used in order to guarantee that the important pieces of physics and mechanics are included in it.

1.2 Aim and scope of the chapter

The aim of this chapter is to present the main elements needed to carry out a proper hydromechanical analysis of petroleum reservoirs. This is achieved dividing the chapter into four parts. Initially, we discuss the fault zones as an engineering structure. We present some terminology related to the subject and discuss two currently accepted fault zone architecture models. With that we establish the geometrical boundaries of a fault zone that will help, later in the chapter, to justify the modeling approach for fault reactivation. In the sequence, we discuss the hydraulic properties of a fault zone. Results from synthetic fault material tests in the laboratory, tests in the field and indirect methods are presented to establish the fault zone permeability.

Faults are considered either as a single planar feature or as a 3D zone. We conclude the third part by discussing the means to evaluate the mechanical properties of faults. In the fourth part we discuss the evaluation of fault reactivation. A short review is presented and, in the sequence, we introduce a methodology for coupling the hydromechanical effects onto the behavior of faulted zones. We present the results of the methodology when applied to a synthetic case in order to show the potential of the method. Finally, some comments are offered.

2 FAULT ZONE AS AN ENGINEERING STRUCTURE

2.1 Generalities

Faults are important geological structures, which are created by tectonic movements and control relative displacements between areas of the earth crust. Faults may be present as either conduits or barriers of fluid flow within the earth. The importance of faults in different areas of geosciences and engineering is reflected in the enormous amount of literature dealing with the subject of fault behavior, *e.g.*, textbooks (Ramsay & Huber, 1983), major literature reviews (Wibberley *et al.*, 2008; Faulkner *et al.*, 2010), specialty conferences and workshops and large research projects both for large faults (ICDP, 1996; IODP, 2003), and reservoir faults (Carstens, 2005).

Faults vary in size and in state of activity. Very large faults do exist at plate boundaries and are the location of earthquakes. Average size faults are boundaries of reservoirs and serve the purpose of reservoir seal. Smaller faults, sometimes subseismic, are ever present within reservoirs. Fault zones are developed over geological time and their composition is defined by a complex combination of strain field, rock types. As suggested by Faulkner *et al.* (2010), flow properties and mechanical behavior of faults are highly dependent upon the fault zone composition. The changes that might occur to the fault structure and its consequence to an engineering problem (reservoir development and eventually some CO_2 sequestration) under an engineering time scale are the main interest of the present paper.

2.2 Fault characterization

The characterization of faults is not a simple issue. The major source of information is the outcropped section of the fault or fault trace at the surface. Seismic sections provide much information about location and geometry of the sections at-depth but are limited in use when it comes to finer details. At-depth observations are also made through electric logs and core samples both obtained during drilling. Examples (by no means the most important ones) of comprehensive fault characterization programs are Lima (2008) for outcropped faults and Ito *et al.* (2007) and Zoback *et al.* (2011) for deep, large faults. Zoback *et al.* (2011) describe in details the results of investigations of a section of San Andreas Fault that include logging and coring. Next, we present a short survey of some ideas regarding fault elements as recorded in some important papers.

2.3 Fault morphology

Faults are 3D geological features even though they are portrayed frequently as planar elements. Their size, composition and structure are a function of the rocks and of the movements. The complexity of a fault zone may be idealized as an inner core, region that concentrates the strains, and a surrounding region, fractured, called the damage zone. Two simple models are represented in Figure 1. Faulkner *et al.* (2010) review extensively some field cases illustrating both the sizes and occurrences of each model.

2.4 Fault thickness

According to Al-Busafi (2005), the thickness of a fault is defined as the separation between the external slip surfaces, in case of more than one being present. Data compilation of faults in outcrops suggests that there is a quasi-linear relationship between the fault slippage and fault thickness, (t_f) (Manzocchi *et al.*, 1999). Figure 2

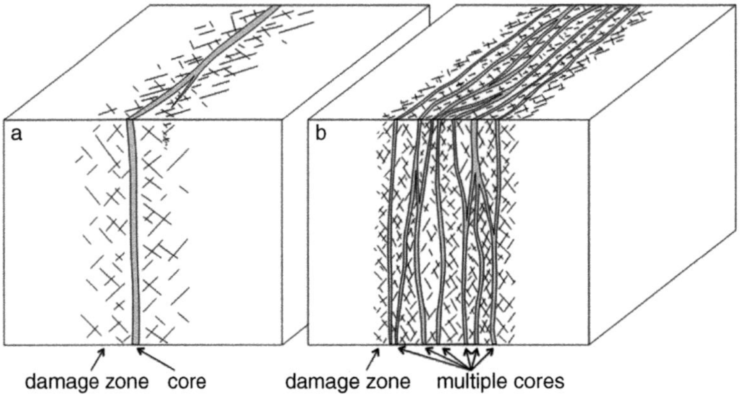

Figure 1 Two models of fault zone structures showing (a) a single high-strain core surrounded by a fractured damage zone (after Chester & Logan, 1986) and (b) multiple cores model, where many strands of high-strain material enclose lenses of fractured protolith (after Faulkner *et al.*, 2003).

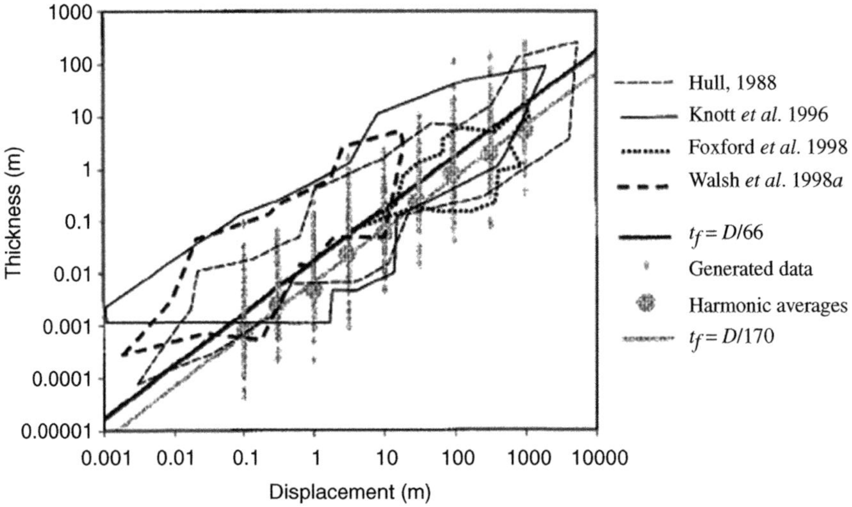

Figure 2 Fault thickness as a function of fault displacement (adapted from Manzocchi *et al.*, 1999).

summarizes the data obtained from Hull (1988), data from faults in the Nubian sandstone at West Sinai (Knott *et al.*, 1996), Moab fault in SE Utah (Foxford *et al.*, 1998) and faults in the sandstone/shale sequence in Westphalian, Lancashire, UK (Walsh *et al.*, 1998).

The average relationship for the dependence of thickness on the displacement is represented in Figure 2 by the line that corresponds to the relation $t_f = d/66$. For small displacement faults this relationship tends to underestimate the thickness of the fault. Very small displacement faults are not considered as a fault zone but as a plane. Another alternative is to use the harmonic average that generates the relationship $t_f = d/170$.

3 HYDRAULIC PROPERTIES OF FAULT ZONES

3.1 Generalities

Fault core is the region that concentrates strains. Several researchers describe the composition of this core as a function of the adjacent rocks and strain level. Faulkner *et al.* (2010) discuss the different core characteristics as a function of the rock types. This section discusses the issue of how to estimate the fault permeability when making predictions. There are two issues to be discussed. Normally, this is how the problem is defined in the literature. First there is the question of defining if the fault is a flow barrier or not. In this case the problem is treated as to define the fault sealing potential. Second, there is the question of defining the actual value of the fault permeability and its variation with the strain field imposed upon the fault during reservoir production.

Faults separate reservoir blocks and they may act as impermeable barriers. The existence of a pressure differential between adjacent reservoir compartments may suggest that the fault does not convey fluid being impermeable. Pressure difference across faults serves as an indicator of fault permeability.

3.2 Fault sealing potential

The seal provided by faults is the main factor that controls hydrocarbon accumulation and it may have a significant influence on the overall behavior of the reservoir during production (Jones & Hillis, 2003). Færseth *et al.* (2007) say that in spite of the large amount of published papers on sealing capacity of faults, there is a considerable uncertainty on the relationship between fault architecture and sealing capacity. Seismic limitations are responsible for the introduction of uncertainties with respect to the factors that have significant influence on the sealing capacity of faults. A better seismic resolution may be able to detect multiple zones with the fault region and also identify zones of different deformations (Wibberley *et al.*, 2008).

3.3 Permeability of fault zones

The determination of the permeability of the fault zones is linked, in the last years, to the analysis of the sealing capacity of the faults in sequences of sandstones/shales as described by Bouvier *et al.* (1989), Gibson (1994), Fristard *et al.* (1997) and Yielding *et al.* (1997). These analyses, normally used by the oil industry, have been carried out using the *Shale Gouge Ratio* (SGR). It is important to notice that, in general, shales are considered to behave as a capillary seal and a barrier of low permeability due to its small pore size.

Harris *et al.* (2002) carried out studies using dynamic pressure data in faults present in the fields of *Strathspey* and *Gullfaks* showing that there exists a relationship between the composition of the material of the fault core and its permeability. The authors also noticed too the existence of a relationship between SGR and the drop in dynamic pressure across the fault. Manzocchi *et al.* (1999) associated values of SGR with values of permeability using permeability of fault gouges at different concentrations of clays. The authors considered a relationship between the volumetric fraction of clay (Vsh)

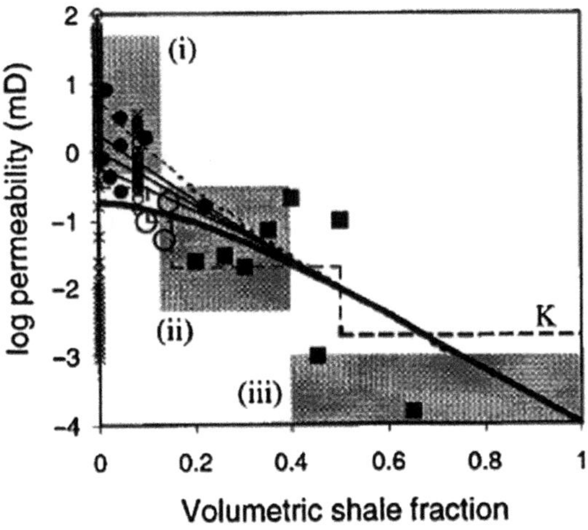

Figure 3 Permeability (log scale) as a function of volumetric clay content (Manzocchi et *al.*, 1999).

and the permeability of the fault using different set of data (Antonellini & Aydin, 1994; Knai, 1996; Gibson, 1998; Ottesen Ellevset *et al.*, 1998).

In Figure 3, the data plotted as full circles refer to cataclastic deformation bands, the open circles refer to deformation bands and the full squares represent clay gouges. The zone marked as (i) represent the cataclastic deformation bands, zone (ii) fault rocks phyllosilicate and zone (iii) *shale smear.* Line K represents average values based on laboratory tests on samples from the Heidrun Field used in flow simulations (Knai, 1996).

As seen from Figure 3, there is a general trend indication a reduction in permeability with the increase in clay/shale content in the formation, sometimes with great dispersion.

Equation 1 represents the relationship proposed by Manzocchi *et al.* (1999) relating permeability with fault displacement, d, and SGR. The lines plotted in Figure 3 correspond to values of d equal to 1 mm, 10 cm, 1 m, 10 m and 1 km, from top to bottom

$$\log k_f = -4SGR - 0.25\log(d) \cdot (1 - SGR)^5 \tag{1}$$

In Equation 1, k_f is the fault permeability in mD, d is the shear displacement of the fault and SGR is the Shale Gauge Ratio. Al-Busafi (2005) says that in Equation 1, the first term represents the reduction of permeability due to the shale smear along the fault plane and that the second term reflects the reduction in permeability due to lithologies of low clay content. Manzocchi *et al.* (1999) say that the influence of the shear displacement, d, in the permeability of the fault is small for low clay contents. Therefore, equation 1 is not indicated for very low SGR.

3.4 Direct measurements of permeability of fault zones

Direct measurement of fault permeability is precluded by the fact the reservoir faults cannot be accessed from surface. In the specialized literature it may be found with

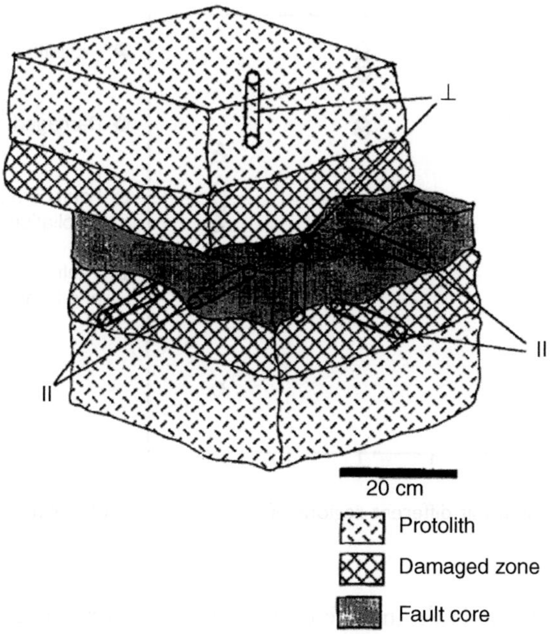

Figure 4 Location of samples within fault zone (adapted from Evans *et al.*, 1997).

greater frequency papers related with the determination of the relationship between fault architecture and flow properties. Next, studies related to the properties of faults will be discussed.

Many studies have been carried out in order to determine the change in fault permeability with the stress field (Teufel, 1987; Caine *et al.*, 1996; Evans *et al.*, 1997; Bolton *et al.*, 1999; Zhang & Cox, 2000; Boutareaud *et al.*, 2008). Evans *et al.* (1997) studied the permeability of a fault zone considering three types of samples, all of them from the East Folk Fault in Wyoming, USA. The samples come from the fault core, the second from the damage zone and the third from the rock itself (see Figure 4).

The tests were carried out under confining pressures between 2 MPa and 50 MPa. The results of permeability for a confining pressure of 3.4 MPa, for the three regions mentioned before, are indicated in Figure 5. As it may be observed, samples from the intact rock showed permeability between 10^{-18} and 10^{-17} m^2, whereas the damage zone present values between 10^{-16} and 10^{-14} m^2 and the fault core showed results between 10^{-20} and 10^{-17} m^2. The results are consistent with the in situ investigation on the flow of fluids through faults present on crystalline rocks.

4 MECHANICAL PROPERTIES OF FAULT ZONES

4.1 Generalities

Faults have been modeled as a single, planar slip surface or as a thick zone. In any case, there is the need to establish a constitutive law that describes the relationship between

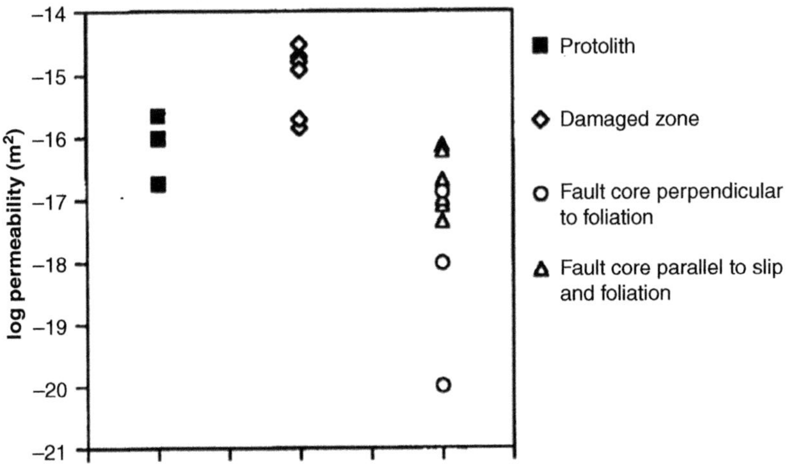

Figure 5 Permeability values at different regions of the fault zone and direction (adapted from Evans *et al.*, 1997).

stresses acting upon the fault and the corresponding strains. The evaluation of such a constitutive law is not an easy task, mainly when these structures are deep as it is the case of faults within reservoirs.

The use of core samples is a possibility to determine such a constitutive law. There are not much samples available but there is the possibility of defining the strength and stress-strain behavior. Considering the complexity of the fault zone (as indicated in Figure 1) one should expect a great variation in the material properties along the strike and down dip. For a practical application one should consider the possibility of working with average values for the material behavior. This practice allows the study of the influence of fault material composition and mechanical properties.

4.2 Shear strength parameters

Early studies on the behavior of rock discontinuities indicate that friction and roughness of the planes control its shear strength (Patton, 1966; Barton, 1973, 1974, 1976). Byerlee (1978), based on a large number of experimental results, suggests that friction between two rock surfaces is not much dependent of the rock type and that the shear strength depends on the normal stress level. Equations 2 and 3 (τ is the shear stress and σ is the normal effective stress) are suggested as shear strength envelopes for normal stress below 200 MPa and for normal stress between 200 MPa and 2000 MPa, respectively.

$$\tau = 0.85\sigma \tag{2}$$

$$\tau = 0.5 + 0.6\sigma \tag{3}$$

Most faults are not single surfaces in contact with each other and may be seen as filled discontinuities. In this case, as also discussed by Barton (1973) and Byerlee (1978), the shear strength depends upon the strength properties of the filling material. Zoback

(2007) uses the friction coefficient of 0.60 for the shear strength of faults when evaluating the risk of fault slipping.

Indirect evaluation of shear strength is possible using the in situ stress field, magnitude and direction, and pore fluid pressure and failure condition as to back calculate the shear strength parameters. Normally, the strength criterion is based upon Mohr-Coulomb and cohesionless material. Many investigators have used this methodology and the obtained values are, sometimes, consistent with Byerlee (1978) friction parameters, based upon the consideration of brittle zone. A compilation of crustal stress measurements by Townend & Zoback (2000) has used Byerlee's law, with a friction coefficient of 0.60, in order to constraint in situ stresses with great success. Townend & Zoback (2001) suggest that in situ stress measurements in crystalline rocks indicate that stress magnitudes are consistent with value of friction coefficients between 0.6 and 1.0 for faults. Faulkner *et al.* (2010) reviewed a large number of cases that indicate that some faults may slip under low friction coefficients, much less than those predicted by Byerlee (1978). They refer to these faults as weak faults.

The geomechanical properties of a fault zone are the result of a combination of factors such as lithology, microstructure, deformation rate, fluid pressure and diagenetic processes (Dewhurst & Jones, 2003). A proper evaluation of the influence of all these factors is possible only when using controlled laboratory experiments on samples from fault zones. Lockner *et al.* (2011) report experiments carried out on samples from San Andreas Fault System when values of friction as low as 0.15 were obtained. Chen *et al.* (2006) present data on shear strength experiments conducted under the Taiwan Chelungpu Fault Drilling Project (TCDP). More recent data on shear tests carried out on core samples of fault gouge are reviewed by Carpenter *et al.* (2016).

5 ANALYSIS OF FAULT REACTIVATION

5.1 Review

The slip tendency of a fault may be verified comparing the shear stress that acts onto the plan that composes the fault with its shear strength (Jaeger & Cook, 1969; Morris *et al.*, 1996). In this approach, the magnitudes and orientations of the principal stresses (in situ total stresses) and pore fluid pressure are known or assumed. The stresses are projected onto the fault plane in order to determine: (i) the effective normal stress (σ_n), (ii) the shear stress magnitude (τ) and (iii) the shear stress direction on the surface. Thus, the magnitude of the slip tendency (τ/σ_n) can be determined, besides the possible direction of the movement, indicated by the direction of the shear stress. This methodology does not consider the eventual changes in total stresses within the reservoir during production. The maximum available pore fluid pressure in order to avoid triggering fault slip may be evaluated as well (Moraes, 2004; Nacht, 2010).

Soltanzadeh & Hawkes (2008) present the use of a semi-analytical method called theory of inclusion in order to evaluate the stress variation due to production and injection processes in hydrocarbon reservoirs in the analysis of fault reactivation. Details about the theory of inclusion and its application in different reservoir geometries can be found in Eshelby (1957, 1959) and Soltanzadeh & Hawkes (2008), respectively.

Regarding the failure of geological media, Sibson (1977) sets the limiting conditions for the case of failure governed by an elastic-frictional mechanism. Sibson (2003) discusses the conditions of occurrence of brittle failure in the rock formation as well as for faults and also presents a study of a brittle failure mode associated to two tectonic schemes (extensional and compressional) in a reservoir environment. A more complete analysis of brittle failure mechanisms can be found in Sibson (1977, 2000, 2003).

Jones & Hillis (2003) proposed a probabilistic approach to assess the risk of loss of sealing capacity of the fault, taking into account the combination of risks associated with juxtaposition seals, membrane seals and fault reactivation, as well as the quantification of uncertainty about each of the variables. The main difference of this methodology for the widely employed methods is the consideration of seal rupture risk due to fault reactivation.

Rutqvist *et al.* (2007) define the maximum injection pressure as the pressure limit that will not cause undesirable damage to the reservoir formation. However, as mentioned above, an estimation of suitable injection pressure that does not result in the reactivation of faults requires knowledge of the stresses acting on the fault. In this context, the works carried out by Streit *et al.* (2004) and Rutqvist *et al.* (2007) exemplify an analytical way of expressing the potential of fault reactivation due to the fluid pressure development in the porous medium and also point out that analytical approaches have limitations, especially when confronted with numerical analyzes and observations made in depleted hydrocarbon reservoirs which indicate that the in situ stress state does not remain constant during the injection of fluid. The stress state may vary in time and space, controlled by changes in pore pressure and temperature. Thus, in order to consider effects such temperature, mechanical and chemical processes, without the limitations of analytical approaches, many efforts have been made to create robust numerical tools to assist geoscientists in solving problems related to disposal of radioactive waste in geological environments, geothermal energy extraction, enhanced oil and gas recovery, natural gas storage and recently CO_2 storage.

Settari & Mourits (1994) pioneered the development and practical use of coupled flow-stress solution for reservoir engineering and evaluation of compaction and subsidence. Fault reactivation evaluation needs the calculation of stresses changes and that can only be properly done if coupled schemes are used. Several authors have discussed the development of both pseudo-coupled and fully coupled schemes.

Rutqvist *et al.* (2007, 2008) performed numerical simulations of fault reactivation considering CO_2 injection in rock masses. To carry out the analyses the authors use coupling numerical scheme between two programs: the flow simulator TOUGH and the stress analysis program FLAC. According to Rutqvist *et al.* (2002), the great advantage of TOUGH-FLAC coupling is that both have been tested and are widely used in their respective areas. One of the main advantages of using numerical approaches is related to the flexibility on the treatment of geological features such as faults using planes or zones. In addition, the discretization of the fault as a zone allows the use of constitutive models for determining its mechanical behavior.

In terms of the different ways to simulate the mechanical behavior, Guimarães *et al.* (2009) presented a comparative study between two constitutive laws in order to determine the maximum injection pressure that leads to the reactivation of a fault zone in an oil reservoir. Adopting a more realistic stratigraphic model, Ducellier *et al.* (2011) also report a study on the impact of CO_2 injection employing a fault zone model. Examples of parametric studies considering the geometric variables of faults, variability of properties as well as different numerical techniques to evaluate the fault behavior can be found in Zhang *et al.* (2008, 2009), Cappa *et al.* (2010) and Souza *et al.* (2014).

Regarding numerical techniques for treatment of discontinuities, such as faults and fractures in geomechanics, a wide range of works can be cited: Goodman *et al.* (1967), Ghaboussi *et al.* (1973), Pande & Sharma (1979), Desai *et al.* (1984), Beer (1985), Wan *et al.* (1990), Day & Potts (1994), Sluys (1997), Sluys & Berends (1998) and Belytschko *et al.* (2009). Even though there is a possibility of simulating fault as a single plane we will concentrate only in the modeling of a fault as a fault zone. Next we describe a robust methodology to perform stress-strain analyses of rocks within an oil reservoir through a two-way coupled flow-stress scheme.

5.2 Flow-stress coupling

Studies conducted by the Computational Geomechanics Group – ATHENA/ GTEP – PUC-Rio showed substantial influence of geomechanical effects on the history of fluid pressures during the development of a reservoir, and also on other aspects such as subsidence and compaction. The coupling methodology developed consists of a two-way partial coupling scheme. The flow variables (pore pressure and saturation of the phases) and the stress variables (displacement field, stress and strain state) are calculated separately and sequentially, using a conventional reservoir simulator and a stress analysis program, respectively. Coupling parameters are exchanged at each time step until reaching the convergence. The quality of this methodology was ensured by the rigorous development of a coupling parameter, which approximates the geomechanical response to the fully coupled behavior (Biot, 1941). Next, the methodology is briefly described and only the equations of the flow problem and the stress analysis problem are shown. For more details about the development of the formulation see Inoue & Fontoura (2009a).

The flow equation can be obtained by combining the mass conservation equation and the Darcy's law. The law of mass conservation is a material-balance equation written for a component in a control volume. In hydrocarbon reservoirs, a porous medium can contain one, two and three fluid phases. Equations 4 and 5 present the governing flow equations (single-phase) for the conventional reservoir simulation and the fully coupled scheme, respectively. In the conventional reservoir simulation, the porosity is related to pore pressure through the rock compressibility using a linear relation, and in the fully coupled scheme, the porosity equation is composed of four components that contribute to the fluid accumulation term. In these equations, the terms are as follows: ϕ is the porosity, p is the pore pressure, t is the time, k is the permeability, μ is the viscosity, c_f is the fluid compressibility, c_s is the solid matrix compressibility, c_r is the rock compressibility, α is the Biot's coefficient and ε_v is the bulk volumetric strain.

$$\left(c_f\phi^0 + c_r\phi^0\right)\frac{\partial p}{\partial t} - \frac{k}{\mu}\nabla^2 p = 0 \tag{4}$$

$$\left[c_f\phi^0 + c_S(\alpha - \phi^0)\right]\frac{\partial p}{\partial t} - \frac{k}{\mu}\nabla^2 p = -\alpha\frac{\partial \varepsilon_v}{\partial t} \tag{5}$$

The formulation of the geomechanical problem takes into account the equilibrium equations, stress-strain-displacement equations, rock-flow interaction and the boundary conditions. The governing equation of the geomechanical problem may be written as indicated in Equation 6, where G is the shear modulus, u is the nodal displacement and v is the Poisson ratio.

$$G\nabla^2 \mathrm{u} + \frac{G}{1-2v}\nabla\nabla \cdot \mathrm{u} = \alpha \nabla p \tag{6}$$

Inoue & Fontoura (2009a, b) described the methodology used herein for the coupling between flow and stress problem. The coupling is achieved through a convenient approximation between of the flow equation of the conventional reservoir simulation and the flow equation of the fully coupled scheme. In this methodology, the effect of solids compressibility is removed from the fully coupled scheme and the effect of volumetric strain of the porous medium is added to conventional reservoir simulation.

The parameters responsible for the coupling, which honor the fully coupling equation, are the porosity ϕ and the pseudo-compressibility c_p, presented in Equations (7) and (8), respectively. These parameters are updated every iteration through the coupled analysis.

$$\phi = \phi^o + \alpha\left(\varepsilon_v - \varepsilon_v^o\right) + \frac{1}{Q}(p - p^o) \tag{7}$$

$$c_p = \frac{\varepsilon_v^{n+1} - \varepsilon_v^n}{\varphi^o\left(p_i^{n+1} - p_i^n\right)} \tag{8}$$

Furthermore, the partial coupling between the stress analysis program and the conventional reservoir simulator is reached using a staggered procedure, implemented in a C++ code (Inoue & Fontoura, 2009b).

5.3 Geometric model

According to the structural geology literature, ideally, a numerical model that aims modeling fault reactivation and leakage should consider its real geomorphology. Nevertheless, a realistic representation of these structures can face several numerical difficulties related to the mesh/grid generation, once refined discretization will be necessary to model them properly.

5.4 Numerical approach for fault reactivation

As can be observed in the works available in the technical literature, evaluating numerically the fault reactivation process is not an easy task, moreover, there is not a single and well-known methodology. Besides that, there is another issue to be analyzed, once faults can be a plane or zone with thickness (Davis & Reynolds, 1996). Thus, the definition of numerical technique to analyze this process properly must take into account the real geomorphology of the fault (Righetto, 2012).

In this work, the stress analysis software (ABAQUS®, 2010) evaluates the fault reactivation mechanism considering a fault as a zone with thickness pre-established.

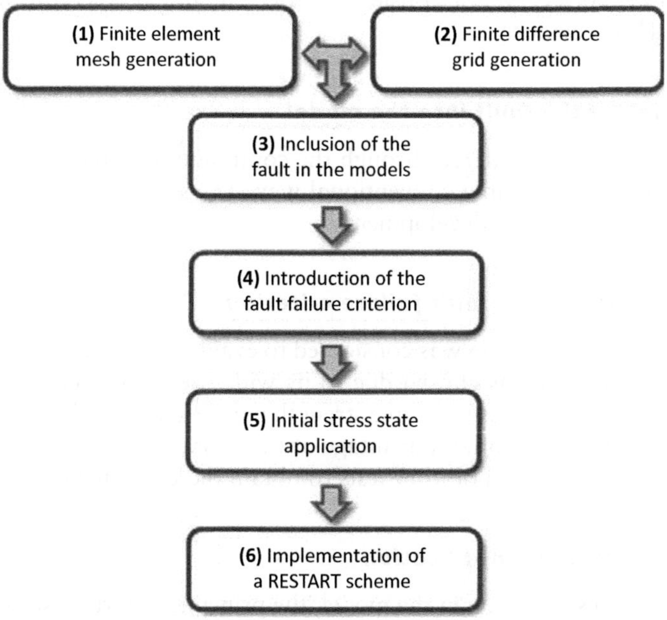

Figure 6 Flowchart with the steps for implementation of the fault reactivation model.

Thereby, this region can be modeled with a continuum approach combined with constitutive models, for instance Mohr-Coulomb. Indeed, Mohr-Coulomb yield criterion was adopted to evaluate the failure of material constituting the fault zone. In the reservoir simulator (ECLIPSE®, 2006), the fault zone was considered also, however, its core was held sealant. All hydraulic characteristics such as permeability, flow rates and rock compressibility were defined in the flow simulator. It should be mentioned that the grid of finite difference method (used in the flow simulation) was coincident with the mesh of finite element method (used in the stress analyses).

The flowchart shown in Figure 6 presents the main steps for the implementation of a fault reactivation model in the partial hydromechanical code. More details about the employed models can be found in Righetto *et al.* (2013).

The six steps shown in the Figure 6 will be presented in detail in the sequence. We will emphasize only the characteristics related to fault zone modeling.

5.4.1 Finite element mesh generation

To generate the finite element mesh composed by reservoir and surrounding rocks, we have used the software PATRAN® (2007) and ABAQUS® (2010).

5.4.2 Finite difference grid generation

The finite difference grid was generated using the flow simulator ECLIPSE® (2006). It should be mentioned that in the flow simulator, only the reservoir grid is generated.

The surrounding rocks are not considered in conventional reservoir simulation, as already discussed previously.

5.4.3 Inclusion of the fault into the model

As the fault zone was considered through the continuum approach, its discretization (grid and mesh) was made in a conventional way. The sealant core of fault zone was defined in this step of model development.

5.4.4 Introduction of the fault failure criterion

Mohr-Coulomb yield criterion was considered to evaluate the failure of the fault zone. This constitutive model was chosen due to its wide use in geotechnical engineering. Besides that, this model needs just three strength parameters, which are: friction angle, cohesion and dilatancy. Another advantage of the Mohr-Coulomb criterion is related to the use of conventional laboratory tests to obtain these parameters.

5.4.5 Initial stress state application

To apply the initial stress state in the model, the principle of virtual stress equilibrium (Herwanger & Koutsabeloulis, 2011) was used. The steps are:

a) The displacements in x, y, z directions were prevented in all nodes of the model;
b) The litho-static stresses were applied in all elements of the model;
c) ABAQUS® (2010) was run with the imposed boundary conditions (a) and the imposed initial conditions (b);
d) As a result, ABAQUS® (2010) calculated the reaction forces in all nodes;
e) Before the beginning of hydromechanical analyses, the reaction forces (d) were applied as nodal forces aiming to guarantee the equilibrium, *i.e.*, null values of displacements and strains.

5.4.6 Implementation of a restart scheme

In each time step of the flow-stress analysis, the coupling code uses an iterative scheme to reach the equilibrium. Thus, a *Restart* scheme was implemented in the stress analysis program, since the Mohr-Coulomb yield criterion needs to be updated due to the stress history.

5.5 Fault failure criterion

As mentioned in the previously sections, the Mohr-Coulomb yield criterion was adopted (see Figure 7) to perform the fault zone reactivation analyses. Thus, ABAQUS® was used in the geomechanical analyses, and the equations available in its theoretical manual will be presented next.

The yield function is defined, properly, according to the three stress invariants (Desai & Siriwardane, 1984). The first stress invariant (p), named equivalent pressure stress, can be defined as in Equation 9.

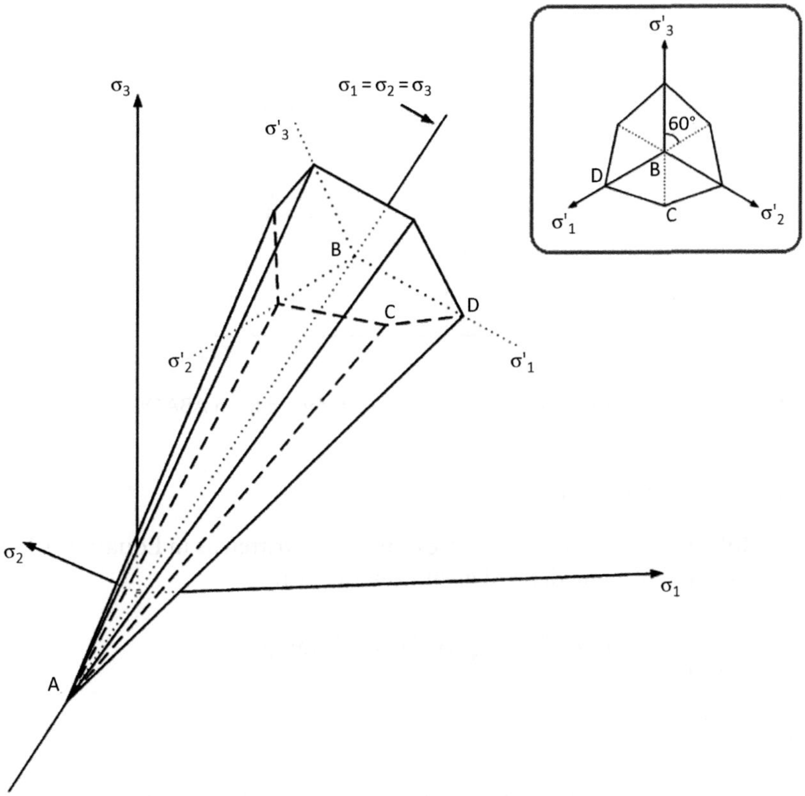

Figure 7 Mohr–Coulomb failure criterion: pyramidal surface in principal stress space and, in a detail, the cross-section in the octahedral plane (Labuz & Zang, 2012).

$$p = -\frac{1}{3} trace(\sigma) \tag{9}$$

The second stress invariant (q), named Mises equivalent stress, can be defined as Equation 10 where S and J_{2D}, are the deviatoric stress and second deviatoric stress invariant, respectively.

$$q = \sqrt{\frac{3}{2}(\boldsymbol{S}:\boldsymbol{S})} = \sqrt{3 \cdot J_{2D}} \tag{10}$$

The third stress invariant (r) can be defined as in Equation 11, where J_{3D} is the third deviatoric stress invariant.

$$r = \sqrt[3]{\frac{9}{2}(\boldsymbol{S} \cdot \boldsymbol{S} : \boldsymbol{S})} = \mathrm{J}_{3D} \tag{11}$$

Therefore, the Mohr-Coulomb yield function (f) can be written as in Equation 12 where ϕ and c are, respectively, the friction angle and cohesion of the geomaterial.

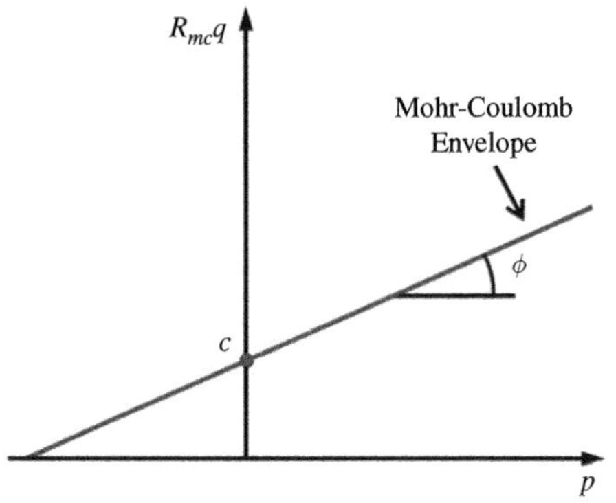

Figure 8 Mohr-Coulomb surface in the meridional plane (adapted from ABAQUS, 2010).

$$f = R_{mc} \cdot q - p \cdot \tan\phi - c = 0 \tag{12}$$

R_{mc} is the Mohr-Coulomb deviatoric stress measure written as in Equation 13, where θ is the deviatoric polar angle defined as in Equation 14.

$$R_{mc} = \frac{1}{\sqrt{3} \cdot \cos\phi} sen\left(\theta + \frac{\pi}{3}\right) + \frac{1}{3}\cos\left(\theta + \frac{\pi}{3}\right) \cdot \tan\phi \tag{13}$$

$$\cos(3\theta) = \left(\frac{r}{q}\right)^3 \tag{14}$$

After the variables definition, the Mohr-Coulomb yield criterion can be verified through the failure envelope as shown in Figure 8.

The stress path resulting from the hydromechanical simulations can be plotted together with the failure envelope to verify the fault zone reactivation.

6 CASE STUDY OF FAULT REACTIVATION

We present next the results of an elasto-plastic analysis of reactivation of a synthetic reservoir case. The right side of Figure 9 shows the face of fault zone to be analyzed in relation to the right portion of the complete model. The portion of the model that contains the reservoir was highlighted in front view on the left side of Figure 9, in order to facilitate the visualization of the region where the failure process was initiated in the coming figures.

Figure 10 illustrates the yielding development in the region of the fault zone, in terms of plastic deformation. Figure 10 (a) indicates the results corresponding to an injection time of 1 day, without evidences of plastic deformation. Figure 10 (b) shows the results corresponding to and injection time of 2163 days, where the onset of the material failure in the fault zone can be observed. Figure 10 (c) shows the results corresponding to an injection time of 4000 days, where the collapse extends throughout fault zone.

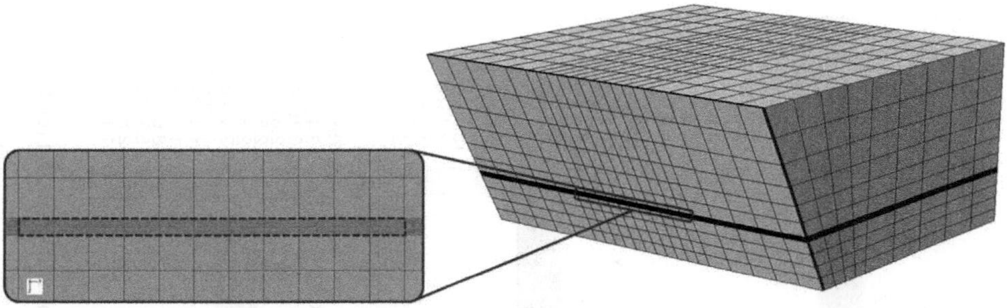

Figure 9 Indication of the fault zone section within the reservoir.

It can be seen that the beginning of yielding occurred in an isolated form at the ends of fault zone, spreading throughout the extent of the fault zone as the injection proceeded, with greater intensity at its ends. At the end of analysis, it can be observed that the whole fault zone presents plastic deformations, thereby emphasizing the generalized rupture of the material that composes it. Furthermore, when comparing the evolution of the yielding with time, the progressive nature, both in the horizontal and vertical direction, of the rupture process is evident.

The three stress invariants along the analysis time were evaluated through Equations 9 to11. From these three stress invariants, the Mohr-Coulomb yielding function can be evaluated and its variation in time with the injection process may be monitored. The yield function value is almost equals to zero when the rupture of the material of the fault zone does occur, as shown in Figure 11.

It was also verified that the yield criterion was met correctly through the Mohr-Coulomb failure envelope and the stress path in the meridional plane, as shown in Figure 12. When the yield function (Equation 12) approaches zero, the stress path touches the Mohr-Coulomb envelope, indicating the collapse. It was observed that for 2163 days, the yielding function has an almost zero value, indicating that the criterion was reached in the material composing the fault zone. This fact corroborates Figure 12, since the trajectory of the envelope touches Mohr-Coulomb envelope after 2163 days of analysis.

Using the two-way partial coupling it was found that the pore pressure value at initial yielding was equal to 50.9 MPa. It is observed that the geomechanical effects caused by injection were significant not only in the reservoir, but also in the surrounding rocks, indicating the importance to its consideration in coupled analyses involving reactivation of faults. The reactivation process can be affected in terms of pressure to be reached to cause yielding of the material of zone fault.

7 SUMMARY

Faults are very important geological structures for the field of petroleum reservoir development. The extraction and injection plan must be carefully planned in order to avoid undesired deformation within and near the reservoir. This chapter discussed the relevance of faults for petroleum reservoir studies. An important point is

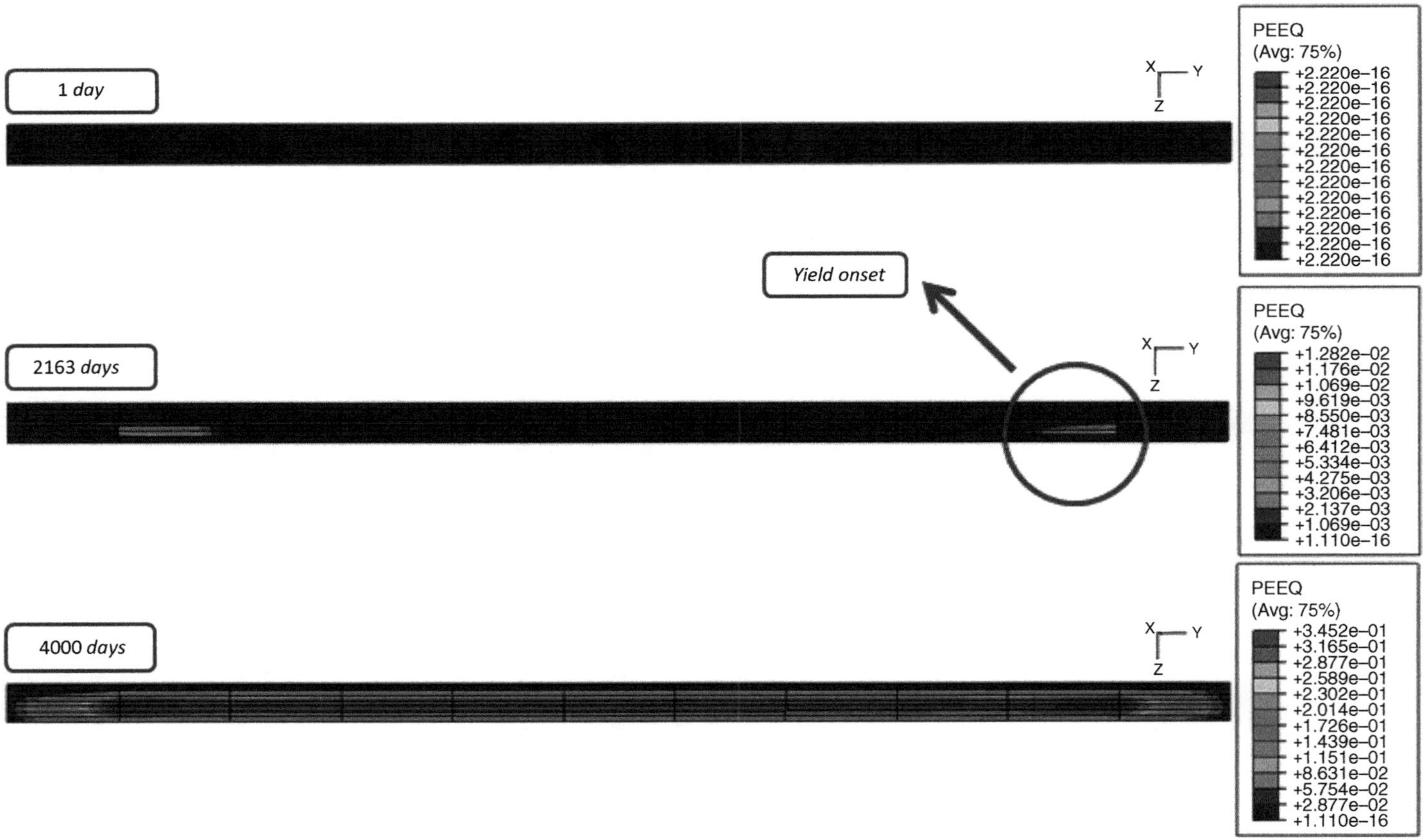

Figure 10 Evolution of plastic deformations throughout the fault zone along the reservoir for injection times of (a) 1 day; (b) 2163 days and (c) 4000 days.

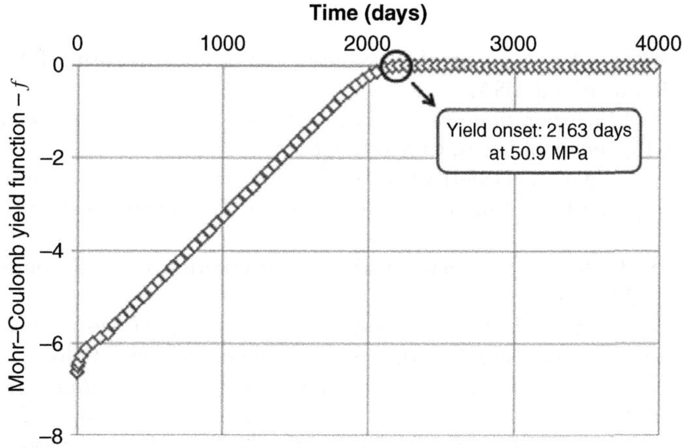

Figure 11 Yield function variation along the time injection.

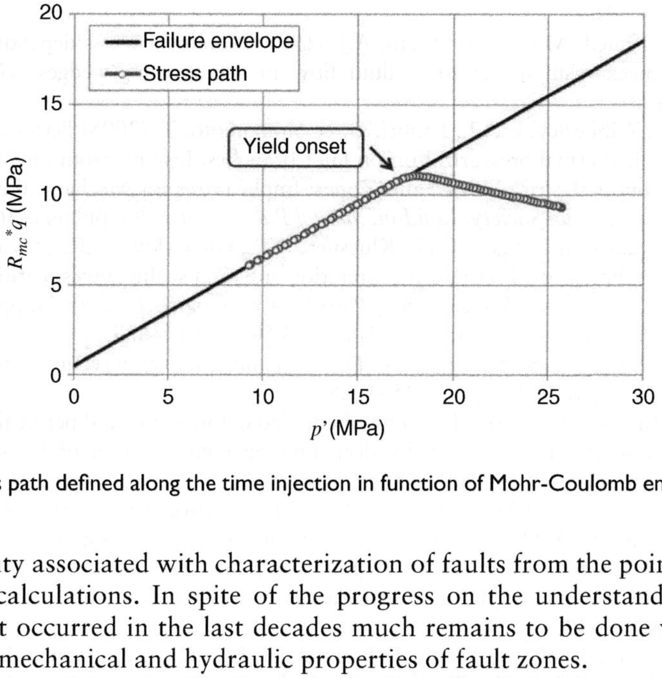

Figure 12 Stress path defined along the time injection in function of Mohr-Coulomb envelope.

the uncertainty associated with characterization of faults from the point of view of engineering calculations. In spite of the progress on the understanding of fault behavior that occurred in the last decades much remains to be done with respect to obtaining mechanical and hydraulic properties of fault zones.

The evaluation of the risk of fault reactivation must be done using coupled analysis tools. The paper presented a methodology of hydromechanical analysis based upon explicit, two-way flow-stress solution. This method is accurate and converges to the fully coupled solution. Faster tools must be made available in order to transform these analyses into normal practice procedures. Very important is the evaluation of the injected volume instead of the injection pressure since it takes some time before a critical pore pressure occurs near the fault.

REFERENCES

Abaqus (2010) *ABAQUS user's manual.* Dassault Systèmes, Simulia Corp. ABAQUS© vs. 6.10, Providence, Rohde Island, USA.

Al-Busafi, B. (2005) *Incorporation of fault rock properties into production simulation models.* Ph.D. Thesis, University of Leeds, United Kingdom.

Antonellini, A. & Aydin, A. (1994) Effect of faulting on fluid flow in porous sandstones: Petrophysical properties. *American Association of Petroleum Geologists Bulletin* 78, pp. 181–201.

Barton, N.R. (1973) Review of a new shear strength criterion for rock joints. *Engineering Geology* 7, pp. 287–332.

Barton, N.R. (1974) *A Review of the Shear Strength of Filled Discontinuities in Rock.* Norwegian Geotech, Inst. Publ. No. 105, Norwegian Geotech. Inst, Oslo

Barton, N.R. (1976) The shear strength of rock and rock joints. *International Journal of Rock Mechanics and Mining Sciences & Geomechanics Abstracts*, 13(10), pp. 1–24.

Beer, G. (1985) An isoparametric joint/interface element for finite element analysis. *International Journal of Numerical Methods in Engineering* 21, pp. 585–600.

Belytschko, T.B., Gracie, R. & Ventura, G. (2009) A review of XFEM/GFEM for material modelling. *Modelling and Simulation in Materials Science and Engineering*, 17(4), pp. 1–24.

Biot, M.A. (1941) General theory of three-dimensional consolidation. *Journal of Applied Physics* 12(2), pp. 155–164.

Bolton, A.J., Clennell, M.B. & Maltman, A.J. (1999) Nonlinear stress dependence of permeability: A mechanism for episodic fluid flow in accretionary wedges. *Geology*, 27(3), pp. 239–242.

Boutareaud, S., Wibberley, C.A.J., Fabbri, O. & Shimamoto, T. (2008) Permeability structure and co-seismic thermal pressurization on fault branches: Insights from the Usukidani fault, Japan. The Internal Structure of Fault Zones: Implications for Mechanical and Fluid-Flow Properties. *Geological Society, London, Special Publications* 299, pp. 341–361.

Bouvier, J.D., Kaars-Sijpesteigen, C.H., Kluesner, D.F., Onyejekwe, C.C. & Vander Pal, R.C. (1989) Three-dimensional seismic interpretation and fault sealing investigations. Num River Field, Nigeria. *American Association of Petroleum Geologists Bulletin* 73, pp. 1397–1414.

Byerlee, J.D. (1978) The friction of rocks. *Paleogh* 116, pp. 615–625.

Caine, J.S., Evans, J.P. & Foster, C.G. (1996) Fault zone architecture and permeability structures. *Geology*, 24(11), pp. 1025–1028.

Cappa, F. & Rutqvist, J. (2010) Modelling of coupled deformation and permeability evolution during fault reactivation induced by deep underground injection of CO_2. *International Journal of Greenhouse Gas Control* 5, pp. 336–346.

Carpenter, B.M., Saffer, D.M. & Marone, C. (2016) Frictional properties of the active San Andreas Fault at SAFOD: Implications for fault strength and slip behavior. *Journal of Geophysical Research: Solid Earth* 120, pp. 5273–5289.

Carstens, H. (2005) Understanding fault facies improves reservoir modelling. *GEO Expro*, February, pp. 27–30.

Chen, N., Louis, L., Tembe, S., Wong, T.-F., Lockner, D., Morrow, C., Song, S.S., Kuo, L.W., David, C., Robion, P. & Zu, W. (2006) Fault rheology: Constraints from laboratory measurements on core samples from deep drilling into fault zone. CIG Workshop.

Chester, F.M. & Logan, J.M. (1986) Implications for mechanical-properties of brittle faults from observations of the Punchbowl fault zone, California. *Pure and Applied Geophysics*, 124(1–2), pp. 79–106

Davis, G.H. & Reynolds, F.J. (1996) *Structural geology of rocks and regions.* 2nd ed. New York: John Wiley & Sons, Inc.

Day, R.A. & Potts, D.M. (1994) Zero thickness interface elements–numerical stability and application. *International Journal of Numerical and Analytical Methods in Geomechanics*, 18(10), pp. 689–708.

Desai, C.S. & Siriwardane, H.J. (1984) *Constitutive laws for engineering materials with emphasis on geologic materials*. New Jersey: Prentice-Hall.

Desai, C.S., Zaman, M.M., Lightner, J.G. & Siriwardane, H.J. (1984). Thin-layer element for interfaces and joints. *International Journal of Numerical and Analytical Methods in Geomechanics*, 8(1), pp. 19–43.

Dewhurst, D.N. & Jones, R.M. (2003) Influence of physical and diagenetic processes on fault geomechanics and reactivation. *Journal of Geochemical Exploration*, 78–79, pp. 153–157.

Donnelly, L.J. (2009) A review of international cases of fault reactivation during mining, subsidence and fluid abstraction. *Quarterly Journal of Engineering Geology and Hydrogeology* 42, pp. 73–94

Ducellier, A., Seyedi, D. & Foerster, E. (2011) A coupled hydromechanical fault model for the study of the integrity and safety of geological storage of CO_2. *Energy Procedia* 4, pp. 5138–5145.

ECLIPSE (2006) *Reservoir Simulator User's Manual*. Schlumberger, Houston, Texas (USA).

Eshelby, J.D. (1957) The determination of the elastic field of an ellipsoidal inclusion and related problems. *Proceedings of the Royal Society of London*, A241 (1226), pp. 326–396.

Eshelby, J.D. (1959) The elastic field outside an ellipsoidal inclusion. *Proceedings of the Royal Society of London*, A252 (1271), pp. 561–569.

Evans, J.P., Forster, C.B. & Goddard, J.V. (1997) Permeability of fault-related rocks, and implications for hydraulic structure of fault zones. *Journal of Structural Geology* 19, pp. 1393–1404.

Færseth, R.B., Johnsen, E. & Sperrevik, S. (2007) Methodology for risking fault seal capacity: Implications of fault zone architecture. *American Association of Petroleum Geologists Bulletin*, 91(9), pp. 1231–1246.

Faulkner, D.R., Jackson, C.A.L., Lunn, R.J., Schlische, R.W., Shipton, Z.K., Wibberley, C.A.J. & Withjack, M.O. (2010) A review of recent developments concerning the structure, mechanics and fluid flow properties of fault zones. *Journal of Structural Geology* 32, pp. 1557–1575.

Faulkner, D.R., Lewis, A.C. & Rutter, E.H. (2003) On the internal structure and mechanics of large-slip fault zones: Field observations of the Carboneras fault in southeastern Spain. *Tectonophysics* 367, pp. 235–251.

Foxford, K.A., Walsh, J.J., Watterson, J., Garden, I.R., Guscott, S.C. & Burley, S. D. (1998) Structure and content of the Moab Fault Zone, Utah, USA and its implication for fault seal predictions. *In*: Jones, G., Fisher, Q. J. & Knipe, R. J. (eds) *Faulting, fault sealing and fluid flow in hydrocarbon reservoirs*. Geological Society, London, Special Publications 147, pp. 87–103.

Fristard, T., Groth, G., Yielding, G., & Freeman, B. (1997) Quantitative fault seal prediction: A case study from Oseberg Syd. *In*: *Hidrocarbon seals: Importance for exploration and production*. Amsterdam: Elsevier Science, 107–124. (Norwegian Petroleum Society. Special Publication 7).

Ghaboussi, J., Wilson, E.L. & Isenberg, J. (1973) Finite element for rock joints and interfaces. *ASCE Journal of Soil Mechanics and Foundations Division* 99, pp. 833–848.

Gibson, R.G. (1994) Fault zones seals in siliciclastic strata of the Columbus Basin, offshore Trinidad. *American Association of Petroleum Geologists Bulletin* 78, pp. 1372–1385.

Gibson, R.G. (1998) Physical character and fluid-flow properties of sandstone derived fault gouge. *In*: Coward, M. P., Johnson, H. & Daltaban, T. (eds) *Structural geology in reservoir characterization*. Geological Society, London, Special Publications 127, pp. 83–97.

Goodman, R.E., Taylor, R.L. & Brekke, T.L. (1968) A model for the mechanics of jointed rock. *ASCE Journal of Soil Mechanics and Foundations Division* 94, pp. 637–659.

Guimarães, L.J.N., Gomes, I.F. & Fernandes, J.P.V. (2009) *Influence of mechanical constitutive model on the coupled hydro-geomechanical analysis of fault reactivation*. Reservoir Simulation Symposium. Society of Petroleum Engineers.

Harris, D., Yielding, G., Levine, P., Maxwell, G., Rose, P.T. & Nell, P. (2002) Using shale gouge ratio (SGR) to model faults as transmissibility barriers in reservoirs: an example from the Strathspey Field, North Sea. *Petroleum Geoscience* 82, pp. 167–176.

Healy, J.H., Rubey, W.W., Griggs, D.T. & Raleigh, C.B. (1968) The Denver earthquakes: Disposal of waste fluids by injection into a deep well has triggered earthquakes near Denver, Colorado. *Science* 161, pp. 1301–1310.

Herwanger, J.V. & Koutsabeloulis N. (2011) *Seismic geomechanics: How to build and calibrate geomechanical models using 3D and 4D seismic data*. EAGE Publications, Houten (The Netherlands).

Hull, J. (1988) Thickness–displacement relationships for deformation zones. *Journal of Structural Geology* 10, pp. 431–435.

ICDP (1996). *International Continental Drilling Program*. www.icdp-online.org

Inoue N. & Fontoura S.A.B. (2009a) Answers to some questions about the coupling between fluid flow and rock deformation in oil reservoirs. *In SPE/EAGE Reservoir Characterization and Simulation Conference Proceedings*. SPE Paper Number 125760-MS.

Inoue, N. & Fontoura, S.A.B. (2009b) *Explicit coupling between flow and geomechanical simulators*. Proceedings of the International Conference on Computational Methods for Coupled Problems in Science and Engineering. Ischia Island, Italy.

IODP (2003). *International Ocean Discovery Program*. www.iodp.org.

Ito, H., Behrmann, J., Hickman, S., Tobin, H. & Kimura, G. (2007) Abstracts and Report from the IODP/ICDP Workshop on Fault Zone Drilling, Miyazaki Japan. *Scientific Drilling*. Special Issue No 1, Reports on Deep Earth Sampling and Monitoring.

Jaeger, J.C. and Cook, N.G.W. (1971) *Fundamentals of rock mechanics*. London: Chapman and Hall.

Jones, R. & Hillis, R. (2003) An integrated, quantitative approach to assessing fault-seal risk. *American Association of Petroleum Geologists Bulletin*, 87(3), pp. 507–524

Knai, T.A. (1996) Faults impact on fluid flow in the Heidrum Field. *In*: *Faulting, fault sealing and fluid flow in hydrocarbon reservoirs*. University of Leeds, 75.

Knott, S.D., Beach, A., Brockband, P.J., Brown, J.L., McCallum, J.E. & Weldon, A.I. (1996) Spatial and mechanical controls on normal fault populations. *Journal of Structural Geology* 18, pp. 359–372.

Labuz, J.F. & Zang, A. (2012) ISRM suggested method: Mohr–Coulomb failure criterion. *Rock Mechanics and Rock Engineering* 45, pp. 975–979.

Lima, C.C. (2008) *Structural Geology and Brittle Deformation: Class notes of Field Trip* (In Portuguese), Petrobras.

Lockner, D., Morrow, C., Moore, D. & Hickman, S. (2011) Low strength of deep San Andreas fault gouge from SAFOD core. *Nature* 472, pp. 82–85.

Manzocchi, T., Walsh, J.J., Nell, P. & Yielding, G. (1999) Fault transmissibility multipliers for flow simulation models. *Petroleum Geoscience* 5, pp. 53–63.

Moraes, A. 2004 *Mechanical behavior of fault zones*. DSc Thesis (In Portuguese), Institute of Geoscience, Federal University of Rio de Janeiro.

Morris, A., Ferril, D.A. & Henderson, D.B. (1996) Slip tendency analysis and fault reactivation. *Geology* 24, pp. 275–278.

Nacht, P.K., de Oliveira, M.F.F., Roehl, D.M. & Costa, A.M. (2010) Investigation of geological fault reactivation and opening. *Mecánica Computacional* 29, pp. 8687–8697.

Ottesen Ellevset, S., Knippe, R.J., Olsen, T.S., Fisher, Q.T. & Jones, G. (1998) Fault controlled communication in the Sleipner Vest Field, Norwegian Continental Shelf: Detailed, quantitative input for reservoir simulation and well planning. *In*: Jones, G., Fisher, Q. J. & Knipe, R. J.

(eds) *Faulting, fault sealing and fluid flow in hydrocarbon reservoirs*. Geological Society, London, Special Publications 147, pp. 283–297.

PATRAN (2007) *Release Guide*. MSC Software Corporation, Santa Ana, California, USA.

Pande, G.N. & Sharma, K.G. (1979) On joint/interface elements and associated problems of numerical ill conditioning. *International Journal of Numerical and Analytical Methods in Geomechanics*, 3(3), pp. 293–300.

Patton, F.D. (1966) Multiple modes of shear failure in rock. *Proceedings of the 1st Congress of International Society for Rock Mechanics*, Lisbon 1, pp. 509–513.

Raleigh, C.B., Healy, J.H. & Bredehoeft, J.D. (1976) An experiment in earthquake control at Rangeley, Colorado. *Science* 191, pp. 1230–1237.

Ramsay, J.G. & Hubber, M.I. (1987) *The techniques of modern structural geology, vol. 2: Folds and fractures*. Academic Press, New York (USA).

Righetto, G.L. (2012) *Hydromechanical simulation of fault re-activation in petroleum reservoirs: Approaches by contact interactions and plasticity* (in Portuguese). M.Sc. Thesis – Department of Civil Engineering, Pontifical Catholic University of Rio de Janeiro.

Righetto, G.L., Lautenschläger, C.E.R., Inoue, N. & Fontoura, S.A.B. (2013) *Analysis of the hydromechanical behavior of fault zones in petroleum reservoirs*. Rock Mechanics for Resources, Energy and Environment, EUROCK 2013, CRC Press, pp. 947–954.

Rutqvist, J., Birkholzer, J.T., Cappa, F. & Tsang, C.F. (2007) Estimating maximum sustainable injection pressure during geological sequestration of CO_2 using coupled fluid flow and geomechanical fault-slip analysis. *Energy Conversion and Management* 48, pp. 1798–1807.

Rutqvist, J., Birkholzer, J.T. & Tsang, C.F. (2008) Coupled reservoir geomechanical analysis of the potential for tensile and shear failure associated with CO_2 injection in multilayered reservoir-caprock systems. *International Journal of Rock Mechanics and Mining Sciences* 45, pp. 132–143.

Rutqvist, J., Wu, Y. S., Tsang, C.F. & Bodvarsson, G. (2002) A modeling approach for analysis of coupled multiphase fluid flow, heat transfer, and deformation in fractured porous rock. *International Journal of Rock Mechanics and Mining Sciences* 39, pp. 429–442.

Sanz, P.F., Lele, S.P., Searles, K.H., Hsu, S.-Y., Garzon, J.L., Burdette, J.A., Kline, W.E., Dale, B.A. & Hector, P.D. (2015) Geomechanical analysis to evaluate production-induced fault reactivation at Groningen gas field. Paper SPE 174942-MS, *SPE Annual Technical Conference and Exhibition*, Houston, Texas.

Settari, A. & Mourits, M. (1994) Coupling of geomechanics and reservoir simulation models. *Computer Methods and Advances in Geomechanics*, Siriwardane, H. J. & Zanan, M. M. (Eds), Balkema, Rotterdam, pp. 2151–2158.

Sibson, R.H. (1997) Fault rocks and fault mechanisms. *Journal of Geological Society of London*, 133, pp. 191–213.

Sibson, R.H. (2000) Fluid involvement in normal faulting. *Journal of Geodynamics* 29, pp. 469–499.

Sibson, R.H. (2003) Brittle-failure controls on maximum sustainable overpressure in different tectonic regimes. *American Association of Petroleum Geologists Bulletin*, 87(6), pp. 901–908.

Sluys, L.J. (1997) Discontinuous modeling of shear banding. In: Computational plasticity – Fundamentals and applications, ed. by Owen, D. R. J., Oñate, E. & Hinton, E. (International Center for Numerical Methods in Engineering, Barcelona), pp. 735–744.

Sluys, L.J. & Berends, A.H. (1998) Discontinuous failure analysis for mode-I and mode-II localization problems. *International Journal of Solids Structures*, 35(31–32), pp. 4257–4274.

Soltanzadeh, H. & Hawkes, C.D. (2008) Semi-analytical models for stress change and fault reactivation induced by reservoir production and injection. *Journal of Petroleum Science and Engineering* 60, pp. 71–85.

Souza, A.L.S, Souza, J.A.B., Meurer, G.B. Naveira, V.P, Frydman, M. & Pastor, J. (2015) Integrated 3D geomechanics and reservoir simulation optimize performance, avoid fault reactivation. *World Oil*, April, pp. 55–58.

Streit, J.E. & Hillis, R.R. (2004) Estimating fault stability and sustainable fluid pressures for underground storage of CO_2 in porous rock. *Energy* 29, pp. 1445–1456.

Suckale, J. (2010) Moderate-to-large seismicity induced by hydrocarbon production. *The Leading Edge*, 29(3), pp. 310–319.

Teufel, L.W. (1987) Permeability changes during shear deformation of fractured rock. *28th United States Symposium on Rock Mechanics*, pp. 473–480.

Townend, J. & Zoback, M.D. (2000) How faulting keeps the crust strong. *Geology*, 28(5), pp. 399–402.

Walsh, J.J., Watterson, J., Heath, A.E. & Childs, C. (1998) Representation and scaling of faults in fluid flow models. *Petroleum Geoscience* 4, pp. 241–251.

Wan, R.G., Chan, D.H. & Morgenstern, N.R. (1990) A finite element method for the analysis of shear bands in geomaterials. *Finite Elements in Analysis and Design* 7, pp. 129–143.

Wibberley, C.A.J., Yielding, G. & Di Toro, G. (2008) Recent advances in the understanding of fault zone internal structure: A review. In: Wibberley, C.A.J., Kurz, W., Imber, J., Holdsworth, R.E. & Collettini, C. (eds) *The Internal Structure of Fault Zones: Implications for Mechanical and Fluid-Flow Properties*. Geological Society, London, Special Publications 299, pp. 5–33.

Wiprut, D. & Zoback, M.D. (2000) Fault reactivation and fluid flow along a previously dormant normal fault in the northern North Sea. *Geology* 28, pp. 595–598.

Yielding, G., Freeman, B. & Needham, T. (1997) Quantitative fault seal prediction. *American Association of Petroleum Geologists Bulletin* 81, pp. 897–917.

Zhang, S. & Cox, S.F. (2000) Enhancement of fluid permeability during shear deformation of a synthetic mud. *Journal of Structural Geology* 22, pp. 1385–1393.

Zhang, Y., Gartrell, A., Underschultz, J.R. & Dewhurst, D.N. (2009) Numerical modeling of strain localization and fluid flow during extensional fault reactivation: implications for hydrocarbon preservation. *Journal of Structural Geology* 31, pp. 315–327.

Zhang, Y., Schaubs, P.M., Zhao, C., Ord, A., Hobbs, B.E. & Barnicoat, A.C. (2008) Fault-related dilation, permeability enhancement, fluid flow and mineral precipitation patterns: numerical models. The Internal Structure of Fault Zones: Implications for Mechanical and Fluid-Flow Properties. *Geological Society, London, Special Publications* 299, pp. 239–255.

Zoback, M.D. & Harjes, H.P. (1997) Injection induced earthquakes and crustal stress at 9 km depth at the KTB deep drilling site, Germany. *Journal of Geophysical Research* 102, pp. 18,477–18,491.

Zoback, M.D. & Townend, J. (2001) Implications of hydrostatic pore pressure and high crustal strength for the deformation of interpolate lithosphere. *Tectonophysics* 336, pp. 19–30.

Zoback, M.D. (2007) *Reservoir geomechanics*. Cambridge University Press, 449p.

Zoback, M.D., Hickman, S. & Ellsworth, W. (2011) Scientific drilling into the San Andreas fault zone – An overview of SAFOD's first five years. *Scientific Drilling*, No 11, pp. 14–28.

Thermo-/Hydro-Mechanics in Gas Storage, Loading and Radioactive Waste Disposal

Chapter 18

Advanced technology of LNG storage in lined rock caverns

E.S. Park[1], Y.B. Jung[1], S.K. Chung[2], D.H. Lee[3] & T.K. Kim[3]
[1]Underground Space Department, Korea Institute of Geoscience & Mineral Resources, Daejeon, Korea
[2]Department of Energy & Resources Engineering, Chonnam National University, Gwangju, Korea
[3]Infra Center of Excellence, SK E&C, Seoul, Korea

Abstract: The challenges in storing Liquefied Natural Gas (LNG) in lined rock caverns are outlined. Many attempts have been made in the past to store LNG underground in unlined containment, though without success. A new concept for storing LNG in lined rock caverns has been developed to provide a safe and cost-effective solution. It consists of protecting the host rock against the extremely low temperature and providing a liquid- and gas-tight liner. In order to verify the technical feasibility of this storage concept, a pilot plant was constructed for storing LNG and has been in operation since January 2004, though it has now been decommissioned. The overall monitored results from the pilot operations confirmed that the construction and operation of underground LNG storage in lined rock caverns are technically feasible. Underground LNG storage systems in lined rock caverns can be realized in some countries which have suffered from the shortage of storage capacity of LNG and seasonal extreme variation of domestic demand, and where industries are already developed and free remaining areas of land are small and expensive.

1 INTRODUCTION

The demand for natural gas in Korea has grown rapidly since the 1980s, and Korea has grown to become the world's largest Liquefied Natural Gas (LNG) importing nation nowadays. However, it is difficult to handle the supply–demand adjustments for LNG due to factors such as seasonal variations in the domestic demand, discordance among import patterns, limits of storage facilities, and so on. Accordingly, it is very important to secure large LNG storage facilities and to stabilize LNG supply management on a long-term basis (Chung *et al.*, 2006). Therefore, Korea currently operates four LNG terminals and a nationwide pipeline network of over 4,240 km in order to ensure a stable supply for the nation.

As Korea imports natural gas in the liquid phase by ships, the refrigerated storage of natural gas is more economical than its conventional storage in the form of gas. Therefore, in-ground and above-ground storage, using insulated steel tanks near the coast, is generally used for storing LNG because LNG is transported by sea and is handled in gas terminals (Cha *et al.*, 2006, 2007). Nevertheless, these types of storage tanks may cause problems in cases that require a large land area for storage.

Based on the experience of underground storage for crude oil and various types of hydrocarbons, an underground storage system was thought to be a more economical way to store LNG. Many attempts have therefore been made in the past to store LNG underground in unlined containment, though without success (Anderson, 1989; Dahlström & Evans, 2002). One of the most significant problems related to the underground storage of cryogenic material is the leakage of liquid and gas from the containment system to the rock mass, caused by tensile failures due to shrinkage of the rock mass around the caverns (Monsen & Barton, 2001). The failures of underground storage caverns were due to thermal shock stresses generating cracks in the host rock mass. The thermal cracks induced by extremely low temperatures in the rock mass contributed to the deterioration of the operational efficiency. Gas leakage and increased heat flux between the ground and storage occurred (Dahlström, 1992; Glamheden & Lindblom, 2002).

The way to prevent a hard rock mass from cracking at LNG boiling temperatures (–162 °C) is to locate the unlined storage cavern deep enough below ground level, so that the geostatic stresses counterbalance the tensile stresses caused by cooling. The necessary depth – from 500 to 1,000 m – varies with rock type, which renders this unlined cavern storage concept very expensive (Amantini & Chanfreau, 2004; Amantini *et al.*, 2005).

A new concept of storing LNG in a lined rock cavern (LRC) with a containment system that can overcome these problems has been developed by Geostock, SK E&C and SN Technigaz with the help of Korea Institute of Geoscience and Mineral Resources (KIGAM). To demonstrate the technical feasibility of this concept, a pilot plant was constructed at the KIGAM in the Daejeon Science Complex in 2003, which was operated for storing liquid nitrogen (LN2, Boiling temperature: –196 °C) from January 2004 to the end of 2004, and is now decommissioned.

2 THEORETICAL STUDIES IN BEHAVIORS OF ROCK MASS UNDER CRYOGENIC CONDITIONS

2.1 Stress changes related to LNG storage

The concepts for underground gas storage differ from each other according to the confinement and sealing principle applied. Gas under atmospheric or low pressure can only be stored in a liquid state, thus applying refrigeration up to –160 °C is necessary. In this case emphasis is placed on thermal insulation and sealing for the liquid and/or the low pressure vapor. The containment – as a structural element – is subjected to very little loading, due to low pressure (Kovári, 1993).

In prior attempts to store LNG in unlined rock caverns, the criterion is mainly concerned with the opening of existing natural fractures on the rock surface. Such opening may cause the leakage of stored liquid gas through fractures, resulting in an excessive boil-off rate of gas and consequent failure of the storage. In the case of a lined rock cavern, since the containment system and concrete lining prevent the existing fractures from being in direct contact with the cryogenic liquefied gas, the opening of such fractures is allowed to some extent. Thus, the criteria for evaluating the thermo-mechanical stability of a lined rock cavern are concerned with the initiation of new fractures and the propagation of existing fractures within a considered temperature range. While the criterion for preventing the initiation of new fractures

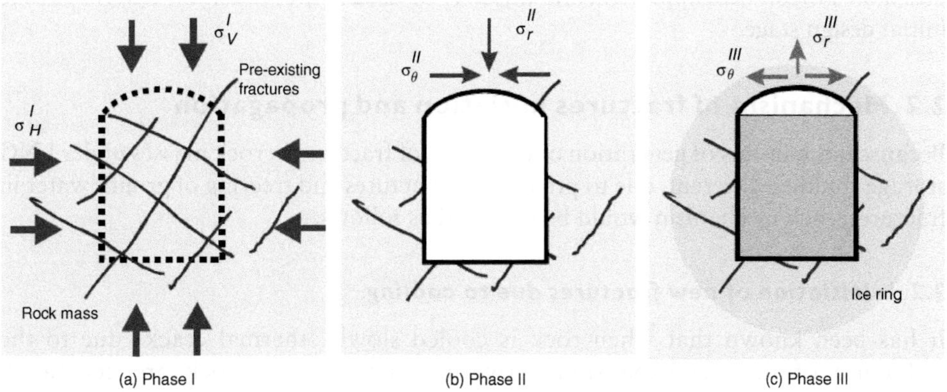

Figure 1 Schematic diagram of involved stress fields in the rock mass during the construction and operation of an underground LNG storage cavern (Lee *et al.*, 2006a).

in rock mass is based on the conventional strength concept, propagation of the existing cracks and fractures should be dealt with using the fracture mechanics concept because the major factor for the assessment of stability in this case is the stress concentration at the crack tip.

When such a lined rock cavern is used for LNG storage, its thermo-mechanical stability depends on the initial *in situ* rock stress (Phase I), stress concentrations in the vicinity of the cavern due to excavation (Phase II), and thermally-induced stress during operation of the cryogenic storage (Phase III), as illustrated in Figure 1. In Figure 1, σ^{I}, σ^{II}, and σ^{III} denote *in situ* rock stress field before excavation (σ^{I}_{V}, σ^{I}_{H}: vertical and horizontal stresses), redistributed stress field after excavation (σ^{II}_{r}, σ^{II}_{θ}: radial and tangential stresses) and thermally-induced stress field during operation of cryogenic storage (σ^{III}_{r}, σ^{III}_{θ}: radial and tangential stresses), respectively. In prior attempts to store LNG in an unlined rock cavern, the criterion is concerned with opening of existing natural fractures on the rock surface. The temperature at which the opening of existing fractures occurs coincides with the point at which the tangential component (σ^{III}_{θ}) of the induced thermal stress during Phase III becomes zero, indicating transition of the stress from compression to tension. At this temperature, the arching effect around the cavern disappears and the existing rock blocks become unstable (Lee *et al.*, 2006a).

The new criteria for evaluating the thermo-mechanical stability of a lined rock cavern are concerned with both the initiation of new fractures and the propagation of existing fractures. Initiation and propagation of such fractures will cause instability of the rock mass around the cavern and, in extreme cases, it may accompany leakage of stored gas, leading to ultimate failure of the system. Generally, new fractures are initiated when the tangential rock stress during Phase III becomes tensile and exceeds the tensile strength of rock. However, since the tensile strength of rock varies with temperature and the amount of thermal shock, such variation should be taken into account when establishing the criteria for preventing the initiation of new fractures in the rock mass. The temperature at which new fractures are initiated can be obtained based on a theoretical approach, and this may be used as an evaluation criterion at the initial design stage.

2.2 Mechanism of fractures initiation and propagation

Because mechanisms of generation or extension of fractures in rock masses under LNG storage could be different, due to pre-existing fractures and freezing of groundwater in fractures, each mechanism would be reviewed as follows.

2.2.1 Initiation of new fractures due to cooling

It has been known that when rock is cooled slowly, thermal cracks due to the difference of thermal expansion between components of rock are generated, although the temperature gradient is not steep. From the results of AE (acoustic emission) experiments on granite by Shell, the generation of micro-cracks started at about -50 °C (Dahlström, 1992). Therefore, a micro-fracture could be initiated when a thermal stress exceeds the tensile strength of a rock. However, because there is an effect of compensating tensile stresses with thermal stresses when initial stresses (compression) are present, the fracture generation due to thermal shock would be suppressed.

According to Goodall (1989), a condition for the successful operation of chilled gas storage can be expressed below in Equation 1.

$$\textit{In situ}\ \text{stress} + \text{tensile strength} > \text{thermal stress} \quad (1)$$

In situations with a significant internal storage pressure, this will contribute to the right hand side of the equation; however, in the case of chilled storage the contribution from the internal gas pressure (0.1–0.3 bar) is negligible. In the same way, water pressure caused by water curtains or by naturally high groundwater will act as a reducing factor on the *in situ* stress situation, in other words destabilizing the equilibrium.

It is suggested that fractures are not initiated when thermal stresses due to cooling are smaller than the minimum compressive stresses in the surrounding rock mass, as shown in Figure 2. The tensile strength of the rock mass is excluded because it is generally assumed that – for the purpose of numerical simulation – rock mass has no tensile strength. So it is thought that this criterion could be considered a conservative one; further studies are needed to verify and modify it.

2.2.2 Propagation of existing fractures due to cooling

In the case of a propagation mechanism of existing fractures, the presence of fractures in the rock mass is the point to be specially considered before the start of cooling. In general, it has been regarded that, as numerous fractures exist in rock masses, the propagation of existing fractures could easily occur compared to the initiation of new fractures in fresh rock during cooling. This means that fracturing could be initiated under lower thermally-induced stresses, which are induced because the mechanical characteristics of existing fractures are weaker than those of fresh rock. Groundwater generally flows through existing fractures in the rock mass and it would stay in the fractures locally if rock masses around

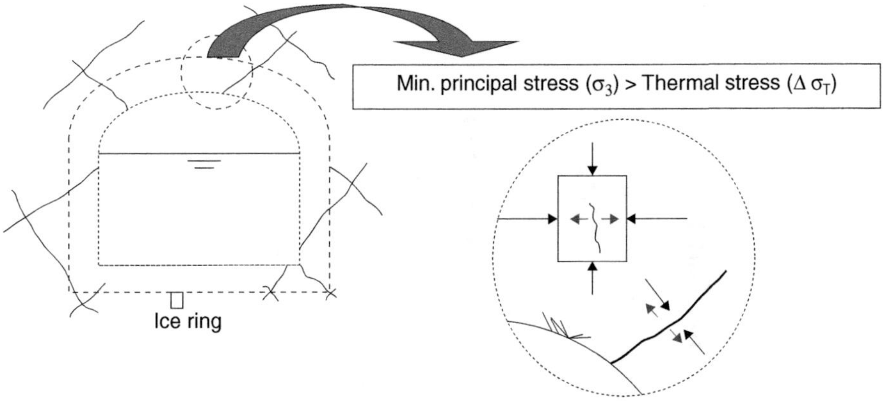

Figure 2 Criterion of suppression of a new fracture initiation (Park, 2006).

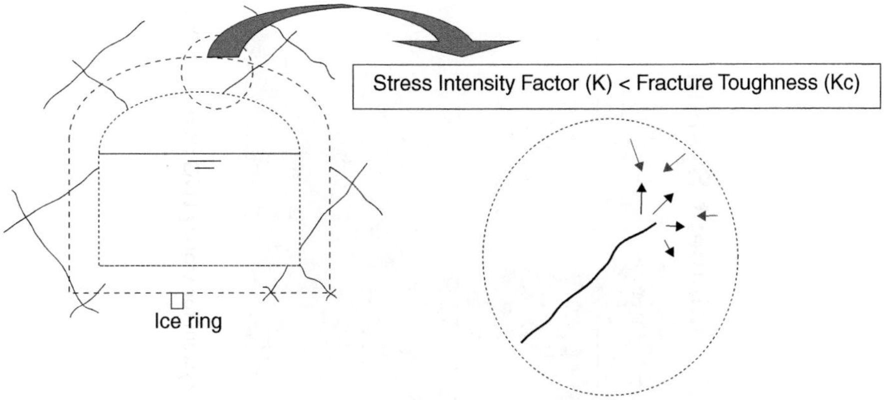

Figure 3 Criterion of suppression of an existing fracture propagation (Park, 2006).

caverns were drained badly during the construction of the caverns. In this case groundwater could be frozen during LNG storage and frost heaving pressures would be created in the existing fractures. This would then affect the expansion of fractures in rock masses.

The propagation mechanism of an existing fracture can be explained through relations between the stress intensity factor (K) and fracture toughness (Kc) at a crack tip, as shown in Figure 3.

3 THE BASIC CONCEPT OF A LINED ROCK CAVERN FOR LNG STORAGE

The basic concept of LNG storage in membrane-lined rock caverns is a combination of well-proven technologies and a new concept named 'formation of ice ring' as shown in Figure 4.

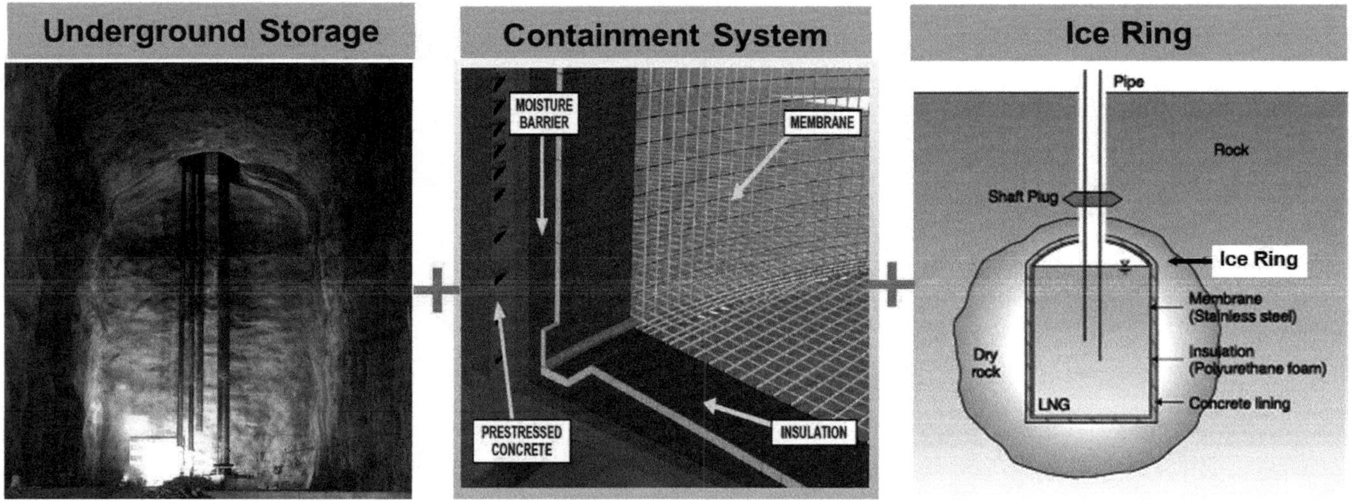

Figure 4 The three basic concepts of the lined rock cavern system for storing LNG (Park *et al.*, 2011).

3.1 Underground rock cavern

The underground storage technology, which proved reliable, was strongly developed. This was related to its intrinsic qualities but also because of the innovations and technical progress with which it was regularly improved. Underground storage facilities provide a lot of particular advantages: no exposure to threats of natural disasters such as earthquake, tsunami and typhoon; increased security against the man-made disasters of fire, war and terrorism; minimized land lease costs; and environmental friendliness.

Most of all, the major advantage of underground storage is its economic feasibility. As the economic feasibility of the underground storage is realised at large scales, the cavern geometry is optimized at the maximum possible cross section with a sufficient degree of structural stability. Currently, a large capacity of storage cavern units, with more than several hundreds of thousands of cubic metres' space, for a broad range type of hydrocarbon products and crude oil, has been constructed and operated successfully, especially in Korea. And Korea has accumulated technical know-how in this field, based on more than 30 years of experience in design, construction and operation of underground storage facilities (Kim, 2006).

3.2 Membrane containment system

The membrane containment system, which was introduced in 1962 on a prototype ship, has since been successfully used in several LNG carriers and storage tanks. This membrane system provides the proper thermal protection to the surrounding rock mass, preventing excessive stresses and crack formation, and reducing boil-off to level comparable with conventional LNG storage tanks. Also it provides a liquid- and gas-tight liner using a corrugated stainless steel membrane fixed onto insulating panels and a concrete lining. The thermal characteristics and thickness of the insulation is designed in order to achieve an allowable minimum temperature in the rock mass for the designed life of the storage, and a boil-off rate of less than 0.1% per day. The modular structure of the containment system makes it very flexible, improving the construction of, and adaptation to, cavern geometry. The thickness of insulating panels can be adapted to increase thermal efficiency and the nearly unstressed membrane permits very large scale projects.

3.3 Ice ring and drainage system

In the case of groundwater intrusion into the containment system, the integrity of the containment system could be destroyed because of the volume expansion of the groundwater during the freezing process. So a new groundwater control system was introduced to drain water around the cavern and prevent hydrostatic pressure acting against the containment system. And it can be used to balance the migration of the 0 °C isotherm by applying water with a specific temperature for water infiltration. The groundwater drainage system is composed of a series of boreholes drilled from the surface and/or dedicated drainage galleries installed around the cavern. Reduction of high water seepage in the cavern will generally be needed. This is achieved by systematic grouting works during excavation.

Although the ice ring is supposed to play an important part as a secondary barrier against the leakage of stored LNG, the location and thickness of the ice ring are also

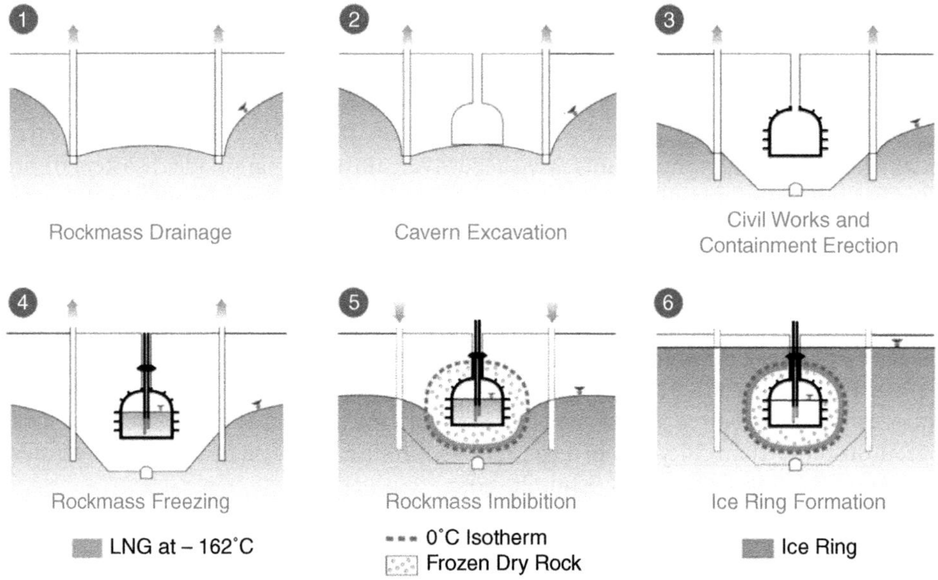

Figure 5 Schematic diagram of the formation of the ice ring (Park *et al.*, 2011).

important factors for the stable storage of LNG in a lined rock cavern. The formation of the ice ring is explained at each stage of the construction of a lined rock cavern (Figure 5).

Groundwater is temporarily removed from the rock surrounding the cavern during the first phase of construction. This preliminary desaturation of the host rock mass aims at preventing unacceptable hydrostatic pore pressure and ice formation behind the cavern lining. After the excavation of cavern and the installation of the containment system are complete, LNG will be stored in the cavern and the propagation of a cold front starts from the cavern at the same time. When the cold front has advanced far enough from the cavern wall, drainage can be stopped to allow groundwater progressively to rise up and quickly form a thick ring of ice around the cavern. After the ice ring is completely formed, operation of the drainage system is stopped. The drainage period during LNG storage will last several months or years, depending on the thermal properties of rock masses and hydrogeological characteristics of the site.

4 UNDERGROUND LNG PILOT PROJECT

4.1 General

From 2003 to 2005, the concept of underground storage of LNG has been validated with the design, construction and operation of a pilot plant in Korea, storing liquid nitrogen at –196 °C. The LNG pilot plant is located in Daejeon, about 200 km south of Seoul, in an existing cavern which is within the KIGAM research facilities. The previous room for cold food storage was enlarged to test the overall performance of a lined rock cavern for LNG storage. A lot of instrumentation was installed to survey both operational parameters and the thermo-hydro-mechanical behavior of rock masses. The key factors considered during design, construction and operation stage of underground LNG storage are as follows.

- Physical and mechanical properties of the rock mass at low temperatures.
- Occurrence and propagation of fractures in the rock mass.
- Stress change by induced thermal stress.
- Calculation of heat flux (conduction + convection).
- Convection rate in rock joints.
- Monitoring of thermal stress and displacement.
- Monitoring of temperature change.
- Resistance to thermal shock during cooling.
- Frost heaving at low depth.

4.2 Geological condition and pilot cavern

Through experience from a former study of underground cold (–25 °C) and chilled (0 °C) storage, lots of information about the thermo-mechanical characteristics of rock masses at low temperature was obtained. However, because a hydrogeological investigation was not included in the previous study, various hydraulic tests were conducted, including single packer tests, straddle packer tests and hydraulic communication tests at specific intervals, in order to evaluate the hydraulic properties of the rock mass around the cavern (Cha *et al.*, 2006).

From the site investigation using the mapping of joints after the local enlargement of the cavern and rock mass classification from about seven boreholes, the estimated results are shown in Figure 6. The orientations (dip/dip direction) of the three major joint sets observed during the site investigation were 60/209, 40/171, and 29/331. Although open joints were seemingly scattered around the pilot cavern, there was a slight concentration of open joints around the major joint sets. The joint (30/154) passing diagonally through the roof of the pilot cavern and carrying a large inflow of water was unique among the joint sets at the site. Even though this joint was not part of any major joint sets, and joints of this type seemed not to occur frequently on this site, its presence showed hydrogeologically important behavior in the pilot cavern (Cha *et al.*, 2007).

The base rock consists of Jurassic biotitic granite intruding Pre-Cambrian gneiss, which has a rock quality designation (RQD) of 80–86 and the most frequent Q-value of 12.5. The Q-value requires no supports or minor unreinforced shotcrete (about 40 mm) and bolt according to support categories of the Q-system. Therefore, it is sensible to adapt rock bolting to stabilize the main fractures of the existing cavern and ensure stability of possible fracture position.

Access to the pilot cavern was provided through an existing horizontal tunnel, and the experimental cavern roof lies at a depth of about 20 m below the ground surface. In order to complete the containment system, a concrete wall closes the entrance of the cavern. The south concrete wall of the cavern is exposed to the entrance tunnel, which is not the typical case for full-scale facilities. Additionally, a platform above the entrance of the cavern was made to install instruments, a manhole and piping. The internal dimensions of the completed pilot plant have a sectional dimension of 3.5 m × 3.5 m, a length of 10 m, amounting to a working volume of 110 m^3 (Figure 7).

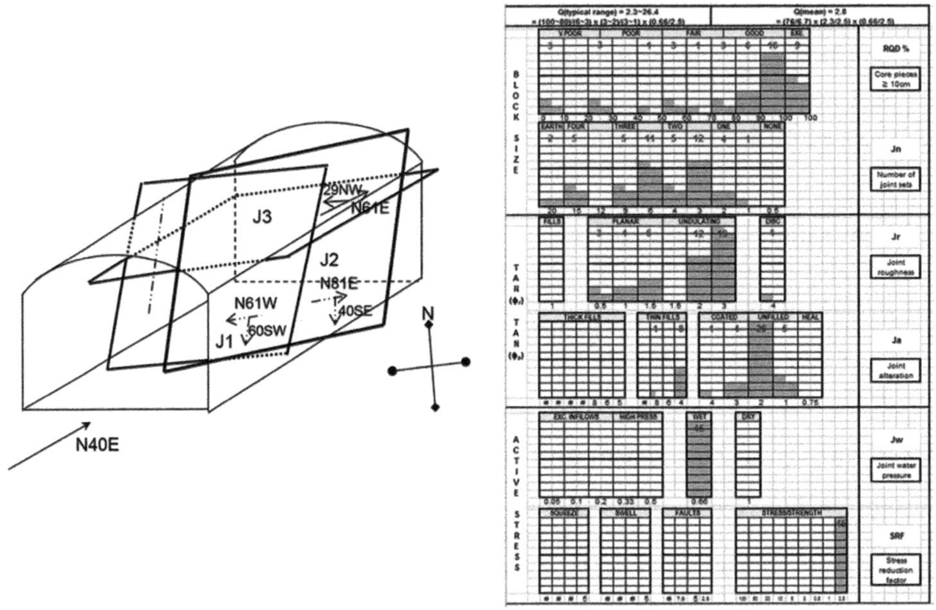

Figure 6 Major joints and Q-values estimated from the site investigation.

Figure 7 A bird's-eye view and cross section of the pilot cavern (Park *et al.*, 2010).

4.3 Drainage system

The purpose of the drainage system is to reduce water entry into the cavern during concrete casting, to reduce humidity percolating through the concrete during the installation of the containment system, to drain the rock mass and maintain a low water saturation degree in the surrounding rock mass during the cooling phase, and to control the resaturation of the host rock mass once the frozen area around the cavern reaches a suitable thickness.

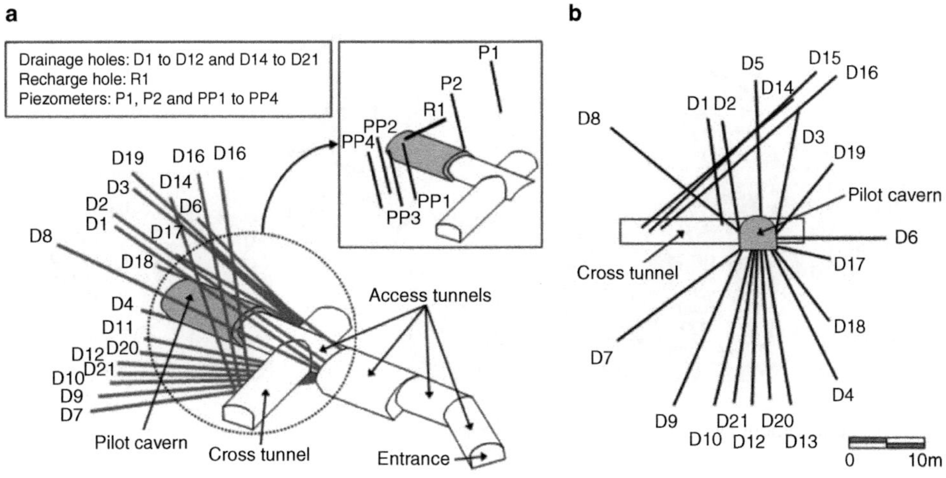

Figure 8 Schematic diagram of the pilot cavern: (a) oblique view of drainage holes (D series), recharge hole (R1), and piezometers (P and PP series), and (b) front view of drainage holes (Cha *et al.*, 2007).

Before the cavern was excavated, 21 boreholes from the tunnel drifts were drilled for drainage near the cavern walls. The diameter of the drainage boreholes was 76 mm. A schematic diagram of drainage holes (D1–D12 and D14–D21), recharge hole (R1), and piezometers (P1, P2 and PP1–PP4) is shown in Figure 8. The major group of drainage holes – that is sub-parallel to the cavern axis – was drilled at the entrance to the cavern. The drainage holes directly cover joint set 1 (60/209). Other drainage holes that were drilled at an angle to the cavern axis were drilled to penetrate joint set 5 (62/265) in the chamber perpendicularly. Other joint sets were intersected at an angle by neighboring boreholes (Cha *et al.*, 2007).

Also, an acceptance test for the cavern was performed by using an interference test after full installation of the concrete structure. The drainage system was used for the first months of operation before the rock was frozen, in order to avoid water intrusion in the containment, and to prevent hydrostatic pressure acting against the containment. Finally, the water table was lowered to 8 m below the cavern floor.

4.4 Containment system and construction works

The containment system, which is used for underground lined rock caverns, is similar to the one used and improved by SN Technigaz since 1962 for the membrane type of LNG storage tanks and LNG carriers. For the Daejeon LNG pilot cavern, insulating panels made of foam were sandwiched between plywood sheets and were bonded on the concrete using load-bearing mastic. The insulation panel thickness is 300 mm to ensure that the rock temperature will not fall below –50 °C after 30 years and the boil-off rate will stay at an acceptable limit. Finally, a 1.2 mm thick stainless steel corrugated membrane, attached to the insulation panel, provides gas-tightness at a low temperature. All the surfaces (*e.g.* bottom, walls, chambers and vault) are

Figure 9 Containment system used for the pilot test.

covered with concrete lining, insulating panels and the stainless steel membrane sheets. The containment system is composed of several layers, from rock to LNG, as shown in Figure 9.

The transition structure between the rock and the containment system consists of a reinforced concrete cast put in place between formwork and the rock, with injection holes for contact grouting. The interface cavern/access tunnel is constructed of a reinforced concrete cast put in place between two frameworks, including a closing wall and a platform used for access, piping and instrumentation. The thickness of the polyurethane insulation panel was 10 cm, and reinforced concrete barriers of 20 cm thickness were formed between the rock and the containment system. The piping for LNG loading/unloading and cavern instrumentation passes through a shaft sealed with a concrete plug embedded in the rock mass and supporting the weight of the whole lines.

4.5 Geotechnical monitoring system

There were several geotechnical issues to be solved from the geotechnical monitoring undertaken during the operation of the pilot storage as follows.

1) Can the monitored temperature distribution in the rock be estimated by the numerical calculation at the design stage?
2) Are the direction and the magnitude of rock displacement compatible with the containment system?
3) Is the opening of rock joints tolerable even at a low temperature?
4) Can the cavern reinforcement maintain its stability with the rock at a very low temperature?

Such questions have been solved from the analysis of the monitoring results. Figure 10 presents the general arrangement of the instruments in a cross section and a plan view (Chung *et al.*, 2007b).

The geotechnical monitoring system was provided for measuring the temperature, thermo-mechanical and groundwater responses of the rock and concrete during the

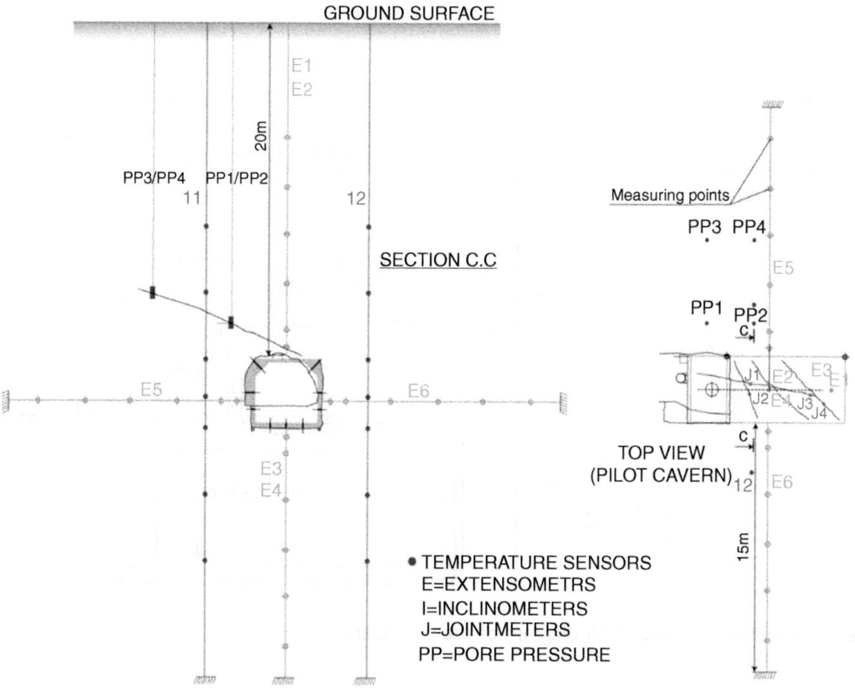

Figure 10 Arrangement of geotechnical instruments around the pilot cavern (Chung *et al.*, 2007b).

pilot test. The pilot cavern and its surrounding rock mass were equipped with a comprehensive set of geotechnical instruments, which allows monitoring of temperature profiles and temperature-induced displacements in rock around the cavern; the opening of rock joints on the cavern surface; the load on the installed rock bolts; settlement of the ground surface; pore pressure distribution in the rock mass; and the variation of ground water level. All instruments are equipped with thermal sensors to measure the temperature at the installed depth.

Moreover, numerous parameters such as level, temperatures, pressure and boil-off rate for the containment system were monitored during the operation. The behavior of surrounding rocks and the containment system were recorded during three successive phases of operation: a) the first six months' duration, during which a full level is maintained by filling with LN2 in order to compensate for loss of boil-off, b) the second six months, during which no more filling is performed and so allowing the cavern to empty naturally, and c) the third six months, during which the empty cavern is heated up until the ambient temperature is reached and the pilot cavern is decommissioned (Chung *et al.*, 2007a).

4.6 Operation of the pilot plant

Process, equipment, monitoring and control during storage operation are similar to above-ground tanks. For safety and practical reasons, LN2 is used instead of LNG.

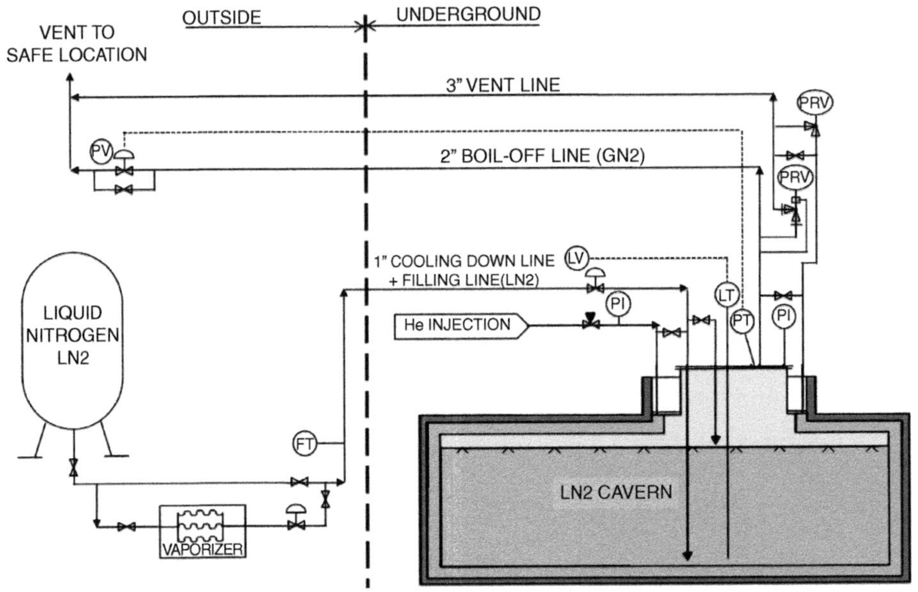

Figure 11 Schematic process flow diagram of the pilot plant.

The cryogenic pilot plant was operated from a laboratory room located in the access tunnel. It houses the Data Control System (DCS) for the process of the cryogenic pilot and for the rock monitoring. An LN2 plant is installed outside on a fenced site, containing one LN2 storage tank, a vaporizer, and associated control systems in order to make the cryogenic pilot inert, cool it down, fill it completely with LN2, and make up to compensate for the gaseous N2 boil-off.

These operations use two insulated pipes from the dedicated site to the cryogenic pilot cavern, which pass through the access tunnel and the special local enlargement in the cavern roof to go over the plug and penetrate inside the cavern. The additional line is necessary to release overpressure if this happens. Some instruments installed in the containment system and cavern permit following the main parameters (inner pressure and temperature, LN2 level, containment deformation) and to operate the pilot (Figure 11).

After several commissioning tests, the operation started on 10 January 2004 with a 2.6 m level of LN2, and the drainage system was stopped gradually between 10 June 2004 and 8 July 2004. The filling pipeline was closed on 10 July, and the storage has been completely empty since 12 August, 2004.

5 MAJOR RESULTS OF THE PILOT PROJECT

The results and data collected during a one-year operation of this pilot plant allowed us to validate the design, to set up criteria and engineering methods and to refine construction methodologies applicable to any type of LNG terminal.

5.1 Drainage system and efficiency test

An efficient drainage system is important in a lined LNG storage cavern. The drainage system in the pilot cavern studied here was designed to take into account the fracture and hydrogeological characteristics of the site. The orientation of drainage holes was limited by the geometry of the pilot cavern and the galleries (Cha *et al.*, 2007). The 21-borehole drainage system has proved its efficiency during the successive construction and operation periods and kept the rock mass desaturated during the cooling period. Efficiency tests performed at this stage showed that almost 99% of the natural water flowing toward the pilot cavern could be drained back by the drainage system. The resaturation of the rock mass was performed by stopping the drainage progressively. Finally, the drainage system was allowed to drain the host rock mass again during the thawing progress of the rock mass (Figure 12).

The rock drainage system worked very efficiently for dewatering the rock mass around the lined pilot cavern used for underground LNG storage. The drainage system drained 97% of inflow around the pilot cavern, even under heavy rainfall conditions and artificial recharge. The tests showed that drainage efficiency depends on the number, spacing and orientation of the drainage holes.

5.2 Temperature changes during cooling and thawing

Figure 13 shows the temperature profiles around the cavern during cooling and thawing stages. After six months of cooling, the 0° C isotherm lay at 4.4 m from the floor, 4.0 m from the side and 3.2 m from the roof (Figure 13a).

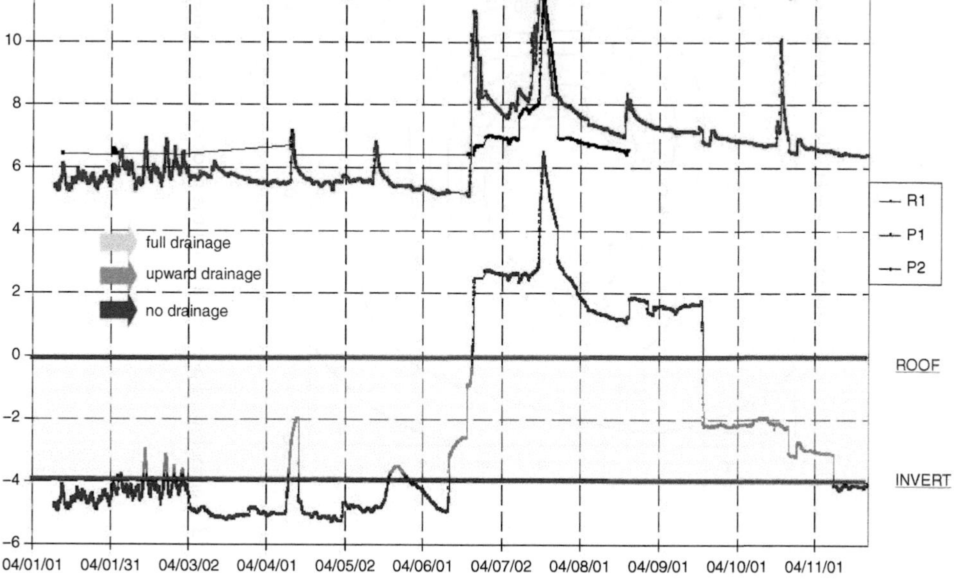

Figure 12 Piezometric variations by operation of the drainage system of the pilot cavern.

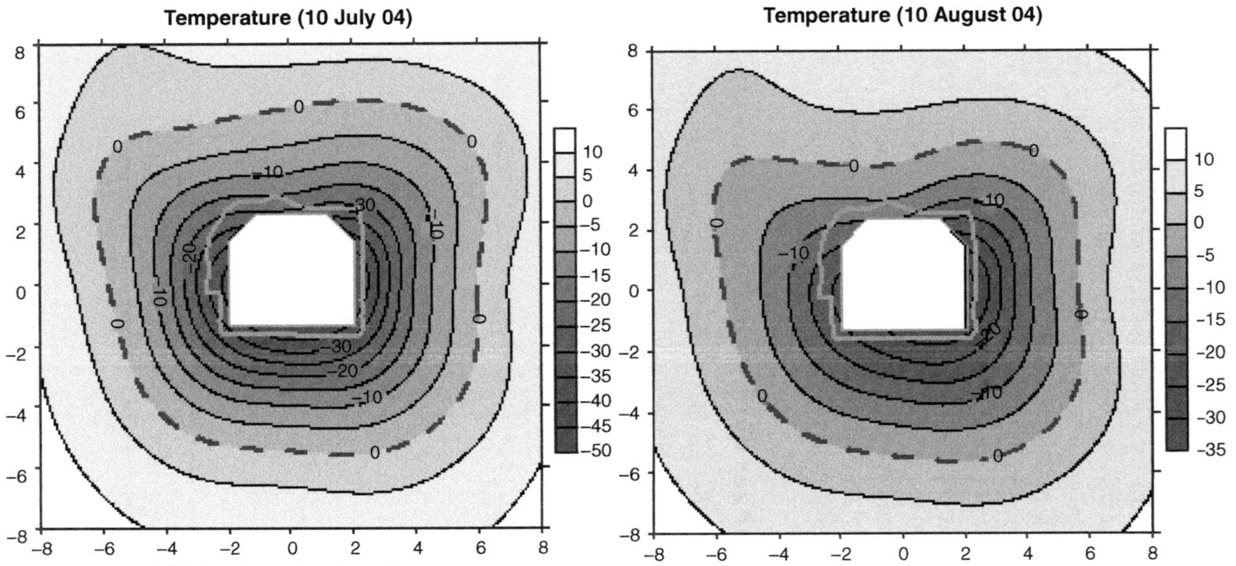

(a) After 6 months of cooling.

(b) After 1 month of thawing.

Figure 13 Temperature profiles around the pilot cavern. Dashed line indicates 0 °C isotherm.

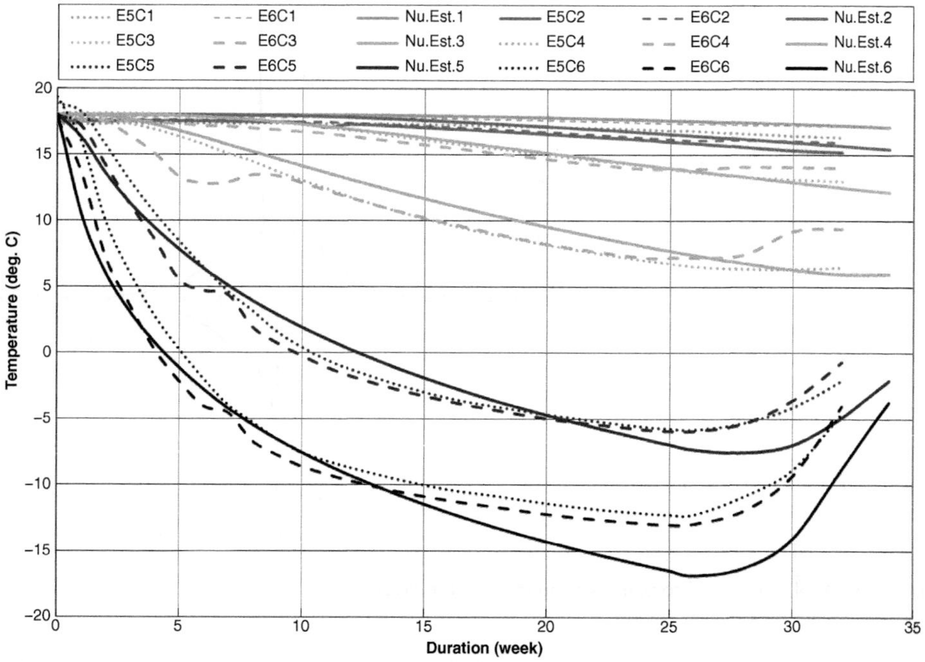

Figure 14 Comparison between measured and predicted temperatures of rock mass around the side wall of the pilot cavern (dotted line: measured temperature, solid line: predicted temperature).

Even after one month of thawing, the 0 °C isotherm maintained a similar profile at the side and the bottom rock of the cavern. However, in the roof rock the distance was reduced to 1.82 m from the roof of the cavern (Figure 13b). The cooling of the rock mass had occurred after operation according to the specified insulation system, and the cold front propagates toward the inner rock mass. The cooling rate of the rock above the roof was slower than that of the other sides because the temperature inside the storage where it was submerged with LN2 was kept constant at –196 °C, and the temperature on the upper part of the storage, filled with boiled-off gas, was about –100 to –120 °C.

Thermal responses of the rock mass under a very low temperature of about –30 °C around the rock cavern, with an absence of water, could be well predicted by numerical models such as FLAC2D code (Figure 14).

5.3 Thermal induced displacement

Thermal stress-induced displacements occurred toward the inner rock mass, which is favorable with regard to stability aspects of the cavern. Displacements are of relatively low amplitude – within 3–5 mm – corresponding to about 0.2 % of the cavern's radius. Data recorded at the end of the thawing period show that a small displacement (about 1 mm) is expected due to local rearrangement of the rock mass. Conventional rock

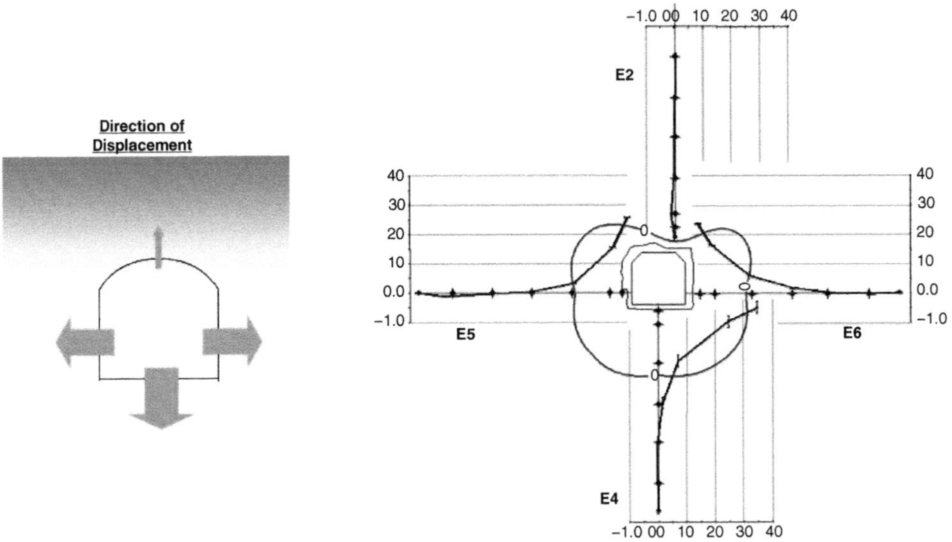

Figure 15 Displacement of rock mass around the pilot cavern.

reinforcements, such as rock bolts and shotcrete, remained effective at very low temperatures during the pilot operation (Figure 15).

5.4 Boil-off gas ratio

In order to evaluate the amount of boil-off gas any approach based on temperature or flux data from the field can be used, or numerical calculation. For simplicity, the boil-off gas rate (BOR) during the operation of LNG storage can be calculated using Equation 2 and temperature data (Lee *et al.*, 2006b):

$$\mathrm{BOR} = 100 \cdot K_i(T_i - T_{LNG}) \cdot (1/L_{LNG} \cdot \rho_{LNG}) \cdot S_f(\%/\mathrm{day}) \tag{1}$$

where

K_i is the coefficient of insulation property
S_f is the shape factor
T_i is the contact temperature at the concrete–insulation interface (°C)
T_{LNG} is the temperature of the stored LNG (°C)
L_{LNG} is the latent heat of vaporization of the LNG (J/kg)
ρ_{LNG} is volumetric weight of the LNG (kg/m^3).

When temperature evolutions at several points are compared with predicted results of numerical heat transfer as shown in Figure 14, the two trends show good agreement during the whole period of operation. Also, the real monitored data from the BOG test was compared with the predicted ones from theoretical and numerical data as shown in Figure 16. The monitored BOR lies in between

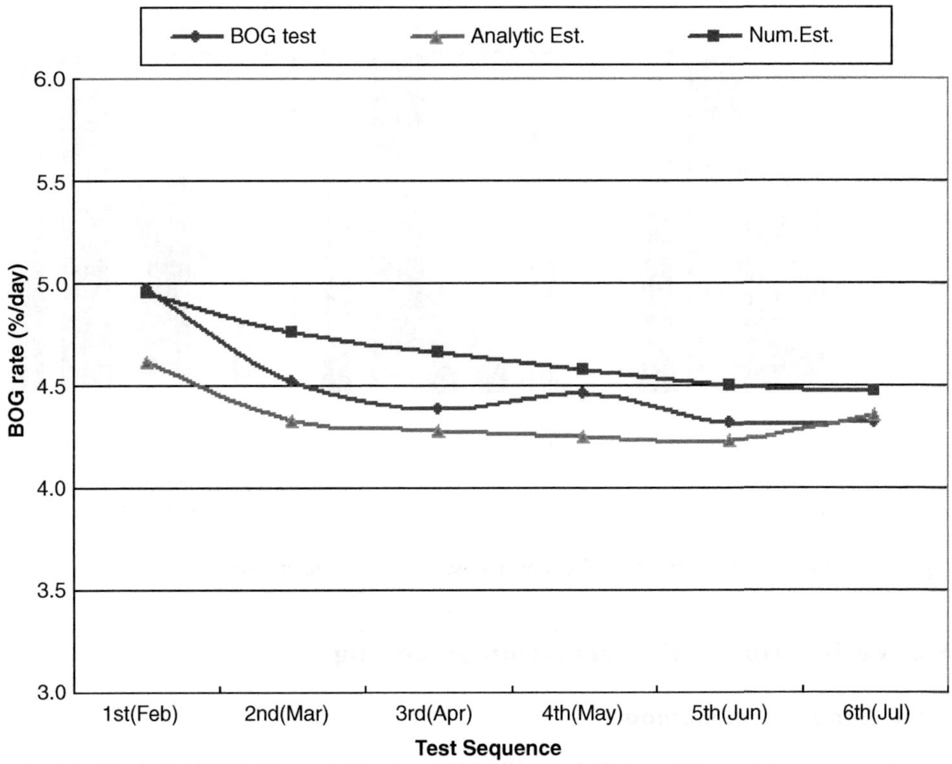

Figure 16 Boil-off gas ratio that occurred during operation of the pilot plant, and a comparison with the plots of two estimated rates.

the theoretical and numerical values. Therefore, the BOR from an underground LNG storage cavern can be estimated by numerical and theoretical methods, as shown in Figure 16.

5.5 Decommissioning

After the completion of the scheduled operations, the cavern and containment system were dismantled in order to judge the validity and safety of the proposed concept (Figure 17). The main findings of the pilot plant are as follows.

- The drainage system was capable of desaturating and resaturating the fissured rock mass.
- There was no damage to the rock mass.
- Temperature variation and boil-off rate estimates were similar to predictions.
- There was no degradation of the containment system efficiency.

Overall, the results from the pilot test confirm that both the construction and operation of underground LNG storage in a lined rock cavern are technically feasible.

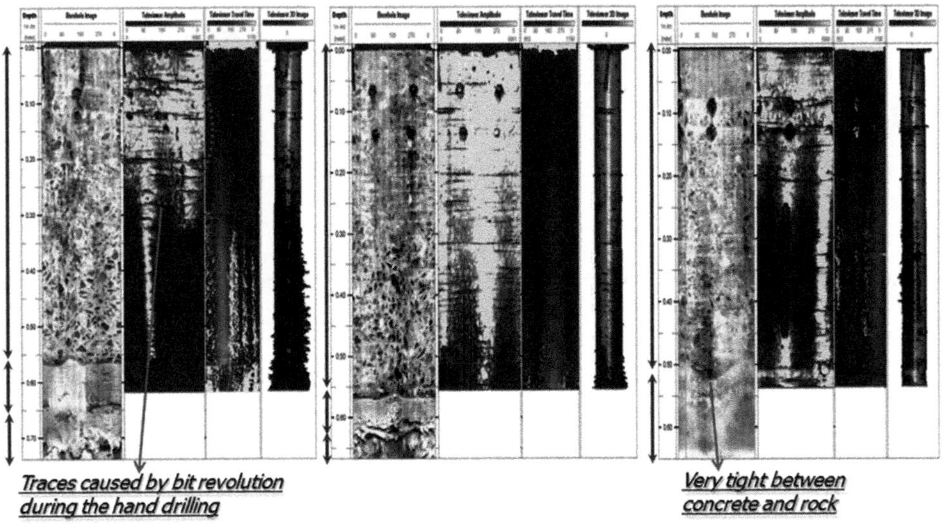

Figure 17 Results of crack inspection after decommissioning of the pilot plant.

5.6 Verification of the formation of ice ring

5.6.1 Geophysical method

It is a very important task to verify the formation of the ice ring and to recognize other changes in ground conditions related to the very low temperature in the storage cavern. To achieve this, a borehole radar survey was used to detect and describe the formation of the ice ring, and a 3D resistivity survey was applied to monitor the changes in ground conditions resulting from the cryogenic environment caused by the stored LN2, whose temperature is −196 °C.

The first set of measurements of 3D resistivity data, which is Phase I of the monitoring survey, was collected on 30 December 2003, just before the plant started operation. On 16 March 2004, after the plant had been operating for about two months, we acquired the second set of measurements, in Phase II. Phase III data collection took place on 14 May 2004, which is about two months after acquiring the Phase II data. This monitoring survey continued until the end of 2004 (Figure 18).

With these surveys, we have shown that there was a major change in hydrogeological conditions in the subsurface due to the cryogenic environment in the LN2 storage cavern. This change in ground conditions was successfully imaged by the 3-D resistivity method, which has proven to be very effective in resolving small changes in ground conditions (Yi *et al.*, 2005; Kim *et al.*, 2007).

5.6.2 Hydro-thermal coupled analysis

In order to investigate relevant mechanisms such as the propagation of the cold front, and the migration of water and formation of ice in the host rock mass, 2-D and 3-D hydro-thermal coupled numerical models were used successfully.

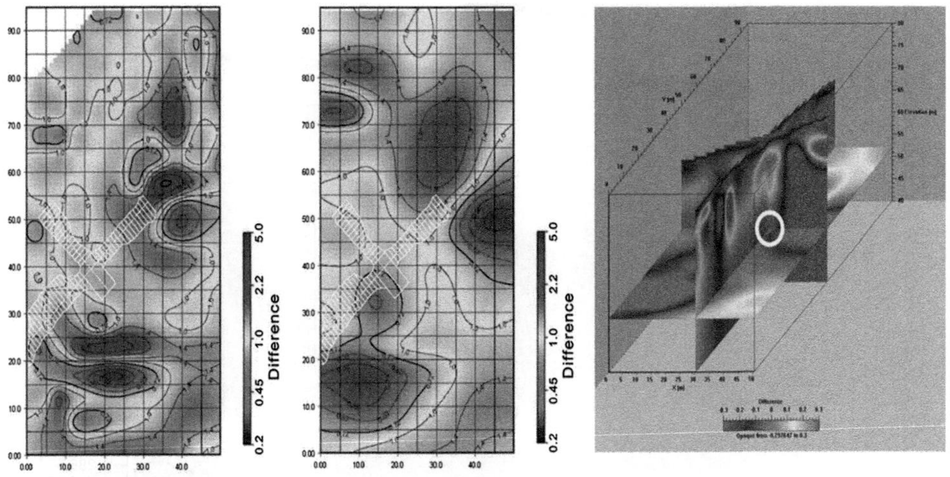

Figure 18 Change in resistivity distribution between Phase III and Phase II. A horizontal slice was made at the elevation (a) above, and (b) at, the level of the storage cavern; (c) shows a 3-D fence diagram of resistivity change between Phase III and Phase II (Yi *et al.*, 2005).

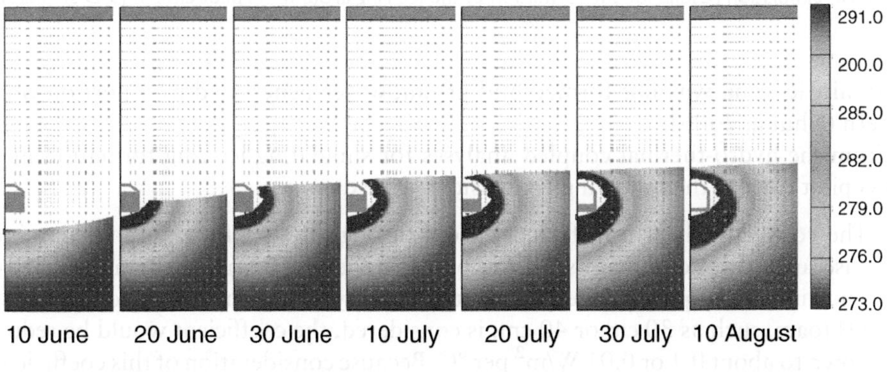

Figure 19 Formation of the ice ring and the groundwater temperature around the pilot cavern (2-D simulation).

The processes of ice ring formation with effective porosity of 3% are shown in Figure 19, together with the temperature distribution of groundwater, respectively. The ice ring in the rock mass below the cavern is thicker than the other regions. The dry zone inside the ice ring is obtained with thickness of maximum 2 m. Until August 10th, ice ring is fully formed in the surrounding rock. However, ice ring contacts partially the corner between wall and floor. It is found from comparison with the early interpretation of the geophysical campaign data that the real process of ice ring formation is very similar to the result from two dimensional simulations (Figure 18). Figure 20 shows the processes of ice ring formation by three dimensional simulations when the access tunnel is considered. Due to concrete wall exposed to the

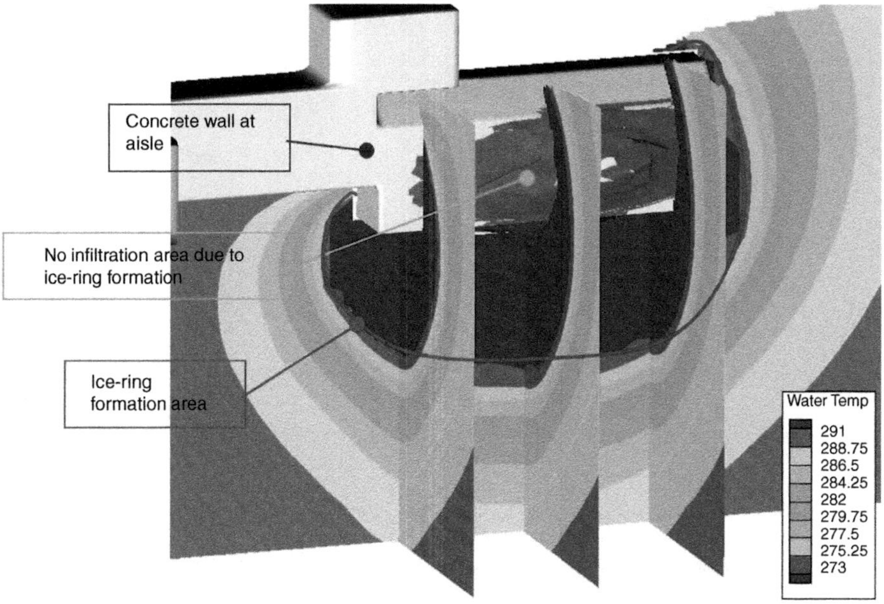

Figure 20 Simulation results for the formation of the ice ring around the pilot cavern (3-D simulation).

open air, ice ring was not formed and groundwater penetrated to a portion of the cavern (Chung, 2006).

From the hydro-thermal coupled analyses for simulating ice ring formation in the LNG pilot cavern, the following conclusions were drawn.

- The convective heat transfer coefficient through the roof membrane containing gaseous sky was obtained as 3 W/m^2 per °C in the containment system with a 10 cm thick insulation panel. When the containment system with thicker non-reinforced PU foam, such as 30 cm or 40 cm, is considered, the coefficient would be reduced down to about 0.1 or 0.01 W/m^2 per °C. Because consideration of this coefficient is very important for the rock mass above the cavern, the probable range of the coefficient in a full-scale cavern model with a standard insulation specification should be obtained by using an appropriate numerical scheme for design purposes.
- The temperature dependency of the input properties must be considered for appropriate modelling of the cryogenic environment. Particularly when the temperature range of the simulation is wide, the variation of properties has to be taken into account.
- Numerical simulation with the thermal properties of dry rock, obtained by laboratory tests, could effectively estimate the real temperature profiles in a rock mass around a cavern.
- The capability of hydro-thermal modelling of ice ring formation has been verified by a comparison of numerical results with *in situ* measurement data. Therefore, a similar approach could be extended to the simulation of ice ring formation in a full-scale LNG storage cavern.

- By controlling the groundwater drainage system, the ice ring can be formed easily based on the assumption that the average distance the 0 °C isotherm reaches is 3–4 m from the cavern wall in a rock mass with the hydraulic conductivity of 10^{-7} to 10^{-6} m/s.

6 AN EXAMPLE OF A COMMERCIAL UNDERGROUND LNG STORAGE SYSTEM

This new concept of LNG storage in membrane-lined rock caverns is valid for any type of exporting or receiving terminal, but also for peak-shaving use or large capacity stockpile storage. Figure 21 shows a full-scale model for an underground LNG storage system including above-ground facilities. Layout of the underground storage system and the number of storage caverns can be varied with the required storage capacity of LNG. Geometry of the storage cavern should be adapted to the local geological and hydrogeological conditions of the LNG terminal facilities. A wide variety of different geometries can be considered. The general cross section of storage gallery is comparable to the unlined underground storage technology, with horse-shoeshaped galleries. The dimension of the gallery is typically 30 m height by 20 m width, which can be adapted to the prevailing rock conditions. The length of cavern can be from 150 m up to 270 m, depending upon the unit capacity per gallery.

The drainage system of the lined underground LNG storage cavern is composed of a drainage tunnel excavated beneath the cavern and drain holes drilled on the rock surface of the drainage tunnel (Figure 22). After the access tunnel for constructing the

Figure 21 Bird's-eye view of a full-scale model for underground LNG storage (Storage capacity of 140,000 m^3).

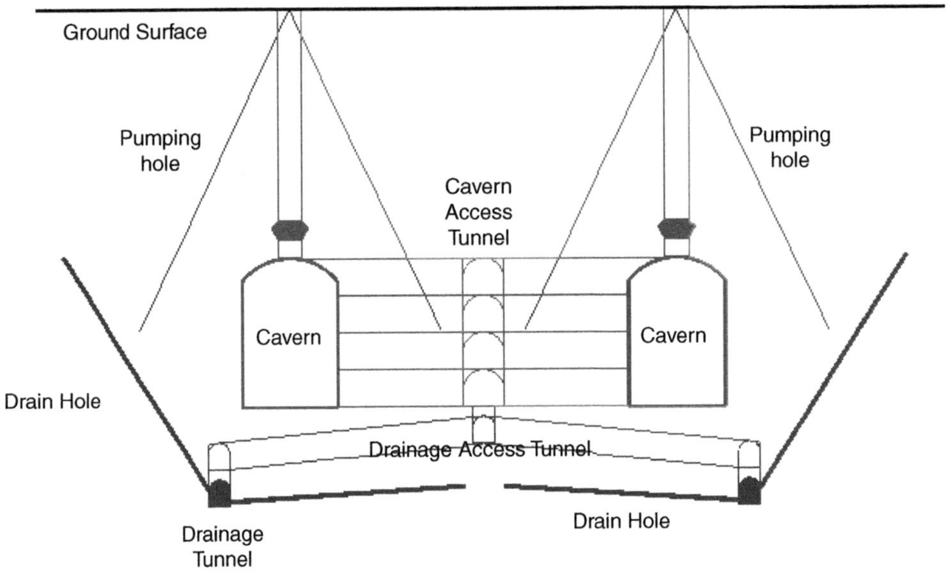

Figure 22 Typical cross section of an underground LNG storage system and its drainage system.

caverns reaches to the cavern bottom, a drainage access tunnel is excavated to the depth of the drainage tunnel bottom. And then drainage tunnels are excavated in parallel to the cavern with a little longer length than that. On the surface of the drainage tunnel, several drain holes should be drilled upward and arranged with the same spacing. Drain holes outside the cavern boundary are inclined with a steep slope and those within the cavern boundary have a gentle slope. The reason why drain holes are installed in an upward direction is that the drainage system is designed to let the groundwater be drained naturally from the rock mass by gravity. That is to say, the groundwater within the rock mass flows into the drain holes along the joint-fracture channels connected to those holes and is collected into the drainage tunnel only by gravity, without pumping. In order to sufficiently desaturate the rock mass around the cavern, the position and horizontal spacing of drain holes should be designed efficiently.

The equipment of a storage cavern is similar to that of a conventional LNG terminal, allowing it to operate in exactly the same way as an above-ground tank (Figure 23). The internal piping arrangement is also similar to that of a membrane LNG storage tank. All pipes are embedded in a concrete structure near the bottom of the operation shaft. Inside the cavern the pipes are joined together by braces forming a pipe tower designed to permit differential pipe contraction.

7 ADVANTAGES AND ECONOMICS OF UNDERGROUND LNG STORAGE

Storing LNG underground provides advantages that cannot be obtained with tank technologies. These are listed below.

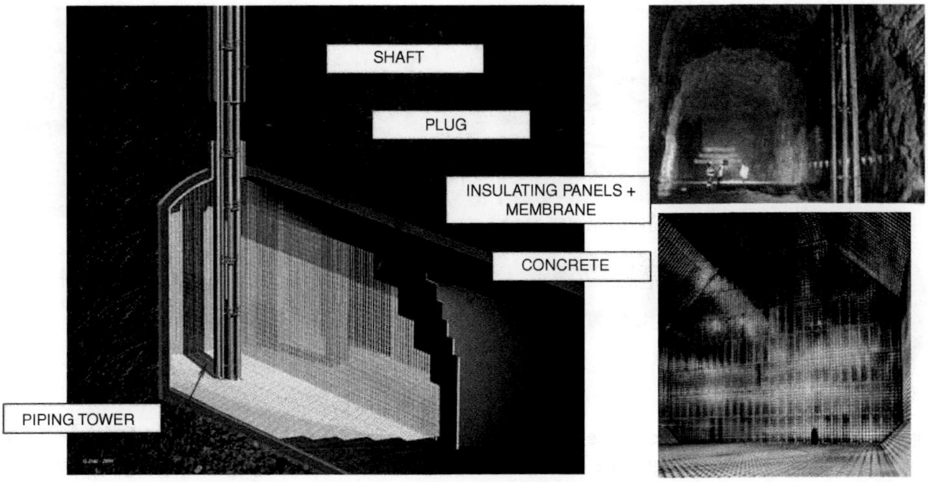

Figure 23 Typical cavern equipment arrangement (with pictures of LPG cavern storage and membrane tank of LNG carrier).

- **Cost effectiveness:** Investment and operating costs are quite attractive as compared with conventional surface tanks. One of the reasons is that underground storage minimizes the total space required for LNG terminals and can therefore represent a huge cost saving, especially in industrial areas.
- **Safety and security:** Owing to its depth, its multi-containment barrier (steel membrane, concrete, frozen rock) and the safety of the systems connecting the storage to the surface, the storage facility is very well protected from outside events such as accidents, sabotage or terrorism acts and has a lower vulnerability to natural events such as earthquakes.
- **Environmental protection and acceptability:** Underground installations are discreet and the visual impact is reduced. Therefore, there is no need of large reclaimed areas and there are fewer earthworks at ground level. This involves a better acceptability by the population located nearby the storage.
- **Use of land space:** Land above ground can be used better because of the minimization of total space required for the LNG terminal (Figure 24). Due to the fact that LNG storage is about 50 m underground, this can represent a huge cost saving, especially in coastal areas where industries are already developed and free remaining areas are small and expensive. It is also the case in areas whose topography needs expensive reclaimed land.

The comparative cost estimate between above-ground and cavern storage concerns only the storage itself and its equipment. The construction costs of varying stored volume from 200,000 kl to 1,000,000 kl with increments of 200,000 kl were evaluated and compared with reference to price as of March 2006 in Korea. It does not take into account the substantial cost saving which could be made, in the case of the cavern storage, on safety equipment (impounding basin, peripheral retention wall, firefighting systems, etc.) and possibly for the piping length and terminal surface area reduction.

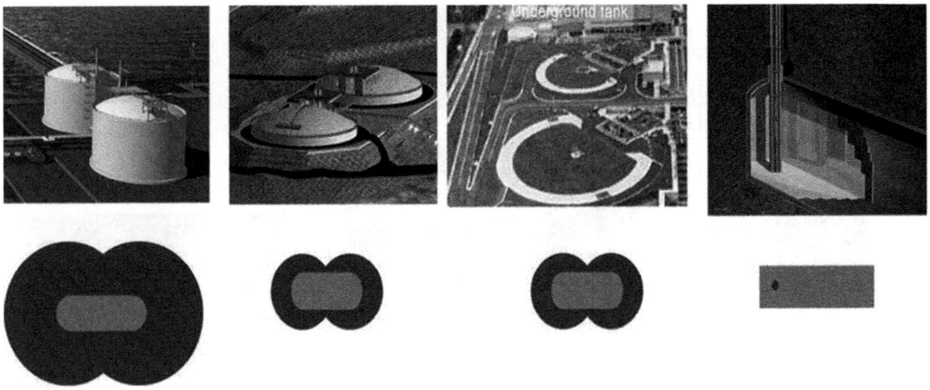

Figure 24 Required area at grade depending on storage type for a storage facility of 320,000 m^3 capacity; illustrated left to right are over-ground, in-ground and underground (light gray means the storage space, dark gray means the required above-ground space).

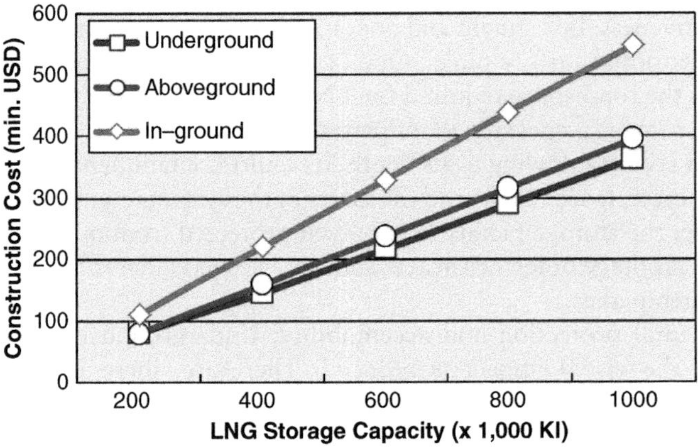

Figure 25 Capital expenditure (CAPEX) evaluation and comparison between above-ground, in-ground and underground storage facilities (Korean context).

In all cases, the underground LNG storage system is the most economic over the stored volume of 300,000 kl, and at 400,000 kl. The construction cost for an underground LNG storage system can be more economical by 8% and 34% compared to an above-ground and an in-ground tank, respectively (Figure 25).

Moreover, operational costs for underground storage units are highly competitive as compared to above-ground and in-ground tanks, as systems like slab heating or fire water are not necessary, or can be tremendously reduced. Based on a Korean reference which has been implemented on crude oil storage by the Korea National Oil Company, operational costs of underground storage are 63% less than those for above-ground.

8 SUMMARY

Based on the experience of underground storage for crude oil and various types of hydrocarbons, an underground storage system was thought to be a more economical way to store LNG. An innovative method of LNG storage in lined rock caverns has been developed to provide a safe and cost-effective solution. To demonstrate the technical feasibility of this method, a pilot plant was constructed within the KIGAM area at Daejeon and was operated from January 2004 to the end of 2004. From the construction and operation of this Daejeon pilot plant, the technical feasibility of an underground lined rock storage system has been well proved. Also the real scale applicability of it has been evaluated by the results from the successful operation of the pilot plant.

As compared with the conventional above-ground and in-ground storage tanks, the use of a lined rock cavern LNG storage system at the LNG terminals can be a more economical way with respect to CAPEX and OPEX. In addition, it also has the advantage of safety, security and environmental acceptability against the conventional tanks.

Underground LNG storage systems in lined rock caverns can be realised in due course in countries which have suffered from a shortage of storage capacity for LNG, which have extreme seasonal variation in domestic demand, and where industries are already developed and free remaining areas of land are small and expensive.

REFERENCES

Amantini, E. & Chanfreau, E. (2004) Development and construction of a pilot lined cavern for LNG underground storage. In: *14th International Conferences and Exhibition on Liquefied Natural Gas*, Doha, Quatar, PO-33.

Amantini, E., Chanfreau, E. & Kim, H.Y. (2005) The LNG storage in lined rock cavern: pilot cavern project in Daejeon, South Korea. *Proceedings, GASTECH 2005: 21st International Conference and Exhibition for the LNG, LPG and Natural Gas Industries*, Bilbao, Spain, pp. 1–16.

Anderson, U.H. (1989) Steel lined rock caverns. In: *Storage of Gases in Rock Caverns*, Nilsen, Olsen (Eds.), Balkema, Rotterdam, pp. 1–10.

Cha, S.S., Lee J.Y., Lee, D.H., Amatini, E. & Lee K.K. (2006) Engineering characterization of hydraulic properties in a pilot rock cavern for underground LNG storage. *Engineering Geology* 84, pp. 229–243.

Cha, S.S., Lee, K.K., Bae, G.O., Lee, D.H. & Gatelier, N. (2007) Analysis of rock drainage and cooling experiments for underground cryogenic LNG storage. *Engineering Geology* 93, pp. 117–129.

Chung, S.K. (2006) Thermo-mechanical behavior of rock masses around underground LNG storage cavern. *Keynote Lectures in Rock Mechanics in Underground Construction – 4th Asian Rock Mechanics Symposium*, Singapore, pp. 19–28.

Chung, S.K, Park, E.S. & Han K.C. (2006) Feasibility study of underground LNG storage system in rock cavern. *Tunnel and Underground Space* 16 (4), pp. 296–306 (In Korean).

Chung, S.K., Park, E.S. & Han, K.C. (2007a) Feasibility study of underground LNG storage system in rock cavern. In: *11th ACUUS International Conference*, Kaliampakos, D. & Benardos, A. (Eds.), NTUA Press, Athene, pp. 501–506.

Chung, S.K., Park, E.S., Kim, H.Y., Lee, H.S. & Lee, D.H. (2007b) Geotechnical monitoring of a pilot cavern for underground LNG storage. In: *11th International Congress on Rock Mechanics*, Sousa, Olalla & Grossman (Eds.), Taylor & Francis, Lisbon, pp. 1245–1248.

Dalström, L.O. (1992) *Rock mechanical consequences of refrigeration – a study based on a pilot scale rock cavern*. PhD thesis, Chalmers University of Technology, Gothenburg, Sweden.

Dalström, L.O. & Evans, J. (2002) Underground storage of petroleum and natural gas. In: *17th WPC*, Portland Press Ltd., London, pp. 128–129.

Glamheden, R. & Lindblom, U. (2002) Thermal and mechanical behaviour of refrigerated caverns in hard rock. *Tunnelling and Underground Space Technology* 17 (4), pp. 341–353.

Goodall, D.C. (1989) Prospects for LNG storage in unlined caverns. *Proceedings of the International Conference on Storage of Gases in Rock Caverns*, Trondheim, Balkema, pp. 237–243.

Jeon, Y.S., Park, E.S., Chung, S.K., Lee, D.H. & Kim, H.Y. (2006) Numerical simulation of fracture mechanisms for rock masses under low temperature conditions. *Tunnelling and Underground Space Technology* 21(3–4), pp. 470–471.

KIGAM. (2003) *Development of Base Technology for Underground LNG Storage and the Analysis on the Operation Results of Pilot Plant*. Research report by KIGAM submitted for SKEC (in Korean).

KIGAM. (2004) *Development of Base Technology for Underground LNG Storage and the Analysis on the Operation Results of Pilot Plant-2004*. Research report by KIGAM submitted for SKEC (in Korean).

KIGAM. (2005) *Development of Base Technology for Underground LNG Storage and the Analysis on the Operation Results of Pilot Plant-2005*. Research report by KIGAM submitted for SKEC (in Korean).

KIGAM. (2006) *Development of Base Technology for Underground LNG Storage and the Analysis on the Operation Results of Pilot Plant-2006*. Research report by KIGAM submitted for SKEC (in Korean).

Kim, H.Y. (2006) 25 years experience of underground hydrocarbon storage in Korea and overstressed problems in large rock cavern. [Lecture] *Japan Society of Civil Engineers & Japanese Society for Rock Mechanics*, 12th January.

Kim, J.H., Park, S.G., Yi, M.J., Son J.S. & Cho, S.J. (2007) Borehole radar investigations for locating ice ring formed by cryogenic condition in an underground cavern. *Journal of Applied Geophysics* 62, pp. 204–214.

Kovári, K. (1993) Basic consideration on storage of compressed natural gas in rock chambers. *Rock Mechanics and Rock Engineering* 26 (1), pp. 1–27.

Lee, D.H., Kim H.Y., Gatelier, N. & Amantini, E. (2003) Numerical study on the estimation of the temperature profile and thermo-mechanical behaviour in rock around the Taejon LNG Pilot Cavern. In: *International Symposium on the Fusion Technology of Geosystem Engineering, Rock Engineering and Geophysical Exploration*, Seoul, Korea, pp. 233–237.

Lee, D.H., Lee, H.S., Kim, H.Y. & Gatelier, N. (2005) Measurements and analysis of rock mass responses around a Pilot Lined Rock Cavern for LNG underground storage. In: *Proceedings of EUROCK 2005*, Konecny, P. (Ed.), Balkema, Brno, pp. 287–292.

Lee, H.S., Lee, D.H., Kim, H.Y. & Choi, Y.T. (2006a) Design criteria for thermo-mechanical stability of rock mass around lined rock cavern for underground LNG storage. *Tunnelling & Underground Space Technology* 21, p. 337.

Lee, H.S., Lee, D.H., Jeong, W.C., Song, Y.W., Kim, H.Y., Park, E.S. & Chung, S.Y. (2006b) Heat transfer and boil-off gas analysis around underground LNG storage cavern. In: *ISRM International Symposium 2006 & 4th ARMS*, Singapore, p. 180.

Monsen, K. & Barton, N. (2001) A numerical study of cryogenic storage in underground excavations with emphasis on the rock joint response. *International Journal of Rock mechanics and Mining Sciences* 38 (7), pp. 1035–1045.

Park, E.S. (2006) Thermo-mechanical consideration for the cryogenic storage. In: *International Workshop on the Underground Storage Facilities in Conjunction with the 4th ARMS*, Singapore, pp. 81–98.

Park, E.S., Chung, S.K., Lee, D.H. & Kim, H.Y. (2007a) Experience of operating an underground LNG storage pilot cavern. In: *11th ACUUS Conference*, Athens, Greece, pp. 157–162.
Park, E.S., Chung, S.K., Lee, H.S., Lee, D.H. & Kim, H.Y. (2007b) Design and operation of a pilot plant for underground LNG storage. In: *Proceedings of 1st Canada-US Rock Mechanics Symposium*, Eberhardt, E., Stead, D. & Morrison, T. (Eds.), Taylor & Francis, Vancouver, pp. 1221–1226.
Park, E.S., Jung, Y.B., Song, W.K., Lee, D.H. & Chung, S.K. (2010) Pilot study on the underground lined rock cavern for LNG storage. *Engineering Geology* 116, pp. 44–52.
Park, E.S., Chung, S.K., Kim, H.Y. & Lee, D.H. (2011) The new way to store LNG in lined rock caverns. In: *The2011 World Congress on Advances in Structural Engineering and Mechanics (ASEM'11)*, Seoul, Korea, pp. 3951–3956.
Yi, M.J., Kim, J.H., Park, S.G. & Son, J.S. (2005) Investigation of ground condition change due to cryogenic conditions in an underground LNG pilot plant. *Exploration Geophysics* 36, pp. 67–72.

Chapter 19

Hydromechanical properties of sedimentary rock under injection of supercritical carbon dioxide

A. Arsyad[1], Y. Mitani[2] & T. Babadagli[3]

[1]Department of Civil Engineering, Faculty of Engineering, Hasanuddin University, Makassar, Indonesia

[2]Department of Civil and Structural Engineering, Graduate School of Engineering, Kyushu University, Fukuoka, Japan

[3]Department of Civil and Environmental Engineering, School of Mining and Petroleum Engineering, University of Alberta, Edmonton, Alberta, Canada

1 INTRODUCTION OF CARBON DIOXIDE GEOLOGICAL STORAGE

It has been widely believed that CO_2 emissions into the atmosphere have increased steadily and become a major contributing factor to global warming over the past several hundred years. This is mainly attributed to burning coal, oil and natural gas for electrical generation, transportation, industrial and domestic needs. Growing CO_2 concentration in the atmosphere will disrupt global climate, which in turn causes a rise in the sea levels, floods in lowered level areas and damage to the ecosystem. Therefore, immediate solutions are urgently needed to reduce CO_2 emission into the atmosphere.

Possible solutions include efficient production and use of energy; exploration of non-fossil fuel energies such as solar power, wind energy, biomass; and development of technologies of disposing CO_2 emission such as CO_2 ocean storage, CO_2 mineral carbonation, and carbon capture and geological storage (CCGS). CCGS is considered the most promising option to reduce atmospheric CO_2 emission among all due to large storage capacity expected to deal with increasing anthropogenic CO_2 emissions, and its readiness for being applied due to similarity to CO_2 injection for enhanced oil recovery (EOR) experienced in the petroleum industry.

IPCC (2005) defined the CCGS as a process in which CO_2 emission produced by large stationery sources such as industrial plants and power stations is captured and compressed to be supercritical CO_2 and then transported via pipelines to suitable geological formations, such as unmineable coal beds, deep saline aquifers, and depleted oil and gas reservoirs (Figure 1). A number of developed countries have investigated the CCGS option since the technology would be able to reduce CO_2 emission from large stationary resources such as coal and gas power plants. By implementing the CCGS, coal and gas can still be used as main energy supply with less CO_2 emission. CCGS is also expected to play important role in the acceleration of the development and infrastructure of CO_2-free hydrogen based transportation system (Benson, 2004). The CCGS will be utilized to reduce CO_2 emission from gasification projects, which converts fossil fuels to be hydrogen for the need of transportation fuel.

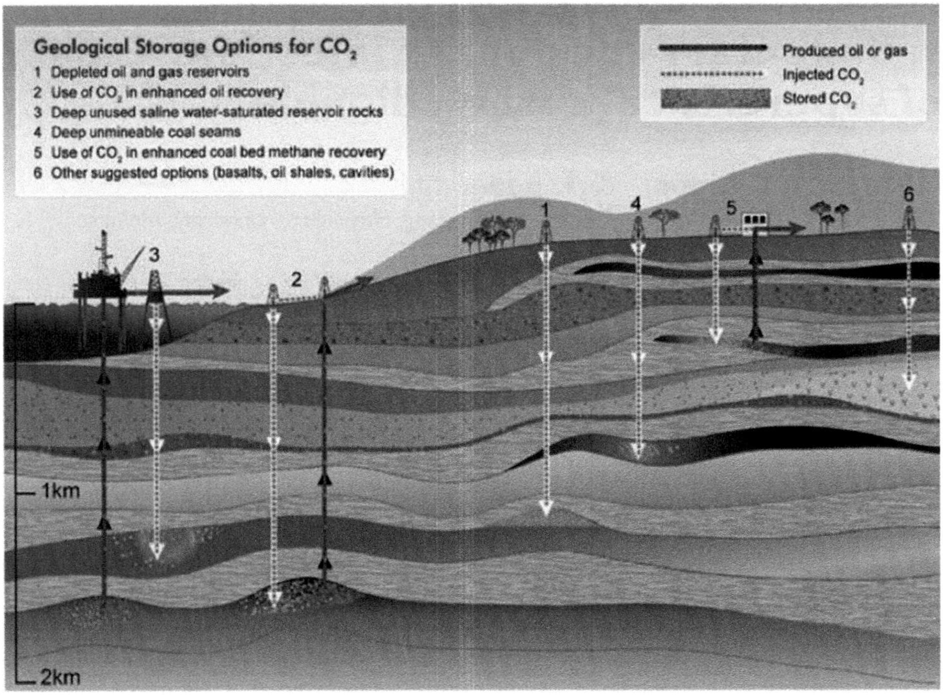

Figure 1 Carbon capture and geological storage (after Cook, 1999).

The most feasible geological formation for CCGS is depleted oil and gas reservoirs due to well-described geological conditions and available volume to be filled with CO_2. This application, however, has certain limitations. First of all oil and gas reservoirs are unequally distributed around the world. Secondly, it will take a long time for those reservoirs to be depleted and ready for CO_2 storage, and vast pipelines distributions are needed due to the fact that the location of the sources of CO_2 emission could be far away from the field for CO_2 storage (Benson, 2004). For those reasons, deep sedimentary basins are considered as an alternative option for geological storage of CO_2.

Sedimentary basins are formed by gradual deposition and compaction of sediments eroded from mountains. As a result, they generally consist of alternating layers of coarse sediments (sandstone) and fine textured sediments (clay, shale and evaporites). The sandstone layers with high permeability will provide storage for CO_2, while the shale layers with low permeability will act as a barrier to prevent CO_2 migration to potable groundwater sources and even to the surface.

The mechanism of disposing CO_2 in sedimentary basins was explained by IPCC (2005). Physical and geochemical processes will be very critical in trapping CO_2 permanently under a thick layer and low permeable seal. The storage mechanisms, as shown in Figure 2, are described as stratigraphical and structural trapping in which folded and fracture rocks can act as permeability barriers in certain circumstances (Salvi *et al.*, 2008); hydrodynamic trapping where process of dissolving CO_2 into rock formation saturated water as CO_2 flowing driven by differential pressure and buoyancy

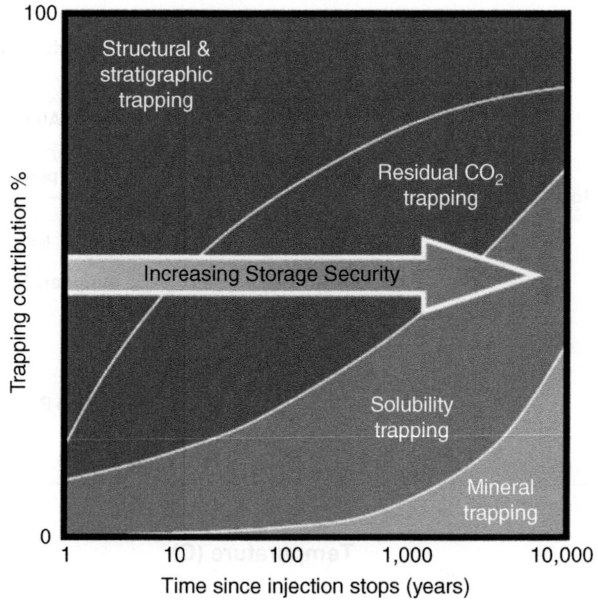

Figure 2 Trapping Mechanisms of geological CO_2 storage (after IPCC, 2005).

effect (Villarasa *et al.*, 2010); and geochemical trapping which is the reaction between dissolved CO_2 and rock mineral, forming ionic species. CO_2 converts into stable carbonate minerals (Gunter *et al.*, 1993).

In this work, we approached storage capacity of sedimentary rocks problem experimentally and numerically. In the experiments, sandstone samples obtained from a specific region were used and flow tests were performed. Then, a numerical model study was performed to clarify the multiphase flow characteristics (relative permeabilities) of CO_2 storage process.

2 EXPERIMENTAL MEASUREMENT OF STORAGE CAPACITY OF ROCK FOR CO_2 IN DEEP SEDIMENTARY BASINS

A number of studies assessed CO_2 storage capacities of sedimentary structures using various methodologies considering different trapping mechanism. These efforts, in turn, just produced widely varying estimates with inconsistency and unreliability (Bachu *et al.*, 2007). The reliability of the methodology depends on the scale and resolution of the assessment undertaken to estimate storage capacity (Bradshaw *et al.*, 2007; Bachu *et al.*, 2007). Effective storage should be a function of reservoir permeability, rock relative permeability to formation water and CO_2, the nature and geometry of reservoirs. Juanes *et al.* (2006) suggested that storage capacity of CO_2 is the volume of irreducible CO_2 saturating rock volume after being invaded by water in such flow reversal:

$$V_{CO_2t} = \Delta V_{trap} \phi S_{irCO_2} \tag{1}$$

where Vtrap is rock volume saturated by irreducible CO_2 (Sir_{CO_2}).

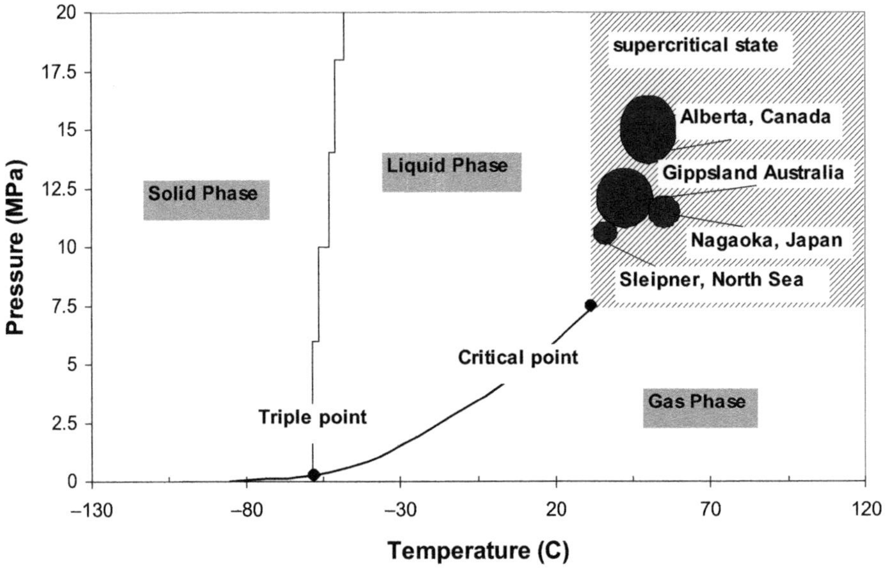

Figure 3 CO_2 phases in the existing CO_2 injection fields (After Sasaki *et al.*, 2008).

Irreducible CO_2 saturation can be determined based on the actual CO_2 saturation at flow reversal and the hysteretic path of the relative permeabilities of CO_2-brine systems for a given aquifer rock. This kind of storage capacity is time dependent as the plume of CO_2 spreads and migrates. Therefore, it needs periodical evaluation as it will vary as the injected CO_2 continues to migrate. Porosity and relative permeability must be obtained from a laboratory experiment, while the irreducible CO_2 saturation can be determined through numerical simulations (Kumar *et al.*, 2005; Juanes *et al.*, 2006).

An experimental system was developed to measure those parameters in this study and relative permeability, irreducible saturation and specific storage of a cored sedimentary rock were determined by injecting supercritical CO_2 into a core obtained from a sedimentary rock. The phase of CO_2 governs its multi-flow characteristics in the brine reservoir. It should be noted that, in the existing field scale CO_2 injection projects, the physical property of CO_2 is mostly in supercritical phase, which is a liquid-like gas (Figure 3). To generate and maintain the phase of CO_2 at high temperature and high pressure, the experimental system was equipped with temperatures, pressure and flow controllers.

2.1 Experimental system

The main difficulty in creating reservoir conditions with high pressure and temperature is the vulnerability of the experimental system to unstable temperature associated with seasonal weather and heat induced by the apparatus. Such conditions will affect the physical property of CO_2, leading to inaccurate measurements. Therefore, the external and internal lab temperatures were controlled by constructing a greenhouse chamber with thermostatic controller (Figures 4 and 5). The apparatus was installed to control temperature using the following instruments:

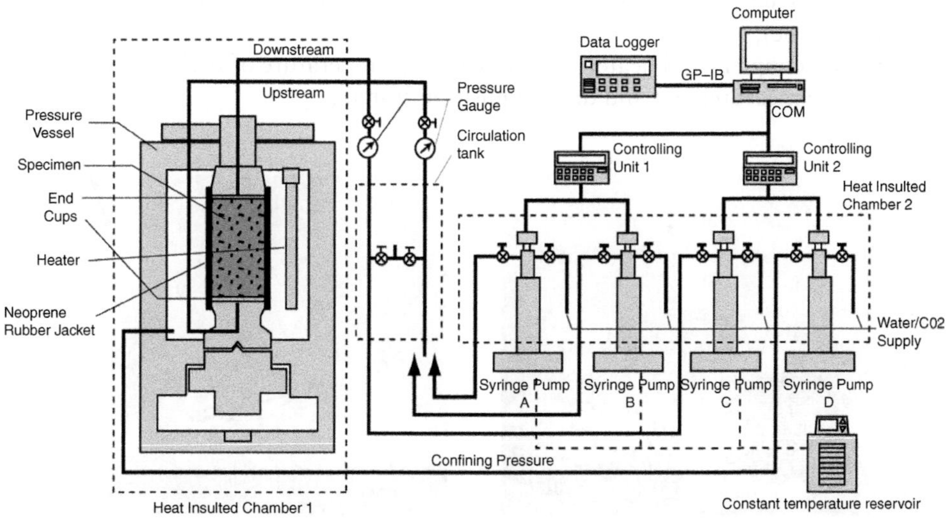

Figure 4 Schematic diagram of experimental system.

Figure 5 Greenhouse chamber.

(1) Thermostatic controller (Figure 6a),
(2) hemathermal circulation tank which controls temperature of water circulating to syringe pumps at desired temperature (Figure 6b),
(3) cylinder jackets (Figure 6c),
(4) water bath of syringe pipes (Figure 6d),

(5) constant temperature water tank controlling temperature in the water bath (Figure 6e),
(6) thermocouplers and heater bars controlling temperature of rock specimen,
(7) remote measurement and data acquisition system.

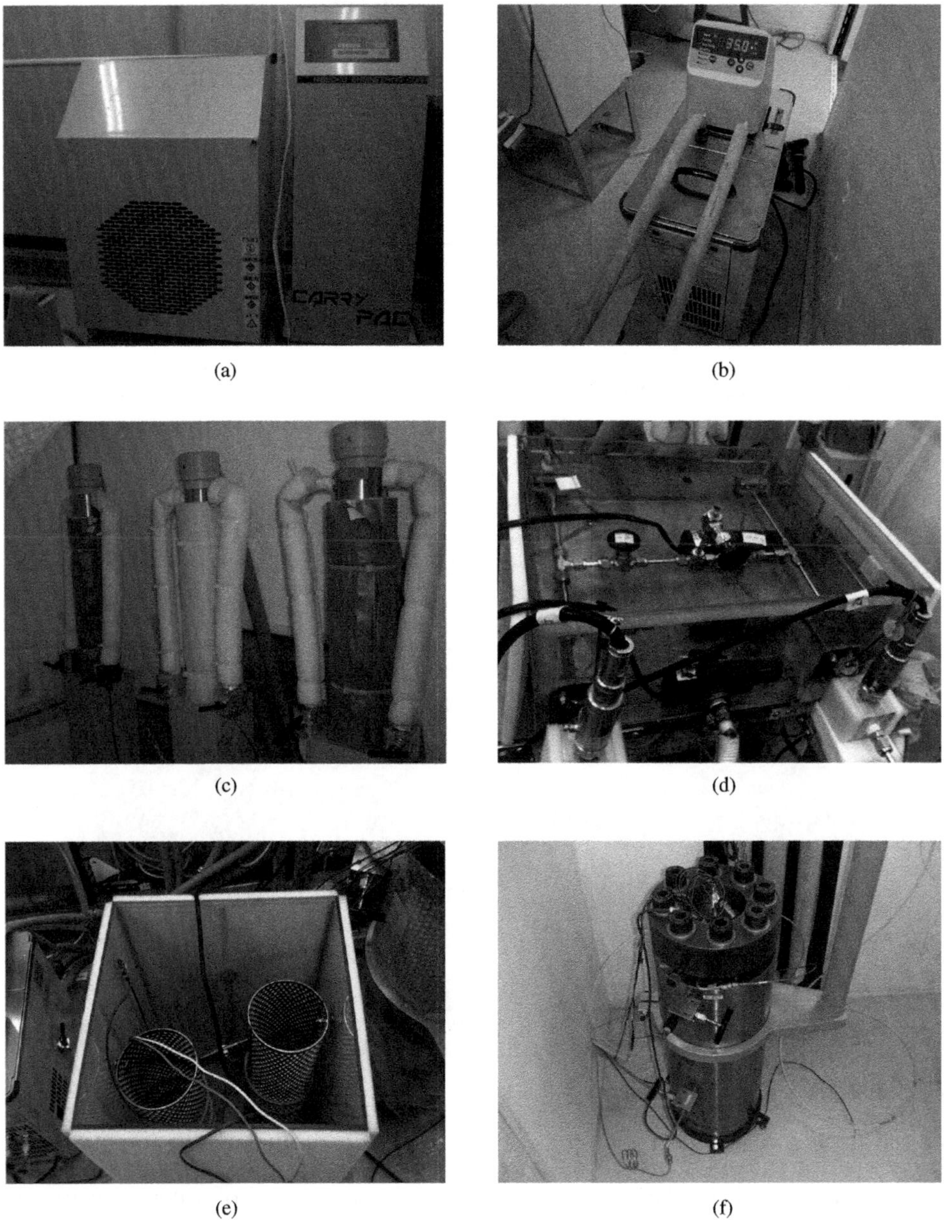

Figure 6 Temperature controller: (a) Thermostatic controller, (b) hemathermal circulation tank, (c) cylinder jacket of syringe pumps, (d) water bath of syringe pumps, (e) water tank for circulating water bath at desired temperature, (f) tri-axial chamber.

Precise measurement is indispensable to obtain accurate and reliable. Therefore, , pressure vessel controlling the confining pressure applied onto the specimen, which is placed in the tri-axial chamber, was installed (Figure 6f). The pressure vessel is also capable to work with rock specimen with a size of 10 cm heights × 10 cm diameters and the maximum confining pressure that it can generate is 100 MPa. Syringe pumps were used to generate pressure or control the flow rate depending on the experimental mode. At a constant pressure mode, the syringe pump can load pressure from 6.9 kPa up to 69 MPa constantly. At the constant flow mode, the pump can generate flow from 1.67 × 10–7 cm3/s up to 0.83 cm3/s. Pressure gauges (manufactured by Research Institute Tokyo) that can measure hydraulic pressure with a resolution of 50 cmH2O were also installed to measure the hydraulic pressure at the upstream and downstream sides of the specimen.

2.2 Rock specimen

The rock specimen used in this study is Ainoura Sandstone obtained from Nagasaki Prefecture, Japan. The cores were obtained from a block with dimensions of 300 × 300 × 150 mm. Using a horizontal milling machine, the cores were re-shaped on the top and bottom edge with surface grinding. The final dimensions of the rock specimen were 50 mm in diameter and 100 mm in height. This conforms to the ISRM standard that the height of the rock specimen should be twice its diameter. Strain gauge devices were installed on the rock specimen to measure lateral and longitudinal strain (Figure 7). However, as fluid leakage may occur due to this installation, silicon was coated in 5 mm thickness on the opening points of the rubber sleeve cover of the rock specimen. Then, the rock specimen was placed in a triaxial test container.

Experiments were conducted to measure the permeability and storage capacity of CO_2. It should be noted that low permeable rock (Ainoura sandstone) was selected in

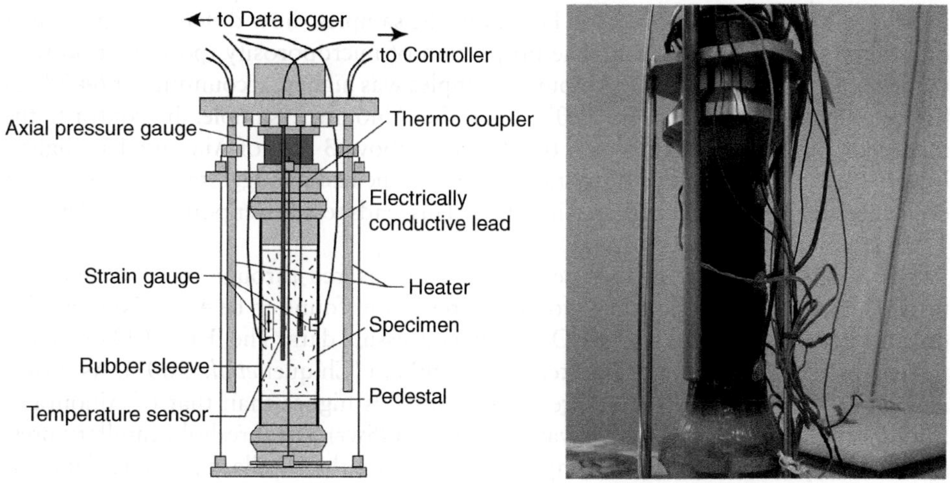

Figure 7 Rock specimen and temperature controller in tri-axial chamber.

Figure 8 Ainoura sandstone samples used in the experiments are 50 mm in diameter and 100 mm in height.

this study due to its high trapping capability, which is suitable for CO_2 geological storage. In case of permeability test method, constant flow rate method was chosen rather than constant pressure. This is because it is easy to fix flow rate injection rather than constant pressure in field scale applications.

The pore size characteristics of the specimens were measured using a mercury-porosimetry. Two specimens of the Ainoura 1 and Ainoura 2 (Figure 8) with slightly different pore characteristics were used. Figure 9 presents the pore-throat size distribution of the specimens. Both rock samples exhibited a bi-modal pore size distribution indicating heterogeneous porosity. However, the samples show differences in number of pores smaller than 1micron. The proportion of microporosity (pore fraction with diameter less than 1 μm) in the Ainoura 1 samples was higher, accounting for 64.7% of the total pores, while this is 51.07% for the Ainoura 2 sample. In contrast, the macroporosity (pore fraction with the diameter above 3 μm) of Ainoura 1 is slightly lower (13.1%) compared to Ainoura 2 (19.6%). The results suggested that Ainoura 1 contained finer grain matrix than Ainoura 2. This leads to a lower porosity for Ainoura 1 (0.126) than Ainoura 2 (0.154).

The capillary pressure of the specimens was measured through mercury injection tests. The interfacial tension (IFT) for air-mercury was found to be 485 mN/m but this data must be converted to water-CO_2 capillary pressure data. The IFT of CO_2-water at the experimental condition was taken as 32.1 mN/m (Chiquet *et al.*, 2007). As shown in Figure 10, the capillary pressure of Ainoura 1 is higher than that of Ainoura 2. In order to determine irreducible water saturation (Swr), the threshold capillary pressure of the specimens (P0), and shape parameter (m), the capillary pressure data was matched to the capillary pressure computed using Van Genuchten equation (1980).

Table 1 Pore characteristics of the tested Ainoura sandstone samples.

Specimens	*% Microporosity*	*% Mesoporosity*	*% Macroporosity*	*Median pore size (μm)*	*porosity*	*IFT (mN/m)*	P_0 *(kPa)*
Ainoura 1	64.7	22.1	13.1	1	12.6	32.1	750
Ainoura 2	51.06	29.4	19.6	1.2	15.46	32.1	25

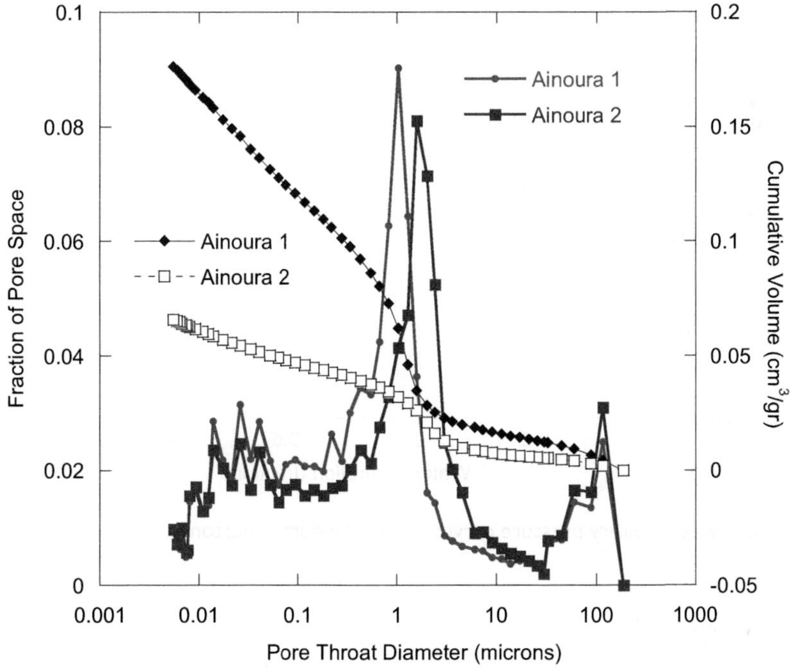

Figure 9 Pore throat-size distribution of the specimens.

The three parameters (Swr, Po, m) for Ainoura 1 and Ainoura 2 specimens were obtained as 0.45, 25 kPa, 0.61, and 0.45, 750 kPa, 0.68, respectively.

2.3 Procedure of the experimental test

Temperature and pressure of the experimental system were adjusted to create reservoir conditions. Pore pressure and confining pressure were set at 10 MPa and 20 MPa, respectively, while temperature was maintained at 20°C. Pure water was injected into the specimen fully saturated with the same fluid previously, to generate 10 MPa pore pressure. This condition was maintained for at least 24 hours for checking any leakage in the system. If the system showed no leakage, the experiment was continued by CO_2 injection. Then, the temperature was increased up to 35°C by setting the temperatures at the syringe pumps, pipes and pressure vessel up to 35°C, 36°C, and 38°C correspondingly. After that, purified water was injected into the fully water-saturated specimen at a constant flow rate of 3 μl/min (Figure 11).

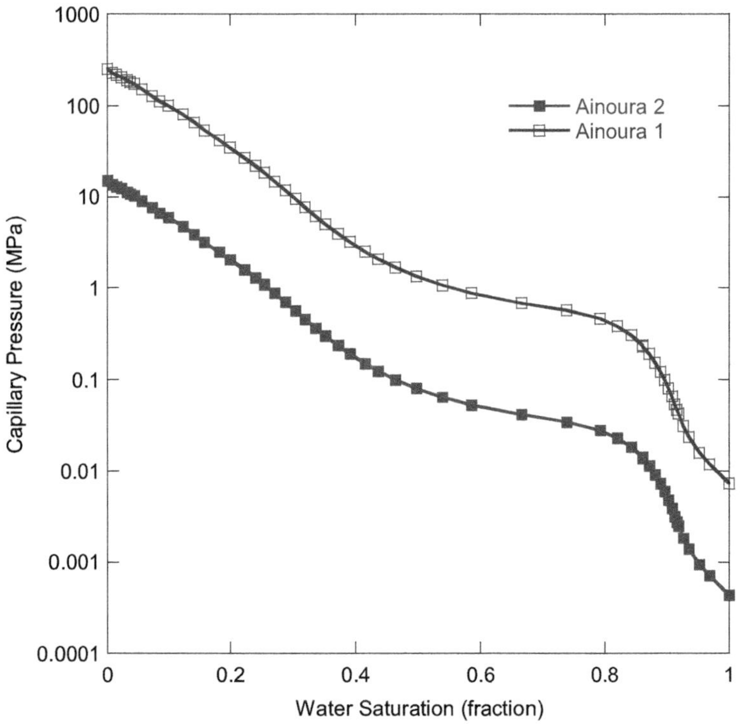

Figure 10 CO_2-water capillary pressure curves for the Ainoura Sandstones.

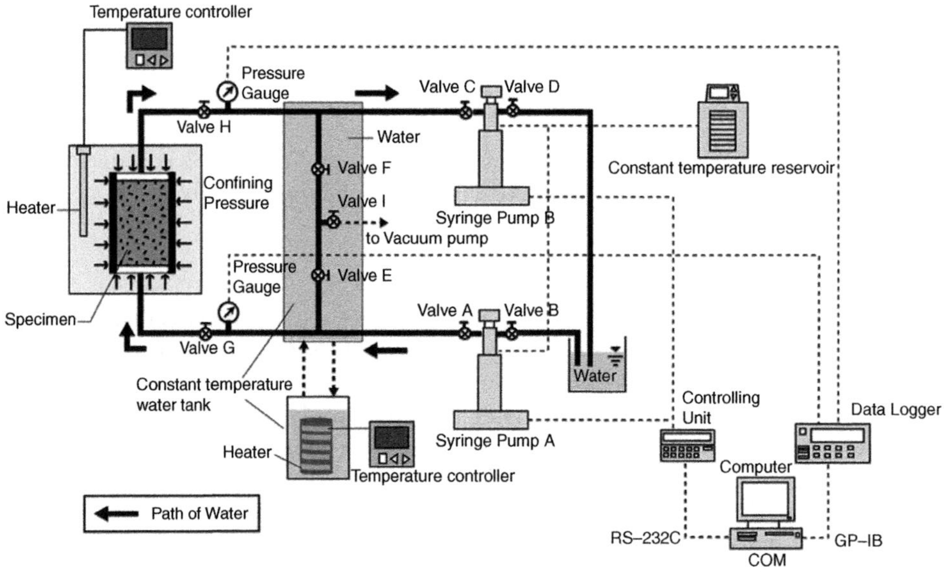

Figure 11 Schematic apparatus of water injection test.

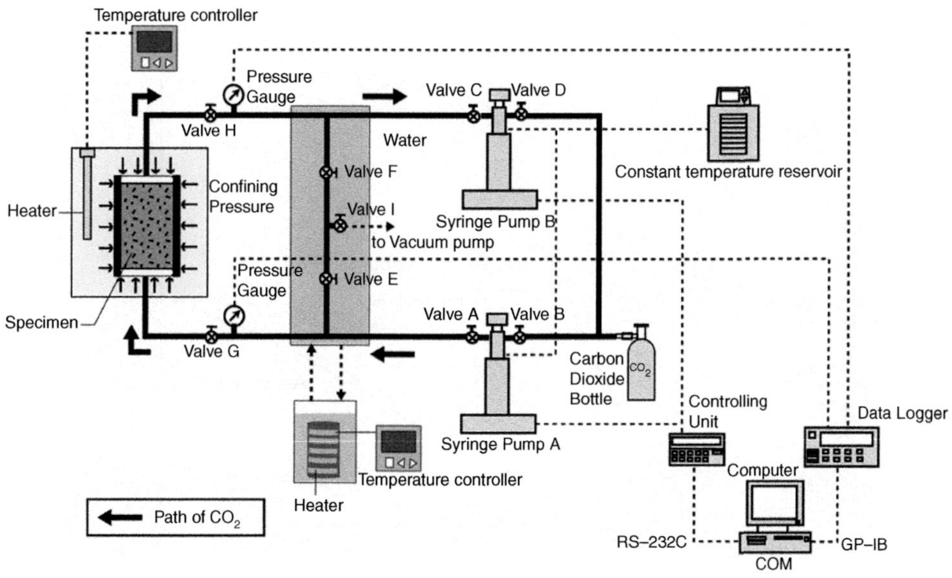

Figure 12 Schematic apparatus of CO_2 injection test.

The pressures at the upstream and downstream gauges were monitored throughout the process. At a steady state, differential pressure was used to determine the intrinsic permeability (K) of the specimens using the Darcy's law. After injection of water, the pressure in the upstream pump was set back to 10 MPa. The water in the upstream pump was discharged and replaced by CO_2. CO_2 was injected into the specimen in the same flow rate (3 μl/min) (Figure 12). The pressures in the upstream and downstream, including the longitudinal and lateral strains of the sample, were continuously measured.

2.4 Experimental results and analysis

During the injection of CO_2 into the specimen, the generated hydraulic pressures in the upstream and downstream were measured. Figure 13 shows that the injection of CO_2 increased the hydraulic pressure both in the downstream and upstream of the specimen. During this process, three distinct stages in the differential pressure between the upstream and downstream are observed (Figure 14).

At the first stage, the differential pressure increased transiently and stabilized at a certain level. The similarity in the results obtained from the previous permeability test with water injection indicated that this stage is the period of the displaced water flow. At the second stage, the differential pressure suddenly increased before it stabilized over certain times. This stage suggested that the injected CO_2 has already penetrated into the bottom of the specimen and began displacing the saturated water out. A drop in the downstream pressure associated with the capillary pressure led to an increase in the differential pressure. The effect capillary pressure is a result of the specimen pores retaining the saturated water until the injected CO_2 pressure exceeded

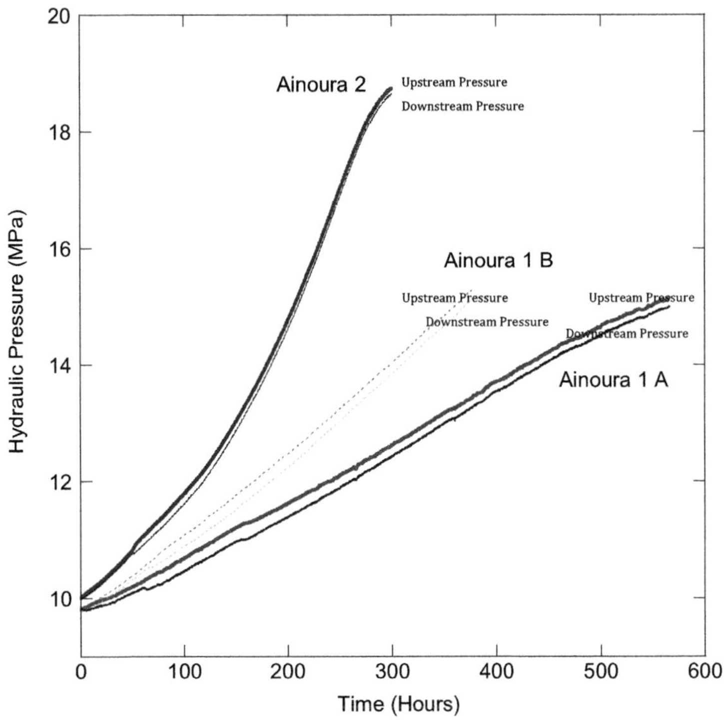

Figure 13 Hydraulic pressures generated in the upstream and downstream of the specimen by the injection of supercritical CO_2.

the pore-water holding pressure. This phenomenon is what Richardson *et al.* (1952) and Dana & Skoczylas (2002) suggested as capillary end effect or capillary pressure effect.

At the third stage, the differential pressure slowly decreased since the injected CO_2 was able to break through. Such stepwise decrease in the differential pressure was observed at this stage implying that the process of CO_2-water displacement occurred in a fingering manner (bypass) rather than frontal sweep. This is consistent with what Bennion & Bachu (2005) suggested as a characteristic of flow in bi-modal pore systems. Indeed, dominating fraction of micropores generated relatively high capillary pressure that would become a barrier for CO_2 to flow. This led to a considerably timely process of CO_2 flow, indicating the capability of Ainoura sandstone in effectively retaining CO_2.

3 ANALYTICAL MODEL

The mathematical model of flow pump permeability test was described as one-dimensional transient flow of a compressible fluid through a saturated porous and compressible medium and numerical solutions were provided. This model combines the principle of fluid mass in a deformable matrix and Darcy's law for laminar flow

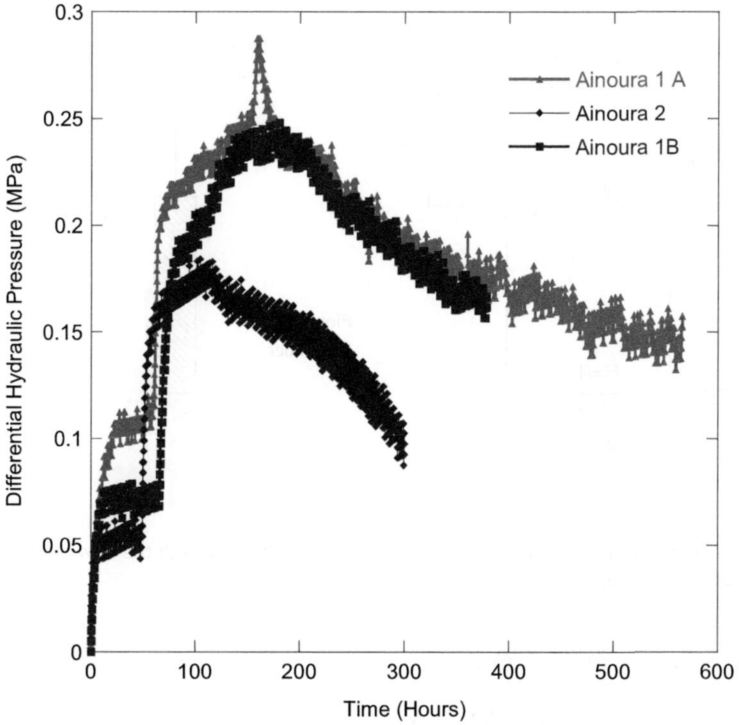

Figure 14 Measured differential pressure during the injection of supercritical CO_2 to the specimen.

through a hydraulic isotropic matrix (Zhang *et al.*, 2000). In order to describe a two phase flow drainage displacement, the boundary condition of the model was changed. The input flow rate in the specimen was assumed to be equal to the total flow rate of the displacing non-wetting fluid and the flow rate of the displaced wetting fluid, at any time minus the volume absorbed within the compressible flow pump test system per unit time interval. The system includes the entire space of the flow pump cylinder, the space in the lower pedestal, and the tubing connecting the flow pump to the test cell. The schematic diagram and boundary conditions associated with the modified mathematical model are depicted in Figure 15.

The governing equation:

$$\frac{\partial^2 H_n}{\partial z^2} - \frac{S_s}{k}\frac{\partial H_n}{\partial t} = 0 \tag{2}$$

Initial condition:

$$H(z,0) = 0 \quad 0 \le z \le L \tag{3}$$

Boundary conditions:

$$z = 0, \quad H(0,t) = 0 \quad t \ge 0 \tag{4}$$

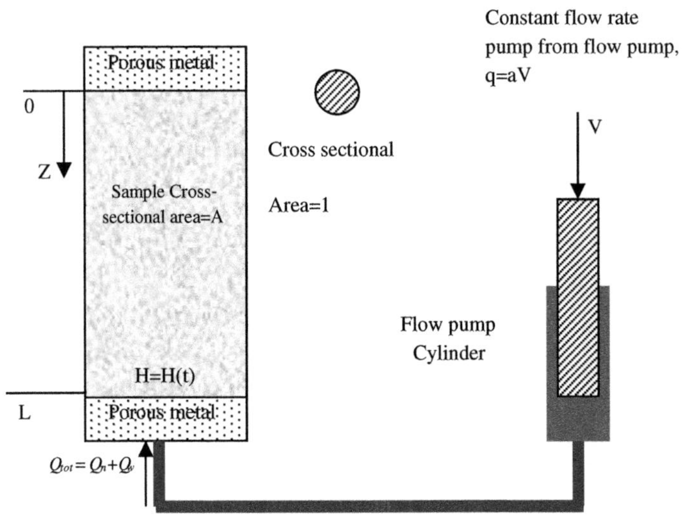

Figure 15 Schematic diagram and the boundary conditions associated with the flow pump permeability. test arrangement.

$$z = L, \quad Q(t) = Q_n(t) + Q_w(t) \tag{5}$$

$$Q(t) = \left(\frac{k_{rn}}{\mu_n}\frac{dH_n}{dz}\rho_n + \frac{k_{rw}}{\mu_w}\frac{dH_w}{dz}\rho_w\right)gKA \tag{6}$$

Due to $\frac{dH_w}{dz} = \frac{dH_n}{dz} - \frac{dH_c}{dz}$, the Equations 3–4a can described as following:

$$Q(t) = \left(\frac{k_{rn}}{\mu_n}\rho_n + \frac{k_{rw}}{\mu_w}\rho_w\right)\frac{dH_n}{dz}gKA - \frac{k_{rw}}{\mu_w}\rho_w\frac{dH_c}{dz}gKA \quad t > 0$$

Therefore the non-wetting fluid pressure gradient becomes

$$\frac{dH_n}{dz} = \frac{Q(t)}{\left(\frac{k_{rn}}{\mu_n}\rho_n + \frac{k_{rw}}{\mu_w}\rho_w\right)gKA} + \frac{\frac{k_{rw}}{\mu_w}\rho_w\frac{dH_c}{dz}}{\left(\frac{k_{rn}}{\mu_n}\rho_n + \frac{k_{rw}}{\mu_w}\rho_w\right)}$$

In which

$$\int_0^t Q(t)dt \int_0^t q\,dt - C_e H_n(L,t)$$

$$\therefore Q(t) = q - C_e\frac{dH_n(L,t)}{dt}$$

Therefore, the complete analytical solution is

$$h(z,t) = \frac{qL + \left(gKA\frac{k_{rw}}{\mu_w}\rho_w L\frac{dH_c}{dz}\right)}{KAg\left(\frac{k_{rn}}{\mu_n}\rho_n + \frac{k_{rw}}{\mu_w}\rho_w\right)} \times \left\{ \frac{z}{L} - 2\sum_{n=1}^{\infty} \frac{\exp\left(-\frac{K\left(\frac{K_{rn}}{\mu_n}\rho_n + \frac{k_{rw}}{\mu_w}\rho_w\right)}{S_s}\beta_n^2 t\right)\sin(\beta_n z)}{L\delta\beta_n\cos(\beta_n L)\left[L\left(\beta_n^2 + \frac{1}{\delta^2}\right) + \frac{1}{\delta}\right]} \right\} \tag{7}$$

where $\delta = \frac{C_e}{AS_s}$, and β_n are the roots of the following equation

$$\tan\phi = \frac{1}{\delta\beta^2}\left(k_{rw} + k_{rn}\frac{\mu_w\rho_n}{\mu_n\rho_w}\right)$$

The roots can be obtained using several numerical methods including Golden-Section method (GSM), the Bi-Sectional Method or Newton Raphson Method, etc. (Carslaw & Jaeger, 1959).

The hydraulic gradient distribution within the specimen can be further derived by differentiating the Equation 7 with respect to the variable z:

$$i_n(z,t) = \frac{qL + \left(gKA\frac{k_{rw}}{\mu_w}\rho_w L\frac{dH}{dz}\right)}{KAg\left(\frac{k_m}{\mu_n}\rho_n + \frac{k_{rw}}{\mu_w}\rho_w\right)} \times \left\{ \frac{1}{L} - 2\sum_{n=1}^{\infty} \frac{\exp\left[-\frac{K}{S_{sn}}\left(\frac{k_{rn}}{\mu_n}\rho_n + \frac{k_{rw}}{\mu_n}\rho_w\right)\beta_n^2 t\right]\cos(\beta_n z)}{L\delta\beta_n\cos(\beta_n L)\left[L\left(\beta_n^2 + \frac{1}{\delta^2}\right) + \frac{1}{\delta}\right]} \right\} \tag{8}$$

where

H = hydraulic pressure, MPa
H_w = hydraulic pressure of water, MPa
H_n = pressure of CO_2, MPa
H_c = capillary pressure, MPa
z = vertical distance along the specimen, cm,
t = time from the start of the experiment, s,
S_s = specimen's specific storage, 1/Pa
K = intrinsic permeability of the specimen, cm^2,
k_{rw} = relative permeability of water, fraction
k_{rn} = relative permeability of CO_2, fraction

L = the length of the specimen, cm,
μ_w = dynamic viscosity of water, $Pa.s$
μ_{nw} = dynamic viscosity of CO_2, $Pa.s$
ρ_w = density of water, gr/cm^3,
ρ_n = density of CO_2, gr/cm^3,
A = the cross-sectional area of the specimen, cm^2,
$Q\ (t)$ = flow in the specimen at time t, cm^3/s,
q = CO_2 flow rate into the upstream of the specimen at time t, cm^3/s,
C_e = storage capacity of the flow pump system, *i.e.*, the change in volume of the permeating fluid in upstream permeating system per unit change in hydraulic head, cm^3/cmH_2O
g = gravity acceleration, cm/s^2

Since the analytical solution of Equation 8 is impossible to obtain, history curve matching was employed and the unknown parameters of *krw*, *krn*, *Ss*, and *Ce* were determined. History curve matching is commonly applied to analyze the experimental data of water-oil unsteady state drainage displacement in petroleum engineering. However, history matching with the four unknown parameters will be time consuming. Therefore, the four parameters were reduced to three by measuring the parameter Ce as upstream pump compressibility from the experimental test of supercritical CO_2 injection into a dummy specimen where the condition of the experiment is similar to the experimental condition of Ainoura sandstone used to inject CO_2.

Figure 16 presents the flow chart of the numerical analysis. The initial values of krw, krn, and Ss were input into Equation 8 to establish the "first guess" of theoretical pressure gradient data at time t, i*(L,t). Meanwhile, the corresponding pressure gradient data at time t, i(L,t), was obtained from the experimental test. Then, the experimental and theoretical pressure gradient data were matched. Once they matched, the krw, krn, and Ss were obtained. The parameters were used to estimate CO_2-water saturation in the specimen during the injection by using volumetric continuity equations. The capillary pressure was also computed by using Van Genuchten (1980) equation with the parameters of m, P0, and Swr.

As a result, the relative permeabilities and specific storage (krw, krn, Ss) including capillary pressure (Hc) were obtained. However, the obtained parameters krw, krn, Ss must be refined due to capillary pressure parameter was not included in the first fitting. Therefore, the capillary pressure (Hc) was input into Equation 8 with the predetermined krw, krn and Ss, to generate refined theoretical pressure gradient data i*(L,t). Again, the theoretical pressure gradient data was fitted into the experimental pressure gradient data. The process was iterative until the theoretical and experimental pressure gradient data matched with minimal errors. Otherwise, the input values of krw, krn and Ss must be alternated and the process restarted from the beginning (Figure 16).

3.1 Applicability of the numerical analysis

Using the procedure explained above and the data obtained from the experiments, CO_2-water relative permeabilities and specific storage were determined. As previously mentioned, the differential pressures across the Ainoura specimen cannot achieve a steady state within the time period allowed during the tests. It means that the complete data for pressure including a steady state flow regime was not obtained experimentally.

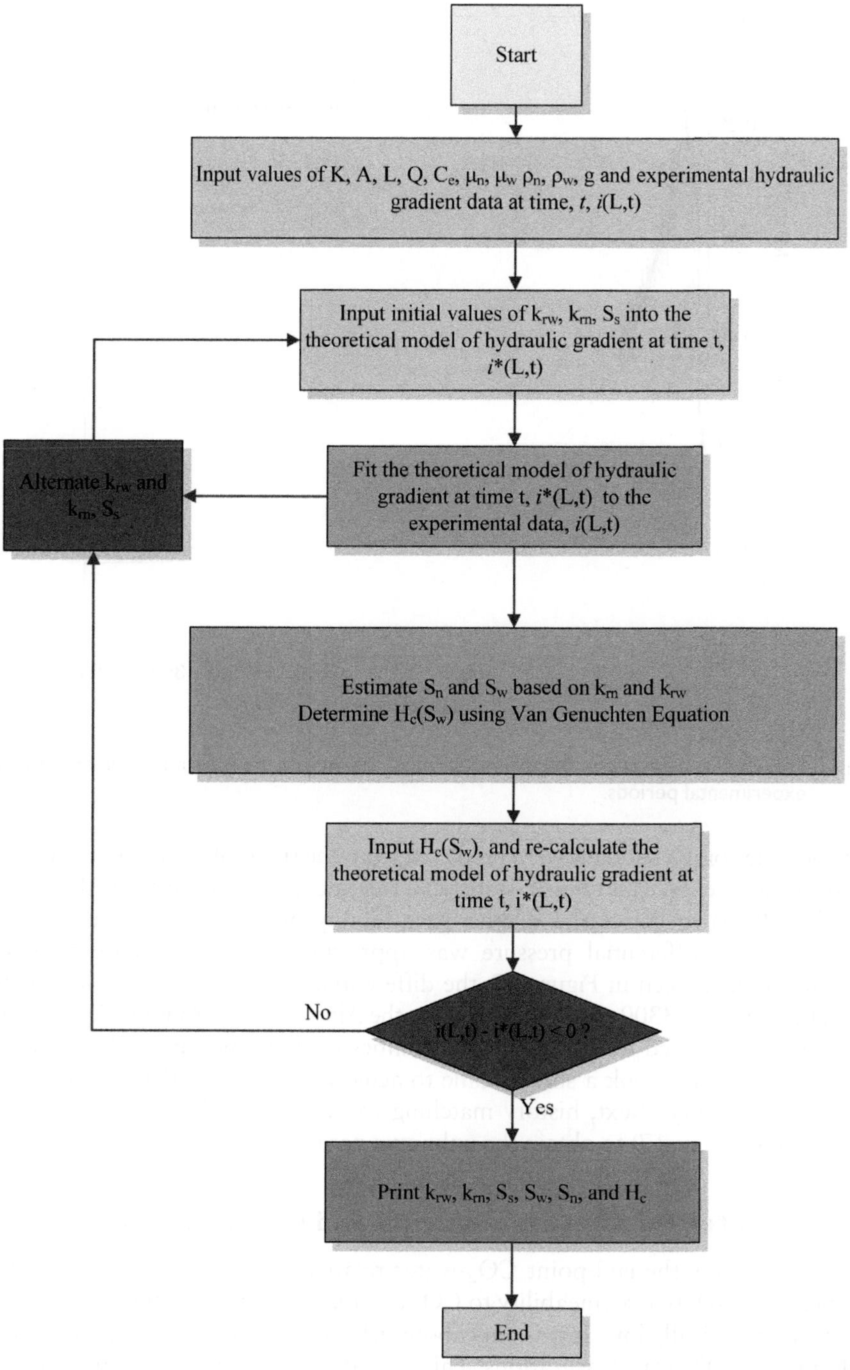

Figure 16 Flow chart of the determination of the unknown parameters krw, krn, and Ss.

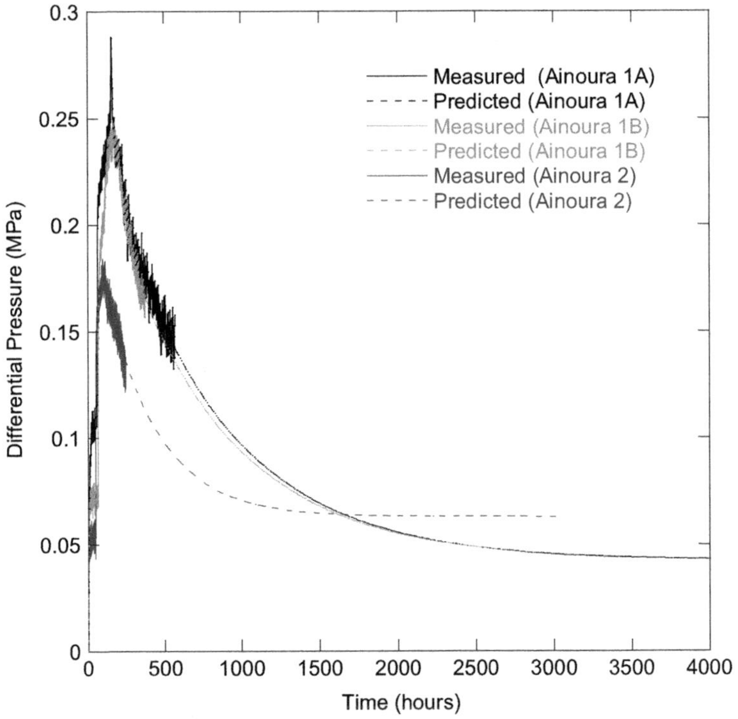

Figure 17 Predicted steady state of the differential pressures across the Ainoura specimens beyond the experimental periods.

Therefore, the completed data were predicted from the trend of the obtained data using statistical approaches. This type of -statistical- treatment was also performed by Bennion & Bachu (2005) to derive the relative permeability of sandstones. In this way, the tendency of the differential pressure was approximated by two-phase exponential decay function. As seen in Figure 17, the differential pressure would achieve a steady state at the period of 3300 and 1600 hours for the Ainoura 1 and Ainoura 2, respectively. This may be attributed to different permeabilities of these specimens. Ainoura 2 with a higher permeability took a shorter time to achieve steady state than Ainoura 1 with a lower permeability. Next, history matching to the measured pressure gradient was implemented (Figure 17) to obtain the unknown parameters, krn, krw, and Ss.

3.2 CO_2-water relative permeability and specific storage

Figure 18 presents the end-point CO_2-water relative permeabilities for the Ainoura specimens. The relative permeability to CO_2 at irreducible water saturation was found to be characteristically low. It was only about 0.15 of the water relative permeability at conditions of 100% water saturations. This indicates a lower displacement efficiency of the saturated water by the injected CO_2 in the Ainoura Sandstones. This may be related to the heterogeneous nature of the Ainoura Sandstones. In this type of bi-modal pore system, CO_2 may channel bypassing water, resulting in non-uniform CO_2 flow as also

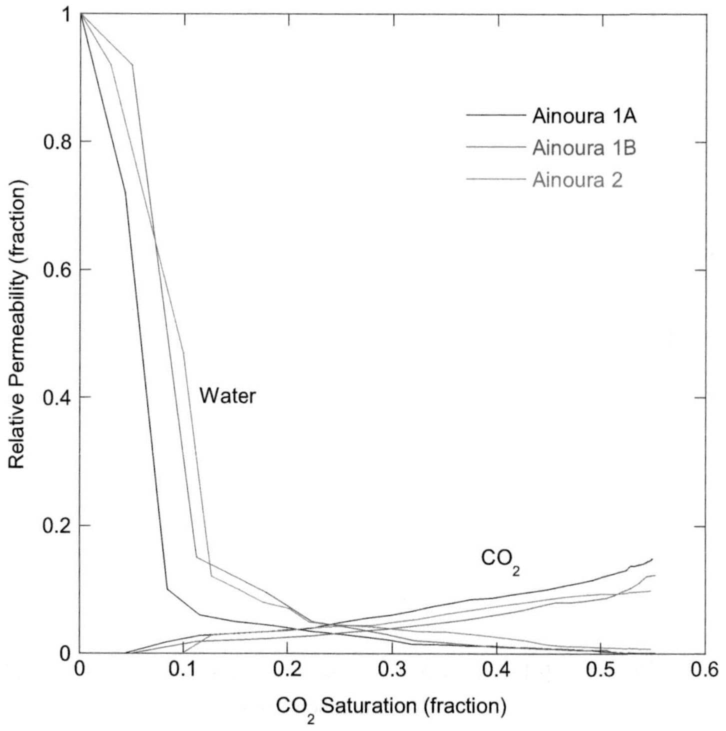

Figure 18 End-point relative permeabilities of the Ainoura specimens versus CO_2 saturation.

observed by Bennion & Bachu (2006). They concluded that pore size distribution critically controls CO_2-water relative permeabilities. Other factor that might be contributing to low relative permeability of CO_2 is the capillary pressure effect. By having large fraction of micropores (more than 50%), the Ainoura specimens yielded relatively higher capillary pressure (Figure 19). Therefore, its irreducible water saturation was still higher (estimated to be ~45%).

Figure 20 presents the change of specific storage of the Ainoura specimens with increasing CO_2 saturations. It was observed that the injection of CO_2 could enlarge the specific storage of the specimen. The specific storage increased by about 0.0004, 0.0003, and 0.0005 1/Pa for Ainoura 1A, 1B and 2, respectively. This can be seen from a transient increase of specific storage as well the significant increase of volumetric strain of the specimen. The increase of specific storage and volumetric strain in the same period, suggested that the change of specific storage is more pronounced as mechanical response rather than just hydraulic process.

4 GEOMECHANICAL PROPERTIES

Compared to numerical ones, laboratory studies associated with hydromechanical aspect of CO_2 injection is still fewer. Li *et al.* (2006) performed a triaxial acoustic emission measurement to monitor the failure mechanism of a rock fracture injected

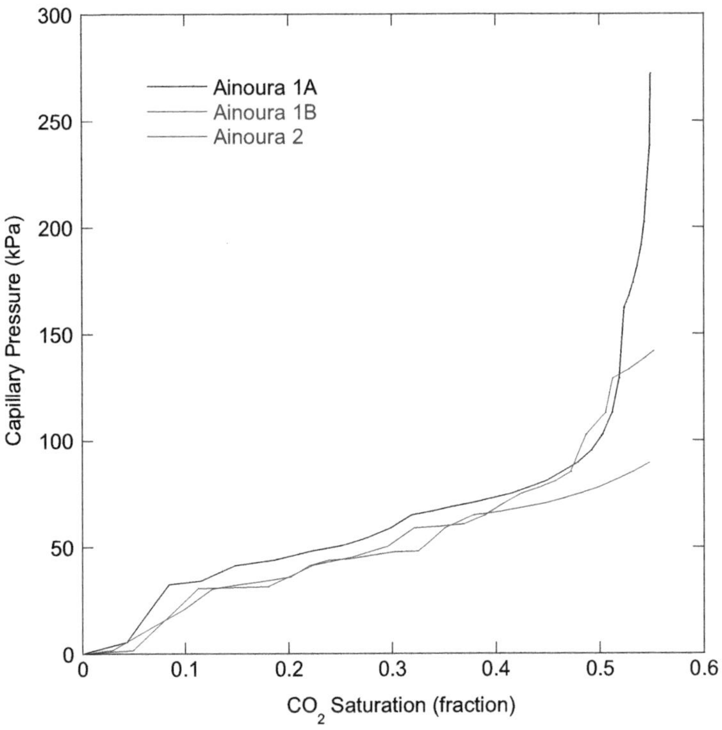

Figure 19 Capillary pressure generated during the injection of CO_2 to the specimens of Ainoura.

with CO_2. They developed a finite element numerical simulation scheme to analyze the abrupt failure process of the rock during a two-phase flow. They found that, during the injection of CO_2 into the rock, the pore pressure is dissipated while the effective stress is quickly dropped, leading to an abrupt failure of the rock.

Other works mainly focused on fracture initiation of rock under tri-axial testing with an analysis of two phase flow (water and air) at a range of confining pressure. A study by Indraratna & Ranjith (2001) concluded that the decrease in two-phase flow rates was due to the closure of fractures in rocks. For further clarification of these effects, more laboratory studies are needed. One specific issue is to develop a new empirical model (or modify existing CO_2-rock hydromechanical models), particularly incorporating the failure criterion of the rock under representative natural reservoir conditions (Shukla *et al.*, 2010). The new empirical model would improve numerical simulation models used to analyze the mechanics of CO_2 transport and storage at the field scale.

4.1 Geomechanical analysis based on poroelastic constants dependent stress

The analysis of geomechanical behavior of the specimen injected with CO_2 can be approached from the mechanism of the interaction of interstitial fluid and porous rock based on linear poroelasticity theory of Biot (1941). Fluid flow will affect the mechanical

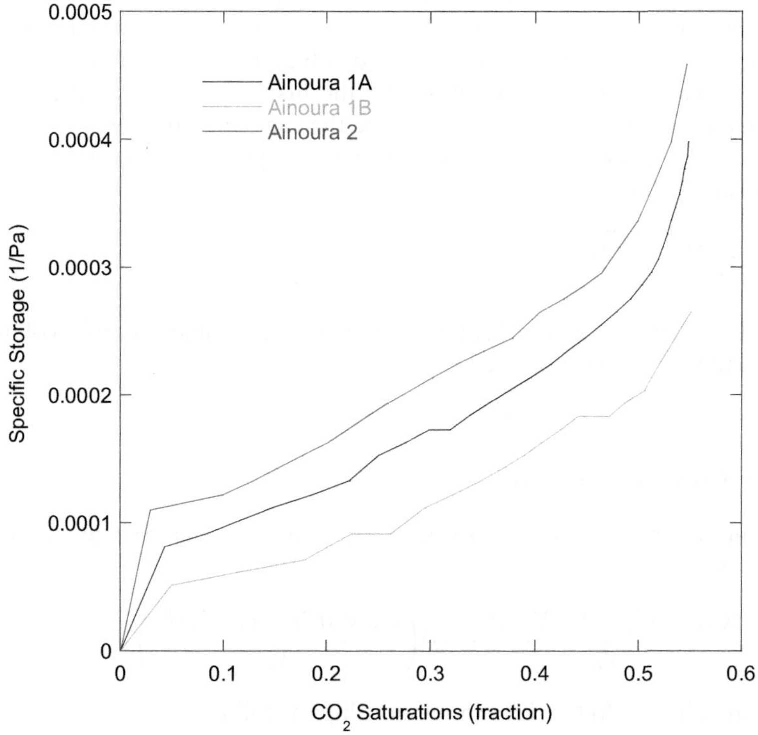

Figure 20 Specific storages of the Ainoura specimens versus CO_2 saturations.

response of rock (Detournay & Cheng, 1993). Compression on rock increases pore pressure, and the increase of pore pressure induces dilatancy of rock. We, in this study, performed linear poroelasticity based analysis to examine hydromechanical behavior of the specimen. Such a method was also used to measure poroelastic constants in addition to hydraulic conductivity and specific storage of tight rocks injected with water using the transient pulse decay test in the past (Hart, 2000; Hart & Wang, 2001).

Rock specimen is modeled as a porous body subjected to internal pore pressure and external confining pressure. Four different compressibilities are subjected on the specimen (Zimmerman, 1991; Jaeger *et al.*, 2007) as follows:

$$C_{bc} = -\frac{1}{V_b}\left(\frac{\partial V_b}{\partial P_c}\right)_{p_p} \quad (9)$$

$$C_{pc} = -\frac{1}{V_p}\left(\frac{\partial V_p}{\partial P_c}\right)_{p_p} \quad (10)$$

$$C_{bp} = \frac{1}{V_b}\left(\frac{\partial V_b}{\partial P_p}\right)_{p_c} \quad (11)$$

$$C_{pp} = \frac{1}{V_p}\left(\frac{\partial V_p}{\partial P_P}\right)_{p_c} \quad (12)$$

where Cbc and Cpc are confining pressure related bulk and pore compressibility; Cbp and Cpp are pore pressure related bulk and pore compressibility; Pp and Pc are fluid pore pressure and confining pressure applied; Vb and Vp are bulk and pore volume, respectively. There are two subscripts with the first subscript denoting the relevant volume change and the second one indicating the changing pressure.

The relationship between porosity change and pore volume and bulk volume changes is defined as

$$d\phi = d\left(\frac{V_p}{V_b}\right) = \frac{dV_p}{V_b} - \phi\frac{dV_b}{V_b} \tag{13}$$

Based on Equations 9–12, and 12, the change of pore volume and bulk volume under loading condition can be described as

$$dV_p = C_{pp}V_p dP_p - C_{pc}V_p dP_c \tag{14}$$

$$dV_b = C_{bb}V_b dP_p - C_{bc}V_b dP_c \tag{15}$$

Substituting Equations 14 and 15 into Equation 13, the change of porosity is expressed as

$$d\phi = \frac{(C_{pp}V_p dP_p - C_{pc}V_P dP_c)}{V_b} - \phi\left(\frac{C_{bp}V_b dP_P - C_{bc}V_b dP_c}{V_b}\right) \tag{16}$$

The compressibilities follow certain relationships as follow

$$C_{bc} = C_{bp} + C_m \tag{17}$$

$$C_{pc} = C_{pp} + C_m \tag{18}$$

$$C_{bp} = \phi C_{pc} \tag{19}$$

where Cm is rock matrix compressibility.

Using the relationships among compressibilities, the change of porosity is expressed as follow:

$$d\phi = C_{bp}(dP_p - dP_c) - \phi(C_{bp} + C_m)(dP_p - dP_c) \tag{20}$$

Bulk volumetric strain, εb, which is defined as the comparison of the increment of bulk volume under loading condition with initial bulk volume, can be defined as:

$$d\varepsilon_b = \frac{dV_b}{V_b} = \frac{C_{bp}V_b dP_p - C_{bc}V_p dP_c}{V_b} = C_{bp}dP_p - (C_{bp} + C_m)\phi\, dP_c \tag{21}$$

As the confining pressure was set constantly in the experiment, the bulk volumetric strain, and the porosity changes can be written as:

$$d\varepsilon_b = C_{bp}dP_p \tag{22}$$

$$d\phi = d\varepsilon_b - \phi(C_{bp} + C_m)dP_p \tag{23}$$

4.1.1 Mean stress

The mean stress is defined from the principal stress as:

$$\sigma'_M = \frac{1}{3}(\sigma'_1 + \sigma'_2 + \sigma'_3) \tag{24}$$

The principal stresses (with tension positive) are calculated as:

$$\sigma'_1 = \sigma_1 + \alpha P_p \tag{25}$$

$$\sigma'_2 = \sigma_2 + \alpha P_p \tag{26}$$

$$\sigma'_3 = \sigma_3 + \alpha P_p \tag{27}$$

Where α is Biot's effective stress parameter (Biot, 1941).

4.2 Relationship between porosity and permeability

The relationship of stress to permeability was investigated by a number of researchers for petroleum reservoir engineering applications (Fatt & Davis, 1952; Thomas & Ward, 1972; Jones & Owens, 1980; Yale, 1984; Kilmer *et al.*, 1987; Morita *et al.*, 1984; Keaney *et al.*, 1998; Han & Dusseault, 2003). In general, the relationships between stress and permeability are empirical derived from a curve fitting analysis of experimental data (Jones & Owens, 1980; Jones, 1998) and no distinctive relationship could be established for a specific rock (Davies & Davies, 2001; Jamveit & Yardley, 1997; Fatt & Davis, 1952). Two equations were employed to determine permeability changes based on porosity changes: popular simplicity of the Carman-Kozeny model as follows:

$$k = \frac{\phi^3}{5(1-\phi)^2 S^2} \tag{28}$$

where specific area, S, derived from $S = \sqrt{\frac{\phi_i^3}{5(1-\phi_i)^2 K_i}}$, ϕi and Ki are porosity and permeability under initial conditions, respectively. Then the exponential function of Davies and Davies (1999) model can be used to represent a correlation between permeability and porosity:

$$K = K_0 \exp\left[22.2\left(\frac{\phi}{\phi_0} - 1\right)\right] \mathrm{K} = \mathrm{K}_0 \exp\left[22.2\left(\frac{\phi}{\phi_0} - 1\right)\right] \tag{29}$$

where K_0 is the initial stress permeability.

4.3 Geomechanical response of the specimens under injection of CO_2

During the injection of CO_2 into the specimen, the generated hydraulic pressures in the upstream and downstream including the longitudinal and lateral strains of the specimens were measured. Overall, the injection of CO_2 has increased the hydraulic

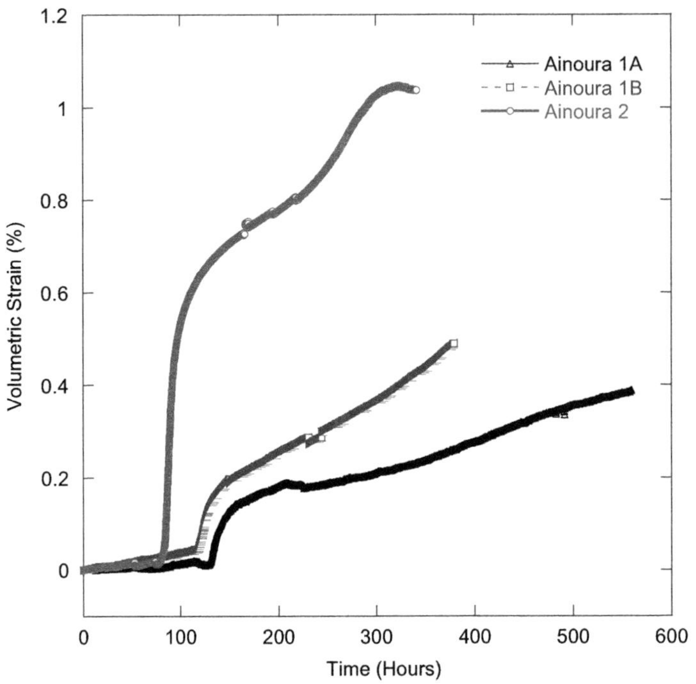

Figure 21 Measured volumetric strain of the Ainoura specimens during CO_2 injection.

pressure in the downstream and upstream of the specimen. The differential pressure between the upstream and the downstream consistently exhibited such three patterns, suggesting a three phases of CO_2 flowing through the specimen (Figure 14). In the first phase, the differential pressure increased transiently and stabilized at a certain level. Relative stable of the longitudinal and lateral strains of the specimens observed as shown in Figures 21 and 22 also proved this indication. In the second phase, the differential pressure suddenly increased again achieving higher level before it stabilized over certain times.

In addition, the longitudinal and lateral strain of the specimen increased shortly after the increase in the differential pressure. The second phase was the starting period of the increasing the specimen strains. The negative direction of the increasing strains indicated expansion of the specimen occurred as the pore pressure increased driven by CO_2 injection. In the third phase, the differential pressure slowly decreased since the injected CO_2 was able to break through the specimen.

4.3.1 Change of bulk compressibility

The bulk compressibility of the specimen was determined based on the volumetric strain and pore pressure of the specimen measured in the experiment. The matrix compressibility (Cm) of the specimens was estimated at 2.54 × 10–5/MPa for typical sandstone (Zimmerman, 1991). Figure 23 presents the bulk compressibilities of the specimens. It was observed that transient increase of bulk compressibility was found at

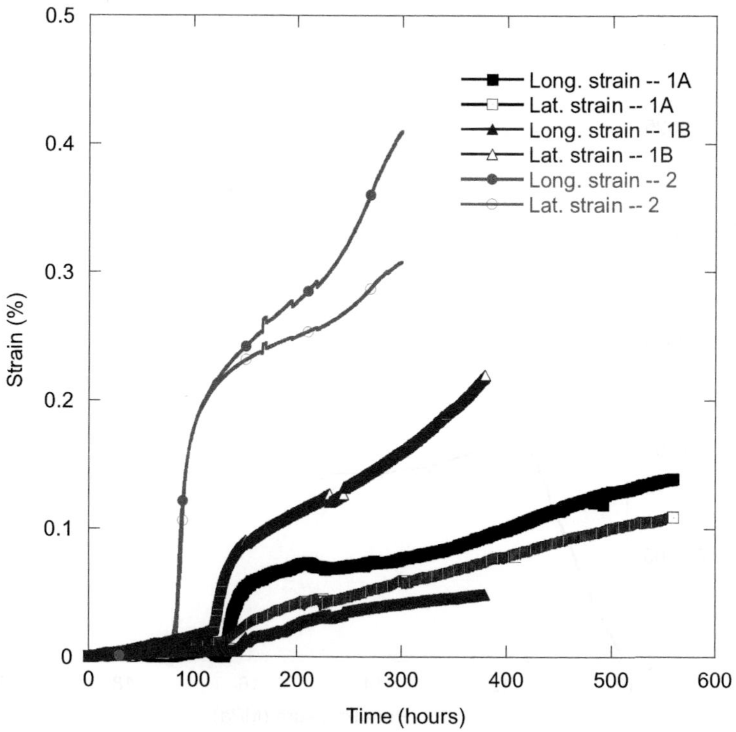

Figure 22 Measured volumetric strain of the Ainoura specimens during CO_2 injection.

the beginning of CO_2 injection. This corresponded to the transition from the displaced incompressible water flow to the displacing compressible CO_2, in the specimen pores. After this period, the bulk compressibility of the specimen decreased with increasing pore pressure. Above a certain pressure, the bulk compressibility reached a plateau that is independent of the pore pressure. Figure 23 also shows that Ainoura 2 has a larger bulk compressibility than Ainoura 1. This is probably due to the higher fraction of macropores in Ainoura 2, which results in more flow of CO_2 inducing higher pore pressure generated.

4.3.2 Effect of pressure margin on volumetric strain

The injection of CO_2 into the rock specimen increased its pore pressure and volumetric strain. As the experiments were constantly set at 20 MPa confining pressure, only the pore pressure increased from the 10 MPa initial pressure. If the pressure margin is defined as the gap pressure of the pore pressure to the confining pressure, the pressure margin decreased during the injection. The pressure margin was analyzed in this study since it is a considerable parameter that might cause hydraulic fracturing. The initiation of hydraulic fracturing will occur when the pore pressure equals the confining pressure (Jaeger *et al.*, 2007). Figure 24 illustrates the relationship between the pressure margin and the volumetric strains

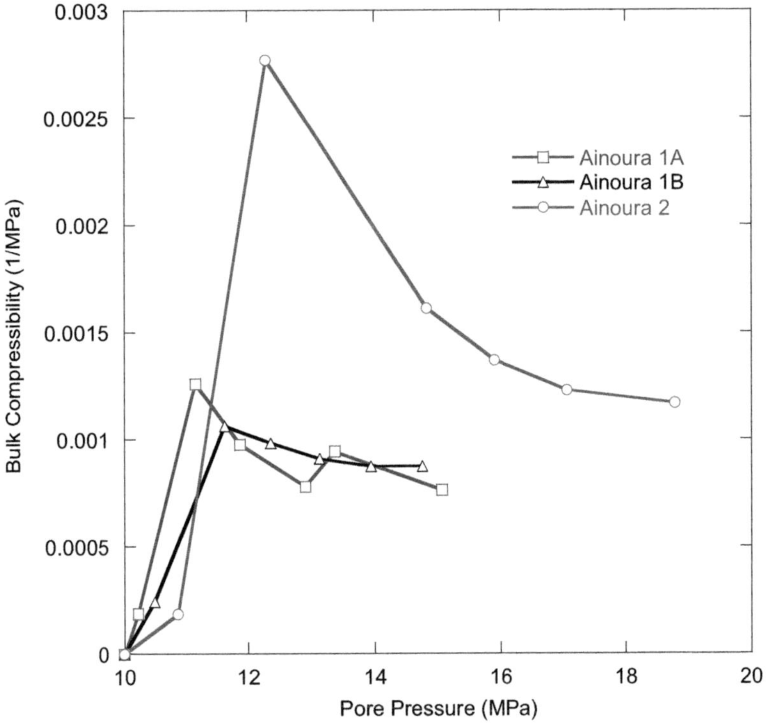

Figure 23 Bulk compressibility measured during CO_2 injection to the specimens.

measured during the experiments. As seen, the pressure margin increased as the volumetric strain increased. Beyond a certain pressure margin, the volumetric strains increased significantly. The transient increase of volumetric strain occurs at the transition of the incompressible water flow to the compressible CO_2 flow in the specimen pores, as observed in the second phase of the experiment. After that, the CO_2 did break through the specimen, generating a higher increase of the volumetric strain.

Given the trend of curves in Figure 24, the flow of CO_2 would generate a significant increase in volumetric strain when the pressure margins were above –9 MPa and –8 MPa for the Ainoura 2 and 1 samples, respectively. This means that the increased volumetric strain of the higher porosity specimen would occur slower than that of the lower porosity specimen. However, in the case of the magnitude of the strains generated, the specimen with higher porosity yielded a larger volumetric strain compared to a lower porosity sample. As a result, the generated pore pressure in the higher porosity specimen took a shorter time to reach the confining pressure level.

The results suggest that the lower porosity of Ainoura sandstones could be beneficial in creating a higher specific storage for CO_2 but would generate lower deformation. It is noted that the lower deformation observed was induced by the injection at the very low flow rate applied in the experimental test. The very low flow rate was selected to mimic laminar flow in deep underground.

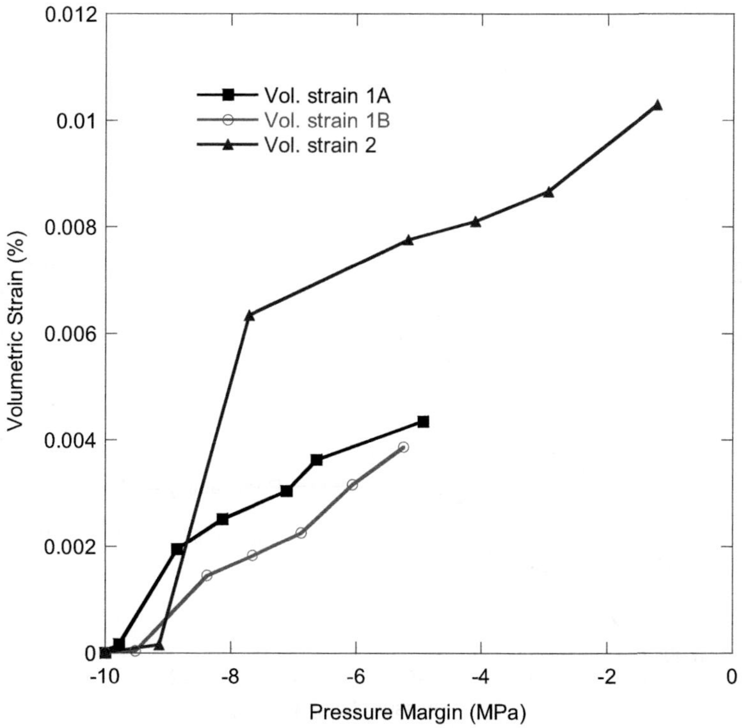

Figure 24 Pressure margin of pore pressure to confining pressure versus volumetric of the specimens.

4.3.3 Permeability evolution of the specimens during CO_2 injection

The injection of CO_2 into the specimen resulted in the increase in porosity. As shown in Figure 25, the porosity increased by about 3% and 5% for the Ainoura 1 and 2 samples, respectively. As a result, their permeability also increased by a factor of two to three with respect to initial permeability. In particular, for Ainoura 2 with higher porosity, the increase in permeability is clearly shown beyond a critical pressure of −8 MPa. On the other hand, for the Ainoura 1A and 1B samples with a lower porosity, this value was measured to be −9 MPa. The period of permeability increase corresponds to the third phase when CO_2 flowed through the rock sample with some fraction of irreducible water. The results confirmed our suggestion that the third phase observed in the experiment is the period for the increase in the volumetric strain yielded by the significant flow of CO_2. This led to the onset of dilatancy of the specimen.

The differential pressure dropped at this period as a result of the specimen dilatancy. The transient increase in permeability and specific storage was attributed to the nucleation and the growth of microcracks. The initiation of microcracking is generally assumed to coincide with the onset of dilatancy (Heiland, 2003). This observation is consistent with Keaney *et al.*'s (1998) results on the deformation of Tennessee sandstone during the transient pulse permeability measurements combined with a triaxial deformation apparatus. Similarly, Zoback & Byerlee (1975) observed that there is

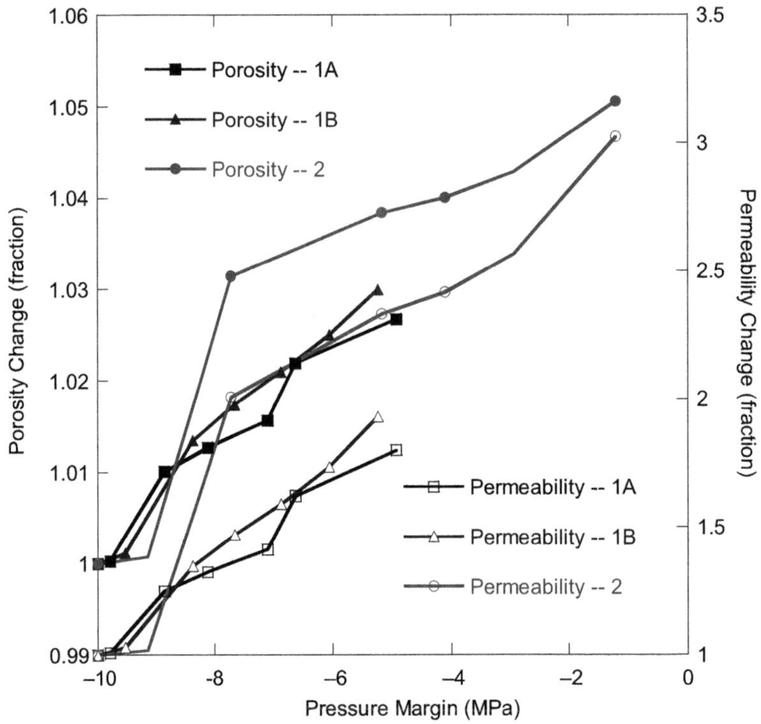

Figure 25 Porosity and permeability change of the specimens with increasing pressure margin of pore pressure to confining pressure.

a strong relationship between the onset of dilatancy and an increase in the permeability of the crystalline rocks. Regarding the failure of the specimens, it was observed that the peak strengths were unachieved due to a continuous increase in the volumetric strain with the increase in total stress at the end of the experiment. The experiment, eventually, had to be ended since the increasing pore pressure generated by injection was expected to exceed the confining pressure and that may have resulted in breaking the silicon on the rubber sleeves covering the rock sample. Nonetheless, based on the increase of the permeability by a factor of two to three to the initial value, which is still lower than the factor of 17 observed by Keaney *et al.* (1998), brittle failure might not yet have been taking place on the specimens at the end of the injection period. The increase in permeability was small, similar to the mechanical behavior of the sandstone with a porosity of about 14% as described by Zhu & Wong (1997).

5 SUMMARY

Based on the observed differential pressures across the cores, it was observed that there are three regimes of CO_2 flow in a sandstone rock sample. The first stage is the flow of the displaced water out of the specimen. The second stage is the flow of the injected CO_2 ending with its breakthrough. The third stage is the flow of CO_2 through the specimen, achieving a steady state. Very slow process of the CO_2-water displacement in the specimen is due to very low hydraulic gradient employed in the injection and the profound effect of capillary pressure in low permeability rocks.

The Ainoura sandstone was observed to have lower CO_2-water displacement efficiency. This is indicated by low relative permeability to CO_2; only 0.15 of the relative permeability of water at 100% water saturation. The average storage capacity of the Ainoura sandstones for supercritical CO_2 is 3.74x10–4 1/Pa at the experimental conditions applied. Ainoura sandstones were effective in retaining the flow of supercritical CO_2, indicated by considerable time needed for CO_2 to migrate through the core sample. The new experimental system of flow pump permeability method with the developed numerical analysis will contribute reliable measurement of relative permeability and specific storage using a standardized geotechnical laboratory method.

The injection of CO_2 into the Ainoura sandstones resulted in the increase in the volumetric strains of the samples. Given by the direction of strains, the sandstones expanded during the injection. The expansion of the Ainoura sandstones is due to the decrease in effective pressure as the pore pressure induced by the injection increases and the confining pressure was set to be constant. The expansion initiated when the pressure margin between the pore pressure and the confining pressure was –9 MPa and –8 MPa for the Ainoura 1 and 2 samples, respectively. The porosity of Ainoura 1 and 2 also changed by 3% and 5%, respectively, during CO_2 injection. This led to an increase in their permeabilities by a factor of two and three. The onset of dilatancy of the sandstone occurred beyond a critical CO_2 saturation. This value was observed to be 13% at the pore pressure above 60% of the confining pressure for the case of a very low flow rate. The results suggested that the failure mechanism did not take place at the end of the experiment as the peak strength of the specimens was unachieved at the condition where the pore pressure is still below the confining pressure.

REFERENCES

Bachu, S., Bonijoly, D., Bradshaw, J., Burrus, R., Holloway, S., Christensen, N.P., Mathiassen, O.M., 2007. CO_2 storage capacity estimation: Methodology and gaps, International Journal of Greenhouse Gas Control 1, pp. 430–443.

Bennion, D.B., Bachu, S., 2005. Relative permeability characteristics for supercritical CO_2 displacing water in a variety of potential sequestration zones in the western Canada Sedimentary Basin, SPE Annual Technical Conference and Exhibition, October 9–12, 2005, Dallas, Texas, SPE 99547.

Bennion, D.B., Bachu, S., 2006. The Impact of Interfacial Tension and Pore Size Distribution/ Capillary Pressure Character on CO_2 Relative Permeability at Reservoir Conditions in CO_2 – Brine Systems, SPE/DOE Symposium on Improved Oil Recovery, April 22–26, 2006, Tulsa, Oklahoma, USA SPE 99325.

Benson, S. M., 2004. Workshop proceeding of the 10–50 solution Technologies and Policies for a Low-Carbon Future, March 25–26, 2004, Washington DC, USA.

Biot, M.A., 1941. General theory of three-dimensional consolidation, Journal of Applied Physics 12, pp. 155–164.

Bradshaw, J., Bachu, S., Bonijoly, D., Burruss, R., Holloway, S., Christensen, N.P., Mathiassen, O.M., 2007. CO_2 storage capacity estimation: issues and development of standards. International Journal of Greenhouse Gas Control 1, pp. 62–68.

Carslaw, H.S., Jaeger, J.C., 1959. Conduction of heat in solids, Oxford University Press, p. 510.
Cook, P.J., 1999. Sustainability and nonrenewable resources. Environmental Geosciences 6(4), pp. 185–190.
Chiquet, P., Daridon, J., Broseta, D., Thibeau, S., 2007. CO_2 /Water interfacial tensions under pressure and temperature conditions of CO_2 geological storage, Energy Conversion and Management 48(3), pp. 736–744.
Dana, E., Skoczylas, F., 2002. Experimental study of two-phase flow in three sandstones. I. Measuring relative permeabilities during two-phase steady-state experiments, International Journal of Multiphase Flow 28, pp.1719–1736.
Davies, J.P., Davies, D.K., 1999. Stress-dependent permeability: Characterization, and modeling. Society of Petroleum Engineers, SPE paper no 56813.
Detournay, E., Cheng, A.H.-D., 1993. Fundamental of Poroelasticity in Comprehensive Rock Engineering: Principles, Practice and Projects, Vol. II, Analysis and Design Method, ed. C. Fairhurst, Pergamon Press, pp. 1130171.
Gunter, W.D., Perkins, E.H., McCann, T.J., 1993. Aquifer disposal of CO_2-rich gases: reaction design for added capacity. Energy Conversion and Management 34, pp. 941–948.
Hart, D.J., 2000. Laboratory measurement of poroelastic constants and flow parameters and some associated phenomena. PhD thesis, University of Wisconsin Madison, USA.
Hart, D.J., Wang, H.F., 2001. A single test method for determination of poroelastic constants and flow parameters in rocks with low hydraulic conductivities. International Journal of Rock Mechanics and Mining Sciences 38, pp. 577–583.
Heiland, J., 2003. Permeability of triaxially compressed sandstone: Influence of deformation and strain-rate on permeability. Pure Application of Geophysics 160, pp. 889–908.
Indraratna B., Ranjith, P.G., 2001. Laboratory measurement of two-phase flow parameters in rock joints based on high pressure triaxial testing. Journal of Geotechnical and Geoenvironmental Engineering 127(6), pp. 530–542.
IPCC, 2005. IPCC special report on carbon dioxide capture and storage, Metz B, Davidson O, de Connick H, Loos. M., and Meyer (eds), Cambridge University Press, New York, USA, pp. 195–276.
Jaeger, J.C., Cook, N.G.W., Zimmerman, R.W., 2007. Fundamentals of Rock Mechanics, Blackwell Publishing. Victoria Australia.
Juanes, R., Spiteri, E.J., Orr jr., F.M., Blunt, M.J., 2006. Impact of relative permeability hysteresis on geological CO_2 storage. Water Resources Research, 42, p. W12418.
Keaney, G.M.J., Meredith, P.G., Murrell, S.A.F., 1998. Laboratory Study of permeability evolution in a tight sandstone under non-hydrostatic stress conditions, SPE Conference Paper, 47265-MS.
Kumar, A., Ozah, R., Noh, M., Pope, G.A., Bryant, S., Sepehrnoori, K., Lake, L.W. 2005. Reservoir Simulation of CO_2 Storage in Deep Saline Aquifers. SPE J., 10(3), pp. 336–348. SPE-89343-PA.
Richardson, J.G., Kerver, J.K., Hafford, J.A., Osoba, J.S., 1952. Laboratory determination of relative permeability, Petroleum Transactions AIME 195, pp. 187–196.
Sasaki, K., Fujii, T., Niibori, Y., Ito, T., and Hashida, T., 2008. Numerical simulation of supercritical CO_2 injection into subsurface rock masses, Energy Conversion and Management 49, pp. 54–61.
Salvi, S., Quattrocchi, F., Angelone, M., Brunori, C.A., Billi, A., Buongiorno, F., Doumaz, F., Funiciello, R., Guerra, M., Lombardi, S., Mele, G., Pizzino, L., Salvini, F., 2008. A multidisciplinary approach to earthquake research: Implementation of a Geochemical Geographic Information System for the Gargano site, Southern Italy. Natural Hazard, 20(1), 225–278.
Shukla, R., Ranjith, P., Haque, A., Choi, X., 2010. A review of studies on CO_2 sequestration and caprock integrity. Fuel 89, pp. 2651–2664.

Van Genuchten, M. Th., 1980. A closed-form equation for predicting the hydraulic conductivity of unsaturated soils. Soil Science American Journal 44, p. 892
Villarasa, V., Bolster, D., Olivella, S., Carrera, J., 2010. Coupled hydromechanical modeling of CO_2 sequestration in deep saline aquifers. International Journal of Greenhouse Gas Control 4, 910–919.
Zimmerman, R.W., 1991. Compressibility of sandstones. Developments in Petroleum Science, Vol. 29. Elsevier, New York.
Zhang, M., Takahashi, M., Morin, R.H., Esaki, T., 2000. Evaluation and application of the transient-pulse technique for determining the hydraulic properties of low-permeability rocks—Part 2: Experimental application. Geotechnical Testing Journal 23, pp. 91–99.
Zhu, W., Wong, T.F., 1997. The transition from brittle faulting to cataclastic flow: Permeability evolution. Journal of Geophysics Research B 102, pp. 3027–3041.
Zoback, M.D., Byerlee, J.D., 1975. The effect of microcrack dilatancy on the permeability of Westerly Granite. Journal of Geophysics Research, 80(5), pp. 752–755.

Chapter 20

Hydro-mechanical coupling of rock joints during normal and shear loading

M. Sharifzadeh[1], *S.A. Mehrishal*[2], *Y. Mitani*[2] *& T. Esaki*[3]

[1]*Western Australian School of Mines (WASM), Curtin University, Kalgoorlie, Australia*
[2]*Department of Mining & Metallurgical Engineering, Amirkabir University of Technology, Tehran, Iran*
[3]*Department of Civil and Structural Engineering, Kyushu University, Kyushu, Japan*

Abstract: Several Hydro-Mechanical coupling tests were conducted on artificial granitic joint samples at Kyushu University in Japan. The experiments consist of three major stages: first, joint geometrical properties were determined. Then, hydro-mechanical properties of rock joints were investigated under normal and shear modes. Finally, hydro-mechanical coupling properties of rock joint were modeled and compared with experimental results.

To determine hydro-mechanical coupling properties of rock joint, a new test apparatus with different flow systems was designed and developed to perform test under high hydraulic head and low hydraulic gradient. Test results showed that during normal closure – hydraulic conductivity test, joint conductivity decreased with increasing normal stress. Increasing in inlet water pressure caused slight increasing of hydraulic conductivity. Hydraulic conductivity decreased at the onset of shear and then increased rapidly and almost remained constant in residual shear region.

To model hydro-mechanical coupling properties, joint aperture distribution data in normal and shear modes were used. Based on the values of each element in joint, an original flow model was developed which applied the effective aperture instead of nominal aperture. Results showed that the new flow simulation gave close results to experimental findings, especially in residual shear region.

1 INTRODUCTION

Understanding water flow through a rock mass is a fundamental issue in many areas of rock engineering, such as hydrogeology, civil engineering, reservoir engineering, underground waste disposal and foundations in fractured rock masses. The flow through a discontinuous rock mass can be divided into flow through discontinuities and flow through the rock matrix. Rock matrix permeability is often much lower than the joint permeability. Therefore it can be assumed that the flow is governed only by discontinuities, and the first step in understanding rock mass conductivity is comprehension of single rock joint conductivity.

Fluid flow through single rock joint depends on fluid and joint geometry parameters. Fluid parameters consist of the density, viscosity, and velocity; whereas, joint parameters consist of joint surface roughness, variable aperture and contact areas distribution which can be referred to as joint geometrical properties (Sharifzadeh, 2005).

The flow through a rock fracture is governed by the Navier–Stokes equations, which is difficult to solve due to a set of three coupled non-linear equations. In case of a fracture bounded by smooth parallel walls, these former equations can be highly simplified and lead to the cubic law, which is still used in the literature in the rock joints context due to its simplicity even if deviations from experimental data due to joint roughness and surface deformability have been observed. Several attempts have been undertaken to improve the cubic law introducing roughness parameters or reducing the value of the hydraulic apertures. These corrections have not been that efficient and the main effect is a diminution of the total flow rate but the description of the flow anisotropy is poor. Reynolds equation which is obtained by considerations of orders of magnitude is more tractable than the Navier–Stokes equations and more accurate than the cubic law. So far, the equation is considered to describe properly the flow anisotropy within a rock joint (Giacomini *et al.*, 2008).

In most research, to characterize the joint hydraulic parameters under various mechanical and flow conditions, several special hydro-mechanical testing apparatuses were developed. Generally, shear–flow coupling test apparatuses can be divided into two categories that consist of radial flow and linear or one-dimensional flow. In the radial flow testing apparatus, water is injected into the joint through hole which is located in the center of the joint and flows around the specimen, with free outlet conditions. In the linear or one-dimensional testing apparatus, water injected is from one side of the specimen and flows through joint void spaces and pours out from the opposite side of the specimen. In this system, joint sides should be sealed to conduct linear flow and prevent water leakage to joint sides (Sharifzadeh, 2005).

2 TYPES OF COUPLINGS IN ROCK ENGINEERING WITH EMPHASIS ON HYDROMECHANICAL COUPLING

The recent concerns about various deep underground utilities, such as geological isolation of chemical and radioactive wastes, have caused great interest in the field of rock engineering. For the appropriate solution of the problem, it is important to understand the coupled behavior of rock joints contained in the rock mass. The coupled behavior means interaction between joint deformation and propagation, hydraulic conductivity and storativity, as well as thermal and chemical properties of rock joints, under various physical and chemical environments. For example, the hydraulic properties of a rock joint are affected by the stresses, the resultant deformation and geometrical factors of joint surfaces. Rock joints are deformed through closure, dilation and shear. The normal and shear stress acting on a joint may close or open the aperture due to contraction or dilatancy. Consequently, the hydraulic properties vary due to the changes of aperture (Esaki *et al.*, 1999).

There are at least five fundamental physical coupling processes consisting of; fluid flow, thermal, mechanical, chemical and biological processes. These can be explained as follows (Tsang, 1987):

- *Mechanical effects (M)*: Coupled processes that involved rock mass stress perturbations of up to few MPa, corresponding to an overburden depth of a few to several hundreds of meters. Mechanical processes include joint closure and dilation due to normal and shear loading and fracturing.

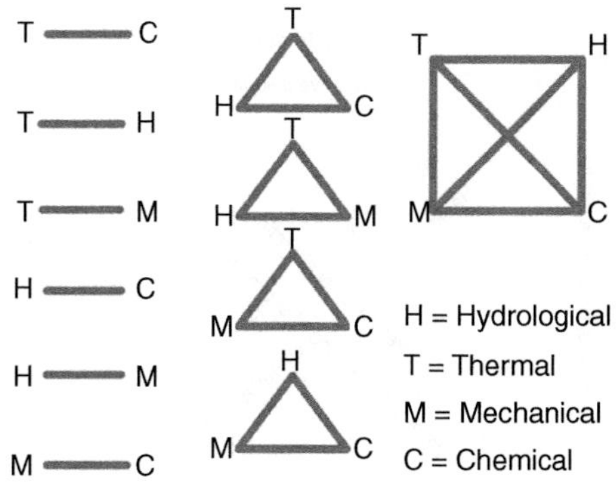

Figure 1 Types of coupling processes between T-H-M-C (modified after Tsang, 1987).

- *Hydrologic flow (H)*: Gas and water can transport contaminants from waste sites to human and ecological receptors. The processes are fluid flow and tracer transport through rock joints and transient pore fluid pressure.
- *Thermal heating and cooling (T)*: Processes include changes of temperature and presence of transient temperature gradient.
- *Chemical effects (C)*: Processes such as fracture healing, mineral dissolution or precipitation, weathering-induced weakening, or contaminant mobilization and transport each have implications for the long-term performance of engineering projects in rock.
- *Microbial effects and radiological effects*: Microbial and radiological processes may each affect rock chemistry on a local scale.

Considering of above mentioned processes, there are eleven types of possible couplings. As shown in Figure 1, some of important couplings in rock fracture are illustrated as followings:

1. *Hydro-mechanical coupling (H-M)*: Stress-flow is a dominant process in most of natural phenomenon and lots of theoretical and experimental studies were performed for this coupling in comparison to others. The joint aperture changes and effect on fluid flow through rock joint.
2. *Thermo-hydraulic (T-H)*: Temperature gradient due to presence of a heat source in the rock formation, and raising the buoyancy flow in joint system plays a major role in thermal conductivity that is important in geothermal energy study.
3. *Hydro-chemical (H-C)*: Change of fluid flow due to solute rock interaction which has changed fluid properties or rock permeability.
4. *Thermo-hydro-mechanical (T-H-M)*: Effect of changing temperature and pore fluid pressure on mechanical changes of the rock fractures, like HDR (Hot Dry Rock).
5. *Thermo*-hydro-*mechanical and chemical (T-H-M-C)*: When dissolution or precipitation processes are also influenced by temperature gradient present, there are four coupling ways as shown in Figure 2.

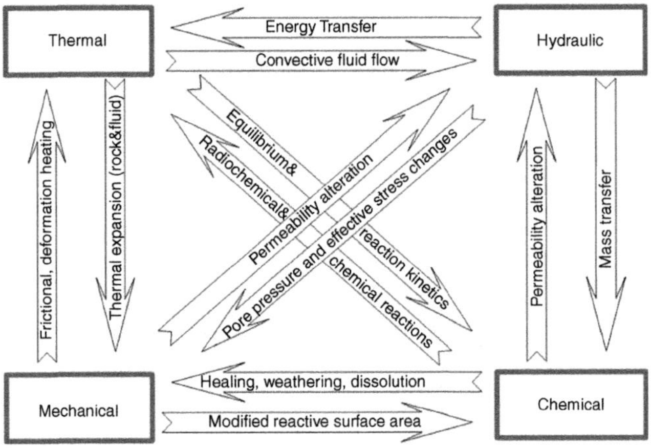

Figure 2 Effect of coupled T-H-M-C processes on rock mass behavior (Modified after Yow & Hunt, 2002).

Coupling process occurrence over a wide range of time and length scales is illustrated by Yow & Hunt (2002). Figure 3 compares the temporal and spatial scales of experiments, engineered structures and natural phenomena involving coupled processes in rock. The time scale reflects the typical duration or life cycle of an experiment or project, while the length scale gives a sense of geometric size. Figure 3 shows that most rock properties, processes measurement in experiments and tests that are orders of magnitude smaller and of much shorter duration than engineering projects. Another aspect of Figure 3 is the scale-dependence of various processes. The line labeled Molecular Diffusion represents the characteristics of distance for a dissolved molecule that will be diffused in water over a specified time using the relationship in which diffusion distance scales as the square root of the product of diffusivity and time. Another line represents

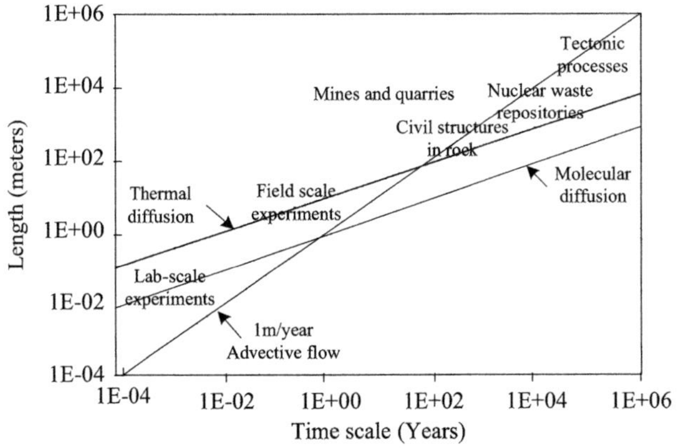

Figure 3 Typical temporal and spatial scales of experiments, engineered structures and natural phenomena involving coupling processes in rock (Modified after Yow & Hunt, 2002).

thermal diffusion, which aligns more closely with projects ranging from field scale experiments to nuclear waste repositories. This suggests that thermal energy transfer is likely to have a significant influence on scales comparable to the size and duration of a wider range of projects. Finally, a line representing advective flow at a velocity of one meter per year helps to delineate the possible importance of water flow.

In the few past decades, hydro-mechanical coupling study has received more attention than the other kinds of coupling study. This kind of coupling occurs in many natural processes and engineering constructions and more popular in various applications in civil, mining, and environmental engineering.

Rock masses naturally are under various loads, deformations and hydraulic conditions. Therefore, comprehension of stress – deformation – permeability history of rock mass can efficiently help us to simulate its behavior. In order to achieve the conceptual understanding of hydro-mechanical coupling in rock joint it is necessary to understand the joint hydraulic, mechanical and geometrical characteristics.

3 ROCK JOINT GEOMETRY DETERMINATION

A rock fracture is composed of two opposite surfaces of usually very complex topography due to numerous asperities of different sizes and shapes. This morphological feature is called the roughness of the rock fracture and is the major reason behind the complexity in mechanical and hydraulic behavior of the rock fractures. The characterization of the surface roughness remains one of the most important and challenging aspects in the study of rock fractures (Jing & Hudson, 2004).

Laboratory research during the past 10 years has made many critical links between the geometrical characteristics of fractures and their hydro-mechanical behavior under normal stress. Understanding the relationship between fluid flow and geometry for rock fractures in shear is more complicated than for under only normal stress because of the need to account for the shearing effect and creation of damage zones.

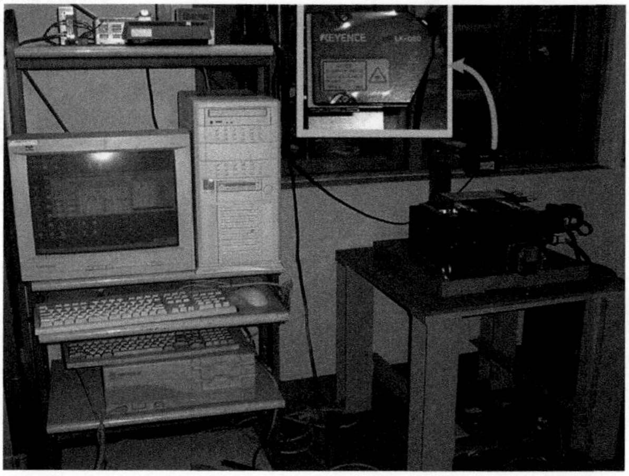

Figure 4 Designed laser scanner system.

3.1 Surface roughness measurements and characterization

Rock joint surface roughness plays a major role in joint mechanical and hydraulic behavior and the precise measurement of rough surface topography is a key to understand these behaviors of joints specially during shearing. In practice, the roughness of rock fractures is measured by means of contact profilometry, Laser profilometry, photogrammetry and advanced topometric scanner at laboratory and/or in-situ scales.

One of the precise measurements of the asperities height is based on using of laser beam (Figures 4 and 5). In this method, the distance between rough surface and laser gauge is measured as Z- elevation by a laser displacement sensor head with zero reference position and the intervals of measurement in X- and Y- directions are set by the positioning table. Firstly, the asperities data are measured in X-direction. Then the table moves in Y-direction with defined intervals. The process is repeated until completion of the whole surface measurement.

Then, using Geographic Information System (GIS) technology along with specifically developed computer programs it is possible to measure asperity heights, process data and evaluate possible deviations and inclination correction to obtain spatial positioning and matching of joint surfaces. It should be noted that, based on surface asperity conditions, each element could be an asperity or several elements may form one asperity. An important intrinsic error from laser measurement is due to the differences of laser light reflection from dark and bright minerals on surfaces. The error points in both upper and lower surfaces are corrected by using the average of neighbor point's elevations. Consequently, each surface is defined as a collection of asperities with different heights, slope angles, aspects and positions. Using GIS three dimensional

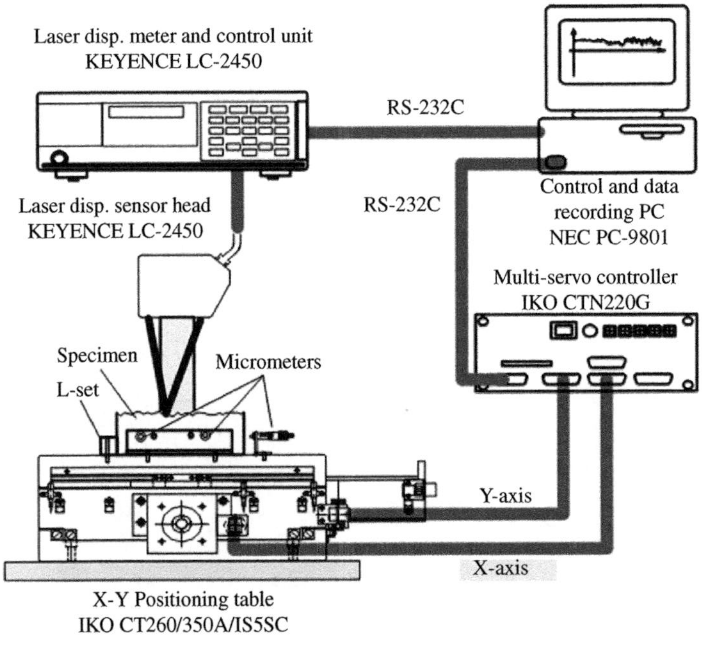

Figure 5 Schematic view of designed laser scanner system (not to scale).

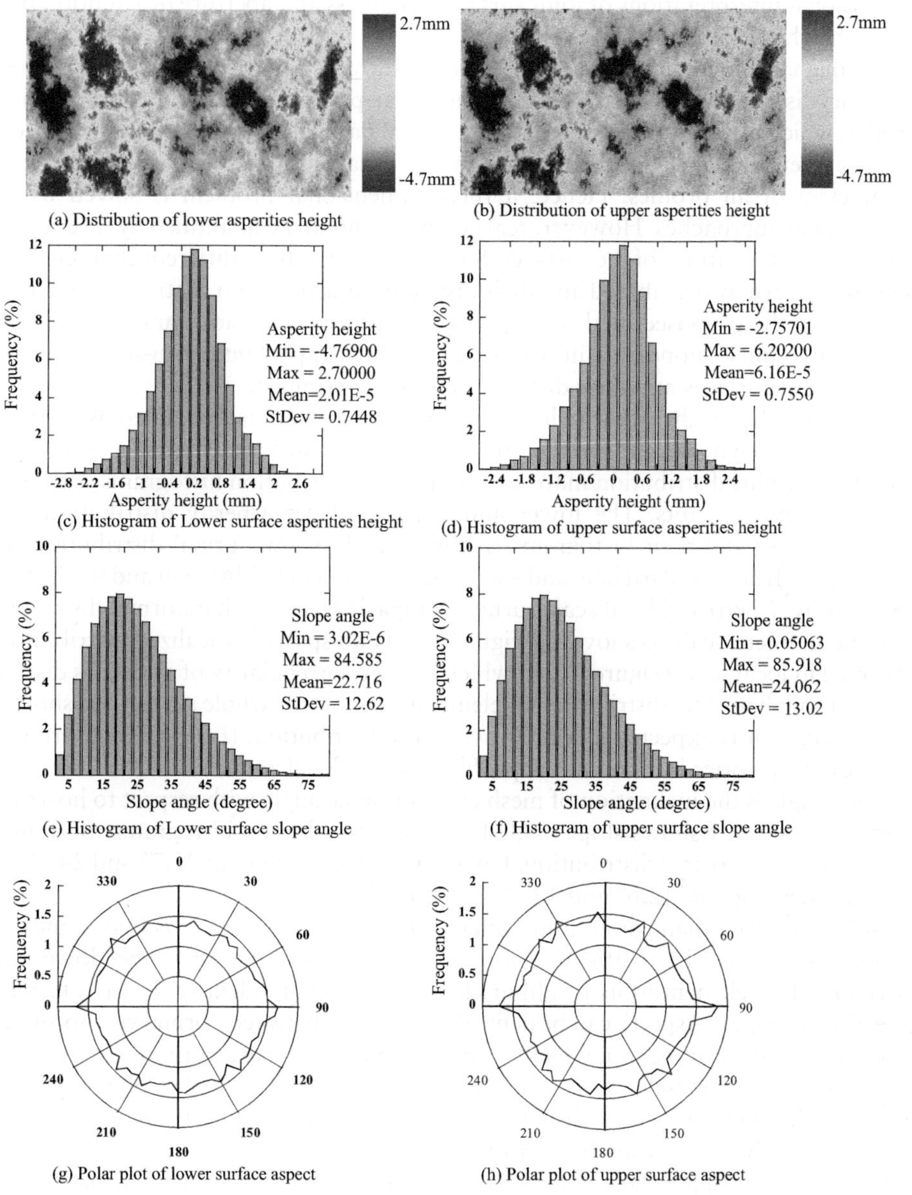

Figure 6 Joint lower and upper surfaces asperities height, slope angle and aspect distribution (Sharifzadeh, 2005).

analysis tools, these characteristics could be analyzed. Each surface is studied with three scales consisting: (i) Element or micro-scale, (ii) local area scale (or a few square centimeters), (iii) laboratory scale.

Micro-scale characteristics are: elements heights, slope angles and aspects determined accurately by using GIS (Figure 6); Therefore it is possible to give three-dimensional interpretations of joint surface roughness and aperture distribution based on the height of local roughness elements.

Although the actual morphology of joint surfaces is three-dimensional, surface roughness is most often described by some linear parameters calculated in individual profiles which are obtained along one-directional parallel lines. Indeed, the parameters describing entire surface of the fracture can be obtained by calculating the average of parameters of all profiles. Hence, a three dimensional problem is solved by two dimensional approaches. However, real geometry should be quantified by three dimensional characterizations of the surfaces. Surface asperities have different characteristics: they are not evenly distributed and their distribution depends on spatial position of the surfaces. In the proposed method, asperity height, asperities slope angle and orientation or direction of slopes, are illustrated as the three important micro-scale characteristics of joint surfaces and their distributions are determined.

To obtain the height of each element, mean surface data is calculated as base level and elements height are measured with respect to mean surface. Figure 6 a, d shows the asperity's heights distribution maps and frequency distribution histograms with statistical calculation results. The lower and upper surfaces asperity distribution maps (Figure 6 a, b) and their histograms (Figure 6 c, d), show normal distribution with mean height from standard line, and standard deviation of 0.7448 mm and 0.7550 mm, respectively. Micro and local scale surface comparison shows that normal distribution is found in micro scale as shown in Figure 6 c, d and spatially localized distribution is observed in local scale (Figure 7 a, b) which indicates irregularity of asperities distribution. Although normal distribution of elements' heights for whole surface are shown in Figure 6 c, d, it is expected a regular asperities distribution, (Figure 7 a, b) irregular changes of asperities is found in the profiles (Spatial localization).

Slope angle is the inclination of mesh element plane angle with respect to horizontal plane. Figure 6 e, f shows slope angle frequency distribution histograms. Both histograms show log-normal distribution, having mean slope angles of 22.7° and 24.1°, and standard deviation of 12.6° and 13.0°, respectively.

Surface asperity plane orientation referred to as "Aspect" and it is one of the most important characteristics affecting joint shear behavior. Aspect is the down slope direction of an element to its neighbor elements and could identify the orientation or direction of slope. "Aspect" can be defined as the angle between normal vectors on each triangle from shear direction, which is a projection of plane orientation with respect to shear direction (Grasselli *et al.*, 2002). Figure 6 g, h show polar plot of aspect's direction that indicates equal distribution of aspect over joint surfaces. Thus, equally dilation is likely to occur in all shear directions.

In order to evaluate the capability of surface morphology measurement and data modification method, joint's upper and lower surface characteristics are compared to each other. Comparison between pairs of upper and lower surface asperity heights, slope angles and aspects direction distribution as shown in Figure 6 indicate close similarity between two surfaces which means high matching of joint surfaces. Thus, it could be concluded that the applied data measurement technique (surface topography) is successful to overcome referencing difficulties which encountered in former researches and achieved matching between joint surfaces sing surface topography method.

Therefore, a three dimensional interpretation of joint surfaces is presented. This analysis can illustrate the aperture distribution during the shear. Each surface can be

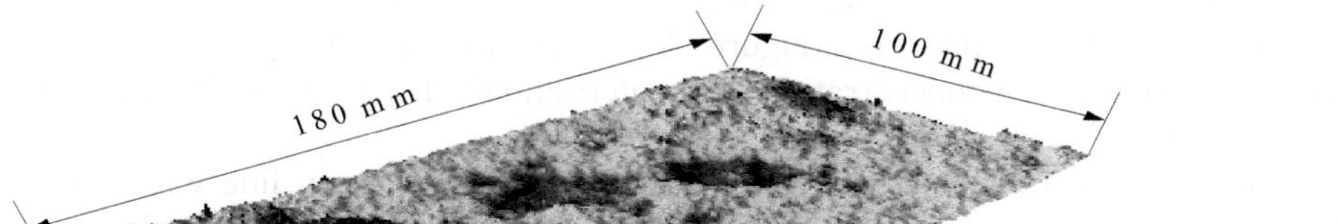

a) Three dimensional view and Profiles in x- and y- direction on upper surface

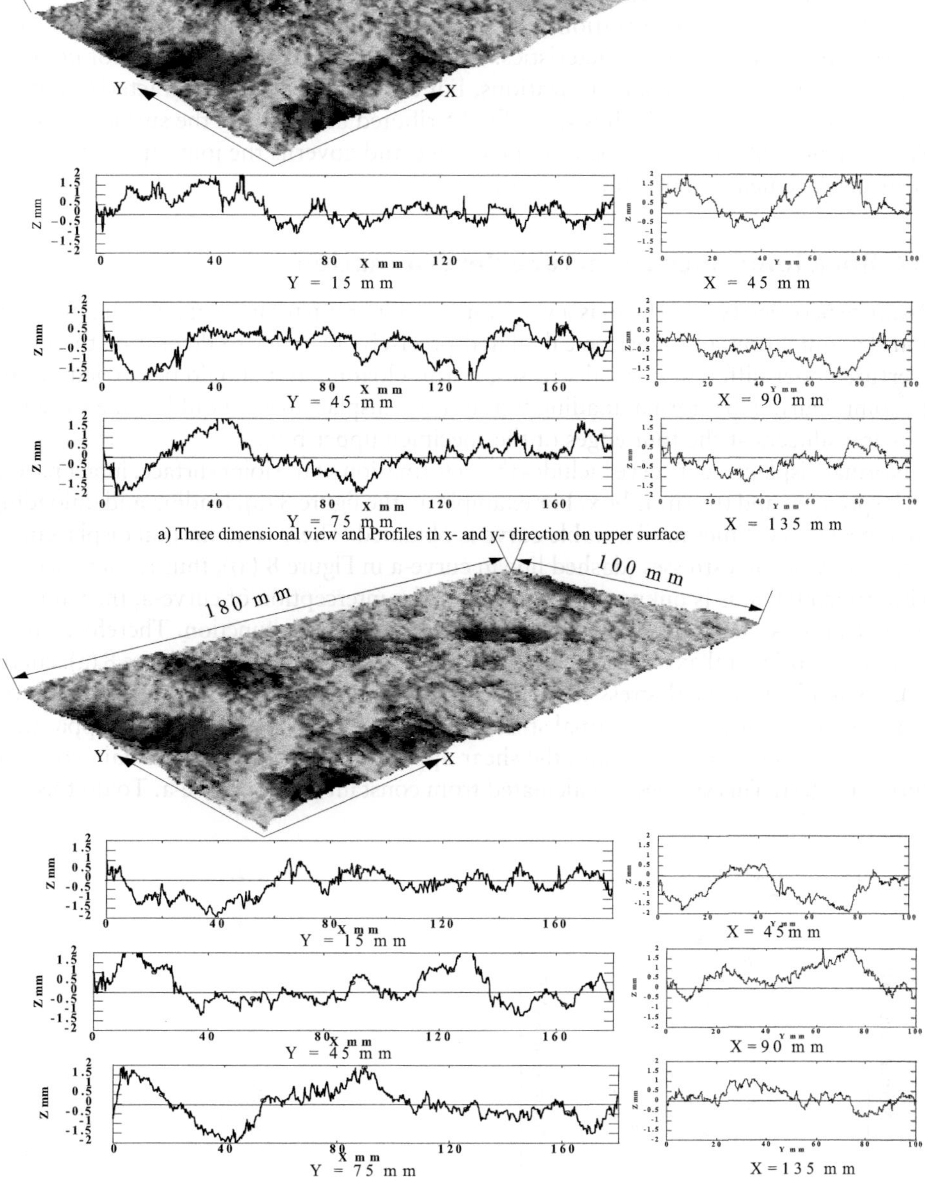

a) Three dimensional view and profiles in x- and y- direction on lower surface

Figure 7 Upper and lower surfaces three dimensional view along with profiles in x- and y- direction to show surface roughness irregularity and elements concentration on local areas over the joint surfaces (Sharifzadeh, 2005).

defined as a collection of asperities with different heights, slopes, aspects and statistical values. The surface contains a collection of elevated and depressed elements having their own statistical quantities (Figure 7), which are different from each other. The difference between concentration parameters and their spatial distribution verifies surface irregularity which can be seen in Figure 7.

On the other hand, if the changes of asperity height trace in line will give two dimensional profiles, which show roughness and undulation. However, in case of three dimensional characterizations, the spatial distribution of asperities concentration is assumed with different characteristics. Therefore, the joint surface behavior not only depends on micro scale characterizations, but also on local scale characterizations of asperity concentrations which is spatially distributed unevenly on the surfaces, thus, it plays a major role on the shape of joint surface and governs the joint aperture during shear displacement (Sharifzadeh, 2005).

3.2 Rock joint initial aperture determination

Distance between two surfaces is a very important parameter in joint aperture distribution determination and must be calculated precisely. In order to determine the initial aperture under different normal stresses, normal loading test should be carried out on the joint. During the normal loading test, normal displacement should be measured by four transducers at the four edges of the specimen upper box.

Normal displacement curve includes the deformation of the joint surface, intact part of rock specimen and the shear box. For example, in the figure 8 (a), loading and unloading are repeated few times until a stable curve is obtained. There is no normal displacement data at low normal stresses (dashed line in curve-a in Figure 8 (a)), thus the intersection with horizontal axis is unknown. To calculate the interception of curve-a, the value for normal stresses: 3MPa is selected and fitted with hyperbolic function. Therefore, interception on horizontal axis is determined and curve-a shifted to origin (Figure 8 (a): curve-a0). As applied, normal stress is increased, and the normal stiffness remains almost constant and represents the normal stiffness of the intact rock and the shear apparatus.

The stiffness of intact rock and the shear apparatus are assumed to remain constant during the test. This stiffness is calculated from constant part of curve-a. To do this, the

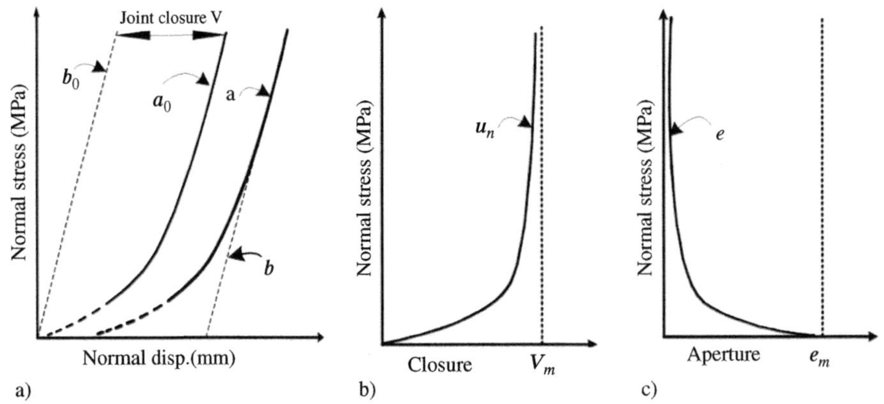

Figure 8 Procedure for joint initial aperture determination from normal loading test (Sharifzadeh, 2005).

data larger than 8MPa is selected and fitted by a line (Figure 8 (a): curve-b) and similarly curve-b is shifted to the origin (Figure 8 (a): curve-b0). After translating curve-a and curve-b to the origin, the normal deformation curve of the rock joint is obtained (Figure 8 (b)). The closure is approximated by a hyperbolic function of Bandis *et al.* (1983) as;

$$\sigma_n = \frac{u_n}{a - bu_n} \tag{1}$$

Therefore maximum closure V_m is obtained as follows;

$$V_m = \lim_{\sigma_n \to \infty} u_n = \frac{a}{b} \tag{2}$$

Where σ_n is the normal stress, u_n is the normal deformation, V_m is the joint maximum closure, a and b are sample coefficients. Initial aperture is calculated by taking the difference of normal deformation from the maximum closure. The joint aperture in proportion to normal stress is obtained as follows (Figure 8 (c)).

$$e_m = V_m - u_n = \frac{V_m}{1 + b\sigma_n} \tag{3}$$

Where e_m is the initial mechanical aperture or distance between the two surfaces at rest (in mm) and σ_n is the normal stress. For this study, Equation 3 becomes as follows (Figure 9):

$$E = \frac{0.099}{1 + 1.14\sigma_n} \tag{4}$$

The initial aperture varies from 0.0607mm at 1MPa to 0.0087mm at 10MPa. Thus, with increasing normal stress up to 4MPa, aperture decreases rapidly, only after it follows a gradual decrease up to 10MPa. This tend indicates that with increasing deformation and contact area, it becomes more difficult to obtain deformation.

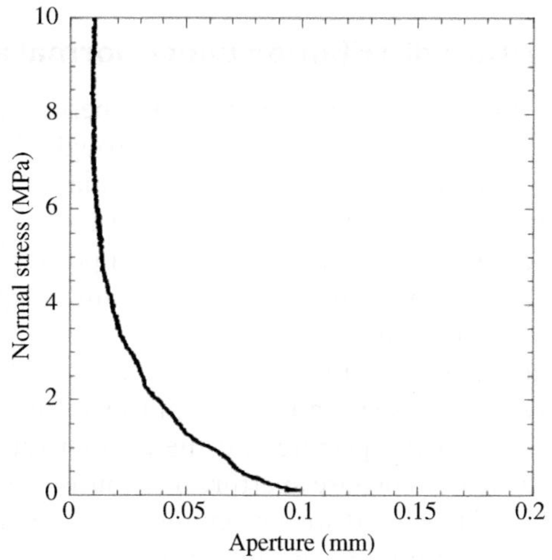

Figure 9 Initial aperture versus normal stress for study case (Sharifzadeh *et al.*, 2004).

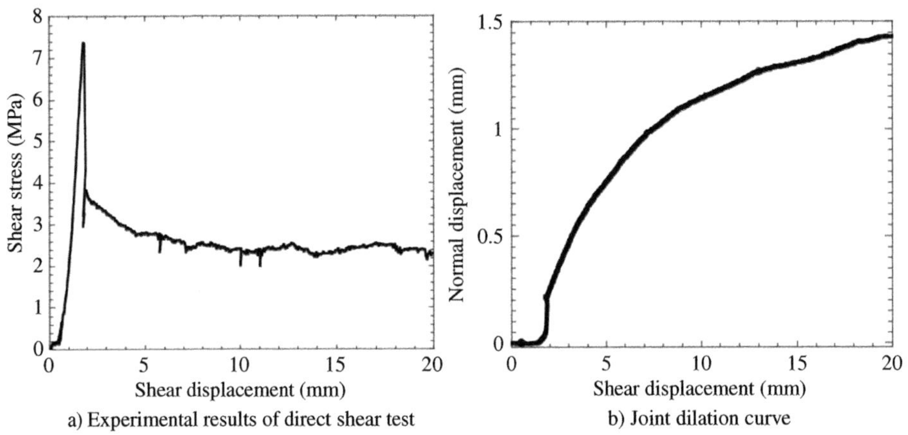

Figure 10 Shear test results under 3MPa of normal stress, showing the changes of shear stress and normal displacement (dilation) versus shear displacement.

In order to study the changes of aperture at different shear displacements, changes of dilation during direct shear test should be evaluated. Changes of shear stress and normal displacement with respect to shear displacement are shown in Figure 10 (a), (b). In this figures, a small contraction is observed up to 1.5mm of shear because of the interlocking of the asperities of upper and lower surfaces. This is followed by a sudden increase in normal displacement (dilation) and shear stress up to peak value. After that, shear stress rapidly decreases to residual stress and remains almost constant (Figure 10 (a)).

The changes of normal displacement during shear are usually used to determine the aperture at different shear displacements. As a result, the distance between the joint surfaces for aperture distribution analysis is determined from initial aperture according to normal stresses and changes of aperture at different shear displacements. These data will be used as a part of input data for aperture distribution analysis.

3.3 Rock joints aperture distribution under normal and shear load

To obtain the aperture distribution using the surface asperities height data, the lower surface data is kept fixed and the upper surface data is brought down in very small steps to form aperture. At each step, the mean aperture is calculated and compared with the mean distance between surfaces for related case. When the mean numerical aperture become equals to the mean experimental aperture, the upper surface moving is stopped and the distance between opposite asperities with 0.2 mm of square cell size at upper and lower surfaces is determined as aperture.

Cells with minus or zero aperture are assumed as contact. ASCII Grid format is used for calculating aperture at each step and Geographical Information System (GIS) is used for visualization. The mean aperture and the contact ratio are also calculated. The general procedure to achieve precise aperture distribution using joint surfaces data is illustrated in Figure 11. Aperture distribution under 1, 3, 5, and 10MPa of normal stresses are represented in Figure 12, and under 5MPa of normal stress, and shear displacements 1, 2, 5, 10 and 20 mm are shown in Figure 13.

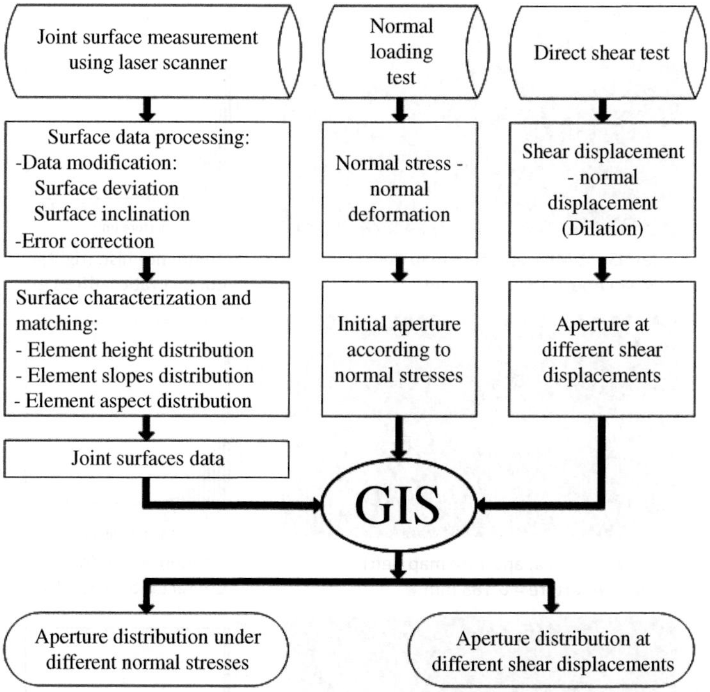

Figure 11 Procedure for determination of aperture distribution.

Figure 12 (a)–(d) shows a high rate of matching between two surfaces under different normal stresses (1–10MPa) without any shear displacement. Increasing normal stress caused an increase in the contact ratio and a decrease of mean aperture, and the whole surfaces come to closer contact. Figure 13 (c) shows aperture distribution before shear and Figure. 13 (a)–(e) shows the change of aperture distribution at different shear displacements which proves that very well matched surfaces become unmatched with proceeding shear displacements due to the rapid increase of aperture and decrease in the contact ratio. Up to one millimeter of shear displacement, homogenous distribution of aperture and contact points are found, but with increasing shear displacement, spatially localized inhomogeneous distribution is observed.

Comparison of aperture frequency histogram before and after peak shear shows that, before peak shear the aperture follows the Poisson distribution; however, after peak shear aperture becomes more similar to normal distribution especially at large shear displacement.

3.3.1 Aperture distribution at different normal stresses

Aperture distribution varies with normal stresses and could be calculated and visualized using the above mentioned method. Figure 14 shows the aperture distribution map (left), its frequency distribution, the mean aperture and contact ratio at 1, 3, 5 and 10MPa of normal stresses. In Figure 11 (a), (d), contact ratio increases from 84.8 to 98.4 percent, while mean aperture decreases from 43.97 to 7.29 microns for normal stresses increasing from 1 to 10MPa, which shows high rate of matching between two

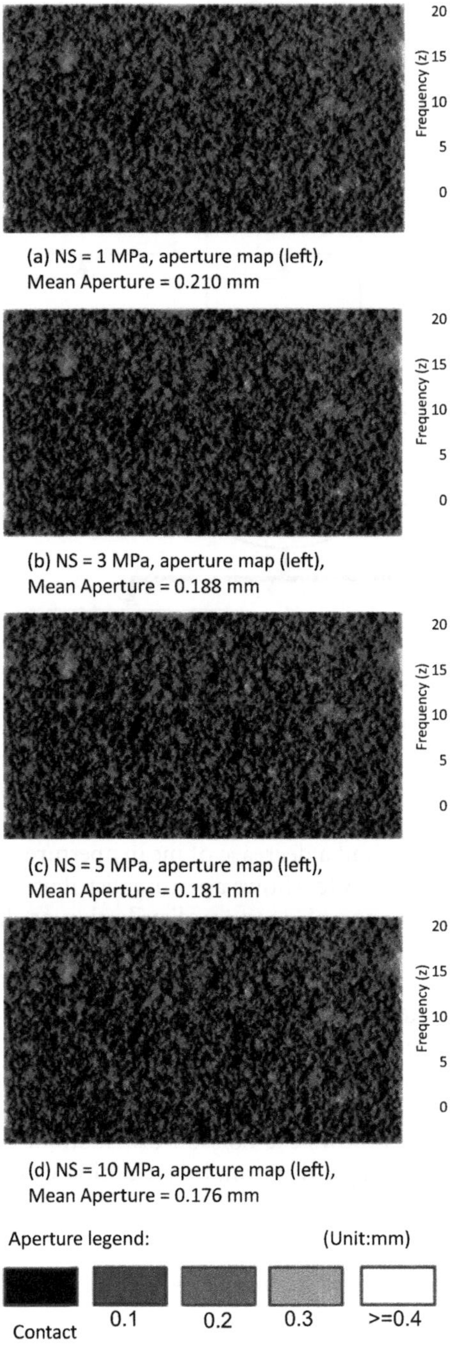

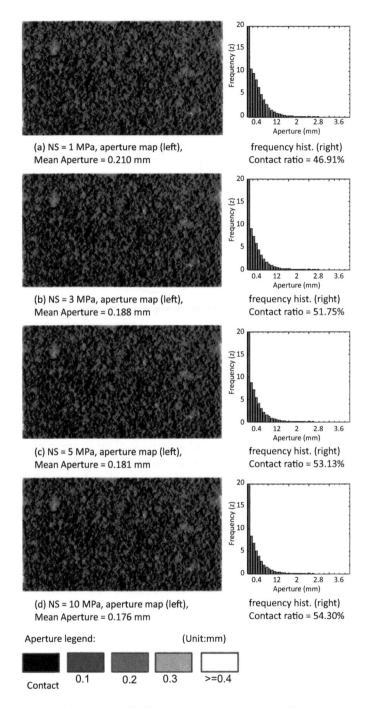

Figure 12 Aperture distribution map (left) and frequency distribution (histogram-right) with mean aperture and percent of contact ratio under different normal stresses.

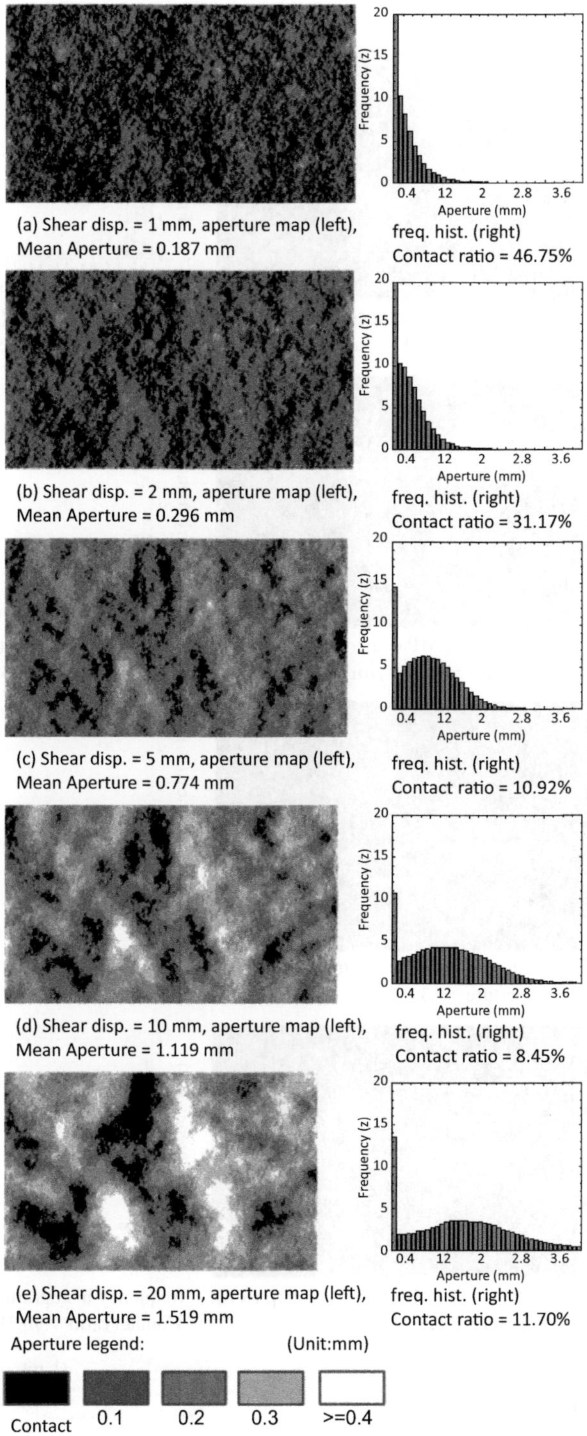

Figure 13 Aperture distribution map (left) and frequency (freq.) histogram (hist.) (right) with mean aperture and percent of contact ratio at different shear displacements under 5MPa of constant normal stress.

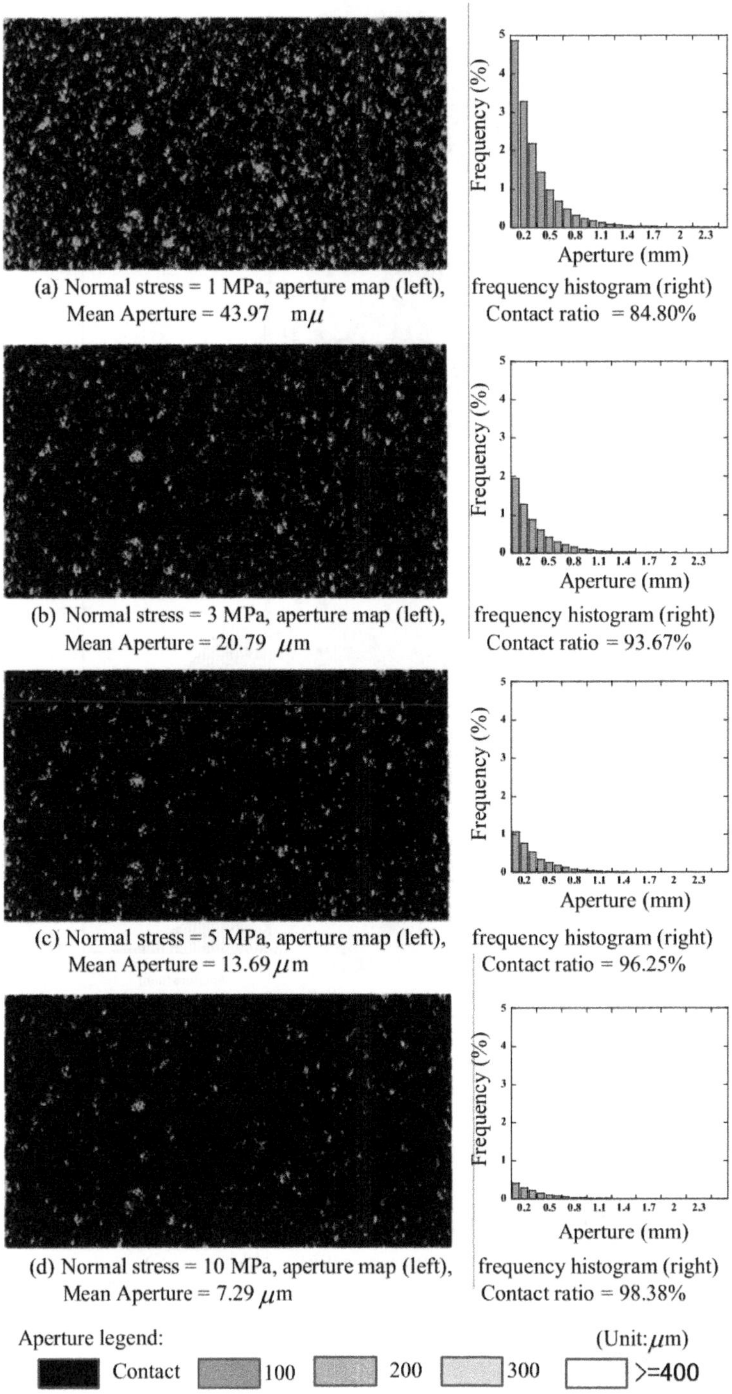

Figure 14 Aperture distribution map (left) and frequency distribution (histogram-right) with mean aperture and percent of contact ratio under different normal stresses.

surfaces and approves our surface measurement and processing technique and specify the correct aperture determination method.

The frequency distribution histogram of the aperture is presented in Figure 14 (right). Both surfaces asperities heights follow Gaussian distributions. However, the aperture frequency distribution under normal load is found similar to Poison or Log-normal distribution. Increasing of normal stresses causes drop in aperture, and increase in the contact ratio, thus results to negative change (shifting to left) in aperture frequency distribution indicating that whole surfaces cometo close contact and therefore high matching between two surfaces is found. At high normal stresses, contact ratio is almost 100% and joint is completely closed which indicate that joint behaves as intact rock. However, a small value of aperture is still remaining.

3.3.2 Aperture distribution at different shear displacements

Changes of aperture distribution during shear are illustrated by a three dimensional surface characterization based on results of shear experiments. Figure 14 shows

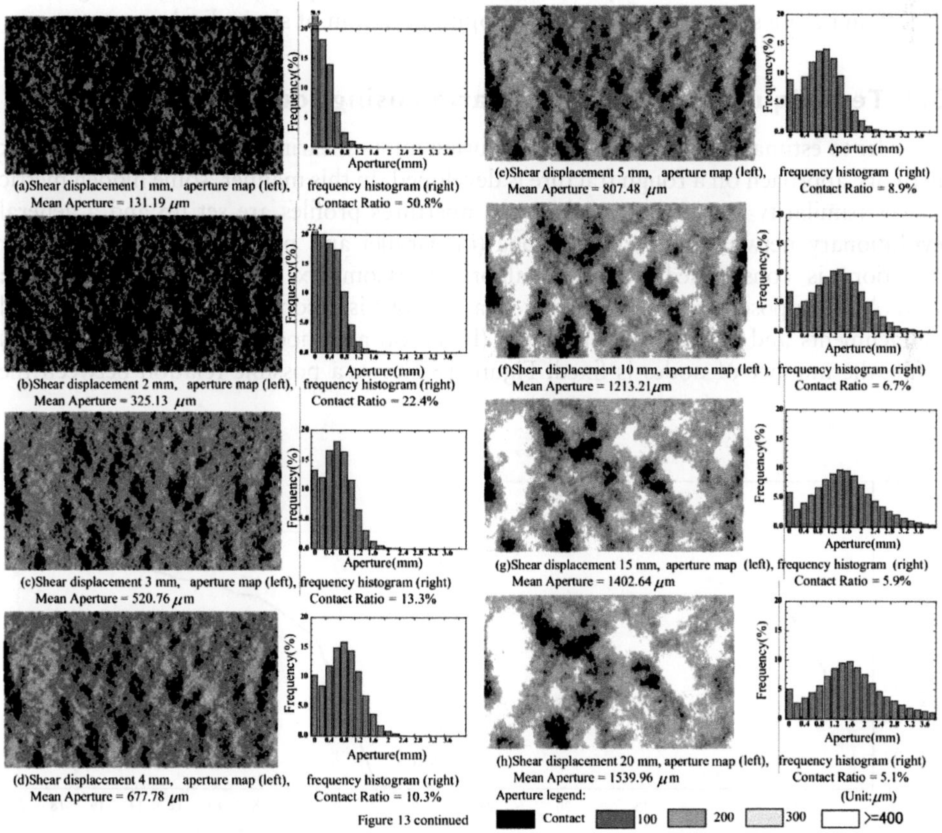

Figure 15 Aperture distribution map (left) and aperture frequency distribution (histogram-right) with mean aperture and percent of contact ratio at different shear displacements (under 3MPa of normal stress).

aperture distribution map (left) and frequency distribution (right) before shear and Figure 15 shows the change of aperture distribution map (left) and aperture frequency distribution (right) during different shear displacements.

Comparison of the aperture maps with aperture frequency distribution before shear show that contact elements are distributed equally because of well matching between two surfaces. With increasing shear, at initial sliding some asperities leave contact, thus contact ratio decreased and mean aperture increased greatly, but aperture and contact pattern still show an even distribution (Figure 15a-b). At critical point near the peak shear, an increment of shear causes simultaneous shearing of all asperities in contact. Consequently, a sudden change in aperture distribution occurs due to un-matching between two surfaces. At this stage, micro scale asperities loose contact and localized areas remain in contact. Aperture and contact are spatially localized; as a result, the ratio of the contact area decreases and aperture increases rapidly.

Finally with increasing shear displacement, dilation is controlled by next asperities in shear and consequently un-matching increases between the surfaces. Distribution of contact area spatially localized on certain asperities and mean aperture and contact ratio shows slight changes. Actually the contact ratio varies from 93.6% before shear to 5.1% at 20mm of shear displacement and also the aperture values change from 20.79μm before shear to 1.54mm (539.96μm) at 20mm of shear displacement.

3.4 Techniques of dilation evaluation using roughness data

In order to estimate the value of dilation during joints shearing displacement a complex network approach on a rough fracture is developed. In this manner, some hidden metric spaces (similarity measurements) between apertures profiles are set up and a general evolutionary network in two directions (in parallel and perpendicular to the shear direction) is constructed. Also, an algorithm (Complex Networks on Apertures: CONA) is proposed in which evolving of a network is accomplished using preferential de-tachments and at-tachments of edges (based on a competition and game manner) while the number of nodes is fixed. Figure 16 shows a possible relation between the

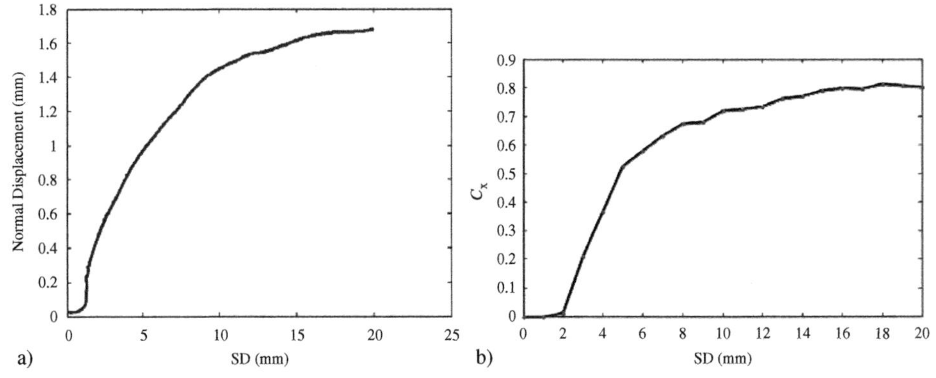

Figure 16 (a) Joint normal displacement–shear displacement, (b) clustering coefficient shear displacement (Ghaffari et *al.*, 2010).

complex networks properties and dilation properties of a rock joint which is under a constant normal stress and successive shear displacements (SD) (Ghaffari *et al.*, 2010).

Concept of Composite Surface is another way to continuously evaluate dilation during shear. In this method the dilation is defined as the average point-to-point distance between two rock joint surfaces perpendicular to a selected plane. Surface geometry is the most significant factor affecting the dilation behavior or mechanical aperture of joints. Thus through definition of composite surface, the effect of joint surface geometry on dilation behavior was investigated.

Accumulative composite surface is obtained through summation of asperities height values corresponding to the upper and lower surfaces of the joint, in different shearing steps. The lower surface is fixed and the upper surface move in shear direction step by step and the composite surface will recalculate for each shearing step.

The joint dilation is a function of joint asperities when the normal stress is low. Thus, joint dilation behavior during shearing (D_i) could be determined by Equation 5.

$$D_i = (Z_{Max(i)} - Z_{Max(0)}) \times (1 - \frac{\sigma_n}{\sigma_c}) \tag{5}$$

Where σ_n is the normal stress (MPa), σ_c is the unconfined compressive strength (MPa) and, $Z_{max(i)}$ is the maximum composite asperity height of the composite surface at each shearing step ($_{i=0}$ for initial step).

The dilation behavior (or normal behavior) of the joint is mainly controlled by first order asperities (or undulation). Therefore, using the multi-scale wave decomposition method based on wavelet theory the joint's surface geometry was decomposed and roughnesses were separated from undulations. The dilation–shear displacement curves resulted from both simulation and laboratory tests are demonstrated in Figure 17. In this figure D=0 indicates that original scanned roughness without decomposition are used for modeling and D=3 means application of three stage decomposition of roughness using wavelet (Mehrishal & Sharifzadeh, 2013).

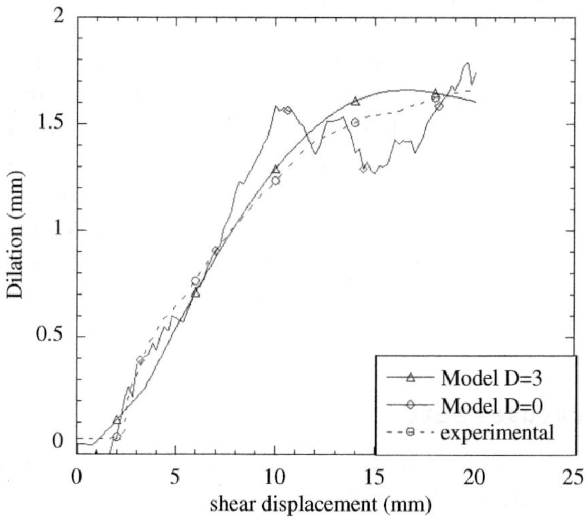

Figure 17 Comparison of dilation behavior resulted from the proposed model, before and after decomposition, and laboratory tests (Mehrishal & Sharifzadeh, 2013).

4 HYDRO-MECHANICAL COUPLING TEST

Considerable experimental works have been investigated the validity of theoretical flow laws derived from the classical parallel plate model and employed a corrective factor adjusting the relation between the mechanical and hydraulic aperture. In addition, a number of analytical studies have also been performed focusing on the variable-aperture nature of the rock joints in flow calculations. Most of the studies on coupled properties were carried out for matched joints under normal stresses evaluation of coupled properties during and/or after shearing was rarely investigated.

It has been examined theoretically that dilatancy and asperity degradation can modify the flow characteristics. However, dilatancy and asperity degradation caused by shearing and the resulting changes in hydraulic aperture and conductivity are much more complicated and little experimental work exists on the effect of shear deformation on hydraulic conductivity of rock joints due to the difficulties of tests and the development of a suitable testing apparatus was necessary to obtain the coupled properties (Esaki *et al.*, 1999).

Makurat *et al.* (1990) developed a coupled shear conductivity test apparatus with a special cell which could apply biaxial stress and pointed out that conductivity increases by several orders of magnitudes during shear. However, the testing apparatus had limited shear displacement ranges and could not apply constant normal stress. Teufel's coupling tests (1987), which were performed by using a triaxial compressive cell, showed that hydraulic conductivity across joint decreases with increasing shear deformation in soft rock because of localized deformation along the fractures and the evolution of a gouge zone. Mohanty *et al.* (1994) investigated the fluid flow through a natural rough joint as a result of both normal and shear deformations, using modified direct shear testing apparatus. The results showed that the hydraulic properties of the joint can change by up to a factor of 3 under shear deformation.

However, the numbers of measurements of hydraulic conductivity were too small to make a quantitative assessment of changes of properties during shear. Therefore, a new coupled shear-flow testing apparatus designed and developed to study the effect of joint dilatancy and shear deformation on hydraulic conductivity. This technique can perform coupled shear-flow tests on an artificial or natural joint sample under constant normal loads and variable shear deformations. The relation between mechanical and hydraulic property variations during shear- flow coupling test was also investigated. Using this device, several hydro-mechanical tests were performed under normal and shear modes with different water inlet pressures and outlet backpressures which at each case, flow tests were performed at different normal stresses and shear displacements, to observe how joint hydraulic conductivity varies with normal and shear displacement (Yow & Hunt, 2002).

4.1 Test apparatuses setup

Usually shear-flow coupling test apparatuses have three major units namely, the mechanical testing unit, the hydraulic testing unit (inlet water supply and outlet water collection), and the control and data acquisition unit. Details of each unit are briefly given here.

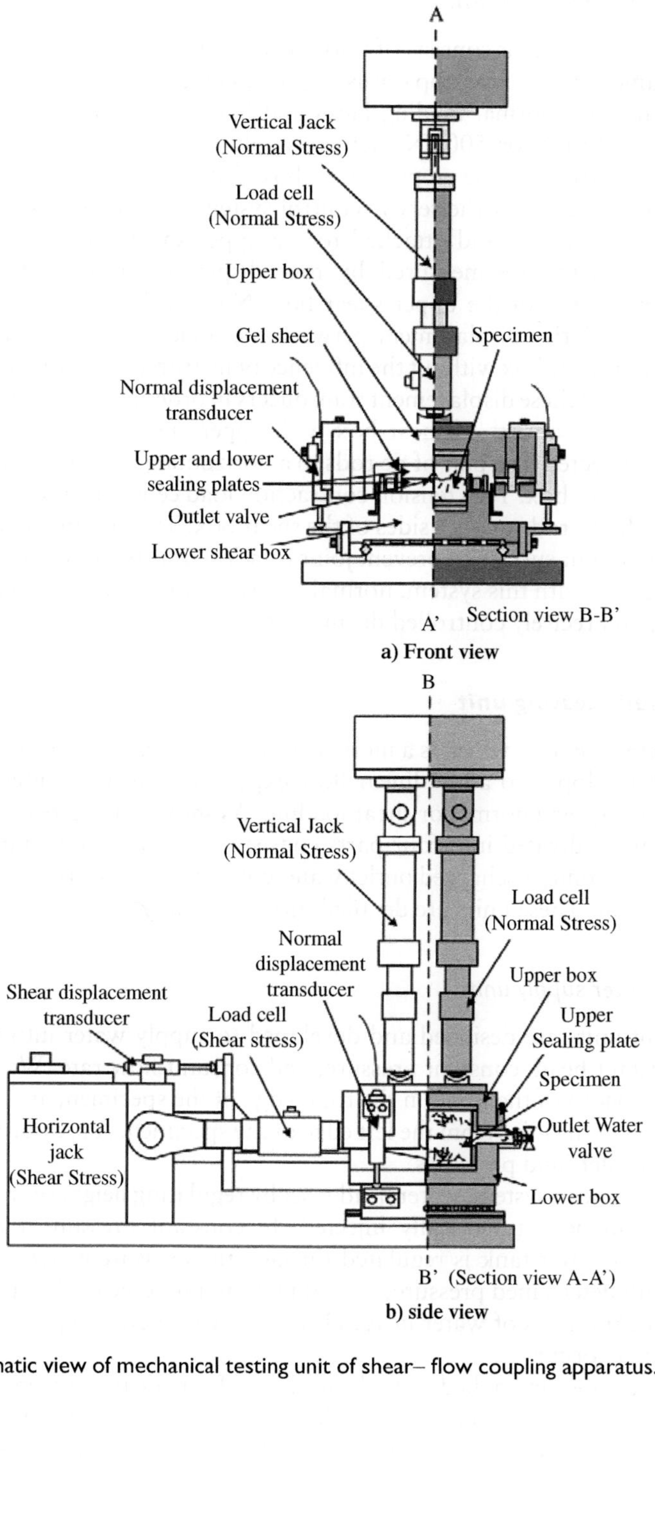

Figure 18 Schematic view of mechanical testing unit of shear– flow coupling apparatus.

4.1.1 Mechanical testing unit

The mechanical testing unit consists of normal and shear loading units. The schematic view of mechanical parts of the apparatus is given in Figure 18. This apparatus has two normal tie rods with normal loading jacks and load cells (tension/contraction dual types, capacity 200 KN or 500KN each) to apply low to high normal stresses and control the inclination of the upper shear box. The normal loading jacks can be controlled independently by the servo– control system. The tie rods are connected by pins to the loading frame and attached to the upper shear box at the pinned end. Normal displacements are measured by four displacement transducers, each one attached to one corner of the upper shear box. Normal displacements are precisely measured as the relative displacement between the upper shear box and target plate fixed to the lower shear box without the influence of instrument deformation. A roller is installed at the tip of these displacement transducers in order to measure it during shear.

Direct shear unit consists of shear box with upper and lower parts, in which the upper part is connected by a pair of tie rods to a horizontal jack. The joints are sheared by moving the lower box. Two tension contraction load cells used for measuring shear load are set in the tie rods to both sides of the shear box. The horizontal jacks are set at the same level as joint surface to prevent joint rotation or produce momentum on joint surface. Therefore, with this system, normal and shear load can be directly applied on joint surface and precisely controlled during test.

4.1.2 Hydraulic testing unit

The hydraulictesting unit, which is a modification of the mentioned mechanical unit, is designed and developed to allow linear flow experiments to be conducted while the rock joint is undergoing normal or shear loading. As shown in Figure19, the hydraulic testing unit can be divided into four parts consisting of inlet water supply units into joint, joint sealing unit, discharged outlet water collection and measurement units, and back pressure unit. These units are detailed further hereafter:

4.1.2.1 Inlet water supply units

Three different units are designed and developed to supply water into the joint consisting of constant head, constant pressure, and constant flow rate. Water is supplied over the entire width of the joint on the right edge of the specimen, and collected over the entire width of the joint from the left edge of the specimen. A pressure gauge is used to measure the inlet fluid pressures.

In the constant head system, water head is set by regulating height of water tank from which water can be automatically injected. In constant pressure, the quantity of pressure on inlet water tank is regulated through the pressure gauge, and water can be injected with determined pressure, according to required head. The pressure can be applied up to 250meters of water head. However, with increasing water head turbulence in flow may occur.

Therefore a balance in applied water head should be made to overcome practical test problem and avoid entering flow in turbulent regime. In constant flow system, water with constant flow rate is supplied to joint and can be adjusted through flow pump

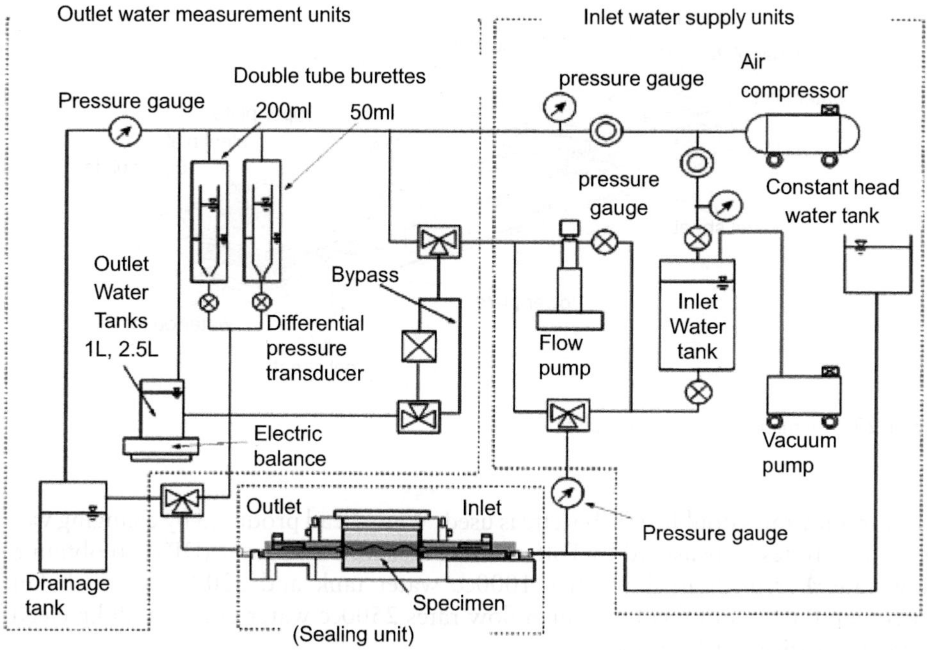

Figure 19 Schematic view of hydraulic testing unit.

settings. The flow pump can supply very low flow rate from 10–5 cm3/s to 10–1 cm3/s. Therefore water can be supplied into joint by different ways depending on joint boundary condition to achieve results close to natural condition.

4.1.2.2 Sealing unit

The hydraulic conductivity testing unit has been designed and developed to allow linear flow test. The mechanism of sealing the joint is shown in Figure 20. A special rubber material, called 'gel sheet', is placed on all sides of the jointed specimen to prevent water leakage during shearing. This rubber material is attached to the steel bars, which are used to push the side of jointed specimen when the hydraulic conductivity test is conducted. The small accumulated space for water injection is covered with rubber plate, lower plate and two side steel bars. The contacted parts of this plate are glued to the special rubber material. With high quality of sealing unit, it is possible to measure low conductivity with high precision.

4.1.2.3 Outlet water collection and measurement units

Water is collected after passing through joint discharged from specimen to outlet system by sealed pipe. Depending on the outlet water quantity, two units consisting of double tube burette and water tank with four devices (50 cc burette, 200 cc burette, 1000 cc and, 2500 cc water tanks) with two electric balances are designed and developed to measure very low to very high amount of discharged flow rate through joint.

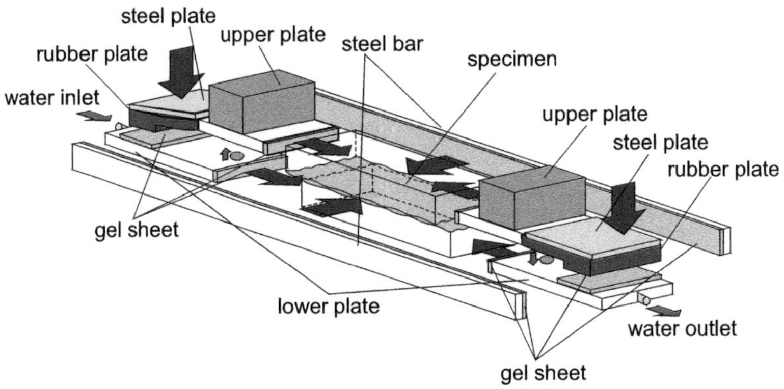

Figure 20 Schematic view of sealing units.

For low flow rates double tube burette is used. Water head produced by changing water level in burettes is measured by Differential Pressure Transducer (DPT) to obtain the flow rate. For medium flow rates 1000cc water tank and 3200g electric balance (precision 0.01g) and finally for high flow rates 2500cc water tank and 20kg electric balance (precision 0.1g) is used.

4.1.2.4 Back pressure unit

One of the major features of present flow system is its capability to apply back pressure from outlet water. This is more similar to natural conditions which are able to apply different heads on both sides of the joint. Back pressure can be applied through pressure gauge on outlet units such as burettes, outlet tank and drainage tank as shown in Figure 19. Likewise, back pressure can be regulated manually and checked by both dial gauge and the data logger.

4.1.3 Control and data acquisition units

Normal and shear loadings are controlled by two hydraulic servo-control systems. Therefore, a computer controlled system is installed in the testing equipment, which was originally developed using LabView 5.1 (National Instrument Corp.). The apparatus, mechanical normal and shear loading are reproduced by digital closed loop control with electrical and hydraulic servos as shown in Figure 21.

In this system, the shear load jack is controlled by shear displacement and two normal load jacks are controlled by normal load or normal displacement. Therefore, shear test can be carried out under constant normal load, constant normal displacement and constant normal stiffness conditions. Normal load and shear load are applied by a closed loop control system using a computer. During measurement, it has monitoring function, while the graph is visually displayed for all data during the test. The data is recorded for all sensors for fixed time, fixed displacement and load intervals.

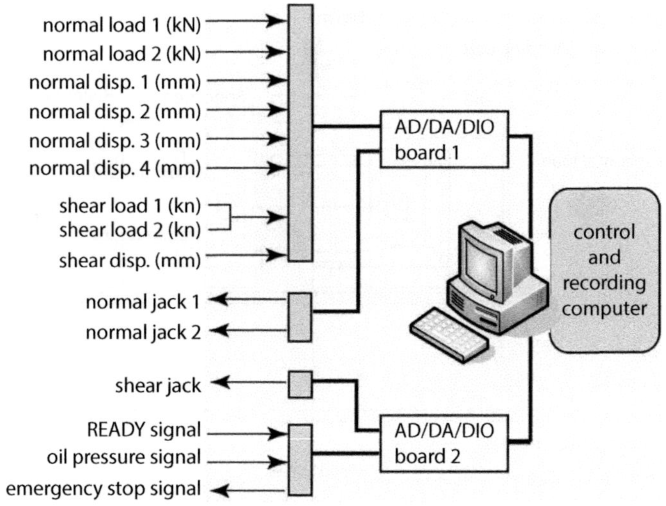

Figure 21 Digital closed loop control system by electrical and hydraulic servos.

Inlet water pressure and back pressure are controlled manually and checked with both dial gauge and pressure gauge (in data logger) with 2 kgf/cm^2 measurement capacity and 1.5 $^{mV}/_{V}$ (PW-2, Tokyo sokki Co.). The variation of water level in burettes is measured by DPT and recorded in data logger within defined time intervals. The data from inlet and outlet pressure gauges, DPT, and room temperature is recorded by data logger. For outlet water, the flow rate is measured by electric balance. For this purpose, two electric balances with small or big scale are individually connected to a computer and the flow quantity with defined time intervals is transferred to the computer. Flow rate is calculated simultaneously and flow rate–time curve is drawn on screen and saved on a personal computer.

4.2 Specimen preparation

Both of artificial and natural rock joint can be tested using the direct shear test apparatus. Usually an artificial rock joint is used in order to raise the reproducibility. An apparatus is developed for creating artificial joint as shown in Figure 22. Firstly; an intact rock block is set in a steel guide box. For steady creating joint in a specimen, after applying a certain normal load (10MPa) to intact rock specimen, a constant horizontal load for splitting is applied through a pair of wedges. Then the normal load is gradually decreased during fracturing while the horizontal load is kept constant. The wedge penetrates into the specimen, and the split is extended by tensile failure. Thus the specimen is fractured smoothly in a stable manner without causing violent vibrations and crashing. Furthermore, a couple of saw-cut slits (width 1 mm, depth 4 mm) are made in the front and backside of the specimen. Steel plates of 6mm thickness are attached to these faces in order to prevent the failure of the edge by shear loading. Finally, a shear plane with 180 mm in length, by 100 mm in width is obtained for the test.

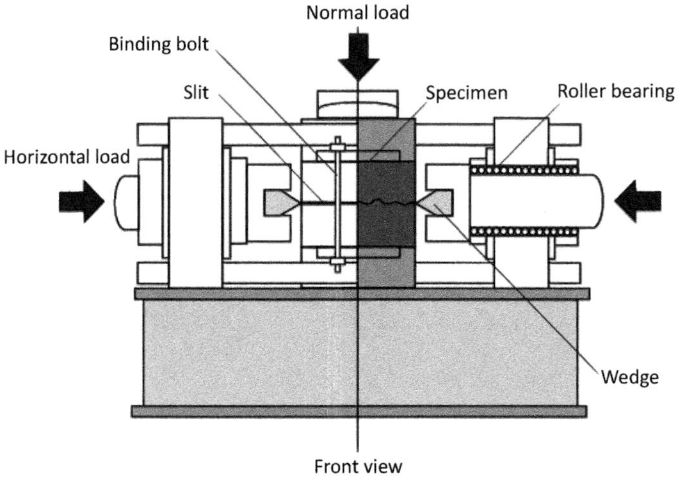

Figure 22 Artificial joint sample creation apparatus.

4.3 Normal stress-normal deformation-hydraulic conductivity

Normal loading up to 10MPa and unloading until 1kN are performed several times to obtain joint matching and determination of joint closure behavior. Normal loading test results are used for initial aperture calculation under different normal stresses which are shown in Figure 23 (a). By increasing normal stress, Joint's initial aperture decreases rapidly from 0.14 mm (at zero normal stress)to 0.03 mm (at 4MPa of the normal stresses). Flow tests with 0.3 kgf/cm^2 (3 m) and 0.6 kgf/cm^2 (6 m) of inlet water pressure are performed at 1, 3,5,7,10MPa of normal stresses. Figure 23 (b) shows the variation of hydraulic conductivity with increasing normal stress, revealing the reduction of hydraulic conductivity with increasing the normal stress. Comparison of Figure 23 (a) and Figure 23 (b) shows that both initial aperture and hydraulic conductivity decrease rapidly up to 4MPa of normal stresses. Then even if initial aperture remains almost constant, hydraulic conductivity decreases. This can be due to residual flow at high normal stresses.

4.4 Shear stress-shear displacement-hydraulic conductivity

The hydro-mechanical coupling during shear relates shear stress, normal displacement, and hydraulic conductivity with shear displacement. The shear stress – dilation – conductivity with respect to shear displacement under 5MPa of constant normal stresses is shown in Figure 24. In this figure, at first shear stress shows a quick rise up to 8.3MPa at 1.75 mm of shear displacement, followed by a gradual decrease in softening region up to 3mm of shear displacement and finally falls to residual shear stress about 4MPa. During residual region, shear displacement is continued with the stick–slipphenomena. Subsequently, normal displacement (dilation) during shear is shown in Figure 24 (a), indicating an initial slight contraction, followed by rapid increase in dilation, which is finally pursued by gradual increase in residual region. The maximum rate of increase of dilation occurs at about the peak shear stress, after

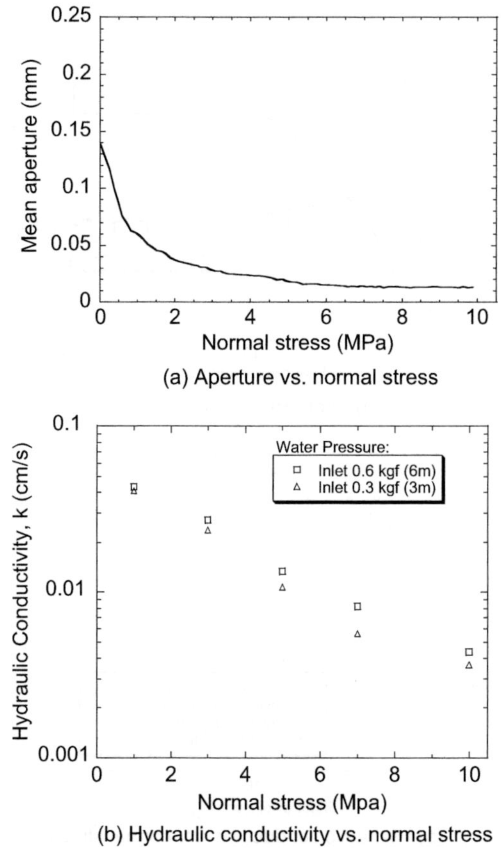

Figure 23 Changes of joint hydraulic conductivity and aperture with respect to normal stress for different inflow water pressures.

that the rate decreased steadily in residual shear. Final normal displacement is 1.35 mm at 20 mm of shear displacement.

As shown in Figure 24 (b), hydraulic conductivity is determined for different inlet water pressures and outlet back pressures. In pre-peak shear region, normal displacement shows slight contraction which causes interlocking between asperities. Hydraulic conductivity also decreases about 0.8 orders of magnitude in this region which is due to good joint matching and reduction of flow pathways. At about peak shear displacement with overriding asperities on each other, dilation increases which causes joint unmatching, increasing aperture, and opening flow pathways resulting in rapid increase in hydraulic conductivity up to 3 orders of magnitude. With the increasing of shear displacement in residual shear region, even though dilation increases, hydraulic conductivity remains almost constant. It could be due to significant increases of aperture and reduction of joint roughness effect of flow path and changing the flow condition. Therefore, hydraulic conductivity during shear is highly affected by joint surface and aperture, matching, contact points and flow path.

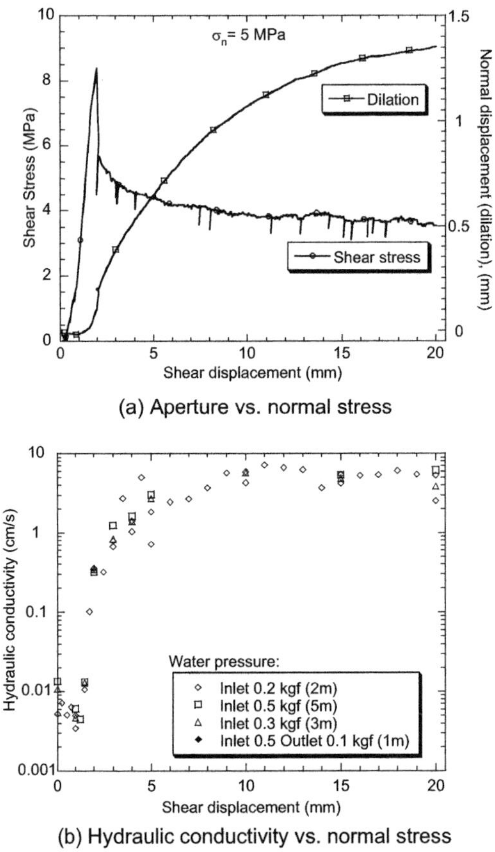

Figure 24 Variation of joint shear stress, normal displacement and hydraulic conductivity, with respect to shear displacement under 5MPa of normal stress.

There is no significant discrimination between hydraulic conductivity with different water inlet pressures. However, the resulting hydraulic conductivity with applying back pressure is almost lower than other cases. It can be concluded that it is difficult to explain the difference between hydraulic conductivity results with different inlet pressures during shear process, while it can be clearly understood from hydraulic conductivity at different normal stresses.

5 FLOW SIMULATION THROUGH THE ROCK JOINT

5.1 Principals of mathematical formulation and computer code

To simulate fluid flow through rock joint, the joint is assumed as a grid mesh for which, aperture in each element of mesh is denoted by d(x,y). Each element in joint has its own transmissivity based on its aperture. Transmissivity of the elements in contact is assumed as transmissivity of intact rock (which is about 10^{-9} cm/s). The procedure for flow simulation using aperture distribution of joint is shown in Figure 25.

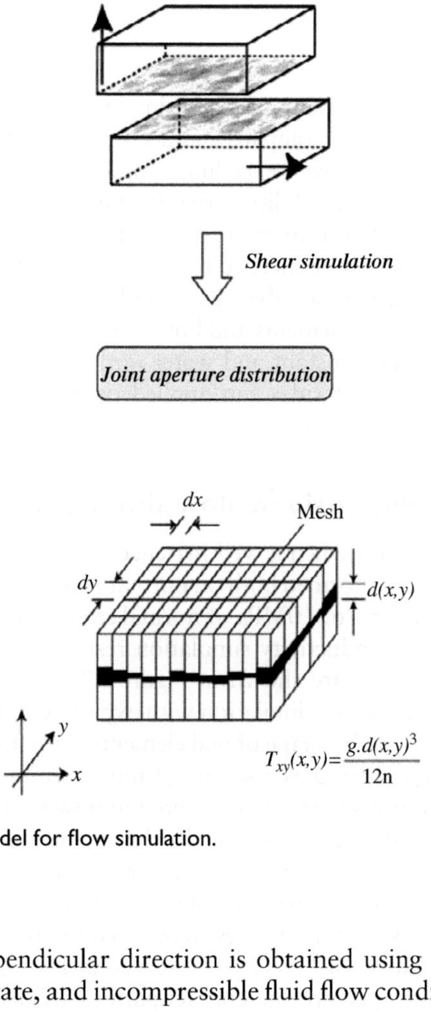

Figure 25 Joint aperture model for flow simulation.

Flow velocity in perpendicular direction is obtained using basic equation of flow under laminar, steady state, and incompressible fluid flow conditions. A basic equation is as follows.

$$\nabla\left(T_{xy}(x,y)\cdot\nabla h\,(x,y)\right)=0 \tag{6}$$

Where $T_{xy}(x,y)$ is water transmissivity, and h (x,y) is water head. The transmissivity of each element in joint mesh grid is calculated using the basic cubic law as follows:

$$T_{xy}(x,y)=\frac{g\cdot d(x,y)^3}{12v} \tag{7}$$

where g is gravity acceleration, d(x,y) is aperture in element coordinated (x,y), and v is the coefficient of kinematic viscosity. The kinematic viscosity of water is set to the values of experimental findings, for instance at 20°C, it is 1.004×10^{-6} m^2/sec. Substituting Equation 7 for Equation 6:

$$\nabla\left(\frac{g\cdot d(x,y)^3}{12v}\cdot\nabla h\,(x,y)\right)=0, \tag{8}$$

Equation 8 is solved by finite element method using underground water infiltration style analysis code FE-FLOW (WASY company), and simulated in this section. FE-FLOW is used because of its high-speed solver, excelling in operating in the input of data and the output of the result, etc. Its compatibility with GIS makes it suitable for flow simulation. Flow simulation is performed using aperture distribution data. Different mesh sizes result in different computer running times needed for simulation. It was found that simulation with one-millimeter mesh size is suitable. Two kinds of boundary conditions are used. The schematic view of the joint elements and boundary conditions is shown in Figure 26. The water inlet head can be constant and water outlet head can be free (atmospheric pressure) and zero flux at joint sides can applied or the same values obtained from experiments are applied for water inlet and outlet heads.

5.2 Results of two-dimensional flow simulation

To understand flow condition when a well-matched joint undergoes normal and shear loading, flow simulation is performed both on normal and shear modes. Flow simulation during normal loading under 1 to 10MPa of normal stresses and water inlet head of 6m are performed. Hydraulic conductivity simulation results for all normal stresses compared with experimental data are shown in Figure 27 and it reveals a big difference between them. The discrepancy of simulation with experimental results can be explained considering the differences in the shape of real elements compared to simulated elements shape. Each element is considered as a square (1mm×1mm in size), however asperities shape in real rock joint is not square as considered in flow simulation. However, in real tests, flow passes through such voids. Especially during normal loading, contact ratio is very high and causes more deviation of simulation results from experimental values. In spite of the discrepancy between experimental and simulation results, general trends of results indicate that increasing normal stress tends to decrease flow rate.

A brief result of simulation of Case 2 is presented in Figure 28. These figures show that before peak shear the aperture and contact pattern are distributed homogenously.

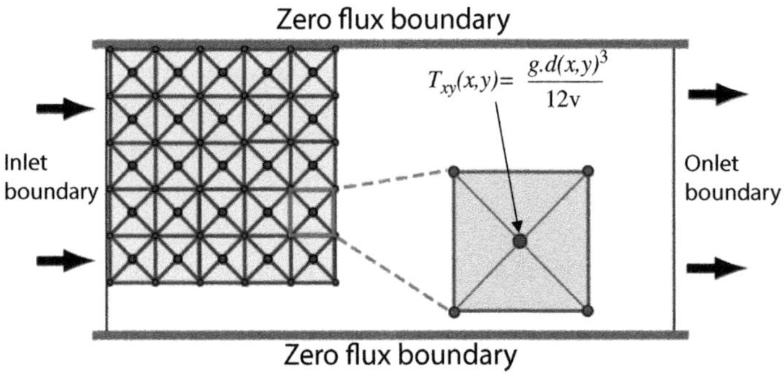

Figure 26 Modeling finite element mesh and boundary conditions.

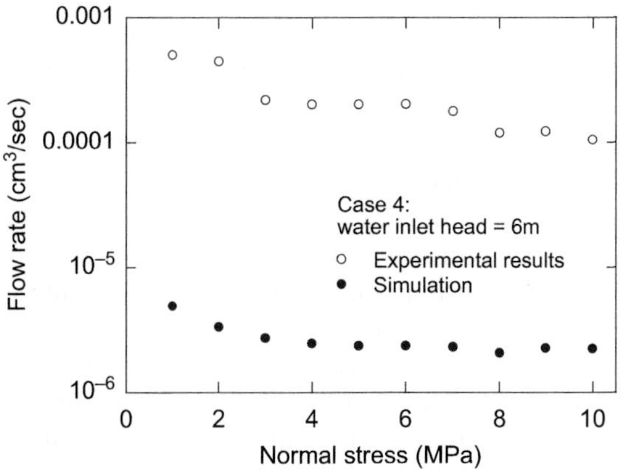

Figure 27 Comparison of flow rate between experimental results and simulation results with 6m of water inlet head for Case 4.

Water head value is also distributed uniformly in proportion to distance from the joint inlet side, and uniform flow occurs over the joint void spaces. After peak shear, the ratio of contact decreases and shear is localized to spatial concentration of elevated asperities; hydraulic head decreases and it is also localized around contact areas. Flow velocity field shows different behavior at pre-peak shear compared to after peak shear displacement. During pre-peak shear, flow velocity is low and well distributed over

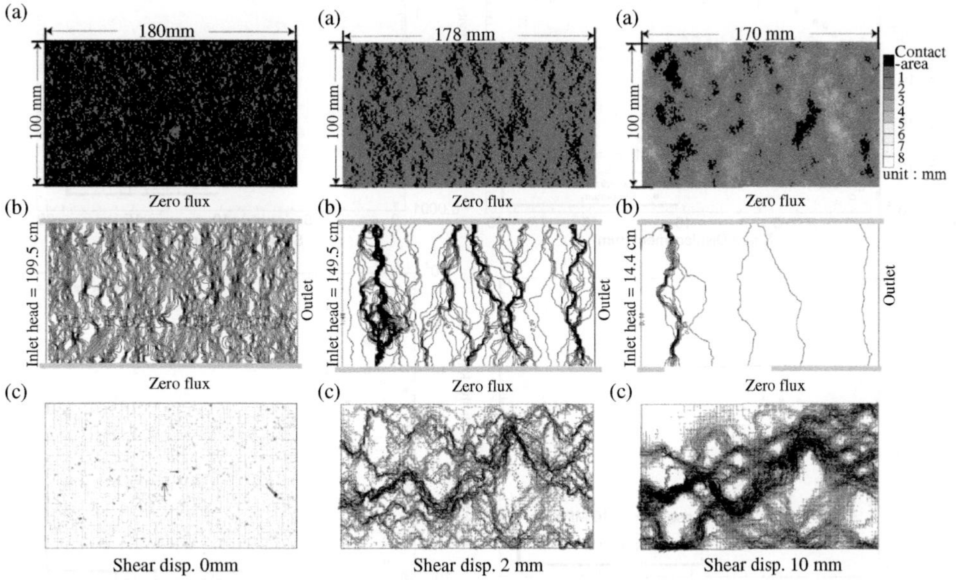

Figure 28 Flow simulation results at shear displacement = 0 mm, 2 mm and 10 mm, under 5MPa of normal stresses, (a) Aperture Distribution, (b) Contour of Hydraulic head, (c) Velocity Vector (Mitani *et al.*, 2003).

joint surface. After peak shear, fluid flow through multiple pathways and is diverted to channels around contact areas, and its velocity in channels is higher than the other areas. Flow path is dictated by void accumulations, which are not linear. Therefore, fluid flows through the tortuous paths. Thus, flow channeling and tortuosity can be at 10 mm of shear displacement. It can be concluded that the flow does not always obey the cubic law at all shear displacements.

Following the head contour lines after peak shear shows several local concentrations of head contour lines, which indicate cyclic strips of contact areas over joint which may be due to local matching and un-matching of joint surfaces during shear. Water velocity field maps during shear show that before peak shear, velocity field concentrated along aperture near to contact areas. After peak shear, water velocity is distributed over all joint surfaces. This is due to joint dilation and continuity of flow pathways. Flow paths are governed by spatially localized contact and aperture distribution patterns, which cause tortuosity and flow channeling. Turbulent flow may also locally occur at channeled flow area.

Hydraulic conductivity is calculated using flow rate obtained from simulation for Case 1 to Case 3. The simulation results for each case at all shear displacements are shown in Figure 29. Comparison of the results of hydraulic conductivity achieved from experiments with flow simulation shows that hydraulic conductivity decreases one

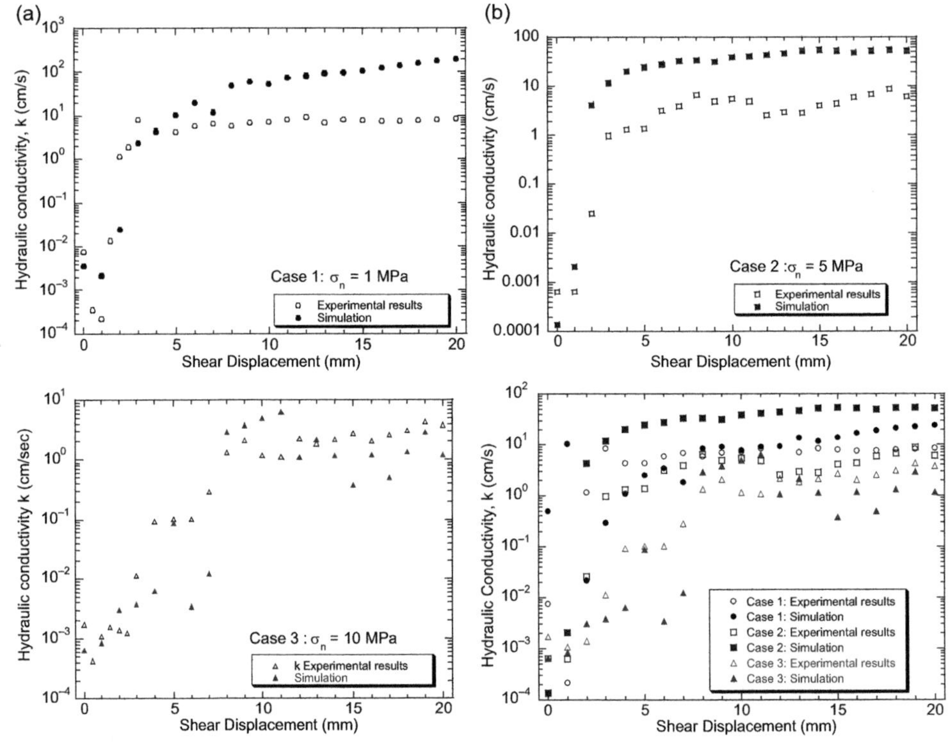

Figure 29 Comparison of hydraulic conductivity between experimental and simulation results for Case 1, Case 2 and Case 3.

order of magnitude before peak shear and increases rapidly about three orders of magnitude after a few millimeters of shear displacement corresponding to peak shear displacements. Subsequently hydraulic conductivity increases gradually up to about 10 mm of shear displacements, which corresponds to post-peak softening region. Finally, it remains almost constant in residual shear region up to 20mm of shear displacements. Comparison of Case 1, Case 2, and Case 3 in Figure 30 indicates that joint hydraulic conductivity descends form Case 1 to Case 3 because of increasing normal stress and reduction of joint aperture.

Overall comparisons of simulation with experimental results show that flow simulation results are relatively close to experimental results. However, about one order of magnitude difference can be seen between experimental results and simulation. Following the water velocity maps show that velocity distribution inside joint varies dramatically from point to point, although overall flow rate seems to be constant. In addition, on one hand, the effect of parameters such as differences between the shape

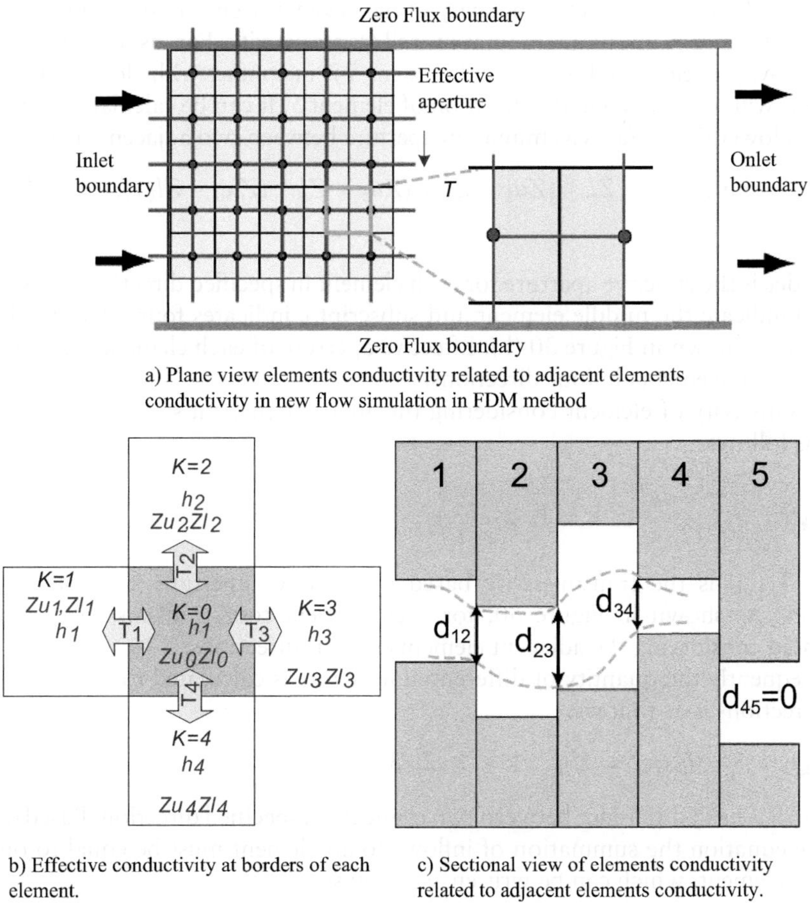

Figure 30 Schematic view of flow simulation considering the conductivity of each element in relation with conductivity of adjacent elements.

of real and simulated asperities, flow channeling, tortuosity, and three-dimensionality which are not considered in simulation may be cause of difference. On the other hand, similarity of simulation results compared to experimental data does not mean that all effective parameters are considered in simulation.

5.3 Flow simulation using effective aperture

In flow simulation using FEM method, surface is divided into grid mesh, and transmissivity of each element in mesh is calculated individually. By assembling the transmissivity of all elements using FEM, the water flow rate, head and velocity distribution can be obtained as described earlier. One of the major problems in FEM method is considering the transmissivity of elements separately. To improve previous simulation, new flow simulation method is developed using the concept of the effective aperture and transmissivity of each element considering the effect of four adjacent elements in four directions using three-dimensional aperture distribution. The basic formulation of the newly developed flow simulation is described in the following paragraphs.

In general, not only the transmissivity of the element itself but also transmissivity of adjacent elements is important in joint overall transmissivity. For instance, in Figure 30, transmissivity of element 4 is high and element 5 is in contact with element 4. It means that flow will not happen in the direction of element 5. It can be calculated using upper (Z_u) and lower (Z_l)surfaces as minimum aperture between two adjacent elements:

$$de_k = \min\{(Zu_0 - Zl_0), (Zu_k - Zl_k), (Zu_k - Zl_0), (Zu_0 - Zl_k)\}, \quad \mathrm{k} = 1,\ 2,\ 3, \tag{9}$$

Where de_k is the effective aperture for each element in specified direction, the subscript zero ($_0$) indicate the middle element and subscript $_k$ indicates four adjacent elements elevation as shown in Figure 30 (b). Effective aperture of each element is calculated in four adjacent elements in four directions.

Transmissivity of element considering the adjacent element's conditions are calculated as follows:

$$T_k(i,j) = \frac{g \times de_k^3}{12 \cdot v}, \quad \mathrm{k} = 1,\ 2,\ 3,\ 4 \tag{10}$$

Where $T_k(i,j)$ is the transmissivity based on effective aperture (de_k) between two elements. As shown in Figure 30, for each element four transmissivity values are calculated considering the adjacent elements in four directions.

Subsequently the quantity of differential flow rate is calculated for each element in four directions is as follows:

$$\Delta Q_k(i,j) = T_k(i,j) \times \Delta h_k, \quad \mathrm{k} = 1,\ 2,\ 3,\ 4 \tag{11}$$

Where Δh is head difference between two elements in specified direction. Based on mass balance equation the summation of inflows to an element must be equal to outflows from an element, which can be written as follows:

$$\sum_{k=1}^{4} \Delta Q_k(i,j) = \sum_{k=1}^{4} T_k(i,j) \times \Delta h_k = 0, \quad \mathrm{k} = 1,\ 2,\ 3,\ 4 \tag{12}$$

By engaging the concept of Finite Difference Method (FDM) the water head at each element is calculated using following equation.

$$\left[\left(T(i,j,1)+T(i,j,2)+T(i,j,3)+T(i,j,4)\right)\cdot h(i,j)\right]-\left[T(i,j,2)\cdot h(i-1,j)\right]$$
$$-\left[T(i,j,2)\cdot h(i,j-1)\right]-\left[T(i,j,3)\cdot h(i+1,j)\right]-\left[T(i,j,4)\cdot h(i,j+1)\right]=0 \qquad (13)$$

The in-house developed computer code is used to solve the above equations. Equation 13 is a linear equation for each element in four directions. The system of equations is solved by Successive Over Relaxation (SOR) method, and output is stored in computer. The accuracy of calculation closely depends on the value of assumed allowable error (0.00001). Basically, inflow water to joint must be equal to outflow water from joint. However, considering allowable error in SOR method of matrix solution makes some differences between inflow and outflow. It is set to make some balance between reasonable program running time and reliable flow simulation. Flow simulation is performed during shear for Case 4 using this method and results are compared with experimental results and FEM simulation results, Figure 31.

With comparison of both FEM and new flow simulation technique results, it is found that new method gives close results to experimental data, especially in residual shear region. The benefit of new technique compared to two-dimensional FEM is its capability to consider the effective aperture.

The major reasons for differences between experimental and simulation results are: (i) the real experiment is in three dimension, however simulation is performed under plane (two dimensional) conditions, (ii) water flows through tortuous paths and is diverted to channels which are not considered in simulation, and (iii) in both simulation

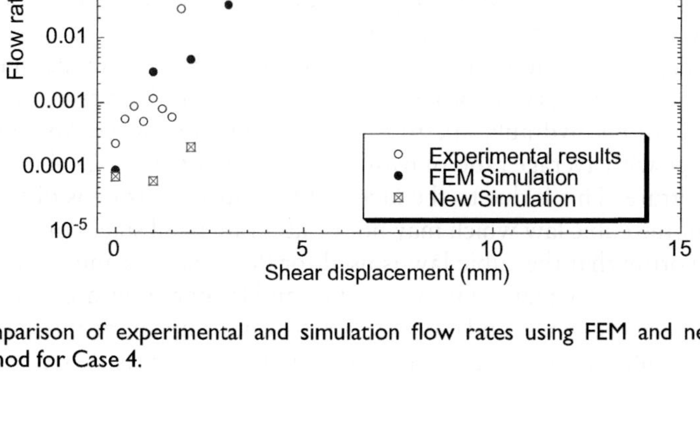

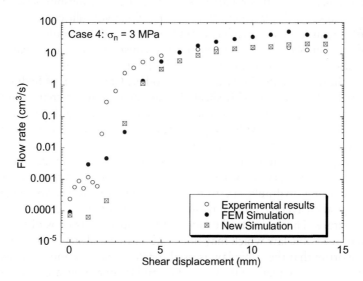

Figure 31 Comparison of experimental and simulation flow rates using FEM and new simulation method for Case 4.

methods elements are considered in square shape, but in reality they may posse other shapes such as circle, ellipse and irregular.

6 DISCUSSION

Although hydro-mechanical coupling properties of rock joints are very important, it is technically difficult to perform laboratory tests coinciding to real field conditions. A new testing technique is designed and developed for joint hydro-mechanical testing, which enables us to study the effect of joint hydraulic properties under normal loading and large shear displacements. Hydraulic testing is instrumented for different flow units, allowing measurements in a wide range of hydraulic conductivity, and testing with low to high hydraulic gradient. This apparatus is used for several coupled hydro-mechanical tests with different water inlet pressures and outlet backpressures.

Test results show that during normal loading joint mechanical aperture and simultaneously hydraulic conductivity decreases with increasing normal stresses. Hydraulic conductivity changes with the increasing of normal load, approximately similar to joint mechanical aperture behavior.

Once the artificial joint is created, it is almost impossible to be closed by mechanical loading, which can be clearly seen in existing flow rate even at high normal stresses. The relationship between hydraulic conductivity and normal stresses of the joint with applying different inlet water pressures shows that increasing in inlet water pressure causes a slight increase in hydraulic conductivity.

During the shear process, shear stress versus shear displacement behavior shows that shear stress reaches a peak at about 2 mm of shear displacement then falls to constant residual shear stresses. Normal displacement versus shear displacement results shows a small contraction then there is a rapid dilation at peak shear stress. With proceeding shear, dilation increases with a relatively slow rate. The relationship between hydraulic conductivity and shear displacement shows that at first hydraulic conductivity decreases one order of magnitude and then it has a rapid increase of about 3 orders of magnitude up to 7 mm of shear displacement. It almost remains constant in residual shear region up to 20mm of shear displacement. In other words, joint dilation behavior and hydraulic conductivity shows intimate similarity to each other. Correlation of hydraulic conductivity and hydraulic aperture with shear displacement shows a trend similar to joint dilation behavior. In general, increases of water inlet pressure slightly increases hydraulic conductivity. Results of tests with applying backpressure show lower values for hydraulic conductivity in comparison to cases with similar conditions and without backpressure. Regarding the relationship between mechanical and hydraulic aperture during shear, both apertures vary in proportion to each other. However, at some critical mechanical apertures, hydraulic aperture changes in different ways. Aperture ratio varies in a wide range with average ratio of hydraulic aperture of about 2 – 4 times larger than hydraulic aperture. This is due to the fact that hydraulic aperture is obtained by back-calculation of the cubic law which may not be accurate in all stages of shear.

It is noteworthy that the cubic law is used for flow analysis and many assumptions such as parallel, smooth plates without contact and laminar flow are used. However, in real joint, surface roughness and contact area are quite important and may result in the deviation of results from the cubic law. Flow in a rock joint is very complex and it is difficult to apply parallel plate model specifically for evaluation of the conductivity of a rock joint. The main reason is the variable flow path in rock joint according to the flow conditions.

7 SUMMARY

The procedure of this research to achieve joint hydro-mechanical properties is illustrated in Figure 32. The specimen was prepared and the asperity heights of the upper and lower surfaces of the joint were measured by a 3-D laser scanning system before the test. These height data were given for each mesh as input data. By overturning the upper surface and overlaying these surfaces, the initial aperture distribution was defined. In hydro-mechanical test, first normal loading-flow tests were conducted and joint normal deformation and flow rate were recorded. Then the shear test with small increments of shear displacement along with flow tests were carried out. During the test, joint shear displacement-normal deformation (dilation)-flow rate were recorded. By combining the joint normal deformation and surface data aperture distribution at each steps were obtained and used for description of the flow mechanism. Aperture distribution was determined and visualized under different stresses and displacement conditions. Variation of aperture distribution during normal loading and shear process were quantified by using both aperture distribution map and aperture frequency distribution.

In this chapter, different parts of hydro-mechanical coupling test apparatus were described in detail and the test results using this apparatus were indicated. Experimental tests were performed under various normal loading conditions and different load and flow conditions, to evaluate the shear-flow coupling properties of rock joint. Finally, the relationships between mechanical and hydraulic properties of joint were analyzed in detail and an original flow model was developed which applies

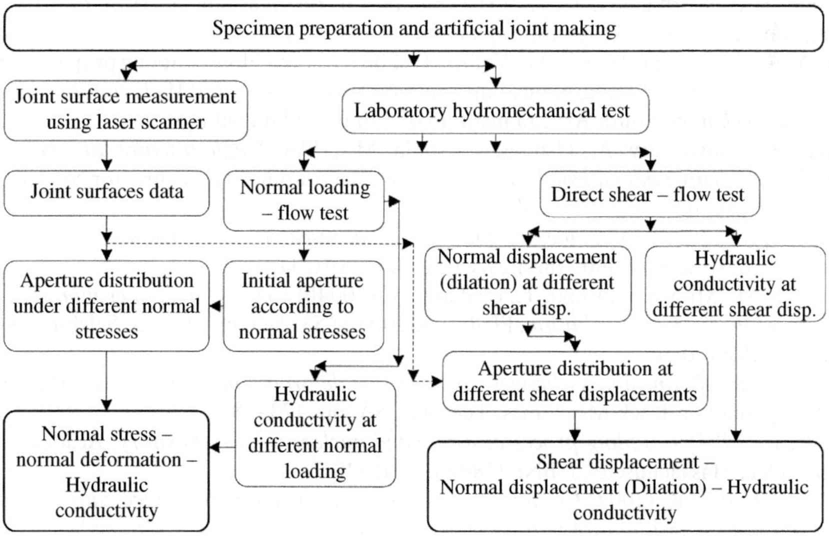

Figure 32 Procedure for hydro-mechanical test with different water head flow system.

the effective aperture instead of nominal aperture. Results show that the new flow simulation gives close results to experimental findings, especially in residual shear region. Effect of mesh size on flow rate revealed that with increasing mesh size, flow rate converges and reaches to almost same results at about 1mm mesh size after which, flow rate remains constant with increasing shear.

REFERENCES

Bandis, S., Lumsden, C., Barton N. R. (1983) Fundamentals of rock joint deformation. *International Journal of Rock Mechanics and Mining Science & Geomechanics Abstracts*, 20, pp. 249–268.

Esaki, T., Dua, S., Mitani, Y., Ikusadaa, K., Jing, L. (1999) Development of a shear-flow test apparatus and determination of coupled properties for a single rock joint. *International Journal of Rock Mechanics and Mining Sciences*, 36, pp. 641–650.

Ghaffari, H., Sharifzadeh, M., Fall, M. (2010) Analysis of aperture evolution in a rock joint using a complex network approach. *International Journal of Rock Mechanics & Mining Sciences* 47, pp. 17–29.

Giacomini, A., Buzzi, O., Ferreroa, A. M., Migliazzaa, M. (2008) Numerical study of flow anisotropy within a single natural rock joint. *International Journal of Rock Mechanics & Mining Sciences*, 45, pp. 47–58.

Grasselli, G., Wirth, J., Egger, P. (2002) Quantitative three-dimensional description of a rough surface and parameter evolution with shearing. *International Journal of Rock Mechanics & Mining Sciences*, 39, pp. 789–800.

Jing, L., Hudson, J. (2004) Fundamentals of the hydro-mechanical behavior of rock fractures: Roughness characterization and experimental aspects. *International Journal of Rock Mechanics & Mining Sciences*, 41, No. 3, Cd-Rom, Elsevier Ltd.

Makurat, A., Barton, N., Rad, N., Bandis, S. (1990) Joint conductivity variation due to normal and shear deformation. Proceedings of the International Symposium on Rock Joints, Balkema, Leon, Norway, pp. 535–540.

Mehrishal, S.A., Sharifzadeh, M. (2013) Evaluation of the hydraulic aperture of a rock joint using wavelet theory. *Geosystem Engineering*, 16:1, pp. 119–127.

Mitani, Y., Esaki, T., Sharifzadeh, M., Vallier, F. (2003) Shear – Flow coupling properties of rock joint and its modeling Geographical Information System (GIS). *ISRM–Technology roadmap for rock mechanics*, South African institute of mining and metallurgy.

Mohanty, S., Chowdhury, A., Hsiung, S., Ahola, M. (1994) *Single fracture flow behavior of Apache Leap Tuff under normal and shear loads*. San Antonio, TX: Center for Nuclear Waste Regulatory Analyses. CNWRA 94–024.

Sharifzadeh, M. (2005) *Experimental and theoretical research on hydro-mechanical coupling properties of rock joint*. PhD thesis, Kyushu University, Japan.

Sharifzadeh, M., Mitani, Y., Esaki, T., Urakawa, F. (2004b) *Development of evaluation method of rock joint aperture distribution*. EUROCK 2004 & 53rd Geomechanics, Salzburg, Austria, Balkema, Rotterdam.

Teufel, L. (1987) *Permeability changes during shear deformation of fractured rock*. In: 28th US Symposium on Rock Mechanics, Tucson, USA, pp. 473–480.

Tsang, C. F. (1987) *Coupling processes associated nuclear waste repository*. Academic Press, Cambridge, Massachusetts, ISBN: 978-0-12-701620-7.

Yow, J. L., Hunt, J. R. (2002) *Coupling processes in rock mass performance with emphasis on nuclear waste isolation*. *International Journal of Rock Mechanics & Mining Sciences*, 39:1, pp. 143–150.

Chapter 21

Thermo-hydro-mechanical couplings in radioactive waste disposal

A. Millard
CEA/DM2S/SEMT/LM2S, Gif sur Yvette, France

Abstract: This chapter presents some of the main issues related to the thermo-hydro-mechanical couplings which can be encountered in the framework of radioactive waste disposal problems and their modeling. First, the different kinds of radioactive wastes are briefly reviewed, and the deep underground disposal solution, based on a multi-barrier concept, is detailed. With regards to the rock mechanics issue, the main processes which can occur at short and long term are explained. Then, the theoretical formulation underlying the thermo-hydro-mechanical studies are fully detailed and some indications on numerical implementation are given. In addition, a specific hydro-mechanical formulation related to a well identified fracture is exposed. Finally, in order to appraise the influence of the various thermo-hydro-mechanical couplings, the study of the short term behavior of a generic high level wastes repository is proposed as an illustration of the previous developments.

1 INTRODUCTION

Since the discovery of radioactivity at the beginning of the 20th century, there has been an increasing use of radioactive substances and nuclear energy throughout the world. Like many others, this activity produces wastes which must be properly handled, in accordance with the general rules regarding the treatment of wastes (NEA, 2010). In fact, nuclear wastes are radioactive substances for which no further use is presently envisaged. These radioactive substances contain radionuclides, which can be artificial as cesium 137 or natural as radium 226. They emit radiations and therefore present a specific risk for human health. As such, they must be carefully managed from the location where they are produced until their final destination (IAEA, 1995).

The major part of the radioactive wastes is produced by nuclear power as well as military facilities, and for a minor part by hospitals, research centers and some industries using radioactive substances.

The radioactive wastes are characterized on one hand by the radionuclides that they contain, their emitted radiations (alpha, beta, gamma), and their activity, which is defined by the number of atoms nuclei that decay spontaneously per unit time (measured in Becquerel), and on the other hand by their radioactive period defined as the time required for the activity of a radionuclide to decrease by a factor 2. As an illustration, the period of Cesium 137 is 30 years, whereas the period of radium 226 is 1600 years. Some radionuclides, like iode 129, have a period of some million years (IAEA, 1994).

These characteristics enable a classification of the radioactive wastes, as well as the level of containment and isolation which must be observed. The radioactive wastes are classified between very low level (VLLW), low level (LLW), intermediate level (ILW) and high level activity (HLW). The intermediate and high level wastes are mostly the residual wastes obtained after reprocessing of spent fuel and some installations from nuclear power plants.

The radioactive wastes are considered as short lived if they contain only radionuclides having a period smaller than 31 years, and they are considered as long lived if they contain a significant amount of radionuclides having a period greater than 31 years.

In many countries, the low level wastes are stored in a near surface disposal, implemented at the ground level (surface) or below the ground level at a depth of tens of meters (sub-surface) (NEA, 1987; IAEA, 1999). The wastes are emplaced either in purposely constructed structures, or in excavated caverns. In the first case, the protection from radiations is ensured by a few meters thick cover of soil, whereas in the second case, it is ensured by the above ground layer. For the high level wastes, the contemplated storage solution consists in burying them in underground stable geological formations (NEA, 1999; Chapman *et al.*, 2003). The protection is then ensured by the thick rock layer, as well as by some specific engineered barrier between the rock and the wastes packages. In the following, the stress will be laid only upon this kind of deep geological disposal, since it involves challenging problems in rock mechanics.

2 THE DEEP UNDERGROUND WASTE DISPOSAL SOLUTION

In order to guarantee the safety of radioactive wastes storage over a very long period, it is of utmost importance to confine and isolate the wastes as long as possible so that their radioactivity has sufficiently diminished to be no longer harmful to living organisms. For this purpose, it is necessary firstly to limit the release of the radioactive substances from the wastes packages and secondly, to delay as much as possible their migration to the biosphere, once they have been released. These conditions are ensured by the multibarrier concept, which consists in interposing several barriers between the wastes and the biosphere (European Commission, 2004; Stief, 1987). The first barrier is the packaging of the wastes, made for example of steel or copper canisters, the second barrier is an engineered barrier, made for example of bentonite, disposed around the wastes packages in order to fill all possible voids between the packages and the rock, and the third barrier is the geological layer surrounding the disposal. Since the long term performance and safety of the disposal rely upon this geological layer, it must be sufficiently thick, which implies a deep disposal.

In practice, the geological layer is selected according to various criteria which are not only technical but also economic, social and environmental (AkEnd, 2002). From a rock mechanics point of view, the stability of the layer with regards to possible perturbations such as seismic events or glaciation is of prime importance. Then, a thick layer is required to limit erosion effects as well as possible human intrusion. In addition, the hydrology and the geochemistry of the layer, as well as its mechanical resistance are important factors. In particular, water circulations must be as low as possible, favoring layers with very low permeability as well as low hydraulic gradients.

Up to now, most of the studies around the world have focused on three principal geological formations, which are the granite, the sedimentary layers made of clay or argillite, and the salt (US DOE, 2011; US DOE, 2014). Granite is a stable material which has a high mechanical resistance, a good thermal conductivity and a very low permeability. However, the presence of fractures may significantly alter these properties, and must therefore be taken into account properly in modeling. The argillite is a sedimentary hard rock which also has a very low permeability. Moreover, it has a good ability to retain radionuclides (Norris *et al.*, 2014). In the case of rock salt, there is practically no groundwater flow, and the viscoplastic properties of the salt (Munson, 1997) induce a progressive self-sealing of the damage caused by excavation (Winterle *et al.*, 2012).

3 THE MAIN ISSUES TO BE SOLVED IN ROCK MECHANICS

3.1 Multi-scales problem

In a nuclear waste underground repository, many complex physico-chemical processes can take place, depending on the type of host rock (Tsang, 1987). As an example, in the base case evolution of a repository, the first perturbation in the rock will be induced by the excavation of the storage facility, which will last for some tenths of years. The initial in situ stress distributions will be modified, as well as the underground water pressure and flow, and also, to a lower degree, the initial geothermal temperature distribution. The galleries might be ventilated, thus causing the desaturation of the host rock. After emplacement of the wastes canisters, the heat released by the wastes will generate thermal stresses as well as a water over-pressure close to the storage. With time, the rock will resaturate and some water will come in contact with the canisters, leading to corrosion and finally to a leak of radionuclides toward the biosphere. Moreover, these processes are accompanied by some possible chemical reactions between the various interplaying materials.

From a methodological point of view, it is possible to distinguish three spatial scales when studying a nuclear waste repository. The smallest scale is the scale of the canister itself, at which specific aspects such as heat release, corrosion (Duquette *et al.*, 2008) and release of radionuclides can be investigated, the results of which will serve as source terms for the upper scale studies. Apart from particular problems such as the mechanical design of the canisters with respect to a potential pressure from the engineered barrier or from the host rock, rock mechanics problems are more related to the upper scale which is the scale of the disposal cells. At this scale, often labeled as 'near field', important gradients are occurring, and thermo-hydro-mechanical as well chemical processes are highly coupled. Moreover, because of the degradation of the rock induced by the excavation of the boreholes, the material properties such as the intrinsic permeability or the mechanical resistance are strongly heterogeneous, even though they were quite homogeneous in the initial state. Finally, the last scale is the scale of the whole repository, in the surrounding geological medium. It is known as the far field. At this scale, the gradients are less important, and the main issue is related to the transport of the radionuclides from the storage to the biosphere. Nevertheless, some rock mechanics questions may be raised such as the influence of regional faults on the evolution of the storage (Andra, 2005).

Concerning the time scales, different scales corresponding to different phases of the repository can be distinguished. The first time scale corresponds to the excavation period, before the emplacement of the canisters. In this phase, coupled hydro-mechanical processes are dominant, with highly non linear phenomena such as creep, initiation and propagation of damage in the rock, as well as desaturation. The stresses and pore pressure obtained at the end of this phase will constitute the initial conditions for the following phase. This phase is associated with the emplacement of the canisters, which engenders a thermal transient evolution over some hundreds of years, as well as a longer lasting hydraulic evolution. Finally, the transport of released radionuclides must be studied over a period of some millions of years.

3.2 THM couplings

One of the main difficulties in the prediction of the evolution in space and in time of a radioactive waste disposal is to account for all the relevant processes as well as their couplings (Tsang, 1987; Nguyen, 1995). According to the various phases in the disposal lifetime, some processes or some couplings might be negligible, but in the general situation, they must be accounted for. The processes and couplings are examined hereafter.

3.2.1 Mechanical aspects

The rock mass may contain a certain amount of water. Therefore, under various loadings, the stresses which develop in the rock, and satisfy the momentum equation, are partly attributable to the solid and the fluid components, leading to the concept of effective stresses for the sole solid component. In addition, the temperature variations engender thermal strains which contribute to the total strains of the rock mass. Moreover, some mechanical properties such as the elasticity modulus or the cohesion may depend upon the temperature as well as the water content. Consequently, the resolution of the mechanical problem requires the knowledge of the state variables of the thermal and hydraulic problems.

3.2.2 Hydraulic aspects

The hydraulic problem is more or less complex, depending on whether the rock is saturated or not. In the simplified case of a saturated rock, the flow of liquid water is not only due to the pore pressure gradient, but also to the porosity variations induced by the rock deformation, and to the water expansion due to the temperature variation. Moreover, some hydraulic properties such as the intrinsic permeability or the water viscosity may depend on the strains and temperature.

3.2.3 Thermal aspects

The evolution of the temperature in the rock is, among other phenomena, due to a simultaneous diffusive flow of heat under temperature gradients, as well as an advective flow by the fluid movement. Here also, some thermal properties such as the thermal conductivity may depend on the temperature and water content.

Consequently, the prediction of the radioactive waste disposal evolution requires the knowledge of the thermo-hydro-mechanical variables at any time, which implies the resolution of a coupled system of partial differential equations. These equations will be detailed in the following paragraph.

4 THEORETICAL FORMULATION OF THM COUPLINGS

The theoretical formulation which is exposed below is based on a macroscopic description of the rock mass, considered as a porous medium (Bower *et al.*, 1997; Nguyen, 1995; Munoz, 2006; Lewis *et al.*, 1987). At this scale, the material is supposed to be continuous and homogeneous. This porous medium is composed of the solid skeleton, which is taken as reference to describe the deformation of the medium. Therefore, the displacement field is attached to the material particles of the solid skeleton. The skeleton is itself composed of solid grains and voids. The occluded porosity composed of non-connected voids will be assimilated to the skeleton, and therefore not be considered explicitly. The open porosity of the rock is defined as the set of the connected voids which enable fluid transfers within the porous medium. In the case of the radioactive wastes problem, the fluids present in the pores are initially the liquid water, if the rock is saturated, and in some zones of the disposal which may be desaturated, liquid water and gas made of a mixture of dry air and vapor. Some other gases, such as hydrogen, may be generated in the disposal, as a result from the interaction between the pore water and the engineering materials present in the disposal. They will not be considered in the following.

The general theoretical formulation is based on two types of equations, on one hand the balance equations and on the other hand the constitutive equations. The balance equations of a given variable give the variation of the quantity of this variable attached to one constituent (solid or fluid) of the porous medium, during an infinitesimal time increment dt. The calculus of this derivative involves the absolute velocity of the constituent. Since the solid skeleton is used as the kinematical reference for the porous medium, all the derivatives are finally expressed in terms of the solid skeleton velocity, and consequently for the fluids, in terms of their velocity relative to the solid skeleton.

In addition, the hypothesis of small displacements for the skeleton is often assumed, and verified in many practical situations. By considering an arbitrary representative elementary volume of rock, the balance equations can then be obtained in a local form. In the following, the presentation will be restricted to this form, which is well suited for numerical implementation for example in finite element codes. More details can be found for example in Lewis *et al.* (1987).

4.1 Balance equations

The balance equations concerning the mass, the momentum and the energy are first presented here. They can be written either for the various constituents (skeleton, water and air) or, more frequently, for the various phases (solid, liquid and gas) of the porous medium.

4.1.1 Solid mass balance

Because of the chemistry of the pore water, the solid mass of the skeleton may be altered, by dissolution or precipitation. Therefore, the general formulation of the solid mass balance equation must incorporate a term corresponding to the rate of solid mass released from the skeleton:

$$\frac{\partial m_s}{\partial t} = -\dot{m}_d \tag{1}$$

where m_s is the solid mass per unit volume of porous medium, and $\dot{m}_d$ is the rate of solid mass chemically released. By introducing the intrinsic density ρ_s of the solid constituent, and the porosity φ of the porous medium, it is possible to express m_s as:

$$m_s = (1 - \varphi)\rho_s \tag{2}$$

which leads finally to the following equation:

$$\frac{\partial[(1 - \varphi)\rho_s]}{\partial t} = -\dot{m}_d \tag{3}$$

4.1.2 Water mass balance

The water may be present in the porous medium in liquid and gas phases. A current practice consists in writing the mass balance equations for each of these two phases and in summing the two equations in order to eliminate the exchange term which exists between the two phases.

The variations of the liquid water mass in the porous medium are due to the water movement in the porosity and to the phase change between liquid water and vapor. Thus, the liquid water mass balance can be written:

$$\frac{\partial m_l}{\partial t} + div\left(m_l \underline{V}_l\right) = -\dot{m}_{l \to v} \tag{4}$$

where $\underline{V}_l$ is the relative velocity of the liquid water with respect to the solid skeleton (the transport by the skeleton being here neglected), and $\dot{m}_{l \to v}$ is the rate of liquid water mass which is evaporated, by unit volume of porous medium. The liquid water mass itself is given by:

$$m_l = \varphi S_l \rho_l \tag{5}$$

where S_l is the saturation degree in liquid water and ρ_l the intrinsic density of the liquid water. In a similar way, the vapor mass balance can be written:

$$\frac{\partial m_v}{\partial t} + div\left(m_v \underline{V}_v\right) = \dot{m}_{l \to v} \tag{6}$$

where $\underline{V}_v$ is the relative velocity of the vapor with respect to the solid skeleton and the vapor mass m_v is given by:

$$m_v = \varphi S_g \rho_v \tag{7}$$

In this equation ρ_v is the intrinsic density of vapor, and S_g the saturation degree in gas, which satisfies:

$$S_l + S_g = 1 \tag{8}$$

4.1.3 Air mass balance

The gas which is present in the pores is a mixture, supposed here ideal, of vapor and dry air. The balance of dry air mass m_a can be simply written:

$$\frac{\partial m_a}{\partial t} + div\left(m_a \underline{V}_a\right) = 0 \tag{9}$$

where $\underline{V}_a$ is the relative velocity of dry air with respect to the skeleton.

4.1.4 Gas relative velocity

As a complement to the vapor and dry air velocities, it can be more convenient to introduce the gas relative velocity since it enables to distinguish between the flow of the gaseous mixture on one hand, and the diffusion of dry air and vapor within this mixture on the other hand. For this purpose, the molar fractions of vapor c_v and dry air c_a are first considered:

$$c_v = \frac{p_v}{p_g} \quad c_a = \frac{p_a}{p_g} \tag{10}$$

where p_v et p_a are respectively the partial pressures of vapor and dry air in the gaseous mixture. The pressure p_g of the mixture is given by the sum:

$$p_g = p_v + p_a \tag{11}$$

The relative velocity $\underline{V}_g$ of the gaseous mixture is then introduced as a mean of the velocities of the constituents weighted by their molar concentration:

$$\underline{V}_g = c_v \underline{V}_v + c_a \underline{V}_a \tag{12}$$

4.1.5 Energy balance

A generally admitted hypothesis supposes that the various phases which are present in the porous medium are in local thermal equilibrium, thus having the same temperature T. Consequently, only one energy balance equation must be written for the medium. Various forms of this equation can be found in the literature, which differ by some simplifications such as neglecting the heat source associated with energy dissipation, as it is the case here:

$$(\rho C)_e \frac{\partial T}{\partial t} + \left(m_l C_l \underline{V}_l + m_g C_g \underline{V}_g\right).\underline{\nabla} T + div \underline{q} = -\Delta H_{l\to v} \dot{m}_{l\to v} \tag{13}$$

In this equation, $(\rho C)_e$ is the effective heat capacity of the porous medium, C_l and C_g are respectively the specific heats of the liquid water and of the gas, $\underline{q}$ is the heat flux supplied by conduction, and $\Delta H_{l\to v}$ is the evaporation enthalpy of liquid water. Note that in case of some endothermic or exothermic chemical reactions within the medium, the corresponding enthalpies should be also incorporated in the energy balance.

The effective heat capacity of the porous medium is determined from the densities, the volume fractions and the specific heats of the individual phases:

$$(\rho C)_e = m_s C_s + m_l C_l + m_v C_v + m_a C_a \tag{14}$$

4.1.6 Momentum balance

The momentum balance equations are written here for the porous medium in quasi-static conditions, neglecting the inertia effects. The stresses $\underline{\underline{\sigma}}$ are thus solution of:

$$div\,\underline{\underline{\sigma}} + \underline{f} = \underline{0} \tag{15}$$

where $\underline{f}$ stands for the body forces applied to the porous medium.

4.2 Constitutive equations

Since a large number of state variables have been introduced to describe the coupled thermo-hydro-mechanical processes in the rock, it is necessary to also supply a large number of constitutive equations in order to obtain a well posed problem. In this presentation, the classical mechanical constitutive equations of the rock skeleton will be omitted, since they are well known, and the stress will be laid upon the thermo-hydraulic aspects of the behavior.

4.2.1 Liquid water mass flux

The liquid water mass flux $m_l\underline{V}_l$ is due to the gradient of the liquid pressure p_l within the porous medium. It is given by the generalized Darcy's law:

$$m_l\underline{V}_l = -\rho_l \frac{Kk_{rl}(S_l)}{\eta_l(T)} \underline{\nabla} p_l \tag{16}$$

where K is the intrinsic permeability of the rock, which depends in general of the porosity, k_{rl} is the relative permeability to liquid water, which depends on the degree of saturation in liquid water, and η_l is the dynamic viscosity of water, which depends on temperature.

4.2.2 Vapor and dry air mass fluxes

The vapor mass flux $m_v\underline{V}_v$ is resulting on one hand from the gas flow within the porous network, and on the other hand from the diffusion of vapor within the gaseous mixture. The flow of the gas is due to the gradient of the gas pressure p_g in the rock, and, like the flow of the liquid water, it obeys the generalized Darcy's law. The diffusion of the vapor in the gas is given by Fick's law. Altogether, the vapor mass flux can be written as:

$$m_v\underline{V}_v = -\rho_v \frac{K\,k_{rg}(S_g)}{\eta_g(T)} \underline{\nabla} p_g - \rho_v \frac{p_g}{p_v} \tau\varphi S_g D(p_g, T) \underline{\nabla} c_v \tag{17}$$

where k_{rg} is the relative permeability to gas, which depends on the degree of saturation in gas, η_g is the dynamic viscosity of gas, which depends on the gas pressure and temperature. τ is the tortuosity and D is the diffusion coefficient of the porous medium.

Following the same reasoning, an analogous expression can be easily established for the dry air mass flux:

$$m_a \underline{V}_a = -\rho_a \frac{K\, k_{rg}(S_g)}{\eta_g(T)} \underline{\nabla} p_g - \rho_a \frac{p_g}{p_a} \tau \varphi S_g D(p_g, T) \underline{\nabla} c_a \tag{18}$$

4.2.3 Conduction heat flux

The conduction heat flux $\underline{q}$ is classically given by Fourier's law:

$$\underline{q} = -\lambda(S_l, T) \underline{\nabla} T \tag{19}$$

where λ is the thermal conductivity of the rock, which depends in general of the degree of saturation and of the temperature.

4.2.4 Water retention curve

The water retention curve, or similarly the sorption-desorption isotherms, refer to the relationship between the water content, or the degree of saturation in liquid water, and the relative humidity or the capillary pressure within the rock. The capillary pressure p_c is defined as the difference between the gas and liquid water pressures:

$$p_c = p_g - p_l \tag{20}$$

Experimentally, the measured isotherms are different when the porous medium is being wetted (sorption) or dried (desorption). However, there are nowadays very few reliable models to describe this hysteretic behavior, and therefore a unique relationship is mostly used. In particular, the Van Genuchten approximation (Van Genuchten, 1980) is often considered for granite as well as argillite:

$$p_c(S_l) = a\left(S_l^{-b} - 1\right)^{\left(1-\frac{1}{b}\right)} \tag{21}$$

The material parameters a and b which are usually calibrated on a given isotherm corresponding to a given temperature, must be temperature dependent in the more general THM case:

$$p_c(S_l, T) = a(T)\left(S_l^{-b(T)} - 1\right)^{\left(1-\frac{1}{b(T)}\right)} \tag{22}$$

Another approach consists in modifying the relation between the degree of saturation and the capillary pressure obtained for example at ambient temperature T_0, by using the dependence of the surface tension of water with temperature $\sigma(T)$:

$$S_l(p_c, T) = S_l(p_c, T_0) \frac{\sigma(T)}{\sigma(T_0)} \tag{23}$$

4.2.5 *Thermal expansion*

The thermal expansion of the rock is classically given by the rate of the thermal strain tensor:

$$\dot{\underline{\underline{\varepsilon}}}^{th} = \alpha_{th}(T)\dot{T}\underline{\underline{I}} \tag{24}$$

where the thermal expansion coefficient α_{th} depends on temperature, and $\underline{\underline{I}}$ stands for the unit second order tensor.

4.3 Numerical implementation

The numerical solution of the previous partial differential equations is achieved by means of the finite elements method (Lewis *et al.*, 1987). The primary unknowns of the problem are the displacement field, two fluid pressure fields (for example the liquid pressure and the gas pressure, or the capillary pressure and the gas pressure), and the temperature field. The interpolation functions are quadratic for the displacements and linear for the fluid pressures and the temperature. The final discretized equations can be solved either in a monolithic way, or in a partitioned way.

4.4 Richards' simplified formulation

In many practical situations of thermo-hydro-mechanical analysis, it is possible to simplify the above formulation, by assuming, following Richards (Richards, 1931), that the gas pressure remains constant, equal to the atmospheric pressure. In that case, only one balance equation is written for the total mass of water, and the liquid water pressure is chosen as single fluid primary unknown. It is however still possible to account for the diffusion of vapor within the gas, by simply relating the vapor pressure gradient to the temperature and liquid pressure gradients by using Kelvin's law.

5 PARTICULAR TREATMENT OF FRACTURES

In hard rocks such as granite, there exists a certain number of existing fractures, which play an important role on the thermo-hydro-mechanical behavior of the rock mass. In view of their specific geometry, it is useful to adopt a specific representation of these fractures by two surfaces. This approach, based on the use of joint or interface finite elements, is well known in rock mechanics modeling and it is developed here in the simplified case of hydro-mechanical couplings.

5.1 Interface finite elements

The interface finite elements are defined by two matching surfaces, and have a zero thickness. The strains are computed from the relative displacements of these two surfaces, and the stresses are the three components of the stress vector along the direction normal to these surfaces. For the hydraulic aspects, it is desirable to model not only a longitudinal flow along the fracture, but also some transverse fluid

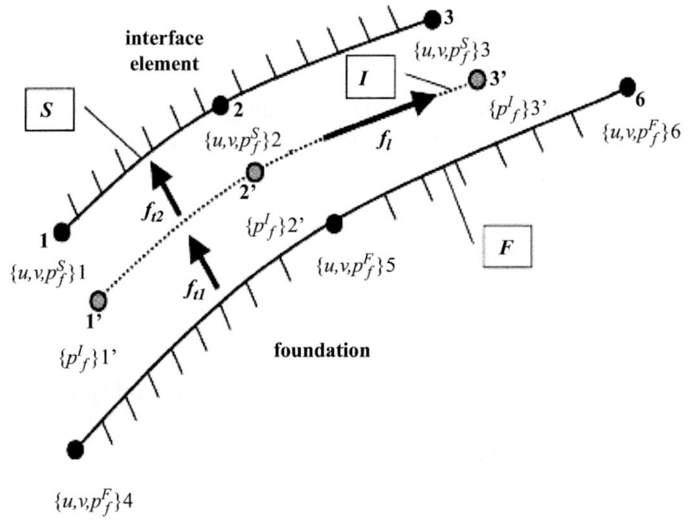

Figure 1 Bidimensional interface finite element for hydro-mechanical analysis of fractures, as proposed by Guiducci *et al.* (2002).

exchanges with the neighboring porous medium. The corresponding finite elements have been proposed for example by Guiducci *et al.* (2002) (see Figure 1).

The previous balance equations need to be reformulated to take into account the special features of the interface elements.

5.2 Balance equations for interface finite elements

In the hydro-mechanical behavior of fractures, two different apertures are generally distinguished, the geometrical one e, and the hydraulic one e_h which is often related to the geometric one by means of an empirical relation involving the fracture roughness.

Restricting the presentation to the bidimensional case for the sake of simplicity, the fluid mass balance written for an infinitesimal length of fracture $d\xi$, is given by the following equation:

$$\frac{\partial(\rho_f e_h)}{\partial t} + \frac{\partial(\rho_f e_h V_f)}{\partial \xi} = m^S_{\to I} + m^F_{\to I} \tag{25}$$

where V_f stands for the longitudinal fluid velocity along the fracture, and $m^S_{\to I}$ et $m^F_{\to I}$ are the transverse mass fluxes coming respectively through the faces S and F of the adjacent porous rock. The longitudinal velocity is given by Darcy's law:

$$V_f = -\frac{K_l}{\eta_l}\frac{\partial p_l^I}{\partial \xi} \tag{26}$$

where K_l is the longitudinal intrinsic permeability of the fracture, which is often calculated from the hydraulic aperture by the following expression derived form a Poiseuille type flow assumption:

$$K_l = \frac{e_h^2}{12} \tag{27}$$

The transverse mass fluxes are given by an analogous expression which involves, instead of the pressure gradient, the pressure differences between one porous rock face and the fracture:

$$m_{\to I}^{S} = \rho_f \frac{K_t}{\eta_l} (p_l^S - p_l^I) \tag{28}$$

$$m_{\to I}^{F} = \rho_f \frac{K_t}{\eta_l} (p_l^F - p_l^I) \tag{29}$$

In these expressions, K_t is the transverse intrinsic permeability of the fracture.As mentioned above, the strains are determined from the relative displacements of the fracture surfaces. Orienting the normal vector from face F to face S, the strains can be written:

$$\underline{\varepsilon} = \begin{bmatrix} v_S - v_F \\ u_S - u_F \end{bmatrix} \tag{30}$$

The first component is along the normal direction, and corresponds to opening-closing movements, whereas the second component, along the tangential direction, corresponds to shearing movements of the fracture. The jump of normal and shear stresses across the fracture obey the momentum balance equation:

$$[\underline{\sigma}]_F^S + e\underline{f} = \underline{0} \tag{31}$$

The formulation can be easily extended to include unsaturated as well as thermal aspects.

6 EXAMPLE OF APPLICATION

As illustration of the previous theoretical formulation, a study performed in the framework of the DECOVALEX international project (Jing *et al.*, 1995; Stephansson *et al.*, 2001) is now presented (Millard & Rejeb, 2003). It is related to a typical three-dimensional pattern of a HLW repository in granite, located at a depth of 1000 m, in order to evaluate the importance of the various THM couplings. This hypothetical case is inspired from a Japanese concept (Figure 2). Since the engineered barrier plays a major role on the flows inside the repository, the study has focused on the near field response, over the first hundred years, during which the thermal paroxysm is reached.

6.1 Repository design and near-field modeling

In the particular repository design which is considered here, the waste canisters are emplaced in vertical boreholes which are sunk in the floor of parallel drifts. Because of symmetry and periodicity considerations, the modeling at the scale of the canisters, in

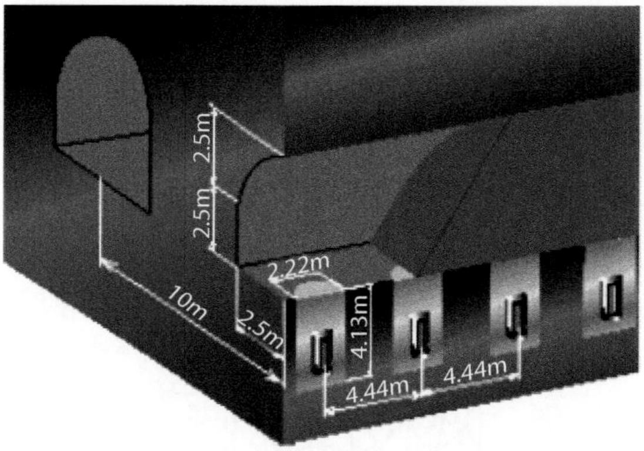

Figure 2 Design of the hypothetic case based on a Japanese HLW disposal concept.

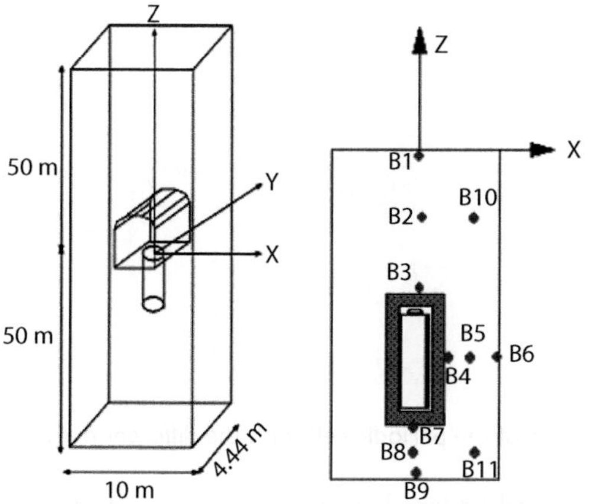

Figure 3 The periodic cell considered for near field evaluation of THM couplings, and some points of interest close to the canister.

order to evaluate the importance of the THM couplings, can be restricted to a single canister with the surrounding rock, as shown on Figure 3.

6.1.1 Modeling setup

The computational domain to be meshed can be limited to only a quarter of the basic periodic cell, thanks to symmetry considerations. Four principal zones, associated to four different materials, can be identified in this domain, as sketched on Figure 4: the overpack *(yellow)*, the buffer *(blue)*, the backfill *(red)* and the rock mass *(green)*. As outlined before, the mesh is composed of quadratic finite elements, with different interpolation functions for the displacements, and for the fluid pressure and temperature.

At the altitude of the repository, in its initial condition, the rock mass is supposed to be fully saturated, and the liquid water pressure is equal to the hydrostatic pressure. The mechanical total stresses are on one hand a vertical stress $\sigma_v = -25.6$ MPa,

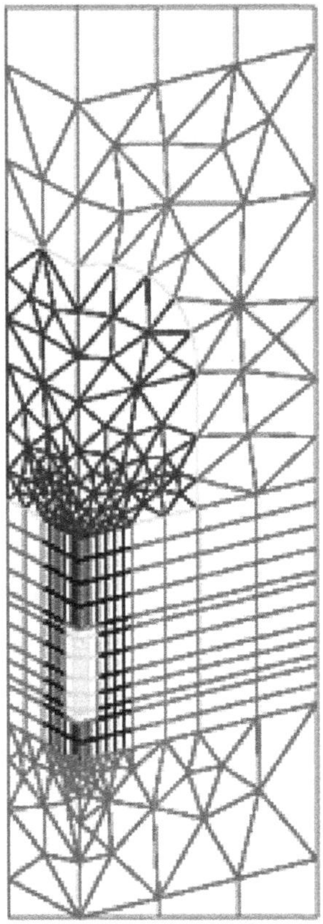

Figure 4 Mesh of one quarter of the periodic cell, with the different materials considered.

corresponding to the overburden dead weight, and on the other hand two isotropic horizontal stresses equal to 1.5 σ_V. Finally, the temperature due to the geothermal gradient is 45°C.

For the mechanical boundary conditions, the displacements which are orthogonal to the lateral limits of the model are set to zero, as well as the vertical displacement at the bottom of the model. A normal pressure corresponding to the overburden load is applied at the surface of the model. For the thermal and hydraulic boundary conditions, the respective thermal and water flows are prevented on the lateral limits of the model, whereas on the top and bottom boundaries, the temperature is prescribed equal to 45°C, and the water pressure is set equal to the in situ hydrostatic pressure.

In order to model the waste disposal in a realistic way, the various sequences of the disposal lifetime are simulated. Thus, the initial sequence consists in setting the initial in situ conditions in the rock mass. Then the excavation of the drift and of the pitch are simulated, at constant temperature, leading to a transient evolution of the stresses and pore pressure in the rock mass. The simulation runs until a new steady state is obtained. In order to start from a nominal geometry of the drift and the pitch before the emplacement of the canister, the displacements corresponding to this steady state are reset to zero. Finally, in the last sequence, the canister, as well as the installation of the engineered barrier, the over-pack, the buffer and the backfill are simultaneously introduced. The heated phase is limited here to the short term period, where the couplings are of major importance. Therefore, the calculation is run over a hundred-year period.

6.1.2 *Material properties*

Although the considered repository is based on a Japanese concept, working in the framework of the Decovalex project enabled to dispose of rock mass properties taken from Canadian granite data (Nguyen *et al.*, 2003). Representative values obtained from laboratory tests on intact rock, have been taken for the simulation:

Young's modulus (30000 MPa), Poisson's ratio (0.3), thermal expansion (8.21 10^{-6} /°C), thermal conductivity (2.7 W/m/°C), density (2300 kg/m^3) and specific heat (833 J/kg/°C). The Biot's coefficient, which is used in the hydro-mechanical couplings, was not measured and has been taken as 1. Because of the importance of the intrinsic permeability K, three different values have been considered in the simulations: 10^{-17}, 10^{-18} (noted in the following as the reference case) and 10^{-19} m^2. In addition, the variations of the porosity φ induced by the thermal and mechanical loadings may affect the intrinsic permeability, according to the following law:

$$K(\varphi) = 2.186\ 10^{-10}\varphi^3 - 5.185\ 10^{-18}, \text{ where K is in m}^2 \tag{32}$$

The buffer (bentonite) and backfill (mixture of bentonite and aggregates) properties have been selected according to the Japanese concept, and calibrated from available laboratory test results (Jing *et al.*, 1999). Following a usual practice, the capillary pressure curve, giving the saturation degree in terms of the capillary pressure, as well as the dependence of the liquid permeability with regards to the degree of saturation, have been approximated by the Van Genuchten relationships.

6.2 Evaluation of the couplings

In order to appraise the importance of the various couplings, three other simulations, denoted respectively TH, TM, and HM, were performed in addition to the reference THM analysis.

In the TH simulations, several coupling effects disappear. Concerning first the hydraulic aspects, in the absence of strains, the skeleton porosity remains constant, and therefore, the rock permeability remains constant. Moreover, the time variation of the porosity is no more involved in the water mass balance, and the mechanical stresses do not influence any more the water pressure field. Then for the thermal aspects, the thermal conductivity as well as the heat capacity vary only with the saturation S_l.

Similarly in the TM simulations, decouplings occur in both equations. In the mechanical equations, the water pressure has no effect on the stresses, and the bentonite does not develop any swelling pressure. In the thermal equations, the thermal conductivity and the heat capacity are constant.

Finally in the HM simulations, a constant temperature, equal to 45°C is maintained leading to zero thermal strains. As a consequence, the porosity variations are only attributable to the mechanical loads and to variations in water pressure. Moreover, there is no thermal gradient-related water transfer, and water viscosity remains constant.

6.3 Simulations results

6.3.1 Results for the THM reference case ($K=10^{-18}$ m^2)

Among the very many results given by the fully coupled THM simulations, a selection of the most illustrative and interesting ones is presented below. A first result, presented on Figure 5 (left), is related to the importance of the porosity variations. On the figure, the red zones correspond to an increase of the porosity, whereas the blue zones correspond to a decrease. As can be seen, the increase can be attributed to the expansion of the rock toward the inside of the cavities, whereas the decrease is due to an increase of mechanical compressive stresses since the vertical load is supported by a narrower zone than prior to excavation.

Traction failure at the wall of the cavity is now investigated. For hard rock such as granite, the maximum principal effective stress gives a good indication on the possibility of failure, associated to a traction loading beyond the maximal traction strength. Figure 5 (right) displays this maximum principal effective stress, with the classical mechanical sign convention of positive traction stresses. The maximum observed here, about 6.7 MPa, is lower than the traction resistance strength of the rock, as measured in laboratory experiments (R_t = 11 MPa), which means that no traction failure is to be expected after the excavation.

The short term evolution of the temperature and resaturation of the backfill and the buffer is illustrated on Figure 6, which shows the temperature field and the degree of saturation field after one year. It can be seen that the backfill re-saturates faster than the

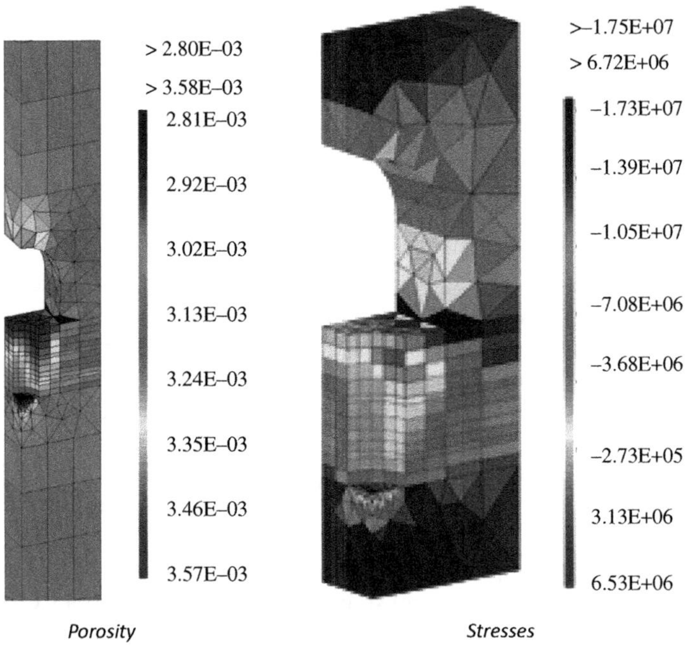

Figure 5 Maps of porosity and maximum principal effective stresses in the rock mass after excavation (zoom around the cavities).

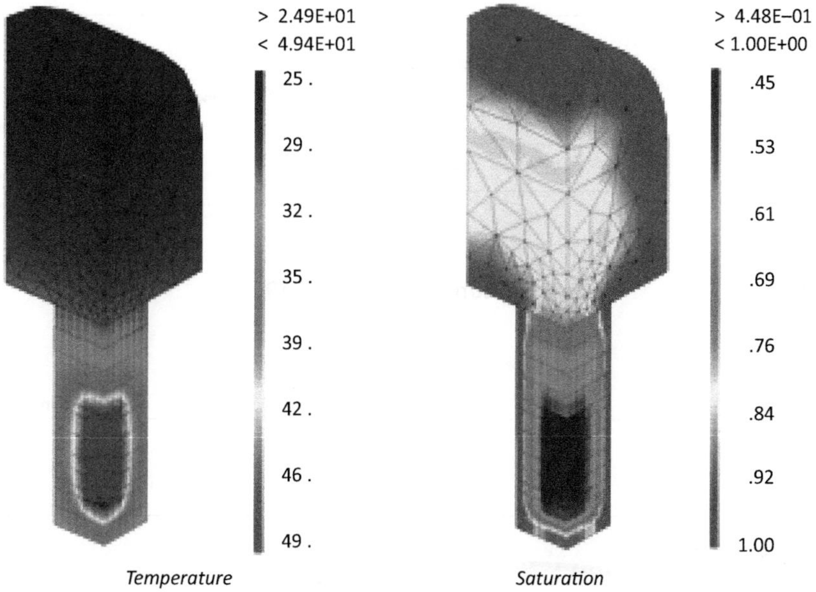

Figure 6 Iso-temperature and iso-saturation in the buffer and backfill after one year (ref. case).

buffer. This is partly due to the fact that the initial degree of saturation is higher in the backfill ($S_l = 0.81$) than in the buffer ($S_l = 0.62$). Moreover, the temperature in the buffer is higher, because of the vicinity of the canister, which tends to push the saturated zone farther. A full resaturation of the buffer and the backfill is obtained after 40 years.

6.3.2 Evaluation of the coupling effects

For the purpose of the comparison to the reference case, two particular points of the buffer material have been chosen where different behavior patterns may be expected because of the different boundary and environmental conditions. The first point is B4 (see Figure 2), at the contact point with the over-pack, which will be subjected to a high temperature. The second point is B6, which lies at the interface between the buffer and the rock-mass, and where water can be directly supplied by the rock. As in the previous paragraph, only a selection of the results will now be presented. In order to evaluate the influence of the couplings, the evolutions of the temperature, the water pore pressure and the horizontal total stress σ_{xx} are compared on Figures 7 to 12. The results are first presented and discussed for point B4. Moreover, for the different quantities, the results have been plotted on each Figure for the lowest rock mass permeability ($K=10^{-19}$ m^2) and the highest ($K=10^{-17}$ m^2).

6.3.3 Comments on results at point B4

Being at the interface with the over-pack, point B4 is subjected to a higher temperature increase, and consequently to a localized drying of the buffer. The consequences of these high thermal and hydraulic gradients, and therefore the importance of the

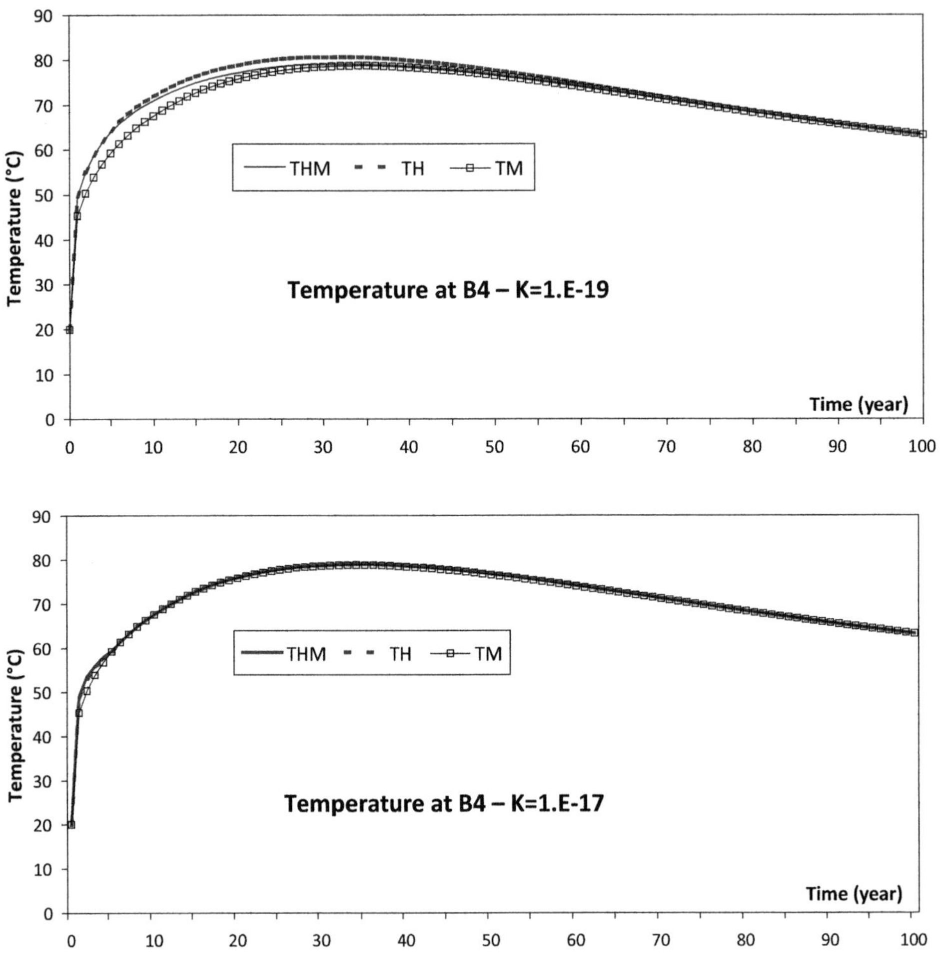

Figure 7 Temperature evolution at point B4.

couplings, are more significant in the short term, as well as for a low rock mass permeability. In particular, for the low permeability $K=10^{-19}$ m^2, the lower and upper curves of Figure 8 are referring to the TM case (constant degree of saturation) and to the TH case respectively (larger de-saturation) while the THM case is closer to the upper bound at the beginning due to drying and closer to the lower bound at the end, once re-saturated.

In the same figure (Figure 8), the drying effect associated with the temperature increase is clearly visible. As can be expected, for a low rock mass permeability, the suction increase is more pronounced. On the contrary, for a high rock mass permeability, the water is supplied from the rock fairly fast, and the M couplings only have little effect. For a low rock mass permeability, in the TH case, re-saturation is not achieved at point B4 after 100 years (Figure 8). Moreover, it is interesting to note that the difference between the THM and TH cases is essentially due to the porosity

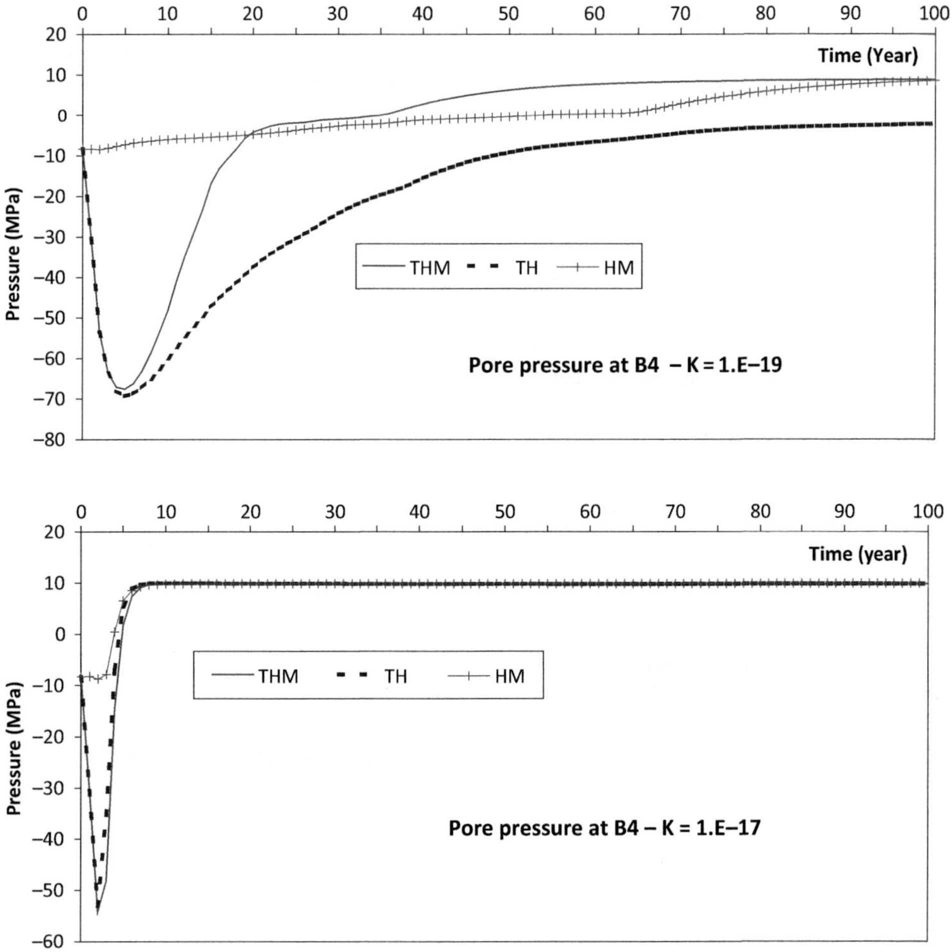

Figure 8 Pore pressure evolution at point B4.

variation induced by the mechanical couplings, which thus have a strong influence on the re-saturation time. This is not the case for the high rock mass permeability.

Another interesting feature linked to the low rock mass permeability concerns the effect of the pore water thermal dilatation. Indeed, as shown on Figure 9, this thermal dilatation leads to an increase of pore pressure which can only dissipate slowly, and which causes a supplement of total horizontal compressive stress σ_{xx}., associated with the confinement due to lateral symmetry boundary conditions.

6.3.4 Comments on results at point B6

Contrary to point B4, point B6 is located sufficiently far from the heat source, and consequently, undergoes lower thermal gradients. For the range of permeabilities considered the couplings have very little effect on temperature (Figure 10).

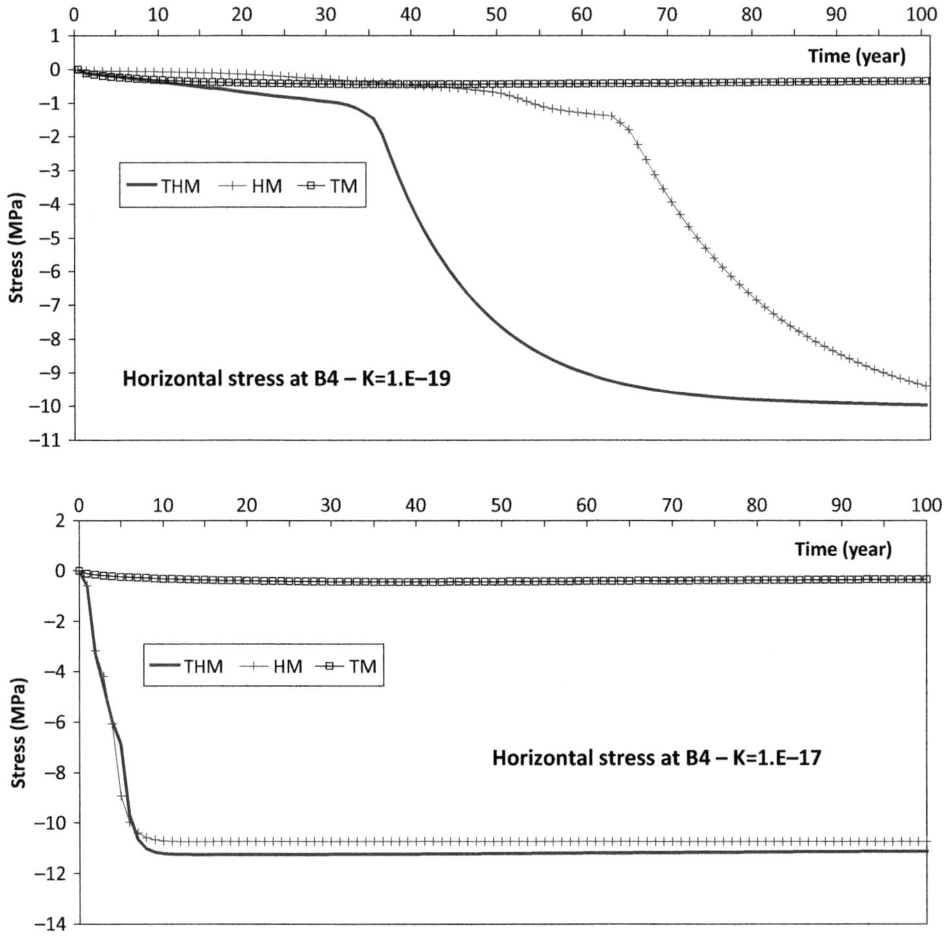

Figure 9 Horizontal total stress σ_{xx} evolution at point B4.

Concerning the pore pressure evolution, the buffer water content at point B6 depends both on one hand on the water supply from the rock mass and on the other hand on the drying from the canister. Therefore, for the high permeability case, the rock mass provides water fast enough to the buffer to prevent any drying, while for the low permeability case, the pressure decrease at the early time is attributable to the drying. As expected, it is less pronounced at point B6 than at point B4. The rebuilding of the hydrostatic water pore pressure starts at about 30 years for the THM case, and about 60 years for the HM case, while the buffer is still unsaturated after 100 years in the TH case.

Finally, the horizontal stress σ_{xx} evolutions at point B6 are very similar to those obtained at point B4, as shown in Figure 12.

The results presented here have been confirmed by those of six research teams involved in the DECOVALEX project (Millard *et al.*, 2003). The predictions of temperature, degree of saturation, pore pressure and total stresses were fairly close to

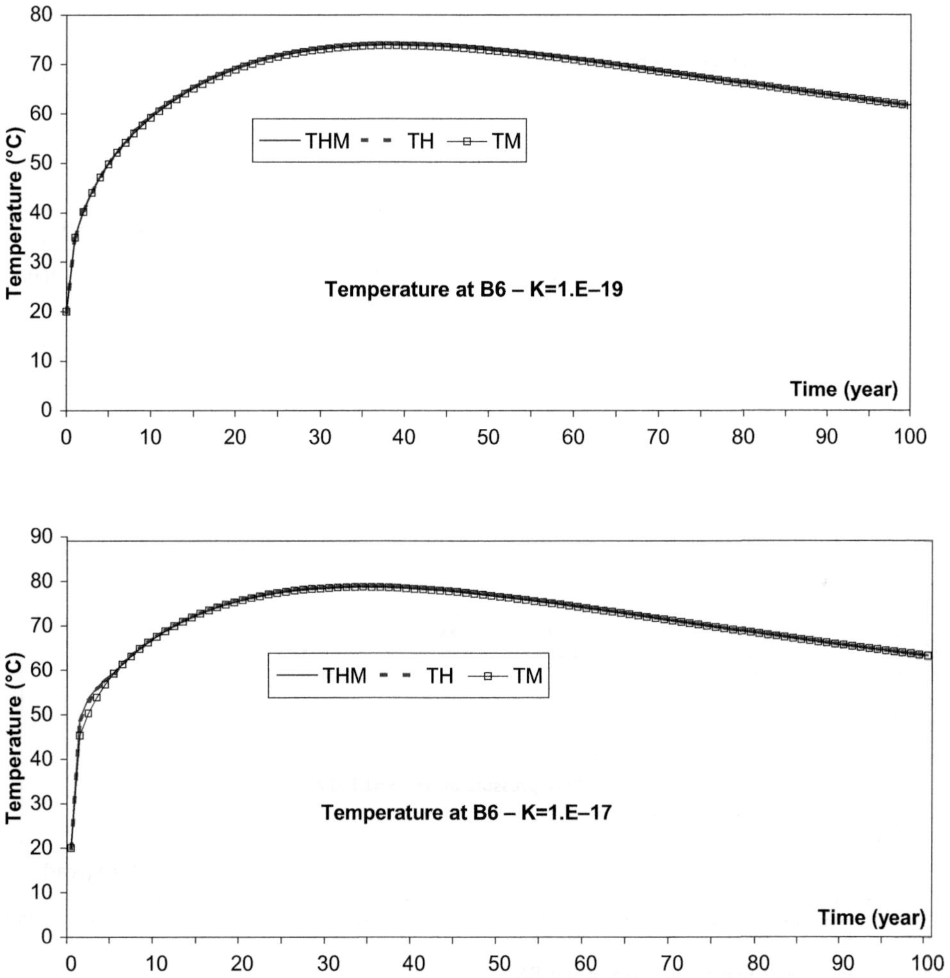

Figure 10 Temperature evolution at point B6.

each other. All the research teams assessed the couplings at the same importance. However, the predicted re-saturation time varied significantly between teams.

7 CONCLUSION

The main thermo-hydro-mechanical processes arising in a deep underground geological disposal of radioactive wastes have been described in details in this chapter. The theoretical formulation underlying these processes has been detailed for the case of an un-fractured rock mass, as well as for a well identified fracture. It has then been illustrated on the case of a generic disposal, in order to appraise the importance of the various THM couplings on the main results. The following general trends have been derived from this study:

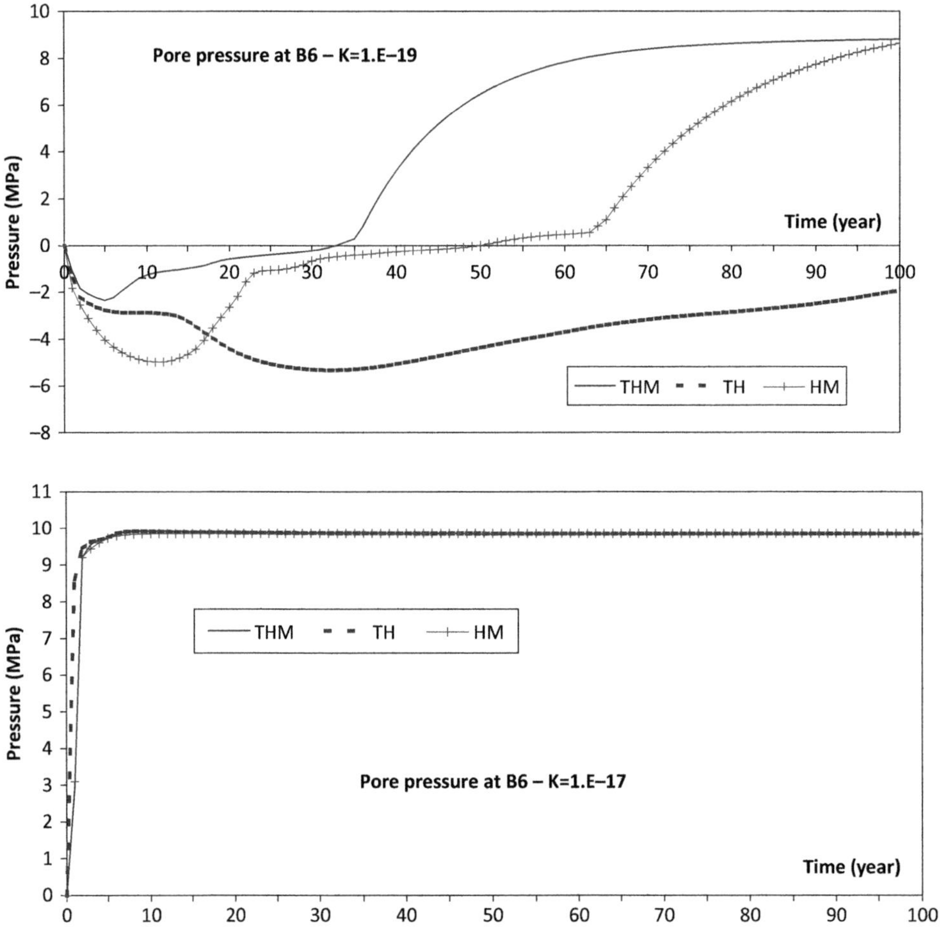

Figure 11 Pore pressure evolution at point B6.

- H and M have a limited effect on the T results during the re-saturation phase.
- During re-saturation, the effect of water pore pressure on the stresses predominates over the effect of skeleton thermal expansion.
- Porosity variations have a significant effect, in particular for low rock mass permeabilities.

These calculations confirm that initial rock mass permeability is a key parameter; it plays a major role in re-saturation time. The following remarks can be made about its influence:

- The effects of the couplings on T, H and M are amplified by low rock mass permeability.
- Suction close to the canisters may be important for low permeability.

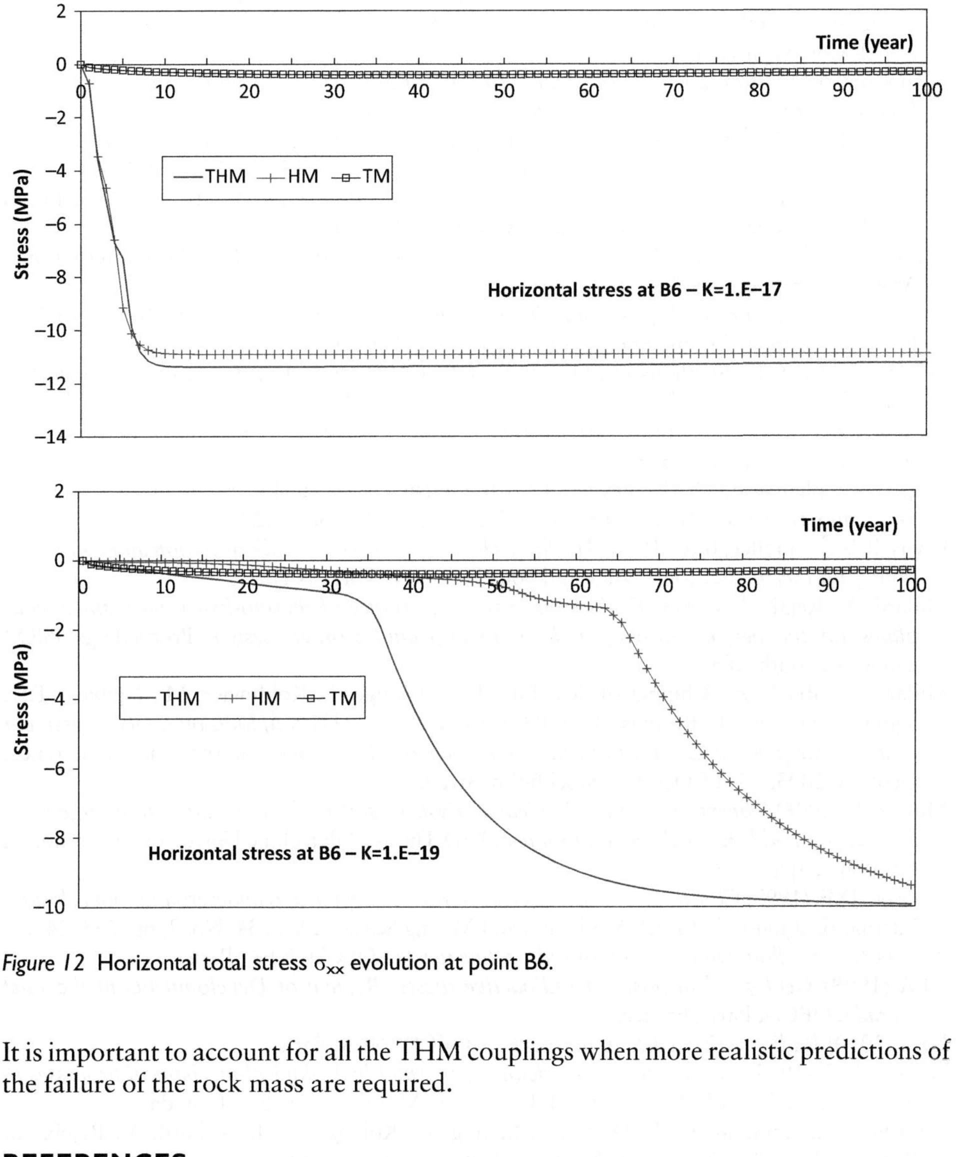

Figure 12 Horizontal total stress σ_{xx} evolution at point B6.

It is important to account for all the THM couplings when more realistic predictions of the failure of the rock mass are required.

REFERENCES

AkEnd (2002) *Site Selection Procedure for Repository Sites. Recommendations of the AkEnd – Committee on a Site Selection Procedure for Repository Sites*, W & S Druck GmbH, Köln.

Andra (2005) *Dossier 2005 Granite, Safety analysis of a geological repository*, Report Series, Andra editor, Chatenay-Malabry, France.

Bower, K.M., Zyvoloski, G. (1997) *A numerical model for thermo-hydro-mechanical coupling in fractured rock*. International Journal of Rock Mechanics and Mining Sciences, Vol. 34, No. 8, pp. 1201–1211.

Chapman, N.A., McCombie C. (2003), *Principles and standards for the disposal of long-lived radioactive wastes*, Elsevier/Pergamon, Oxford, UK.

Duquette, D., Latanision, R., Di Bella, C., Kirstein, B. (2008) *Corrosion issues related to disposal of high-level nuclear waste in the Yucca mountain repository. Peer reviewers perspective*, NACE Corrosion Conference, 2008.

European Commission (2004) *Geological Disposal of Radioactive Wastes Produced by Nuclear Power ... from concept to implementation*, EUR 21224.

Guiducci, C., Pellegrino, A., Radu, J.P., Collin, F., Charlier, R. (2002) *Numerical modeling of hydro-mechanical fracture behavior*, Proceedings NUMOG VIII, pp. 293–299.

IAEA (1994) *Classification of radioactive waste – IAEA safety guide*. Safety Series No. 111.G 111, International Atomic Energy Agency, Vienna, Austria.

IAEA (1995) *Principles of radioactive waste management*. Safety Series No.111-F, International Atomic Energy Agency, Vienna, Austria.

IAEA (1999) *Near surface disposal of radioactive waste*. IAEA Safety Standards Series No. WS-R-1, International Atomic Energy Agency, Vienna, Austria.

Jing, L., Tsang, C.F., Stephansson, O. (1995), DECOVALEX – *An international co-operative research project on mathematical models of coupled THM Processes for safety analysis of radioactive waste repositories. International Journal of Rock Mechanics and Mining Sciences*, Vol. 32, No. 5, pp. 389–398.

Jing, L., Stephansson, O., Börgesson, L., Chijimatsu, M., Kautsky, F., Tsang, C.F. (1999) *Decovalex II project, Technical report – Task 2C*, SKI Report 99:23.

Lewis, R.W., Schrefler, B.A. (1987) *The finite element method in the deformation and consolidation of porous media*, John Wiley and Sons Inc., New York, NY.

Millard, A., Rejab, A. (2003) *Evaluation of the importance of thermo-hydro-mechanical couplings on the performance of a deep underground storage design*. Proceedings ISRM Congress, South Africa.

Millard, A., Rejeb, A., Chijimatsu, M., Jing, L., De Jonge, J., Kohlmeier, M., Nguyen, T.S., Sigita, Y., Souley, M., Rutqvist, J. (2003) *Evaluation of THM couplings on safety assessment of a nuclear fuel waste repository in a homogeneous hard rock*. International Conference, GeoProc 2003, 13–15 October, Stockholm, Sweden.

Munoz, J. (2006) *Thermo-hydro-mechanical analysis of soft rock, application to a large scale heating test and large scale ventilation test*. PhD Thesis, Polytechnic University of Catalonia, Catalonia, Spain.

Munson, D.E. (1997) *Constitutive model of creep in rock salt applied to underground room closure*. International Journal of Rock Mechanics and Mining Sciences, Vol. 34, No. 2, pp. 233–247.

NEA (1987) *Shallow land disposal of radioactive waste*,. OECD/ NEA, Paris.

NEA (1999) *Geological disposal of radioactive waste, Review of Developments in the Last Decade*, OECD, Paris, France.

NEA (2010) *Radioaxctive waste in perspective*, OECD/ NEA, Paris.

Nguyen, T.S. (1995) *Computational modeling of thermal-hydrological-mechanical processes in geological media*. PhD Thesis, McGill University, Montreal, Quebec, Canada.

Nguyen, T.S., Chijimatsu, M., De Jonge, J., Jing, L., Kohlmeier, M., Millard, A., Rejeb, A., Rutqvist, J., Souley, M., Sugita, Y. (2003) *Implications of coupled thermo-hydro-mechanical processes on the safety of a hypothetical nuclear fuel waste repository*. International conference, GeoProc 2003, 13–15 October, Stockholm, Sweden.

Norris, S., Bruno, J., Cathelineau, M., Delage, P., Fairhurst, C., Gaucher, E.C., Hohn, E.H., Kalimichev, A., Lalieux, P., Sellin, P. (2014) *Clay in natural and engineered barriers for radioactive waste confinement*. Geological society special publication n° 400, London.

Richards, L.A. (1931) *Capillary conduction of liquids in porous mediums. Physics*, Vol. 1, pp. 318–333.

Stephansson, O., Tsang, T.C., Kautsky, F. (2001) DECOVALEX II. *Foreword*. International Journal of Rock Mechanics and Mining Sciences & Geomechanics Abstracts, Vol. 38, No. 1, pp. 1–5.

Stief, K. (1987) *The multi-barrier concept – A German approach*. International Symposium on Process, Technology and Environmental Impact of Sanitary Landfill, 19–23 October 1987, Sardinia, Cagliari, Italy.

Tsang, C.F. (1987) *Coupled processes associated with nuclear waste repositories*, Academic Press, Orlando, USA.

US DOE (2011) *Basis for Identification of Disposal Options for R&D for Spent Nuclear Fuel and High-Level Waste*, FCRD-USED-2011-000071.

US DOE (2014) *Assessment of Disposal Options for DOE-Managed High-Level Radioactive Waste and Spent Nuclear Fuel, Evaluation of Options for Permanent Geologic Disposal of Spent Nuclear Fuel and High-Level Radioactive Waste*, Volume I, Sandia National Laboratories, USA.

Van Genuchten, M.T. (1980) *A closed-form equation for predicting the hydraulic conductivity of unsaturated soils*. Soil Science Society of America Journal, Vol. 44, No. 5, pp. 892–898.

Winterle, J., Ofoegbu, G., Pabalan, R., Manepally, C., Mintz, T., Pearcy, E., Smart, K., McMurry, J., Pauline, R., Fedors, R. (2012) *Geologic disposal of high-level radioactive wastes in salt formations*, U.S. Nuclear Regulatory Commission Contract NRC-02-07-006.

The five-volume set *Rock Mechanics and Engineering* consists of the following volumes

Volume 1: Principles
ISBN: 978-1-138-02759-6 (Hardback)
ISBN: 978-1-315-36426-1 (eBook)

Volume 2: Laboratory and Field Testing
ISBN: 978-1-138-02760-2 (Hardback)
ISBN: 978-1-315-36425-4 (eBook)

Volume 3: Analysis, Modeling and Design
ISBN: 978-1-138-02761-9 (Hardback)
ISBN: 978-1-315-36424-7 (eBook)

Volume 4: Excavation, Support and Monitoring
ISBN: 978-1-138-02762-6 (Hardback)
ISBN: 978-1-315-36423-0 (eBook)

Volume 5: Surface and Underground Projects
ISBN: 978-1-138-02763-3 (Hardback)
ISBN: 978-1-315-36422-3 (eBook)

9781032652276